ELECTRONIC ENGINEERING

ELECTRONIC ENGINEERING

THIRD EDITION

Charles L. Alley
Kenneth W. Atwood

Electrical Engineering Department
University of Utah

JOHN WILEY & SONS
New York   London   Sydney   Toronto

JUL   1994

Library of Congress Cataloging in Publication Data

Alley, Charles L
  Electronic engineering.

  1. Electronics. I. Atwood, Kenneth W., joint
author. II. Title.
TK7815.A46 1973   621.381   72–8520
ISBN 0–471–02450–3

Printed in the United States of America

10 9 8 7 6 5 4 3 2 1

To Mildred and Ruth
whose patience and understanding
made this book possible.

Preface

The rapidly changing field of electronics requires a continual reevaluation of the content of an electronics textbook. Also, improved methods of presenting familiar subjects are constantly being sought by a resourceful teacher. For these reasons, we felt that a new edition would be very beneficial to those teachers who strive to keep their material up-to-date. Also, the practicing engineer requires reference material which is current.

The first edition unified the treatment of vacuum tubes and transistors. In addition, the transient as well as the steady-state dc and ac behavior of electronic circuits was given. The second edition brought this material up-to-date and included some of the newer devices which had been developed.

This third edition keeps the unified treatment of tubes and transistors, but the emphasis has been shifted in keeping with the modern uses of the devices. Accordingly, the vacuum tube treatment has been substantially reduced and the semiconductor treatment has been expanded to include MOSFETs (both single- and dual-gate) and some of the other newer devices. In addition to these changes, new chapters on *integrated circuits* (ICs), both linear and digital, have been included. Once these ICs have been introduced, typical uses of ICs are presented in the following chapters so the student can realize the impact these devices are having in the electronics industry. Material on active filters and digital logic has also been added.

This book is primarily concerned with electronic devices and circuits. However, basic field theory is given in Chapter 1 and basic semiconductor physics is briefly presented in Chapter 2. Students with adequate backgrounds in these areas could skip these chapters.

Chapters 3–6 introduce the basic electronic devices and their characteristics. Typical circuits and applications using these devices are also included. Various methods of coupling amplifier stages with emphasis on gain and bandwidth are considered in Chapters 7–10. Chapter 11 discusses power amplifiers with emphasis on class B operation. Small-scale electronic systems such as multistage amplifiers, feedback systems, linear integrated circuits, active filters, and power supplies are introduced in Chapters 12–17. Communication circuits and systems are discussed in Chapters 18–21. Finally, basic pulse and digital circuits are considered in Chapters 22 and 23.

The material in Chapters 3–7 is basic to the understanding of the material in the following chapters. However, the remaining chapters or sections may be selected according to the needs of the students.

The material in this third edition has been presented to Electrical Engineering classes at the University of Utah with gratifying results. We are indebted to these students for their helpful suggestions. We also wish to extend our sincere thanks to Mrs. Ruth Eichers and Mrs. Doris Bartsch of the secretarial staff of the Electrical Engineering Department, under the able direction of Mrs. Marian Swenson, for their able help in preparing the manuscript.

We hope you enjoy this new edition.

Salt Lake City   *Charles L. Alley*
                                                    Kenneth W. Atwood

Contents

A LIST OF SYMBOLS AND ABBREVIATIONS USED IN THIS TEXT

Standard Prefixes and Abbreviations for Decimal Multipliers

Multiplier	Prefix	Abbreviation
10^{12}	tera	T
10^9	giga	G
10^6	mega	M
10^3	kilo	k
10^2	hecto	h
10	deka	da
10^{-1}	deci	d
10^{-2}	centi	c
10^{-3}	milli	m
10^{-6}	micro	μ
10^{-9}	nano	n
10^{-12}	pico	p
10^{-15}	femto	f
10^{-18}	atto	a

A	Amperes
A	Cross-sectional area in square meters
A	A constant (may have subscripts)
A'	A constant
Å	Angstrom
AM	Amplitude modulation
AF	Audio frequency
a	Acceleration in meters/second^2
ac	Alternating current
α	Common-base current transfer ratio of a transistor
B	A constant
B	The bandwidth of an amplifier
B	Bel (a unit to express power gain) (Eq. 12.20)
b	The ratio of actual coupling to critical coupling in a transformer

β	Common-emitter current transfer ratio of a transistor
β	Feedback factor (may have subscripts to indicate voltage or current feedback)
C	Coulomb
C	Capacitance in farads
C	The gap-energy temperature coefficient
C	A constant
°C	Temperature in degrees Celsius (centigrade)
CW	Continuous wave
CMRR	Common-mode-rejection ratio (Eq. 10.3)
c	The velocity of light (2.998×10^8 meters/sec)
cm	Centimeter
D	A constant
D_n	Diffusion constant for electrons (see Table 3.1)
D_p	Diffusion constant for holes (see Table 3.1)
d	The derivative of a variable
d	Distance in meters
dB	Decibels (Eq. 12.21)
dc	Direct current
Δ	A small change (Δt = a small change in time)
E	Energy
e	The base of the natural logarithm (e = 2.7182818)
eV	Electron volts (a unit of energy)
$\mathscr{E}$	Electric field intensity in volts/meter
ε	Dielectric constant
F	Farad
F	A function (usually in terms of s)
F	Force in newtons
FET	Field effect transistor
FM	Frequency modulation
F_a	Figure-of-merit of an amplifier
f	Frequency in cycles per second or hertz
f_L	Low cutoff frequency of a multistage amplifier in Hz
f_H	High cutoff frequency of a multistage amplifier in Hz
f_α	The alpha cutoff frequency of a transistor in Hz
f_β	The beta cutoff frequency of a transistor in Hz
f_1	Low cutoff frequency of a single-stage amplifier in Hz
f_2	High cutoff frequency of a single-stage amplifier in Hz

f_{1f}	Low cutoff frequency of an amplifier with feedback in Hz
f_{2f}	High cutoff frequency of an amplifier with feedback in Hz
G	Conductance in mhos $(G = 1/R)$
G_i, G_p, G_v	Gain of a single-stage amplifier as a function of frequency (subscripts indicate current, power, or voltage)
G_a	Gain of a multistage amplifier as a function of frequency
GaAs	Gallium arsenide
Ge	Germanium
g	The gravitational constant
g	An equivalent conductance (not an actual resistor)
g_d	The drain admittance of a FET or MOSFET
g_g	The gate conductance of a FET or MOSFET
g_m	Transconductance of a tube or transistor
H	Henry
H	A constant
Hg	Mercury
Hz	Hertz (cycles per second)
h	Height in meters
h	The symbol for the hybrid parameters (must have subscripts, see page 114)
I	Current in amperes (subscripts as listed in Table 4.1)
I_{CO}	The saturation current across the reverse biased collector-base junction
I_{EO}	The saturation current across the reverse biased emitter-base junction
I_I	The diode injection current
I_S	The diode saturation current
IGFET	Insulated-gate field-effect transistor
IF	Intermediate frequency
i	Current in amperes (subscripts as listed in Table 4.1)
IC	Integrated circuit
J	Current density in amperes/meter^2
j	$\sqrt{-1}$
K_i, K_p, K_v	The mid-frequency gain (subscripts indicate current, voltage, or power)
K	A constant
$°K$	Degrees Kelvin

K_{vf}, K_{if}	The mid-frequency gain of an amplifier with feedback (subscripts indicate current or voltage gain)
k	Boltzmann's constant (1.38×10^{-23} joule/$^\circ$K)
k	The coefficient of coupling in a transformer
k_c	The critical coefficient of coupling on a transformer
L	Length in meters
L	Inductance in henries
l	A distance of length in meters
$\ln$	Logarithm to the base e
$\log$	Logarithm to the base 10
M	The modulation index of an amplitude modulated wave
M	Mutual inductance in henries
MOSFET	Metal-oxide semiconductor field-effect transistor
m	Mass in kilograms
M_f	The modulation index for a frequency modulated signal
M_p	The modulation index for a phase modulated signal
μ	Carrier mobility (see Table 2.1)
μ	Voltage amplification factor for a vacuum tube
N	The numerator of a function
N_a	Acceptor atom density
N_d	Donor atom density
n	An integer
n	The free electron density in a material
n	The turns ratio in a transformer
n	The total number of turns on a single coil
n_1	The number of turns on the primary of a transformer
n_2	The number of turns on the secondary of a transformer
ω_c	The radian frequency of the carrier signal in an amplitude modulated signal
ω_m	The radian frequency of the modulating signal in an amplitude modulated signal
Ω	Ohms
ω	Frequency in radians per second $= 2\pi f$
ω_1	The low-frequency cutoff frequency of a single stage amplifier (Eq. 7.20 or 7.40)
ω_2	The high-frequency cutoff frequency of a single stage amplifier
ω_E	The frequency where $R_E = 1/\omega C_E$

ω_H	The high half-power frequency of a multistage amplifier
ω_L	The low-half-power frequency of a multistage amplifier
ω_n	The natural frequency of a circuit
ω_o	The resonant frequency of a circuit
ω_α	The alpha cutoff frequency of a transistor
ω_β	The beta cutoff frequency of a transistor
ω_τ	The frequency where $\beta = 1$ in a transistor
P_d	The power dissipation capacity of a device (in watts)
p	The hole density in a material
π	Pi ($\pi = 3.1415927$)
Q	Charge on a body in coulombs
Q	The figure-of-merit of a tuned circuit
Q_o	The figure-of-merit of a coil
q	The charge on an electron (16.019×10^{-10} coulombs)
R	Resistance in ohms (may have subscripts)
R_i	The input resistance to an amplifier
R_{it}	The input resistance to a transistor or tube
R_L	The dc load on a transistor or tube
R_l	The ac load on a transistor or tube
R_o	The output resistance of an amplifier
R_{ot}	The output resistance of a transistor or tube
R_s	The internal resistance of a voltage driving source
R_{sh}	The total shunt resistance in an amplifier
RF	Radio frequency
r	An equivalent resistance (not an actual resistor)
R_{ser}	The resistance in series with an inductance
R_{par}	The resistance in parallel with an inductance
ρ	Resistivity of a material
S_I	The current stability factor
Si	Silicon
Sn	Tin
SNR	Signal-to-noise ratio
SCR	Silicon controlled rectifier
s	Distance
s	Laplace operator ($s = \sigma + j\omega$)
σ	The conductivity of a material
σ	The real component of the Laplace operator

xx   **List of Symbols and Abbreviations**

T	Temperature (usually in °K but may be °C)
T_a	Ambient temperature in °K or °C
T_j	The junction temperature in a diode or transistor (usually in °C)
t	Time in seconds
t'	Thickness in meters
τ	The time constant of a circuit (example $\tau = RC$)
θ	A phase angle (may be given in degrees or radians)
Θ_T	Thermal resistance in °K/watt or °C/watt
V	Potential difference in volts (subscripts as listed in Table 4.1)
V_P	The pinch-off voltage in a FET
v	Potential difference in volts (subscripts as listed in Table 4.1)
v	The velocity in meters/second
$\bar{v}$	The average drift velocity of a carrier
W	Watts
W	Energy in joules
W_g	The gap energy of a semiconductor (see Table 2.2)
X_C	Capacitive reactance in ohms
X_L	Inductive reactance in ohms
x	A distance in a given direction (along the x-axis)
Y	An admittance in mhos (may be complex)
Y_i	The input admittance of an amplifier
Y_o	The output admittance of an amplifier
Y_{if}	The input admittances of an amplifier with feedback
Y_{of}	The output admittance of an amplifier with feedback
y	An equivalent admittance (not composed of actual R, L, C elements)
y	A distance in a given direction (parallel to the y-axis)
y	The symbol for y-parameters (must have subscripts)
Z	An impedance in ohms (may be complex)
Z_i	The input impedance of an amplifier
Z_{if}	The input impedance of an amplifier with feedback
Z_{oa}	The output impedance of an amplifier
Z_{ot}	The output impedance of a transistor or tube

z	An equivalent impedance (not composed of actual R, L, C, elements)
z	The symbol for z-parameters (must have subscripts)
ζ	The damping ratio

1

Introduction

During the twentieth century, technological developments have occurred at a phenomenal rate. The electronic technology has not only contributed to this development but has itself experienced fantastic growth. Thus we find that radio, television, radar, stereo, automatic control systems, and computers are familiar items to most members of our society. While these devices are widely used, the general public considers them too complicated to be understood by anyone but an expert. Actually, most electronic devices can be reduced to a group of interconnected basic electronic building blocks or circuits. It is not too difficult to design electronic equipment if one is well acquainted with the characteristics of these basic blocks or units. In fact, many of these basic electrical circuits are now available in compact packages known as *integrated circuits.* Much of this text is devoted to the analysis and design of these basic blocks or circuits, however, we also consider the problems encountered when the discrete circuits or the integrated circuits are interconnected.

1.1 TYPICAL ELECTRONIC SYSTEMS

The engineer usually begins the design of an electronic device by laying out a *block diagram.* This block diagram is then used as a guide while he designs the circuits represented by the various blocks in the system.

To illustrate this concept, let us consider the block diagram of an audio amplifier. Assume the input signals will be from a microphone or from a phonograph pickup. The output signal must be large enough to drive (or activate) a loudspeaker. A typical block diagram for this device might be sketched as shown in Fig. 1.1.

The electrical signal from a microphone is usually much smaller than the electrical signal from a phono pickup. Therefore, the preamplifier is used to amplify the signal from the microphone until it is about as large as

1

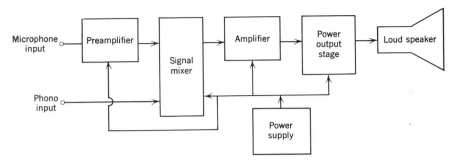

FIGURE 1.1
The block diagram of an amplifier.

the signal from the phono pickup. The signal mixer combines the two signals (microphone and phonograph) and is usually designed so an adjustment of one signal will not effect the other signal. The combined signals are then sent through an amplifier to increase the signal amplitude. This amplified signal is then used to drive the power output stage. This power output stage must produce enough electrical signal power to drive the loudspeaker at the required signal level. The power supply furnishes enough dc power to operate all of the amplifier stages. In a simple portable circuit, this power supply may be a single battery. However, in a high-quality, high-power amplifier, this power supply may be quite a sophisticated and complex circuit.

The block diagram for a simple radio receiver is given in Fig. 1.2. The radio signal is received by the antenna and amplified in the radio frequency (RF) amplifier. This RF amplifier is tuned to amplify only one radio signal

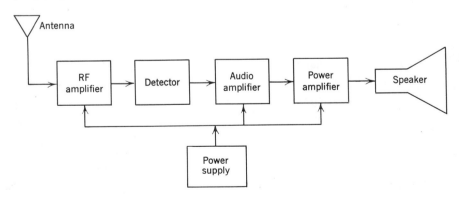

FIGURE 1.2
The block diagram of a radio receiver.

and to reject all the other radio signals which may be present. The detector converts the radio signal to an electrical audio frequency signal. This audio signal is then amplified in the audio amplifier. The amplified audio signal is used to drive the power amplifier stage which furnishes enough power to drive the loudspeaker. Again, a power supply is required to furnish dc power to the various amplifiers in the radio. While this receiver is very simple, more complex circuits are usually used. The block diagrams of some of these more complicated receivers are given in later chapters in this text.

When the engineer has drawn the block diagram of the system and has determined the required specifications for this system, he is then ready to design a circuit for each of the blocks. Some of these circuits may be available as integrated circuit modules. In these cases, the entire block (or in some cases, several blocks) may be available in a single, tiny package.

In order to use integrated circuits or to design the various stages of a system, one must understand how these circuits operate. Although most of this book is devoted to the understanding of electronic devices and circuits, as previously mentioned, these devices cannot be well understood without some insight into the characteristics of semi-conductor materials. The authors hope that Chapter 2 will provide at least this minimal required insight. Also, the basic laws which govern the flow of electric charges must be understood. Therefore, these basic electrical laws are briefly treated in the remainder of this chapter for the benefit of those who need to review them.

1.2   BEHAVIOR OF AN ELECTRON IN AN ELECTRIC FIELD

Electronics is defined as "the study of the properties and behavior of electrons under all conditions, especially with reference to technical and industrial applications."[1] Some of the electron properties are well known:

$$\text{mass} = 9.1066 \times 10^{-31} \text{ kilograms}$$
$$\text{charge} = 1.6019 \times 10^{-19} \text{ coulombs}$$

Some electron properties, such as size and shape, are unknown.

Other properties of an electron are somewhat paradoxical. As an example, electrons act as if they were small particles in some experiments; whereas in other experiments, electrons act as if they were waves. The nature of the experiment determines which property will be the more pronounced. However, as far as the electrical engineer is concerned, the electron *usually* behaves as if it were a small particle of matter.

---

[1] *Britannia World Language Dictionary*, Charles E. Funk, editor, Funk and Wagnalls Company, New York, N.Y.

Since electrons obey certain basic laws, their behavior can be predicted and controlled quite accurately. In fact, the entire science of electronics is based on man's ability to predict and control the movements of electrons. Thus, if an electron is assumed to be a small, charged particle, the basic laws of physics can be applied. One law states that the force on an object is equal to the time rate of change in momentum of the object. This law is written as follows:

$$\mathbf{F} = \frac{d(m\mathbf{v})}{dt} \tag{1.1}$$

where

$\mathbf{F}$ = the force, newtons
m = the mass, kg
$\mathbf{v}$ = the velocity, m/sec

Now, if the mass is assumed to be constant,[2] Eq. 1.1 can be reduced to the familiar form

$$\mathbf{F} = m \frac{d\mathbf{v}}{dt} = m\mathbf{a} \tag{1.2}$$

where $\mathbf{a}$ is the acceleration[3] in m/sec.[2] The equation for the electrostatic force on a charged body in an electric field is

$$\mathbf{F} = Q\mathscr{E} \tag{1.3}$$

where Q is the charge on the body in coulombs and $\mathscr{E}$ is the electric field intensity at the position of the charge in volts per meter. A combination of Eqs. 1.2 and 1.3 gives

$$Q\mathscr{E} = m\mathbf{a} \tag{1.4}$$

Since acceleration is the time rate of change of velocity, Eq. 1.4 can be written as

$$\frac{Q\mathscr{E}}{m} = \frac{d\mathbf{v}}{dt} \tag{1.5}$$

or

$$\mathbf{v} = \int \frac{Q\mathscr{E}}{m} dt \tag{1.6}$$

where $\mathbf{v}$ is the velocity in m/sec. Equation 1.6 will apply in all cases when mass is constant. The importance of this generality can be realized if we note that in some cases the electric field and perhaps even the charge are

[2] The more general case where mass is not a constant will be treated later in this section.
[3] A symbol set in boldface type indicates a vector quantity.

each a function of time. However, a simpler equation can be found if the charged body is in a uniform electric field that does not change appreciably while the electron is in this field. This condition does not restrict the use of this simpler equation to fields which do not change with time. Because of its small mass and relatively large charge, the electron travels at very high velocities. Consequently, as long as the $\mathscr{E}$ field does not change appreciably during the time an electron is in the field, the approximate formula will be valid. Under these conditions, the $\mathscr{E}$ term of the equation may be treated as a constant. If the charge on the charged body is constant and if the mass is assumed to be constant, Eq. 1.6 becomes

$$\mathbf{v} = \frac{Q\mathscr{E}}{m} \int dt \tag{1.7}$$

or

$$\mathbf{v} = \frac{Q\mathscr{E}}{m} t + \mathbf{v}_o \tag{1.8}$$

where $\mathbf{v}_o$ is the constant of integration and in this case is the initial velocity of the charged body at time $t = 0$.

Velocity is the time rate of change of distance, so Eq. 1.8 can be written as

$$\frac{d\mathbf{s}}{dt} = \frac{Q\mathscr{E}t}{m} + \mathbf{v}_o \tag{1.9}$$

or

$$\mathbf{s} = \int \frac{Q\mathscr{E}t}{m} dt + \int \mathbf{v}_o \, dt \tag{1.10}$$

where $\mathbf{s}$ is the distance in meters. Again, assuming Q, $\mathscr{E}$, and m are constant, Eq. 1.10 becomes

$$\mathbf{s} = \frac{Q\mathscr{E}}{2m} t^2 + \mathbf{v}_o t + \mathbf{s}_o \tag{1.11}$$

where $\mathbf{s}_o$ is the constant of integration or initial displacement at time $t = 0$.

Equations 1.4, 1.8, and 1.11 are very similar to the three corresponding equations of motion for a body of mass m in a gravitational field. In the electrical system, the product $Q\mathscr{E}$ assumes the role of the gravitational force.[4] As a consequence, a charged body in an electric field "falls" like a body in a gravitational field.

---

[4] In the foregoing derivation, the effect of gravity was ignored. Since the electrostatic force on an electron is usually very much greater than the gravitational force, very little error is introduced by ignoring the gravitational force. If large bodies with small charges are considered, both gravitational and electrostatic forces must be considered.

When a charged body falls through an electric field, the charged body loses potential energy as it gains kinetic energy. From the law of conservation of energy, the kinetic energy gained is equal to the potential energy lost. Then

$$QV = \tfrac{1}{2}mv^2 \tag{1.12}$$

or

$$v = \left(\frac{2QV}{m}\right)^{1/2} \tag{1.13}$$

where

Q = the charge on the charged body, coulombs
v = the velocity, m/sec, after falling through a potential difference V
V = the potential difference through which the charged body has fallen, volts
m = the mass of the body, kg

This equation will, of course, hold for any charged particle. If the numerical values for Q and m are substituted into Eq. 1.13, the equation for an electron becomes

$$v = 5.93 \times 10^5 (V)^{1/2} \text{ m/sec} \tag{1.14}$$

It is interesting to note in passing that if an electron has fallen through only 3×10^{-7} V, the electron will be traveling at approximately the speed of sound in air at sea level. The very large velocities encountered in work with electronics can thus be visualized.

PROBLEM 1.1   Through what potential must an electron be accelerated to reach a velocity of 18,000 mph which is required to place a satellite into orbit?

The foregoing equations are based on the assumption that the mass of the charged particles remains constant. However, when large velocities are encountered, the mass of an object changes. The manner in which the mass changes is now considered.

From Eq. 1.1

$$F = \frac{d(mv)}{dt} \tag{1.1}$$

If mass is assumed to be constant, Eq. 1.2 results. However, if mass is *not* assumed to be constant, Eq. 1.1 must be written as

$$F = m\frac{dv}{dt} + v\frac{dm}{dt} \tag{1.15}$$

Also, force multiplied by distance in this derivation is equal to kinetic energy W. Consequently, Eq. 1.15 can be modified to

$$dW = F \, ds = m \frac{dv}{dt} \, ds + v \frac{dm}{dt} \, ds \qquad (1.16)$$

Furthermore, since ds/dt is equal to v, Eq. 1.16 can be written as

$$dW = mv \, dv + v^2 \, dm \qquad (1.17)$$

Einstein has shown that mass and energy are related by the equation

$$E = c^2 m \qquad (1.18)$$

where E is the total energy, including mass energy, and c is the velocity of light (2.998×10^8 m/sec). The derivative of Eq. 1.18 may be taken to yield

$$dE = c^2 \, dm \qquad (1.19)$$

This change of energy (dE) is equal to the dW in Eq. 1.17. If this value of energy (dE) is substituted into Eq. 1.17,

$$c^2 \, dm = mv \, dv + v^2 \, dm$$

or

$$(c^2 - v^2) \, dm = mv \, dv \qquad (1.20)$$

This equation can be rearranged to yield

$$\frac{dm}{m} = \frac{v}{c^2 - v^2} \, dv \qquad (1.21)$$

If the mass of the object is m_o when the velocity is zero, the mass m at velocity v can be found by integrating Eq. 1.21 as shown below.

$$\int_{m_o}^{m} \frac{1}{m} \, dm = -\frac{1}{2} \int_{o}^{v} \frac{-2v}{c^2 - v^2} \, dv \qquad (1.22)$$

Carrying out the integration,

$$\ln m - \ln m_o = -\tfrac{1}{2} \ln(c^2 - v^2) + \tfrac{1}{2} \ln c^2 \qquad (1.23)$$

which can be written as

$$\ln m = \ln \left(\frac{m_o c}{(c^2 - v^2)^{1/2}} \right) \qquad (1.24)$$

or

$$m = \frac{m_o}{[1 - (v^2/c^2)]^{1/2}} \qquad (1.25)$$

The relationship between mass and velocity has now been achieved. Other useful equations may be established by inserting the value of m from Eq. 1.25 into Eq. 1.17. Then,

$$dW = F \, ds = \frac{m_o v}{[1 - (v^2/c^2)]^{1/2}} \, dv + v^2 \, d\left\{\frac{m_o}{[1 - (v^2/c^2)]^{1/2}}\right\} \qquad (1.26)$$

However, F is given in Eq. 1.3 as $Q \mathcal{E}$, hence, Eq. 1.26 can be written as

$$Q\mathcal{E} \, ds = \frac{m_o \, cv}{(c^2 - v^2)^{1/2}} \, dv + v^2 \frac{m_o \, cv}{(c^2 - v^2)^{3/2}} \, dv \qquad (1.27)$$

When the integral is taken of both sides of the equation,

$$Q \int_o^s \mathcal{E} \, ds = m_o c \int_o^v \frac{v}{(c^2 - v^2)^{1/2}} \, dv + m_o c \int_o^v \frac{v^3}{(c^2 - v^2)^{3/2}} \, dv \qquad (1.28)$$

Since $\int_o^s \mathcal{E} \, ds$ is the voltage V through which the charged particle has fallen, Eq. 1.28 becomes

$$QV = m_o c \left[-(c^2 - v^2)^{1/2} \right]_o^v + m_o c \left[2(c^2 - v^2)^{1/2} + \frac{v^2}{(c^2 - v^2)^{1/2}} \right]_o^v \qquad (1.29)$$

which reduces to

$$QV = m_o c^2 \left[\frac{1}{(1 - v^2/c^2)^{1/2}} - 1 \right] \qquad (1.30)$$

or

$$QV = (m - m_o)c^2$$

Equation 1.30 can be solved for v. After some manipulation,

$$v = c \left[1 - \frac{1}{\left(1 + V \dfrac{Q}{m_o c^2} \right)^2} \right]^{1/2} \qquad (1.31)$$

When the numerical values for c, Q, and m_o are inserted in Eq. 1.31, the velocity of an electron is

$$v = c \left[1 - \frac{1}{(1 + 1.966 \times 10^{-6} V)^2} \right]^{1/2} \text{ m/sec} \qquad (1.32)$$

This value of v can be inserted into Eq. 1.25 to yield

$$m = m_o(1 + 1.966 \times 10^{-6} V) \qquad (1.33)$$

A plot of velocity versus electron energy[5] is shown in Fig. 1.3. Figure 1.3 may be divided into three regions. Region 1 extends from 0 to about 40,000 eV. In this region, m is approximately equal to m_o, and Eq. 1.14 yields the velocity. In region 2, both mass and velocity change simultaneously. This region 2 extends from about 40,000 eV to approximately 3,000,000 eV. Equations 1.25 or 1.33 must be used to determine the mass, and Eq. 1.32 can be used to find the velocity in region 2. Region 3 extends from 3,000,000 eV upward. In region 3, the velocity is almost equal to the velocity of light and the mass can be calculated from Eq. 1.33. Of course, Eqs. 1.32 and 1.33 are applicable in any region, but accurate values are quite difficult to evaluate in region 1.

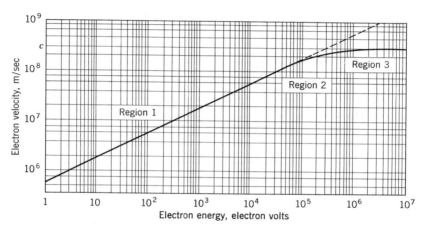

FIGURE 1.3
Velocity versus energy of an electron as found from Eq. 1.32.

PROBLEM 1.2   What is the velocity and mass of an electron which has been accelerated through $10,000V$?
Answer: $v = 5.93 \times 10^7$ m/sec $m = 1.02m_o$.

PROBLEM 1.3   What velocity must an electron have in order to exhibit a mass equal to the mass of a proton at rest? What potential must this electron have fallen through in order to achieve this velocity?
Mass of a proton = 1832 mass of an electron.
Answer: $V = 9.32 \times 10^8$ V, $v = 2.997 \times 10^8$ m/sec.

An example will help us visualize the use of the equations just derived. In this example, a cathode-ray tube will be used. The cathode-ray tube

---

[5] This electron energy is shown as electron-volts where an electron volt is the amount of energy an electron receives when accelerated through one volt.

is used in oscilloscopes, TV sets, radar systems, sonar units, curve tracers, spectrum analyzers, computer displays, hospital heart monitors, and a multitude of other uses. At present there is no solid-state device which offers serious competition for the cathode-ray tube. Thus, the electronics student should become familiar with these devices.

Figure 1.4 shows the principal parts of a cathode-ray tube. The electron gun (see Fig. 1.5) produces a narrow beam of electrons which have been

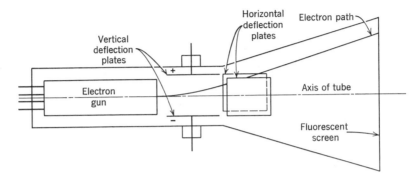

FIGURE 1.4
A cathode-ray tube.

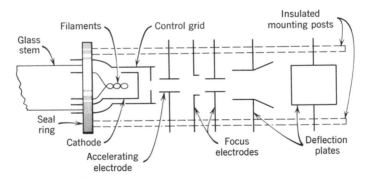

FIGURE 1.5
A typical electron gun.

accelerated through a potential difference of several thousand volts. As a result, the electrons are traveling at a high velocity. These electrons pass between the vertical deflection plates. If a potential difference exists between these deflection plates, the electrons will be attracted toward the positive plate and repelled by the negative plate. Thus the beam of electrons is deflected from the axis of the tube an amount proportional to the voltage difference between the deflection plates. Two sets of deflection plates are provided to obtain both vertical and horizontal deflection of the beam. After being deflected, the electron beam continues on and impinges on a fluorescent screen which emits visible light when struck by an electron. The cathode-ray tube thus allows the visual examination of a voltage waveform.

Example 1.1 A cathode-ray tube has the dimensions as shown in Fig. 1.6. The electrons (in the electron beam) have been accelerated through a

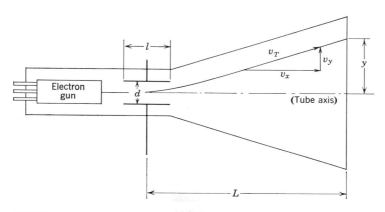

FIGURE 1.6
Configuration for Example 1.1.

voltage V_x when they enter the space between the deflection plates. The velocity in the x direction (along the axis of the tube) is, therefore,

$$v_x = \left(\frac{2qV_x}{m}\right)^{1/2} \tag{1.34}$$

where q is the charge on an electron (16.019×10^{-20} coulombs).[6] The distance traveled by an electron is equal to the velocity multiplied by the time. Hence the time a given electron will be between the deflection plates is given by

$$t = \frac{l}{v_x} = \left(\frac{m}{2qV_x}\right)^{1/2} l$$

The component of velocity at right angles to the axis of the tube v_y can be found by using the relationship

$$v_y = a_y t$$

[6] Most cathode-ray tubes have accelerating voltages, V_x of 18,000 V or less. Consequently, the mass can be assumed to be constant.

The acceleration in the y direction a_y can be found from the relationship

$$F_y = ma_y$$

But the force in the y direction is

$$F_y = -q\mathscr{E}_y$$

The electric field intensity can be found from

$$-\mathscr{E}_y = \frac{V_y}{d}$$

where d is the distance between the deflection plates in meters and V_y is the potential difference between the vertical deflection plates. This relationship assumes the electric field between the deflecting plates to be constant. This is a good approximation if the dimensions of the plates are large in comparison with the distance between them. When the foregoing substitutions are made for a_y and t, the formula for v_y becomes

$$v_y = \frac{qV_y l}{md}\left(\frac{m}{2qV_x}\right)^{1/2}$$

where v_y is the electron velocity in the y direction after passing through the deflection plates. This equation can be simplified to the relationship

$$v_y = \frac{lV_y}{d}\left(\frac{q}{2mV_x}\right)^{1/2} \tag{1.35}$$

Now, if we assume that after leaving the deflection plates, the velocities in the x and y directions are constant, we can write

$$\frac{l_y}{v_y} = \frac{l_x}{v_x}$$

where l_y is the distance the electron has traveled in the y direction and l_x is the corresponding distance in the x direction. In terms of Fig. 1.6, the foregoing equation becomes

$$\frac{y}{v_y} = \frac{L}{v_x}$$

or

$$y = L\frac{v_y}{v_x} \tag{1.36}$$

where L and y are defined in Fig. 1.6. When the expressions for v_x and v_y (Eq. 1.34 and Eq. 1.35) are substituted into this equation

$$y = L\, \frac{(lV_y/d)(q/2mV_x)^{1/2}}{(2qV_x/m)^{1/2}}$$

or

$$y = \frac{LlV_y}{2dV_x} \tag{1.37}$$

This equation is approximate, since certain simplifications were made in the derivation. For example, the fringing effect of the flux field at the edges of the plates was ignored. When this fringing is included, the equation is about the same as Eq. 1.37, except l is slightly longer than the deflection plates.

The reader may question the validity of Eq. 1.36, which infers that the electron has no velocity in the y direction until the electron reaches the center of the deflection plates. On reaching the center of the deflection plates, however, the electron suddenly acquires the full velocity v_y. This, of course, is certainly not the way the electron would behave. The electron actually follows a parabolic path while between the two deflection plates. However, if tangents are drawn to the parabolic curve where this curve enters and leaves the area between the two deflection plates, these tangents will intersect at the center of the deflection plates as shown in Fig. 1.6. There is, therefore, no error introduced due to the simplification of Eq. 1.36.

PROBLEM 1.4 The sensitivity of a cathode-ray tube is defined as the ratio of deflection distance (of the electron beam on the screen) to the deflecting voltage applied to the deflection plates. What is the sensitivity in cm/v of the cathode-ray tube shown in Fig. 1.6 if $d = 1$ cm, $l = 4$ cm, and $L = 10$ in.? Assume the screen is at the same potential as the point where the electron leaves the area between the deflection plates. The electron has been accelerated through 2000 V when it enters the space between the deflection plates.
Answer: 0.0254 cm/V.

PROBLEM 1.5 How far from the axis of the tube in Prob. 1.6 will an electron be when striking the screen if a potential of 10 V is applied between the deflection plates? If a particle with the same charge as an electron but with a mass 800 times as large as an electron is used in this tube, how far from the axis of the tube will this particle strike? (Assume 10 V between deflection plates.) Explain. What would be the effect of increasing the charge by a factor of 10?

1.3   THE OSCILLOSCOPE

The oscilloscope is the most useful single piece of electronic test equipment that has been developed. The block diagram for a simple oscilloscope is shown in Fig. 1.7.

The sawtooth voltage generator produces an output voltage signal as shown in Fig. 1.8. Since this voltage increases linearly with time, the electron beam sweeps linearly across the face of the cathode-ray tube in a horizontal direction. Thus, if a signal voltage is applied to the vertical input, this signal is amplified and applied to the vertical deflection plates of the cathode-ray tube. Consequently, the electron beam is deflected vertically in accordance to the vertical input signal and simultaneously moves linearly in the horizontal direction. Thus, a plot of the vertical signal versus time

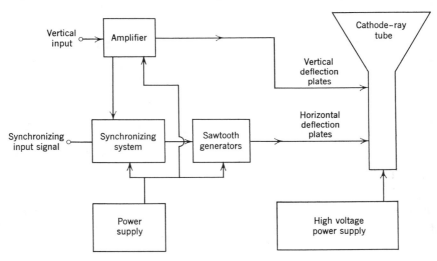

FIGURE 1.7
A block diagram of an oscilloscope.

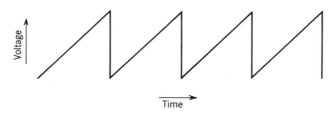

FIGURE 1.8
A sawtooth voltage waveform.

is traced on the face of the cathode-ray tube. As soon as one horizontal trace is completed, the beam returns to make the next trace. The synchronizing system is used to start each trace at the same voltage level on the vertical signal. Thus, if the vertical signal is a repeating function, each trace will follow the same path as the preceding trace. Hence, a fixed plot appears on the face of the cathode-ray tube. The cathode-ray tube requires one power supply to furnish the high voltage required to produce a bright trace. The amplifier, synchronizing system, and sawtooth generator use a lower dc voltage that is supplied from a separate power supply.

PROBLEM 1.6   An electron enters the space between two deflection plates with a velocity of 13.27×10^6 m/sec. If the configuration of the deflection plates is as given in Fig. 1.9, how far from the input end of the deflection plate will the electron strike the deflection plate?

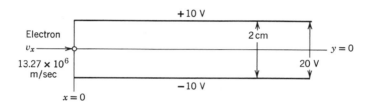

FIGURE 1.9
The configuration for Problem 1.8.

PROBLEM 1.7   How far will the electron of Fig. 1.9 have traveled in the y direction when it has traveled 10 cm in the x direction?

PROBLEM 1.8   The cross-section of a cathode-ray tube is as shown in Fig. 1.10. Answer the following questions.

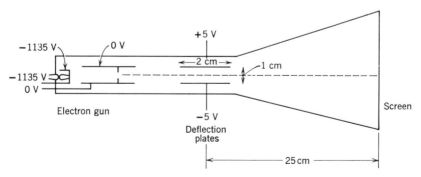

FIGURE 1.10
The configuration for Problem 1.8.

    a.   What is the electron velocity on leaving the electron gun?

    b.   How long is an electron between the deflection plates?

    c.   What is the electric field intensity between the deflection plates?

    d.   What is the velocity component in the vertical direction of an electron when it leaves the deflection plates?

    e.   How far is an electron beam deflected in the vertical direction when it arrives at the screen?

PROBLEM 1.9   When X rays and gamma rays are absorbed by an electron, the energy contained in the X ray is converted to kinetic energy. Determine the velocity of an electron which is excited by a 30,000 eV X ray.

Answer: 1.644×10^8 m/sec.

PROBLEM 1.10 Atomic bombs convert mass to energy. How much mass is converted to energy in the explosion of an atomic bomb which is rated equal to a 20 kiloton TNT explosion? Assume that the heat of combustion of 1 gram of TNT is 1,000 calories. (This size atomic bomb is known as a "nominal" size.)

2

Semiconductors

As the name implies, *semiconductors* are those materials which are neither good conductors nor good insulators. In fact, the resistivity, or conductivity, of a semiconductor can be readily changed by either changing its temperature or by adding small amounts of impurity to the pure, or intrinsic, material. The *thermistor* is a temperature-sensitive resistor made from a semiconductor material. Thermistors are often used as the temperature-sensing elements in automatic temperature controls, electronic thermometers, and so on.

The immediate goal of this chapter may appear to be the study of thermistors. However, some understanding of semiconductor materials is essential to the comprehension of the operating principles of many intriguing semiconductor devices, such as diodes, transistors, and silicon-controlled rectifiers. Therefore, the broader purpose of this chapter is to provide that insight into semiconductor materials which is essential to the pursuit of a satisfying study of semiconductor devices and circuits. The in-depth study of semiconductor materials and junction theory is properly the content of an entire course. Such a course usually either precedes or follows an electronics *circuits course*, which is the main content of this book. If your background includes a semiconductor theory course, you may wish to proceed immediately to Chapter 3, or Chapter 4, depending upon your background.

2.1 CONDUCTORS, INSULATORS, AND SEMICONDUCTORS

The chemical properties of an atom depend almost entirely on the number of electrons in the outer shell which is known as the valence shell. The atoms which have four valence electrons, such as carbon, silicon, and germanium are known as tetravalent atoms and are of special interest to us because they include the semiconductors. Most elements combine chemically

17

with other elements so that the outer electron shells contain eight electrons. It is possible for the tetravalent atoms to achieve this desired configuration by each atom sharing its valence electrons with its four adjacent neighbors. In this configuration, the pure tetravalent material forms a crystal. While the actual crystal is, of course, a three-dimensional structure, we can represent the structure schematically as shown in Fig. 2.1. The nucleus and the filled inner shells are represented by the circle with the +4 which is the net positive charge of this group. The valence electrons are represented by the negative signs and the bonds which result from the reduced energy and which hold the crystal together are represented by the curved lines. These bonds are known as *covalent bonds*. In this crystalline form, carbon is known as diamond.

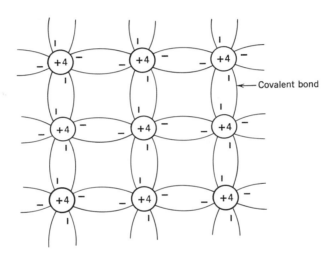

FIGURE 2.1
Two-dimensional representation of covalent bonds in a tetra-valent crystal.

An electrical conductor is a material which will permit electric charge to flow (current) when a potential difference is applied to the material. Therefore, conduction relies on the availability of charge carriers, such as electrons, which can move freely about in the material. A good conductor has many free electrons, but a semiconductor has comparatively few electrons at normal temperatures. Of course, an insulator does not have free electrons. At $0°K$ (absolute zero) all the electrons in a semiconductor crystal are bound by the covalent bonds as shown in Fig. 2.1. Therefore, a semiconductor is actually an insulator at $T = 0°K$.

A semiconductor can become a fairly good conductor, however, if its temperature is raised high enough to give a sufficient number of electrons enough energy to break their covalent bonds and drift freely through the crystal. This electron liberation process compares roughly with the liberation of a rocket from the earth by transferring the energy of the rocket fuel, first primarily to kinetic energy and then, as the rocket gains altitude, to potential energy until the rocket gains an altitude equal to many diameters of the earth. The gravitational pull of the earth is then negligible and the rocket is free to drift through the solar system.

The tetravalent atoms such as carbon, silicon, germanium, and tin do not form crystals of equal conductivity at a given temperature, because the energy required to break a covalent bond depends upon the closeness of the atomic spacing in the crystal. The smaller the atom, the closer the spacing, and the greater the energy required to break the covalent bonds. Figure 2.2

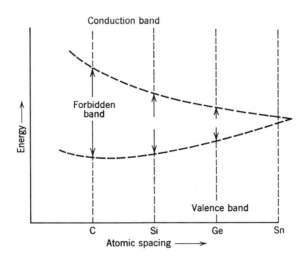

FIGURE 2.2
Energy bands of tetravalent crystals as a function of lattice spacing.

shows the relative atomic spacing and the relative energy required to break a covalent bond for carbon, silicon, germanium, and tin. Compared with carbon, silicon has one more filled shell, germanium has two more, and tin has three more filled shells between the valence shell and the nucleus; therefore, the atom size and spacing increases in that order. The energy represented by the forbidden band is the minimum energy required by a valence (band) electron to break a covalent bond and become a conduction (band) electron. Smaller amounts of energy cannot free the electron from the valence band. Therefore, electrons can pass through the forbidden band but cannot remain there.

At room temperature, germanium is neither a good insulator nor a good conductor; hence, it is a semiconductor.

PROBLEM 2.1   Of those tetravalent crystals shown in Fig. 2.2, which would you expect to be:

    a.   The best conductor?
    b.   The best insulator?
    c.   The best semiconductors?

Diamond is an insulator, and tin is a fairly good conductor at normal room temperatures. At very high temperatures, assuming the crystal to remain intact, they all become conductors, while at $0°K$, they all become insulators. Photons, or light energy, and strong electric fields may also break covalent bonds and elevate electrons to the conduction band, thus increasing the conductivity of the material.

2.2   CONDUCTION IN A SEMICONDUCTOR

As mentioned, heat or other sources of energy, such as light, cause valence band electrons to break their covalent bonds and become free electrons in the conduction band, thus producing modest conductivity in a semiconductor at normal temperatures. As the electron leaves the valence band, it creates a missing covalent bond which is known as a *hole*. This hole permits charge movement and, hence, conduction by the valence electrons as illustrated by Fig. 2.3. A nearby valence band electron can fill a hole and thus create another hole with practically no exchange of energy. If an observer could see electrons and holes, he would notice that the holes move rather than the electrons. The hole acts as a positive charge since the atom where the hole is located is missing an electron and has a net positive charge of 1.6×10^{-19} coul. Thus, electric conduction is caused by both free electrons and free holes, which are known as *charge carriers*. Unfortunately, the detailed mechanism by which hole conducting occurs is not as simple as indicated by the above discussion. However, this concept is sufficiently accurate for our purposes.

The conduction due to the electrons in the conduction band is a different process than the conduction due to the holes left in the valence band. In the intrinsic or pure semiconductor material, there are as many holes as there are free electrons because the free electron leaves a missing covalent bond or hole. An analogy (which is attributed to Shockley)

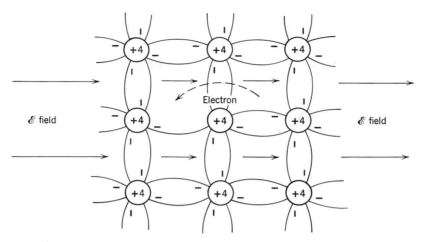

FIGURE 2.3
Illustration of conduction resulting from broken covalent bonds or holes.

might be used to illustrate the conduction processes. A parking garage with two floors has the lower completely filled with automobiles and the upper floor completely empty. Under these conditions, there can be no movement of automobiles on either floor. However, if one automobile is elevated from the lower to the upper floor, there can be motion of auto-mobiles on each floor. The auto on the upper floor may move freely over comparatively large distances. In contrast, the motion on the lower floor is accomplished by moving one car at a time into the available space. Hence, an observer near the ceiling of the first floor would see the "hole" move rather than the automobiles. The hole would have less mobility than the auto on the upper floor. Nevertheless, both would contribute to the total motion. Similarly, a free electron in a semiconductor has greater mobility, or ease of movement, than a hole.

We now know that electrical conductivity depends upon free charge carriers in a material and that semiconductors have two different charge carrier mechanisms known as electrons and holes. Our next goal is to determine the conductance (or resistance) of a piece of semiconductor material. But before that can be done, we need to know how many of each type of carrier there are (or the carrier densities) in the material, and how the conductivity is related to the carrier densities. Since the latter problem is the easiest, it will be considered first.

As you probably know,

$$\text{Resistance } R = \frac{V}{I} = \rho \frac{l}{A} \qquad (2.1)$$

where ρ is the resistivity of the material, l is the length, and A is the cross-sectional area of the conductor.

Similarly,

$$\text{Conductance } G = \frac{I}{V} = \sigma \frac{A}{l} \qquad (2.2)$$

where $\sigma = 1/\rho$ is the conductivity of the material.

Then, from Eq. 2.2,

$$\sigma = \frac{I/A}{V/l} = \frac{J}{\mathscr{E}} \qquad (2.3)$$

where $I/A = J$ is the current per unit cross-sectional area, or *current density*. The term $V/l = \mathscr{E}$ is known as the electric field intensity. The electric field intensity $\mathscr{E}$ is constant in a uniform conductor of constant cross-section (area A). Equation 2.3 is therefore Ohms law on a *per unit* basis.

Since current density is the rate of flow of charge through a unit cross-sectional area,

$$J = qn\bar{v} \qquad (2.4)$$

where n is the number of carriers (electrons for example) per cubic meter, q is the charge per carrier 1.6×10^{-19} coul for electrons or holes, and $\bar{v}$ is the average drift velocity of the carriers due to the electric field. Substituting the expression for J in Eq. 2.4 into Eq. 2.3,

$$\sigma = \frac{qn\bar{v}}{\mathscr{E}} \qquad (2.5)$$

One problem remains. Neither the drift velocity $\bar{v}$ or the electric field $\mathscr{E}$ are usually known. Fortunately, however, their ratio $\bar{v}/\mathscr{E}$ is known and is constant at a given temperature for a given material. This constant is known as the carrier mobility and has been given the symbol μ. Thus,

$$\mu = \frac{\bar{v}}{\mathscr{E}} \qquad (2.6)$$

As previously mentioned, the mobility of electrons is greater than the mobility of holes.

Since the total conductivity is the sum of the conductivities due to electrons and holes,

$$\sigma = qn\mu_n + qp\mu_p = q(n\mu_n + p\mu_p) \qquad (2.7)$$

where

  n is free electron density
  p is free hole density
  μ_n is electron mobility
  μ_p is hole mobility

The mobilities of electrons and holes in both silicon and germanium at 300°K are given in Table 2.1. As the reader may suspect, these mobilities are temperature dependent. The approximate temperature dependence for each mobility is also given in Table 2.1.

TABLE 2.1
Carrier Mobilities

Material	Mobility at 300°K $(cm^2/volt\text{-}sec)$	Approximate Mobility at a given Temperature
Ge		
Free electrons	3900	$4.9 \times 10^7 T^{-1.66}$ (100–300°K)
Holes	1900	$1.05 \times 10^9 T^{-2.33}$(125–300°K)
Si		
Free electrons	1350	$2.1 \times 10^9 T^{-2.5}$ (160–400°K)
Holes	480	$2.3 \times 10^9 T^{-2.7}$ (150–400°K)

† E. M. Conwell, *Proc. IRE*, **46** (June 1958), pp. 1281–1300.

Example 2.1   The density of free electrons in pure, or *intrinsic*, germanium at 300°K (room temperature) is about 2.4×10^{19} electrons per cubic meter. Let us determine the resistance of a bar of germanium which is 1 millimeter by 2 millimeters and 1 centimeter long.

  The conductivity (from Eq. 2.7) is $\sigma = q(n\mu_n + p\mu_p) = 1.6 \times 10^{-19}$ ($2.4 \times 10^{19} \times .39 + 2.4 \times 10^{19} \times 0.19) = 2.22$ mho/meter. The resistivity is $\rho = 1/\sigma = 1/2.22 = 0.45$ ohm-meter at 300°K. The resistance of the germanium bar is $R = \rho l/A = 0.45 \times 0.01/(0.001 \times 0.002) = 2{,}250 \ \Omega$.

PROBLEM 2.2   If the density of free electrons in pure silicon at room temperature is 1.7×10^{16} electrons/m³, find the resistance of the bar given in Example 2.1 if it is made of silicon.
Answer: 10 MΩ.

2.3   CHARGE CARRIER DENSITY IN A SEMICONDUCTOR

  The next problem to be treated is that of determining the free carrier density as a function of temperature and forbidden band energy in a semiconductor. A formal derivation, or even an informal one, is beyond the scope of this work. However, an analogy will be used which will make plausible the carrier density formula to be presented later.

As pointed out in Chapter 1, charge in an electric field behaves very much like mass in a gravitational field. Therefore, the air molecules, or atmosphere, above the earth may be used as a model to predict the density of electrons in the conduction of a semiconductor. Figure 2.4 is a sketch of the surface of the earth, in which an instrumented balloon is released vertically to measure the density of the atmosphere or air molecules.

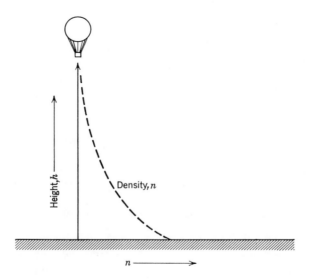

FIGURE 2.4
An arrangement for measuring air density as a function of altitude or height.

As a beginning point, we should recall that the atmosphere exists only because of its temperature which provides kinetic energy and mobility to the air molecules. At 0°K, there would be no atmosphere, only a thin film of solid oxygen and nitrogen on the surface of the earth. As the temperature is increased, the air molecules gain sufficient kinetic energy, which can be exchanged for potential energy, to place them high above the earth. If the temperature becomes sufficiently high, many molecules can reach escape velocity and go into outer space. This escape compares with thermionic emission where the electron is ejected from the heated surface of the emitting material.

Boltzmann has shown mathematically that if the temperature of a gas, such as the atmosphere, is constant, the density of the gas is an exponential function of the altitude h. His equation follows:

$$n = A \, e^{-(mgh/kT)} \tag{2.8}$$

where n is the number of molecules per cubic meter, A is a constant, m is the mass of a molecule, g is the gravitational constant, and k is Boltzmann's constant which relates temperature T to its equivalent energy. Note that mgh is the potential energy $w(h)$ of a molecule at altitude h. Therefore, Eq. 2.8 can be written as

$$n = A \, e^{-(w(h)/kT)} \tag{2.9}$$

This exponential density distribution, which is known as the Boltzmann distribution, has been theoretically and experimentally verified. However, the atmospheric temperature and composition vary with altitude, so the Boltzmann distribution given by Eq. 2.9, is more easily applicable to semiconductors than it is to the atmosphere.

Note two things about the ideal atmospheric model.

1. The Boltzmann equation gives the density at any altitude h providing the density at any given reference $h = h_0$ is known. The reference may be at sea level, on a mountain top, or at the center of the earth, assuming a hole could be dug to the center of the earth, and the temperature is uniform. For example, if the density of air is 5×10^{25} molecules per cubic meter at sea level, the constant $A = 5 \times 10^{25}$ if the reference for h ($h = 0$) is at sea level.

2. The Boltzmann distribution predicts the density only if the molecules are not restricted or forbidden for some reason. For example, the density of air molecules in Moffat tunnel in Colorado may be 4×10^{24} molecules per cubic meter, but the air density ten meters above the tunnel inside the mountain is zero if the soil and rocks inside the mountain completely excludes the air.

When the gas molecule analogy is applied to the free electron density problem in a semiconductor, the availability of occupiable electron energy levels or states must be considered and must weigh, or multiply, the Boltzmann density distribution function. For example, electrons cannot exist in the forbidden band. Also, as previously mentioned, each electron energy level or shell in an atom will accept only a limited number of electrons. When these factors are included and an integration is performed over the range of conduction band energy levels, the total number of electrons in the conduction band can be obtained for a unit volume of the material as follows

$$n_i = A \, T^{3/2} \, e^{-(W_g/2kT)} \tag{2.10}$$

where

    T is temperature in °K

    k is the Boltzmann constant $= 1.38 \times 10^{-23}$ joule/°K

    W_g is the forbidden band or *gap* energy (the energy required to break a covalent band) in joules

    A is a constant.

The gap energy W_g decreases as the temperature increases, because the crystal lattice spacing increases due to thermal expansion. This atomic spacing, and hence, W_g, is a linear function of temperature over the useful temperature range. Thus W_g can be written as

$$W_g = W_{g0}(1 - CT) \tag{2.11}$$

where

    C is the gap-energy temperature coefficient (2.8×10^{-4} for silicon and 3.9×10^{-4} for germanium)

    W_{g0} is the gap energy at 0°K

Then, Eq. 2.10 can be written

$$n_i = AT^{3/2}e^{-W_{g0}(1 - CT)/2kT}$$
$$= Ae^{(W_{g0}C/2k)}T^{3/2}e^{-(W_{g0}/2kT)} \tag{2.12}$$

But $e^{W_{g0}C/2k}$ is a constant, so it can be combined with A to give a new constant A'. Therefore,

$$n_i = A'T^{3/2} e^{-(W_{g0}/2kT)} \tag{2.13}$$

The constant A' can be either derived theoretically or determined experimentally. Commonly accepted values of W_{g0} and A' for silicon and germanium are given in Table 2.2. Since each free electron leaves a missing, or broken, covalent bond in the crystal lattice, the hole density p_i is equal to the free electron density n_i in the intrinsic crystal, and thus Eq. 2.13 can also be used to determine free hole density in the intrinsic crystal. Since the gap energies are given in electron volts, these values must be multiplied by the electronic charge $q = 1.6 \times 10^{-19}$ to convert them to joules.

TABLE 2.2
Constant A' and Gap Energy at 0°K

Material	W_{g0} electron volts	Constant A'
Silicon	1.20	3.88×10^{22}
Germanium	0.782	1.76×10^{22}

The free electron density can now be calculated for temperatures near 300°K (approximately room temperature), which is the normal operating range. For example, the density of free electrons in pure, or intrinsic, silicon at 300°K is 1.7×10^{16} electrons/m³. The conductivity of pure silicon at 300°K is 0.5×10^{-3} mho/m, and the resistivity is 2000 ohm-meter.

The free-carrier density, n_i or p_i in germanium and silicon, is plotted as a function of temperature in Fig. 2.5. Note the exponential shape of the curves.

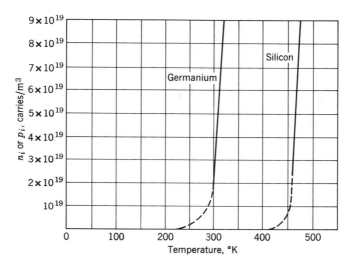

FIGURE 2.5
Free carrier density, n_i or p_i, in germanium and silicon.

PROBLEM 2.3   Using the value of resistivity determined for germanium (0.47 ohm-meter) at 300°K, what is the approximate ratio of the resistivity of silicon to that of germanium at this temperature?
Answer: 4250.

As one would expect, free carriers are continually being generated in a semiconductor, and their density would continually increase if recombinations did not also occur. A *recombination* is the process of filling a hole with a free electron, thus eliminating both. Since the opportunities for recombination are proportional to the product of the number of free electrons and the number of holes, the recombination rate is proportional to this product pn. Thus, at a given temperature, the recombination rate must equal the generation rate so the free carrier density remains essentially constant at the value predicted by Eq. 2.13. Therefore, if the temperature of an intrinsic semiconductor were increased enough to cause the rate of charge carrier generation to double, the carrier density would increase until the rate of recombination would also double. However, each carrier density would increase by only a factor of $\sqrt{2}$.

The variation of resistivity of a semiconductor with temperature is utilized in a temperature-sensitive resistor known as a *thermistor*.

PROBLEM 2.4   If a thermistor is made of intrinsic (pure) silicon, calculate its resistivity at $T = 400°K$.

Answer: 7.5 ohm-m.

PROBLEM 2.5   By what factor is the resistance of the silicon thermistor decreased if its temperature is raised from $300°K$ to $400°K$?

Answer: 272.

In addition to its resistivity, the resistance of a thermistor depends upon its length and cross-sectional area.

2.4   DOPED SEMICONDUCTORS

The conductivity of a semiconductor may be greatly increased if small amounts of specific impurities are introduced into the crystal. For example, if a few pentavalent atoms, such as arsenic or antimony, which have five valence electrons, are added to each million semiconductor atoms, the impurity atoms form covalent bonds with their neighbors as shown in Fig. 2.6. The pentavalent atom, after furnishing four electrons for the covalent bonds, has an extra valence electron which will not fit into the lattice arrangement. This extra electron is loosely bound to the atom at $0°K$ and requires very little thermal energy to become a free electron. In fact, as shown by Fig. 2.7, this extra electron is located at the level W_d and is about 0.01 eV below the conduction band. Therefore, nearly all the extra electrons gain sufficient additional energy to become free electrons in the conduction band when the crystal temperature is above about $50°K$. Observe that each impurity atom, essentially, provides a free electron *without* creating a hole. Thus, the pentavalent atom is known as a *donor* because it donates, or provides a free electron, and the semiconductor is said to be *doped* or *extrinsic*. Also note that the donor provides a *fixed* positive charge in the crystal lattice and the crystal is electrically neutral. The donor-doped crystal is known as *n*-type crystal because most of the charge *carriers* are negative.

Example 2.2   The effectiveness of the donor impurity in increasing the conductivity of a semiconductor will be emphasized by an example. Let us consider a silicon crystal which has approximately 1.7×10^{16} free

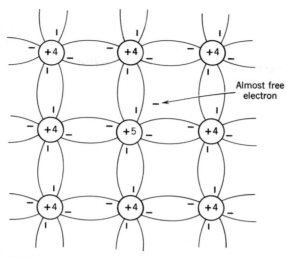

FIGURE 2.6
Semiconductor crystal with a donor impurity.

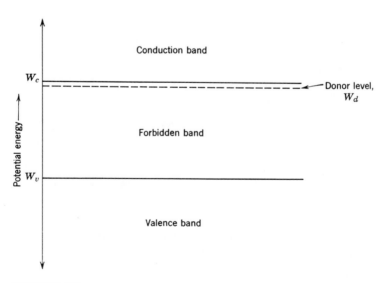

FIGURE 2.7
Energy level diagram for an *n*-doped semiconductor.

29

electrons which result from thermally broken covalent bonds, per cubic meter at 300°K, according to our previous calculation. Then let us assume that one part arsenic per million parts silicon was added when the crystal was formed. Since there are approximately 5×10^{28} silicon atoms per cubic meter, the density of arsenic atoms is 5×10^{22} atoms/m³. Then assuming all the donors to be activated (electrons are free), the conductivity of the doped silicon is 960 mho/m, which is 2×10^6 times as high as the 300°K conductivity of intrinsic silicon which we calculated to be 5×10^{-4} mho/m.

PROBLEM 2.6   Verify that the conductivity of the doped silicon is as given in Example 2.2.

As noted, the loosely bound electrons of the donor atoms become activated at about 50°K, and for the 5×10^{22} atoms/m³ doping concentration of the preceding example, the thermally generated carriers are few compared with the donors until the temperature reaches about 450°C. Fig. 2.8 is a sketch of the free electron and hole densities in this doped silicon as functions of temperature.

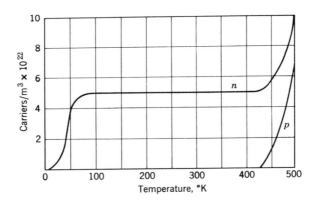

FIGURE 2.8
Charge carrier density in a donor-doped semiconductor as a function of temperature.

In a donor-doped crystal, the free electrons are known as the *majority* carriers, and the holes are known as the *minority* carriers. The majority carrier density n_n is the sum of the density due to the donors, N_d and the intrinsic density n_i. Thus,

$$n_n = N_d + n_i \tag{2.14}$$

The conductivity is

$$\sigma = q(\mu_n n_n + \mu_p p_n) \tag{2.15}$$

where p_n is the minority carrier density in the n-type crystal.

When the density of donor atoms N_d is large in comparison with the intrinsic carrier density n_i, as in Example 2.2, the total carrier density, at normal temperatures, is essentially equal to the doping density N_d. Then Eq. 2.15 can be simplified to

$$\sigma \simeq q\mu_n N_d \tag{2.16}$$

The minority carrier density is much lower than you may have suspected. As previously discussed, the rate of carrier recombination is proportional to the product of the free electrons and holes, but the rate of carrier generation is dependent only on the temperature. Therefore, when additional free electrons are introduced into the crystal by doping, the rate of recombination increases until the population of holes is reduced to the point where the recombination rate again equals the rate of thermal generation. For example, if the doping suddenly increased the free electron density by a factor of 100, the recombination rate would initially increase by a factor of 100 but would then decrease as the hole population decreases until the normal rate of recombination, which is equal to the rate of thermal generation, is reached. This normal rate is reached when the hole density reaches one percent of its value for the intrinsic material. An analogy of the reduction of minority carrier density may be the reduction of free-girl density in San Diego after the U.S. Navy comes ashore.

Since the rate of carrier generation is the same for both intrinsic and doped semiconductors, assuming the temperature to be the same, it should be apparent from the above discussion that the steady-state recombination rates must be the same. Therefore, since the recombination rate is proportional to the product of the free electrons and holes, the following relationship holds.

$$p_n n_n = p_i n_i = p_i^2 = n_i^2 \tag{2.17}$$

Therefore, in a donor-doped crystal, the minority carrier density p_n may be written

$$p_n = \frac{p_i^2}{n_n} = \frac{n_i^2}{n_n} \tag{2.18}$$

Since n_n is very nearly equal to the doping concentration, or density, N_d

$$p_n = \frac{n_i^2}{N_d} = \frac{p_i^2}{N_d} \tag{2.19}$$

PROBLEM 2.7   In Example 2.2, where the doping concentration was assumed to be 5×10^{22} donors/m³ and free electron density in the intrinsic crystal is 1.7×10^{16}, determine the minority carrier (or hole) density in the doped silicon.

Answer: 5.8×10^9 holes/m³.

P-doped semiconductors can be produced by adding a trace of trivalent material such as indium or gallium to the molten tetravalent germanium or silicon. Then, as the crystal forms, each impurity atom fits into the crystal structure but lacks one electron in forming covalent bonds with its neighbors (see Fig. 2.9). Therefore, each trivalent atom provides a hole in the

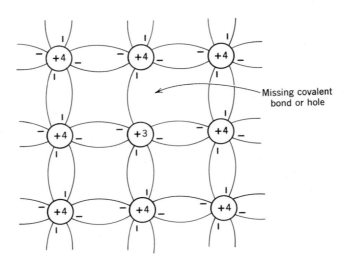

FIGURE 2.9
An acceptor-doped semiconductor crystal.

semiconductor crystal. At temperatures above about 50°K, the valence electrons have enough energy to transfer from a neighboring atom to fill a hole. Therefore, the hole moves freely through the crystal while a fixed negative charge remains with the trivalent atom. This atom is therefore known as an acceptor, since it accepts an electron from its neighbors, and the symbol for acceptor doping concentration or density is N_a. The trivalent-doped crystal is called p doped because most of the charge carriers are holes which behave as positive charges.

Since very little energy is needed to move an electron from the valence shell of a neighboring tetravalent atom into the hole produced by an acceptor, the acceptor level is very near the valence band as shown by the

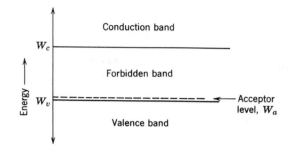

FIGURE 2.10
Energy level diagram showing the acceptor level.

energy level diagram of Fig. 2.10. Thus, as noted, at temperatures above about 50°K, the holes created by the acceptors circulate freely and behave as free positive charges. Therefore, the probability that the acceptor holes are filled at normal temperatures is about the same as the probability that the valence band holes are filled. Thus at normal doping levels of a few parts per million, the holes associated with an acceptor atom are nearly all filled and holes are moving at random through the crystal.

From the preceding discussion, the following relationships can be written for a p-doped semiconductor

$$p_p \simeq N_a \qquad (2.20)$$

$$n_p = \frac{n_i^2}{N_a} = \frac{p_i^2}{N_a} \qquad (2.21)$$

$$\sigma = q N_a \mu_p \qquad (2.22)$$

Since the conductivity of a semiconductor, especially silicon, can be controlled over such wide limits by doping, a complete circuit (including resistors, capacitors, and conductors) can be made in a single piece or chip of material by controlling the area and concentration of doping. In the following chapters you will learn how diodes and transistors are formed from p-doped and n-doped semiconductors and thus may be an integral part of the circuit formed in the single or monolithic chip.

PROBLEM 2.8 If you were making an integrated or monolithic circuit, how would you dope for:

    a. A conductor?
    b. An insulator?
    c. A resistor?
    d. A capacitor?

If a semiconductor is doped with equal concentrations of donor and acceptor atoms, the free electrons from the donors fill the holes created by the acceptors, and the material behaves as though it were intrinsic. Also, a semiconductor may be doped with unequal concentrations of donors and acceptors. Then the effective doping is the difference between the two concentrations and has the characteristic of the higher concentration.

2.5 GALLIUM ARSENIDE SEMICONDUCTORS

A semiconductor crystal can be formed from a compound composed of trivalent and pentavalent atoms. Gallium arsenide is the prime example of this type of compound and the only one in common use at the time of this writing. The gallium arsenide forms a crystal very similar to the silicon or germanium crystal except the gallium and arsenic atoms alternate in the crystal lattice so that each gallium atom is surrounded by four arsenic atoms and each arsenic atom is surrounded by four gallium atoms. Covalent bonds are formed as in a germanium or silicon crystal. The extra electrons from the pentavalent arsenic atoms fill the holes produced by the trivalent gallium atoms so that the crystal has the same general properties as an intrinsic semiconductor composed of tetravalent atoms.

The gallium arsenide (GaAs) crystal can be p doped by adding small amounts of group II atoms, such as zinc, with two valence electrons. These atoms replace the trivalent gallium atoms and provide an extra hole in addition to accepting the extra electron from an arsenic atom neighbor. Also, the crystal can be n doped by adding a small quantity of group VI atoms, such as selenium, with six valence electrons. These atoms replace arsenic atoms and provide a free electron in addition to donating an electron to fill the hole of a neighboring gallium atom.

The GaAs crystal is used to make semiconductor devices which are superior to either germanium or silicon in some respects. These advantages arise from the following characteristics of GaAs.

1.   The forbidden band, or gap, energy is 1.40 eV at 25°C compared with about 1.1 eV for silicon and 0.67 eV for germanium. This higher gap energy provides satisfactory performance of GaAs devices up to about 300°C with present technology, and this limit could be increased to 400°C or above if some aging problems can be overcome. The upper temperature limit for silicon devices is about 200°C, and germanium is useful up to about 100°C. The intrinsic GaAs crystal is a much better insulator at normal temperatures than silicon or germanium because of the higher gap energy. This characteristic is desirable for making high-quality integrated circuits and low-loss varactor diodes which are discussed in Chapter 3.

2. GaAs has much higher electron mobility than either silicon or germanium. Therefore, the upper frequency limit of a transistor, which is proportional to charge carrier mobility, can be much higher if the transistor is made of GaAs. The theoretical electron mobility in a pure, perfect GaAs crystal is 1.1 m^2/V-sec which is about three times as high as that of germanium and six times as high as electron mobility in silicon. Present refining processes can produce GaAs with about 0.8 m^2/V-sec electron mobility which is about twice that of germanium.

3. Electrons which return from the conduction band to the valence band transform their potential energy into electromagnetic radiation. In a GaAs crystal this radiation is in the visible spectrum (light). In contrast, the radiation from silicon and germanium is in the infrared (or heat) region. Therefore, GaAs can be used in low-voltage, solid-state display devices. An example of such a device is the light diode which is discussed in Chapter 3.

GaAs has other interesting properties which are useful in micro-wave devices which will not be discussed in this book. Some other semiconductor compounds with interesting properties have been produced but are not yet available commercially. These compounds, including GaAs, will undoubtedly become increasingly important in the world of semiconductors as improved techniques of refining and metallurgy are developed.

As a summary of the characteristics of intrinsic and doped semiconductors, the conductivity of intrinsic and doped germanium, silicon, and gallium arsenide is sketched as a function of temperature in Fig. 2.11. Note that the conductivity scale is logarithmic. Also note that the conductivity of the doped semiconductor *decreases* slightly as temperature increases, over the moderate

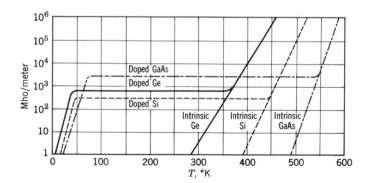

FIGURE 2.11
Conductivity of intrinsic and doped (10^{22} donors/m^3) Ge, Si, and GaAs as a function of temperature.

temperature range, because of the reduced mean-free-path of the carriers between collisions. Thus the carrier mobility μ decreases somewhat as temperature increases. This effect occurs in nearly all conductors and results in the positive temperature coefficient of resistance. The n-type doping was assumed to be 10^{22} donors/m³ for all materials.

TABLE 2.3
Some Fundamental Properties of Germanium, Silicon, and GaAs

Specific Property		Silicon	Germanium	GaAs
Atomic number		14	32	
Atomic weight		28.06	72.6	
Density (25°C) kg/m³		2.33×10^3	5.33×10^3	5.32×10^3
Melting point, °C		1420	936	1238
Relative dielectric constant $\varepsilon/\varepsilon_v$		12	16	11
Intrinsic carrier density n_i (300°K)/m³		1.7×10^{16}	2.4×10^{19}	1.4×10^{12}
Gap energy, W_{go}, electron volts:	0°K	1.2	0.782	1.63
	300°K	1.12	0.67	1.40
Carrier mobility (300°K) m²/V-sec	μ_n	0.135	0.39	1.10
	μ_p	0.05	0.19	0.05
Diffusion constant (300°K) m²/sec	D_n	33.8×10^{-4}	98.8×10^{-4}	43×10^{-3}
	D_p	13×10^{-4}	46.6×10^{-4}	13×10^{-4}

PROBLEM 2.9   Calculate the end-to-end resistance of a rectangular bar of intrinsic GaAs crystal at 300°K if $w = 2$ mm, $t = 1$ mm, and $l = 1$ cm. *Answer:* $R = 1.94 \times 10^{10}\Omega$.

PROBLEM 2.10   There are 4.43×10^{22} atoms per cc in gallium arsenide at 300°K. If donor-type atoms are added until the total number of free electrons is equal to $2 n_i$, what will be the free hole density and what will be the ratio of intrinsic atoms to donor atoms? *Answer:* $p_i = 0.5 n_i$; 3.16×10^{16}.

PROBLEM 2.11   Would impurity atoms (either donors or acceptors) either increase or decrease the rate of change of resistance with temperature? Why?

PROBLEM 2.12   Assume that you are developing an electronic fever thermometer which uses a thermistor in a bridge circuit to sense the temperature. You want the resistance of one leg of the bridge to decrease by only 10 percent as the temperature rises from 95°F to 105°F. A fixed resistor can be connected either in series or in parallel with the thermistor, of course. If the thermistor is made of intrinsic germanium, determine the resistances of both the thermistor and the fixed resistor if the total resistance of the combination is to be 10 kΩ at 95°F.

PROBLEM 2.13   You intend to mass produce the thermometer of Problem 2.12 and recognize that you could possibly save the cost of the fixed resistor by doping the germanium to produce the desired resistance change. Can this technique be used? If so, determine the approximate n-type doping density required.

3

Diodes

A diode is an electronic device that readily passes current in one direction but does not pass appreciable current in the opposite direction. Such a device is formed when a piece of n-type semiconductor is connected to a piece of p-type semiconductor. In actual production, a single crystal of semiconductor is formed with half of the crystal doped with acceptor impurities and the other half of the crystal doped with donor impurities. Other types of diodes are constructed by placing an electron emitter (cathode) and an electron collector (plate) in a vacuum or in a controlled gaseous atmosphere.

Diodes are used in a number of electronic applications such as converting ac to dc, detecting radio signals, performing computer logic, and other useful functions. In this chapter we will study the basic characteristics of diodes and examine some of their uses.

3.1 THE p-n JUNCTION

The basic mechanism of a p-n junction is illustrated in Fig. 3.1. Assume two pieces of semiconductor (one p- and one n-type material) exist as shown in Fig. 3.1a. The p material contains a certain density (N_a) of acceptor $(+3$ valence) atoms. These atoms are represented in the diagram by the negative signs with the circles around them. These atoms are fixed at a given location in the crystal structure. Associated with each acceptor atom is a free hole which is represented by a positive sign in the diagram. Of course, some of the intrinsic atoms have lost electrons due to thermal agitation. These thermal electrons and the holes they produced are represented, respectively, by the negative sign in the square and the positive sign in the square. Of course thermally generated carriers are indistinguishable from carriers produced by doping. Thus, the total density of majority carriers (p_p) is equal to the density of acceptor atoms (N_a) plus the density

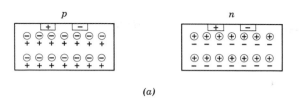

(a)

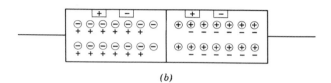

(b)

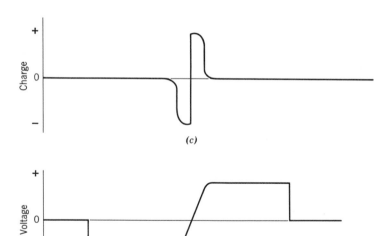

(c)

(d)

FIGURE 3.1
The p-n junction : (a) **representation of p- and n-type material; (b) representation of p-n junction; (c) charge distribution of the p-n junction; (d) potential distribution of the p-n junction.**

of thermal holes (p_i). As we have already noted, since $N_a \gg p_i$, $p_p \simeq N_a$. The density of minority carriers (n_p) are due to the thermally generated electrons, but as noted in Eq. 2.21, $n_p \ll n_i$.

The n material (Fig. 3.1a) contains a given density (N_d) of donor $(+5$ valence) atoms. The donor atoms (fixed in the crystal structure) are represented in the diagram by the positive signs with the circles around them. The associated free electrons are represented by negative signs. Again, the thermally generated electrons and holes are represented by the negative and positive signs in the squares. In the n material, the density of majority carriers (n_n) is equal to $(N_d + n_i)$. However, since $N_d \gg n_i$, $n_n \simeq N_d$. The p (or minority) carriers are due entirely to the thermally generated carriers. However, from Eq. 2.19 we note that $p_n \ll p_i$ if $N_d \gg p_i$. Of course, both pieces of crystal are electrically neutral.

When the p-type and n-type crystals are joined together, assuming their crystal structure to be continuous, a charge redistribution occurs as shown in Fig. 3.1b. Some of the free electrons from the n material migrate across the junction and combine with the free holes in the p material. Similarly, free holes from the p material migrate across the junction and combine with free electrons in the n material. Therefore, as a result of the redistribution, the p material acquires a net negative charge and the n material acquires a net positive charge as shown in Fig. 3.1c. This electric charge produces an electric field and a potential difference between the two types of material as shown in Fig. 3.1d. As previously mentioned, the diode manufacturing process does not actually employ the joining technique but usually employs selective doping of a continuous crystal.

The process by which the charges cross the junction is known as *diffusion*. This process may be visualized if only the action of the electrons from the n piece of material is considered. When the two pieces of material are joined, there is a concentration of free electrons in the n material but essentially none in the p material. The random motion of the electrons will allow some electrons to pass from the n to the p region. Because there are fewer electrons in the p region than in the n region, fewer electrons will tend to pass from the p to the n region. If no other forces exist, the diffusion process will continue until the concentration of electrons is uniform throughout the material. This process is the same process which occurs when two containers of dissimilar pure gases are joined. Eventually, both containers will contain a uniform mixture of both gases. However, because of the charges on the electrons, these electrons are attracted toward the positively charged n region. Similarly, the holes are attracted toward the negatively charged p region. Hence, an electric field is established (by the diffusion process), which inhibits the diffusion process.

Since diffusion is a common process, mathematical relationships for this process have been established. In simple terms, the rate at which electrons (gas atoms, etc.) will cross a given reference plane of unit cross-sectional area is proportional to the distribution gradient (the rate at which the density of electrons changes with distance) of the electrons. This relationship can be expressed mathematically as

$$\frac{dn}{dt} = D_n \frac{dn}{dx} \tag{3.1}$$

where dn/dt is the rate at which electrons will cross a given unit cross-sectional area, D_n is the constant of proportionality called the *diffusion constant*, and dn/dx is the distribution gradient.

Since electrical current is the rate at which charge passes a given reference, we can use Eq. 3.1 to find the diffusion current density. Thus,

$$J_n = q \frac{dn}{dt} = q D_n \frac{dn}{dx} \tag{3.2}$$

where J_n = current density due to electrons in A/m^2
q = the charge of an electron in C
D_n = the diffusion constant for electrons in m^2/sec

The diffusion constants for electrons in silicon, germanium, and gallium arsenide are given in Table 3.1.

TABLE 3.1
Diffusion Constants at 300°K in m^2/sec

	Ge	Si	GaAs
D_n	98.8×10^{-4}	33.8×10^{-4}	174×10^{-4}
D_p	46.8×10^{-4}	13.0×10^{-4}	17.4×10^{-4}

Holes also diffuse in a semiconductor. Thus, an equation similar to Eq. 3.2 can be developed for current flow due to hole diffusion.

$$J_p = q D_p \frac{dp}{dx} \tag{3.3}$$

where J_p = current density due to holes in A/m^2
D_p = the diffusion constant for holes in m^2/sec
dp/dx = the hole distribution gradiant.

Again, the diffusion constants for holes in silicon, germanium, and gallium arsenide are listed in Table 3.1. Of course, the diffusion constant is a function of material and temperature as well as the type of carrier.

In the actual *n-p* junction, as stated previously, both types of semi-conductive materials are neutral before the junction is made. After the junction is made, the *n* material loses electrons due to diffusion and gains holes.

However, as the electrons enter the *p* region, recombination with the numerous holes occurs. In fact, the hole population next to the junction is said to be "depleted," since practically all these holes are filled with electrons. The original electrically neutral nature of this region is replaced with negative ions wherever a $+3$ valence atom exists. Similarly, any holes which "travel" (or are *injected*) into the *n* material are immediately removed by recombination with the free electrons near the junction. In the *n* material, the free electrons are depleted and positive ions occur wherever a $+5$ valence atom exists. As a result of this action (the *p* material becomes negative and the *n* material becomes positive), an electric field or potential hill is established at the junction. In the steady-state condition, this field is just strong enough to inhibit the diffusion action of the electrons from the *n* material and the holes from the *p* material.

The potential hill at the junction is caused by the diffusion of the majority carriers, while the minority carriers which are swept across the junction tend to reduce the height of the potential hill. As a result, an equilbrium condition is reached in which the flow of minority carriers is equal to the flow of majority carriers. The current resulting from the diffusion of the majority carriers is known as the *injection current* or *majority current* and the current resulting from the minority carriers is known as the *saturation current* or *minority current*.

As previously mentioned, the charge distribution of Fig. 3.1c causes a potential difference across the *p-n* junction (Fig. 3.1d). This potential difference is a few tenths of a volt at room temperature. It seems possible that a current would flow in an external circuit if a conductor were connected to the open ends of the *p-n* combination since they are at different potentials. This supposition is *not* true, because the contact difference of potential at the junctions of the crystal and the external conductor causes the total emf around the closed circuit to be zero, and no current will flow. For example, if a conductor is brought into contact with an *n* crystal, the net diffusion of electrons will be from the crystal to the conductor until the crystal becomes positive with respect to the conductor and equilibrium is reached. Of course, the opposite is true if the conductor is connected to a *p*-type crystal.

The directions of the injection current I_I and the saturation current I_S are shown in Fig. 3.2b. In addition, the lengths of the current arrows indicate the relative magnitude of these currents. The net current flow in this case is zero, $I_I = I_S$. The majority carriers must have enough kinetic

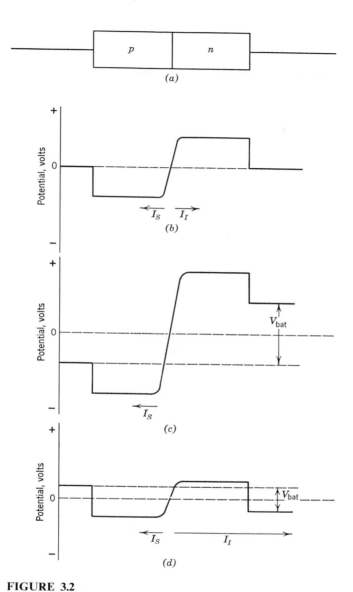

FIGURE 3.2
The effect of an external voltage on the potential distribution of a
p-n junction: (a) the p-n junction; (b) potential distribution with no
external battery; (c) potential distribution with a reverse-bias external
battery; (d) potential distribution with a forward-bias external battery.

energy to "climb" the potential "hill," whereas all minority carriers just "slide down" this potential hill with no kinetic energy required. In fact, the minority carriers gain kinetic energy in traversing the junction.

If an external battery is connected with the positive terminal to the n-type material and the negative terminal to the p-type material, the potential difference across the junction will increase as shown in Fig. 3.2c. The height of the potential "hill" is increased by the amount of the external battery voltage. Then only those majority carriers which have a very large amount of kinetic energy can "climb" the potential "hill." As a consequence, the injection current I_I is essentially eliminated as shown in Fig. 3.2c. Since the minority carriers still require no energy, the current I_S remains the same as in Fig. 3.2b. The external current flow is, therefore, mainly due to the thermally generated minority carriers of the semiconductor. This current I_S is known as the *reverse current* of the diode.

If the external battery is connected with the negative terminal to the n-type material and the positive terminal to the p material, the potential difference across the junction is decreased as shown in Fig. 3.2d. The potential "hill" is reduced by the magnitude of the external battery voltage, neglecting the IR drop in the crystals. As a result, a large number of the majority carriers are able to cross the junction. Therefore, the injection current I_I is greatly increased as shown in Fig. 3.2d. Since the minority carriers can still traverse the junction with no loss of energy (these carriers still gain kinetic energy but not as much as in Fig. 3.2b and Fig. 3.2c), the current I_S does not change. The *forward current* of the diode $I_I - I_S$ is, therefore, quite large. In fact, because of the large number of majority carriers as compared to the minority carriers, the forward current is usually thousands of times larger than the reverse current.

The diode symbol is shown in Fig. 3.3. The arrow points in the direction of positive current flow when forward bias is applied to the diode.

3.2 THE DIODE EQUATION

A relationship between the voltage across a diode and the current through it is quite easily derived. The Boltzmann distribution function has already been discussed in Chapter 2. In fact, Eq. 2.9 (repeated here for your convenience) gave the Boltzmann distribution

$$n = Ae^{-wh/kT} \tag{2.9}$$

for a gas. In this equation, the term wh represents the weight times the height or the potential energy of the gas. In electrical terms, potential energy is equal to qV where q is the charge on a carrier and V is the electrical potential above the reference value. Thus, if n_p is the density of

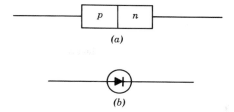

FIGURE 3.3
The p-n diode : (a) the physical representa-
tion of a diode; (b) the diode symbol.

electrons in the n material which have enough energy to surmount the potential hill and diffuse into the p material, we can write a relationship between n_p and temperature.

$$n_p = n_n e^{-qV_h/kT} \tag{3.4}$$

where n_n = the reference density of electrons in the n material $\simeq N_d$
V_h = the difference of potential or height of the potential hill between the n and p material
k = Boltzmann's constant
T = the temperature in °K.

Similarly, the density of holes p_n in the p material which have enough energy to cross into the n material is

$$p_n = p_p e^{-qV_h/kT} \tag{3.5}$$

Of course, p_p is essentially equal to N_a of the p material.

The injection current I_I is proportional to the total number of majority carriers which have enough energy to pass through the potential barrier V_h. Consequently, if K is the proportionality constant and $n_n = N_d$ and $p_p = N_a$, I_I is

$$I_I = K(n_p + p_n) = K(N_d + N_a)e^{-qV_h/kT} \tag{3.6}$$

But $K(N_d + N_a)$ is another constant K', since the doping concentrations are constant for a given diode. Then

$$I_I = K'e^{-qV_h/kT} \tag{3.7}$$

The constant K' can be evaluated in terms of the saturation current I_S, since the injection current is of equal magnitude but opposite direction to I_S when the externally applied voltage is zero. In this zero bias condition, we will let V_h be V_{ho}. Then,

$$I_I = -I_S = K'e^{-qV_{ho}/kT} \tag{3.8}$$

and

$$K' = -I_S e^{qV_{ho}/kT}$$

Substituting this value for K' into Eq. 3.4,

$$I_I = -I_S e^{q(V_{ho} - V_h)/kT} \tag{3.9}$$

But $V_{ho} - V_h = V$, the bias voltage or change in barrier height resulting from the external voltage source. Thus,

$$I_I = -I_S e^{qV/kT} \tag{3.10}$$

The diode current I is the algebraic sum of the injection current I_I and the saturation current I_S. Therefore,

$$I = I_I + I_S = -I_S(e^{qV/kT} - 1) \tag{3.11}$$

At normal temperatures, around $300°K$, the factor $q/kT = 38.7$. This number is usually rounded to 40, and used in Eq. 3.11 to obtain a simple equation which is adequately accurate for normal temperatures. Making this approximation,

$$I = -I_S(e^{40V} - 1) \tag{3.12}$$

If V is about 0.1 V or more negative, I is approximately equal to I_S. If V is 0.1 V or more positive, the exponential term of Eq. 3.12 is large in comparison with unity, and I is approximately equal to the injection current.

The maximum rated current for a given diode is determined by the heat dissipation qualities of the mounting system and the cross-sectional area of the diode. In addition, the type of material used in the diode has a bearing on the maximum current rating.

PROBLEM 3.1   A typical value of saturation current for a modest size silicon diode is $I_S = -10^{-8}$ A. Sketch a current versus voltage curve for this silicon diode by calculating values of diode current for forward-bias voltages of 0.1 V, 0.2 V, 0.4 V, and 0.6 V, respectively, and for reverse bias voltages of -0.1 V and -1.0 V.

3.3   SEMICONDUCTOR DIODE CHARACTERISTIC CURVES

A sketch of current versus voltage as given by Eq. 3.12 is shown as the solid line in Fig. 3.4. The plot of current versus voltage for an actual diode is similar to the expected curve, but considerable departure exists for large values of both negative and positive bias voltage. This departure is shown by the dashed line in Fig. 3.4. The departure in the reverse-bias region from a to b is partially due to leakage along the

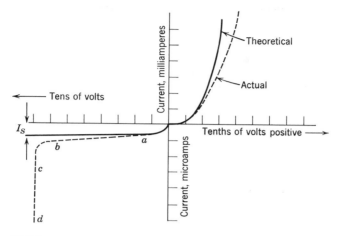

FIGURE 3.4
Characteristic curve of a typical germanium *p-n* junction diode.

surface of the junction. Other leakage paths exist through the mounting material which protects the junction. As a result, this departure is known as the *leakage* component.

At point *c* of Fig. 3.4, a new effect known as the *breakdown* of the crystal is noted. In this "breakdown" region, high currents may be passed, and they are limited only by the resistance in the external circuit. These high currents may generate enough heat to destroy the junction. Consequently, most diodes have a maximum reverse voltage rating. This maximum reverse voltage depends somewhat on the temperature of the diode. The "breakdown" mechanism is investigated in considerable detail in Section 3.10.

The departure from the theoretical curve in the forward bias region is caused by the *IR* drop in the doped crystal as previously mentioned. Thus, the actual diode behaves as the circuit shown in Fig. 3.5. In this

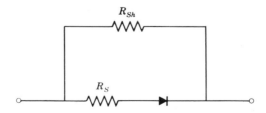

FIGURE 3.5
An equivalent circuit for a junction diode.

48   **Diodes**

figure, the diode has the characteristics given by Eq. 3.12. The resistor R_{Sh} accounts for the shunt leakage around the diode and is normally in the order of 10^5 Ω or higher. The resistor R_S represents the resistance in series with the junction. Typical values for R_S are in order of 10 Ω or less. The diode symbol represents a theoretical diode.

The ability of a diode to pass current in one direction and block current from flowing in the other direction is very useful. Let us consider a simple example.

PROBLEM 3.2 Assume that you own a 1961 model automobile. This car contains a dc generator and has a relay which connects the generator to the battery when the generator voltage is greater than the battery voltage. When the generator voltage drops below the battery voltage, the relay should open and disconnect the battery from the generator. Unfortunately the relay occasionally becomes stuck and does not open properly. When this happens, the battery rapidly discharges through the generator when the generator stops, causing damage to both the battery and the generator in addition to discharging the battery. This problem can be solved by connecting a diode in series with the battery and generator. The generator has a rating of 30 A so a 1N1183 diode which has a current rating of 35 A is suitable. Assume that you buy a 1N1183 and make the following measurements in the lab.

Forward Bias $V = 0.75$ V, $I = 30$ A
Reverse Bias $V = 12$ V, $I = 0.5$ mA
Reverse Bias $V = 0.1$ V, $I = 0.1$ mA

a. Draw the circuit diagram of the battery charging circuit, including the diode.

b. If the open circuit battery voltage is 12 V and the battery contains 0.1 Ω internal impedance, what must be the generator voltage in order to deliver 30 A to the battery?

c. The battery has a 60 ampere-hour capacity. If the battery were completely charged, how long could the car sit unused before the battery becomes discharged?

d. Determine the equivalent circuit parameters R_{Sh} and R_S for this diode (Fig. 3.5).

e. What must the generator voltage be when 20 A are delivered to the battery?

PROBLEM 3.3 Assume that 1 μA of current flows when 1 V reverse bias is applied to a junction diode. Calculate the approximate forward current when 0.256 V forward bias is applied to the *junction*. If the diode has

10 Ω internal resistance, what is the magnitude of external voltage required in order to obtain 0.256 V across the junction?

Answer: $V = 0.476$ V.

3.4   TEMPERATURE AND SEMICONDUCTOR MATERIAL EFFECTS ON THE CHARACTERISTIC CURVE

Equation 3.11 shows that the diode current is a function of temperature, which appears in the denominator of the exponent. However, the variation of the saturation current with temperature is much greater than the variation of the exponential term. The saturation current is proportional to the number of minority carriers which are swept across the p-n junction previously discussed and the number of minority carriers swept across the junction is proportional to the minority-carrier density in the material under consideration. Using the relationship $pn = n_i{}^2$ (see Eq. 2.17),

$$p_n = \frac{n_i{}^2}{n_n} \simeq \frac{n_i{}^2}{N_d} \tag{3.13}$$

and

$$n_p = \frac{n_i{}^2}{p_p} \simeq \frac{n_i{}^2}{N_a} \tag{3.14}$$

Then

$$J_s \simeq C(p_n + n_p) = Cn_i{}^2\left(\frac{1}{N_d} + \frac{1}{N_a}\right) = C'n_i{}^2 \tag{3.15}$$

where C and C' are constants. Using Eq. 2.10, we see that

$$pn = n_i{}^2 = B^2 T^3 e^{-W_g/kT} \tag{3.16}$$

The rate of change of J_s with temperature may be obtained by differentiating J_s with respect to temperature.

$$\frac{d(J_s)}{dT} = \frac{d(C'B^2T^3e^{-W_g/kT})}{dT} = C'B^2\left(\frac{3}{T} + \frac{W_g}{kT^2}\right)T^3e^{-W_g/kT} \tag{3.17}$$

The fractional increase of J_s per °K can be determined at a given temperature T by dividing Eq. 3.17 by J_s. Then

$$\frac{1}{J_s}\left(\frac{dJ_s}{dT}\right) = \frac{3}{T} + \frac{W_g}{kT^2} \tag{3.18}$$

For germanium at 300°K, the fractional increase of J_s per °K is

$$\frac{1}{J_s}\left(\frac{dJ_s}{dT}\right)\bigg|_{300°K} = \frac{3}{300} + 39\left(\frac{0.68}{300}\right) \simeq 0.1 \tag{3.19}$$

For silicon at 300°K, the fractional increase per °K is

$$\frac{1}{J_s}\left(\frac{dJ_s}{dT}\right)\bigg|_{300°K} = \frac{3}{300} + 39\left(\frac{1.1}{300}\right) \simeq 0.16 \qquad (3.20)$$

This derivation is valid for either the intrinsic or extrinsic (p- or n-type) material. Equation 3.19 shows that the saturation current in germanium near 300°K increases approximately 10 per cent for each degree K, but this rate decreases as the temperature increases. Therefore, the saturation current approximately doubles for each 10°K increase in temperature. Equation 3.20 shows that the saturation current in silicon doubles for approximately each 6°K increase. However, the saturation current of a silicon diode is usually much smaller than the leakage current. The saturation component is the temperature-sensitive component of reverse current. Consequently, the reverse current of an actual silicon diode will not double for a temperature increase of less than 10°K. Now, remember that intrinsic germanium has a much higher concentration of charge carriers than intrinsic silicon. Since the minority carriers are proportional to the intrinsic carriers squared (Eq. 3.14 or Eq. 3.13), silicon has a much lower concentration of minority carriers than germanium. Consequently, I_S is much higher in germanium diodes than in silicon diodes. As a result, germanium begins to be useless as a diode material at 100°C. In contrast, silicon does not generally begin to degrade too badly until temperatures of 200°C or higher are encountered. Gallium arsenide is usable to temperatures of 500°C.

The effect of increased temperature on the characteristic curve of a p-n junction diode is shown in Fig. 3.6.

In the diode equation (Eq. 3.12), we note that as I_S decreases, V must increase to produce the same diode current. Since I_S for the silicon diode is much less than I_S for the germanium diode, a higher forward voltage drop exists across the silicon diode than across a germanium diode with a similar current flowing. This difference is illustrated in Fig. 3.7.

PROBLEM 3.4 (a) If a germanium diode has a saturation current I_S of 10 μA at room temperature (300°K), what will the saturation current be at 400°K? (b) A silicon diode has a saturation current I_S of 0.01 μA at room temperature. What will the saturation current of this diode at 400°K? Assume the rate of increase to be constant at the 300°K value.

PROBLEM 3.5 The diodes of Prob. 3.4 have 0.2 V forward bias across the junction. (a) Find the current at room temperature if the diode is germanium. (b) Find the current at 400°K if the diode is germanium. (c) Find the current at room temperature if the diode is silicon. (d) Find the current at 400°K if the diode is silicon.

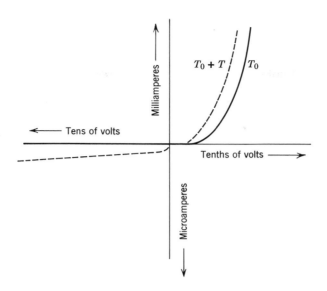

FIGURE 3.6
The effect of increased temperature on the characteristic curve of a *p-n* junction diode.

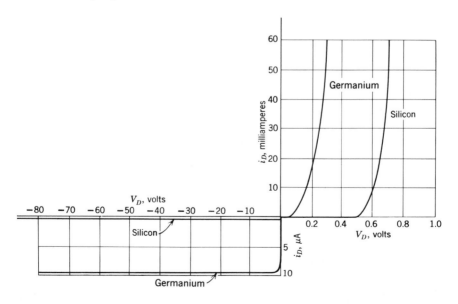

FIGURE 3.7
A comparison of silicon and germanium diode characteristics.

Answer: (*a*) 23 mA; (*b*) 3.43 A; (*c*) 23 μA; (*d*) 3.43 mA (if I_S doubles every 10°C).

3.5 GRAPHICAL SOLUTION OF DIODE CIRCUITS

We now have sufficient background to consider some typical diode applications. Two basic methods of analysis and design will be described. One approach (described in Section 3.6) uses an *equivalent circuit* for the diode. The other approach, which will be used in this section, is a *graphical method.*

In the graphical method of solution, the current-versus-voltage characteristics of the diode are expressed in graphical form. These curves may be obtained from a transistor curve tracer. These plots of current versus voltage are known as *characteristic curves*. The characteristic curves furnished by the manufacturer represent *average* diodes of a given type. However, if very accurate curves are required (or if the curves are not given), the characteristic curve for the particular diode in question can be plotted from measured voltages and currents. If a *curve tracer* and a polaroid camera are available, accurate plots can be obtained quickly and easily.

If a circuit is connected as shown in Fig. 3.8, the voltage equation can be written as

$$v_I = v_D + i_D R_L \tag{3.21}$$

where v_I = the input voltage
v_D = the voltage across the diode
i_D = the diode current.

This equation can also be written as

$$i_D = \frac{v_I}{R_L} - \frac{1}{R_L} v_D \tag{3.22}$$

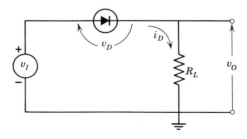

FIGURE 3.8
A diode clipping circuit.

From analytical geometry, the equation for a straight line is

$$y = b + mx \qquad (3.23)$$

Note that Eq. 3.22 has this same form where i_D represents y and v_D represents x. Thus, Eq. 3.22 represents a straight line in the v_D and i_D plane with an intersection on the $i_D = 0$ line at $v_D = v_I$. The intersection on the i_D axis occurs at $i_D = v_I/R_L$. The slope of this line, known as a *load line*, is $-1/R_L$. Now, since the characteristic curve of the diode is also plotted on the v_D and i_D plane, the intersection of the diode curve and the load line provides the solution of the circuit current and voltages. To help clarify this concept, let us consider an example

Example 3.1   A circuit is connected as shown in Fig. 3.8. If $R_L = 5\ \Omega$ and v_I has the waveform shown in Fig. 3.9, let us determine the waveform of v_O. The characteristic curves for the diode are given in Fig. 3.10. When

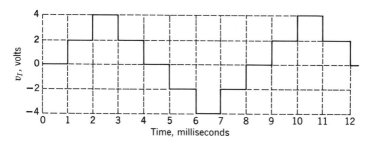

FIGURE 3.9
A plot of v_I for Example 3.1.

v_I is 0 V, no current flows through the circuit. Hence, v_O is 0 if v_I is 0. If v_I is 2 V, we can use Eq. 3.22 to draw a load line on the characteristics of Fig. 3.10. As noted before, this load line is a straight line. Thus, if we can locate two points on the load line, we can draw in the rest of the line. Now, from Eq. 3.22 if $i_D = 0$, $v_D = v_I = 2$ V. Similarly, if $v_D = 0$, $i_D = v_I/R_L = 2/5 = 0.4$ A. These two points $i_D = 0$, $v_D = 2$ V and $v_D = 0$, $i_D = 0.4$ A are joined by the load line. The intersection of the load line and the characteristic curve (point A) is the actual point of operation. Thus, $v_D = 0.5$ V and $I_D = 0.3$ A. Then, the voltage drop v_O across R_L is $(v_I - v_D) = 2 - 0.5$ V or we could also use the relationship $v_I = i_D R_L = 0.3 \times 5 = 1.5$ V. A similar construction for $v_I = 4$ V produces the load line from $i_D = 0$ A, $v_D = 4$ V to $v_D = 0$, $i_D = 0.8$ A. The solution in this case is $v_D = 0.75$ V and $v_I = 4 - 0.75 = 3.25$ V.

When $v_I = -2$ V, the load line is drawn from $v_D = -2$ V, $i_D = 0$ A to $v_D = 0$ V, $i_D = -0.4$ A. However, the intersection for this load line and the

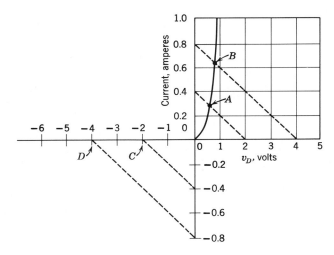

FIGURE 3.10
The characteristics of the diode in Example 3.1.

characteristic curve occurs when $i_D = 0$ (point C in Fig. 3.10). Thus, $v_O = 0$ V. Similarly, when $v_I = -4$ V, $v_O = 0$ V. Thus, a plot of v_O is as shown in Fig. 3.11. (Note that since the slope is equal to $-R_L$, all of the load lines are parallel.)

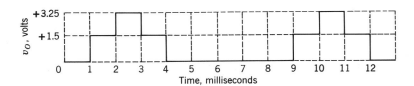

FIGURE 3.11
A plot of v_O for Example 3.1.

The foregoing circuit is known as a *clipper circuit*, since the negative portion of the input voltage waveform was clipped off. If the polarity of the diode is reversed, the positive portion of the input wave will be clipped.

PROBLEM 3.6   Repeat Example 3.2 if R_L is changed to 10 Ω.

PROBLEM 3.7   The diode of Example 3.1 is connected as shown in Fig. 3.8. Determine the value of R_L if $v_O = 4.6$ V when $v_I = 5$ V.

Answer: $R_L \simeq 20$ Ω.

3.6   DIODE EQUIVALENT CIRCUITS

As mentioned in Section 3.5, an equivalent circuit can be used for the analysis and design of diode circuits. In this approach, the diode is replaced by linear circuit elements (resistors, capacitors, batteries, etc.) and switches, the combination of which has *approximately* the same current-versus-voltage characteristics as the diode. Thus, we construct a *model* which electrically performs like the actual device. With the equivalent circuit used in place of the diode, standard circuit analysis can be used to determine the circuit performance.

To illustrate the process for obtaining a suitable model, consider the diode characteristic curve shown in Fig. 3.12*b*. This characteristic curve is approximated (for the range of 0 V to 0.35 V) quite closely by the dashed line 0*A* of Fig. 3.12*b*. This dashed line is a graphical representation of a linear relationship between voltage and current. Since a resistor has the same linear relationship, the diode can be approximated (as least in

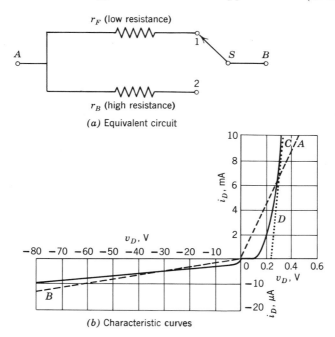

(*a*) Equivalent circuit

(*b*) Characteristic curves

FIGURE 3.12
The characteristic curve of a diode and an equivalent circuit for this diode.

the voltage range of 0 to 0.35 V) by a resistor. In Fig. 3.12b, 6 mA of current flows when v_D is 0.28 V. Hence, the line $0A$ represents a resistance of $0.28/.0006 \simeq 47$ Ω. This dashed line is known as the *dc forward resistance* of the diode (at 6 mA) and is given the symbol r_F. If small voltage fluctuations (in the order 0.1 V or so) occur around a given location on the characteristic curve, a second type of equivalent resistance can be found. Then, the *change* of diode voltage with diode current becomes the important relationship. Thus, if i_D is the instantaneous value of diode current and v_D is the instantaneous value of diode voltage,

$$r_f = \frac{\Delta v_D}{\Delta i_D} \tag{3.24a}$$

or in the limit

$$r_f = \frac{dv_D}{di_D} \tag{3.24b}$$

where r_f is known as the *small-signal* or *incremental forward resistance*. Thus, if voltage on the diode of Fig. 3.12 varies about the $v_D = 0.28$ V value, the incremental forward resistance would have the slope given by the dotted line CD of Fig. 3.12b. In this case, $r_f = \Delta v_D/\Delta i_D = 0.1$ V$/0.010 = 10$ Ω when $v_D \simeq 0.28$ V.

If the internal ohmic resistance of the diode (R_S in Fig. 3.5) is ignored, there is another way to calculate r_f. Then, the diode equation (Eq. 3.11) expresses the relationship between diode current and diode voltage.

$$I = -I_S(e^{qV/kT} - 1) \tag{3.11}$$

Since dI/dV is the incremental conductance (the inverse of incremental resistance), let us take the derivative of Eq. 3.11.

$$\frac{dI}{dV} = -I_S\left(\frac{q}{kT}\right)(e^{qV/kT}) \tag{3.25}$$

Let us add and subtract ($I_S q/kT$) from this equation.

$$\frac{dI}{dV} = \frac{q}{kT}[-I_S e^{qV/kT} + I_S - I_S] \tag{3.26}$$

or

$$\frac{dI}{dV} = \frac{q}{kT}[-I_S(e^{qV/kT} - 1) - I_S] \tag{3.27}$$

Note that one term inside the bracket is now equal to I (Eq. 3.11). Thus, the incremental conductance, g_d, is

$$g_d = \frac{dI}{dV} = \frac{q}{kT}(I - I_S) \qquad (3.28)$$

However, if we have very much forward bias, $I \gg I_S$. Then from Eq. 3.28, the forward incremental conductance g_f is

$$g_f = \frac{dI}{dV} \simeq \frac{q}{kT} I \qquad (3.29)$$

The term $g_f = 1/r_f$. Therefore, from Eq. 3.29

$$r_f \simeq \frac{kT}{qI} \qquad (3.30)$$

But as we have already noted, $q/kT \simeq 40$ at room temperature, so $kT/q \simeq 1/40 = 25 \times 10^{-3}$. Thus, Eq. 3.30 can also be written as

$$r_f = \frac{25 \times 10^{-3}}{I} = \frac{25}{I \text{ (in mA)}} \ (\Omega) \qquad (3.31)$$

Note that if Eq. 3.31 is used for a diode current of 6 mA, the value of r_f will be 4.17 Ω. From the characteristic curve (Fig. 3.12b) we found $r_f = 10 \ \Omega$. Consequently, the internal resistance (R_S in Fig. 3.5) of this diode must be equal to $10 - 4.7 = 5.83 \ \Omega$.

Thus far, we have only considered the diode with forward bias applied. If reverse bias is applied, the characteristic curve can be approximated by the dashed line $0B$ in Fig. 3.12b. This dashed line represents a resistance which is known as the *back-biased resistance* r_B and for this diode, $r_B = v_D/i_D = -65 \text{ V}/-10^{-5} = 6.5 \text{ M}\Omega$.

From the foregoing discussion, an equivalent circuit for a diode can be drawn as shown in Fig. 3.12a. The switch S is in position 1 if the diode is forward biased and in position 2 if the diode is reverse biased. Manufacturers often list the values of r_F and r_B (or data from which these values can be determined) for their diodes. In many applications, diodes can be considered as simple switches ($r_F = 0$ and $r_B = \infty$) which are ON when forward bias is applied and OFF when the diodes are reverse biased. Under these conditions, we refer to the diode as an *ideal diode*.

Examination of Fig. 3.12b indicates that the dotted line approximates the actual characteristic curve much better than the dashed line. However, if we use the dotted curve, i_D must be zero when v_D is still slightly positive. If silicon diodes are used, this offset is more pronounced. Thus, the characteristic curve of a silicon diode appears as the solid curve in Fig. 3.13b. An equivalent circuit which approximates this electrical behavior is shown in Fig. 3.13a. In this equivalent circuit, the diode symbol in the square represents an ideal diode (this ideal diode replaces the switch in

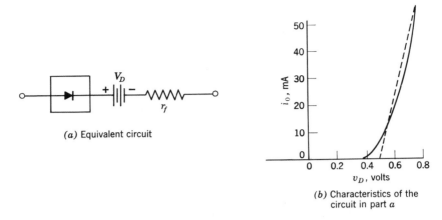

(a) Equivalent circuit

(b) Characteristics of the
circuit in part a

FIGURE 3.13
A more accurate equivalent circuit for a diode and its characteristic curve.

Fig. 3.12a). The battery V_D keeps the ideal diode reverse biased until the potential V_D is reached. Then, when the voltage across the diode exceeds V_D, the diode becomes forward biased, and current flows through the diode. The dashed line in Fig. 3.13b represents the characteristics of this equivalent circuit. Note that V_D is about 0.5 V for this particular diode. (Germanium diodes have a V_D voltage of about 0.1 V.) Of course, r_f is found from Eq. 3.24a in this configuration. In Fig. 3.13b, the value of $r_f = \Delta v / \Delta i =$ 0.2 V/40 mA = 5 Ω. We have omitted r_B in Fig. 3.13a. If r_B is small enough to require inclusion, it would be connected in parallel across the ideal diode.

The equivalent circuit shown in Fig. 3.13a is more accurate than the equivalent circuit shown in Fig. 3.12a but does require the additional voltage source v_D to obtain this increased accuracy. The engineer must determine how much accuracy he requires and then develop models or equivalent circuits which will produce this accuracy.

A few simple examples may help illustrate the usefulness of the equivalent circuits.

Example 3.2   A diode with the characteristic curve shown in Fig. 3.12 is connected as shown in Fig. 3.14. Let us determine the current which flows through R_L.

We have already determined values for r_F (47 Ω) and r_B (6.5 MΩ), but we should check the validity of these values. The curve given by r_F was quite accurate up to $i_D \simeq 10$ mA. We note that if the diode in Fig. 3.14 is replaced by a short circuit, the current will not rise above 100 V/10,000 Ω = 10 mA. Consequently, the equivalent circuit will be valid for our use.

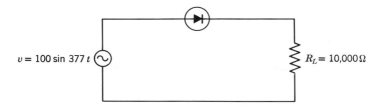

FIGURE 3.14
The circuit for Example 3.2.

The voltage v is positive for t from 0 to $1/120$ sec. Thus, the diode is forward biased and the equivalent circuit during this period of time is as shown in Fig. 3.15a. The current i in the circuit is given by the relationship

$$i = \frac{100}{10,047} \sin 377\, t \qquad (3.32)$$

For t between $1/120$ and $1/60$ seconds, the voltage v is negative and the diode is reverse biased. Under these conditions, the equivalent circuit given in Fig. 3.15b is valid. Then, i is

$$i = \frac{100}{6,510,000} \sin 377\, t \qquad (3.33)$$

Now, since the $\sin 377\, t$ is positive for t greater than $1/60$ sec but less than $3/120$ sec, Eq. 3.32 will also apply for this time interval. In fact, Eq. 3.32 applies for $t > n/60$ but less than $(2n + 1)/120$. By similar reasoning, Eq. 3.33 is valid for t greater than $(2n + 1)/120$ but less than $(n + 1)/60$ where n is 0, 1, 2, 3, 4, and so forth.

A plot of the current in R_L can be made using Eq. 3.32 and Eq. 3.33. This plot is shown in Fig. 3.16. Again, the negative portion of the input signal has been clipped off.

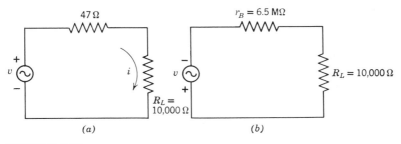

FIGURE 3.15
The equivalent circuit for Fig. 3.13: (a) the circuit when v is positive; (b) the circuit when v is negative.

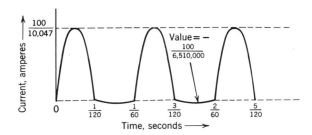

FIGURE 3.16
The current output of the circuit in Fig. 3.15.

The average current can be found by integrating the total current in one cycle and dividing by the time for one cycle. Therefore, if the reverse current is assumed to be negligible,

$$I_{ave} = \frac{\int_0^{1/120}(100/10{,}047)\sin 377t\; dt}{1/60} \tag{3.34}$$

or

$$I_{ave} = \frac{100}{(10{,}047)\pi} \tag{3.35}$$

The example just considered is known as a *half-wave rectifier*. The rectifier circuits are used in power supplies which convert ac to dc power. In this example, negligible error would be introduced by assuming an ideal diode ($r_F = 0$ and $r_B = \infty$).

PROBLEM 3.8   Repeat Example 3.2 if $R_L = 200\;\Omega$ and $v = 2\sin 377\,t$. How much error would be introduced by assuming an ideal diode in this problem? Under what conditions can you assume an ideal diode?

Now, let us consider a further example.

Example 3.3   A diode is connected in a circuit as shown in Fig. 3.17. The diode has a forward resistance r_F of 100 Ω and a reverse resistance r_B of 1,000,000 Ω. If a voltage with the waveform shown in Fig. 3.17 is applied to the input, plot the output voltage v_o.

The capacitor in the circuit is assumed to be completely uncharged at the time $t = 0$. Then, at time $t = 0+$, the circuit would be as shown in Fig. 3.18a. This circuit is a simple series RC circuit whose loop equation can be written as

$$10 = r_F i + \frac{1}{C}\int i\; dt \tag{3.36}$$

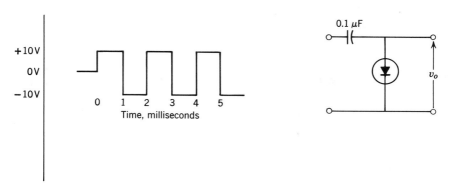

FIGURE 3.17
A diode clamper circuit.

Solving this equation for i,

$$i = \frac{10}{r_F} e^{-t/r_F C} \tag{3.37}$$

The term $r_F C$ is known as the time constant of the circuit and is usually given the symbol τ.

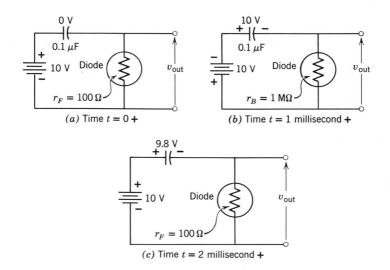

FIGURE 3.18
The equivalent circuits for Fig. 3.17.

Since v_o is equal to i times r_F,

$$v_o = 10e^{-t/\tau} \qquad (3.38)$$

Thus at time $t = 0$, the output voltage is 10 V, but this output voltage decays rapidly (with a time constant of 10 μsec) to essentially zero. This portion of the output voltage is plotted (for 1 msec $> t > 0$) in Fig. 3.19.

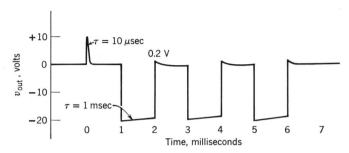

FIGURE 3.19
A plot of v_{out} for Fig. 3.17.

At time $t = 1$ msec, the capacitor is charged to essentially 10 V when the input voltage shifts to become -10 V. Consequently, the equivalent circuit is as shown in Fig. 3.18b, where

$$v_o = -20e^{-t'/r_B C} \qquad (3.39)$$

where t' has the value of 0 at the time $t = 1$ msec. Thus, the output voltage at time $t = 1$ msec is -20 V and decays toward 0 potential. However, the time constant $r_B C$ is large (10^{-1} sec) compared to one msec (the time during which this circuit is valid), so the capacitor does not discharge appreciably before the input voltage reverses again. Actually,

$$v_o = -20e^{-0.01} = -19.8 \text{ V} \qquad (3.40)$$

for the output voltage at the end of this portion of the cycle. This voltage variation is plotted in Fig. 3.19.

Since the capacitor discharged only 0.2 V during the period from the time $t = 1$ msec to $t = 2$ msec, a charge of 9.8 V remains on the capacitor. Hence, at time $t = 2$ msec, the equivalent circuit is as shown in Fig. 3.18c. In this case,

$$v_o = 0.2e^{-t''/r_F C} \qquad (3.41)$$

where t'' has the value of zero at time $t = 2$ msec. As in the circuit in Fig. 3.18a, the capacitor quickly charges to 10 V again. From this point on, the action of the circuit is repetitive.

The foregoing circuit is known as a *clamper circuit*. In effect, this circuit has "clamped" the most positive portion of the input waveform to a value of 0 V. Many modifications of this circuit are used in various pulse circuit applications and TV receivers.

PROBLEM 3.9 Repeat Example 3.3 if a 110,000 Ω resistor is placed in parallel with the diode.

PROBLEM 3.10 A circuit is connected as shown in Fig. 3.20. If $r_F = 10 \ \Omega$ and $r_B = 1 \ M\Omega$, construct a plot of v_o.

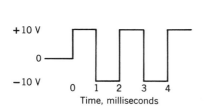
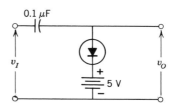

FIGURE 3.20
The configuration for Problem 3.10.

3.7 FULL-WAVE RECTIFIER CIRCUITS

The circuit considered in Example 3.2 removed the negative portion of an ac signal and was called a half-wave rectifier circuit. The circuit for a *full-wave rectifier* is given in Fig. 3.21. This circuit is frequently used to obtain dc power from an ac power line. Electronic amplifiers and switching circuits as well as many other electronic devices require this type of power supply.

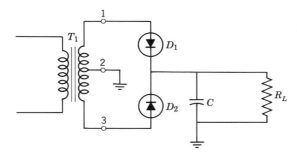

FIGURE 3.21
A full-wave rectifier circuit.

64 Diodes

Let us analyze the circuit shown in Fig. 3.21. Assume the diodes have low forward resistance and very high back resistance. Let time $t = 0$ when the terminal 1 of transformer T_1 is at the maximum positive potential. Since diode D_2 has a reverse voltage, very little current will flow through side 3 of the transformer. In fact, since the reverse current is so small, the reverse current can be ignored without introducing any appreciable error. Since side 1 is positive with respect to ground, a current will flow through diode D_1. This current will charge capacitor C as shown in Fig. 3.22 (curve A to B) and also cause current to flow through R_L. If the forward resistance of the diode and the resistance of the transformer are ignored, the capacitor will charge to the peak value of the voltage across $1/2$ of the transformer secondary. After the peak value of voltage is past, the voltage at point 1 will decrease as shown by the dashed line B to C of Fig. 3.22. As soon as the voltage at point 1 becomes less than the charge on the capacitor (time t_1 in Fig. 3.22), the diode D_1 is biased in the reverse direction and the diode becomes an open circuit. The charge on the capacitor C starts to leak off through the resistor R_L as soon as D_1 is cut off. At point C in Fig. 3.22, the voltage at point 1 (Fig. 3.21) becomes zero. Then the voltage at point 1 becomes negative, and the voltage at point 3 becomes positive as shown by curve $CDEF$ in Fig. 3.22. At time t_2, the voltage at point 3 (Fig. 3.21) is equal to the voltage on the capacitor. After time t_2, the capacitor is charged through diode D_2 to the peak value of the transformer secondary voltage (point E, Fig. 3.22). The cycle then repeats.

A rigorous analysis of the full-wave rectifier with filter capacitor can be developed.[1] However, considerable simplification results when certain

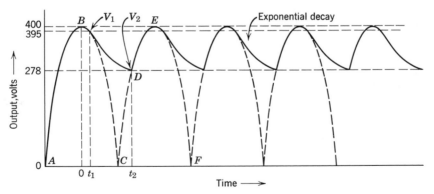

FIGURE 3.22
Output voltage waveform of the circuit of Fig. 3.21.

[1] *Electronic Engineering*, Second Edition, by C. L. Alley and K. W. Atwood, John Wiley and Sons, Inc., New York, 1966, pp. 100–103.

approximations, which will be discussed, are allowable. This simplified method produces results accurate enough for most engineering applications. Whenever a capacitor is used in a rectifier system to improve the rectification efficiency and reduce the ac component or *ripple* in the load, the capacitance value is usually chosen so that the ripple voltage is small in comparison with the dc component of load voltage. (The actual ripple voltage usually has a triangular waveform but is assumed to be essentially a sinusoid in this simplified method.) Under this condition, the time constant of the load resistance and filter capacitance must be long compared with the period T of the input voltage (Fig. 3.23). Then the capacitor (and load) voltage decreases almost linearly at the initial discharge rate $V_{max}/R_L C$. This initial slope would reduce the load voltage to zero at $t = R_L C$ if it were allowed to continue.

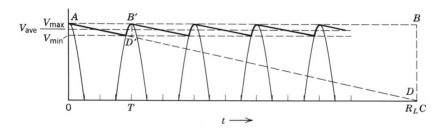

FIGURE 3.23
Constructions used in the approximate solution.

If semiconductor diodes are used as rectifiers, the maximum voltage across the load is approximately equal to the peak input voltage, since the forward drop across the diode is approximately one volt. Then an approximate relationship between the filter capacitance and the ripple voltage can easily be obtained by the following procedure.

　　1.　Assume that the load voltage decreases linearly from $t = 0$ until $t = T$, and then the capacitor is instantly recharged to V_{max} and so on. Then triangle $AB'D'$ is similar to triangle ABD and

$$\frac{T}{R_L C} = \frac{B'D'}{V_{max}} \tag{3.42}$$

　　2.　$B'D'$ is the peak-to-peak ripple voltage which may be approximated from the specified rms ripple voltage. $B'D' \simeq 2\sqrt{2}\, v_{ripple}$.
　　3.　The average or dc load voltage is obtained by subtracting the peak ripple voltage $B'D'/2$ from V_{max}, or, more often, the maximum

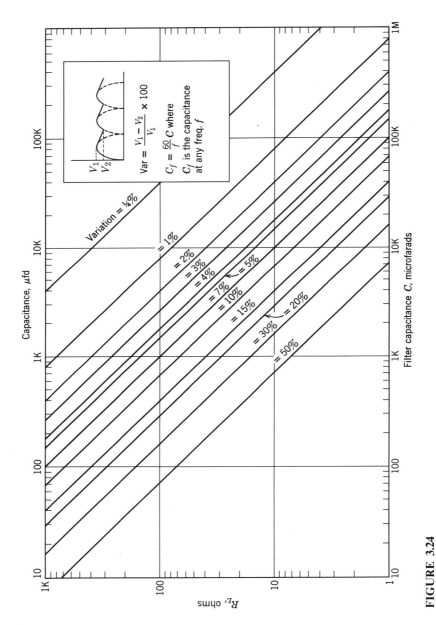

FIGURE 3.24

A nomograph to help determine the size of filter capacitor in a full-wave rectifier for 60 Hz.

input voltage is obtained by adding the peak ripple voltage to the specified dc load voltage.

4. The minimum value of load resistance, which is the worst case, may be obtained by dividing the dc load voltage by the maximum specified load current.

5. The required value of capacitance can be obtained from the specifications for the power supply and with the aid of Eq. 3.42.

$$C = \frac{T V_{max}}{2\sqrt{2} \, v_{ripple} \, R_L} \qquad (3.43)$$

When a full-wave rectifier is used, the discharge period is approximately $T/2$ instead of T. Therefore, the required filter capacitance is reduced by a factor of two.

A computer solution of the full-wave rectifier is presented in nomograph form in Fig. 3.24. This nomograph is very useful when determining the ripple factor as a function of $\omega \mathbf{R}_L C$. This particular nomograph assumes that the diode voltage drop is negligible and the resistance in the transformer winding is zero. A nomograph which provides for finite values of these parameters is given in Chapter 17.

PROBLEM 3.11 Design a full-wave power supply with a capacitor filter that will provide 30 V dc at 1 A into a resistive load. The permissible rms ripple voltage is 5 percent of the dc load voltage. Use silicon rectifiers and determine the filter capacitance as well as the rms voltage of the transformer secondary. The primary power is 115 V 60 Hz.

3.8 JUNCTION CAPACITANCE

A p-n junction and a charged capacitor are similar. As previously noted, the stored charge in the region of the junction results from the removal of free electrons from the n region, which leaves fixed positive donors. Similarly, the filling of the missing covalent bonds of the acceptor atoms in the p material produces fixed negative charges. The removal of the free or mobile carriers near the junction produces a *depletion* region which supports fixed excess charges and an electric field (Figs. 3.1 and 3.25). Some relationships between the barrier potential and the depletion width, the junction capacitance and the barrier potential, the maximum field intensity and the doping concentrations, and so on will be developed with the aid of Fig. 3.25, representing an abrupt junction *planar* diode of unit cross-sectional area. The excess positive charge density in the n-doped crystal is approximately equal to qN_d and penetrates a distance L_n (Fig. 3.25b). Similarly, the excess negative charge density in the p-doped crystal

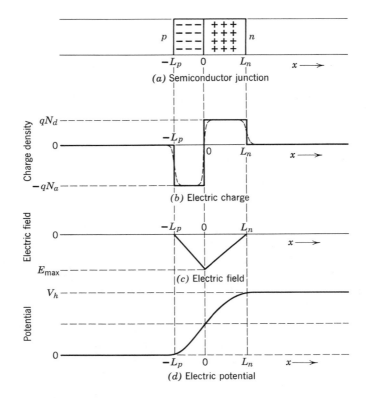

FIGURE 3.25
**The electric charge, field, and potential relationships in the depletion
region associated with a p-n junction.**

is $-qN_a$ and it penetrates a distance L_p. As indicated by the dashed lines
in Fig. 3.25, the excess-charge regions do not terminate abruptly, but little
error will be introduced by assuming abrupt termination at the effective
distances L_n and L_p from the junction. Conservation of charge requires that
the total charge be zero. Then

$$qN_d L_n = qN_a L_p \qquad (3.44)$$

The electric field intensity (Fig. 3.25c) may be determined from Gauss'
law, which states that the total electric flux DA passing through a given
closed surface with area A is equal to the coulomb charge enclosed by the
surface. Imagine that a ZY plane is located through the depletion region
to the left of the junction. The total charge to the left (and to the right) of
the plane is $-qN_a(L_p + x)$. All the electric flux lines originate on positive

charges to the right of the plane and terminate on the negative charges to the left of the plane. Then, since the cross-sectional area is unity,

$$\mathscr{E} = \frac{D}{\epsilon} = -\frac{qN_a(L_p + x)}{\epsilon} \tag{3.45}$$

Since the imaginary plane is to the left of the junction, x is negative and the electric field reduces to zero at $x = -L_p$. The maximum field exists at $x = 0$ and is given by

$$\mathscr{E}_{max} = -\frac{qN_a L_p}{\epsilon} = -\frac{qN_d L_n}{\epsilon} \tag{3.46}$$

The electric potential, using the p material to the left of L_p as the reference, is

$$V_h = -\int_{-L_p}^{L_n} \mathscr{E} \, dx \tag{3.47}$$

Since the potential difference V_h across the barrier is equal to the negative of the area under the $\mathscr{E}$ curve, inspection of Fig. 3.25c shows that

$$V_h = \frac{\mathscr{E}_{max}(L_p + L_n)}{2} = \frac{qN_a L_p(L_p + L_n)}{2\epsilon} \tag{3.48}$$

But from Eq. 3.44, $L_n = N_a L_p/N_d$ and

$$V_h = \frac{qN_a L_p{}^2(1 + N_a/N_d)}{2\epsilon} \tag{3.49}$$

Therefore,

$$L_p = \left[\frac{2\epsilon V_h}{qN_a(1 + N_a/N_d)} \right]^{1/2} \tag{3.50}$$

A similar solution for L_n yields

$$L_n = \left[\frac{2\epsilon V_h}{qN_d(1 + N_d/N_a)} \right]^{1/2} \tag{3.51}$$

Note that the depth of charge penetration, L_p or L_n, is proportional to the square root of the total barrier potential V_h and is roughly inversely proportional to the square root of the doping concentrations.

A concentration of charge exists at the junction and also a potential difference appears across the junction. Whenever these conditions exist, a capacitance also exists. The usual definition of capacitance C is charge Q, divided by voltage V. However, this junction capacitance is a function of the junction voltage and is therefore nonlinear. Thus an incremental capacitance may be defined as

$$C = \frac{dQ}{dV} \tag{3.52}$$

For the junction, the charge and voltage are both functions of the distance L_p or L_n. Therefore, Eq. 3.52 may be written as

$$C = \frac{dQ/dL_p}{dV_h/dL_p} \tag{3.53}$$

A voltage increase dV_h will cause an increased depth of penetration dL_p to the left of L_p and an increased depth of penetration dL_n to the right of L_n. But, dQ for the increase to the left of L_p is

$$dQ = qN_a \, dL_p \tag{3.54}$$

or

$$\frac{dQ}{dL_p} = qN_a \tag{3.55}$$

We can take the derivative of Eq. 3.49 to obtain

$$\frac{dV_h}{dL_p} = \frac{2qL_p N_a}{2\epsilon} \left[1 + \frac{N_a}{N_d} \right] \tag{3.56}$$

The value of L_p from Eq. 3.50 is substituted into Eq. 3.56 to yield

$$\frac{dV_h}{dL_p} = \left[\frac{2qN_a\left(1 + \dfrac{N_a}{N_d}\right)V_h}{\epsilon} \right]^{1/2} \tag{3.57}$$

Now, substitution of Eqs. 3.55 and 3.57 into Eq. 3.53 yields

$$C = \frac{qN_a}{\left[\dfrac{2qN_a(1 + N_a/N_d)V_h}{\epsilon}\right]^{1/2}} = \left[\frac{qN_a \epsilon}{2(1 + N_a/N_d)V_h}\right]^{1/2} \text{ farad/m}^2 \tag{3.58}$$

Equation 3.58 can be written as

$$C \simeq KV_h^{-1/2} \tag{3.59}$$

where K is a constant.

PROBLEM 3.12   A given silicon diode has $N_d = 10^{16}$ atoms/cm³ and $N_a = 10^{17}$ atoms/cm³. Determine the effective length of the depletion region on each side of the planar junction in a silicon diode with 20 V reverse bias applied. Assume the zero bias barrier voltage $V_{ho} \simeq 0.65$ V at this doping concentration. $\epsilon = 1.06 \times 10^{-10}$ F/m for silicon.

Answer: $L_p = 1.58 \times 10^{-5}$ cm, $L_n = 1.58 \times 10^{-4}$ cm.

PROBLEM 3.13   What is the incremental junction capacitance of the diode of Problem 3.12 if the cross-sectional area of the junction is 10^{-6} m²?

The junction capacitance may be troublesome in some applications. For example, let us consider the use of a diode to *detect* radio signals. The radio transmitter propagates a radio frequency signal which varies in amplitude according to the music or voice signal which is being transmitted. (These basic concepts will be treated in much greater detail in Chapters 16 and 17.) Thus, a radio frequency (RF) voltage wave which carries an audio signal would appear as shown in Fig. 3.26*a*. If this signal is applied to a diode in series with a load resistor and capacitive filter, as shown in Fig. 3.26*b*, the RF signal is rectified and converted to an audio signal (or the audio signal is *detected*) as shown in Fig. 3.26*c*. The reactance of the capacitor should be small in comparison with the load resistance for the RF signal so this RF component is shunted around the load resistor. On the other hand, the reactance of the capacitor at the audio frequency should be high compared with the load resistance so this audio component is *not* shunted to ground. In actual radio systems, the radio frequency carrier is

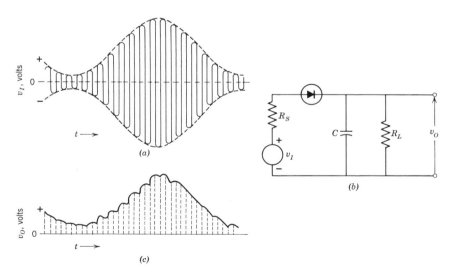

FIGURE 3.26
An RF detector circuit: (*a*) the input voltage waveform; (*b*) the detector circuit; (*c*) the output voltage waveform.

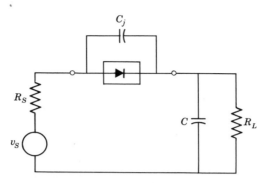

FIGURE 3.27
The detector circuit of Fig. 3.26b with the diode
replaced by a high-frequency equivalent circuit.

much higher than the audio signal so the RF ripples on the audio signal are not nearly as pronounced as those shown in Fig. 3.26. Radio detectors are discussed with much more rigor in Chapter 20.

Now, back to the capacitive effects of the detecting diode. Figure 3.27 shows the detector circuit of Fig. 3.26b with the actual diode replaced by an equivalent circuit which includes an ideal diode in parallel with the junction capacitance C_j. The resistive elements of the diode equivalent circuit and the dc offset voltage are not shown because they are assumed to be negligible. The junction capacitance C_j has essentially no effect during the forward-bias half cycles because it is essentially shorted out by the low resistance of the forward-biased junction. However, during the reverse-bias half cycles, the capacitor C_j permits current to flow in the reverse direction through the load. If this reverse current approaches the magnitude of the forward current, the diode becomes ineffective as a rectifier. Consequently, low-capacitance (this usually means small cross-sectional junction area) diodes must be used for high-frequency signals. In general, the higher the frequency of operation, the smaller the permitted capacitance of the diode. In fact, some diodes are made which utilize the contact potential between a piece of semiconductor material and a fine metal wire (called a cat's whisker because it is so fine) as the potential hill of the diode. Since these diodes have a very small junction area (the size of the cross-section of the wire) and consequently a very low capacitance, they can be used for detecting signals to 10^{10} Hz or higher. The crystal sets used in the very early days of radio had this type of detector.

PROBLEM 3.14 A circuit is connected as shown in Fig. 3.26. The frequency of the RF signal is 1 MHz and the frequency of the audio signal is 5,000 Hz. Let $R_L = R_S = 1,000 \ \Omega$. If X_C is 10,000 Ω for the audio signal, what is the value of C? With this size of capacitance, what is the impedance of X_C for the RF signal? What maximum value of junction capacitance can be tolerated if the reverse current is not to exceed 10 percent of the forward current? Assume C_j constant.

While junction capacitance may be undesirable in some applications, this same capacitance can actually be very useful in other applications. We have a capacitance which can be controlled by a voltage. For example, resonant circuits are used to select one radio signal from the numerous signals received by the antenna. Thus, if a circuit is connected as shown in Fig. 3.28, the resonant frequency of the tuned circuit can be controlled

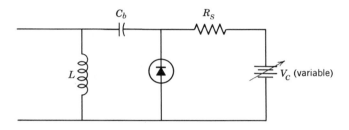

FIGURE 3.28
A voltage-controlled resonant circuit.

by adjusting the voltage V_C. The capacitor C_b is used to block the dc from flowing through L. This capacitor is large enough to be effectively a short circuit for the ac signal. The diode acts as the capacitor for the resonant circuit. This diode is reverse biased so it draws no appreciable current from V_C. Consequently, R_S can be very large (so it doesn't load down the resonant circuit) and still produce a negligible dc voltage drop for V_C. Special diodes which are constructed to be used as variable capacitors are known as *varactor diodes*.

PROBLEM 3.15 A circuit is connected as shown in Fig. 3.28. If the diode capacitance is given by Eq. 3.59 and the resonant frequency $\omega_0 = 1/\sqrt{LC}$, derive the relationship between the resonant frequency f_0 and voltage V_C.

3.9 DIFFUSION CAPACITANCE

In addition to the junction capacitance just considered, semiconductor diodes also exhibit an effect known as *diffusion capacitance*. When no external potential is applied, a profile of n carriers across the junction would appear as shown by the dashed curve in Fig. 3.29a. The relationship

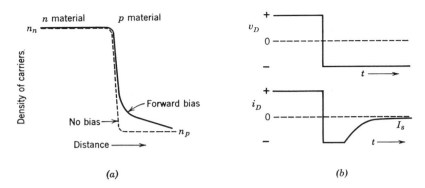

FIGURE 3.29
The diffusion capacitive effect : (a) n-carrier profile ; (b) effect of diffusion capacitance.

between the n_n and n_p carriers is given by Eq. 3.4 (repeated here for convenience),

$$n_p = n_n e^{-qV_h/kT} \tag{3.4}$$

Note that when forward bias is applied, the height of the potential hill is reduced so the n carriers which are injected from the n to the p material must increase. (Of course the p carriers in the n material also increase.) Thus, the minority carrier density near the junction increases as the forward bias potential increases. This effect is illustrated by the solid curve in Fig. 3.29a. The additional minority carriers which result from forward bias (represented by the area between the two curves in Fig. 3.29a) are often referred to as *stored charges*.

If the forward bias across the diode is suddenly reversed (Fig. 3.29b), there are a relatively large number of minority carriers near the junction. Consequently, these minority carriers diffuse to the junction and are swept back across the junction by the electric field due to the reverse-bias potential. As a result, the stored charges produce a reverse current through the diode as shown in Fig. 3.29b. In addition to the minority carriers which are swept back across the junction, some of the stored charges are also lost by recombination. Therefore, the stored charges are rapidly depleted and the reverse current across the diode decreases to the normal saturation current I_S. Some diodes called *hot-carrier* diodes or *fast* diodes have comparatively few stored charges. Other diodes are purposely constructed so they have a large stored charge but drop to the I_S value very rapidly when the stored charge is depleted. These diodes are known as *snap diodes* and are used for frequency multiplication applications.

PROBLEM 3.16 A snap diode is connected in series with a high-frequency sinusoidal voltage generator. As a result of the stored charge, the diode current waveform has the value shown in Fig. 3.30. Use a Fourier analysis to determine the magnitude of third harmonic current in this waveform. Would it be possible to obtain a third harmonic voltage from this current waveform? Explain.

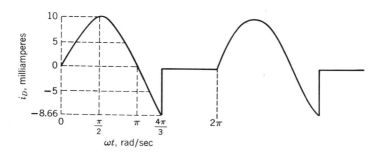

FIGURE 3.30
The current through a snap diode.

3.10 JUNCTION BREAKDOWN

The breakdown effect shown at point c in Fig. 3.4 will now be considered. One theory attributes the rather abrupt current increase to the high potential gradient (or electric field) which exists at the junction. According to this theory, the high electric field is able to disrupt the covalent bonds and therefore greatly increases the minority carriers. This effect is known as *Zener* breakdown. Another accepted theory for the voltage breakdown is the *avalanche* breakdown. This theory was founded by Townsend while he was studying the behavior of gases subject to electron bombardment. Accordingly, this theory was originally applied to gaseous conduction but applies just as well to the semiconductor.

According to the avalanche theory, a few carriers are generated in the intrinsic semiconductor material owing to thermal action, as previously discussed. These carriers are accelerated by the high electric field near the junction until high velocities are acquired. A carrier with sufficient energy can produce an electron-hole pair when this carrier collides with a neutral atom. The new carriers so produced are free to be accelerated and, in turn, to produce additional carriers. The origin of the term *avalanche* can be seen by the foregoing explanation.

Since the electrons are much more mobile than the holes, most of the carriers are produced by electron collisions. The electrons have many

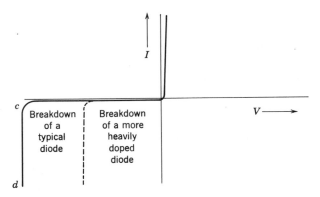

FIGURE 3.31
The effect of impurities on the breakdown potential of *p-n*
junction.

random collisions as they travel through the semiconductor. In order for
the avalanche effect to manifest itself, the electrons must obtain sufficient
energy in traveling one mean free path to produce ionization of the atoms
in the semiconductor. Hence, the electrons must have a kinetic energy equal
to or greater than the gap energy of the semiconductor for an avalanche
to be produced.

The junction breakdown potential is a function of the impurity concentrations of the semiconductor and actually involves both breakdown
mechanisms. As can be noted by Eqs. 3.50 and 3.51, the depletion width
varies inversely with the impurity concentration. Hence, as the impurity
concentration increases, the depletion region becomes thinner. As the
depletion region becomes thinner, the electric field intensity becomes higher
for a given junction voltage. Higher electric field potentials produce higher
electron energies per mean free path. Therefore, a heavily doped *p-n*
junction will have a lower breakdown potential than a relatively lightly
doped *p-n* junction. This effect is illustrated in Fig. 3.31.

Many diodes are designed and constructed for operation on the avalanche
portion of the characteristic curve. These diodes are known as *Zener diodes*
or *reference diodes*. These diodes operate in a region (from *c* to *d* on the
curve of Fig. 3.31) where the voltage is essentially independent of current.
The uses of these diodes are discussed in Section 3.11.

PROBLEM 3.17   Zener breakdown will occur in a silicon diode if the
electric field intensity exceeds about 10^8 V/m. A given silicon diode has
$N_a = N_d = 10^{18}$ atoms/cm^3. Will Zener breakdown occur with 5.0 V reverse
bias applied to this diode? $V_{ho} \simeq 0.9$ V, $\epsilon = 1.06 \times 10^{-10}$ F/m. At what
voltage will a Zener breakdown occur?

3.11 REFERENCE DIODES

Reference diodes are available with breakdown voltages ranging from about 3 V to well over 100 V. The avalanche appears to be the primary breakdown mechanism in diodes with reference voltages above about 7 V because the breakdown voltages of these diodes increase with temperature. As temperature increases, the atoms in the crystal structure vibrate about their 0°K position. The higher the temperature, the greater the amplitude of vibration. Thus, the atoms effectively occupy a larger volume. Consequently, the probability of a free electron-atom collision increases with temperature. Thus, increased temperature means a reduced mean free path for the free electrons. As a result, a higher voltage will be required to produce breakdown at a higher temperature. This type of behavior produces a positive *temperature coefficient*. The temperature coefficient is normally given in millivolts/°C and indicates the change in breakdown voltage which accompanies a one degree increase in temperature.

In constrast to the foregoing behavior, diodes with breakdown voltages less than about 6 V have a negative temperature coefficient, which indicates that the Zener breakdown mechanism is predominant in this range. The increased kinetic energy of the valence electrons would aid the high field in producing carriers and thus cause a negative temperature coefficient.

Unfortunately, the reference voltage increases slightly as the current increases. The important parameters specified for a reference diode are:

1. Breakdown potential
2. Temperature coefficient
3. Power rating
4. Incremental resistance, r_z

If a series of similar diodes but with different breakdown potentials are examined, additional insight can be gained. Thus, a plot of temperature coefficient versus breakdown potential is given in Fig. 3.32. The transition from negative to positive temperature coefficient occurs at 6 V (for $I_z = 40$ mA) for this series. A plot of incremental resistance versus breakdown potential is shown in Fig. 3.33. Note that a minimum of resistance also occurs near the 6 V value. Note that if cost is not an important factor, two 6 V reference diodes connected in series would give better performance than one 12 V reference diode. The temperature coefficient of the combination would be approximately zero, and the combined resistance of two 6 V diodes is less than the resistance of a single 12 V diode.

The equivalent circuit for a reference diode is a battery. If the diode becomes forward biased, it behaves the same as a regular diode. Consequently, an equivalent circuit for a reference diode may be as shown

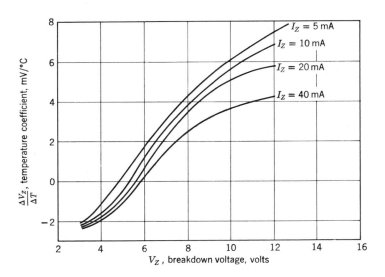

FIGURE 3.32
Temperature coefficient versus breakdown voltage: IN746 series.
(Courtesy of Texas Instruments, Inc.)

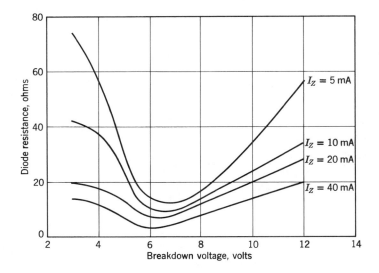

FIGURE 3.33
Incremental diode resistance r_Z versus breakdown voltage : IN746 series.
(Courtesy of Texas Instruments, Inc.)

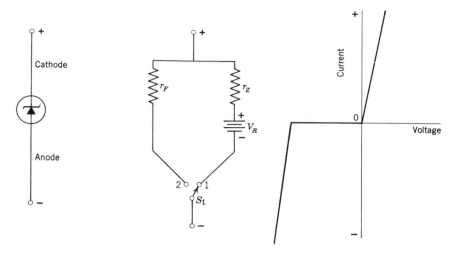

FIGURE 3.34
The equivalent circuit for a reference diode.

in Fig. 3.34. The current-versus-voltage characteristics for this circuit are also shown. The switch S_1 is in position 1 when a positive voltage is applied to the anode and switches to position 2 when a negative voltage is applied to the anode. r_Z is the diode resistance noted in Fig. 3.33.

Reference diodes can be used as voltage regulators. An example will be used to illustrate this use.

Example 3.4 A reference diode is connected as shown in Fig. 3.35. The characteristic curve of the diode is given in Fig. 3.36. The minimum value of V_I is 18 V.

The voltage across the reference diode is approximately 10 V as noted from Fig. 3.36. Consequently, the current through the resistor R_L is

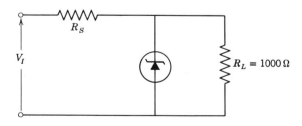

FIGURE 3.35
A voltage-regulating circuit.

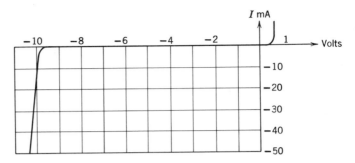

FIGURE 3.36
Characteristics of a typical 500 mW 10 V reference diode.

10 V/1,000 Ω = 10 mA. In order to maintain the load voltage essentially constant, the diode current should not drop below approximately 3 mA as seen in Fig. 3.36. If the minimum input voltage V_I is 18 V, the value of R_S should be approximately $(18 - 10)$ V/$(10 + 3)$ mA = 615 Ω.

The reference diode can dissipate up to 500 mW. Then, the maximum current through the diode is 500 mW/10 V = 50 mA. Under these conditions, the current through R_S will be $(50 + 10)$ mA. The input voltage will then be $10 + R_S(60$ mA$) = 10 + 615 \times .06 = 46.9$ V. From Fig. 3.36 we note that the voltage across the diode (and therefore across R_L) is about 10.3 V when the diode current is 50 mA. Thus, while the input voltage rises from 18 to 46.9 V, the voltage across the load (R_L) only changes from 10 to 10.3 V.

PROBLEM 3.18   Determine the values of the elements in the equivalent circuit (Fig. 3.34) for the diode whose characteristics are given in Fig. 3.36.

PROBLEM 3.19   (a) A voltage regulator circuit is connected as shown in Fig. 3.35. The voltage $V_I = 18$ V and $R_S = 615$ Ω. Determine the minimum and maximum values of R_L and the maximum load voltage variation if the diode has the characteristics given in Fig. 3.36. (b) Repeat (a) using a 400 mW IN746 series diode (Fig. 3.34). What is the maximum temperature coefficient?

3.12   OTHER TYPES OF SEMICONDUCTOR DIODES

A diode may be so heavily doped that the depletion region experiences Zener breakdown with only V_{ho} (no external bias) across the junction. The characteristics of this diode are shown in Fig. 3.37. This diode is known as a *backward* diode because it conducts more readily in the reverse-bias direction than in the forward-bias direction. For example, the diode of

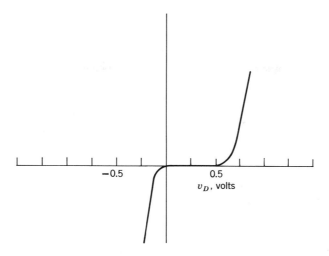

FIGURE 3.37
Characteristics of a backward diode.

Fig. 3.37 would be an effective rectifier of small ac voltages providing the peak amplitude does not exceed approximately 0.5 V.

If a diode is more heavily doped than a backward diode, the depletion region is in breakdown even with small amounts of forward bias applied. Thus, as the forward bias is increased, the current rises rapidly until the potential hill is reduced to the breakdown voltage, as shown at point B in Fig. 3.38. The current then falls rapidly until current again begins to rise due to normal majority carrier injection through the depletion region. This diode, known as a *tunnel diode*, or Esaki diode after its inventor, is useful because of its negative resistance, or conductance, which occurs between points B and C on the characteristic curve. (An increase of voltage causes a decrease of current.) This characteristic may be used in conjunction with a tuned circuit to produce very high-frequency oscillations. Doping densities in tunnel diodes are of the order of 10^{20} doping atoms/cm^3.

As previously mentioned, light (or photons) can give sufficient energy to a bound electron to break the covalent bond and produce a free electron and a hole. Conversely, an electron can fall from the conduction band into a hole and give up its energy in the form of a photon, or light. However, the momentum and energy relationships in silicon and germanium are such that the electron gives up its energy as heat when it returns from the conduction band to the valence band. On the other hand, the electron in a gallium arsenide crystal *does* produce a photon when it returns from the

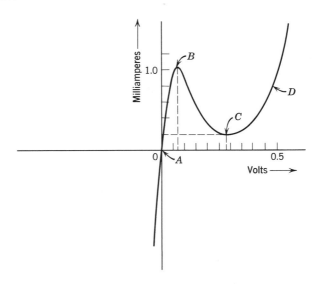

FIGURE 3.38
The characteristic curve of a typical tunnel diode.

conduction band to fill a hole in the valence band. This event does not occur often enough to produce a useful light intensity in the intrinsic crystal. However, gallium arsenide is used to make light-emitting diodes. When forward bias is applied to diodes, large numbers of electrons are injected from the n material to the p material. Most of these electrons combine with the holes in the p material. Since these holes are at the valence band energy level, photons are given off as the electrons combine with these holes. The light intensity is proportional to the rate of recombination of electrons, and thus is proportional to the diode current. The GaAs diode emits wavelengths in the near infrared region. To produce visible light, a mixture of GaAs and gallium phosphide must be used. The gallium phosphide absorbs the GaAs radiation and emits light in the visible spectrum.

A photo diode is about the inverse of a light diode. Reverse bias is applied to the photo diode, and the saturation (reverse) current is controlled by the light intensity which shines on the diode and generates electron-hole pairs.

PROBLEM 3.20   A battery with a voltage V_B is connected in series with a 5 Ω resistor, a 5 mh inductance, and 0.05 μF capacitance.

   a.   Determine the voltage across the inductance as a function of time if the circuit is connected together at time $t = 0$.
   b.   Add a negative resistance of 5 Ω in series with the circuit in part

a and then find the voltage across the inductance as a function of time.

    c.   What use does this problem suggest for a tunnel diode?

PROBLEM 3.21   An ac voltage generator has a rms open terminal voltage of 0.01 V and an internal resistance of 10 Ω. The generator is connected to a resistive load of 100 Ω.

    a.   Find the voltage across the load, the current through the load, and the power delivered to the load.

    b.   Add a negative resistance of 105 Ω in series with the circuit in part *a*. Find the voltage across the load, the current through the load, and the power delivered to the load under these conditions.

    c.   What use does this problem suggest for a tunnel diode?

3.13 TUBE-TYPE DIODES

The high-vacuum diode is constructed as shown in Fig. 3.39. The important components of this vacuum diode are: a heater wire, an electron-emitter surface, and an element known as a plate. This entire structure is placed in a glass or metal envelope. Most of the air in the envelope is removed so the active elements are surrounded by an effective vacuum. It is impossible to remove all of the air in the envelope, but the pressure inside the envelope is reduced to the vicinity of 10^{-6} mm Hg (atmospheric pressure is 760 mm Hg).

The heater wire is a piece of resistance wire and is heated by an electric current which flows through the wire. In some vacuum tubes, the heater

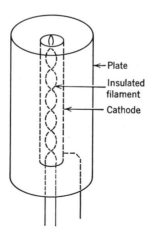

FIGURE 3.39
The high-vacuum diode.

wire and the electron-emitter surface are combined. In this case, the hot heater wire or *filament* emits the electrons. This type of emitter is known as a *filament-type emitter*. Other tubes use the heater wire or filament to heat a metal sleeve (Fig. 3.39). The metal sleeve, which is usually coated with a good thermal electron emitter, such as the rare-earth oxides, is known as a cathode. The cathode is usually electrically insulated from the heater.

The heated cathode emits electrons into the vacuum. If the plate is more positive than the cathode, these electrons are attracted to the plate. As a consequence, current flows through the tube. If the plate is more negative than the cathode, the electrons are repelled back to the cathode and no current flows through the tube. The action is similar to the *p-n* junction action except the only carriers are electrons. A second difference exists because the plate cannot emit electrons at normal temperatures so no reverse current flows in the vacuum tube. In fact, since the electrons in actual tubes are emitted from the cathode with a finite velocity, some current flows in the forward direction for zero and even slightly negative plate voltages.

If vacuum diodes are operated at the manufacturers' recommended filament voltage, a cloud of free electrons will be formed near the cathode surface. As a positive potential v_P is applied to the plate, these electrons will proceed to the plate and produce a plate current i_P. The relationship between plate current and plate voltage is

$$i_P = K v_P^{3/2} \tag{3.60}$$

where K is a constant and is determined by the diode geometry. The characteristic curve of a 5U4 vacuum diode is given in Fig. 3.40.

The semiconductor diode has replaced the vacuum diodes in most applications. The semiconductor diodes are much more rugged than the vacuum tubes (some vacuum tubes have a glass envelope). In addition, the vacuum diodes require a separate voltage source to heat the filaments. However, the main disadvantage of the vacuum diode is the high-voltage drop across the tube for a given plate current.

The voltage drop across the tube can be reduced by placing a small amount of a pure gas (mercury vapor, for example) in the previously evacuated envelope. In these *gas diodes*, the voltage drop across the tube is essentially equal to the ionizing potential of the gas (about 15 V for mercury vapor). The voltage drop is almost independent of current flow in the tube, as in the reference diode. However, even this 15 V drop is much greater than the typical voltage drop across a semiconductor diode (usually 1 V or less). The graphical symbols for directly and indirectly heated diodes are given in Figs. 3.41a and 3.41b, respectively. The symbol for the gas

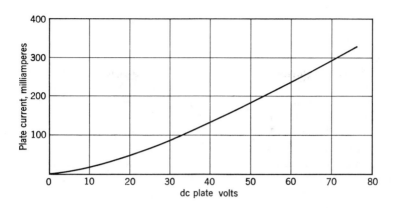

FIGURE 3.40
The characteristic curve of one plate in a 5U4 tube with 5.0 V across the filaments.

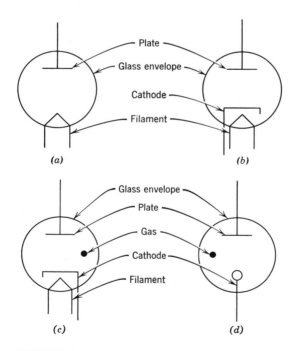

FIGURE 3.41
Symbols for vacuum diodes; (*a*) filament-type emitter; (*b*) cathode-type emitter; (*c*) gas diode; (*d*) voltage regulator tube.

rectifier tube is shown in Fig. 3.41c. Many tubes (intended for rectifier applications) contain two plates so they can be used for full-wave rectifiers (see Fig. 3.42).

A class of gas tubes known as *voltage regulator tubes* are available with voltage breakdown ratings between 75 V and 150 V. The voltage regulator tubes do not have a heated cathode as indicated by the symbol in Fig. 3.41d. These tubes do contain a small amount of radioactive material to produce a slight ionization of the gas. When sufficient voltage is present, more ions are produced by collisions between electrons and gas molecules. In addition, the impact of the positive ions on the cathode produce secondary electrons. As noted before, the potential across the tube remains essentially constant over fairly wide variations of diode current. The type of gas in the tube determines the ionization potential and consequently the voltage across the tube.

In order to develop more insight into the relative characteristics of the various types of diodes, consider the following example.

Example 3.5   A circuit is to be connected as shown in Fig. 3.42. The desired peak current to the load is 150 mA. Three types of diodes are available.

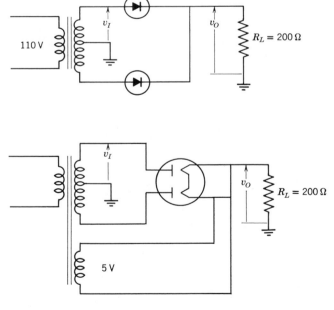

FIGURE 3.42
The configuration for Example 3.5.

1. A IN4004 semiconductor diode
2. A 5U4 vacuum diode
3. An 83 mercury vapor diode

Let us compare the relative merits of these three diodes.

The characteristics of the IN4004 state that the typical voltage drop across the diode is 0.8 V when $i_D = 150$ mA. The peak voltage across the load is $iR_L = 0.15 \times 200 = 30$ V. Thus, the required peak value of v_I is 30.8 V. The peak power lost in the IN4008 is 0.15 A $\times$ 0.8 V = 0.12 W.

The 83 gas diode contains mercury vapor and has about 15 V drop across the tube when it is conducting. In addition, the filaments require 5 V and 3 A for proper heating. Thus, the required peak value of $v_I = 30 + 15 = 45$ V for the 83 tube. The peak power to be lost in the diode is 0.15 A $\times$ 15 V + 5 V $\times$ 3 A = 2.25 + 15 = 17.5 W. (Most of this power loss occurs in the filament.

The 5U4 vacuum diode requires 5 V and 3 A to heat the filament, and the plate characteristics are given in Fig. 3.40. From the plate characteristics, we note that the voltage drop across the tube is 44 V when the current is 150 mA. Consequently, the required peak value of $v_I = 30 + 44 = 74$ V for the 5U4 tube. The peak power lost in the tube is 0.15 A $\times$ 44 V + 5 V $\times$ 3 A = 6.6 + 15 = 21.6 W.

No wonder the semiconductor diodes have essentially replaced the tube diodes!

PROBLEM 3.22   Repeat Example 3.2 if one section of a 5U4 diode is used instead of the semiconductor diode.

PROBLEM 3.23   A germanium diode (at room temperature) has doping concentrations such that $N_a = 10^{17}/\text{cm}^3$ and $N_d = 10^{15}/\text{cm}^3$.

    a. What are the values of n_n and p_p?
    b. What are the values of n_p and p_n?
    c. What is the magnitude of the potential V_{ho}?
    d. What are the lengths of L_p and L_n?
    e. What is the maximum electric field intensity in the depletion region?

PROBLEM 3.24   A silicon diode (at room temperature) has the following physical parameters.

$$N_a = 10^{17}/\text{cm}^3$$
$$N_d = 10^{15}/\text{cm}^3$$
$$\text{Cross-sectional area} = 0.01 \text{ cm}^2$$

    a. Determine the junction capacitance of this diode when no external voltage is applied.

b.   Make a plot of capacitance versus external voltage for this diode if the external voltage were to vary from −25 to 0 V.

PROBLEM 3.25   A junction diode has 1 V reverse bias applied. The measured current is 0.1 μA. What current would flow ($T = 300°K$) if 10 V forward bias is applied? The internal resistance of the diode is 5 Ω.

PROBLEM 3.26   A circuit is connected as shown in Fig. 3.43*b*. Plot the output voltage waveform if the input voltage has the form shown in Fig. 3.43*a*, and the characteristics of the diode are as shown in Fig. 3.43*c*.

PROBLEM 3.27   A circuit is connected as shown in Fig. 3.44. Plot the steady-state output waveform if the input signal has the form shown.

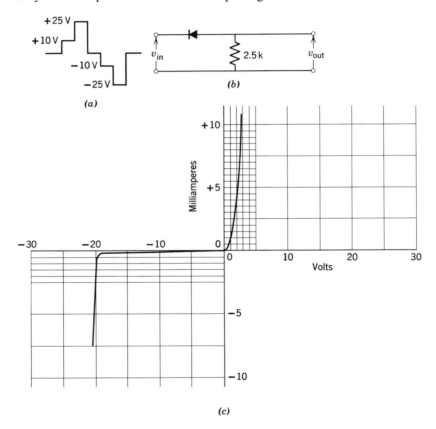

FIGURE 3.43
Information required by Problem 3.26 : (*a*) input voltage; (*b*) circuit configuration; (*c*) characteristic curves.

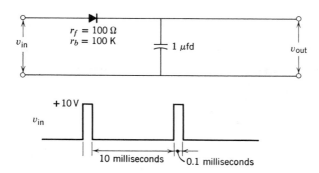

FIGURE 3.44
Information for Problem 3.27.

PROBLEM 3.28 Design a silicon Zener diode which will break down at an applied voltage of 6.0 V. Assume breakdown will occur in silicon if the electric field intensity exceeds 10^8 V/m. The data in Table 2.2 may be useful.

PROBLEM 3.29 Vacuum tubes and transistors require dc potentials for proper operation. However, in many instances, it is desirable to use the 115 V ac as a source. The specifications for a typical power supply for vacuum tubes and the specifications for a typical transistor supply are given below.

Supply for Vacuum Tubes	Transistor Supply
Input	Input
115 V 60 Hz ac	115 V 60 Hz ac
Output	Output
dc 300 V @ 100 mA Ripple 2% max ac 6.3 V @ 2 A	dc 30 V @ 3 A Ripple 2% max

a. Design a power supply to meet each of these specifications.

b. Give the circuit diagrams and determine the value of each component in the diagram.

c. Determine the approximate cost of each power supply.

PROBLEM 3.30 Diodes can be used to direct current to certain elements in a circuit but prohibit current from other circuit elements. A diode configuration of this type is commonly known as a diode matrix. Diode matrices find wide use in computer and logic circuits. To investigate this method of control, consider the following situation.

A railroad dispatcher wishes to set the track turnouts in his yard so a train will follow a given path. Each turnout is constructed as shown in Fig. 3.45a. When power is applied to solenoid S_{1L}, this solenoid pulls the turnout to direct the train to the left track. When power is applied to

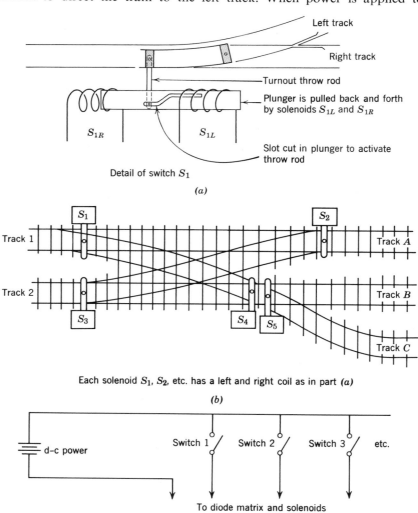

Detail of switch S_1

(a)

Each solenoid S_1, S_2, etc. has a left and right coil as in part (a)

(b)

(c)

FIGURE 3.45
The configuration for Problem 3.30.

solenoid S_{1R}, the turnout directs the train to the right track. A simple track layout is shown in Fig. 3.45b. The dispatcher would like to close switch 1 (Fig. 3.45c) and direct dc power to the proper solenoids so a train will be directed from track 1 to track A. Develop a diode circuit which will direct the train as follows:

From track 1 to track A when switch 1 is closed.
From track 1 to track B when switch 2 is closed.
From track 1 to track C when switch 3 is closed.
From track 2 to track A when switch 4 is closed.
From track 2 to track B when switch 5 is closed.
From track 2 to track C when switch 6 is closed.

PROBLEM 3.31   A silicon diode is tested in the laboratory. The following measurements are made at room temperature.

Forward bias $V = 1.1$ V, $I = 1.0$ A
Reverse bias $V = 1.0$ V, $I = 0.02$ μA
Reverse bias $V = 400$ V, $I = 0.1$ μA

Determine the values of R_{Sh} and R_S (Fig. 3.5) for this diode.

4

Common-Base
Transistor Amplifiers

Transistors are used in a wide variety of applications, including television, automatic control, satellite instrumentation, and medical electronics. The ability of the transistor to amplify electrical signals accounts for its wide use.

The amplifier is actually an energy converter. The input signal merely controls the current that flows from the power supply or battery. Thus the energy from the power supply is converted by the amplifier to signal energy.

4.1 JUNCTION TRANSISTORS

When a semiconductor is arranged so that it has two p-n junctions as shown in Fig. 4.1a, it is known as a *p-n-p junction transistor* or *bipolar transistor*. The electrical potential in the transistor as a function of distance along the axis is shown in Fig. 4.1b. Note that this potential is the same as expected from two diode junctions. Fortunately, the electrical performance of a transistor is very different from the electrical performance of two diodes connected in a similar configuration. To illustrate this electrical behavior, consider a transistor connected as shown in Fig. 4.1c. The potential profile has now been altered as shown in Fig. 4.1d. The left-hand junction is forward biased, the potential hill is lowered, and carriers are injected into the n region which is known as the *base*. This base is made very thin, so nearly all the injected carriers diffuse across the base and are accelerated across the right-hand junction into the region known as the *collector*. The left-hand region is known as the *emitter* because it provides, or emits, the injected carriers. The current which flows out the base lead results from recombinations in the base. Therefore the base is lightly doped, in addition to being thin, in order to minimize recombinations. The ratio

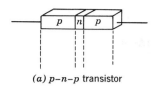

(a) p–n–p transistor

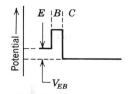

(d) Idealized potential of *(c)*

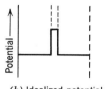

(b) Idealized potential with
no bias applied

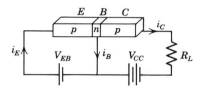

(e) Load resistance and reverse–biased
collector–base junction

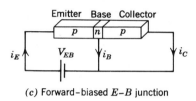

(c) Forward–biased *E–B* junction

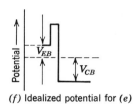

(f) Idealized potential for *(e)*

FIGURE 4.1
Potentials and currents in a *p-n-p* **transistor.**

of carriers flowing into the collector to those injected across the emitter-base junction is known as α (alpha) which ranges from about 0.90 to 0.998 in modern transistors.

The transistor is useless as an amplifier unless a load resistance, such as a headset or relay, is placed in the collector, or output, circuit. But a resistance in the collector circuit of Fig. 4.1c would cause the collector terminal to become positive with respect to the base and would thus forward bias the collector-base junction. The injection current resulting from this bias would tend to cancel the initial collector current. However, if a battery is placed in the collector circuit as shown in Fig. 4.1e, the collector-base junction will remain reverse biased, providing the voltage drop across the load is less than the collector supply voltage V_{CC}. The reverse bias across the collector-base junction increases the height of the potential hill, as shown in Fig. 4.1f. Therefore the electrons, or charge carriers, gain more kinetic energy as they are accelerated from the base into the collector region. This energy is dissipated as heat and thus causes the temperature of the transistor to rise.

From Fig. 4.1e, we note that the supply voltage V_{CC}, and also the voltage across the load resistor, may be very large in comparison with the input voltage across the forward-biased, emitter-base junction. Therefore, the circuit is an *amplifier* because it has voltage amplification or *gain*.

From the preceding discussion, it would seem that either end of the transistor could be used as the emitter. This would be true for the transistor configuration of Fig. 4.1. However, most transistors are made with collector junctions larger than emitter junctions for improved collector power dissipation and increased α (which will be discussed later) and thus α is higher in the normal, or forward, direction than in the reverse direction.

The circuit diagram for the transistor amplifier of Fig. 4.1e, with an ac signal source added to the input circuit, is shown schematically in Fig. 4.2a.

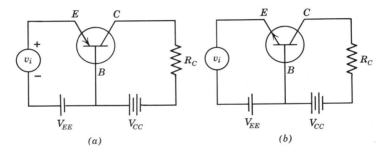

(a)   (b)

FIGURE 4.2
Circuit diagrams for (a) **a** *p-n-p* **transistor amplifier, and** (b) **a** *n-p-n* **transistor amplifier.**

Note that the emitter-base voltage v_{EB} is not the same as the bias battery voltage V_{EE}, but is the sum of V_{EE} and the signal voltage v_i. Fig. 4.2 also introduces the conventional transistor symbol. The *n-p-n* transistor amplifier may be identical to the *p-n-p* amplifier, except all the current and voltage polarities are reversed, as shown in Fig. 4.2b.

At this point a few words should be said about conventional current and voltage directions. Actual current directions and potential polarities were shown in Fig. 4.1. However, IEEE standards require that currents which flow *into* a device be defined as *positive*, and currents which flow *out* of the device be defined as *negative*. Therefore, the collector and base currents which usually flow out of the *p-n-p* transistor are negative currents because they flow in the opposite direction to the conventional positive currents. Also, a potential is considered positive if it is positive with respect to the common terminal or ground, which is the base terminal in Fig. 4.1.

Thus the emitter terminal is positive and the collector terminal is negative in Fig. 4.2*a*. Either + and − signs, or an arrow pointing in the direction of positive potential are used to indicate voltage polarity.

An amplifier may frequently be called upon to amplify ac signals. Capacitors are then used to couple the time-varying signal into the amplifier and out of the amplifier and to block (or prevent) the dc, or *bias*, components from entering the driving source or load as shown in Fig. 4.3.

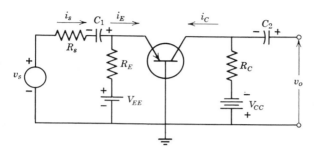

FIGURE 4.3
A common-base ac amplifier using a *p-n-p* transistor.

The capacitors are assumed to be large enough to offer negligible reactance to the signal currents. The input signal voltage v_s varies the voltage across the forward-biased emitter junction and thus causes the emitter current i_E to vary. The collector current, being almost equal to the emitter current, varies in accordance with the emitter current, and thus causes a varying voltage across the collector circuit resistor R_C. This varying, or signal component, is passed through the capacitor C_2 and becomes the output voltage v_0. Note that the emitter circuit resistor R_E should have high resistance compared with the input resistance of the transistor to avoid shunting an appreciable part of the signal current i_s to ground. Then, V_{EE} must be larger than V_{EB} because of the $I_E R_E$ voltage drop across R_E. In order to emphasise these concepts, let us consider an example.

Example 4.1   A transistor is connected as shown in Fig. 4.3. Let us determine proper circuit components if the average or bias current I_E is 2.0 mA.

Since the emitter-to-base junction is a forward-biased *p-n* junction, the dynamic input resistance r_e from emitter to base is the same as the dynamic forward-biased resistance r_f for a diode. Thus, from Eq. 3.31, we have $r_e \simeq 25/I_E$ (in mA) $= 25/2 = 12.5 \ \Omega$. If we choose the resistor R_E so it is 100 times as large as r_e, then only one percent of the signal current i_s will be shunted through R_E. This value of R_E will be $12.5 \times 100 = 1250 \ \Omega$. The average (or dc) voltage drop across R_E is $I_E R_E = 1250 \times .002 = 2.5$ V. If the transistor is silicon, the average voltage from base to emitter may be about 0.5 V. Then, the battery voltage V_{EE} should be 0.5 V + 2.5 V = 3.0 V.

A suitable value for R_C can be determined if we first choose a value for V_{CC}, which must be below the voltage breakdown rating of the collector junction. Let us assume $V_{CC} = -20$ V. Now, if α for this transistor is 0.95, the average collector current I_C will have a value of $I_C = -\alpha I_E$ (the actual collector current is flowing *out* of the collector) $= -.95 \times 2 = -1.9$ mA. Normally, the voltage drop across R_C is chosen to be about one-half of V_{CC}. (This value is chosen so the signal variations of collector voltage can be symmetrical.) As a result, the average voltage drop across R_C should be about 10 V. Then, the value of R_C is 10 V/I_C = 10 V/1.9 mA $\simeq 5.25$ kΩ. The commercially available value of 5 kΩ would probably be used.

Now, if the signal current, i_s, has a peak value of 0.5 mA, the peak signal voltage from emitter to base v_{eb} is (neglecting the one percent loss of signal current through R_E) $i_s r_e = 5 \times 10^{-4} \times 12.5 = 6.25$ mV. The signal current component i_c in the collector circuit is $-\alpha i_s = -.95 \times 0.5$ mA $= -0.475$ mA. When this current flows through R_C, the voltage drop across R_C and therefore the output voltage v_o will have a peak magnitude of $i_c R_C = 4.75 \times 10^{-4} \times 5 \times 10^3 = 2.37$ V. The signal voltage gain of this amplifier is 2.37 V/$6.25 \times 10^{-3} = 380$.

PROBLEM 4.1   The emitter-bias current I_E of Example 4.1 is changed from 2 mA to 1 mA. Repeat Example 4.1 under these conditions.
Answer: $R_E = 2.5$ kΩ, $V_{EE} = 2.98$ V, $V_{CC} = -20$ V, $R_C = 10.5$ kΩ (use 10 kΩ), voltage gain = 380.

In Example 4.1, several different current symbols were used. A few words should be said here about voltage and current symbols. The IEEE standards are based on the system given in Table 4.1.

The double subscripts associated with the voltage symbols (except the bias battery symbols) indicate the terminals between which the voltage is

TABLE 4.1
Standard Voltage and Current Notation

Component	Symbol	Subscripts	Example
dc or average	Capital	Capital	V_{CB}, I_C
s domain or ω domain	Capital	Lower case	V_{cb}, I_c
Total (dc + time varying)	Lower case	Capital	v_{CB}, i_C
Instantaneous time varying	Lower case	Lower case	v_{cb}, i_c
Bias supply voltage	Capital	Double capital	V_{CC}, V_{EE}

measured. For example, V_{BC} is the dc or average value of potential difference between the base and the collector. Therefore, $V_{BC} = -V_{CB}$. Sometimes a single subscript is used when the potential is with respect to ground or the chassis. For example, v_C is the total voltage between the collector and ground (or chassis or other ground reference) at any instant. If the base terminal is grounded, $v_C = v_{CB}$. As previously mentioned, the current flows into the transistor if its sign is positive, whereas a negative sign indicates that current is flowing out of the transistor. Current arrows always indicate the direction of positive current flow, therefore, the current arrowheads always point toward the transistor, or device. Of course, ac or signal components that alternately change direction can also be represented by arrows or polarity signs. The arrows (or polarity signs) indicate the direction of the current or voltage when it is positive.

Now, let us again refer back to Fig. 4.3 and develop some general equations. If $R_E \gg r_e$, the transistor input voltage v_i due to the signal current i_s is

$$v_i = i_s r_e \tag{4.1}$$

Also, the output voltage v_o due to the signal current i_s is

$$v_o = \alpha i_s R_C \tag{4.2}$$

Then the voltage amplification or gain[1] is

$$K_v = \frac{v_o}{v_i} = \frac{\alpha i_s R_C}{i_s r_e} = \frac{\alpha R_C}{r_e} \tag{4.3}$$

PROBLEM 4.2 If the transistor amplifier of Fig. 4.3 has the values given in Example 4.1 ($r_e = 12.5\ \Omega$ and $R_C = 5.0\ k\Omega$) but $\alpha = 0.99$, calculate the voltage gain K_v of the amplifier.

Answer: 396.

PROBLEM 4.3 Draw the circuit diagram for an ac amplifier using an n-p-n transistor in the grounded base configuration. Also determine suitable values for the circuit elements (excluding capacitors) and calculate the voltage gain if $I_E = -1$ mA, $\alpha = 0.98$ and $V_{CC} = 30$ V. Assume $V_{BE} = 0.5$ V. *Answer:* $R_E = 2.5\ k\Omega$ (assuming 100 r_e), $V_{EE} = -3.0$ V, $R_C = 15\ k\Omega$, and $K_v = 588$. (Any suitable set of answers is acceptable.)

Saturation currents due to minority carriers flow across the junctions in transistors as well as diodes. These currents which were designated I_S in the diode are known as I_{CO} for the collector junction and I_{EO} for the emitter

---

[1] The symbol K is used in this text to designate a special type of gain which is not a function of frequency. The more general gain symbol G will be introduced in Chapter 7.

junction in a transistor. Transistor data sheets usually list these currents, plus any leakage currents, as I_{CBO} and I_{EBO}. The first two subscripts designate the terminals between which the current flows, and the third subscript indicates that the third (other) terminal is open. The current I_{CO} or I_{CBO} must be included in the collector and base currents if highly accurate results are to be obtained. The relationships between these currents are shown in Fig. 4.4. Observe that I_E is a negative current while I_C and I_{CO} are positive currents. Therefore,

$$I_C = -\alpha I_E + I_{CO} \tag{4.4}$$

and

$$I_B = -(1 - \alpha)I_E - I_{CO} \tag{4.5}$$

The magnitude of thermal current I_{CO} is usually small in comparison with αI_E but may not be small in comparison with $(1 - \alpha)I_E$. In fact, these currents may be equal at moderate values of I_E in a germanium transistor, and the base current will then be zero. For example, if a germanium transistor with $\alpha = 0.98$ has $I_{CO} = 10 \ \mu A$, the value of I_E which will reduce I_B to zero is -0.5 mA.

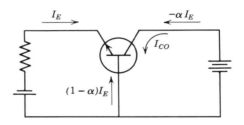

FIGURE 4.4
Bias currents and the collector saturation (or thermal) current in an *n-p-n* **transistor.**

PROBLEM 4.4   A germanium transistor has $\alpha = 0.99$ and $I_{CO} = 12 \ \mu A$. For what value of I_E will $I_B = 0$?

Answer: -1.2 mA.

4.2   GRAPHICAL ANALYSIS

Improved understanding of the common-base transistor amplifier can be gained from a graphical presentation of the input and output (or collector) characteristics of the transistor. The diagram of a circuit which might be used to obtain the common-base characteristics is given in Fig. 4.5. The batteries and potentiometers which are used as variable bias sources in this circuit are customarily replaced by ac operated power supplies with adjust-

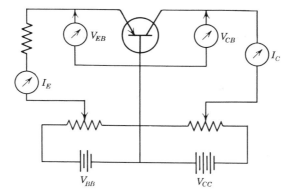

FIGURE 4.5
A circuit for determining transistor characteristics.

able dc output voltages. The voltmeters shown should be highly sensitive, preferably electronic voltmeters, so that their operating currents do not make an appreciable contribution to i_C or i_E.

There are four important variables in the circuit, as indicated by the meters. Some of these variables are held constant while the relationships between others are found. For example, the emitter current may be held constant while the effect of the collector voltage on collector current is determined. When the collector current is plotted as a function of the collector-base voltage (with emitter current held constant), the resulting curve is known as a *collector-characteristic curve*. Each different value of emitter current will yield a different collector-characteristic curve. A set of several curves obtained from several representative values of emitter current is known as a *family* of collector-characteristic curves. A typical family or set of collector characteristics for a *p-n-p* transistor is shown in Fig. 4.6. The negative current indicates that the current is flowing out of the transistor where the reference direction is into the transistor. A set of curves for an *n-p-n* transistor might be identical to the set shown except that the polarities of currents and voltages would be reversed.

From Fig. 4.6, it should be observed that:

1. The collector current is almost equal to the emitter current when reverse bias is applied to the collector junction.

2. The collector current is almost independent of the collector voltage when reverse bias is applied to the collector junction.

3. The collector current is rapidly reduced to zero and then reversed when forward bias is applied to the collector junction. This behavior occurs because the junction, or majority, current across the collector junction opposes the majority current across the emitter junction. The

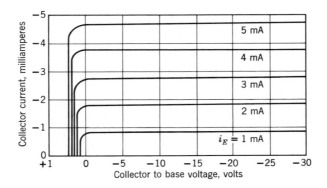

FIGURE 4.6
**A family of collector characteristic curves. (Note the change
of scale for positive collector-to-base voltage.)**

collector current is essentially the algebraic sum of these two majority
currents.

The emitter-base voltage does not appear in the set of collector
characteristics. Therefore, a relationship is needed between emitter current
and emitter-base voltage to determine the input resistance of the transistor.
Experimentation reveals that the collector voltage has a slight influence
on the emitter current. Therefore, a family of curves is needed to completely
define the input characteristics of the transistor. Normally the emitter-base
voltage is plotted as a function of the emitter current with the collector
voltage held constant (Fig. 4.7). A curve is obtained for each different
value of collector voltage. Note that these curves are typical junction-diode

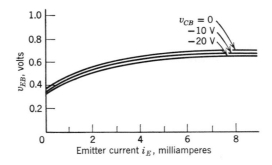

FIGURE 4.7
**Input characteristics of a typical common-base
transistor.**

curves. However, the voltage and current axes have been exchanged when compared with the diode characteristics of Chapter 3 because the emitter current is here considered as the independent variable. Figure 4.7 shows that an increase of reverse collector bias increases the emitter current slightly if v_{EB} is held constant. The reason for this collector voltage dependence will be discussed in Section 4.3.

The characteristic curves[2] given above are known as *static* curves because any point on these curves represents the dc or static current and voltage relationships of the transistor. However, when a signal is applied to a transistor amplifier, most voltages and currents vary with time. Therefore, it may seem that the static curves are of little use in determining the operating characteristics of an amplifier. Fortunately, this assumption is not true because the dynamic characteristics can be obtained simply by drawing a load line on the static collector characteristics in exactly the same manner as for the diode characteristics described in Section 3.5. In this case, the voltage-current relationship in the collector circuit is given by the equation $v_{CB} = V_{CC} - i_C R_C$. Therefore, the load line intercepts the v_{CE} axis ($i_C = 0$) at $v_{CB} = V_{CC}$ and intercepts the i_C axis ($v_{CB} = 0$) at $i_C = V_{CC}/R_C$. This load line is known as a dc load line because the direct current must flow through the collector load resistor R_C. For example, the dc load line for a value of $R_C = 5$ kΩ has been drawn on a set of collector characteristics in Fig. 4.8a for a V_{CC} value of -20 V. The dc load will be considered as the only load at this time.

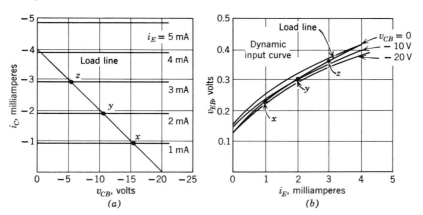

FIGURE 4.8
Transistor characteristics: (a) load line for $R_L = 5$ kΩ; (b) the matching dynamic input characteristics.

[2] The common-base characteristics of Figs. 4.6 and 4.7 are seldom given by the manufacturers but can be readily obtained from a transistor curve tracer.

A dc bias point or *quiescent point* is selected some place along the load line. The transistor operates at this quiescent point when no signal currents are present. If the expected input signals are symmetrical about the quiescent or *q* point, such as sinusoidal signals, a good selection for the *q* point would be near the center of the load line. The *q* point is selected at $i_E = 2.0$ mA in this example.

Values from the load line can be transferred to the input characteristics to produce a curve which is known as a *dynamic* input curve. Thus, point *x* in Fig. 4.8*a* occurs where $v_{CB} = -15$ V and $i_E = 1$ mA. In Fig. 4.8*b*, the corresponding values of v_{CB} and i_C determine point *x* on the dynamic input curve. Enough additional points are transferred from the load line (points *y* and *z*, for example) to the input characteristics until the dynamic input curve can be drawn.

The operating characteristics of the amplifier can be readily determined graphically. In order to illustrate the procedure, we will consider an example.

Example 4.2 The characteristic curves for a transistor are given in Fig. 4.9. If $R_C = 5$ kΩ and $V_{CC} = -20$ V, let us determine the characteristics of this amplifier.

First, the load line and the dynamic input curve are drawn on the transistor characteristic curves as shown. Then, a *q* point is chosen on the load line. In this example, let $i_E = 2$ mA be the required value. Now, the signal current waveform can be plotted at right angles to the load line.

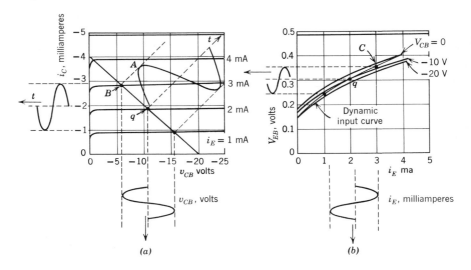

(a)   (b)

FIGURE 4.9
(a) **A load line, and** (b) **dynamic input characteristics for** $R_L = 5$ kΩ.

In this example, we have chosen a signal current of $i_s = 10^{-3} \sin \omega t$. At $\omega t = 0$, the transistor is operating at its quiescent value. From Fig. 4.9a, we note $i_C = -1.9$ mA, $v_{CB} = -10.5$ V. This q point can be transferred to the dynamic input curve where we note $v_{EB} = 0.31$ V. When $\omega t = \pi/2$ radians, the value of i_s is 1 mA. This is represented as point A on the input signal and is projected to the load line as point B. We note that $i_E = 3$ mA, $i_C = -2.85$ mA, $v_{CB} = -6$ V. Point B can be transferred to the dynamic input curve (point C) where we note $v_{BE} = 0.35$ V. Other points can be found in a similar fashion until complete plots of i_C, v_{CB}, i_E, and v_{EB} can be obtained. In this example, the peak-to-peak value of emitter current (Δi_E) is $3 - 1 = 2$ mA. Corresponding to these values, we see the peak-to-peak magnitude of collector current (Δi_C) is $2.85 - 0.95 = 1.9$ mA. The current gain of this amplifier is $\Delta i_C / \Delta i_E = 1.9$ mA/2 mA $= 0.95$ which is essentially α for this transistor. The peak-to-peak magnitude of collector voltage (Δv_{CB}) is $15 - 6 = 9$ V and the peak-to-peak value of emitter voltage (Fig. 4.9b) is $\Delta v_{EB} = 0.35 - 0.24 = 0.11$ V. Therefore, the voltage gain of this amplifier is $\Delta v_{CB} / \Delta v_{EB} = 9$ V/0.11 V $= 82$. The input impedance from the emitter to the base is $R_{i_t} = \Delta v_{EB} / \Delta i_E = 0.11$ V/0.002 A $= 55$ Ω. This input resistance is abnormally high because the input curves were drawn unusually steep so the graphical construction could be more easily visualized. In addition, the output impedance[3] of this amplifier is $Z_o = \Delta v_{CB} / \Delta i_C = 9$ V/1.9 mA $= 4.75$ kΩ.

PROBLEM 4.5 The common-base characteristics of a 2N3903 transistor are given in Fig. I-1 (Appendix I). If this transistor is used in a common-base amplifier with $V_{CC} = 30$ V and $R_C = 10$ kΩ, determine the voltage gain of the amplifier when the peak-to-peak variation of the emitter current is 1.0 mA. R_C is the total load resistance. Draw the load line on the characteristics, choose a suitable quiescent operating point, and graphically determine the voltage gain, the current gain, the input impedance, and the output impedance of this amplifier.

4.3 BASE WIDTH MODULATION

The collector voltage has an influence on the collector and emitter currents because of the dependence of the depletion region width on the collector voltage, as illustrated in Fig. 4.10. In this figure, the vertical line at $x = 0$ represents the emitter junction and the solid vertical line beyond $x = w$ represents the collector junction. Since the base is lightly doped compared with the collector, and since depletion region width is inversely

---

[3] This output impedance is actually R_C in parallel with the output impedance of the transistor itself. The reader will be able to appreciate this concept better as he advances in the book.

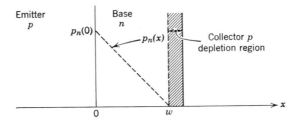

FIGURE 4.10
Diagrammatic sketch showing the effect of depletion
region width and, hence, collector voltage, on base width.

proportional to the doping density, the depletion region extends primarily into the base region. But the injected carriers which diffuse across the base region are swept into the collector as soon as they reach the depletion region, therefore, the effective base width (Fig. 4.10) is w. Now the current flow through the base is primarily by diffusion, as discussed further in Section 4.5. The diffusion equation was developed in Section 3.1 (Eq. 3.3) and is repeated here for hole diffusion.

$$J_p = -qD_p \frac{dp}{dx} \tag{4.6}$$

But the diffusion current through the base region must be continuous, or constant, except for the small loss due to recombinations. Therefore, in the rectangular configuration assumed here, the slope of the $p_n(x)$ curve, dp/dx, must be essentially constant. Then the density of injected carriers $p_n(x)$ must decrease almost linearly from $p_n(0)$ at the emitter junction to essentially zero at the edge of the collector-junction depletion region. Figure 4.10 shows that this slope may be expressed as $p_n(0)/w$. This slope is negative since it is downward. Then, since the diffusion current is proportional to the slope $p_n(0)/w$ and the base width w decreases as the collector voltage increases, the emitter current increases as the collector voltage increases, assuming the emitter-junction voltage remains constant. Alpha (α) also increases with collector voltage because the narrower base region provides less opportunity for carrier recombination. This variation of effective base width with collector-base voltage is known as *base width modulation*.

The effect of collector voltage on α becomes more pronounced as the collector voltage becomes high and approaches avalanche breakdown of the collector junction. Then, carrier multiplication occurs across the collector junction because high-velocity carriers collide with atoms in the crystal and produce additional carriers. A sketch of α as a function of v_{CB} is given in

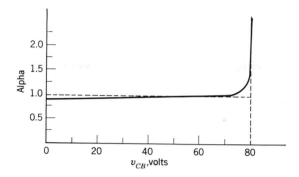

FIGURE 4.11
A sketch of α as a function of v_{CB} for a typical transistor.

Fig. 4.11 for a typical transistor with a collector-junction breakdown voltage of 80 V. Observe that α is unity at a collector voltage considerably below junction breakdown. In the transistor of Fig. 4.11, α is unity at approximately $v_{CB} = 60$ V.

4.4   TRANSISTOR EQUIVALENT CIRCUITS

Equivalent circuits, or models, of the transistor can be devised in the same manner as for the diode. One basic equivalent circuit known as the Ebers-Moll model is shown in Fig. 4.12. The diode symbols may be considered as either actual diodes or ideal depending upon the accuracy desired. The dependent current generator $\alpha_F i_F$ accounts for the current which flows in the collector circuit as a result of the biased emitter-base junction. Similarly, the dependent current generator $\alpha_R i_R$ is the current which flows in the emitter circuits as a result of the biased collector junction. Note that in this model, either junction may be either forward

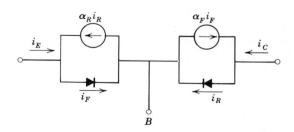

FIGURE 4.12
The Ebers-Moll model of a transistor.

or reverse biased. The collector and emitter currents can be expressed by observation of Fig. 4.12 as

$$i_C = i_R - \alpha_F i_F \tag{4.7}$$

$$i_E = -\alpha_R i_R + i_F \tag{4.8}$$

The currents i_F and i_R are the diode currents which result from the bias voltages across the emitter and collector junctions, respectively. These currents can be expressed by the familiar diode equation form as

$$i_F = I_{ES}(e^{qv_{EB}/kT} - 1) \tag{4.9}$$

$$i_R = I_{CS}(e^{qv_{CB}/kT} - 1) \tag{4.10}$$

These equations can be substituted into Eqs. 4.7 and 4.8 to obtain

$$i_C = I_{CS}(e^{qv_{CB}/kT} - 1) - \alpha_F I_{ES}(e^{qv_{EB}/kT} - 1) \tag{4.11}$$

$$i_E = -\alpha_R I_{CS}(e^{qv_{CB}/kT} - 1) + I_{ES}(e^{qv_{EB}/kT} - 1) \tag{4.12}$$

where I_{CS} and I_{ES} are known as the short-circuit saturation currents of the junctions.

The Ebers-Moll model is used primarily to model switching circuits in which both junctions are often either forward biased (a condition known as *saturation*) or reverse biased (a condition known as *cutoff*).

Whenever a transistor is used as an amplifier, the emitter junction is always forward biased and the collector junction is always reverse biased, as previously discussed. Therefore, simpler and more accurate models will be developed for amplifier use. These amplifier models are called ac equivalent circuits because the dc currents are ignored and only the ac or time-varying signal currents are included. Therefore, the diode saturation currents and the q-point bias currents are not included in the models. Also, the signal currents are assumed to be small in comparison with the bias currents, so the voltage drop across a forward-biased diode can be represented by a linear incremental resistance as was done in Chapter 3. One ac model, known as an *equivalent T* circuit is shown in Fig. 4.13. This is a low-frequency model because it does not include the effects of junction capacitance or diffusion capacitance, which will be included later. The point b' can be imagined as being in the center of the base region. Then the incremental resistance r_e accounts for the voltage-current relationship across the forward-biased emitter junction. This resistance is identical with the incremental resistance r_f of a forward-biased diode previously discussed. Thus, its value at normal temperatures and at a given q-point emitter current I_E may be obtained from the relationship $r_e \simeq 25/I_E$ (mA) Ω. As previously mentioned, the ideal-diode symbols are not included in the diagram because the emitter junction is considered to be forward biased

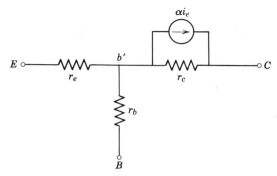

FIGURE 4.13
An equivalent T circuit for a common-base transistor.

and the collector junction is assumed to be reverse biased in amplifier operation.

The incremental resistance r_b represents the effective transverse ohmic resistance in the lightly doped base region. This resistance is typically 100 to 500 Ω in low-power transistors and varies somewhat with collector voltage because the base width varies with the collector voltage. The incremental resistance r_c appears in the same position as the leakage resistance for the reverse-biased collector junction. This resistance was called r_B in the diode model of Chapter 3. However, r_c is smaller than normal values of surface leakage because it primarily accounts for the change in collector current which results from a change in collector voltage, assuming emitter current remains constant, because of base width modulation. Typical values of r_c for a low-power transistor range from one to ten MΩ.

The current generator $\alpha' i_e$ (Fig. 4.13) accounts for the current which diffuses across the base region from the emitter junction as in the Ebers-Moll model. This current source does *not* include the thermal current I_{CO} since the model is a small-signal, ac equivalent circuit as previously discussed. The current $\alpha' i_e$ is not quite equal to αi_e for two reasons. First, $\alpha' i_e$ is slightly larger than i_c since part of $\alpha' i_e$ is shunted through r_c even at $R_L = 0$. Also $\alpha' i_e$ is assumed to be constant while αi_e is a function of v_{CB}, as previously discussed. However, the difference between α' and α is so small, and the value of α' is so difficult to obtain that α is normally used instead of α' in the equivalent T circuit. This practice will be adopted in the work that follows. Again, the small-signal requirement results from the assumption that the incremental components, such as r_e, do not vary over the signal cycle. This is a good approximation only if the signal currents are considerably smaller than the bias currents.

TABLE 4.2
Parameters for a Typical
Low-power Transistor with $I_E = 1$ mA

$\alpha = 0.99$
$r_e = 25\ \Omega$
$r_b = 300\ \Omega$
$r_c = 1\ M\Omega$

Of course, the parameters for the equivalent T circuit vary depending on the type of transistor used. In fact, these incremental parameters even vary for a given transistor if the bias point is changed. However, in order to establish relative magnitudes, the T-circuit parameters for a typical low-power transistor are given in Table 4.2.

Amplifier problems may be solved analytically by using equivalent circuits. Thus, if a transistor is connected as shown in Fig. 4.14a, the equivalent circuit will be as shown in Fig. 4.14b. Note that the transistor has been replaced by its equivalent T configuration. In addition, the batteries have been replaced by short circuits. This is possible because the batteries maintain essentially a constant output potential and consequently are short circuits for the ac or time-varying signals to be considered. Also, the

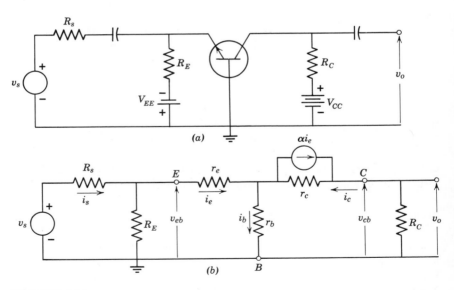

(a)

(b)

FIGURE 4.14
A common-base amplifier: *(a)* actual circuit; *(b)* ac equivalent circuit.

capacitors have been assumed to be sufficiently large to offer very little impedance to the ac signal so these too are replaced by a short circuit.

In order to find the input impedance of the transistor, we note that from Fig. 4.14b the voltage v_{eb} is

$$v_{eb} = i_e r_e + i_b r_b \qquad (4.13)$$

However, $i_b = i_e + i_c$. Normally, the resistance of R_C is much less than r_c. If this condition is true, essentially all of the current from the current generator αi_e flows through R_C. Then, $i_c = -\alpha i_e$ and the equality $i_b = i_e + i_c$ can be written as $i_b = (1 - \alpha)i_e$. When this value of i_b is substituted back into Eq. 4.13, the equation for v_{eb} is

$$v_{eb} = i_e r_e + (1 - \alpha)i_e r_b \qquad (4.14)$$

Since $R_i = v_{eb}/i_e$, we can write

$$R_i = \frac{v_{eb}}{i_e} = r_e + (1 - \alpha)r_b \qquad (4.15)$$

In order to determine the voltage gain, we note that if $R_C \ll r_c$, the voltage v_{cb} which is also equal to v_o is

$$v_o = \alpha i_e R_C \qquad (4.16)$$

and

$$v_{eb} = i_e R_i \qquad (4.17)$$

Then,

$$K_v = \frac{v_o}{v_{eb}} = \frac{\alpha i_e R_C}{i_e R_i} = \frac{\alpha R_C}{R_i} \qquad (4.18)$$

As noted in Example 4.1, the resistor R_E is normally chosen to be much greater than the input impedance to the transistor. If this condition is true, essentially all of the current from the signal generator flows through r_e. Then, $i_s \simeq i_e$. In addition, $v_s = i_e (R_s + R_i)$. Then, the voltage gain v_o/v_s, from source to load, can be written as

$$K_v' = \frac{v_o}{v_s} = \frac{\alpha R_L}{R_s + R_i} \qquad (4.19)$$

To get a feel for typical magnitudes, let us consider another example.

Example 4.3 The transistor whose parameters are listed in Table 4.2 is connected as shown in Fig. 4.14a. Determine the voltage gain of this amplifier if $R_E = 1$ kΩ, $R_s = 25$ Ω, and $R_C = 10$ kΩ.

The equivalent circuit with all of the parameter values is given in Fig. 4.15. From Eq. 4.15, we note the input impedance to the transistor is $R_{it} = r_e + (1 - \alpha)r_b = 25 + (0.01)300 = 28\ \Omega$. Thus, the input impedance to the transistor is $\simeq r_e$. We note that $r_c = 100R_C$ so the simplifying assumptions made in developing Eq. 4.14 and consequently Eq. 4.18 are valid. Therefore, $K_v = \alpha R_C/R_i = 0.99 \times 10^4/28 = 354$. Since $R_E \gg R_i$, Eq. 4.19 is also valid so $K_v = v_o/v_s = 0.99 \times 10^4/(25 + 28) = 187$. Since $R_C \ll r_c$, the current flowing back through r_c is negligible compared to the current flowing through R_C. Consequently, the current gain is $K_i \simeq \alpha i_e/i_e = \alpha = 0.99$. The output impedance of the amplifier is equal to the resistor R_C in parallel with the output resistance of the transistor. Since the output impedance of the transistor is very high in comparison with R_C, the output impedance of the amplifier is approximately R_C. The output impedance of an amplifier will be considered in more detail in a later section of the text.

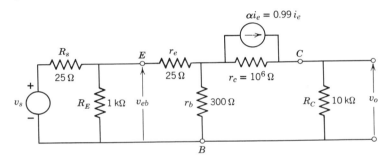

FIGURE 4.15
The equivalent circuit for Example 4.3.

The equivalent T circuit shows, as previously discussed, that the collector current is less than αi_e because of the shunting effect of r_c. However, when R_C is very small in comparison with r_c, the approximation may be made that $i_c = \alpha i_e$ and circuit calculations are simply as noted above. Otherwise, loop or nodal equations, or similar techniques must be used to solve for the currents or voltages in the circuit. If loop equations are used, the solution is made easier by transforming the current generator, αi_e, and its associated internal impedance, r_c, of Fig. 4.13 into a Thevenin's equivalent voltage generator. The current generator and its equivalent voltage generator are shown in Fig. 4.16. Note that the open-circuit voltage of both sources is $(\alpha r_c\, i_e)$ volts and the short-circuit current of both sources is αi_e. Consequently, these two sources *are* equal.

It is standard practice to let $\alpha r_c = r_m$. If this standard practice is followed and the current source of Fig. 4.13 is replaced by its equivalent

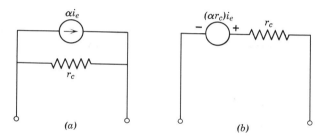

FIGURE 4.16
Equivalent sources for the T configuration : (a) current source;
(b) voltage source.

voltage source, the equivalent T circuit assumes the form shown in Fig. 4.17.
   The standard form for writing the loop equations of Fig. 4.13 or Fig. 4.17
is

$$v_1 = z_{11}i_e + z_{12}i_c \tag{4.20}$$

$$v_2 = z_{21}i_e + z_{22}i_c \tag{4.21}$$

The term z_{11} is known as the self impedance at the input terminals, or
the *input impedance* with the output open ($i_c = 0$). Inspection of Fig. 4.17
reveals that

$$z_{11} = r_e + r_b \tag{4.22}$$

   The term z_{12} is known as the reverse transfer impedance or *reverse
mutual impedance*. This impedance is the ratio of the voltage which appears
at the open-circuited input terminals ($i_e = 0$) to the current which flows
in the output, or collector, circuit. Inspection of Fig. 4.17 shows that

$$z_{12} = r_b \tag{4.23}$$

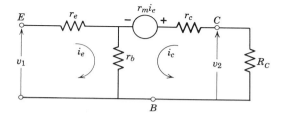

FIGURE 4.17
Equivalent T circuit with a dependent voltage source
in the collector circuit.

Similarly, z_{22} is the self impedance of the output loop, or the output impedance with the input open $(i_e = 0)$. Figure 4.17 shows that

$$z_{22} = r_c + r_b \tag{4.24}$$

Also, z_{21} is the forward transfer, or mutual, impedance which is the ratio of the open-circuit output voltage to the input (emitter) current. Inspection of Fig. 4.17 reveals that this open-circuit output voltage is $i_e r_b + r_m i_e$, so

$$z_{21} = r_b + r_m \tag{4.25}$$

Note that z_{21} is much larger than z_{12} because of the addition of the r_m term in z_{21}. This feature distinguishes an *active* circuit, which is capable of power amplification, from a *passive* circuit in which $z_{21} = z_{12}$.

In order to solve the loop equations 4.20 and 4.21, one needs to note that $v_2 = -i_c R_L$ and then transform this term to the right side of the equation. Then Cramer's rule, or other techniques, can be used to solve the two simultaneous equations.

PROBLEM 4.6   Solve the loop equations 4.20 and 4.21 for the collector current i_c in terms of v_1, and determine the voltage gain if the transistor has the parameters given in Table 4.2 and $R_C = 20$ kΩ. Compare your answer with the approximate value obtained by using Eq. 4.18.

Another transistor equivalent circuit known as a z-parameter circuit is derived directly from the loop equations 4.20 and 4.21. This circuit is shown in Fig. 4.18. Although the elements of this circuit are not identifiable with physical regions in the transistor, these elements can be obtained from the measurements of the terminal currents and voltages. However, the required open-circuit measurements such as the input impedance with the output open, as discussed above, are difficult to make because the measurements need to be made at the desired q point, and thus the desired bias currents must flow in both the input and the output circuits.

An equivalent circuit, known as a y-parameter circuit, is obtained by

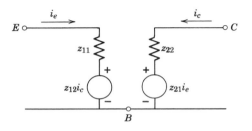

FIGURE 4.18
A z-parameter equivalent circuit.

replacing the dependent voltage generators of the z-parameter circuit with dependent current generators as shown in Fig. 4.19. Note that the input admittance is y_{11} when the output is *shorted* because $v_2 = 0$, and the current generator $y_{12}v_2$ has zero current when $v_2 = 0$. Also, observe that y_{11} is *not* the reciprocal of z_{11} because the measurements are made under different conditions. In fact, the main advantage of the y-parameter circuit is that the short-circuit measurement can be more easily made than the open-circuit measurement because the output is shorted for ac signals when $R_C = 0$. Similarly, y_{22} is the output admittance with the input shorted. The forward transfer admittance y_{21} is the ratio of the short-circuit output current to the input voltage and the reverse transfer admittance y_{12} is the ratio of the short-circuit input current to a voltage v_2 applied to the output.

The basic nodal equations for the y-parameter circuit may be written from observation of Fig. 4.19

$$i_e = y_{11}v_1 + y_{12}v_2 \qquad (4.26)$$

$$i_c = y_{21}v_1 + y_{22}v_2 \qquad (4.27)$$

The y parameters could be obtained in terms of the T parameters or Z parameters, or they could be measured directly, as previously mentioned.

Many different models have been developed for the transistor. You may wonder why we need any more models (or equivalent circuits) since the equivalent T seems to yield satisfactory circuit solutions and gives additional insight into the characteristics of the transistor. In fact, you may wonder why models are necessary at all since graphical analysis will also yield solutions to electronic circuits. This latter question will be answered first, since it is the easiest to answer. Graphical analysis gives no indication concerning the frequency capabilities of a device. Graphical analysis is also cumbersome, and relies on the availability of suitable sets of characteristic curves, which are usually *not* available. Then to the first question, why not stick with one model? You may as well ask, "Why not use only one model automobile—why compacts, sedans, station wagons, and trucks?" The

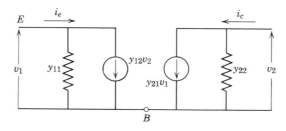

FIGURE 4.19
A y-parameter equivalent circuit.

reasons are similar. One model will not perform all functions well. The equivalent T is a poor high-frequency model. It isn't even a good low-frequency model because it isn't easy to use. It does look like the skeleton of the physical model we have used for a transistor, and therefore, has some educational value in developing better models because its components can be related to the physical parts of a transistor.

At least two additional models are essential to an adequate understanding of semiconductor circuits, and a reasonable dexterity in their use is essential. The first of these is the hybrid or h-parameter circuit which can yield easy *approximate* solutions to circuit problems, providing the high-frequency characteristics are not important. The z-parameter and y-parameter models were used as intermediate steps in the development of the h-parameter model in this chapter. The other popular and very useful model, particularly at high frequencies, is the hybrid-π model, which will be developed in Chapter 7. The y-parameter model is also very useful for high-frequency tuned amplifiers and will be discussed in connection with that application.

4.5   THE h-PARAMETER CIRCUIT

A popular circuit which is a cross between the z-parameter and y-parameter circuits and is therefore called a hybrid or *h-parameter* circuit is shown in Fig. 4.20. The input voltage and output current are assumed to be the dependent variables. Therefore, the basic circuit equations are

$$v_i = h_{ib} i_e + h_{rb} v_o \tag{4.28}$$

$$i_c = h_{fb} i_e + h_{ob} v_o \tag{4.29}$$

The subscript notation has been changed from that given z- and y-parameter circuits to conform to standard transistor form. The second subscript b, designates the common-*base* configuration. The meaning of the first subscript can be seen from the following definitions of the h parameters.

$h_i = $ *input* impedance with the output shorted (v_2 or $v_o = 0$).
$h_r = $ *reverse voltage* transfer ratio with input open (i_1 or $i_e = 0$).
$h_f = $ *forward current* transfer ratio with output shorted ($v_o = 0$).
$h_o = $ *output* admittance with input open ($i_e = 0$).

These parameters are easily obtained in terms of the equivalent T parameters by inspection of the equivalent T circuit of Fig. 4.13 as shown in the following discussion. The input resistance with R_C small compared with r_c was determined and is given by Eq. 4.15. This equation is precisely correct if R_C is zero, in which case $R_i = h_{ib}$ by definition. Then, using Eq. 4.15,

$$h_{ib} = r_e + (1 - \alpha)r_b \tag{4.30}$$

Also, note from Fig. 4.21 that a signal voltage v_c applied to the output

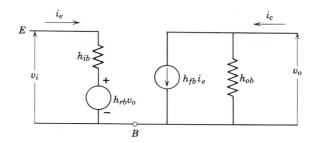

FIGURE 4.20
An *h*-parameter equivalent circuit.

terminals will cause current i to flow through r_c and r_b. With the input open, $i_e = 0$ so i is the only current. Then the voltage appearing in the input circuit is the drop across r_b, as shown in Fig. 4.21. Then

$$h_{rb} = \frac{v_e}{v_c} = \frac{ir_b}{i(r_c + r_b)} = \frac{r_b}{r_c + r_b} \tag{4.31}$$

From observation of Fig. 4.13, or from the definition of α, one can see that

$$h_{fb} = -\alpha \tag{4.32}$$

The negative sign appears because the current generators αi_e and $h_{fb} i_e$ have opposite polarities. Also, from Fig. 4.21, one can see that the output admittance with the input open is

$$h_{ob} = \frac{1}{r_c + r_b} \simeq \frac{1}{r_c} \tag{4.33}$$

One main advantage of the *h*-parameter equivalent circuit is that the *h* parameters may be obtained from the transistor input and collector characteristics previously discussed. Since these curves are easily obtained from a transistor curve tracer, the *h* parameters are readily available. They are also frequently given by the transistor manufacturer.

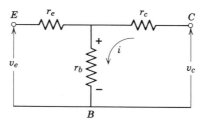

FIGURE 4.21
Equivalent circuit used to determine
$h_{rb} = v_e/v_c$.

The h-parameters are often defined by a set of partial derivatives which are given below as an introduction to the graphical method of obtaining the parameters. Since the partial derivative is the slope of a function at a given point, the incremental h-parameters can be defined as

$$h_{ib} = \frac{\partial v_{EB}}{\partial i_E} \simeq \frac{\Delta v_{EB}}{\Delta i_E}\bigg|_{v_{CB} = \text{constant}} \tag{4.34}$$

Thus, h_{ib} is the slope of the static input characteristic with v_{CB} constant. Of course, the slope at the q point is normally used.

$$h_{ob} = \frac{\partial i_C}{\partial v_{CB}} \simeq \frac{\Delta i_C}{\Delta v_{CB}}\bigg|_{i_E = \text{constant}} \tag{4.35}$$

Therefore, h_{ob} is the slope of the collector characteristic with i_E held constant at the q-point value

$$h_{fb} = \frac{\partial i_C}{\partial i_E} \simeq \frac{\Delta i_C}{\Delta i_E}\bigg|_{v_{CB} = \text{constant}} \tag{4.36}$$

$$h_{rb} = \frac{\partial v_{EB}}{\partial v_{CB}} \simeq \frac{\Delta v_{EB}}{\Delta v_{CB}}\bigg|_{i_E = \text{constant}} \tag{4.37}$$

The curves which will yield h_{fb} and h_{rb} have not yet been drawn but could easily be constructed from the information contained in the collector and input characteristics However, the h_{fb} and h_{rb} parameters can be obtained from the collector and input characteristics directly by using the Δ values, as illustrated by the following example.

Example 4.4   Input and collector characteristics are given in Fig. 4.22 for a typical low-power n-p-n transistor. We will determine the h parameters

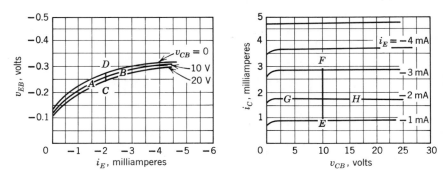

FIGURE 4.22
Illustration of the technique used to obtain the h parameters from characteristic curves.

for this transistor. The (dynamic) input resistance with the output shorted h_{ib} is $\Delta v_{EB}/\Delta i_E$ with v_{CB} held constant (shorted for ac). This is the slope of an input characteristic curve at a given point. The point at which the h parameters are desired is, of course, the q point. In Fig. 4.22, the q point has been selected at $i_E = -2.0$ mA, $v_{CB} = 10$ V. Therefore, h_{ib} at this point is the slope of the line AB which has the value $h_{ib} = .03$ V/1 mA = 30 Ω.

The parameter h_{rb} is $\Delta v_{EB}/\Delta v_{CB}$ with i_E held constant (ac input current = 0). This value of h_{rb} may be determined from the line CD which gives the change in v_{EB} produced by a 20 V change in v_{CB}. The value of h_{rb} at this point is .025 V/20 V = 1.25×10^{-3}.

The forward current transfer ratio with output shorted, h_{fb}, can be determined at the q point from line EF which gives the collector current change $\Delta i_C = 1.9$ mA for an emitter current change $\Delta i_E = 2$ mA with v_{CB} constant (or zero ac voltage). Thus, the value of h_{fb} is -1.9 mA/2.0 mA = $-.95$.

The output admittance with the input open, h_{ob}, is $\Delta i_C/\Delta v_{CB}$ with i_E held constant. Thus, h_{ob} is the slope of the output characteristic curve at the q point or the line GH. This slope is so small it is difficult to determine with satisfactory accuracy. However, h_{ob} for this q point is roughly .01 mA/10 V = 10^{-6} mho.

We will now calculate the voltage gain of the transistor $\Delta v_{CB}/\Delta v_{BE}$ with $R_C = 10$ kΩ, using the h parameters determined from Fig. 4.22. The h-parameter circuit with these values listed is given in Fig. 4.23. At this time we will avoid a solution using simultaneous equations by assuming that $i_e = 1$ mA. Then $v_{cb} = 0.95$ mA/$(10^{-6} + 10^{-4})$ = 9.4 V. Then

$$v_{eb} = h_{ib} i_e + h_{rb} v_{cb} = 0.3 + .0117 = .0417 \text{ V}$$

and the voltage gain is 9.4/.0417 = 225.

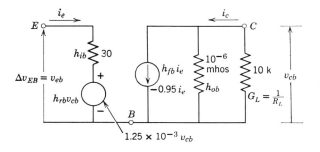

FIGURE 4.23
An *h*-parameter circuit with values determined from Example 4.4

PROBLEM 4.7   A given transistor has $r_e = 15\ \Omega$, $r_b = 200\ \Omega$, $\alpha = .98$ and $r_c = 2\ M\Omega$ at a certain q point. Determine the h parameters, draw an h-parameter equivalent circuit and calculate the voltage gain v_{cb}/v_{eb} if $R_C = 5\ k\Omega$.

The h-parameter circuit, Fig. 4.23, will now be solved generally to obtain the input impedance, voltage gain, and current gain for a general load resistance[2] R_L or load conductance $G_L = 1/R_L$. This solution will follow the procedure used in Example 4.4.

$$v_{cb} = -\frac{h_{fb}\, i_e}{h_{ob} + G_L} \tag{4.38}$$

The negative sign in Eq. 4.38 results from the negative voltage polarity on the collector when the current directions are as shown. Also

$$v_{eb} = h_{ib}\, i_e + h_{rb}\, v_{cb} \tag{4.39}$$

Substituting Eq. 4.38 into Eq. 4.39,

$$v_{eb} = \left(h_{ib} - \frac{h_{rb}\, h_{fb}}{h_{ob} + G_L}\right) i_e \tag{4.40}$$

Then the transistor input impedance is

$$z_{it} = \frac{v_{eb}}{i_e} = h_{ib} - \frac{h_{rb}\, h_{fb}}{h_{ob} + G_L} \tag{4.41}$$

The voltage gain can be obtained from dividing Eq. 4.38 by Eq. 4.40,

$$K_v = \frac{[-h_{fb}/(h_{oe} + G_L)]}{h_{ib} - [h_{fb}\, h_{rb}/(h_{ob} + G_L)]} = -\frac{h_{fb}}{h_{ib}(h_{ob} + G_L) - h_{fb}\, h_{rb}} \tag{4.42}$$

The current i_c through the load is $-v_{cb}\, G_L$, so using Eq. 4.38

$$i_c = \frac{h_{fb}\, i_e\, G_L}{h_{ob} + G_L} \tag{4.43}$$

so the current gain is

$$K_i = \frac{i_c}{i_e} = \frac{h_{fb}\, G_L}{h_{ob} + G_L} \tag{4.44}$$

The output admittance is so small that it can usually be neglected and will therefore not be pursued at this time.

---

[2] Previously we have used R_C as the load on a transistor. This notation is too restrictive, since other circuit elements may parallel R_C and become part of the load.

4.6 DIFFUSION CAPACITANCE AND α CUTOFF FREQUENCY

The response time, or frequency response of a transistor, like a diode, is limited by diffusion capacitance and junction capacitance. The diffusion capacitance results from the carriers injected into the base region as shown in Fig. 4.24. The injected carriers tend to produce an excess charge (positive for a p-n-p) in the base, but the electric field which begins to build up causes electrons to flow through the base connecting lead to

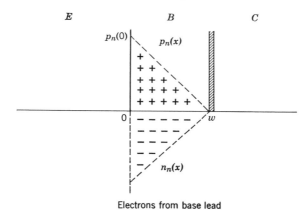

Electrons from base lead

FIGURE 4.24
A representation of stored charge in the base which results in diffusion capacitance.

neutralize the charge. Note that these neutralizing charges do not necessarily combine with the injected carriers. In fact, they do not appreciably increase the recombination rate as long as their density is small in comparison with the base doping concentration. The capacitive current which flows in the emitter-base circuit does not contribute to the collector current. Therefore, as the frequency increases and this capacitive current becomes significant, α decreases.

The dynamic or ac value of the diffusion capacitance can be determined as a function of the effective base width w of the transistor with the aid of Fig. 4.25. If the injected carrier density is increased by Δp from p_o, the stored charge is increased by ΔQ, where ΔQ is the increase in average charge density times the effective volume of the base region. Then, since the average increase in charge density is $q\Delta p/2$,

$$\Delta Q = \frac{q\Delta p w A}{2} \tag{4.45}$$

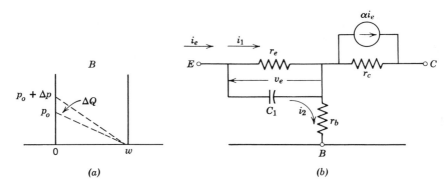

FIGURE 4.25
(a) A sketch to assist in the determination of diffusion capacitance, and (b) a high-frequency equivalent T circuit.

where A is the effective cross-sectional area of the base region. The diffusion capacitance is

$$C_d = \frac{\Delta Q}{\Delta v_E} \tag{4.46}$$

where Δv_E is the change in voltage across the capacitor or the ac voltage v_e shown in Fig. 4.22. But, from this figure, $v_e = i_1 r_e$. Then, using Eq. 4.45,

$$C_d = \frac{q\Delta p w A}{2 i_1 r_e} \tag{4.47}$$

Now, i_1 is the ac component of diffusion current. We can convert current density J (see Eq. 4.6) to current by multiplying J by area A. Then, i_1 is

$$i_1 = AqD_p \frac{\Delta p}{w} \tag{4.48}$$

Substituting this value of i_1 into Eq. 4.47

$$C_d = \frac{w^2}{2D_p r_e} \tag{4.49}$$

The value of diffusion capacitance may be calculated for a uniform base transistor at a given q point if the effective base width is known. For example, a p-n-p silicon transistor with $w = 10^{-4}$ meters and $I_E = 1$ mA has $C_d = (10^{-8})/[2 \times 13 \times 10^{-4}(25)] = 1.54 \times 10^{-7}$ F. The capacitor C_1 in Fig. 4.25 is the diffusion capacitance C_d plus the emitter junction capacitance C_{je} and is therefore somewhat larger than C_d.

Alpha cutoff frequency f_α is defined as the frequency at which α decreases to $0.707\alpha_0$ where α_0 is the low-frequency value of α. This f_α is the frequency at which the capacitive current i_2 is equal to the diffusion current i_1 (Fig. 4.25), since these currents are 90° out of phase and must be added vectorially to obtain i_e. Remember that $\alpha = i_c/i_e$.

Then r_e must equal X_c at f_α or ω_α. Therefore, r_e is

$$r_e = \frac{1}{\omega_\alpha C_1} \qquad (4.50)$$

Thus,

$$\omega_\alpha = 2\pi f_\alpha = \frac{1}{r_e C_1} \qquad (4.51)$$

Note that if C_d is large compared with C_{je}, so the latter can be neglected, the expression for C_d in Eq. 4.49 can be substituted for C_1 in Eq. 4.51. Then

$$\omega_\alpha \simeq \frac{2D_p}{w^2} \qquad (4.52)$$

Thus, the *p-n-p* silicon transistor with $w = 10^{-4}$ meter would have $\omega_\alpha \simeq (2 \times 13 \times 10^{-4})/10^{-8} = 2.6 \times 10^5$ rad/sec and $f_\alpha \simeq 4.14 \times 10^4$ Hz. This value of alpha cutoff frequency is disappointingly low. If the base width w is reduced to 10^{-5} meter, f_α is increased to 4.14×10^6 Hz.

PROBLEM 4.8 If the transistor above were an *n-p-n* silicon with $w = 10^{-5}$ m, what would be the value of f_α?
Answer: 1.08×10^7 Hz.

PROBLEM 4.9 Determine f_α for an *n-p-n* germanium transistor with $w = 10^{-5}$ m.
Answer: 3.15×10^7 Hz.

Gallium arsenide is an especially attractive material for high-frequency transistors because of the very high electron mobility and consequently high value of D_n. Physically, the high carrier mobility and large diffusion constant provide a high value of f_α because the carriers diffuse through the base so quickly that the stored charge, or diffusion capacitance, is relatively small for a given value of I_E.

PROBLEM 4.10 Determine the value of f_α for an *n-p-n* GaAs transistor with an effective base width $w = 10^{-5}$ meters if $D_n = 200 \times 10^{-4}$ m²/sec for GaAs.
Answer: 6.37×10^7 Hz.

PROBLEM 4.11 Determine the common-base *h* parameters approximately for the 2N3903 transistor at the q point $I_C = 3.0$ mA, $V_{CB} = 6.0$ V. Curves are given in Fig. I.1 (Appendix I).

Answer: $h_{ib} = 12 \ \Omega$, $h_{rb} \simeq 0$, $h_{fb} \simeq -.99$, $h_{ob} \simeq 3 \times 10^{-7}$ *mho* (estimate).

PROBLEM 4.12 Assume that you want to build a sound amplifier to eavesdrop on your sister and her boyfriend. You have an 8 Ω (internal resistance) loudspeaker which will serve as a microphone. You find by measurement that moderate sound levels will generate .01 V peaks in the 8 Ω voice coil of this loudspeaker. You also have a 2,000 Ω (internal impedance) headset with which you can do the listening. In addition, you have a 2N3903 transistor (Fig. I-1), a 12 V battery, a 3 V battery, and a wide assortment of resistors and capacitors. Design a common-base amplifier, using these components and determine the peak voltage expected across the headset.

Answer: 1.2 V, if $I_C = 3$ mA.

5

Common-Emitter Amplifiers

The student may have surmised while reading Chapter 4 that a re-arrangement of the input terminals of the common-base amplifier would reduce the input current requirement for a given output current and thus provide higher power gain. Such a configuration would employ the emitter as the common or grounded terminal and the base as the other input terminal as shown in Fig. 5.1.

5.1 CURRENT GAIN OF THE COMMON-EMITTER AMPLIFIER

Since $i_B = (1 - \alpha)i_E$, the input current is reduced by the factor $(1 - \alpha)$, and the input impedance is increased by the factor $1/(1 - \alpha)$ as compared with the common-base configuration, assuming the q point to remain the same. Since the base current is the input current, it is desirable to obtain an expression for the collector, or output, current as a function of the input current. This may be done by using Eq. 4.4, repeated below.

$$i_C = -\alpha i_E + I_{CO} \tag{4.4}$$

But, from Fig. 5.1

$$-i_E = i_C + i_B \tag{5.1}$$

Substituting this value of $-i_E$ into Eq. 4.4,

$$i_C = \alpha(i_C + i_B) + I_{CO} \tag{5.2}$$

123

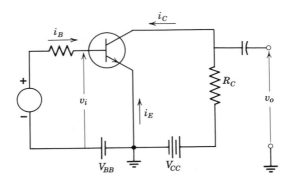

FIGURE 5.1
A common-emitter amplifier.

Then solving for i_C in terms of i_B and I_{CO},

$$i_C = \frac{\alpha}{1 - \alpha} i_B + \frac{I_{CO}}{1 - \alpha} \tag{5.3}$$

The term $\alpha/(1 - \alpha)$ is known as β (beta), the current gain of the common-emitter configuration when the output is shorted. A transistor which has $\alpha = 0.99$, has $\beta = .99/.01 = 99$ and $1/(1 - \alpha) = 1/.01 = 100$. Note that $1/(1 - \alpha) = \beta + 1$, as can be shown by adding $\alpha/(1 - \alpha)$ to $(1 - \alpha)/(1 - \alpha)$. Thus i_C may be written in terms of i_B, i_{CO}, and β, using Eq. 5.3.

$$i_C = \beta i_B + (\beta + 1)I_{CO} \tag{5.4}$$

Perhaps it is surprising that the thermally generated current I_{CO} is multiplied by $(\beta + 1)$ if the base current is zero, or constant. For example, if $i_B = 10 \ \mu A$, $I_{CO} = 10 \ \mu A$ and $\alpha = .99$, $i_C = 99 \times 10 \ \mu A + 100 \times 10 \ \mu A = 1.99 \ mA$.

One way to visualize the control action of the base is to assume the forward bias between the base and emitter to be increased. This increased forward bias causes the height of the potential hill to be reduced and the majority current across the junction to be increased. Most of the carriers which comprise this current diffuse across the thin base region and are swept into the collector. Thus the increase of collector current may be much greater than the increase of base current. The collector current change is almost proportional to the base current change. Therefore, the base current is often assumed to be the control parameter. The thermal current I_{CO} is amplified because it is forced to flow across the emitter junction. As a result, a small forward bias is produced across that junction, and the collector current is increased as though I_{CO} were added to the base current.

The process by which I_{CO} is amplified is shown in Fig. 5.2. With the base circuit open, the current I_{CO} is not free to flow out of the base lead to return to V_{CC}, but must flow across the emitter junction and therefore produce a forward-bias voltage across that junction which in turn causes an additional current of magnitude βI_{CO} to flow in the collector circuit, thus making a total current of $(\beta + 1)I_{CO}$. The I_{CO} causes a similar increase in collector current when the base current is held at a constant nonzero value.

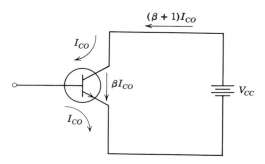

FIGURE 5.2
An illustration of the process by which I_{CO} is multiplied in the common-emitter configuration.

PROBLEM 5.1   A given transistor has $\alpha = .995$ and $I_{CO} = 5\ \mu A$. What is the value of collector current when $i_B = 20\ \mu A$?
Answer: 4.98 mA.

5.2   GRAPHICAL ANALYSIS OF THE COMMON-EMITTER AMPLIFIER

Collector characteristics and input characteristics may be obtained for the common-emitter configuration in a manner similar to that described for the common-base configuration. However, a curve tracer is recommended as the easy and accurate method of obtaining transistor characteristics.

A set of collector characteristics for a typical *n-p-n* germanium transistor is given in Fig. 5.3. Comparison of these curves with those given in Fig. 4.6 for the common-base configuration show the following differences.

1.   The base current i_B is a parameter in the common-emitter set.
2.   The collector current is approximately equal to zero when v_{CE} equals zero, regardless of the value of i_B. The reason for this behavior is that the collector junction has the same forward-bias voltage as the

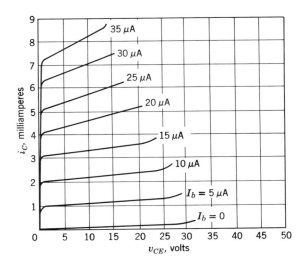

FIGURE 5.3
Collector characteristics for the common-emitter configuration.

emitter junction when $v_{CE} = 0$, as seen from Fig. 5.1 or Fig. 5.2. Therefore, an injection current flows across the collector junction into the base. This current is almost equal in magnitude and opposite in direction to the emitter junction injection current, and therefore the collector current, which is essentially the difference between these two currents, is almost zero. As v_{CE} is increased in the reverse-bias direction, the collector current increases very rapidly until the collector junction bias is reduced to zero. Larger values of v_{CE} apply reverse bias to the collector junction and the collector current is then relatively independent of collector voltage and the collector characteristic is nearly horizontal.

3.   The spacing between curves increases as v_{CE} increases because α and hence β increase with v_{CE}. As shown by the curves, the increase in β is much more pronounced than the increase in α because small changes in α produce large changes in β.

4.   The slope of the collector characteristics are much steeper in the common-emitter set because the change in collector current which results from base-width modulation, like the thermal current I_{CO}, is forced to flow across the emitter junction, if base current is held constant. The change in forward bias thus produced causes β times as much change in collector current as that produced by the initial basewidth modulation. Thus the slope of the output characteristic in the common-emitter set is increased by the factor $(\beta + 1)$.

A set of input characteristics for the transistor of Fig. 5.3 is given in Fig. 5.4a. The curve obtained with $v_{CE} = 0$ is widely spaced from the others because both junctions are forward biased when $v_{CE} = 0$, as previously mentioned, and the majority currents across both junctions store charge in the base. This excess charge results in increased recombination and a larger base current for a given value of forward-emitter bias. Except for switching applications, the transistor does not normally operate in this region of very small collector voltage which is known as the *saturation region*.

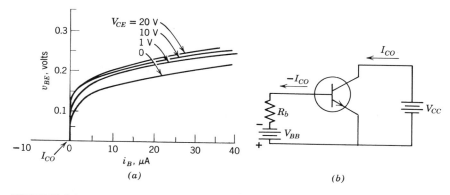

(a)   (b)

FIGURE 5.4
(a) **Common-emitter input characteristics, and** (b) **illustration of reverse-base current required to eliminate the forward bias across the emitter junction.**

Observe from Fig. 5.4a that, except for $v_{CE} = 0$, the emitter junction is forward biased when $i_B = 0$ because I_{CO} flows across the emitter-base junction and produces forward bias. Therefore, as shown in Fig. 5.4b, the emitter junction voltage can be reduced to zero and consequently the collector current reduced to I_{CO} only if a reverse base current equal to I_{CO} flows in the base circuit.

The current gain, voltage gain, and permissible signal levels of a transistor amplifier may be determined graphically, using the characteristic curves, in the manner indicated for the common-base amplifier in Section 4.2. An example will be used to illustrate the procedure.

Example 5.1   Let us analyze an amplifier which is connected as shown in Fig. 5.5. This is an ac amplifier which uses an *n-p-n* germanium transistor with $V_{CC} = 25$ V, collector-circuit load resistor $R_C = 6$ kΩ, and base-bias battery $V_{BB} = 3$ V.

The input and collector characteristics for the transistor of Fig. 5.5 are the same as those given in Figs. 5.3 and 5.4 and are repeated in Fig. 5.6 for

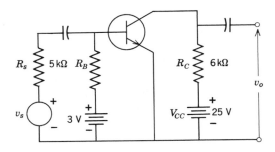

FIGURE 5.5
A circuit diagram for an ac common-emitter amplifier using an *n-p-n* transistor.

convenience. A 6 kΩ load line is drawn on the collector characteristics and
the q point is chosen at $i_B = 10\ \mu$A. From the collector characteristics, we
note that the q-point collector current is 2.2 mA. The base circuit
resistance required to obtain this q point is $R_B = (V_{BB} - v_{BE})/i_B = (3 - 0.2)/(10 \times 10^{-6}) = 280$ kΩ.

Data from along the load line is transferred to the input characteristics to
produce the dynamic input characteristic in the same manner as shown in
Example 4.2. Next, a waveform for the signal voltage v_s is chosen. Let us
assume v_s will produce a base current of the form $i_b = 5 \cos \omega t$ microamperes. Then, the total base current is $i_B = (10 + 5 \cos \omega t)$ microamperes.
One cycle of the input base current is plotted at right angles to the load
line as in Example 4.2. The peak base current is 15 μA and is noted as
point A on both the base and collector characteristics. The minimum base
current is 5 μA and is noted as point B on both characteristic curves.
The variations of v_{BE}, i_C, and v_{CE} are also plotted. The variation of base
current is $\Delta i_B = (15 - 5)\ \mu$A $= 10\ \mu$A. Similarly, $\Delta v_{BE} = 0.22$ V $- 0.175$ V $=$
0.045 V. $\Delta i_C = 3.3$ mA $- 1.1$ mA $= 2.2$ mA. Finally, $\Delta v_{CE} = 18$ V $- 5.5$ V $=$
12.5 V.

The input impedance of the transistor is $Z_{it} = \Delta v_{BE}/\Delta i_B = 0.045/10^{-5} =$
4.5 kΩ. The current gain is $K_i = \Delta i_C/\Delta i_B = 2.2 \times 10^{-3}/10^{-5} = 220$. The
output impedance of this amplifier is $Z_o = \Delta v_{CE}/\Delta i_C = 12.5/(2.2 \times 10^{-3}) =$
5.7 kΩ. Finally, the voltage gain is $K_v = \Delta v_{CE}/\Delta v_{BE} = 12.5$ V$/0.045 = 278$.
There is an interesting feature to be observed in connection with the voltage
gain. Note when $t = 0$, the voltage v_{BE} is at point A (Fig. 5.6) and is the
maximum value of input voltage. In contrast, v_{CE} is at its *minimum* value
when $t = 0$ (point A). Therefore, the output voltage waveform is *inverted* in
relation to the input voltage waveform in the common-emitter configuration. The inversion is normally noted by listing the voltage gain as -278.

The power gain of the amplifier is defined as the signal power output

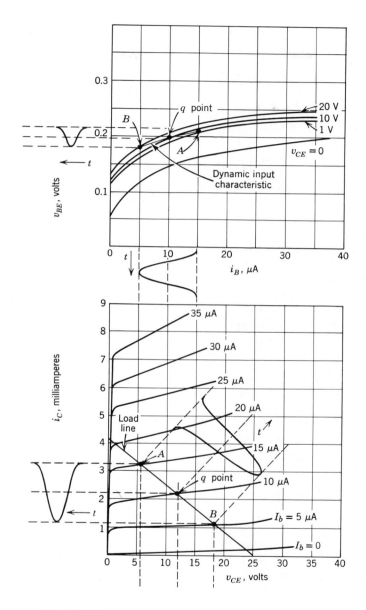

FIGURE 5.6
Input and collector characteristics for the transistor of Fig. 5.5

divided by the signal power input to the amplifier. This power gain K_p is also equal to the voltage gain times the current gain. Therefore, $K_p = K_i \, K_v = 220 \times 278 = 61200$. Obviously, this amplifier produces substantial current, voltage, and power gains.

PROBLEM 5.2   The amplifier of Fig. 5.5 has $R_C = 10 \, \text{k}\Omega$. Other components remain the same except R_B. Draw a load line, choose a suitable q point and determine R_B, K_i, and K_p.

Answer: $R_B \simeq 560 \, \text{k}$, $K_i = 180$, $K_v = 270$, $K_p = 4.8 \times 10^4$.

5.3   COMMON-EMITTER EQUIVALENT CIRCUITS

The common-base equivalent T circuit could be used for the common-emitter configuration by exchanging the input terminals as shown in Fig. 5.7a. However, this equivalent circuit is not convenient for the common-emitter configuration because the current source in the collector circuit is given in terms of i_e, which is not the input current. The circuit of Fig. 5.7b is therefore better. Note that the resistance $r_d = (1 - \alpha)r_c = r_c/(\beta + 1)$ as previously discussed because with $i_b = 0$, the output current which results from base width modulation and change in α flows across the emitter junction and is therefore amplified by the factor $(\beta + 1)$. This reduction in output resistance can be easily shown mathematically by applying a voltage v_{ce} to the output terminals of the circuit of Fig. 5.7a. Then the impedance seen at the collector terminal is this voltage divided by the current flowing into the transistor. Assuming $r_e \ll r_c$, this current is

$$i_c \simeq \frac{v_{ce}}{r_c} + \alpha i_e \tag{5.5}$$

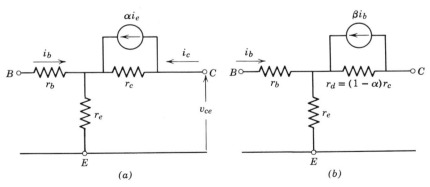

FIGURE 5.7
Equivalent T circuits for the common-emitter configuration.

But with $i_b = 0$ (or if i_B is constant), $i_e = i_c$. Making this substitution for i_e in Eq. 5.5,

$$i_c = \frac{v_{ce}}{r_c(1 - \alpha)} \tag{5.6}$$

Then,

$$r_d = \frac{v_{ce}}{i_c} = r_c(1 - \alpha) \tag{5.7}$$

The *h*-parameter circuit for the common-emitter configuration is identical to the common-base *h*-parameter circuit as shown in Fig. 5.8. The

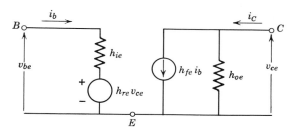

FIGURE 5.8
The common-emitter *h*-parameter circuit.

definitions of the *h*-parameters are the same for the common-emitter configuration as for the common-base, although their values are different as indicated by the second subscript. A more precise definition for h_{ie} than $\Delta v_{BE}/\Delta i_B$ with v_{CE} constant, is the derivative form which follows.

$$h_{ie} = \frac{dv_{BE}}{di_B}\bigg|_{v_{CE}\,=\,\text{constant}} = \text{the input impedance with the output shorted} \tag{5.8}$$

The improved preciseness results because the derivative is the slope of the input curve at the desired point, as discussed in Section 4.4, whereas the ratio $\Delta v_{BE}/\Delta i_B$ is only the approximate slope and the goodness of the approximation depends upon the size of Δv_{BE}. Similarly the other common-emitter *h* parameters can be defined.

$$h_{re} = \frac{dv_{BE}}{dv_{CE}}\bigg|_{i_B\,=\,\text{constant}} = \text{the reverse voltage transfer ratio with the input open} \tag{5.9}$$

$$h_{fe} = \frac{di_C}{di_B}\bigg|_{v_{CE}\,=\,\text{constant}} = \text{the forward current transfer ratio with the output shorted} \tag{5.10}$$

$$h_{oe} = \frac{di_C}{dv_{CE}}\bigg|_{i_B\,=\,\text{constant}} = \text{the output admittance with the input open} \tag{5.11}$$

The slope of the input characteristics will yield h_{ie} and the slope of the output characteristics will give h_{oe} at the prescribed point. Also, from the information given in these sets of curves, additional curves could be plotted for v_{BE} as a function of v_{CE}, with i_B held constant, and i_C could be plotted as a function of i_B with v_{CE} held constant. The slope of these curves would give h_{re} and h_{fe}, respectively, with improved accuracy. However, for most applications, adequate accuracy will be obtained by using the Δ value technique employed in finding the common-base h parameters h_{rb} and h_{fb}. Thus, all the common-emitter h parameters can be determined from the curves of Fig. 5.6 for the transistor represented there.

PROBLEM 5.3   Determine the common-emitter h parameters for the transistor of Fig. 5.6 at the q point $i_B = 10\ \mu A$, $v_{CE} = 10$ V.

Answer: $h_{ie} = 4000\ \Omega$, $h_{re} = 4 \times 10^{-4}$, $h_{fe} = 230$, and $h_{oe} = 25\ \mu mho$.

As previously noted, the input impedance (with output shorted) of the common-emitter transistor is $(\beta + 1)$ times as high as that of the common base. Therefore,

$$h_{ie} = (\beta + 1)h_{ib} \tag{5.12}$$

Also, as previously shown, the output impedance (with input open) of the common-emitter configuration is reduced by the factor $(\beta + 1)$, as compared with the common-base. Thus the admittance h_{oe} is greater than the admittance h_{ob} by the factor $(\beta + 1)$.

$$h_{oe} = (\beta + 1)h_{ob} \tag{5.13}$$

In addition, since $h_{fb} = -\alpha$, $h_{fe} = \alpha/(1 - \alpha)$ and $(\beta + 1) = 1/(1 - \alpha)$. Therefore,

$$h_{fe} = -(\beta + 1)h_{fb} \tag{5.14}$$

As seen above, the common-emitter and common-base h parameters, except h_{re} and h_{rb}, are related by the factor $(\beta + 1)$ or $(h_{fe} + 1)$. In modern transistors, h_{re} and h_{rb} are so small they can usually be neglected at low frequencies without significant error. Therefore, sets of curves are not needed for both common-base and common-emitter configurations. The common-emitter curves are almost universally used because they provide better accuracy in obtaining the parameters.

PROBLEM 5.4   Using the h parameters determined in Problem 5.3, calculate the common-base h parameters for the transistor of Fig. 5.6 at the same q point.

Answer: $h_{ib} = 17.3\ \Omega$, $h_{fb} = -.996$, and $h_{ob} = 1.1 \times 10^{-7}$ mho.

The current gain, voltage gain, input resistance, and output admittance (or impedance) can be determined for the common-emitter transistor amplifier by use of the h parameter equivalent circuit in the same manner as described for the common-base amplifier (Sec. 4.4). The h parameters obtained from Fig. 5.5 are listed on the h parameter circuit of Fig. 5.9.

Since the equivalent circuit configurations are identical, the equations derived for the common-base amplifier are applicable to the common-emitter amplifier by simply replacing the common-base h parameters with

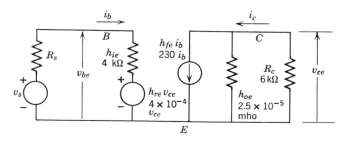

FIGURE 5.9
The h-parameter circuit with values listed for the amplifier of Figs. 5.5 and 5.6.

the common-emitter h parameters. Therefore, from Eq. 4.41, the value of the input impedance of the transistor Z_{it} is

$$Z_{it} = h_{ie} - \frac{h_{re} h_{fe}}{h_{oe} + G_L} \tag{5.15}$$

where G_L is the admittance of the transistor load. In addition, from Eq. 4.42, the value of voltage gain is

$$K_v = \frac{v_{ce}}{v_{be}} = \frac{-h_{fe}}{h_{ie}(h_{oe} + G_L) - h_{fe} h_{re}} \tag{5.16}$$

Finally, from Eq. 4.44, the value of current gain is

$$K_i = \frac{i_c}{i_b} = \frac{h_{fe} G_L}{h_{oe} + G_L} \tag{5.17}$$

The output admittance of the common-base transistor was not determined because of its negligibly small value. However, the output admittance of the common-emitter configuration is much larger than that of the common-base, and a technique for determining the output admittance will now be discussed. First, the driving-source voltage is turned off, but its internal resistance R_s is retained, as shown in Fig. 5.10. Then a voltage v_2 is applied to the output terminals, as shown, and the current i_2 which flows into the output terminals is determined. Thus $i_2 = i_3 + h_{fe} i_b$ or

$$i_2 = v_2 h_{oe} + h_{fe} i_b \tag{5.18}$$

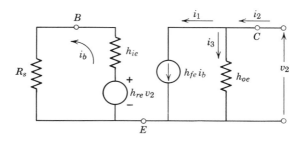

FIGURE 5.10
Circuit used to determine the output admittance of a
transistor in the common-emitter configuration.

The current i_b which flows in the base circuit is caused by the dependent voltage source $h_{re} v_2$ and, as seen from Fig. 5.10, is

$$i_b = - \frac{h_{re} v_2}{h_{ie} + R_s} \tag{5.19}$$

The negative sign appears because i_b flows out of the transistor. Substituting Eq. 5.19 into Eq. 5.18,

$$i_2 = v_2 h_{oe} - \frac{h_{re} h_{fe} v_2}{h_{ie} + R_s} \tag{5.20}$$

Then the output admittance, or the admittance looking into the collector terminal, is

$$Y_{ot} = \frac{i_2}{v_2} = h_{oe} - \frac{h_{re} h_{fe}}{h_{ie} + R_s} \tag{5.21}$$

Example 5.2 We will now use the h parameter equivalent circuit to determine the characteristics of the amplifier of Fig. 5.5, with common-emitter characteristics given in Fig. 5.6. The h parameters for this transistor were determined in Problem 5.3 and are listed below, for convenience, in addition to R_C and R_s.

$h_{ie} = 4 \text{ k}\Omega$	$h_{fe} = 230$	$R_C = 6 \text{ k}\Omega$
$h_{oe} = 2.5 \times 10^{-5}$	$h_{re} = 4 \times 10^{-4}$	$R_s = 5 \text{ k}\Omega$

Better insight is gained by observing the h parameter circuit (Fig. 5.9) rather than mechanically substituting values in the Eqs. 5.15, 5.16, 5.17, and 5.21. Therefore, we find from the equivalent circuit that $v_o = -230$ $i_b/(2.5 \times 10^{-5} + 1.67 \times 10^{-4}) = -1.2 \times 10^6 \, i_b$. Therefore, $v_i = 4 \times 10^3 \, i_b$ $- 4 \times 1.2 \times 10^2 \, i_b = 3.52 \times 10^3 \, i_b$ and $Z_{it} = v_i/i_b = 3.52 \text{ k}\Omega$. The voltage gain $K_v = v_o/v_i = 1.2 \times 10^6 \, i_b/3.52 \times 10^3 \, i_b = 340$. The load current $i_c =$

$v_o G_L = 1.2 \times 10^6 \times 1.67 \times 10^{-4} = 200\ i_b$ and the current gain $K_i = i_c/i_b =$ 200.

Let us assume that the driving-source voltage is turned off and an ac voltage v_o is applied between the collector and the emitter. Then $i_c = v_o(2.5 \times 10^{-5}) + 230\ i_b$. But $i_b = -4 \times 10^{-4}\ v_o/9 \times 10^3\ \Omega$. Then $i_c = 2.5 \times 10^{-5}\ v_o - 230 \times 4.44 \times 10^{-8}\ v_o = 1.5 \times 10^{-5}\ v_o$. Thus the output admittance of the transistor is $Y_{ot} = i_c/v_o = 1.5 \times 10^{-5}$ mhos. The output admittance of the amplifier is the output admittance of the transistor plus the load conductance. For this example, this total *amplifier* output admittance $Y_o = 1.5 \times 10^{-5} + 1.67 \times 10^{-4} = 1.82 \times 10^{-4}$ mho and $R_o = 5.5$ kΩ.

These results agree well (within the accuracy of reading the graphs) with the characteristics determined from the graphical solution, Example 5.1.

PROBLEM 5.5 A given transistor has $h_{ie} = 1$ kΩ, $\beta = 100$, $h_{re} = 10^{-4}$, and $h_{oe} = 2 \times 10^{-5}$ mho at a given q point. Calculate the voltage gain v_{ce}/v_{be}, current gain i_c/i_b, the input resistance and the output admittance of an amplifier which uses this transistor with $R_L = 5$ kΩ and $R_s = 2$ kΩ. *Answer:* $K_v = 476$, $K_i = 91$, $R_{it} = 955\ \Omega$.

Transistor types that have been developed recently have values of h_{re} much smaller than the value 4×10^{-4} determined for the transistor of Fig. 5.6. Consequently, the value h_{re} for these modern transistors can be assumed to be approximately equal to zero over the frequency range below beta cutoff frequency (to be discussed later) where the h parameter circuit is most useful. This assumption greatly simplifies the determination of the voltage gain, input impedance, and output impedance of the transistor in a given circuit configuration.

Example 5.3 Let us consider the transistor amplifier of Example 5.2, but assume that the input characteristics can be adequately represented by a single curve instead of the family of curves shown in Fig. 5.6. Then $h_{re} \simeq 0$. Otherwise, the h parameters have the same values as previously determined.

$$Z_{it} \simeq h_{ie} = 4000\ \Omega$$

The output impedance of the *transistor* is

$$Z_{ot} = 1/h_{oe} = 40\ \text{k}\Omega$$

$$K_v = \frac{h_{fe}}{h_{ie}(h_{oe} + G_L)} = \frac{230}{4000(1.92 \times 10^{-4})} = 300$$

$$K_i = \frac{h_{fe}}{1 + h_{oe} R_L} = 200\ \text{(same as before)}$$

Note that a further simplification can be made if $G_L \gg h_{oe}$, so h_{oe} can be neglected. Then

$$K_v \simeq \frac{h_{fe}}{h_{ie} G_L} = \frac{h_{fe} R_L}{h_{ie}} \tag{5.22}$$

$K_v = 343$ for this example.

$$K_i \simeq h_{fe} = 230$$

PROBLEM 5.6   Calculate the voltage gain v_{ce}/v_{be}, the current gain i_c/i_b, the input resistance and output resistance (of the transistor) of the amplifier of Problem 5.5, assuming $h_{re} \simeq 0$.

Answer: $K_v = 454$, $K_i = 91$, $R_{it} = 1 \text{ k}\Omega$, $R_{ot} = 50 \text{ k}\Omega$.

PROBLEM 5.7   Calculate the voltage gain v_{ce}/v_{be}, current gain, and transistor input resistance of the amplifier of Problem 5.5, assuming $h_{re} \simeq 0$ and h_{oe} negligible compared with G_L.

Answer: $K_v = 500$, $K_i = 100$, $R_{it} = 1 \text{ k}\Omega$.

5.4   TRANSISTOR RATINGS

Essentially the only cause of deterioration and destruction of a semiconductor device is heat, which may melt the solder connections, deteriorate the insulating materials, and change the crystal structure of the semiconductor. Therefore, manufacturers rate their semiconductor products in accordance with their power dissipation capability and maximum permissible temperature. Also, maximum voltage ratings are given and sometimes maximum current is specified. These ratings, and their application to circuit design will be discussed in this section.

The ratings of a typical low-power *n-p-n* germanium transistor follow.

Absolute Maximum Ratings, 25°C

V_{CEO}	30 V
I_C	100 mA
Power dissipation, P_d*	200 mW
Junction temperature, T_j	85°C

* Derate 3.33 mW/°C for ambinent temp-
atures above 25°C.

The safe operating area of this transistor may be marked on the collector characteristics as shown in Fig. 5.11. The transistor operation will be confined to this area if the dc load line remains *below* the maximum

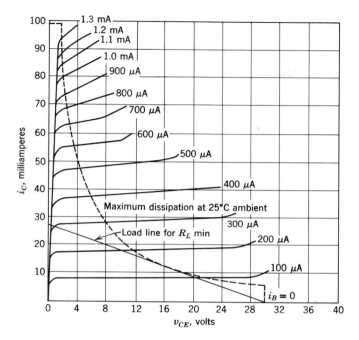

FIGURE 5.11
Safe operating area for a typical 200 mW transistor.

dissipation curve which is drawn through all points where $v_{CE} i_C = P_d$ max. For example, if $v_{CC} = 30$ V, R_C min $= 1.11$ kΩ.

If the ambient, or surrounding, temperature is higher than 25°C, the maximum dissipation rating must be reduced, as specified by the manufacturer. For example, the transistor with the ratings given above must have its maximum dissipation rating reduced 3.3 mW for each degree C of ambient temperature above 25°C. Therefore, the maximum dissipation curve drawn on the collector characteristics should represent the maximum permissible dissipation at the highest expected ambient temperature. For example, if the transistor above is to be enclosed in a metal cabinet and used on the desert in the summer, the ambient temperature may rise to 55°C (131°F). The maximum dissipation rating is then $P_d(\text{max}) = 200 - 3.33 \times 30 = 100$ mW.

PROBLEM 5.8 Draw the maximum dissipation curve for 100 mW dissipation on the collector characteristics of Fig. 5.11. Determine the minimum safe value of the collector load resistance for this dissipation if $V_{CC} = 20$ V.
Answer: $R_C = 1$ kΩ.

Sometimes a derating curve is given instead of a derating factor. A derating curve for the transistor of Fig. 5.11 is given in Fig. 5.12. Observe that the slope of the derating curve is equal to the derating factor. The slope of the curve in Fig. 5.12 (above 25°C) is -200 mW$/60°$C $= -3.33$ mW$/°$C. The negative sign results from the negative slope. The derating factor is usually given as a positive number, however.

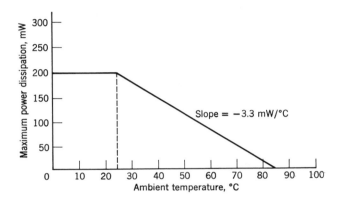

FIGURE 5.12
Derating curve for the transistor of Fig. 5.11.

Most of the power dissipation in a transistor occurs at the collector junction because of the relatively large voltage across that junction. The junction temperature rises above the ambient temperature because of this dissipation. The rise in junction temperature is equal to the power dissipation divided by the derating factor. For example, if the transistor of Fig. 5.11 is operating with $V_{CE} = 10$ volts and $i_C = 10$ mA, the rise in junction temperature above the ambient is $v_{CE}(i_C)$/derating factor $= 100$ mW/3.3 mW per °C $\simeq 30$°C.

Heat flow, which results from a temperature difference, is analogous to current flow, which results from a potential difference. Therefore, a *thermal resistance* θ_T has been defined as the ratio of the temperature rise to the power dissipation, or

$$\theta_T = \frac{\Delta T}{P_d} \tag{5.23}$$

and

$$\Delta T = \theta_T P_d \tag{5.24}$$

Note that if ΔT is replaced by V, θ_T replaced by R, and P_d replaced by I, the foregoing two equations express Ohm's law. These equations show that the thermal resistance is the reciprocal of the derating factor and is therefore the negative reciprocal of the slope of the derating curve. The thermal resistance between the junction and the ambient surroundings of the transistor in Fig. 5.12 is $\theta_T = 1/3.33 = 0.3°\text{C/mW}$.

The junction temperature is the ambient temperature T_a plus the temperature rise due to power dissipation. Therefore, using Eq. 5.24,

$$T_j = T_a + \Delta T = T_a + \theta_T P_d \tag{5.25}$$

For example, if the ambient temperature is 40°C and the transistor of Fig. 5.12 is dissipating 100 mW, the collector junction temperature $T_j = 40°\text{C} + 0.3 \times 100°\text{C} = 70°\text{C}$. Observe and verify that at any point on the derating curve, the ambient temperature plus the corresponding power dissipation times the thermal resistance θ_T gives the maximum permissible junction temperature.

PROBLEM 5.9   A given silicon transistor has 300 mW maximum dissipation rating at $T_a = 25°\text{C}$ and T_j max $= 175°\text{C}$. Determine (a) the derating factor, (b) the thermal resistance and (c) the junction temperature when $T_a = 50°\text{C}$ and the average power dissipation is 100 mW.
Answer: (a) 2 mW/°C, (b) 0.5°C/mW, (c) 100°C.

The maximum current rating of a transistor, if given, usually indicates either the current at which the maximum dissipation curve crosses the saturation voltage, as shown in Fig. 5.11, or the current at which β falls below the minimum specified by the manufacturer.

The maximum voltage rating is not so simply specified for a transistor as for a diode. The transistor may *appear* to break down at the voltage for which $\alpha = 1$, although this voltage may be considerably below the avalanche breakdown voltage as shown in Fig. 4.10. The reason the transistor appears to break down is because the collector current approaches infinity as α approaches unity, as seen by Eq. 5.3, repeated below.

$$i_C = \frac{\alpha}{1 - \alpha} i_B + \frac{I_{CO}}{1 - \alpha} \tag{5.3}$$

Observe that i_C is equal to infinity for $i_B = 0$ or for any positive value of i_B when $\alpha = 1$. The *apparent* breakdown voltage when $i_B = 0$, is known as the sustaining voltage or *maximum* V_{CEO}. The first two subscripts indicate the electrodes to which the voltage is applied and the third subscript indicates the conditions at the third terminal (base) which, in this case, is open. Also observe from Eq. 5.3, however, that finite positive values of collector current may be obtained for values of α greater than unity if the base current is negative. In fact, if $i_B = -I_{CO}$, Eq. 5.3 shows that the collector current

$$i_C = I_{CO}\left(\frac{-\alpha}{1-\alpha} + \frac{1}{1-\alpha}\right) = I_{CO}$$

for any value of α. Of course, I_{CO} increases rapidly because of carrier multiplication as the avalanche breakdown voltage is approached. This avalanche breakdown voltage is known as the maximum V_{CBO}, because there is no current across the emitter junction when the emitter is open and the base current is automatically held to $-I_{CO}$.

When neither the base nor the emitter circuits are open and the base circuit has a finite resistance R as shown in Fig. 5.13a, part of I_{CO} flows as

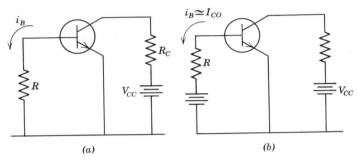

(a)   (b)

FIGURE 5.13
Negative i_B flows through resistance R in the base circuit.

a negative base current i_B and the remaining part of I_{CO} flows across the emitter junction. This negative base current reduces the collector current, as compared with the open base configuration, and causes the apparent breakdown voltage, *maximum* V_{CER} to be higher than maximum V_{CEO}. The improvement in apparent breakdown voltage depends upon the value of R. Maximum improvement occurs when $R = 0$, because this value gives the maximum negative i_B. The apparent breakdown voltage with $R = 0$ is known as maximum V_{CES} where S means the base is shorted to the emitter. The value of base resistance must be specified when the maximum V_{CER} is given.

Not all the I_{CO} flows in the base circuit when $R = 0$ because of the internal resistance r_b. Therefore, the negative base current can be increased and hence the apparent breakdown voltage increased if a reverse-biasing voltage is included in the base circuit as shown in Fig. 5.13b. The apparent breakdown voltage with this reverse voltage applied is known as *maximum* V_{CEX} and is essentially equal to V_{CBO}.

The collector characteristics for a typical *n-p-n* transistor, including the avalanche region, are given in Fig. 5.14a, and the breakdown characteristics of this transistor are given in Fig. 5.14b. The various *maximum* voltage ratings for this transistor are indicated along the voltage axis for a typical transistor.

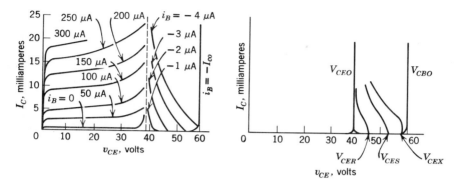

FIGURE 5.14
Voltage breakdown characteristics of a transistor.

5.5   BIASING CIRCUITS

The transistor circuits previously considered have used a bias battery and a current-limiting resistor to provide forward bias to the emitter junction. However, in the common-emitter configuration, the battery V_{BB} which supplies forward bias to the base has the same polarity, with respect to the emitter, as the battery V_{CC} which supplies reverse bias to the collector. Therefore, a single battery, or power supply, can supply the proper bias to both junctions, as shown in Fig. 5.15. Since the dc voltage across the bias

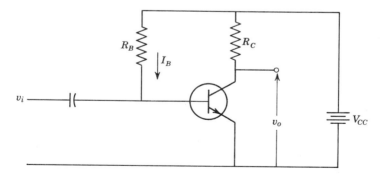

FIGURE 5.15
Fixed bias obtained from a single voltage since V_{CC}.

resistor R_B is $V_{CC} - V_{BE}$, the bias circuit resistance can be determined, using Ohm's law, from the relationship

$$R_B = \frac{V_{CC} - V_{BE}}{I_B} \tag{5.26}$$

Example 5.4   Let us determine R_B if $V_{CC} = 20$ volts, the desired q-point base current I_B is 20 μA and the q-point value of v_{BE} is $V_{BE} = 0.5$ V. The proper value of R_B is

$$\frac{19.5 \text{ V}}{2 \times 10^{-5} \text{ A}} = 9.75 \times 10^5 \simeq 10^6 \ \Omega.$$

Observe that V_{BE} may be neglected in Eq. 5.26 if V_{CC} is large in comparison with V_{BE} (by a factor of 10, at least).

This type of bias is known as *fixed* bias because the base bias current I_B is determined, or *fixed*, almost entirely by the values of R_B and V_{CC}. The idea of I_B being *fixed* may seem good at first thought. However, as noted in Chapter 3, the thermally generated current I_{CO} doubles, approximately, for each 10°C temperature increase. Therefore, the q point shifts up the load line with increasing temperature as shown in Fig. 5.16 for a typical *n-p-n* germanium transistor. The reason for the large increase in i_C, with i_B held constant, is that I_{CO} is forced to flow across the emitter junction and is therefore amplified by the factor $(\beta + 1)$, as previously discussed. At the

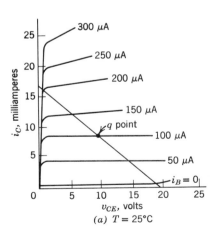
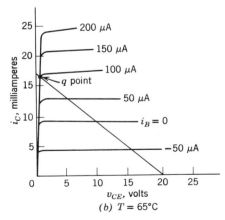

FIGURE 5.16
Collector characteristics of a typical *n-p-n* germanium transistor (*a*) at $T = 25$°C, (*b*) at $T = 65$°C.

65°C temperature, the amplifier of Fig. 5.16*b* is useless because the *q* point is at the upper end of the load line and the transistor is in saturation.

The *q* point shift, or collector current instability, can be reduced considerably if a resistor R_E is inserted in the emitter lead and the base circuit resistor R_B is reduced from the fixed bias value. The reason for this improvement in *q* point stability is shown in Fig. 5.17, where the thermally generated current I_{co} is shown to divide between the base circuit and the emitter circuit. Only that portion, KI_{co}, which flows across the emitter junction is amplified by β, therefore, the collector current which results from the thermal current is $(I_{co} + K\beta I_{co}) = (1 + K\beta)I_{co}$ and K can have values between 1 and 0. Since the shift in *q* point is caused primarily by the increase of I_{co} with temperature, the best stability is obtained when K is 0.

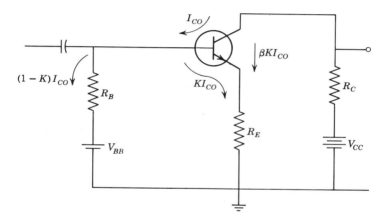

FIGURE 5.17
Circuit showing the division of thermal current I_{co} between the emitter and base circuits.

A current stability factor S_I is defined as the ratio of collector current change to the change in I_{co}, or

$$S_I = \frac{\Delta I_C}{\Delta I_{co}} \tag{5.27}$$

Thus, if I_{co} in Fig. 5.17 is increased by ΔI_{co}, the collector current is increased by $(1 + K\beta)\Delta I_{co}$, and the current stability factor $S_I = (1 + K\beta)$. Note that a small stability factor results in good *q* point stability. Therefore, S_I is actually an *instability* factor. If the base resistor R_B is very large and the emitter resistor R_E is zero, which occurs when fixed bias is used, the value of K is 1 and the value of S_I is $(1 + \beta)$. On the other hand, if R_B is zero and R_E is large, which occurs when the common-base configuration is used, $K = 0$ and $S_I = 1$. Intermediate values of R_E and R_B give values of S_I which lie between one and $(\beta + 1)$. In fact, it can be shown[1] that when S_I is large compared with one, but small compared with $(\beta + 1)$,

$$S_I \simeq R_B/R_E \qquad (5.28)$$

A desirable value for S_I can be determined after a transistor has been chosen for a given application, as illustrated by the following example.

Example 5.5   A given n-p-n germanium transistor is to operate as a common-emitter amplifier with $V_{CC} = 25$ V, and $R_L = 5$ kΩ. The load line is drawn on the collector characteristics in Fig. 5.18. Let us assume that the q point should remain within the limits q_1 and q_2 as the ambient temperature changes from 25°C to 55°C. The transistor ratings are given in Table 5.1 for 25°C ambient temperature. As seen in Fig. 5.18, $\Delta I_C = 1$ mA. The value of I_{co} depends upon the junction temperature, not the ambient, so the maximum junction temperature must be determined from the relationship $T_j = T_a + \theta_T P_d$ where $\theta_T = 0.5$°C/mW. Thus the maximum junction temperature in this application is $T_a + \theta_T \times I_C V_C$ or T_j max $= 55° + .5 \times 3 \times 11 = 71$°C.

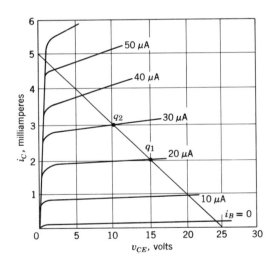

FIGURE 5.18
Characteristics used to visualize the permissible shift in q point.

[1] *Electronic Engineering*, Alley and Atwood, Second Edition, John Wiley and Sons, Inc., New York, New York, p. 195.

TABLE 5.1
Transistor Ratings for Example 5.5

V_{CEO} (max)	30 V
P_d (max)	140 mW
T_j (max)	95°C
I_{CO}	4 μA

Derate 2 mW per °C for ambient temperatures above 25°C.

We have no information from the manufacturer concerning the rate of increase of I_{CO} with temperatures, so will assume that I_{CO} doubles for each 10°C junction temperature increase which is typical for transistors. Then at $T_j = 71$°C, the value of I_{CO} can be determined approximately from Table 5.2. We may estimate I_{CO} at $T_j = 71$°C to be about 100 μA. Then, assuming T_j (min) to be 25°C, $\Delta I_{CO} = 96$ μA and $S_I = \Delta I_C / \Delta I_{CO} = 1/.096 = 10.4$. Note that $T_{j\,min}$ rises above 25°C as the transistor warms up, but if we want the q point to *not* be below the lower limit q_1 during the warm-up period, we must use $T_{j\,min} = T_{a\,min} = 25$°C.

A more accurate method of predicting the maximum value of $I_{CO}(I_{CO2})$, when a reference value I_{CO1} is given, follows.

$$I_{CO_2} = I_{CO_1} 2^{(\Delta T_j / 10)} \qquad (5.29)$$

Using Eq. 5.29, $I_{CO2} = 97$ μA. This value is so near the first estimate that we will not alter our value of S_I.

We now know the required S_I, which is approximately the ratio of R_B to R_E, but we need to determine a suitable value of R_E. R_E is actually part of the dc resistance between the collector and emitter and therefore should be included in the dc resistance and load line. Generally, R_E should be small

TABLE 5.2
Variation of I_{CO}
With
Temperature

T_j(°C)	I_{CO}
25	4 μA
35	8 μA
45	16 μA
55	.32 μA
65	64 μA
75	128 μA

in comparison with R_C to avoid unnecessary power loss in R_E. $R_E =$ 0.2 R_C to 0.25 R_C is usually adequate. In this example we will choose $R_E = 1$ kΩ. R_C will then be 4 kΩ so the total dc load resistance will be 5 kΩ as specified. The value of R_B required to give proper stability is $R_B \simeq S_I R_E = 10.4 \times 1\text{k} = 10.4$ kΩ or approximately 10 kΩ. Since the transistor input resistance is approximately $(\beta + 1)\, r_e = (100)(25)/2 = 1250\ \Omega$, the 10 k$\Omega$ value of R_B would shunt about 12 percent of the ac signal to ground, which is a small price to pay for the good stability achieved. The transistor input resistance would be much higher than $(\beta + 1)\, r_e$ if the capacitor C_E, Fig. 5.19, were not included. This capacitor, which will be discussed later, is known as a bypass capacitor because it shunts the ac signal currents around R_E and thus eliminates the effect of R_E insofar as the input signal is concerned.

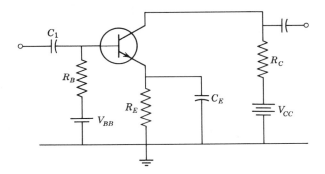

FIGURE 5.19
A stabilized transistor amplifier with the emitter resistor bypassed.

The two batteries shown in Fig. 5.19 are not required for a stabilized bias system. A single battery system is shown in Fig. 5.20a. Observe that a dc Thevenin's equivalent circuit looking into the base bias circuit to the left of the points A and B is as shown in Fig. 5.20b, where R_B is the parallel combination of R_1 and R_2, or

$$R_B = \frac{R_1\, R_2}{R_1 + R_2} \tag{5.30}$$

and the open-circuit voltage, or voltage between points A and B with the base disconnected, is

$$V_{BB} = \frac{R_1}{R_1 + R_2}\, V_{CC} \tag{5.31}$$

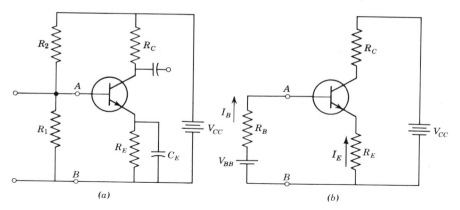

FIGURE 5.20
(a) A stabilized-bias system which uses a single battery, and (b) the dc Thevenin's equivalent of this circuit.

Note that this equivalent circuit is the same as the circuit of Fig. 5.19. The value of V_{BB} can be found by adding the voltage drops around the base circuit at one of the q points, preferably the lower one. Thus

$$V_{BB} = I_B R_B + V_{BE} - I_E R_E \qquad (5.32)$$

Observe that I_E is a negative current since it flows out of the transistor. Therefore, the term $-I_E R_E$ will be positive for an n-p-n transistor. Input characteristics may not be available, in which case V_{BE} may be assumed to be about 0.2 V for a germanium transistor or 0.6 V for a silicon transistor. For example, in the preceding amplifier of Figs, 5.18 and 5.19, $V_B = (20 \times 10^{-6}) \, 10^4 + 0.2 + (1.9 \times 10^{-3}) \, 10^3 = 2.3$ V.

Since V_{CC} and V_{BB} are now known, there are only two unknowns (R_1 and R_2) in equations 5.30 and 5.31. Therefore, these equations can be solved simultaneously for R_1 and R_2 in terms of V_{CC}, V_{BB}, and R_B. This solution yields

$$R_1 = \frac{V_{CC}}{V_{CC} - V_{BB}} R_B \qquad (5.33)$$

$$R_2 = \frac{V_{CC}}{V_{BB}} R_B \qquad (5.34)$$

Therefore, the circuit of Fig. 5.20a can be used for the amplifier with characteristics shown in Fig. 5.18, in which case $R_1 = 25 \times 10.4$ k/(25 − 2.3) = 11.5 kΩ, and $R_2 = 25 \times 10^4$ k/2.3 = 113 kΩ. Let $R_1 = 12$ kΩ and $R_2 = 120$ kΩ, which are stock size 10 percent resistors and would be suitable.

A more general and precise method of designing a bias circuit will be discussed in Section 5.6 which follows.

PROBLEM 5.10   The germanium transistor used in the preceding example, with characteristics given in Fig. 5.18 and Table 5.1, is used in the amplifier circuit given in Fig. 5.20a. This amplifier must operate over the ambient temperature range $T_a = 5°C$ to $T_a = 65°C$ and the q point may vary from $I_C = 1.5$ mA to $I_C = 3.5$ mA. $R_C = 4$ kΩ and $R_E = 1$ kΩ as in the preceding example. Determine suitable values for R_1 and R_2.

Answer: $R_1 = 15$ k, $R_2 = 180$ k.

5.6   STABILIZED BIAS

In the preceding sections, I_{CO} was considered to be the only parameter which varies with temperature. If this were true, the bias stabilization technique discussed in the preceding section would be entirely adequate. However, two parameters in addition to I_{CO} may change with temperature and cause undesirable q point shift. One of these parameters is V_{BE} which is primarily the forward-bias voltage across the emitter junction. The diode equation (Eq. 3.11) shows that the diode current is strongly dependent on the saturation current I_S or I_{EO}, which is an exponential function of temperature, if the junction voltage is held constant. Similarly, if the emitter current is held constant, the junction voltage will vary with temperature, as shown in Fig. 5.21.

When i_E is held constant and the temperature is increased from T_1 to T_2,

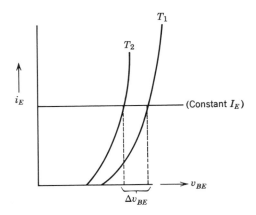

FIGURE 5.21
A sketch showing the effect of temperature on the input characteristics of a transistor.

v_{BE} decreases by the amount Δv_{BE}. The rate of change of v_{BE} with temperature is typically about -2.5 mV per °C. Thus, if a typical silicon transistor has $v_{BE} = 0.55$ V at $i_E = 1$ mA and $T = 25°C$, it would have $v_{BE} = 0.425$ V at $T = 75°C$ and $i_E = 1$ mA.

The other parameter which is temperature sensitive, particularly in a silicon transistor, is β. Figure 5.22 shows how a change in β causes a q point

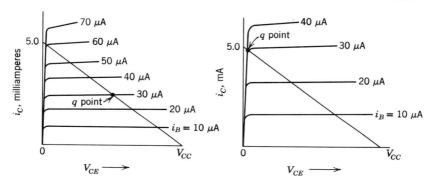

FIGURE 5.22
Collector characteristics showing q-point shift with changing β but i_B constant.

shift, assuming base current is held constant. Generally, β increases with temperature.

The stabilized-bias circuit of Fig. 5.20a can be used to control the q point shift which may result from a combination of changes in I_{CO}, V_{BE} and β as shown below. We will begin with the base circuit voltage equation (Eq. 5.32), repeated below, which was written by inspection of Fig. 5.20b.

$$V_{BB} = I_B R_B + V_{BE} - I_E R_E \qquad (5.32)$$

However, we will specify the q point in terms of I_C, not I_B and I_E. Therefore, the following relationships are needed. First, using Kirchoff's current law, $I_E + I_C + I_B = 0$.

$$-I_E = I_C + I_B \qquad (5.35)$$

Then, using the relationship $I_C = \beta I_B + (\beta + 1)I_{CO}$, which is Eq. 5.4 written for dc values,

$$I_B = \frac{I_C}{\beta} - \frac{\beta + 1}{\beta} I_{CO} \qquad (5.36)$$

Substituting Eq. 5.35 into Eq. 5.32 and then substituting Eq. 5.36 into the result,

$$V_{BB} = \left(\frac{I_C}{\beta} - \frac{\beta + 1}{\beta} I_{CO} \right) R_B + V_{BE} + \left(I_C + \frac{I_C}{\beta} - \frac{\beta + 1}{\beta} I_{CO} \right) R_E \qquad (5.37)$$

This equation can be simplified if β is assumed to be much larger than one, then $(\beta + 1) \simeq \beta$ and I_C is much larger than I_C/β in the right-hand term. With these simplifications, Eq. 5.37 can be written as

$$V_{BB} \simeq \left(\frac{I_C}{\beta} - I_{CO}\right)R_B + V_{BE} + (I_C - I_{CO})R_E \qquad (5.38)$$

Now, the objective is to restrict the q-point values of I_C to a predetermined range ΔI_C between I_{C1} and I_{C2} as shown in Fig. 5.23. At I_{C1} the collector

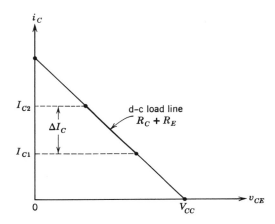

FIGURE 5.23
Confinement of q-point range from I_{C1} to I_{C2}.

junction temperature is minimum, the thermal current I_{CO1} is minimum and β_1 is minimum. Conversely, when the q point is at I_{C2}, the values of T_j, I_{CO2} and β_2 are maximum. To restrict the q point, we will select a suitable value for R_E, discussed in Section 5.5, and then calculate the required value of R_B, which may be determined by first writing Eq. 5.38 with values at I_{C2} substituted, and then writing Eq. 5.38 again with values at I_{C1} substituted. The second equation is then subtracted from the first.

$$\begin{aligned} V_{BB} - V_{BB} = &\left(\frac{I_{C2}}{\beta_2} - I_{CO2}\right)R_b + V_{BE2} + (I_{C2} - I_{CO2})R_E \\ &-\left(\frac{I_{C1}}{\beta_1} - I_{CO1}\right)R_b - V_{BE1} - (I_{C1} - I_{CO1})R_E \end{aligned} \qquad (5.39)$$

If we let $(I_{C2} - I_{C1}) = \Delta I_C$, $(I_{CO2} - I_{CO1}) = \Delta I_{CO}$ and $(V_{BE2} - V_{BE1}) = \Delta V_{BE}$, Eq. 5.39 becomes

$$0 = \left(\frac{I_{C2}}{\beta_2} - \frac{I_{C1}}{\beta_1} - \Delta I_{CO}\right)R_B + \Delta V_{BE} + (\Delta I_C - \Delta I_{CO})R_E \qquad (5.40)$$

Solving Eq. 5.40 explicitly for R_B,

$$R_B = \frac{\Delta V_{BE} + (\Delta I_C - \Delta I_{CO})R_E}{[(I_{C1}/\beta_1) - (I_{C2}/\beta_2)] + \Delta I_{CO}} \tag{5.41}$$

Although Eq. 5.41 was developed for an n-p-n transistor, it is applicable to a p-n-p type because all the signs of the variables I_C, I_{CO} and V_{BE} change. The use of Eq. 5.41 will be illustrated by carrying through a more precise design for the bias circuit of Example 5.5.

Example 5.6 The following values have already been determined or assumed:

$$
\begin{array}{lll}
T_{j1} = 25°C & T_{j2} = 71°C & \Delta T_j = 46°C \\
I_{C1} = 1.9 \text{ mA} & I_{C2} = 2.9 \text{ mA} & \Delta I_C = 1 \text{ mA} \\
I_{CO1} = 4\ \mu A & I_{CO2} = 97\ \mu A & \Delta I_{CO} = 93\ \mu A \\
V_{BE1} = 0.2 \text{ V} & \beta_1 \simeq 95 & R_E = 1 \text{ k}\Omega
\end{array}
$$

We will also assume that β rises to 150 at the maximum junction temperature and that $\Delta V_{BE}/°C = -2.5 \text{ mV}/°C$. Then $\Delta V_{BE} = -2.5 \times 10^{-3}(46) = -.115$ V. Substituting these values into Eq. 5.41, $R_B = 8.5 \text{ k}\Omega$. Note that R_B is less than the value calculated in Example 5.5 because changes in V_{BE} and β were taken into account.

The value of V_{BB} can be calculated by using a known set of parameters at any given q point. The values associated with I_{C1} can be obtained most easily because I_{CO} is usually small compared with I_{C1}/β_1 at this low temperature q point. (See Eq. 5.36 for calculating I_B.) The value of V_{BB} for this example was determined in Example 5.5 to be 2.3 V. Although R_B is now less than calculated in Example 5.5, we will assume $V_{BB} = 2.3$ V is sufficiently accurate for our purposes here. The values of R_1 and R_2 for the circuit of Fig. 5.20a, with $V_{CC} = 25$ V, are $R_1 = 9.3 \text{ k}\Omega$ and $R_2 = 92 \text{ k}\Omega$.

When Eq. 5.41 yields values of R_B which are too low, and therefore seriously shunt the input currents, either a larger value of R_E is needed or a larger value of ΔI_C should be permitted. Perhaps replacement of a germanium transistor with a silicon transistor to reduce ΔI_{CO} may be the best solution. In some instances, negative values of R_B will be obtained from Eq. 5.41. The causes and possible solutions to this problem follow.

1.   The numerator may become negative because ΔV_{BE} is negative and $(\Delta I_C - \Delta I_{CO})R_E$ is either: (a) Negative, or (b) Positive but smaller in magnitude than ΔV_{BE}. In case a, ΔI_{CO} is larger than ΔI_C, and if ΔI_C cannot be increased considerably, a different transistor with lower ΔI_{CO} must be selected. In case b, the same solutions as listed for case a are applicable. In addition, R_E may be increased.

2. The denominator may be negative because $I_{C1}/\beta_1 - I_{C2}/\beta_2$ is negative and has a greater magnitude than ΔI_{CO}. This indicates that for the given temperature range, you have chosen ΔI_C larger than can be obtained with any positive value of R_B. Decreasing ΔI_C will solve the problem. Also, if your value of ΔI_C gives positive but very large values of R_B so that V_B is greater than V_{CC}, R_1 will be negative. The solution to this problem is also to either reduce ΔI_C or arbitrarily reduce R_B to a practical value.

The stabilized-bias circuit is also applicable to a mass-produced amplifier which must accept a specified variation in transistors due to manufacturing tolerances as well as temperature differences.

PROBLEM 5.11   Let us assume that the amplifier design, previously considered in Section 5.5 and continued in this section, is to be mass produced using factory-run transistors with $\beta_{min} = 50$ at $25°C$ and $\beta_{max} = 200$ at $75°C$, $V_{BEmax} = 0.25$ at $25°C$ and $V_{BE\,min} = 0.08$ at $75°C$ (at $I_E = 2$ mA). The value of $I_{CO} = 4.0\ \mu A$ at $25°C$ is the maximum for this type of transistor. Determine suitable values for R_B, V_B, R_1, and R_2 if the q-point shift, V_{CC} and R_E remain the same as previously specified.

Answer: $R_B = 6.3$ kΩ, $V_{BB} = 2.39$ V, $R_1 = 6.95$ k and $R_2 = 66$ k. ($R_1 = 6.8$ k and $R_2 = 68$ k would probably be used.)

Some signal shunting by R_B in the mass-produced amplifier may not be objectionable because the higher gain transistors have higher input impedance (almost proportional to β), and therefore the shunting effect of R_B tends to make the variation (or spread) of amplifier gain much less than the spread of β.

The required value of the emitter-bypass capacitor C_E (Fig. 5.20) will be discussed in Chapter 7. However, a design equation is given below to permit the immediate design of a common-emitter amplifier with stabilized bias.

$$C_E \simeq \frac{h_{fe}}{\omega_1(h_{ie} + R_s)} \tag{5.42}$$

where ω_1 is the lowest signal frequency, in rad/sec, and R_s is the resistance of the driving sourse.

The value of the input coupling capacitor C_1 (Fig. 5.19) is also discussed in Chapter 7. A suitable relationship for determining the value of this capacitance is given below.

$$C_1 = \frac{10}{\omega_1(h_{ie} + R_s)} \tag{5.43}$$

A procedure for designing a stabilized-bias circuit for a specific amplifier is given below.

1. Choose $R_E \simeq 0.2\, R_C$. Use the nearest stock size.
2. Draw a load line for $(R_C + R_E)$ on the collector characteristics or a set of i_C versus v_{CE} coordinates, as a visual aid.
3. Choose I_{C2} and I_{C1} as the upper and lower limits, respectively, of the q-point excursions on the load line.
4. Determine the power dissipation $P_{d2} = v_{CE2} I_{C2}$ at the upper q point and calculate the maximum junction temperature from the relationship $T_{j\,max} = T_{a\,max} + \theta_T P_{d2}$.
5. Calculate I_{CO2} using Eq. 5.29 or, preferably, manufacturer's data.
6. Assuming $T_{j\,min} = T_{a\,min}$, determine I_{CO1} and $\Delta I_{CO} = I_{CO2} - I_{CO1}$.
7. Calculate $\Delta V_{BE} \simeq -2.5 \times 10^{-3}(T_{j\,max} - T_{j\,min})$.
8. Calculate R_B using Eq. 5.41 where $\Delta I_C = I_{C2} - I_{C1}$, β_2 is the maximum expected β at $T_{j\,max}$ and β_1 is the minimum expected β at $T_{j\,min}$. If R_B is negative, make the modifications suggested in the material following Example 5.6.
9. Determine $V_{BB} = I_B R_B + V_{BE} - I_E R_E$ (Eq. 5.32) using all values from the I_{C1} q point, including $I_B = I_{C1}/\beta_1$.
10. Solve for R_1 and R_2 using Eq. 5.33 and Eq. 5.34
11. Calculate C_1 and C_E using Eq. 5.42 and 5.43.

The simpler bias calculation can be made for more relaxed specifications by replacing steps 7 and 8 with the approximate relationship $R_B = S_I R_E$.

Although the stabilized-bias circuit and design equations were developed for the common-emitter configuration, they are applicable to *all* bipolar (or two-junction) transistor configurations. We have assumed, to this point, the common-base configuration always has $R_b = 0$ and $\Delta I_C = \Delta I_{CO}$, but this is not always the case.

A firm basis for determining the maximum permissible ΔI_C will be given in Chapter 7.

PROBLEM 5.12 Assume that you are hired by an electronics firm to design a transistor preamplifier which will amplify the signal from a phonograph pickup to at least a 1.0 V peak level for driving a main (power) amplifier. The phono-pickup generates a 10 mV peak signal, open circuit, and has an internal resistance of 2000 Ω. The preamp will be mass produced and must operate satisfactorily over the ambient temperature range of 0°C to 50°C. The available supply voltage is $V_{CC} = 20$ V. An n-p-n silicon transistor with $I_{CO} = .01\ \mu$A max at 25°C, P_d max = 300 mW at 25°C, and derating = 2 mW/°C above 25°C has been chosen. The max β spread (range) is 80 to 300 over the temperature range desired, and the recommended value of collector circuit resistor R_C is 6.8 kΩ. The amplifier will operate properly if the q-point variation is limited to $I_{C1} = 0.8$ mA and $I_{C2} = 1.8$ mA. V_{BE} max $= 0.50$ V at $I_C = 0.8$ mA, $T = 25°$C. Choose a suitable value of R_E, calculate values for R_1 and R_2, and check the output voltage at both the minimum and maximum q-point values, assuming $h_{ie} \simeq (\beta + 1)r_e$, $h_{re} \simeq 0$, and h_{oe} is very small in comparison with Y_L. The main amplifier has very high input resistance compared with R_C, so R_C is the ac load resistance. Assume that the reactances of coupling and bypass capacitors are negligible. Is the voltage gain adequate?

6

High Input Impedance
Circuits and Devices

Maximum power is transferred from a source to a load if the impedance of the load is equal to the conjugate of the source impedance. Since the output impedances of most devices are essentially resistive at the lower frequencies, the input impedance of the load should be approximately equal to the source impedance for good power transfer. Consequently, the common-emitter amplifier is usually preferred over the common-base configuration because the common emitter has reasonably good match between input and output impedance and, therefore, can be operated in cascade (output of one amplifier or stage provides the input signal for a following amplifier) without the use of transformers. Many signal sources, such as crystal or ceramic microphones and phonograph pickups require load impedances of the order of a megohm or higher, which is much higher than the input impedance of a common-emitter amplifier. Therefore, this chapter will be devoted to the common-collector configuration, or emitter follower, and other high input impedance circuits and devices.

6.1 THE COMMON-COLLECTOR AMPLIFIER

The common-collector amplifier, which is frequently known as an *emitter follower*, has the load in the emitter circuit as shown in Fig. 6.1. The collector is normally connected directly to V_{CC}, which is at ac ground potential; thus, the name *common collector*.

Figure 6.1 shows that the input voltage v_i is equal to the output voltage v_o plus the ac voltage across the forward-biased emitter junction,

155

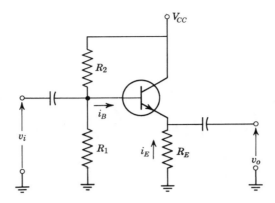

FIGURE 6.1
A common-collector amplifier, or emitter-follower.

provided the reactances of the coupling capacitors are small enough to be neglected. This voltage relationship can be expressed as

$$v_i = v_{be} + v_o \qquad (6.1)$$

Thus, the input voltage is larger than the output voltage, and the voltage gain is less than one. Usually the output voltage v_o is much larger than v_{be}, so the voltage gain is almost one. Thus, the emitter voltage *follows* the base or input voltage very closely, as the name *emitter follower* indicates.

The characteristics of the common-collector configuration can be determined from Fig. 6.2 which shows the common-collector amplifier and

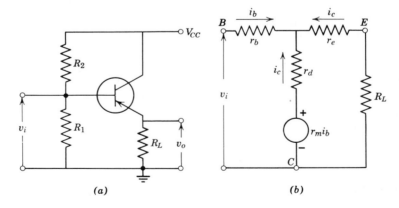

(a)   (b)

FIGURE 6.2
The common-collector transistor amplifier : (a) circuit diagram; (b) equivalent circuit.

its equivalent T circuit. The short-circuit current gain may be determined by letting R_L in Fig. 6.2 approach zero. Then

$$-i_e = i_b + i_c = i_b + h_{fe}i_b \tag{6.2}$$

$$h_{fc} = \frac{i_e}{i_b} = -(1 + h_{fe}) \tag{6.3}$$

Equation 6.3 shows that the forward current amplification factor of the common-collector configuration is slightly greater in magnitude than that of the common-emitter connection. As already noted, the voltage gain is approximately one. Consequently, the power gain is approximately equal to the current gain. As a result, the power gain of the common collector is considerably less than that of the common-emitter configuration.

The input and output impedances of the common-collector configuration may be determined in terms of the equivalent T parameters by writing Kirchhoff's equations for the circuit of Fig. 6.2b. However, it is more convenient to use an h-parameter equivalent circuit as shown in Fig. 6.3a.

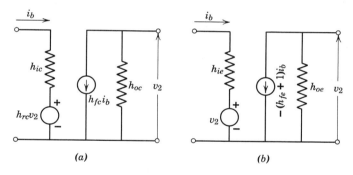

(a)   (b)

FIGURE 6.3
The h-parameter equivalent circuit for the common-collector configuration (a) using common-collector h parameters, and (b) using common-emitter h parameters.

The forward current amplification factor h_{fc} has already been found in terms of the common-emitter current amplification factor (Eq. 6.3). Thus

$$-h_{fc} = h_{fe} + 1 \simeq h_{fe} \tag{6.4}$$

Figure 6.2b shows that with the output shorted, the common-collector configuration is indistinguishable from the common-emitter configuration. Therefore, the low-frequency input resistance with the output shorted is

$$h_{ic} = h_{ie} \tag{6.5}$$

When the input is open, it can be seen from Fig. 6.2b that the output resistance is $(r_e + r_d)$ since i_b, and hence $r_m i_b$, is equal to zero. Then

$$h_{oc} = h_{oe} \qquad (6.6)$$

Figure 6.2b also shows that

$$h_{rc} = \frac{r_d}{r_d + r_e} \simeq 1 \qquad (6.7)$$

Thus it should be observed that the common-collector h parameters are almost identical in magnitude with the common-emitter h parameters, with the exception of h_{rc} which is practically unity. The common-collector h-parameter equivalent circuit is redrawn in Fig. 6.3b, using the common-emitter h parameters.

As noted previously, the general h-parameter equations can be used for any configuration if the proper parameters are used. Thus, Eq. 4.41 or Eq. 5.15 can be used to find the input impedance to the transistor in Fig. 6.2a.

$$R_{it} = h_{ic} - \frac{h_{fc} h_{rc}}{h_{oc} + G_L} \qquad (6.8)$$

where G_L is equal to $1/R_L$. When the equivalent common-emitter parameters (Eqs. 6.3, 6.5, 6.6, and 6.7) are used, this equation becomes

$$R_{it} = h_{ie} + \frac{h_{fe} + 1}{h_{oe} + G_L} \qquad (6.9)$$

Since G_L is normally large in comparison to h_{oe}, Eq. 6.9 can be written as

$$R_{it} \simeq h_{ie} + (h_{fe} + 1)R_L \qquad (6.10)$$

We note from Eq. 6.10 that the input impedance of a high-gain (large h_{fe}) transistor can be much greater (at least two orders of magnitude) than R_L.

The output admittance was determined for the common-emitter configuration in Chapter 5. Thus, from Eq. 5.21, we can write

$$Y_{ot} = h_{oc} - \frac{h_{fc} h_{rc}}{h_{ic} + R_s} \qquad (6.11)$$

where R_s is the impedance of the source or generator which drives the transistor. When the equivalent common-emitter parameters are substituted into Eq. 6.11,

$$Y_{ot} = h_{oe} + \frac{h_{fe} + 1}{h_{ie} + R_s} \qquad (6.12)$$

Observe from Eq. 6.10 and Eq. 6.12 that the emitter follower is an impedance transformer with an impedance ratio equal to $h_{fe} + 1$. Impedances in the output circuit are increased by a factor $h_{fe} + 1$ as viewed from the input terminals. In contrast, impedances in the input circuit are reduced by a factor of $h_{fe} + 1$ when viewed from the output terminals.

An example may help clarify these concepts.

Example 6.1   The transistor used in Example 5.2 is to be used in a common-collector (emitter-follower) configuration, as shown in Fig. 6.2a. The load resistance R_L is the same (6 kΩ) as used in Example 5.2. Let us determine the behavior of this amplifier. The common-emitter h-parameters at the desired q point ($I_C = 2.2$ mA, $V_{CE} = 12$ V) are:

$$h_{ie} = 4 \text{ k}\Omega \qquad h_{oe} = 2.5 \times 10^{-5}$$
$$h_{fe} = 230 \qquad h_{re} = 4 \times 10^{-4}$$

The value of G_L is $1/R_L = 1.67 \times 10^{-4}$ mhos. From Eq. 6.9 we have

$$\begin{aligned} R_{it} &= h_{ie} + (h_{fe} + 1)/(h_{oe} + G_L) \\ &= 4 \text{ k} + (230 + 1)/(2.5 \times 10^{-5} + 1.67 \times 10^{-4}) \\ &= 4 \text{ k} + 1.2 \text{ M} \simeq 1.2 \text{ M}\Omega. \end{aligned}$$

Assume the resistance of the source which drives this stage is $R_s = 5$ kΩ. Then, from Eq. 6.12,

$$\begin{aligned} Y_{ot} &= h_{oe} + (h_{fe} + 1)/(h_{ie} + R_s) \\ &= 2.5 \times 10^{-5} + (230 + 1)/(4 \times 10^3 + 5 \times 10^3) \\ &= 2.5 \times 10^{-5} + 2.56 \times 10^{-2} \simeq 2.56 \times 10^{-2} \text{ mhos.} \end{aligned}$$

Therefore, the output impedance $Z_{ot} = 1/Y_{ot} = 1/(2.56 \times 10^{-2}) = 39$ Ω. We would expect the current gain to be about equal to $h_{fc} = (h_{fe} + 1)$. If Eq. 5.17 is used with the common-collector subscripts, the current gain is

$$\begin{aligned} K_i &= h_{fc} G_L/(h_{oc} + G_L) \\ &= -(h_{fe} + 1)G_L/(h_{oe} + G_L) \\ &= -(230 + 1)(1.67 \times 10^{-4})/(2.5 \times 10^{-5} + 1.67 \times 10^{-4}) \\ &= -201. \end{aligned}$$

In addition, the voltage gain should be about one. If we use Eq. 5.16 with the common-collector subscripts, the voltage gain is

$$\begin{aligned} K_v &= -h_{fc}/[h_{ic}(h_{oc} + G_L) - h_{fc}h_{rc}] \\ &= (h_{fe} + 1)/[h_{ie}(h_{oe} + G_L) + (h_{fe} + 1)] \\ &= 231/[(4 \times 10^3)(2.5 \times 10^{-5} + 1.67 \times 10^{-4}) + 231] \\ &= 231/(0.77 + 231) = 0.997. \end{aligned}$$

The results of Example 6.1 should be compared to the results of Example 5.2 to fully appreciate the relative impedances and gains. Also, notice that the current gain is negative. This means that if the input current flows into the transistor, the output current flows out of the transistor.

The biasing circuit for the emitter follower may be determined in the same manner as described for the common-emitter amplifier in Chapter 5. However, since R_L is in the emitter circuit, the permissible values of base circuit resistance are usually quite large and the resistor R_1 may often be omitted in order to prevent the bias resistance from seriously reducing the input resistance. An example will illustrate the ease with which the bias problem can be handled.

Example 6.2   A transistor is to be connected as shown in Fig. 6.2. The value of V_{CC} is 19 V and R_L is 2 kΩ. The transistor characteristics are given in Fig. 6.4. If R_1 is omitted, let us determine the proper value for R_2.

The load line is drawn on the collector characteristics in Fig. 6.4 for $V_{CC} = 19$ V and $R_L = 2$ kΩ. The q point is chosen at $v_{CE} = 10.5$ V (or at about the center of the load line). Actually, the vertical axis of the collector characteristics should be i_E rather than i_C, but $i_E \simeq i_C$ so the available common-emitter curves are used. We will assume the transistor is silicon with $v_{BE} = 0.5$ V at the q point. As noted, V_{CE} at the q point is 10.5 V. The voltage across R_2 is then $V_{CE} - V_{BE} = 10$ V and with R_1 omitted, the value of R_2 is 10 V/50 μA = 200 kΩ. From Fig. 6.4, the value of $h_{fe} \simeq 100$, and if the value of h_{ie} is 600 Ω, the input resistance of the transistor is 600 Ω + (101) 2 kΩ, which is approximately 200 kΩ. Thus, the input

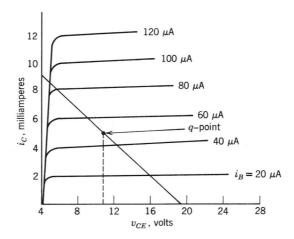

FIGURE 6.4
Collector characteristics and load line for the amplifier of Fig. 6.2.

resistance to the amplifier is approximately 100 kΩ, including the parallel biasing resistor.

The reader may assume that if R_1 is omitted, we will have a very bad current stability factor. This is not necessarily true. To illustrate this point, let us assume a current ΔI_{CO} flows across the collector-base junction. As just noted, the base has two paths to ground. One path is through R_2 and has an impedance of 200 kΩ. The other path is across the emitter-base junction and is also 200 kΩ. Consequently, ΔI_{CO} splits and $\Delta I_{CO}/2$ flows across the emitter-base junction. This current is amplified by h_{fe} so the total change of I_C is $\Delta I_C = \Delta I_{CO} + (h_{fe}\,\Delta I_{CO}/2) = 51\,\Delta I_{CO}$. Thus, $S_I = \Delta I_C/\Delta I_{CO} = 51\,\Delta I_{CO}/\Delta I_{CO} = 51$.

PROBLEM 6.1   The load resistor R_L of the emitter follower of Figs. 6.2 and 6.4 ($V_{CC} \simeq 20$ V) is changed to 5 kΩ. Draw a load line, select a q point, and determine the approximate input resistance of the transistor and the circuit, using a single biasing resistor, R_2.

Answer: 500 kΩ, 250 kΩ if $V_{CB} = 10$ V.

PROBLEM 6.2   A 2N2712 transistor (Appendix I) is used as an emitter follower with $R_L = 5$ kΩ. The driving source resistance $R_s = 3$ kΩ. Using the q point for which the common-emitter h parameters are given, determine the common-collector h parameters. Draw an h-parameter equivalent circuit and use it to calculate the current gain, voltage gain, input impedance, and output impedance of the amplifier.

Answer: $K_i = 183$, $K_v = .975$, $R_{it} = 470$ kΩ, $R_{ot} = 29$ Ω.

6.2   BOOTSTRAPPING TECHNIQUES

The input impedance of the common-collector amplifier is limited by the bias resistors, as discussed in Section 6.1. Also, in the h-parameter circuit, the load resistance R_L is in parallel with h_{oe}. Therefore, the input resistance at the base of the transistor is $h_{ie} + (h_{fe} + 1)/(G_L + h_{oe})$ as given by Eq. 6.9. Thus, the maximum input resistance which is obtained when $R_L = \infty$ or $G_L \simeq 0$ is $h_{ie} + (h_{fe} + 1)/h_{oe}$. But $(h_{fe} + 1)/h_{oe} \simeq (h_{fe} + 1)r_d = r_c$. Therefore, the maximum input impedance of the conventional emitter follower is approximately r_c, which may be a few megohms for a low-current transistor. The bias resistance is in parallel with this input resistance, so the emitter follower may not have sufficiently high input impedance for many applications.

A technique which can be used to increase the input impedance of an emitter follower is known as *bootstrapping*. This word comes from the phrase "lifting oneself by his own bootstraps." The basic principle of bootstrapping is illustrated in the circuit of Fig. 6.5. In this circuit the

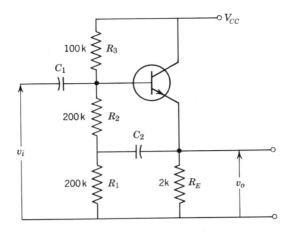

FIGURE 6.5
Bootstrapping used to increase biasing impedance.

capacitor C_2 is a dc blocking capacitor and has negligible reactance at the lowest signal frequency. Therefore, the output voltage v_o is applied at the junction of resistors R_1 and R_2. These resistors replace the single resistor which normally appears between base and ground in a stabilized-bias circuit. But the ac voltage across R_2 is the difference between the input voltage v_i and the output voltage v_o, or $v_{R2} = (v_i - v_o)$. Since v_o is nearly equal to v_i in an emitter follower, there is very little signal voltage across R_2 and, hence, very little signal current through R_2. This current can be determined by dividing the voltage by the resistance, or $i_{R2} = (v_i - v_o)/R_2$. The effective resistance of the series combination of R_1 and R_2, as seen by the input voltage, is the ratio of the input voltage to the current i_{R2}. Then

$$R_{eff} = \frac{v_i}{i_{R2}} = \frac{v_i}{(v_i - v_o)/R_2} = \frac{R_2 v_i}{v_i - v_o} \qquad (6.13)$$

If both the numerator and the denominator of Eq. 6.13 are divided by v_i, and it is recognized that v_o/v_i is the voltage gain K_v, Eq. 6.13 may be rewritten as

$$R_{eff} = \frac{R_2}{1 - K_v} \qquad (6.14)$$

If the transistor in Fig. 6.5 is the same type as in Example 6.1, where K_v was found to be 0.997, the effective resistance of R_1 and R_2 is $2 \times 10^5/0.003 = 6.67 \times 10^7 \, \Omega$.

This high effective resistance is encouraging for the attainment of a high input resistance. The fact that the resistance of R_1 did not appear directly in

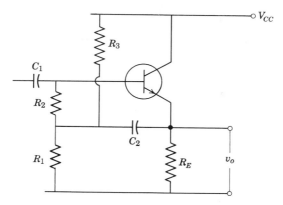

FIGURE 6.6
A circuit which bootstraps both R_2 and R_3.

the effective resistance formula may give the impression that the value of R_1 is unimportant. However, R_1 is in parallel with R_L, so far as the signal is concerned, and should be large in comparison with R_L. Otherwise, the voltage gain will be reduced, with a resulting reduction of effective resistance.

With the effective resistance of R_1 and R_2 increased to many MΩ, the bias resistor R_3 in Fig. 6.5 seriously limits the input resistance. This loading may be eliminated, as shown in Fig. 6.6, by connecting R_3 to the bootstrapping point between R_1 and R_2. From a bias standpoint, only i_B now flows through R_2. Consequently, R_3 must be reduced slightly from its value in the preceding configuration to compensate for the smaller dc voltage drop across R_2. The total bias resistance R_b is, of course, R_2 plus the parallel combination of R_1 and R_3.

The effective resistance of the bias circuit of Fig. 6.6 is probably higher than r_c, which is the limiting value of the input resistance of the transistor, as previously discussed. The value of r_c is inversely proportional to the collector current in a given transistor, therefore, very high values of input impedance may be obtained if the q-point value of collector current is very small. However, the current gain or β should be high at this small value of collector current in order to maintain a high value of $(\beta + 1)R_L$, which, in parallel with r_c, primarily determines the input impedance of the transistor.

PROBLEM 6.3 The amplifier of Fig. 6.6 has $V_{CC} = 20$ V, $h_{ie} = 5.0$ kΩ, $h_{oe} = 10^{-5}$ mho, $h_{fe} = 200$, $R_E = 10$ kΩ, $R_1 = 200$ kΩ, and $R_2 = 50$ kΩ. Determine a value of R_3 that will provide a 10 V dc drop across R_E. Assume $v_{BE} \approx 0.5$ V. Also, determine the input impedance of the amplifier at frequencies for which the capacitors have negligible reactance.

6.3   FIELD-EFFECT TRANSISTORS

The field-effect transistor (FET) is a voltage-controlled semi-conductor which has very high input impedance, particularly at low frequencies, such as audio frequencies. This transistor has only one p–n junction as shown in Fig. 6.7. It is sometimes known as a unipolar field-effect transistor (UNIFET) because it uses only one type of charge carrier. This type of FET configuration is also known as a *junction* FET as contrasted to the insulated gate FETs discussed in Section 6.4.

The schematic representation in Fig. 6.8 will be used to explain the principles of operation of the junction FET. Figure 6.8a shows that a narrow

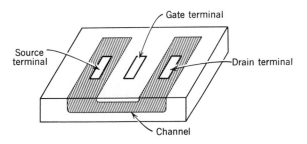

FIGURE 6.7
Typical field-effect transistor structure.

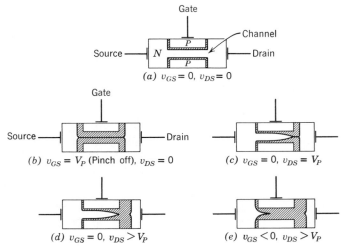

FIGURE 6.8
Schematic representation of the junction FET.

semiconductor channel provides a conducting path between the *source* and the *drain*. This channel may be either an *n*- or *p*-type crystal. The *n*-type is used in this discussion. With no biases applied to the transistor, the channel conductance $G_c = \sigma(wt/l)$ where σ is the conductivity of the crystal and w, t, and l are the width, thickness, and length of the channel, respectively. For example, if $\sigma = 1$ mho/cm, $w = 0.1$ cm, $t = 0.01$ cm and $l = 0.1$ cm, the channel $G_c = .01$ mhos and the channel resistance $R_c = 100\ \Omega$.

If reverse bias is applied between the gate and the source, the depletion region width is increased, the thickness of the channel is decreased, and therefore the conductivity of the channel is decreased. The gate bias required to just reduce the channel thickness to zero, as shown in Fig. 6.8*b* is called the *pinch-off* voltage, V_P.

When the gate-source voltage v_{GS} is zero and the drain is made positive with respect to the source, electrons drift through the channel because of the electric field. The drain current i_D is equal to the drain-source voltage v_{DS} times the channel conductance G_c, providing v_{DS} is very small. However, the positive drain voltage reverse biases the *p-n* junction near the drain end of the channel, and when the drain voltage is increased to the pinch-off voltage, the channel thickness is reduced to zero at a point near the drain end of the channel as shown in Fig. 6.8*c*. The drain current does not stop when the drain voltage reaches pinch-off because a voltage equal to V_P still exists between the pinch-off point and the source, and the resulting electric field along the channel causes the free carriers in the channel to drift from the source to the drain.

As the drain voltage is increased above V_P, the depletion region thickness is increased between the drain and the gate, as shown in Fig. 6.8*d*. In fact, the additional drain voltage is absorbed by the increased field in the wider pinched-off region, and the electric field between the original pinch-off point and the source remains essentially unchanged. Therefore, the channel current and, hence, the drain current remains essentially unchanged. The carriers which arrive at the pinch-off point are swept through the depletion region in the same manner as carriers which are swept from the base into the collector region in a conventional or *bipolar* transistor. Thus, whenever v_{DS} is higher than the pinch-off voltage V_P, the drain current is essentially independent of the drain voltage.

The field-effect transistor normally operates with the drain voltage v_{DS} beyond the pinch-off voltage V_P and reverse bias applied between the gate and the source. The electric field and thus the drain current in the channel is then controlled by the gate voltage v_{GS}. The effect of the gate voltage v_{GS} on the channel conductance is shown in Fig. 6.8*e*. The channel thickness is reduced as a result of the reverse gate bias. The drain current is essentially independent of the drain voltage whenever the sum of the magnitudes of the

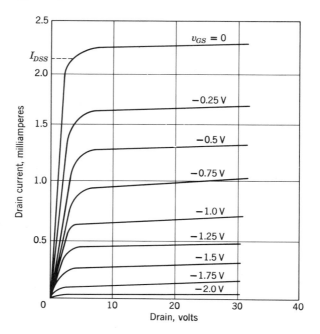

FIGURE 6.9
Drain characteristics of a typical *n*-channel FET.

drain voltage and the reverse-bias gate voltage exceeds the pinch-off voltage.

The drain characteristics of a typical *n*-channel FET are shown in Fig. 6.9. The drain current which flows when $v_{GS} = 0$ and $v_{DS} = V_P$ is known as I_{DSS}, which means saturated drain current with input shorted. The transistor of Fig. 6.9 has $I_{DSS} = 2.2$ mA. Observe that the slope of the drain characteristic curves are quite flat in the normal operating range. This means the output resistance (equivalent to $1/h_{oe}$ in a bipolar transistor) is very high. Avalanche breakdown occurs at the junction whenever the drain-gate voltage exceeds a given value.

The input characteristics of a typical *n*-channel FET are shown in Fig. 6.10. The gate current is the saturation current of the reverse-biased junction. This current is of the order of 10^{-9} A or 1 nano-amp for a low-power silicon FET at 25°C. The dynamic input conductance g_g is the slope of the input characteristics. The dynamic input conductance of the FET of Fig. 6.10 is $\Delta i_G/\Delta v_{GS} = 1$ nA/5 V $= 2 \times 10^{-10}$ mho and the dynamic or incremental input resistance is 5×10^9 Ω when the junction is reverse biased. Although the input resistance decreases rapidly as the junction becomes forward biased,

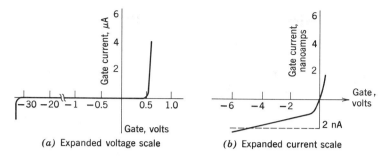

FIGURE 6.10
Input characteristics of a typical *n*-channel FET.

Fig. 6.10*a* shows that the input resistance remains quite high (a megohm or more) in a silicon FET so long as the forward bias does not exceed about 0.25 V at 25°C. Therefore, $v_{GS} = 0$ is a suitable q point for a small-signal amplifier. Note that the input resistance drops abruptly when avalanche breakdown occurs. Avalanche breakdown occurs whenever the algebraic difference between the gate and drain potentials exceeds the avalanche breakdown voltage.

In order to analyze the FET amplifier, we need to develop an equivalent circuit for the FET device. The basic *y*-parameter circuit shown in Fig. 6.11 is the most useful configuration. At low frequencies, the depletion-region capacitances can be neglected and the admittances become conductances. The admittance y_i is the gate-to-source conductance and is normally given the symbol g_g for *gate input conductance*. The gate conductance is found from the relationship

$$g_g = \frac{\Delta i_G}{\Delta v_{GS}}\bigg|_{v_{DS} = K} \tag{6.15}$$

We have already noted this conductance is very low (about 10^{-9} mhos or so). In fact, for many applications g_g can be considered essentially equal to zero which simplifies the calculations a bit.

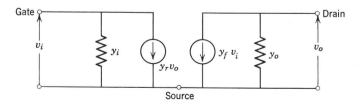

FIGURE 6.11
The *y*-parameter equivalent circuit of a FET.

The parameter y_r is given by the relationship

$$y_r = \frac{\Delta i_G}{\Delta v_{DS}}\bigg|_{v_{GS}=k} \tag{6.16}$$

However, at low frequencies the drain-source voltage has essentially no effect on the gate current so y_r is normally assumed to be zero.

The parameter y_f is called the *forward transconductance* or *mutual conductance* of the transistor and is normally given the symbol g_m. The mutual conductance is found from the relationship

$$g_m = \frac{\Delta i_D}{\Delta v_{GS}}\bigg|_{v_{DS}=k} \tag{6.17}$$

If the transfer characteristics of the transistor are provided as in Fig. 6.12, the value of g_m is the slope of this transfer curve at the required q point. In addition, g_m can be obtained with less accuracy from the drain characteristics of Fig. 6.9 by using the same technique as was used to obtain h_{fe} or h_{re} in the bipolar transistors.

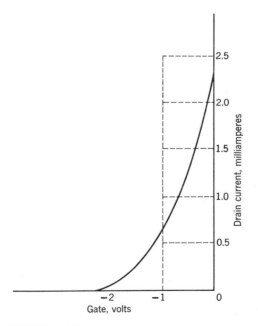

FIGURE 6.12
The transfer characteristics of the FET in Fig. 6.9.

Finally, the parameter y_o is the output or *drain admittance* with the input terminals shorted. This drain admittance is usually given the symbol g_d and is found from the following relationship.

$$g_d = \frac{\Delta i_D}{\Delta v_{DS}}\bigg|_{v_{GS}=k} \tag{6.18}$$

Therefore, g_d is the slope of the drain characteristic (Fig. 6.9) at the desired q point. From this analysis, a low-frequency equivalent circuit for the FET may have the form shown in Fig. 6.13.

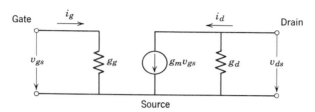

FIGURE 6.13
An equivalent circuit for a FET.

A typical amplifier configuration (corresponding to the common-emitter configuration of a bipolar, *n-p-n* or *p-n-p* type, transistor) is shown in Fig. 6.14a. In Fig. 6.14b, the FET of Fig. 6.14a has been replaced by its equivalent circuit. We have assumed C_S is effectively a short circuit for the ac signals in developing the equivalent circuit. Now if C_{C1} is sufficiently large, $v_{gs} \simeq v_i$, and if C_{C2} is also large, $v_d \simeq v_o$. Then, if $G_D = 1/R_D$,

$$v_o = \frac{-g_m v_i}{g_d + G_D} \tag{6.19}$$

The voltage gain is

$$K_v = \frac{v_o}{v_i} = -\frac{g_m}{g_d + G_D} \tag{6.20}$$

In many amplifiers, r_d is much greater than R_D. Then, the voltage gain is

$$K_v \simeq -g_m R_D \tag{6.21}$$

Of course, Eqs. 6.20 and 6.21 are only valid if the transistor operates on the linear portion of its characteristic curves.

Since y_r in Fig. 6.11 is zero, there is no feedback from the output to the input loop in the equivalent circuit. (This condition is similar to $h_{re} = 0$ for the bipolar transistor.) Then the input impedance of the transistor is

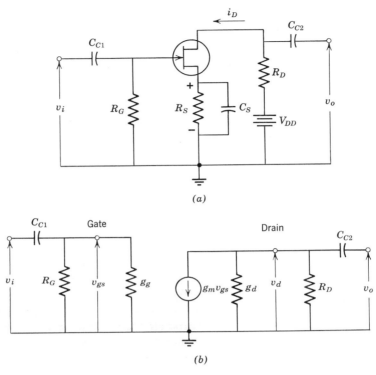

(a)

(b)

FIGURE 6.14
A FET amplifier circuit: (a) the actual circuit; (b) the equivalent circuit.

$R_{it} = 1/g_g = r_g$, and the output impedance of the transistor is $R_{ot} = 1/g_d = r_d$. Since the signal input current to a FET is so small, the current gain is very high but of little importance. However, the current gain can be readily found from the equivalent circuit if it is required.

The resistor R_G in Fig. 6.14a permits the saturation current which flows across the junction to flow to ground and the negative terminal of V_{DD}. Thus, the gate is maintained at approximately ground potential, for dc, which is negative with respect to the source. Usually, not more than 0.1 V or so should be dropped across R_G. If bias at $v_{GS} = 0$ is desired, the source would be returned to ground. However, as already noted, the gate signal should be restricted to a few tenths of a volt in magnitude if $V_{GS} = 0$.

Reverse bias is often desired for the gate of the FET with respect to the source. Under these conditions, the resistor R_S of Fig. 6.14 is used to obtain the proper value of bias potential for the FET. The drain current then flows through this resistor in the direction which will provide reverse bias. If the gate circuit resistance R_G is low enough so the drain-gate saturation current I_{DO} produces negligible voltage across R_G, the gate is maintained at dc ground potential and the bias voltage is $I_D R_S$. Thus, for a desired gate bias voltage V_{GS}, R_S can be obtained from the relationship

$$R_S = \frac{V_{GS}}{I_D} \qquad (6.22)$$

A bypass capacitor C_S across R_S will prevent degeneration and loss of gain for ac signals. This capacitor must bypass the impedance looking in at the FET source terminal in parallel with R_S. This source impedance may be obtained by making a small change in the source voltage Δv_S and noting the change in source current Δi_S as shown in Fig. 6.15. However, the

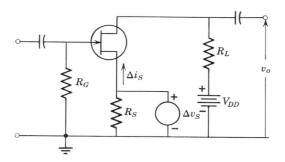

FIGURE 6.15
The configuration used to determine the impedance into the source of a FET.

gate is maintained at ground potential so the source-to-ground voltage v_S is equal to the source-to-gate voltage v_{SG}. Also $\Delta i_S = \Delta i_D = g_m \Delta v_{GS}$, providing g_d is small compared with G_L, which is the usual situation. Therefore, the resistance into the source is

$$r_S = \frac{\Delta v_S}{\Delta i_S} = \frac{\Delta v_{GS}}{\Delta i_D} = \frac{1}{g_m} \qquad (6.23)$$

If the reactance of the bypass capacitor C_S is to be one-tenth of the magnitude of the resistance being bypassed at some low frequency ω_1, then

$$\frac{1}{\omega_1 C_S} = \frac{1}{10(g_m + G_S)} \qquad (6.24)$$

where $G_S = 1/R_S$, and

$$C_S = \frac{10(g_m + G_S)}{\omega_1} \qquad (6.25)$$

The drain current in a FET varies approximately as the square of the gate-source voltage and can be expressed by the following relationship

$$i_D \simeq I_{DSS}\left(1 - \frac{v_{GS}}{V_p}\right)^2 \qquad (6.26)$$

Since $g_m = di_D/dv_{GS}$, g_m can be obtained by differentiating Eq. 6.26,

$$g_m \simeq 2I_{DSS}\left(1 - \frac{v_{GS}}{V_P}\right)\left(-\frac{1}{V_P}\right) \qquad (6.27)$$

$$g_m \simeq -\frac{2I_{DSS}}{V_P}\left(1 - \frac{v_{GS}}{V_P}\right) \qquad (6.28)$$

The value of g_m obtained at zero bias, or $v_{GS} = 0$, is usually given as g_{mo} and can be readily obtained from Eq. 6.28 as

$$g_{mo} = -\frac{2I_{DSS}}{V_P} \qquad (6.29)$$

Observe that g_m or g_{mo} will always be positive since either V_P of I_{DSS} will be a negative quantity. Substituting Eq. 6.29 into Eq. 6.28,

$$g_m \simeq g_{mo}\left(1 - \frac{v_{GS}}{V_P}\right) \qquad (6.30)$$

Thus g_m can be obtained analytically from the data normally given in the manufacturer's data sheets.

The variation of drain current with temperature in a field-effect transistor is determined by two factors. One factor is the temperature variation of depletion-region width, which results from the temperature variation of barrier height $(V_{ho} - V)$, as discussed in Chapter 3, where V_{ho} is the zero-bias barrier height. As previously discussed, the temperature coefficient of this voltage is about $-2.2\ \text{mV}/^\circ\text{C}$, which results in an increased drain current with increased temperature. The other factor is the variation of majority carrier mobility with temperature. This mobility influences the transconductance g_m. As the temperature increases, the carrier mobility, and hence g_m, decrease and tend to compensate for the variation of V_{ho} with temperature. In fact, the proper choice of q point will give essentially zero temperature coefficient of drain current from about -50 to 100°C. The temperature coefficient due to mobility change is about $0.7\%/^\circ\text{C}$. Therefore, the condition for zero temperature coefficient is

$$0.007(-i_D)/^\circ\text{C} = g_m(-0.0022)/^\circ\text{C} \qquad (6.31)$$

$$\frac{i_D}{g_m} = +0.315\ \text{V} \qquad (6.32)$$

Substituting the expression for i_D (Eq. 6.26) and g_m (Eq. 6.28) into Eq. 6.32, the following relationship results.

$$-\frac{V_P}{2}\left(1 - \frac{v_{GS}}{V_P}\right) = +0.315 \text{ V} \qquad (6.33)$$

or

$$v_{GS} = V_P + 0.63 \text{ V} \qquad (6.34)$$

Equation 6.34 shows that zero thermal drift may be achieved if the FET is biased 0.63 V above the pinch-off voltage V_P. Note that V_P and v_{GS} are negative for an n-channel FET. All signs should be reversed for a p-channel FET.

An example will be used to help clarify the design ideas for a FET amplifier.

Example 6.3   Let us design an amplifier using the 2N5457 n-channel FET. The characteristic curves for a typical transistor are given in Fig. 6.16.

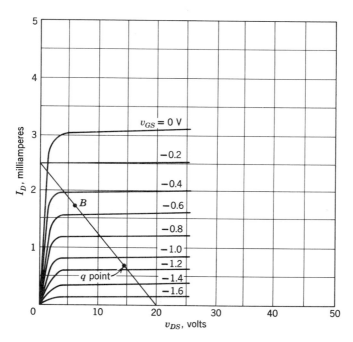

FIGURE 6.16
The drain characteristics for a 2N5457.

In addition, the following electrical data are given.

> Gate-source breakdown voltage, $BV_{GSS} = 25$ V minimum.
> Gate reverse current for $T_A = 25°C$, $I_{GSS} = 1$ nA maximum.
> $T_A = 100°C$, $I_{GSS} = 200$ nA maximum.
> Maximum power dissipation at $T_A = 25°C$, $P_D = 310$ mW maximum.

The circuit for the amplifier is given in Fig. 6.17. Since the gate-source breakdown voltage is 25 V, let us use $V_{DD} = 20$ V.

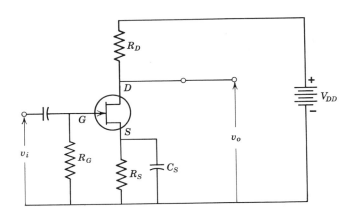

FIGURE 6.17
The amplifier configuration for Example 6.3.

First, let us determine the value of R_G. We would like this resistance to be high, but we do not want an appreciable voltage drop due to the gate reverse current. If the ambient temperature never exceeds 25°C, the value of R_G can be very large. However, if the temperature can rise to 100°C and if we limit the maximum voltage drop across R_G to 0.2 V, the maximum value of R_G is $R_G = 0.2$ V$/I_{GSS} = 0.2$ V$/2 \times 10^{-7}$A $= 1$ MΩ.

From the characteristics in Fig. 6.16, the pinch-off voltage $V_P = -1.8$ V. If we bias the transistor at its zero thermal-drift value, the voltage $v_{GS} = V_P + 0.63 = -1.8 + 0.63 = -1.17$ V. Then, the quiescent drain current (Fig. 6.16) is 0.70 mA. The value of R_S (Eq. 6.22) is $R_S = V_{GS}/I_D = 1.17$ V$/0.70$ mA $= 1.67$ kΩ. The value of g_m can be found from the relationship $g_m = \Delta i_D/\Delta v_{GS}$. From Fig. 6.16, the value of g_m at $v_{GS} = -1.17$ V is $g_m \simeq (0.8 - 0.6)$ mA$/[-1.0 - (-1.2)$ V$] = 0.2$ mA$/0.2$ V $= 1000$ μmhos. Now from Eq. 6.25, the value of $C_S = 10(g_m + G_S)/\omega_1$. If the lowest frequency to be bypassed is $\omega = 200$ rad/sec, $C_S = 10(10^{-3} + 0.60 \times 10^{-3})/200 = 80$ μF. (We would probably use a 100 μF value.)

If the transistor is biased at -1.17 V, a signal swing of -0.63 V would be sufficient to swing the gate to pinch-off ($V_P = -1.8$ V). Therefore, if we provide for a symmetrical signal swing, the maximum signal swing in the positive direction is -1.17 V $+ 0.63$ V $= -0.54$ V. Point B in Fig. 6.16 will provide a signal swing to -0.54 V without getting onto the knee of the characteristic curve. A load line through point B and $V_{DD} = 20$ V has been drawn on Fig. 6.16. The value of this load line is $R_L = 20$ V$/2.5$ mA $= 8.0$ kΩ. Then, since $R_L = R_D + R_S$, the value of $R_D = 8.0$ k $- 1.67$ k $= 6.33$ kΩ. (We would use resistor values of 1.5 kΩ for R_S and 6.8 kΩ for R_D.)

Now, from Fig. 6.16, $r_D \gg R_D$. Therefore, the voltage gain (Eq. 6.21) is $K_v \simeq g_m R_L = 10^{-3} \times 6.8 \times 10^3 = 6.8$. The input impedance is approximately equal to $R_G = 1$ MΩ, and the output impedance is approximately equal to $R_D = 6.8$ kΩ.

This amplifier should be very stable over wide temperature variations.

PROBLEM 6.4   Use the 2N5457 transistor of Example 6.3 to design an amplifier. The supply voltage $V_{DD} = 25$ V and the desired q point is at $v_{DS} = 15$ V and $i_D = 1$ mA. Limit the voltage across R_G to 0.1 V if the temperature rises to 55°C. (Assume the gate reverse current doubles for every 10°C temperature increase.) The lowest frequency to be amplified is 111 rad/sec. What is the voltage gain of your amplifier?

Answer: $R_G = 12.5$ MΩ, $R_S \simeq 900$ Ω, $R_D \simeq 9$ kΩ, $C_D = 100$ μF.

PROBLEM 6.5   A typical crystal microphone has 2 mV peak open-circuit output voltage and an internal capacitance of 10^{-9} F in parallel with 10 MΩ resistance. Design a preamp for this microphone using a 2N5457 FET, assuming the input impedance of the main amplifier to be 100 kΩ. Assume $V_{DD} = 30$ V. The desired frequency response is 30 Hz to 15 kHz. Determine the values of all components and calculate the expected voltage at the input terminals of the main amplifier.

6.4   INSULATED GATE FETs

If extremely high input resistance is desired in a solid-state circuit, insulated gate FETs may be used. These devices are also known as metal-oxide semiconductor (MOS) field-effect transistors (FET) or metal-insulator semiconductor (MIS) field-effect transistors (FET). Usually, they are referred to simply as MOSFETs.

The MOSFETs are typically constructed as shown in Fig. 6.18. Note that the gate is insulated from the semiconductor material (usually silicon) by a very thin layer of oxide insulation (usually silicon dioxide or glass). Consequently, the input resistance is typically in the range of 10^{12} to 10^{14} Ω.

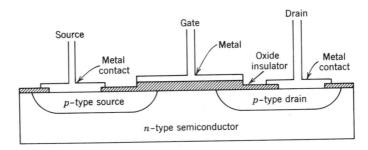

FIGURE 6.18
Construction of a *p*-channel MOSFET.

Under typical operating conditions, the gate is maintained negative. The negative charge on the gate repels the electrons in the *n* material and attracts the holes. If the potential on the gate is sufficiently negative, the number of holes near the surface of the *n* material will exceed the number of electrons. The surface of the *n* material is said to have *inverted* and now behaves as if it were *p* material. Fig. 6.19*a* shows the configuration of a MOSFET with negative gate and zero potential on the source and drain. As the gate becomes more negative, the depth of the channel increases. Hence, the channel width is controlled by the gate potential as in the junction FET.

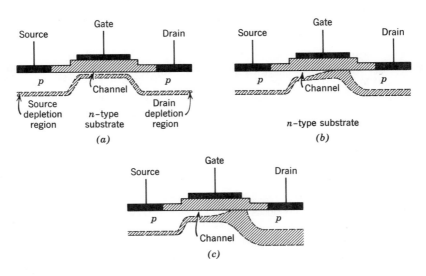

FIGURE 6.19
A MOSFET with applied bias.

When a negative potential is applied to the drain as well as the gate, the channel is distorted as shown in Fig. 6.19*b* and Fig. 6.19*c*, and current flows from the source to the drain.

When drain current flows along the channel, an *IR* drop is developed along the channel. This *IR* drop tends to cancel the field produced by the gate bias. When the cancellation is sufficient to *almost* prohibit the formation of the inversion layer, the channel *pinches off* and the drain current tends to *saturate* at a constant value independent of increased drain voltage. Of course, if there were no channel, no current would flow to the drain. Thus the *pinch-off* or *threshold* voltage, V_{TH}, is the voltage from gate to channel necessary to just produce inversion in the channel.

The characteristics of a typical MOSFET are shown in Fig. 6.20*a*. Note that in contrast to the junction FET, the gate and drain potential have the same polarity. This means that MOSFETs can be direct coupled from the drain of one stage to the gate of the next stage with no isolating capacitors required.

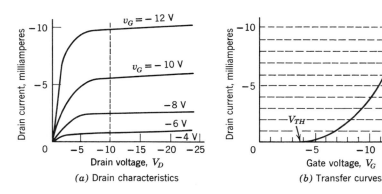

(*a*) Drain characteristics   (*b*) Transfer curves

FIGURE 6.20
Characteristics of a *p*-channel MOSFET.

If the drain voltage is maintained constant (say 10 V in Fig. 6.20*a*) and a plot of drain current versus gate voltage is obtained, this characteristic is known as a transfer curve. The transfer curve of the MOSFET in Fig. 6.20*a* is drawn in Fig. 6.20*b*. In this transistor, the threshold voltage, V_{TH}, is −4 V. The threshold voltage is readily apparent in either set of curves in Fig. 6.20.

These curves illustrate that *the drain current is proportional to the square of the gate-source voltage, the same as in the junction FET*. This square-law effect may be desirable in some applications but can be a handicap in a linear amplifier.

A p-channel MOSFET amplifier is connected as shown in Fig. 6.21. Note that a connection is made to the bulk or substrate (the n material in Fig. 6.18) of the semiconductor as well as to the gate, to the source, and to the drain. In a p-channel MOSFET, the substrate is connected to the positive terminal of the power supply. This condition is necessary to insure the p-n junctions in the device do not become forward biased. (The substrate of an n-channel device is connected to the negative terminal of the power supply.) Of course, if the p-n junctions should become forward biased, the circuit will cease to behave as a MOSFET.

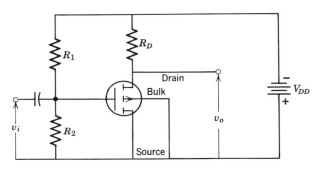

FIGURE 6.21
A MOSFET amplifier.

In the foregoing transistor (Fig. 6.20) the electric field from the gate produces the conduction channel. A higher field increases or *enhances* the channel thickness and consequently the drain current magnitude. A MOSFET operating in this manner is operating in the *enhancement mode*. It is also possible to construct MOSFETs with a channel present when $v_{GS} = 0$. An actual channel can be diffused into the substrate when the gate and source volumes are constructed. Or, by the proper choice of insulating and gate materials, the contact potentials of the different materials plus charges trapped in or near the gate insulator may create a built-in field to produce inversion and a channel in the substrate when the gate voltage is zero. If a channel is present when v_{GS} is zero, the transistor is operating in the *depletion mode*. A junction FET operates in the depletion mode. In this case, the channel thickness is controlled by adjusting the depletion width of the p-n junction.

In order to distinguish between enhancement mode and depletion mode MOSFETs, the channel of a depletion-mode MOSFET is usually represented by a solid line. In contrast, the channel of an enhancement-mode MOSFET is represented by a dashed line as shown in Fig. 6.22.

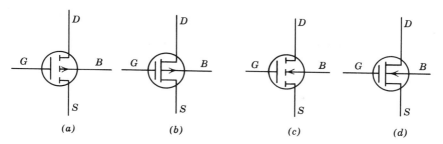

FIGURE 6.22
Symbols for various types of MOSFET transistors; (*a*) *p*-channel enhancement mode; (*b*) *p*-channel depletion mode; (*c*) *n*-channel enhancement mode; (*d*) *n*-channel depletion mode.

An *n*-channel MOSFET would be constructed as shown in Fig. 6.23*a*. The contact potentials of an *n*-channel configuration usually produce a depletion-mode device. (In contrast, most *p*-channel MOSFETs are enhancement-mode devices.) Drain characteristics for a typical *n*-channel MOSFET are given in Fig. 6.23*b*.

A circuit for an *n*-channel depletion-mode transistor is given in Fig. 6.24.

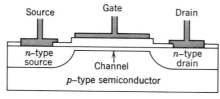

(*a*) *n*-channel configuration

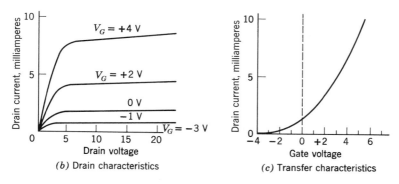

(*b*) Drain characteristics

(*c*) Transfer characteristics

FIGURE 6.23
A typical *n*-channel MOSFET and its characteristics.

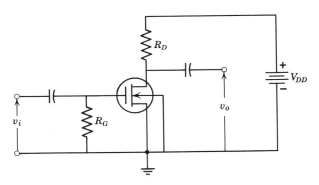

FIGURE 6.24
An *n*-channel MOSFET amplifier.

Since current flows with zero bias, the gate can be returned to the source as shown. Since the gate draws zero bias current, the value of R_G can be very large. Similarly, in Fig. 6.21, the two resistors R_1 and R_2 form a voltage dividing network to obtain the proper gate bias, but both R_1 and R_2 can be very large.

In Fig. 6.21 and Fig. 6.24, the substrate lead is connected to the source. However, it is possible to change the characteristics of the device by applying a different bias to the substrate. To illustrate the effect of substrate bias, three sets of output characteristics for a *p*-channel MOSFET are given in Fig. 6.25. Note that if the gate voltage is maintained constant, the drain current decreases as the substrate (or bulk) voltage increases in the positive direction. The threshold voltage also changes with the substrate voltage. Since drain current control can be achieved by either the gate or the substrate, the gate is sometimes called the *front gate* and the substrate is referred to as the *back gate*. However, the input impedance of the substrate is much less than the input impedance of the gate. In fact, since the substrate to source or drain forms an *n-p* junction, the input impedance of the substrate is in the same order of magnitude as the input impedance of a conventional FET.

An equivalent circuit for a MOSFET with substrate maintained constant (as in Fig. 6.21 or Fig. 6.24) is given in Fig. 6.26. Since the impedance is so high, the effective input impedance is essentially an open circuit at low frequencies as shown. (However, the MOSFET does have an input capacitance in the order of a picofarad or so that must be considered at higher frequencies.) The parameter g_m is defined by the equation

$$g_m = \frac{\Delta i_D}{\Delta v_G}\bigg|_{v_D = \text{constant}} \tag{6.35}$$

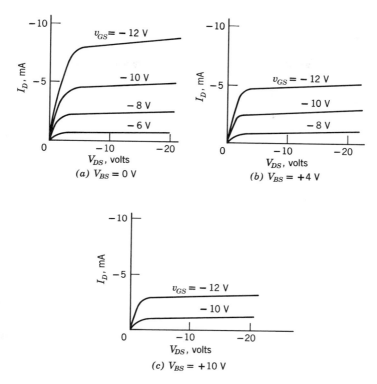

FIGURE 6.25
The effect of substrate bias on the output characteristics of a MOSFET:
V_{GS} = gate voltage, V_{DS} = drain voltage, V_{BS} = substrate voltage.

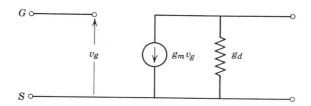

FIGURE 6.26
An equivalent circuit for a MOSFET.

Of course, the value of g_m can be determined from the characteristic curves of the transistor. Typical values of g_m are in the range of 1,000 to 2,000 μmhos. From the curves in Fig. 6.20 or Fig. 6.22, we note that the value of g_m is a function of drain current. The value of g_m may be found from the following equation.

$$g_m = \frac{2i_D}{v_G - v_{TH}} \tag{6.36}$$

where

> i_D is the drain current in amperes
> v_G is the gate voltage in volts
> v_{TH} is the pinch-off voltage in volts

The value of g_D can be found from the characteristic curves by using the relationship

$$g_D = \frac{\Delta i_D}{\Delta v_D}\bigg|_{v_G = \text{constant}} \tag{6.37}$$

PROBLEM 6.6   Determine g_d and g_m for the MOSFET whose characteristics are given in Fig. 6.23. Use both Eq. 6.35 and 6.36 to determine g_m and compare the results. Assume the q point is at $v_G = 0$ V and $v_D = 10$ V. Answer: $g_d \simeq 10$ μmhos, $g_m = 1000$ to 1100 μmhos.

PROBLEM 6.7   Determine g_d and g_m for the MOSFET whose characteristics are given in Fig. 6.20. Assume a q point of $v_G = -10$ V and $v_D = -10$ V.

If both gate and substrate voltages are to be varied, the equivalent circuit of the device becomes more complicated. One approach would be to use a circuit similar to that given in Fig. 6.26, but g_m would be proportional to the substrate voltage, v_{BS}. Thus, g_m in Fig. 6.26 for a p-channel MOSFET would be given as

$$g_m = g_{m1} - D_{m2} v_{BS} \tag{6.38}$$

where

> g_{m1} is the value of g_m where $v_{BS} = 0$
> D_{m2} is the rate at which g_m changes with v_{BS}

An example will be given to clarify this concept.

Example 6.4   Let us draw the equivalent circuit for the transistor whose characteristics are given in Fig. 6.25. We will assume the q point for this transistor is at $v_{GS} = -10$ V, $v_{BS} = +4$ V, $v_{DS} = -10$ V. Note that if

$v_{BS} = 0$ V, the threshold voltage is $v_{TH} = -4$ V and the drain current (for $v_{GS} = -10$ V and $v_{DS} = -10$ V) is -5 mA. From Eq. 6.36, the value of g_m is

$$g_m = \frac{2(-5 \times 10^{-3})}{-10 - (-4)} = \frac{-10^{-2}}{-6} = 1,667 \ \mu\text{mhos}$$

If v_{BS} is $+4$ V, the threshold voltage v_{TH} is -6 V and the drain current for the q point is 2.6 mA. Then, Eq. 6.36 gives a value of g_m as

$$g_m = \frac{2(-2.6 \times 10^{-3})}{-10 - (-6)} = \frac{-5.2 \times 10^{-3}}{-4} = 1,300 \ \mu\text{mhos}$$

When v_{BS} is $+10$ V, the threshold voltage is $v_{TH} = -8$ V and the drain current (for $v_{GS} = -10$ V and $v_{DS} = -10$ V) is -0.8 mA. The value of g_m changes to the following value

$$g_m = \frac{2(-0.8 \times 10^{-3})}{-10 - (-8)} = \frac{-1.6 \times 10^{-3}}{-2} = 800 \ \mu\text{mhos}$$

A plot of g_m versus v_{BS}, as determined from the foregoing information, is given in Fig. 6.27. Note that g_m decreases as v_{BS} increases, thus justifying

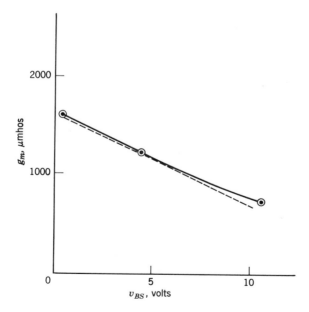

FIGURE 6.27
A plot of g_m versus v_{BS} for the transistor of Fig. 6.24.

the negative sign in Eq. 6.38. (An n-channel MOSFET would require a positive sign in this equation.) The desired q point is at $v_{BS} = 4$ V so a tangent is drawn to the curve in Fig. 6.27 at $v_{BS} = 4$ V. This tangent represents the linear model we are developing and the difference between the straight tangent line and the actual curve represents the error in our model. Note that a change of one volt (along the tangent line) represents a change of 85 micromhos for g_m. Thus the slope of the tangent at the desired operating point determines the value of D_{m2} and the intercept with the $v_{BS} = 0$ axis determines the value of g_{m1}. Hence, Eq. 6.38 would be

$$g_m = (1,600 - 85\, v_{BS}) \text{ micromhos}$$

for the transistor of Fig. 6.25.

The equivalent circuit of the MOSFET in Fig. 6.25 (biased at the q point previously noted) would be as shown in Fig. 6.28. The substrate (or bulk) terminal is included as one of the input terminals. Since this input terminal has a much lower input impedance than the insulated gate, an input resistor r_b is included. The output current is a function of both input circuits as noted.

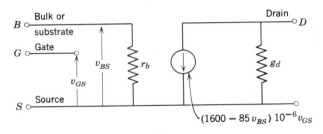

FIGURE 6.28
The equivalent circuit for the MOSFET of Fig. 6.25 when both gates are used for signal inputs.

The MOSFET amplifier with both bulk and gate input signals could be connected as shown in Fig. 6.29. If this configuration is used, the value of v_{BS} should never be more negative than -0.3 V or so (for a silicon transistor). Otherwise the source-to-bulk junction will begin to conduct an appreciable current and the device will no longer act as a MOSFET.

Because of the extremely high input impedance of the MOSFET, one must be very careful when installing or handling them. To more fully understand this statement, consider the situation illustrated in Fig. 6.30. In this figure, a positive charge $(+q)$ is placed near the gate lead of a MOSFET. The positive charge attracts electrons to the end of the gate

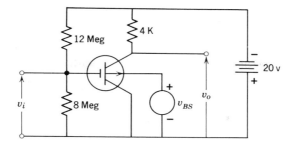

FIGURE 6.29
A MOSFET amplifier with adjustable substrate voltage.

lead. The loss of these electrons creates a positive charge on the gate; a potential is developed across the insulation between the gate and substrate. In fact, the gate, insulation, and substrate is a capacitor, and the voltage across the insulation in a capacitor is

$$V = \frac{Q}{C} \tag{6.39}$$

As mentioned previously, the input capacitance of the gate in a MOSFET is typically in the order of one picofarad. Thus, the voltage across the insulation is

$$V \simeq 10^{12}Q \tag{6.40}$$

Hence, a small charge can produce a large potential across the insulation.

The gate insulating material in a MOSFET is typically 10^{-5} to 2×10^{-5} cm thick. Consequently, gate-to-substrate voltages in the order of 50 V or so will cause breakdown of the insulation. Once breakdown has occurred, the insulating qualities of the insulator are destroyed and the MOSFET is ruined.

The charge shown in Fig. 6.30 can be a static charge on a person's finger or some tool he is using. Thus, *one can destroy a MOSFET without even touching it.* To prevent destruction, most MOSFETs are shipped with their leads twisted so the gate is shorted to the source and/or

FIGURE 6.30
The effect of a static charge near the gate of a MOSFET

substrate. Shorting leads should be clipped to the gate when soldering or installing MOSFETs.

Some MOSFETs are constructed with a reverse-biased junction connected internally to the gate. This junction conducts (as a Zener diode) if the gate voltage exceeds the diode breakdown potential. The diode does protect the gate insulator, but the total input impedance is now equal to the impedance of the reverse-biased diode. Hence the input impedance is about equal to that of a junction FET.

An example will help illustrate one application for a MOSFET.

Example 6.5   A MOSFET transistor has the characteristics given in Fig. 6.25. This transistor is connected as shown in Fig. 6.29. Let us determine the voltage gain of this amplifier if $v_{BS} = 0$ V and if $v_{BS} = 4$ V.

The equivalent circuit for this MOSFET was given in Fig. 6.28. The value of g_d can be found from Fig. 6.25. $g_d = \Delta i_D / \Delta v_D \simeq 0.1$ mA/10 V $= 10^{-5}$ mhos. Then, $r_d = 1/g_d = 1/10^{-5} = 10^5$ Ω. The voltage gain (Eq. 6.20) is $K_v = g_m/(g_d + G_D) = g_m \times 3.85 \times 10^3$. Now, $g_m = (1600 - 85v_{BS})10^{-6}$ and $v_{BS} = 0$, $g_m = 1.6 \times 10^{-3}$. Then, $K_v = 1.6 \times 10^{-3} \times 3.85 \times 10^3 = 6.16$.

In contrast, if $v_{BS} = 4$ V, the value of g_m is $(1600 - 85 \times 4)10^{-6} = 1.26 \times 10^{-3}$ mhos. The voltage gain is now

$$K_v = 1.26 \times 10^{-3} \times 3.85 \times 10^3 = 4.85.$$

The gain of the amplifier in the preceding example is controlled by the voltage on the base. There are many applications where automatic gain control of a circuit is desirable. The MOSFET would perform well in these applications.

PROBLEM 6.8   A circuit is connected as shown in Fig. 6.29. The characteristics of the device are given in Fig. 6.25. Find the equation for v_o if $v_i = 0.1 \cos(10^7 t)$ and $v_{BS} = 4 + 4 \cos(10^3 t)$.

Answer: $v_o = 0.5 \cos 10^7 t - 0.136(\cos 10^3 t)(\cos 10^7 t)$. We have accomplished amplitude modulation!

PROBLEM 6.9   Design a two-stage, direct-coupled MOSFET amplifier which uses a 20 V battery for a power supply. The q point for both amplifiers is to be $v_{DS} = -10$ V and $v_{GS} = -10$ V.

6.5   DUAL INSULATED-GATE FIELD-EFFECT TRANSISTORS

The dual-gate IGFET or MOSFET has two control gates as shown in Fig. 6.31. With this configuration, the output can be controlled by two different signals while the substrate remains grounded. Even if only one

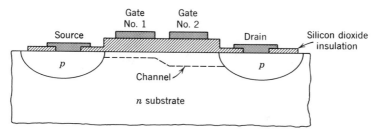

FIGURE 6.31
A dual-gate MOSFET configuration.

signal is to be used, the dual-gate MOSFET has characteristics which make it superior to a single-gate MOSFET for some applications.

As shown in Fig. 6.31, the first gate (gate No. 1) controls the channel near the source while the second gate (gate No. 2) controls the channel near the drain. With this configuration, we have four variables v_{G1S} (the voltage from gate No. 1 to the source), v_{G2S} (the voltage from gate No. 2 to the source), v_{DS} (the voltage from the drain to the source), and i_D (the drain current) to consider. There are several ways we could present this information in graphical form. One presentation could be a series of characteristic curves as was done in Fig. 6.25. One manufacturer presents the characteristics of a 3N200 (an n-channel depletion-mode dual-gate MOSFET) as shown in Fig. 6.32.

The incremental equivalent circuit for a dual-gate MOSFET can be developed by noting (Fig. 6.32) that i_D is dependent on the values of v_{G1S}, v_{G2S}, and v_{DS}. We can express this relationship in the following equation

$$di_D = \frac{\partial i_D}{\partial v_{G1S}} dv_{G1S} + \frac{\partial i_D}{\partial v_{G2S}} dv_{G2S} + \frac{\partial i_D}{\partial v_{DS}} dv_{DS} \qquad (6.41)$$

Notice that $\partial i_D / \partial v_{G1S}$ is the rate at which i_D changes with a change of v_{G1S}. When this rate is multiplied by the actual change of v_{G1S}, we have the actual change in i_D due to the change in v_{G1S}. Similarly, the second term on the right of Eq. 6.41 produces the change in i_D due to the change in v_{G2S}. The third term on the right of Eq. 6.41 determines the change in i_D due to a change in v_{DS}. This same approach could have been used in developing all of the equivalent circuits up to this point. This technique is especially useful as more variables are included in the device.

The relationship given by Eq. 6.41 can be expressed in delta form as

$$\Delta i_D = \frac{\Delta i_D}{\Delta v_{G1S}} \bigg|_{\substack{v_{G2S}=K_1 \\ v_{DS}=K_2}} \Delta v_{G1S} + \frac{\Delta i_D}{\Delta v_{G2S}} \bigg|_{\substack{v_{G1S}=K_3 \\ v_{DS}=K_2}} \Delta v_{G2S} + \frac{\Delta i_D}{\Delta v_{DS}} \bigg|_{\substack{v_{G1S}=K_3 \\ v_{G2S}=K_1}} \Delta v_D \qquad (6.42)$$

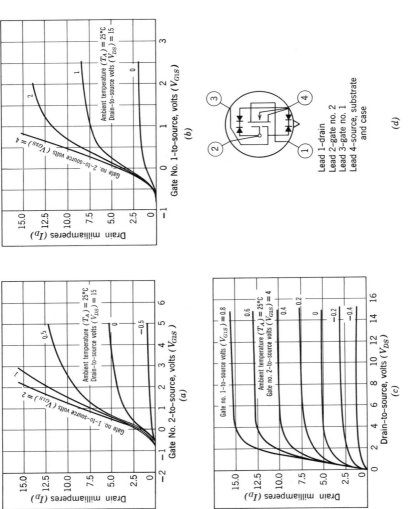

FIGURE 6.32

Characteristics of a 3N200 transistor: (a) i_D versus v_{G2S}; (b) i_D versus v_{G1S}; (c) i_D versus v_{DS}; (d) terminal diagram. (Courtesy of RCA Corporation, New Jersey.)

Electrical characteristics at $T_A = 25°C$ unless otherwise specified

Characteristic	Symbols	Test conditions	Min.	Typ.	Max.	Units
Gate no. 1-to-source cutoff voltage	$V_{G1S(off)}$	$V_{DS} = +15$ V, $I_D = 50$ μA; $V_{G2S} = +4$ V	-0.1	-1	-3	V
Gate no. 2-to-source cutoff voltage	$V_{G2S(off)}$	$V_{DS} = +15$ V, $I_D = 50$ μA; $V_{G1S} = 0$	-0.1	-1	-3	V
Gate no. 1-terminal forward current	I_{G1SSF}	$V_{G1S} = +1$V, $V_{G2S} = V_{DS} = 0$; $T_A = 25°C$ / $T_A = 100°C$	— / —	— / —	50 / 5	nA / μA
Gate no. 1-terminal reverse current	I_{G1SSR}	$V_{G1S} = -6$ V, $V_{G2S} = V_{DS} = 0$; $T_A = 25°C$ / $T_A = 100°C$	— / —	— / —	50 / 5	nA / μA
Gate no. 2-terminal forward current	I_{G2SSF}	$V_{G2S} = +6$ V, $V_{G1S} = V_{DS} = 0$; $T_A = 25°C$ / $T_A = 100°C$	— / —	— / —	50 / 5	nA / μA
Gate no. 2-terminal reverse current	I_{G2SSR}	$V_{G2S} = -6$ V, $V_{G1S} = V_{DS} = 0$; $T_A = 25°C$ / $T_A = 100°C$	— / —	— / —	50 / 5	nA / μA
Zero-bias drain current	I_{DS}	$V_{DS} = +15$ V, $V_{G1S} = 0$; $V_{G2S} = +4$ V	0.5	5.0	12	mA
Forward transconductance (gate no. 1-to-drain)	g_{fs}	$f = 1$ kHz	10,000	15,000	20,000	μmho
Small-signal, short-circuit input capacitance	C_{iss}	$f = 1$ MHz	4.0	6.0	8.5	pF
Small-signal, short-circuit, reverse transfer capacitance (drain-to-gate-no. 1)	C_{rss}	$V_{DS} = +15$ V, $I_D = 10$ mA, $V_{G2S} = +4$ V	0.005	0.02	0.03	pF
Small-signal, short-circuit output capacitance	C_{oss}		—	2.0	—	pF

Now, we can define g_{m1} as

$$g_{m1} \equiv \left.\frac{\Delta i_D}{\Delta v_{G1S}}\right|_{\substack{v_{G2S}=K_1 \\ v_{DS}=K_2}} \tag{6.43}$$

and

$$g_{m2} \equiv \left.\frac{\Delta i_D}{\Delta v_{G2S}}\right|_{\substack{v_{G1S}=K_3 \\ v_{DS}=K_2}} \tag{6.44}$$

Finally, let g_{ds} be defined as

$$g_{ds} \equiv \left.\frac{\Delta i_D}{\Delta v_D}\right|_{\substack{v_{G1S}=K_3 \\ v_{G2S}=K_1}} \tag{6.45}$$

Now, if we let Δi_D be written as i_d, Δv_{G1S} be written as v_{g1s}, Δv_{G2S} be written as v_{g2s}, and Δv_{DS} be written v_{ds}, we can write Eq. 6.42 as

$$i_d = g_{m1}v_{g1s} + g_{m2}v_{g2s} + g_{ds}v_{ds} \tag{6.46}$$

An equivalent circuit which behaves according to Eq. 6.46 is given in Fig. 6.33. Notice that the input impedances of the two gates are assumed

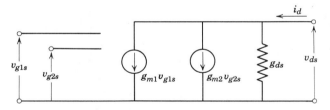

FIGURE 6.33
An equivalent circuit for a dual-gate MOSFET.

to be infinite in this equivalent circuit. Now let us consider an example where we determine the magnitude of the parameters in our equivalent circuit.

Example 6.6 Let us determine the equivalent circuit parameters for a 3N200 at the q point, $V_{G1S} = 1$ V, $V_{G2S} = 1$ V, and $V_{DS} = 6$ V.

The typical characteristics for a 3N200 transistor are given in Fig. 6.32. Notice that a pair of diodes are internally connected back-to-back from each gate to the source and substrate. These diodes break down in the avalanche mode when the reverse voltage across any one of them exceeds approximately 10 V. Consequently, the internal impedance of each gate is very high until the input signal exceeds $+10$ V or drops below -10 V.

Then the diodes break down to protect the thin gate region insulators. The addition of this protection has reduced the input impedance of each gate so that up to 5 μA of current can flow (at 100°C) when 6 V is applied to the gate (see the electrical characteristics in Fig. 6.32). Thus, the input impedance to each gate may be as low as 6 V/5 $\times$ 10^{-6} A = 1.2 MΩ.

The typical drain characteristics (Fig. 6.32c) are not drawn for our particular q point. However, if we assume the value of g_d at $v_{G1S} = 0.8$ V and $v_{G2S} = 4$ V is approximately the same as g_d at our q point, we can take the slope of the $v_{G1S} = 0.8$ V curve at $v_{DS} = 6$ V to obtain g_{ds}. Then $g_{ds} \simeq 0.5$ mA/10 V = 5×10^{-5} mhos.

We note from Fig. 6.32c that the drain current for $v_{DS} = 15$ V is approximately equal to the drain current for $v_{DS} = 6$ V if the gate voltages remain constant. Consequently, the curves of Fig. 6.32a and Fig. 6.32b are also essentially correct for $v_{DS} = 6$ V. Then the value of g_{m1} is equal to the slope of the $v_{G2S} = 1$ V curve (in Fig. 6.32b) at $v_{G1S} = 1$ V. Thus, $g_{m1} = \Delta i_D/\Delta v_{G1S} = 1$ mA/1 V = 10^{-3} mhos. Similarly, the value of g_{m2} is equal to the slope of the $v_{G1S} = 1$ V curve (Fig. 6.32a) at $v_{G2S} = 1$ V. In this case, $g_{m2} \simeq \Delta i_D/\Delta v_{GS} = 11$ mA/1 V = 11×10^{-3} mhos.

Similarly, the value of g_{m2} is equal to the slope of the $v_{G1S} = 1$ V curve (Fig. 6.32a) at $v_{G2S} = 1$ V. In this case, $g_{m2} \simeq \Delta i_D/\Delta v_{GS} = 11$ mA/1 V = 11×10^{-3} mhos.

From this analysis, our equivalent circuit for a typical 3N200 at the required q point is as shown in Fig. 6.34.

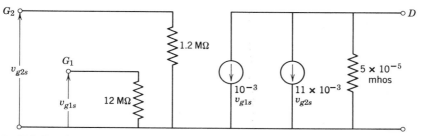

FIGURE 6.34
An equivalent circuit for a 3N200 transistor where
$$v_{G1S} = 1, V, v_{G2S} = 1 \text{ V, and } v_{DS} = 6 \text{ V.}$$

The equivalent circuit we have developed in Example 6.6 (Fig. 6.34) can be used to determine the behavior of a 3N200 with small-signal voltages near the required q point.

If the dual-gate MOSFET is biased on a linear portion of its curve there is very little cross-modulation (the production of frequencies which are not present in the input signals) if the signals are small. Then the output signal is the sum of the input signals. Thus, if we examine the $v_{G1S} = 1$ V curve

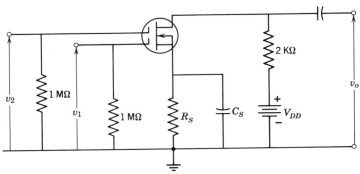

FIGURE 6.35
A dual-gate MOSFET amplifier.

in Fig. 6.32a, we find that the slope is essentially constant (hence g_{m2} is constant) for values of v_{G2S} from about 0.5 V to 1.5 V. Similarly, the $v_{G2S} = 1$ V curve in Fig. 6.32b has essentially a constant slope (g_{m1} is constant) for values of v_{G1S} from about 0.7 V to 1.4 V. Consequently, the equivalent circuit in Fig. 6.34 is valid if the input signals do not exceed these values.

However, if the input signals are large or if the transistor is biased on a nonlinear section of the characteristic curves, g_{m1} may be dependent on v_{G2S}, and g_{m2} may be dependent on v_{G1S}. Thus, the gain of a signal inserted on gate No. 1 could be controlled by the voltage on gate No. 2 or vice versa. Consequently one common use for dual-gate MOSFETs is in circuits which use automatic gain control. In addition, as noted in Chapters 20 and 21, the dual-gate MOSFET can be used in modulation circuits where we require the production of frequency components not present in the original input signals. The fact that one gate can close or cut off the flow of drain current regardless of the signal on the other gate permits the dual-gate MOSFET to also be used in switching circuits.

Finally, the extremely low capacitance (C_{rss} in the characteristics of Fig. 6.32) between gate No. 1 and the drain make the dual-gate MOSFET highly desirable for use in RF circuits. We will elaborate on this aspect of these transistors in Chapters 7 and 9. As a result of the variety of applications for dual-gate MOSFETs, they are an important addition to the solid-state family of devices.

PROBLEM 6.10   A 3N200 is to be operated at the q point $V_{G1S} = 0$ V, $V_{G2S} = 4$ V, and $V_{DS} = 8$ V. Determine the parameters for the equivalent circuit at this q point. In addition, determine the approximate range of voltages on gate No. 1 and gate No. 2 where this equivalent circuit would be valid.

PROBLEM 6.11 The 3N200 is to be operated with gate No. 2 held constant at 4 V. The desired q point occurs where $V_{DS} = 8$ V and $V_{G1S} = 0.2$ V.

  a. Draw the equivalent circuit for this mode of operation and determine the values of each parameter in the equivalent circuit.

  b. Draw the actual circuit for this transistor if $V_{DD} = 16$ V.

PROBLEM 6.12 A 3N200 is to be operated with $V_{G1S} = 0.5$ V. v_D will vary with v_{G2S}, but for all practical purposes the curves of Fig. 6.32a will be essentially correct for v_D from 4 V to 16 V. Therefore, if the load in the drain is a 1 kΩ resistor, sketch the gain versus v_{G2S} curve for this amplifier.

6.6 VACUUM TRIODES

The triode tube was invented in 1906 by Dr. Lee DeForest, who inserted a third element in the form of a wire mesh or screen (called a control grid) between the cathode and the plate of a high-vacuum diode. The resulting configuration is shown in Fig. 6.36.

The operating principle of the triode can be explained with the help of Fig. 6.37 which shows the cross-section of the area between the plate and the cathode. The plate is maintained positive (usually 100 V or so) with respect to the cathode. Consequently, the electrons emitted from the heated cathode are attracted toward the plate. However, the control grid is maintained negative (typically 5 to 10 V) with respect to the cathode.

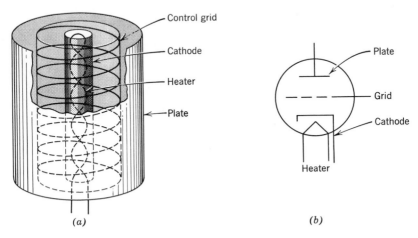

(a)   (b)

FIGURE 6.36
A triode vacuum tube; (a) actual construction; (b) symbol.

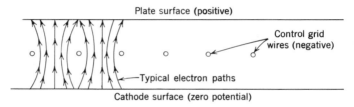

FIGURE 6.37
The action of a triode tube.

Consequently, the electrons are repelled away from the control grid wires. As a result, the electrons follow the typical paths shown in Fig. 6.37. Notice that an electron "channel" exists between each control grid wire. As the control grid becomes more negative, the width of the conducting channel decreases and less current passes to the plate. When the control grid becomes negative enough, the current to the plate may cease to flow and the triode is in the "cutoff" condition. Of course, the amount of current flowing to the plate will be dependent on both the control grid voltage and the plate voltage. Similarly, the required value of grid-cutoff voltage will change if the plate voltage changes.

From the foregoing description, the action of the triode tube is quite similar to the action of the MOSFET transistor. The control grid of the triode and the gate of the MOSFET perform similar electrical control functions. Since the control grid is usually maintained negative with respect to the cathode, the control grid draws essentially no current, so the input impedance of a triode tube is very high. Actually, it is impossible to obtain a perfect vacuum, so some residual gas is left in the tube. As the electrons travel toward the plate, they acquire sufficient kinetic energy to ionize these gas particles. The positive ions thus produced will be attracted toward the negative control grid so some grid current will flow due to these ions. In addition, some of the electrons emitted from the cathode have enough kinetic energy to overcome the negative bias of the grid and may hit the grid. Therefore, the input impedance to the grid is not infinite, but it is very high. The manufacturers usually specify the maximum resistance allowed between the grid and ground, so the voltage drop across this resistance will be negligible.

Of course, characteristic curves for the triode can be constructed in the same fashion as for transistors. The plate characteristics for a typical triode tube are given in Fig. 6.38. Notice that each curve has approximately the same shape as the current-versus-voltage curve of a vacuum diode.

The equivalent circuit for a triode (Fig. 6.39) has the same form as the equivalent circuit for a FET or MOSFET. Since the collector of a transistor

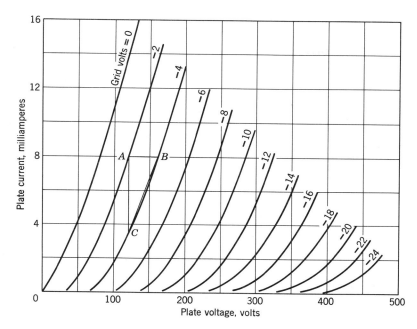

FIGURE 6.38
The characteristic curves for a 6J5 triode.

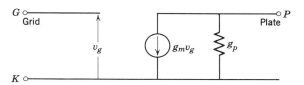

FIGURE 6.39
An equivalent circuit for a triode.

has been labeled C, the cathode of a vacuum tube is given the label K to avoid confusion. The parameters g_m (called the *transconductance*) and g_p (called the *plate conductance*) are defined as

$$g_m = \frac{\Delta i_P}{\Delta v_G}\bigg|_{v_P = \text{constant}}$$

$$(6.47)$$

and

$$g_p = \frac{1}{r_p} = \frac{\Delta i_P}{\Delta v_P}\bigg|_{v_G = \text{constant}}$$

$$(6.48)$$

where r_p is the *plate resistance* of the tube.

In some applications, it is desirable to have a voltage source rather than a current source for the equivalent circuit. Then, Thevenin's theorem can be applied to the circuit shown in Fig. 6.39 to produce the configuration shown in Fig. 6.40. In order for both equivalent circuits to be equal, the open-circuit voltages of both circuits must be equal. Then, from Fig. 6.39,

$$v_o = iR = -v_g \frac{g_m}{g_p} = -v_g g_m r_p \tag{6.49}$$

From Fig. 6.40,

$$v_o = \mu v_g \tag{6.50}$$

If these two circuits are equivalent,

$$\mu v_g = -g_m v_g r_p \tag{6.51}$$

or

$$\mu = -g_m r_p \tag{6.52}$$

The parameter μ is known as the *voltage amplification factor* of the tube. The negative sign appears because the current source and voltage source

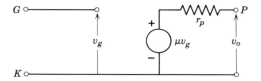

FIGURE 6.40
A voltage-source equivalent circuit for a triode.

are opposite in directions. If partial derivatives are used to express Eqs. 6.47 and 6.48, these derivatives can be used to express Eq. 6.52 as

$$\mu = \frac{\partial i_P}{\partial v_G} \cdot \frac{\partial v_P}{\partial i_P} = \frac{\partial v_P}{\partial v_G} \tag{6.53}$$

or, in terms of delta values, μ is

$$\mu = \frac{\Delta v_P}{\Delta v_G}\bigg|_{i_P = \text{constant}} \tag{6.54}$$

Observe that μ is the ratio of the effectiveness of the grid voltage to the effectiveness of the plate voltage in controlling plate current. The value of the equivalent circuit parameters g_m, μ, and r_p can be found from the tube plate characteristics (Fig. 6.38).

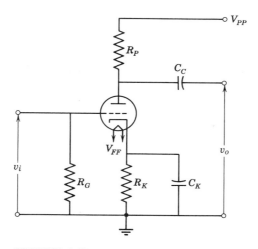

FIGURE 6.41
A typical triode amplifier.

Since the plate requires a positive potential with respect to the cathode and the control grid requires a negative potential with respect to the cathode, the bias arrangement for a triode is similar to the bias arrangement for an n-channel FET. A typical triode amplifier is shown in Fig. 6.41. An example will be given to illustrate the procedure used to design a triode amplifier.

Example 6.7   A 6J5 triode is connected as shown in Fig. 6.41. The characteristics of this tube are given in Fig. 6.38. Let us determine the characteristics of this amplifier if $V_{PP} = 250$ V and the desired q point is at $i_P = 8$ mA and $v_P = 120$ V.

First, the quiescent value of $V_{GK} = -2$ V and $I_P = 8$ mA. Therefore, the value of R_K is $V_{GK}/I_P = 2$ V/8 mA $= 250$ Ω. Let us choose a high value of resistance for R_G so we will not load down the signal source appreciably. The manufacturer suggests (in a tube manual) that the maximum value of grid resistance be 1 MΩ. Therefore, let $R_G = 1$ MΩ.

The load line must pass through the point $v_P = V_{PP}$ when $i_P = 0$ and also through the q point. Thus, the load resistance is $R_L = \Delta v_P/\Delta i_P = (250 - 120)$ V/8 mA $= 130$ V/8 mA $= 16.25$ kΩ. Since R_L is equal to $R_P + R_K$, the plate load resistance should be $R_P = 16$ kΩ. A common stock size of $R_P = 15$ kΩ would be used. We will find how to determine the size of C_K in the next section.

The q point is marked on Fig. 6.38 as point A. The line AC is a $v_P =$ constant line so we will use it to find g_m. $g_m = \Delta i_P/\Delta v_G = 4.5$ mA/2 V $= 2.25 \times 10^{-3}$ mhos. Similarly, line AB is an $i_P =$ constant line so it can be used to find μ. $\mu = \Delta v_P/\Delta v_G = 40$ V/-2 V $= -20$. The value of r_p or g_p can be found by taking the slope of the $v_G = -2$ V curve at the q point. Since the $v_G =$ constant lines are essentially parallel near the q point, let us use line BC to find r_p. $r_p = \Delta v_p/\Delta i_p = 40$ V/4.5 mA $= 8.9$ kΩ.

Since we have used the current-source equivalent circuits in the previous examples, let us use the voltage-source equivalent circuit for this example. Then, the equivalent circuit for the amplifier will have the form shown in Fig. 6.42. The current i_p is

$$i_p = \frac{\mu v_g}{r_p + R_P} \tag{6.55}$$

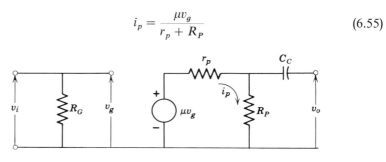

FIGURE 6.42
An equivalent circuit for the amplifier in Fig. 6.41.

Since v_o is $i_p R_P$, we can write

$$v_o = \frac{\mu v_g R_P}{r_p + R_P} \tag{6.56}$$

or

$$K_v = \frac{v_o}{v_g} = \frac{\mu R_P}{r_p + R_P} \tag{6.57}$$

For this example, $K_v = -20 \times 15$ k/(8.9 k $+ 15$ k) $= -12.5$.

The input impedance of the tube is very high, so the input impedance of the amplifier is $\simeq R_G = 1$ MΩ. If a Thevenin's equivalent circuit is drawn for the output loop, the effective output impedance of the amplifier is r_p in parallel with R_P or

$$R_o = \frac{r_p R_P}{r_p + R_P} \tag{6.58}$$

For this example, $R_o = 8.9$ k $\times 15$ k/(8.9 k $+ 15$ k) $= 5.6$ kΩ.

PROBLEM 6.13   Repeat Example 6.7 if the q point is changed to $v_P = 150$ V and $i_P = 4$ mA.

PROBLEM 6.14   Use the current-source equivalent circuit to solve Example 6.7.

6.7   TETRODE VACUUM TUBES

In order to eliminate the troublesome capacitance between the plate and the control grid in a triode, a tetrode vacuum tube was developed in the early 1930's. The tetrode has two grids located between the cathode and plate. The symbol for a tetrode is given in Fig. 6.43a and the circuit for

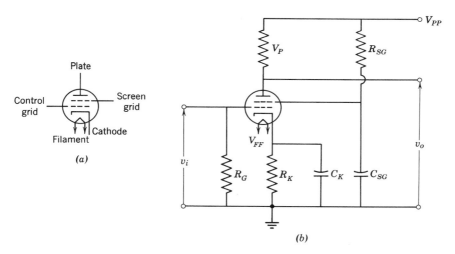

(a)

(b)

FIGURE 6.43
A tetrode amplifier : (a) tetrode symbol; (b) the tetrode amplifier.

a typical tetrode amplifier is given in Fig. 6.43b. In normal operation, the grid nearest the cathode is used as the control grid. The second grid is used to shield the control grid from the electric field of the plate. This second grid is normally maintained at a constant dc potential and is usually called the *screen grid*.

The characteristic curves for a type 24A tetrode are shown in Fig. 6.44. Notice that a change of plate voltage (above 90 V) causes very little change in the plate current. In effect, the electric field in the cathode-control grid region is shielded from the plate by the screen grid. Therefore, the value of r_p for a tetrode is very high. In fact, the plate resistance of the type 24A tube shown in Fig. 6.44 is about 0.5 MΩ.

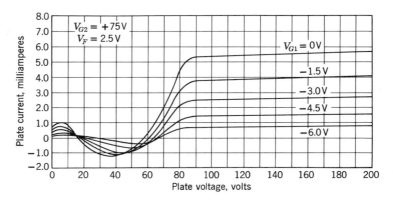

FIGURE 6.44
Characteristic curves of a type 24A tetrode.

The unusual dips in the characteristic curves for plate voltages between 8 and 85 V require further explanation. When the curve of screen current-versus-plate voltage is displayed as well as the curve of plate current versus plate voltage, the reason for this unusual behavior becomes clear. Figure 6.45 illustrates the two curves as well as a curve showing the variation of total cathode current (plate current plus screen current) as a function of plate voltage. The total cathode current is seen to be almost independent of the plate voltage. However, as the plate voltage is increased above 8 V, the plate current begins to decrease, because the kinetic energy required of a primary electron is about 8 to 10 eV (depending on the type of metal) in

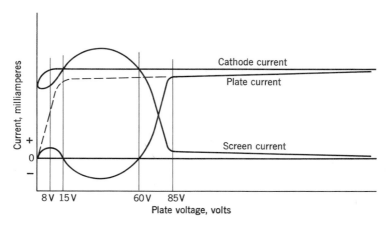

FIGURE 6.45
Plate current and screen current characteristics for a tetrode.

order to release a secondary electron from a metal sheet. Therefore, as the energy of the primary electrons reaches about 8 eV, some secondary electrons are emitted from the plate.

These secondary electrons are released in the region between the plate and the screen grid. Since the screen grid is more positive than the plate, the electrons are attracted to the screen grid. The net current to the plate is equal to the primary current minus the secondary current. Therefore, the total current to the plate decreases. At the same time, the current to the screen grid increases due to the additional secondary electrons. As the energy of the primary electrons increases, the number of secondary electrons also increases. In fact, as the curves of Fig. 6.44 indicate, when the plate voltage is between 15 and 60 V, the number of secondary electrons may be greater than the number of primary electrons. The plate then acts as an *electron emitter*, and the plate current is *reversed*. As the plate potential becomes more positive than the screen potential, some of the secondary electrons are attracted back to the plate. When the plate voltage is 85 V or greater, essentially all of the secondary electrons are returned to the plate.[1] If the secondary electrons were not present, the characteristic curve would have the shape indicated by the dotted curve of Fig. 6.45.

The characteristic curve indicates that the tetrode has a *negative* plate resistance for low values of plate voltage. (An increase of plate voltage causes a decrease in plate current.) This negative resistance characteristic of the tetrode, like the tunnel diode, has been used to produce oscillations. Oscillators will be discussed in subsequent chapters. However, for most amplifier applications, the dip in the characteristic curve is undesirable as this dip seriously reduces the useful operating range of the amplifier. Consequently, the regular tetrode has been largely replaced by the pentode or beam-power tetrode.

PROBLEM 6.15   Sketch v_o as a function of time if the tube of Fig. 6.44 is connected as shown in Fig. 6.43. $R_L = 30$ kΩ and $v_i = -4 \sin \omega t$. Would you recommend operation of this tube under these conditions? Why?

PROBLEM 6.16   Draw the dynamic transfer characteristic for the tube whose characteristic curves are given in Fig. 6.44 if (a) $R_L = 20$ kΩ, (b) $R_L = 40$ kΩ, (c) $R_L = 100$ kΩ. Assume the tube is connected as shown in Fig. 6.43.

[1] About 90 percent of the secondary electrons are emitted with an energy of 10 eV or less. All of the electrons are emitted at different angles. As a result, the component of energy away from the plate is usually less than the net energy of the electron. Consequently, very few electrons have a net energy away from the plate greater than 10 eV.

6.8 PENTODE TUBES

The undesirable secondary emission effects of the tetrode led to the development of the *pentode tube*. The pentode, which has three grids between the cathode and the plate is essentially a tetrode with a grid inserted between the screen grid and the plate. This third grid is called the *suppressor grid*. The physical arrangement and symbol for the pentode tube are shown in Fig. 6.46.

The suppressor grid is usually conected to the cathode. A basic pentode amplifier circuit is shown in Fig. 6.47. As in the tetrode, the field from

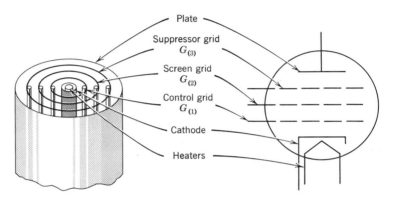

FIGURE 6.46
The pentode construction and graphical symbol.

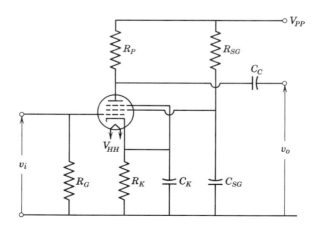

FIGURE 6.47
A typical pentode amplifier circuit.

the positive screen grid causes electrons to move from the cathode toward the screen grid. The screen grid intercepts some of these electrons but most of them continue on toward the plate. The suppressor grid causes a field of low potential to exist between the screen grid and the plate. However, the electrons which were not captured by the screen grid have enough energy to pass through this low-potential field to the plate. A few electrons which travel straight toward the wire of the suppressor grid will either be returned to the screen grid or captured by the suppressor, depending on the energy they possess.

The electrons which arrive at the plate cause the emission of secondary electrons, as in the tetrode. However, the suppressor grid is negative in relation to the plate and returns the secondary electrons to the plate. The suppressor grid, therefore, does not suppress the *emission* of secondary electrons but does suppress the *effect* of the secondary emission.

The characteristic curves of a typical pentode are shown in Fig. 6.48. In this figure, the dip due to the secondary emission is absent. Except for this difference, the curves are very similar to the tetrode curves. The advantage of the pentode over the tetrode is obvious from the characteristic

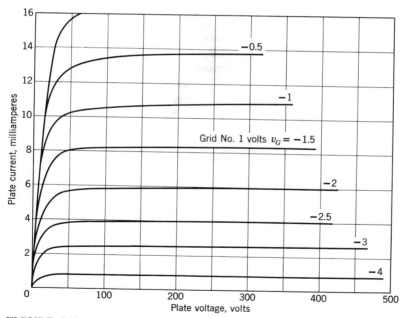

FIGURE 6.48

Characteristic curves of a 6AU6 pentode $v_{G2} = 150$ V. **(Courtesy of Radio Corporation of America.)**

curve. The potential on the plate of the tetrode has to remain at least 10 V higher than the screen grid potential or the characteristic curves are non-linear. In contrast, the plate of the pentode can be reduced far below the screen-grid potential and the operation is still on the linear portion of the characteristic curve. If the plate has a very low voltage, a cloud of electrons will accumulate near the suppressor grid. Consequently, a virtual cathode forms in this region, and the characteristic curves for very low plate voltages have the same form as the vacuum diode. As the voltage increases, all the electrons in the screen-to-plate area are attracted to the plate and a saturation condition exists in which the plate potential has a *very* small effect in plate current. Therefore, the plate resistance of a pentode is very high.

Equivalent circuits can be drawn to represent the tetrode or the pentode. Since the pentode contains one more grid than the tetrode, the equivalent circuit for the pentode will be developed, and modifications which will fit this development to the tetrode should be obvious.

The pentode contains three grids and a plate. Consequently, if the cathode is taken as reference, each of these electrodes should have an effect on the plate current. Then, for small signals, the law of super-position can be used to determine the characteristic equation for the pentode-current.

$$di_P = \frac{\partial i_P}{\partial v_{G1}} dv_{G1} + \frac{\partial i_P}{\partial v_{G2}} dv_{G2} + \frac{\partial i_P}{\partial v_{G3}} dv_{G3} + \frac{\partial i_P}{\partial v_P} dv_P \qquad (6.59)$$

or in delta form as

$$\Delta i_P = g_m \Delta v_{G1} + g_{m2} \Delta v_{G2} + g_{m3} \Delta v_{G3} + \frac{1}{r_p} \Delta v_P \qquad (6.60)$$

where g_m indicates the effectiveness of v_{G1} in controlling the plate current. The parameter g_m can be measured by applying proper potentials to the various grids and plate. Then, note the change of i_P as v_{G1} is changed slightly. All other grid and plate potentials must be maintained constant while this change of plate current and grid 1 voltage occurs. From Eq. 6.59 and Eq. 6.60, g_m is

$$g_m = \frac{\partial i_P}{\partial v_{G1}} \qquad (6.61)$$

Similarly, the rest of the parameters in Eq. 6.60 are as follows

$$g_{m2} = \frac{\partial i_P}{\partial v_{G2}} \qquad (6.62)$$

$$g_{m3} = \frac{\partial i_P}{\partial v_{G3}} \qquad (6.63)$$

$$r_p = \frac{\partial v_P}{\partial i_P} \qquad (6.64)$$

In a typical pentode amplifier circuit, the screen voltage v_{G2} and the suppressor voltage v_{G3} are maintained constant. Consequently, $\Delta v_{G2} = \Delta v_{G3} = 0$ and Eq. 6.60 can be written[2] as

$$\Delta i_P = g_m \,\Delta v_{G1} + \frac{1}{r_p} \Delta v_P \qquad (6.65)$$

This equation is identical to the characteristic equation of the triode. Consequently, the equivalent circuits for the pentode, for the tetrode, and for the triode are the same. For convenience, the two equivalent circuits for the pentode or tetrode are shown in Fig. 6.49. The foregoing derivation applies to the tetrode if g_{m3} is equated to zero or is omitted.

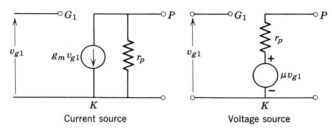

FIGURE 6.49
Equivalent circuits for a pentode or tetrode.

PROBLEM 6.17 Draw the equivalent circuit of a pentode when the voltages of all the grids are permitted to vary.

The characteristics of the pentode are so much improved over those of the tetrode that conventional tetrodes are no longer made. A form of tetrode known as a *beam-power tube* is made, but these special tubes have characteristics very similar to the pentode tube.

The design of a pentode amplifier is very similar to the design of a triode amplifier. However, the cathode current, I_K, of a pentode is equal to the screen current, I_{G2}, plus the plate current I_{P2}. Thus, the cathode resistor R_K is

$$R_K = \frac{V_{GK}}{I_P + I_{G2}} \qquad (6.66)$$

where V_{GK} is the required bias between the control grid and the cathode. For proper bypassing, the reactance of the capacitor C_K should be one-tenth of the resistance across which it is placed. This resistance is R_K in parallel with the input resistance to the cathode of the tube. The input resistance to the cathode of the tube, R_{ik}, can be found from the equivalent circuit

---

[2] By common usage, v_G in a pentode or tetrode tube is understood to be v_{G1}. Hence, the second subscript may be omitted when no confusion will result.

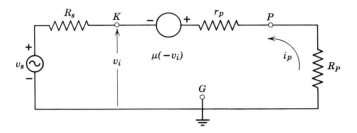

FIGURE 6.50
The circuit for determining R_{ik} for the cathode circuit of a vacuum tube amplifier.

shown in Fig. 6.50. In this equivalent circuit, a signal generator v_s and R_s is connected to the cathode of the tube. A load resistance R_P is connected in the plate circuit. The voltage v_i is the voltage v_{KG} which is equal to $-v_{GK}$. The loop equation is

$$v_i - \mu v_i = -i_p(r_p + R_P) \tag{6.67}$$

or

$$-i_p = \frac{(-\mu + 1)v_i}{r_p + R_P} \tag{6.68}$$

The input impedance is $v_i/-i_p$ or

$$R_{ik} = \frac{r_p + R_P}{-\mu + 1} \tag{6.69}$$

For triode tubes, r_p is about the same order of magnitude as R_P, so both resistances must be included. However, for pentode or beam-power tubes, r_p is usually greater than R_P. In addition, $-\mu$ is much greater than one. Under these conditions, R_{ik} is

$$R_{ik} \simeq \frac{r_p}{-\mu} = 1/g_m \tag{6.70}$$

The value of C_K is given by the relationship

$$\frac{1}{\omega_1 C_K} = \frac{1}{10(G_K + g_m)} \tag{6.71}$$

or

$$C_K = \frac{10(G_K + g_m)}{\omega_1} \tag{6.72}$$

where ω_1 is the lowest radian frequency to be amplified.

The resistance R_{SG} is used to drop the voltage V_{PP} to the proper screen-grid voltage V_{G2}. Then,

$$R_{SG} = \frac{V_{PP} - V_{G2}}{I_{G2}} \tag{6.73}$$

The capacitor C_{SG} should be large enough to bypass R_{SG}. However, R_{SG} has the input impedance to the screen grid in parallel with R_{SG}. Normally, the input impedance to the screen grid is not given and cannot be determined from values that are given. However, this input impedance can be measured in the laboratory. Usually, sufficient accuracy can be obtained by assuming the input impedance to the screen grid is about equal to R_{SG}. Then,

$$\frac{1}{\omega_1 C_{SG}} \simeq \frac{1}{10} \cdot \frac{R_{SG}}{2} \tag{6.74}$$

or

$$C_{SG} \simeq \frac{20}{\omega_1 R_{SG}} \tag{6.75}$$

Example 6.8 A pentode tube is connected as shown in Fig. 6.47. The characteristics of this tube are given in Fig. 6.51. Let us design this amplifier if $V_{PP} = 250$ V and $R_P = 25$ kΩ.

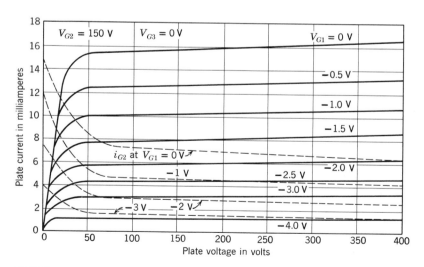

FIGURE 6.51
The characteristics of a 6AU6 pentode.

The load line for $V_{PP} = 250$ V and $R_P = 25$ kΩ is drawn on the characteristic curves. The proper bias appears to be at about $v_{GK} = -2.5$ V. Then, from Fig. 6.51, $I_K = 4.4$ mA. The screen-grid current is given by the dotted lines in Fig. 6.51. Then, $I_{G2} \simeq 2.2$ mA. The value of R_K is 2.5 V/(4.4 + 2.2) mA = 380 Ω. This resistance is so much lower than R_P that the load line need not be redrawn. The value of R_G will be chosen as 1 MΩ.

The characteristics given in Fig. 6.51 are valid if V_{G2} is 150 V. Then, $R_{SG} = (250 - 150)/2.2$ mA $\simeq 45$ kΩ. If the lowest frequency to be amplified is $\omega_1 = 200$ rad/sec, the value of $C_{SG} \simeq 20/(200 \times 45$ k$) = 2.2$ μF.

In order to determine the size of C_K, we must know the value of g_m at the q point. $g_m \simeq \Delta i_P/\Delta v_G = 3$ mA/1 V = 3×10^{-3} mhos. Then, using Eq. 6.72, the value of C_K is $10(3 + 2.63)10^{-3}/200 = 280$ μF.

PROBLEM 6.18 Determine the values of r_p and μ for the pentode tube used in Example 6.8.

Answer: $r_p = 1$ MΩ, $\mu = -3000$

PROBLEM 6.19 Determine the input resistance, the output resistance, and the voltage gain of the amplifier in Example 6.7.

Answer: $R_i = 1$ MΩ, $R_o \simeq 25$ kΩ, $K_v \simeq 75$.

PROBLEM 6.20 Determine the proper value of C_K for the amplifier in Example 6.7 if the lowest frequency to be amplified is 200 rad/sec.

PROBLEM 6.21 Use the 6AU6 pentode to design an amplifier with a voltage gain of 100. Assume V_{PP} is 250 V and the lowest frequency to be amplified is 400 rad/sec.

7

RC Coupled Amplifiers

In previous chapters, our attention has been focused on the principles of operation and the characteristics of semiconductor devices. Simple circuit applications have been given to illustrate the characteristics and usefulness of these devices. Capacitors have been used to provide ac coupling and dc blocking between the particular device and its signal source and load. In addition, capacitors have been used to bypass the ac signal around bias resistors. Equations were given to determine the appropriate values of the capacitors, but in most cases the development of these equations has been deferred to this chapter.

In this chapter, the characteristics of an entire circuit will be considered instead of restricting our attention to the semiconductor device. The circuits to be considered are typical of those used in sound amplifiers, audio amplifiers in communications equipment, video (or picture) amplifiers in television sets, and in many other applications.

7.1 THE MID-FREQUENCY RANGE

All of the amplifiers considered in the previous sections of this book have been connected similarly to that shown in Fig. 7.1. Of course, we have used different configurations (common base and common collector) and different amplifying devices (FETs, MOSFETs, and tubes) but the load was always connected in series with the output terminals, and the output voltage was always measured across the open-circuit output terminals. In actual practice, very few amplifiers have their loads connected directly in the output circuit (such as R_C in Fig. 7.1). Most of the amplifiers have their actual loads connected across the terminals where v_o is noted in Fig. 7.1. In fact, many amplifiers consist of cascaded stages where the output signal from one amplifier stage becomes the input signal for the

209

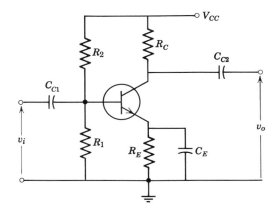

FIGURE 7.1
A typical transistor amplifier.

next amplifier stage as shown in Fig. 7.2. In this configuration, the load for one stage is the input impedance to the next stage.

The configurations shown in Fig. 7.2 are known as *RC* coupled amplifiers. While two configurations are shown, many other combinations of MOSFETs, tubes, bipolar transistors, and FETs are possible. Fortunately, the methods of analysis are straightforward and if understood for one configuration can be readily applied to any other configuration. Consequently, we will use the configuration shown in Fig. 7.2*a* to illustrate the graphical method of solving *RC*-coupled amplifiers.

The term *mid-frequency range* is used to designate the frequency range in which all reactances are negligible and the gain is independent of frequency. (These assumptions have been made in the previous sections on amplifiers.) Thus, the coupling capacitors (C_C in Fig. 7.2) and the bypass capacitors (C_S and C_E in Fig. 7.2) are assumed to have zero reactance in the mid-frequency range. The dc load line in Fig. 7.3*a* is determined by $R_L = R_C + R_E$ since the current must flow through these two resistors. In contrast the ac signal is bypassed through C_E to ground. In addition, since C_{C2} is a short circuit for ac signals, the input impedance to the second amplifier is in parallel with R_C for the ac signal. Thus, the ac impedance (denoted by R_l) is the parallel combination of R_C, R_1, R_2, and R_{it} where R_{it} is the input impedance to transistor T_2 ($R_{it} \simeq h_{ie2}$). Thus, the ac load impedance, R_l, is often much less than the dc load impedance, R_L.

Load lines can and should be drawn on the output characteristics for both dc and ac loads. Both load lines must pass through the *q* point so the

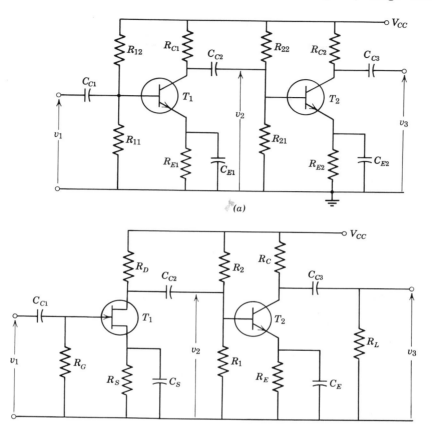

FIGURE 7.2
Typical *RC*-coupled amplifiers: (*a*) bipolar transistor configuration; (*b*) A FET
coupled to an *n-p-n* transistor.

q point is at the intersection of the two load lines. An example will help
clarify this procedure.

Example 7.1 A circuit is connected as shown in Fig. 7.2*a*. The value of
h_{ie} for transistor T_2 is 500 Ω and the parallel combination of R_{21} and
R_{22} is 10 kΩ. The collector characteristics of transistor T_1 are given in
Fig. 7.3. Let us design this circuit so 1 mA peak base signal current will be
delivered to transistor T_2.

The q point must be chosen at a value of I_C greater than 1 mA because
part of the signal collector current will be shunted through R_{C1} and R_b
(the parallel combination of R_{21} and R_{22}). However, if R_{C1} and R_b are

both large in comparison with h_{ie}, most of the signal component of collector current will flow into the base of the following transistor. In order to insure adequate current-drive capability, we will choose the q point at $I_C = 2$ mA, $V_{CE} = 10$ V with $V_{CC} = 20$ V. The dc load line is then drawn through this q point and V_{CC}. This load line establishes the dc load resistance which includes R_E, as previously mentioned. For this example, $R_L = R_{E1} + R_{C1} = 20V/4$ mA $= 5$ kΩ. Since R_C is usually much larger than R_E, we will select $R_{C1} = 4$ kΩ, $R_{E1} = 1$ kΩ. Then, since h_{ie} of the following transistor is 500 Ω and $R_b = 10$ kΩ, the load resistance is the parallel combination of R_{C1}(4 kΩ), h_{ie2} (500 Ω), and R_b (10 kΩ) or $R_l = 425$ Ω. The ac load line can be most easily drawn by using the relationship

$$\Delta v_{CE} = \Delta I_C R_l \tag{7.1}$$

If the Δi_C is taken as the change from the q-point value of i_{C1} to $i_C = 0$, (2 mA) as shown in Fig. 7.3, the value of Δv_{CE} is $2 \times 10^{-3}(425) = 0.85$ V as shown. The v_{CE} axis intercept of the ac load line is thus located at $V_{CEq} + \Delta v_{CE} = 10.85$ V, and the ac load line can be drawn through this intercept and the q point as shown.

The peak signal input current available for the actual load, or second transistor, may be determined after a minimum acceptable value of i_C is chosen. In view of the facts that i_C cannot be reduced below I_{CO} and that the h_{fe} of most transistors decrease rather rapidly with decreasing i_C

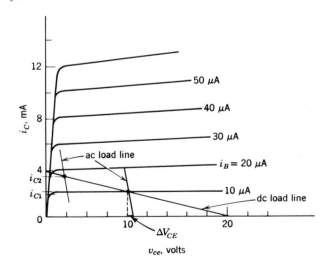

FIGURE 7.3
Q-point selections and load lines on the collector characteristics.

below about 0.1 mA (depending upon the type transistor), we will choose 0.1 mA as the minimum permissible value of i_C. Then the peak signal current available for the following transistor is Δi_C max $R_l/h_{ie} = 1.9$ mA (425 Ω)/500 $\Omega = 1.6$ mA which is well above the required 1 mA.

The permissible excursion of the q point along the dc load line can also be determined from the foregoing information. You may recall that the limits of collector current I_{C1} and I_{C2} must be established before the value of R_b in the stabilized-bias circuit can be determined. We have already determined the minimum value of collector current I_{C1} as 2.0 mA, allowing a generous safety factor. The maximum permissible value of collector current can most easily be determined by recalling that the peak value of output signal voltage is $\Delta I_{C\,max} R_l = 0.85$ V. The maximum q-point value of v_{CE} must be at least this distance to the right of the transistor saturation region to prevent saturation. If the maximum saturation voltage is approximately 1.0 V at the maximum expected value of collector current (6 mA in this example) the minimum q point value of v_{CE} is then $1 + 0.85$ or 1.85 V. The value of 2.0 V used in Fig. 7.3 allows an additional margin of safety. The value of I_{C2} at this q point can be determined from the slope of the dc load line and the desired value of $v_{CE\,min}$. Thus,

$$I_{C2} = \frac{V_{CC} - V_{CE}\,\text{min}}{R_C + R_E} \tag{7.2}$$

For this example, $I_{C2} = (20 - 2)\text{V}/5\ \text{k}\Omega = 3.6$ mA.

Since the input impedance to FET, MOSFET, and tube amplifiers is very high, the ac load impedance is not much lower than the dc load resistance for these devices. However, when a significant difference between ac and dc loads exists, the method outlined in Example 7.1 can be used.

Through the use of equivalent circuits, much greater insight into the behavior of RC-coupled amplifiers can be obtained. To illustrate this idea, consider the various types of RC-coupled amplifiers shown in Fig. 7.4. The corresponding equivalent circuits are given in Fig. 7.5. The coupling capacitors are retained in the equivalent circuits because they will be used in a later section. Note that when parallel resistors are combined to one single resistance, each of these equivalent circuits can be reduced to the one equivalent circuit shown in Fig. 7.6. The current source i is either equal to $h_{fe}i_b$ or $g_m v_g$, depending on the type of device used. Note that the resistance R_o is, in every circuit, equal to the output resistance of the amplifier (not the output resistance of the transistor or tube). In addition, the resistance R_i is the input resistance to the amplifier in the circuit. Again, this input resistance is not the input resistance to the tube or transistor, R_{it}, but is the input resistance to the total amplifier, R_i which includes the bias circuit.

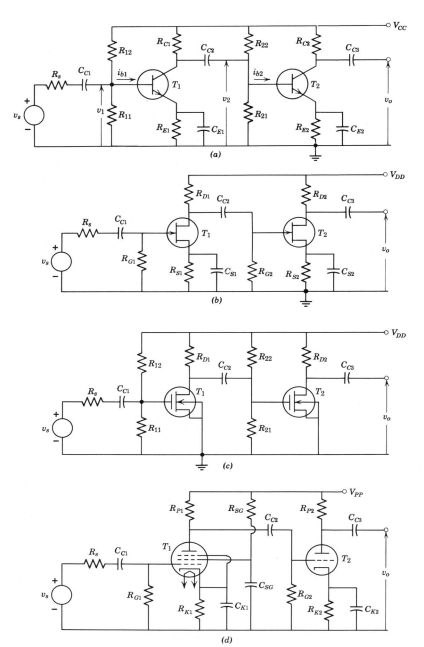

FIGURE 7.4
Several *RC*-coupled amplifiers: (*a*) bipolar transistors; (*b*) FET transistors;
(*c*) MOSFET transistors; (*d*) vacuum tube.

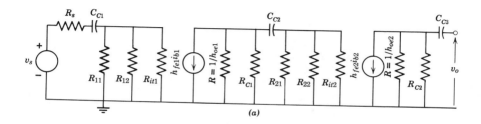

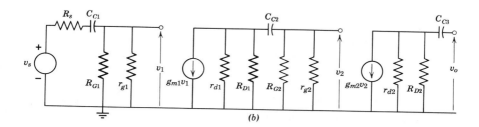

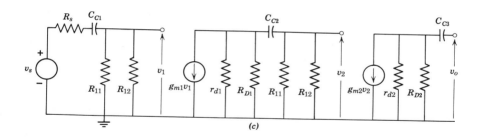

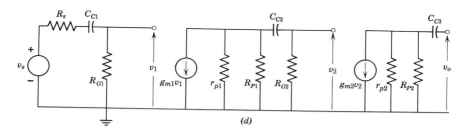

FIGURE 7.5
Equivalent circuits for the amplifiers in Fig. 7.4; (a) bipolar transistors; (b) FET transistors; (c) MOSFET transistors; (d) vacuum tubes.

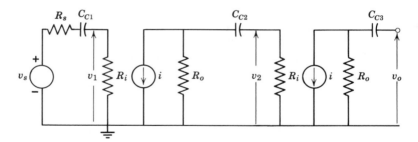

FIGURE 7.6
A simplification of the equivalent circuits shown in Fig. 7.5.

In Fig. 7.5a, the input resistance of the bipolar transistor is given as R_{it}. In RC-coupled amplifiers, the value of R_{it} is usually approximately equal to h_{ie}. This approximation is valid because the term $h_{fe}h_{re}/(h_{oe} + Y_l)$ which appears in the input impedance evaluation is usually very small. First, the parameter h_{re} is usually very small in modern transistors. Secondly, the load Y_l is equal to $1/R_l$ where R_l is the ac load impedance of the transistor. As shown in Example 7.1, this ac load impedance is usually quite small (less than h_{ie} of the following stage). If R_l is small, Y_l will be relatively large. For these two reasons, the term $h_{fe}h_{re}/(h_{oe} + Y_l)$ is usually negligibly small for RC-coupled amplifiers. By similar reasoning, the output conductance G_{ot} of the RC-coupled bipolar transistor is usually approximately equal to h_{oe}. However, if R_{it} or G_{ot} differs significantly from h_{ie} or h_{oe} respectively, the values of R_{it} and G_{ot} should be used in the analysis which follows.

PROBLEM 7.1   Test the validity of the assumption that $R_{it} \simeq h_{ie}$ and $G_{ot} = 1/R_{ot} \simeq h_{oe}$ for an RC coupled transistor. This can be done by assuming that several identical stages are connected in cascade so that the driving source resistance R_s for an inner stage is approximately $1/h_{oe}$ in parallel with R_C. The input resistance of the following stage, which is a predominant part of the load of the inner stage, is assumed to be h_{ie}. Then a more accurate value of input resistance and output admittance of this inner stage can be calculated with the effects of h_{re} included in the calculation. Make this calculation, using an h-parameter model, for an inner stage of a multistage amplifier with all stages having $h_{ie} = 1\ \text{k}\Omega$, $h_{oe} = 4 \times 10^{-5}\ \text{mho}$, $h_{fe} = 100$, $h_{re} = 2 \times 10^{-4}$, $R_B = 10\ \text{k}\Omega$, and $R_C = 5\ \text{k}\Omega$. All bypass and coupling capacitors have negligible reactance. Determine the percent error in the assumptions that (a) $R_{it} = h_{ie}$, (b) $G_{ot} = h_{oe}$.

In order to determine the voltage gain of the first amplifier stage in Fig. 7.4, only the middle loop of Fig. 7.6 need be considered. This center

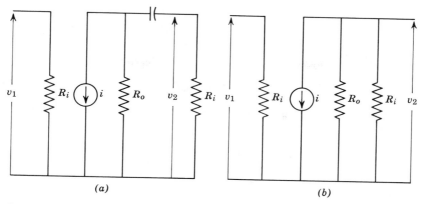

(a)   (b)

FIGURE 7.7
The circuit required to find voltage gain: (a) basic equivalent circuit; (b) mid-frequency equivalent circuit.

loop is redrawn in Fig. 7.7a. In the mid-frequency range, C_{C2} acts as if it were a short circuit. Thus, the equivalent circuit shown in Fig. 7.7a can be reduced to that shown in Fig. 7.7b in the mid-frequency range. Note that the current i must flow through the parallel combination of R_o and R_i. Then,

$$v_2 = -i\frac{R_o R_i}{R_o + R_i} \qquad (7.3)$$

However, the parallel combination of R_o and R_i is approximately equal to the ac load impedance R_l. Actually, the parallel combination of R_o and R_i is $1/h_{oe}$ (or r_d or r_p depending on the active device) in parallel with R_l. Let us define R_{sh} as

$$R_{sh} = \frac{R_o R_i}{R_o + R_i} \qquad (7.4)$$

For most high output impedance devices (bipolar transistors, FETs, MOSFETs, and pentodes), $R_{sh} \simeq R_l$.

If the value of R_{sh} defined by Eq. 7.4 is used in Eq. 7.3, the value of v_2 is

$$v_2 = -iR_{sh} \qquad (7.5)$$

Now, if i is $g_m v_1$, the value of v_2 is

$$v_2 = -g_m v_1 R_{sh} \qquad (7.6)$$

or

$$K_v = \left|\frac{v_2}{v_1}\right| = g_m R_{sh} \qquad (7.7)$$

where K_v is the magnitude of the mid-frequency voltage gain for an amplifier stage. Of course, this same relationship will also give the voltage gain for the configurations shown in Chapter 6 where $R_L = R_l$.

If the current source i is equal to $h_{fe} i_{b1}$, the value of v_2 (Eq. 7.5) becomes

$$v_2 = -h_{fe} i_{b1} R_{sh} \tag{7.8}$$

From Fig. 7.4a, the current i_{b2} is equal to v_2/R_{it2} where R_{it2} is the input impedance ($R_{it} \simeq h_{ie}$) to transistor T_2. Then,

$$i_{b2} = \frac{v_2}{R_{it2}} = \frac{-h_{fe} i_{b1} R_{sh}}{R_{it2}} \tag{7.9}$$

The current gain is

$$K_i = \left| \frac{i_{2b}}{i_{b1}} \right| = \frac{h_{fe} R_{sh}}{R_{it2}} \tag{7.10}$$

where K_i is the magnitude of the current gain for an amplifier stage.

An additional relationship can be found from the relationships $v_2 = i_{b2} R_{it2}$ and $v_1 = i_{b1} R_{it1}$. Then,

$$\frac{v_2}{v_1} = \frac{i_{b2} R_{it2}}{i_{b1} R_{it1}} \tag{7.11}$$

or

$$K_v = K_i \frac{R_{it2}}{R_{it1}} \tag{7.12}$$

While this relationship (Eq. 7.12) was developed for a bipolar transistor amplifier, it is a general relationship and is valid for all amplifiers.

PROBLEM 7.2 Determine K_v and K_i for the first stage of an amplifier which is connected as shown in Fig. 7.4a. Assume: $h_{ie1} = 2.5 \ k$, $h_{fe1} = 200$, $h_{ie2} = 500$ and $R_{l1} = 400 \ \Omega$.
Answer: $K_v = 32$, $K_i = 160$.

An example will be used to illustrate some of the applications of the foregoing principles.

Example 7.2 Let us design an amplifier stage which can be used to couple a crystal microphone to the two-stage transistor amplifier in Example 7.1. The crystal microphone requires a high load resistance (or amplifier input resistance) of the order of one megohm, thus, the FET is a sensible choice for the first, or input, stage of the amplifier. A circuit diagram of the FET

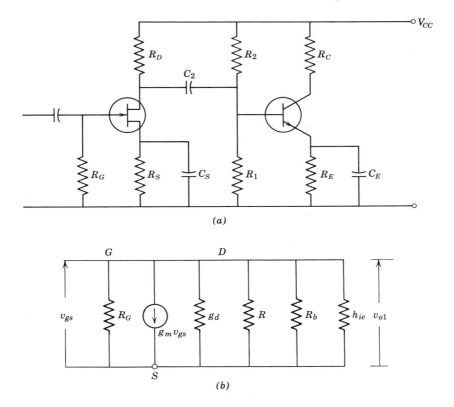

FIGURE 7.8
(a) **A FET coupled to a bipolar transistor**, (b) **a mid-frequency equivalent circuit of the FET stage.**

coupled to the conventional or bipolar transistor is given in Fig. 7.8a and the mid-frequency equivalent circuit of the FET stage is given in Fig. 7.8b.

In Example 7.1 we found that the bipolar transistor of Fig. 7.8 must provide a peak output signal of 0.8 V. The h_{ie} of this transistor at the given q point is 2.5 kΩ. Figure 7.3 shows that an input signal current of 10 μA peak will provide the 0.8 V output. Thus the peak input voltage is $i_b \times h_{ie} = 10^{-5}A \times 2.5 \times 10^3 \, \Omega = 0.025$ V. Also, we chose $R_E = 1$ kΩ, and we will assume that a current stability factor of 10 is suitable for this application and, therefore, $R_b = 10$ kΩ will be used.

A typical output voltage for a crystal microphone is about 3 mV peak at a normal sound pressure input. Therefore, the voltage gain of the FET stage must be 25 mV/3 mV = 8.33.

Since the voltage gain of a FET is $g_m R_{sh}$, and R_{sh} is the parallel combination of $1/g_d$, R_D, R_b, and h_{ie}, we must choose a FET with a suitable combination of g_m and I_{DSS} to provide the required voltage gain. We will assume that the drain resistance $r_d = 1/g_d$ is large in comparison with h_{ie}, but the drain circuit resistor R_D may not be high in comparison with h_{ie} and this resistor cannot be determined until the ·FET and its q point are chosen. A basis for the choice of FET is given in the following relationship which was derived in Chapter 6.

$$g_{mo} = \frac{2I_{DSS}}{V_p} \tag{7.13}$$

Where g_{mo} is the transconductance at $v_{GS} = 0$, $v_{DS} = V_P$, and I_{DSS} is the drain current at this point, as previously defined.

As a first approximation, we will neglect R_D as well as r_d and assume R_{sh} is approximately equal to h_{ie} in parallel with R_b, or approximately 2.0 kΩ. Then, since $K_v = g_m R_{sh}$, the required value of g_{mo} is then greater than $K_v/R_{sh} = 8.33/2k = 4.17 \times 10^{-3}$ mho. Thus, we must look for an FET with g_{mo} about 5×10^{-3} mho, since R_D has not yet been included in R_{sh}. But in order to keep R_D large so it will not seriously reduce R_{sh}, I_{DSS} should be as small as possible. Equation 7.13 shows that a low pinch-off voltage V_P will give a high ratio of g_{mo} to I_{DSS}. The 2N3459 n-channel FET appears promising for this application. It has a typical $I_{DSS} = 2.0$ mA and typical $V_P = 0.8$ V. Then typical $g_{mo} = 5 \times 10^{-3}$ μmhos. A family of typical drain characteristics for this transistor is given in Fig. 7.9. We must choose a q point on the $v_{GS} = 0$ curve in order to obtain the value of $g_m = g_{mo}$. Also, the q point should be well to the right of the knee of the characteristic curve so g_d will be small. A value of $v_{DS} = 5.0$ V is indicated as a suitable value in Fig. 7.9. Using this q point, and $V_{DD} = 20$ V, $R_D = 15$ V/2 mA = 7.5 kΩ.

PROBLEM 7.3 Determine R_{sh} and K_v for the FET amplifier above.
Answer: $R_{sh} = 1.67$ kΩ, $K_v = 8.35$.

PROBLEM 7.4 Two identical bipolar transistors are connected as shown in Fig. 7.4a. The circuit and transistor parameters are:

$$h_{ie} = 1000 \ \Omega \qquad h_{re} \simeq 0$$
$$h_{fe} = 100 \qquad h_{oe} = 10^{-5} \text{ mhos}$$
$$R_C = 1{,}000 \ \Omega \qquad R_b = R_1 \| R_2 = 20 \text{ k}\Omega$$

Determine the values of R_o, R_i, R_{sh}, K_i and K_v for one stage of this amplifier.
Answer: $R_o = 990 \ \Omega$, $R_i = 952 \ \Omega$, $R_{sh} = 486 \ \Omega$, $K_i = K_v = 51$

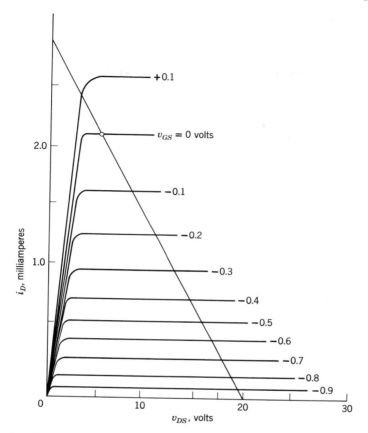

FIGURE 7.9
Drain characteristics of a FET with low pinch-off voltage.

7.2 THE LOW-FREQUENCY RANGE

We have previously assumed that the reactance of a coupling capacitor or a bypass capacitor is negligible. This is true if the frequency is sufficiently high (mid-band frequencies or higher). However, there certainly will be frequencies for which the reactance of these capacitors cannot be neglected. By definition, this range of frequencies is known as the *low-frequency* range.

Let us assume the emitter (or source or cathode) resistor is perfectly bypassed by C_E or (C_S or C_K). Then, the effect of the coupling capacitor C_C can be isolated and studied. In the next section we will study the effect of the emitter (or source or cathode) capacitor.

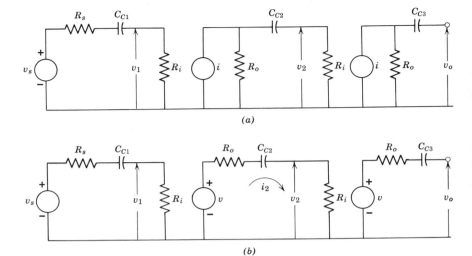

FIGURE 7.10
The low-frequency circuit configurations for the circuits in Fig. 7.5: (a) current source circuits; (b) voltage source circuits.

In Section 7.1 we noted that each of the amplifier configurations in Fig. 7.4 could be reduced to the form given in Fig. 7.6. This general configuration is repeated in Fig. 7.10a for your convenience. When the current generator and R_o are replaced by a Thevenin's equivalent circuit, the configuration would appear as shown in Fig. 7.10b. The new voltage generator in this Thevenin's circuit has a value

$$v = -iR_o \qquad (7.14)$$

The loop equation for the center loop (Fig. 7.10b) in the s domain[1] yields the following equation.

$$I_2 = \frac{V}{(1/sC_{C2}) + R_i + R_o} \qquad (7.15)$$

where I_2 is the Laplace transform of the current in the center loop and V is the Laplace transform of the voltage source v. This equation can be written in a form as

$$I_2 = \frac{sC_{C2}V}{1 + sC_{C2}(R_i + R_o)} \qquad (7.16)$$

---

[1] The reader who is not familiar with the Laplace transform can replace s by $j\omega$ in the above equation. In making this substitution, the general nature of the solution will be lost, but the resulting solution will be valid for steady-state ac signals. Of course, the circuit can also be solved by classical differential equation techniques.

Then

$$V_2 = I_2 R_i = \frac{sC_{C2}VR_i}{1 + sC_{C2}(R_i + R_o)}$$ (7.17)

or

$$V_2 = \frac{VR_i}{(R_i + R_o)} \frac{s}{s + [1/C_{C2}(R_i + R_o)]}$$ (7.18)

When we write Eq. 7.14 in s domain notation and insert this result into Eq. 7.17 in place of V, we have

$$V_2 = -I \frac{R_o R_i}{R_o + R_i} \frac{s}{s + [1/C_{C2}(R_o + R_i)]}$$ (7.19)

Note that the term $R_o R_i/(R_o + R_i)$ is equal to R_{sh}. In addition, let us define ω_1 as

$$\omega_1 = \frac{1}{C_{C2}(R_o + R_i)}$$ (7.20)

Then, Eq. 7.19 can be written as

$$V_2 = -IR_{sh} \frac{s}{s + \omega_1}$$ (7.21)

Now, if I is $g_m V_1$, the value of V_2 is

$$V_2 = -g_m V_1 R_{sh} \frac{s}{s + \omega_1}$$ (7.22)

The voltage gain is equal to V_2/V_1 so the voltage gain is

$$G_v = \frac{V_2}{V_1} = -g_m R_{sh} \frac{s}{s + \omega_1}$$ (7.23)

However, from Eq. 7.7, the term $g_m R_{sh}$ is equal to K_v. Thus,

$$G_v = -K_v \frac{s}{s + \omega_1}$$ (7.24)

The gain K_v is the magnitude of the mid-band gain and is independent of frequency. Consequently, the symbol K_v is used to indicate this gain is a constant. In contrast, since $s = j\omega + \sigma$, the gain G_v is a function of frequency. This general notation, K_v or K_i for a constant gain and G_v or G_i for a frequency dependent gain will be used to differentiate the gains in this text.

If the current I in Eq. 7.21 is equal to $h_{fe}I_1$, the current gain becomes important. Then, from Fig. 7.4a, the current I_{b2} is equal to V_2/R_{it2}. When the value of V_2 given in Eq. 7.21 is used, the value of I_{b2} is

$$I_{b2} = -\frac{IR_{sh}}{R_{it2}}\frac{s}{s + \omega_1} \tag{7.25}$$

And if $I = h_{fe}I_{b1}$,

$$G_i = \frac{I_{b2}}{I_{b1}} = -\frac{h_{fe}R_{sh}}{R_{it2}}\frac{s}{s + \omega_1} \tag{7.26}$$

Using Eq. 7.10, the value of G_i is

$$G_i = -K_i\frac{s}{s + \omega_1} \tag{7.27}$$

Note that either Eq. 7.24 or Eq. 7.27 can be written in the general form

$$G = -K\frac{s}{s + \omega_1} \tag{7.28}$$

where G and K are both either current gains or voltage gains. A pole-zero plot of Eq. 7.28 is given in Fig. 7.11.

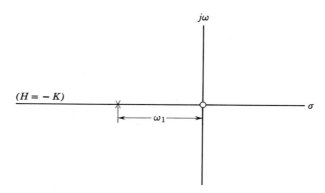

FIGURE 7.11
A pole-zero plot of Eq. 7.28.

When steady-state ac signals are used, $s = j\omega$. Therefore, the ac steady-state gain equation can be written by replacing s in Eq. 7.28 by $j\omega$. Thus,

$$G = -K\frac{j\omega}{j\omega + \omega_1} \tag{7.29}$$

This relationship can be rearranged to have the form

$$G = -K \frac{1}{1 - j(\omega_1/\omega)} \tag{7.30}$$

Now, from Fig. 7.11 or from Eq. 7.30, a plot of gain magnitude, $|G|$, versus ω can be found. In addition, a plot of phase angle (arg G) versus ω can also be developed. The resulting curves are given in Fig. 7.12.

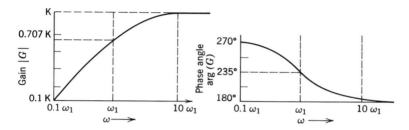

FIGURE 7.12
A plot of gain magnitude versus ω and phase angle versus ω for an RC-coupled amplifier.

At low frequencies the gain is proportional to frequency. However, as ω increases, the gain approaches the mid-band gain K as an asymptote. The frequency where $\omega = \omega_1$ is of particular interest. Notice that Eq. 7.20 can be rearranged to have the form

$$R_o + R_i = \frac{1}{\omega_1 C_{C2}} \tag{7.31}$$

Thus, ω_1 is the frequency at which the reactance of the coupling capacitor (C_{C2} in Fig. 7.10) is equal to the resistance facing the coupling capacitance. (The resistance "facing" a capacitor is the resistance encountered in traversing the circuit from one capacitor plate to the other plate.) We can also note that since $\omega_1 = 1/C_C(R_o + R_i)$, the time constant τ of the circuit in question is equal to $1/\omega_1$.

From Eq. 7.30, the gain magnitude of the amplifier stage at the frequency $\omega = \omega_1$ is 0.707 K and the phase shift is 235°. This frequency at which the gain magnitude is reduced to 0.707 of the mid-band gain is, by definition, known as the *low cutoff frequency* of the amplifier. Since the power delivered to a load is V^2/R_L or $I^2 R_L$, the power delivered to the load of an amplifier at the mid-band frequencies is either $(V_1 K_v)^2/R_L$ or $(I_{b1} K_i)^2 R_L$. In contrast, the power delivered to the load of an amplifier at the frequency $\omega = \omega_1$ is $(0.707 \, V_1 K_v)^2/R_L$ or $(0.707 I_{b1} K_i)^2 R_L$. In either notation, the power at the frequency ω_1 is reduced to one-half the power delivered in the mid-band frequency range. Thus, ω_1 is also known as the *lower half-power frequency.*

In design applications, the lower cutoff frequency will usually be specified and the engineer must determine the required size of the coupling capacitor. A slight rearrangement of the terms in Eq. 7.31 yields the following relationship:

$$C_{C2} = \frac{1}{\omega_1(R_o + R_i)} \tag{7.32}$$

The low-frequency region is defined as the frequency range in which the reactance of the coupling and bypass capacitors cannot be neglected. This reactance can normally be neglected if it is less than one-tenth as large as the associated resistance. Thus, if $X_{C2} = 0.1(R_o + R_i)$, the frequency ω must be $10\omega_1$. From Eq. 7.30, the gain is reduced to $1/\sqrt{1.01}$ of the mid-band gain. This reduction (about 0.5 percent) is normally negligible. Consequently the upper frequency limit of the low-frequency region is considered to be $10\omega_1$. At this frequency, the phase of the output voltage with respect to the input voltage is 5.7° above the mid-frequency value of 180°.

An analysis, similar to that just given, can be developed for the input loop of Fig. 7.4 or Fig. 7.10. In this analysis, we find the mid-frequency gain is

$$K_v = \frac{v_1}{v_s} = \frac{R_i}{R_s + R_i} \tag{7.33}$$

The general gain equation is

$$G_v = K_v \frac{s}{s + \omega_1} \tag{7.34}$$

where

$$\omega_1 = \frac{1}{C_{C1}(R_s + R_i)} \tag{7.35}$$

The gain-versus-frequency curve for Eq. 7.34 is identical to that given in Fig. 7.12. However, the phase angle will have the same shape as that shown in Fig. 7.12, but the angle will shift from 90° as ω approaches zero to 0° as ω becomes large.

One word of caution should be given at this time. Assume that a circuit is connected as shown in Fig. 7.4 and both C_{C1} and C_{C2} are chosen to provide low-frequency cutoff at the same frequency. Then, at this given frequency, ω_1, the voltage v_1 will be 0.707 of the mid-band gain due to capacitor C_{C1}. When the amplified signal passes through capacitor C_{C2}, this signal will again be attenuated. Thus, the signal v_2 at the frequency ω_1 will be 0.707×0.707 (or 0.50) of the mid-band signal. Obviously the lower cutoff frequency where the gain of both amplifier stages is 0.707 of the mid-frequency gain will occur at a frequency above ω_1. One solution would be to make one capacitor ten times the calculated size. Then, the lower frequency will be determined by only one coupling capacitor. A more sophisticated solution to this problem will be given in Chapter 12.

PROBLEM 7.5   The output impedance of a signal generator is 14 kΩ. This signal generator is connected through a 0.5 μF capacitor to a transistor with an h_{ie} of 1 kΩ. If the base biasing resistance R_b is much greater than h_{ie}, calculate the mid-frequency value of the voltage ratio v_{B1} (the voltage at the base of the transistor) to v_s (the open-circuit output voltage of the signal generator). In addition, determine the value of ω_1 for this configuration. *Answer:* $K_v = 0.0667$, $\omega_1 = 133$ rad/sec, $f_1 = 21$ Hz.

PROBLEM 7.6   Two identical bipolar transistors are connected as shown in Fig. 7.4a. The circuit and transistor parameters are:

$$h_{ie} = 1000\ \Omega \qquad\qquad h_{re} \simeq 0$$
$$h_{fe} = 100 \qquad\qquad h_{oe} = 10^{-5}\ \text{mhos}$$
$$R_C = 1000\ \Omega \qquad\qquad R_b = R_1 \parallel R_2 = 20\ \text{k}\Omega$$
$$C_E\ \text{and}\ C_{C1}\ \text{are very large} \qquad C_{C2} = 1\ \mu\text{F}$$

Determine the value of ω_1 for this configuration. (See Problem 7.4.) *Answer:* $\omega_1 = 515$ rad/sec, $f_1 = 82$ Hz.

7.3   THE EFFECT OF THE EMITTER BYPASS CAPACITOR

To this point, the emitter (or source or cathode) bypass capacitor has been assumed to be very large so that its reactance is negligible for any frequency of interest. In fact, the previous equations for determining the size of these capacitors have given values so the bypass capacitor effect would be negligible. Let us now study the effect of the emitter (or source or cathode) bypass capacitors at frequencies where their effect can no longer be ignored.

A circuit is connected as shown in Fig. 7.13a. Normally a capacitor would be connected in series with R_s, but in order to isolate the effect of C_E, we will assume that the reactance of the coupling capacitor is zero. The source and transistor input portion of this circuit will be of primary importance in this analysis. Consequently, let us rearrange this portion of the circuit into a simpler configuration. First, the circuit to the left of

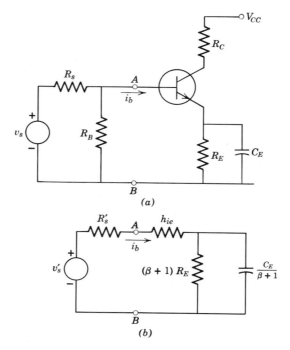

FIGURE 7.13
A circuit for determining the effect of C_E : (a) actual
circuit; (b) equivalent circuit.

points A and B will be replaced by its Thevenin's equivalent configuration,
as shown in Fig. 7.13b. The values of the Thevenin's generator are:

$$R_s' = \frac{R_s R_b}{R_s + R_b} \tag{7.36}$$

$$v_s' = v_s \frac{R_b}{R_s + R_b} \tag{7.37}$$

Our experience with the emitter follower can now be used to complete the
simplified equivalent circuit of Fig. 7.13b. You will recall that the input
impedance to a transistor with a load Z_E in the emitter is $h_{ie} + (\beta + 1)Z_E$.
Thus, the equivalent circuit has the form shown in Fig. 7.13b.

Your first impression may be that the emitter bypass capacitor C_E need
only have small reactance in comparison with R_E (for example 0.1 R_E) for
effective bypassing. This would be a good rule if the impedance seen by
looking into the emitter terminal of the transistor were not in parallel with
R_E. As we found in studying the emitter follower, this impedance into the

emitter (R_{ot} for the emitter follower) is usually much smaller than R_E. Consequently, the impedance into the emitter primarily determines the value of C_E, and the reactance of C_E will be much less than R_E in the frequency range near ω_1. As a result, the effect of R_E may be neglected. If R_E is neglected, the value of I_b in Fig. 7.13b is

$$I_b = \frac{V_s'}{R_s' + h_{ie} + \dfrac{1}{[sC_E/(\beta + 1)]}} \tag{7.38}$$

The terms in Eq. 7.38 can be rearranged to produce the following arrangement.

$$I_b = \frac{V_s'}{R_s' + h_{ie}} \frac{s}{s + [(\beta + 1)/C_E(R_s' + h_{ie})]} \tag{7.39}$$

The first term in Eq. 7.39 gives the value of I_b at the mid-band frequencies and the second term is of the form $s/(s + \omega_1)$ where ω_1 is

$$\omega_1 = \frac{(\beta + 1)}{C_E(R_s' + h_{ie})} \tag{7.40}$$

The value of $1/\omega_1$ is equal to C_E times a resistive term. However, this resistive term, $(R_s' + h_{ie})/(\beta + 1)$, is the input impedance to the transistor from the emitter terminal. If we had not neglected the effect of R_E, the resistive term would have been R_E in parallel with the input impedance to the emitter terminal.

From our previous analysis, the value of I_b and consequently I_c drop to 0.707 of their mid-band values when the reactance of C_E is equal to the resistance facing C_E. This same general statement is true for FET, MOSFET, or tube devices. *The gain drops to 0.707 of its mid-band gain whenever the reactance of C_E (or C_S or C_K) is equal to the resistance facing C_E (or C_S or C_K).*

A sketch of voltage gain as a function of frequency is given in Fig. 7.14 for the frequency range where the emitter bypass capacitor limits the voltage gain. The output is not zero at zero frequency when only the bypass capacitor is considered because the resistor R_E is the maximum limiting value of Z_E. Thus, there is base current at zero frequency. In fact, the voltage gain is easily determined when a resistance R_E is in the emitter circuit and is not bypassed. The analysis of the emitter follower showed that the signal voltage across the emitter resistor is essentially equal to the input voltage if $(\beta + 1)R_E$ is large compared with h_{ie}. Then

$$v_i \simeq i_e R_E \tag{7.41}$$

and

$$v_o = i_c R_l \tag{7.42}$$

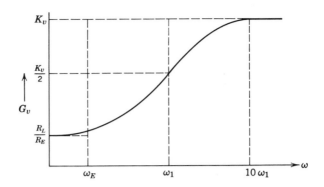

FIGURE 7.14
A sketch of voltage gain as a function of frequency,
considering only the emitter bypass.

where R_l is the total load resistance, including R_C, R_b, and h_{ie} in parallel, if the transistor is used to drive a following transistor. Thus, when the emitter resistor is unbypassed,

$$K_{vu} = \frac{v_o}{v_i} = \frac{i_c R_l}{i_e R_E} = \frac{\alpha R_l}{R_E} \simeq \frac{R_l}{R_E} \qquad (7.43)$$

The main problem in calculating the voltage gain from Eq. 7.43 is determining the value of the load resistance R_l. The following example will help clarify this procedure.

Example 7.3   Two identical transistors are connected as shown in Fig. 7.15. Let us assume that the coupling capacitors are very large and consequently

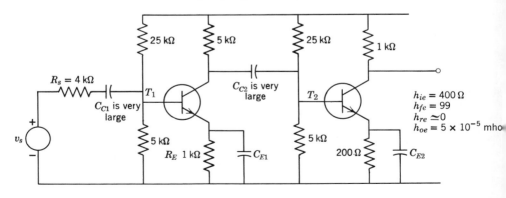

FIGURE 7.15.
The amplifier configuration used for Example 7.3.

have a negligible reactance at all frequencies considered. We wish to design the first stage with ω_1 equal to 100.

The value of R_s' is the parallel combination of R_s (4 kΩ in this example) and R_b (the parallel combination of 5 kΩ and 25 kΩ or 4.17 kΩ). Thus, R_s' is 2.04 kΩ or approximately equal to 2 kΩ. Then, the impedance into the emitter of transistor T_1 is $(h_{ie1} + R_s')/(\beta_1 + 1)$ or $(2k + 0.4k)/(99 + 1) = 24$ Ω. This value of resistance is so much smaller than the value of R_E that the latter may be neglected in calculating the value of C_E. By rearranging Eq. 7.40, we have

$$C_E = \frac{\beta + 1}{\omega_1(R_s' + h_{ie})} \qquad (7.44)$$

Then, $C_E = 1/(100 \times 24) = 416$ μF.

In order to determine the gain of transistor T_1 much below ω_1, the value of R_l must be determined. Since C_C is assumed to have negligible reactance, the input impedance to transistor T_2 becomes part of R_l. However, if the value of C_E in the transistor T_2 circuit is chosen to provide ω_1 of 100 rad/sec, then, at frequencies much below ω_1 the capacitor C_E will be ineffective. Thus, the input impedance to transistor T_2 will be equal to $h_{ie2} + (\beta_2 + 1)R_E$. For this example, the input impedance to T_2 is $400 + (99 + 1)200 = 20.4$ kΩ. This resistance is in parallel with the bias resistors $R_1 \| R_2 = 4.17k$, R_C (5 kΩ in this example) and $1/h_{oe1}$ (20 kΩ). $R_l = 20.4k \| 4.17k \| 5k \| 20k = 1.85$ kΩ. Then, from Eq. 7.43, $K_{vu} = R_l/R_E = 1.85k/1k = 1.85$.

PROBLEM 7.7 Determine the size of C_E in the transistor T_2 circuit in Example 7.3 that will produce $\omega_1 = 200$ rads/sec.

Answer: $C_E = 208$ μF.

PROBLEM 7.8 Assume that all of the capacitors in Fig. 7.15 are very large except C_{C2}. Determine the size of C_{C2} that will produce $\omega_1 = 100$ rad/sec.

Answer: $C_{C2} = 2.3$ μF.

The frequency at which $X_{CE} = R_E$, shown as ω_E in Fig. 7.14, is the frequency below which the emitter bypass capacitor C_E is ineffective and the emitter resistor R_E can be considered to be unbypassed. Since the gain does not drop to zero when ω approaches zero, Eq. 7.39 is not valid at frequencies much less than ω_1. By neglecting R_E, this error occurred. For reasons which will become obvious in Chapter 12, the exact form for I_b should be

$$I_b = \frac{V_s'}{R_s' + h_{ie}} \frac{s + \omega_E}{s + \omega_1} \qquad (7.45)$$

where $\omega_E = 1/R_E C_E$.

We now know how to determine either the coupling capacitance C_C or the emitter bypass capacitance C_E separately to obtain a given low-frequency cutoff ω_1. However, a stage of amplification (including one transistor and its associated components) usually includes both a coupling and a bypass capacitor. One solution is to let one capacitor determine ω_1 and make the reactance of the other capacitor negligibly small at ω_1. Since C_E must be so much larger than C_C (see Example 7.3 and Problem 7.8), size and cost can be minimized by letting C_E determine ω_1 and using a value of C_C which will let X_c be $0.1(R_s' + h_{ie})$ at ω_1. The capacitance of C_C will then be ten times as large as the value which would be used if C_C were to determine ω_1.

Example 7.4   In the amplifier stage including T_2 (Fig. 7.15), we calculated $C_E = 416$ μF (Example 7.4) and $C_C = 2.3$ μF (Problem 7.7) if each individually determined $\omega_1 = 100$ rad/sec. But, if both capacitors are to be used simultaneously and ω_1 is to remain at approximately 100 rad/sec, an economical combination of values is $C_E = 416$ μF and $C_C = 23$ μF. The effect on ω_1 when several stages are included in one amplifier will be discussed in a later chapter.

PROBLEM 7.9   A single transistor amplifier with $R_L = 2$ kΩ is capacitively coupled to a 6 kΩ driving source. If $h_{ie} = 700$ Ω, $\beta = 150$, $R_B = 10$ kΩ, and $R_E = 200$ Ω, determine values of coupling capacitance and emitter bypass capacitance which will provide a low-frequency cutoff, or half-power frequency, $f_1 = 16$ Hz, approximately.
Answer: $C_C = 16$ μF, $C_E = 340$ μF.

As noted previously, the coupling capacitor for a FET amplifier, or in fact any amplifier, can be calculated in precisely the same manner as for the bipolar transistor amplifier. The required coupling capacitance is much smaller, however, for the high input impedance devices. Also, the capacitor leakage currents may become important when devices with extremely high input impedance, such as MOSFETs or vacuum tubes are capacitively coupled.

The FET, MOSFET, or tube bypass capacitor must bypass the source bias resistor R_S (not to be confused with the driving source resistance R_s) in parallel with the resistance looking into the source terminal. This source terminal resistance can be determined, as usual, by applying a signal voltage between the source terminal and ground, and noting the signal current which flows into the source terminal. But, since there is no gate signal while we make this impedance test, the gate potential remains at ground and the source-to-ground voltage is also applied between the source and gate. This signal current which flows is therefore approximately $g_m V_{gs}$, as previously discussed, and the impedance seen at the source terminal is $1/g_m$. Therefore, the source bypass capacitor C_S must bypass the conductance $g_m + G_S$ or

$$C_S = \frac{G_S + g_m}{\omega_1} \tag{7.46}$$

where $G_S = 1/R_S$.

PROBLEM 7.10 A given FET amplifier has $R_G = 5$ MΩ, $R_D = 10$ kΩ and $R_S = 100$ Ω. If $g_m = 4 \times 10^{-3}$ mho and the driving-source resistance is 100 kΩ, determine suitable values for the input coupling capacitor and source bypass capacitor to provide $f_1 = 16$ Hz.
Answer: $C_C = .02$ μF, $C_S = 140$ μF, (or 150 μF).

7.4 TIME OR FLAT-TOP RESPONSE

In Sections 7.2 and 7.3, the input voltage was assumed to be a steady-state sinusoid and the output voltage was plotted as a function of frequency. This plot is usually known as a *frequency response* curve. Frequently, however, the amplifier input is *not* sinusoidal and the output waveform, or the output voltage as a function of *time* is of major interest. For example, the input voltage may be a step voltage, a series of pulses, or a voltage ramp and the waveform, or time response, of the output is desired. The gain equations have been developed in a very general form and can be used to solve for the desired output.

To illustrate the procedure, let us assume that the input waveform is a negative step function with a magnitude of V volts. Then, Eq. 7.24 has the form

$$G_v = \frac{V_o}{V_i} = -K_v \frac{s}{s + \omega_1} \tag{7.24}$$

This equation can be written as

$$V_o = -V_i K_v \frac{s}{s + \omega_1} \tag{7.47}$$

The notation indicates that the input voltage must be expressed in the Laplace form. The Laplace transform for the step function input has the form $V_i = -V/s$. When this substitution is made in Eq. 7.47, the value of V_o becomes

$$V_o = K_v V \frac{1}{s + \omega_1} \tag{7.48}$$

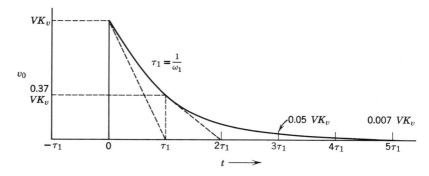

FIGURE 7.16
A sketch of output voltage as a function of time if the input signal is a negative step function.

The inverse Laplace transform of Eq. 7.48 is

$$v_o = VK_v e^{-\omega_1 t} \qquad (7.49)$$

This output voltage is plotted as a function of time in Fig. 7.16. The signal steps (at time $t = 0$) from zero to VK_v volts as expected. However, since the coupling capacitor blocks the dc, the output voltage exponentially drops to zero with a time constant $\tau_1 = 1/\omega_1$. Of course, the output signal is inverted with respect to the input signal because we have assumed the common-emitter configuration, with its inherent polarity reversal, is used. Since Eq. 7.27 is so similar to Eq. 7.24, a step function of input current will also produce an output current with the shape shown in Fig. 7.16.

Actually, if only an emitter (or source or cathode) bypass capacitor exists, the output will not approach zero but will approach the unbypassed gain K_{vu} as shown in Fig. 7.17a. In most actual RC-coupled amplifiers, both the emitter bypass capacitor and the coupling capacitor are present. If ω_1 is

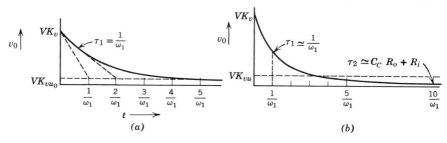

(a)

(b)

FIGURE 7.17
The effect of the emitter bypass capacitor on the signal waveforms: (a) only emitter bypass present; (b) both emitter bypass and coupling capacitor present.

determined by the emitter bypass capacitor and the reactance of the coupling capacitor is negligible at ω_1, the response would be as shown in Fig. 7.17b. In this plot, the total response is the sum of two exponentials. The shape for the first few time constants (to about three time constants) is determined primarily by the emitter circuit with a time constant approximately equal to $1/\omega_1$. After about four time constants (where $\tau = 1/\omega_1$), the shape is determined essentially by the coupling capacitor circuit and has a time constant essentially equal to $C_c(R_i + R_o)$ as the output voltage (or current) approaches zero. Usually, interest centers about the first few time constants (in fact, much less than one time constant in most applications). Consequently, only the effect of the low-frequency determining circuit need be considered, and the response for either coupling capacitor limitation or emitter bypass limitation is assumed to have the form shown in Fig. 7.16.

PROBLEM 7.11   The response of an amplifier with C_E as its only capacitor is given by Eq. 7.45. Prove that the response of i_b has the form shown in Fig. 7.17a if V_s' is a step function of voltage with an amplitude of V volts.

Although the RC-coupled amplifier is not suitable for amplifying dc signals, the information gained from the step function analysis is very useful in predicting the behavior of the amplifier when it is called on to amplify rectangular pulses. A series of rectangular pulses is shown in Fig. 7.18. These

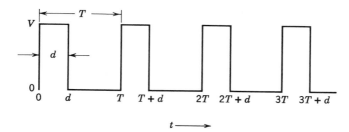

FIGURE 7.18
A series of rectangular pulses.

pulses have amplitude V, pulse duration d, and period (of repetition) T. In the case $d = T/2$, this signal is known as a square wave.

If we assume the excitation to be a series of voltage pulses as illustrated in Fig. 7.18, we may express the time-domain input voltage as the sum of a series of step voltages.

$$v_i = V[u(t) - u(t - d) + u(t - T) - u(t - (T + d)) + u(t - 2T) \cdots] \quad (7.50)$$

The $u(t - a)$ symbolizes a step voltage that occurs at $t = a$ and so on. The Laplace transform of this excitation voltage is

$$V_i = V\left(\frac{1}{s} - \frac{e^{-ds}}{s} + \frac{e^{-Ts}}{s} - \frac{e^{-(T+d)s}}{s} + \frac{e^{-2Ts}}{s} \cdots\right) \qquad (7.51)$$

The s domain output voltage is $V_o = V_i G_v$. Therefore,

$$V_o = -VK_v\left(\frac{1}{s + \omega_1} - \frac{e^{-ds}}{s + \omega_1} + \frac{e^{-Ts}}{s + \omega_1} - \frac{e^{-(T + d)s}}{s + \omega_1} + \frac{e^{-2Ts}}{s + \omega_1} \cdots\right) \qquad (7.52)$$

The time-domain response is obtained from the inverse transform[2]

$$v_o = Vk_v[e^{-\omega_1 t} - u(t - d)e^{-\omega_1(t-d)} + u(t - T)e^{-\omega_1(t-T)}$$
$$- u(t - T - d)e^{-\omega_1(t-T-d)} + u(t - 2T)e^{-\omega_1(t-2T)} \cdots] \qquad (7.53)$$

The time response may be plotted by adding the exponential terms of Eq. 7.53, as illustrated in Fig. 7.19. Since the average steady-state current through the coupling capacitor must be zero, the cross-hatched area labeled A below the reference axis must approach the corresponding area labeled B above the axis after several periods T have elapsed.

Ideally, the output voltage should remain constant during the pulse. However, the output current and voltage decay exponentially as shown because of the charging of the coupling capacitor. The departure of the actual output from the ideal flat response at the termination of the pulse is called *sag*. From a consideration of the first pulse as shown in Fig. 7.19, the sag can be determined by solving Eq. 7.53 at $t = d$. Then

$$\text{Sag} = VK_v - VK_v e^{-\omega_1 d} = K_v V(1 - e^{-\omega_1 d}) \qquad (7.54)$$

The fractional sag, or ratio of sag to the output at the beginning of the pulse, is of greater significance than the amount of sag.

$$\text{Fractional sag} = \frac{\text{Sag}}{K_v V} = 1 - e^{-\omega_1 d} \qquad (7.55)$$

If the pulse duration d is small in comparison with $1/\omega_1 = (R_o + R_i)C_C$, the fractional sag can be determined more easily than by the use of Eq. 7.55. The initial slope of the exponential decay is, using Eq. 7.53,

$$\text{Initial slope} = \frac{d(-K_v V e^{-\omega_1 t})}{dt}\bigg|_{t=0} = -K_v V \omega_1 \qquad (7.56)$$

The exponential sag closely follows the initial slope as long as the pulse duration is small (0.1 or less) in comparison with the time constant $1/\omega_1$.

[2] This time transformation is treated in *Network Analysis*, Second Edition, by M. E. Van Valkenburg, Prentice-Hall, Englewood Cliffs, New Jersey, 1964.

Function   Plot

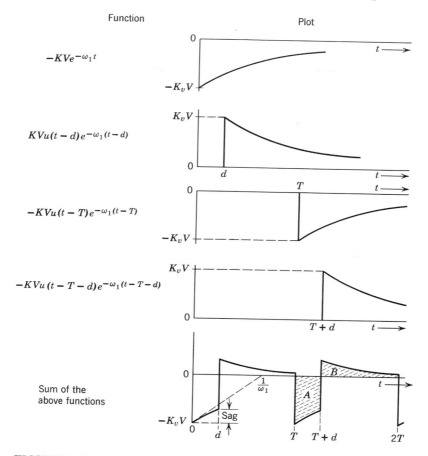

FIGURE 7.19
The time response of an RC-coupled amplifier to a periodic rectangular pulse.

Then the total sag is approximately equal to the pulse duration times the magnitude of the initial slope

$$\text{Sag} \simeq K_v V \omega_1 d \qquad (7.57)$$

and

$$\text{Fractional sag} \simeq \omega_1 d \qquad (7.58)$$

Equation 7.58 points up the desirability of a small value of ω_1 when rectangular pulses are to be amplified.

Figure 7.19 shows that the first pulse after $t = 0$ has greater sag than the succeeding pulses. The sag decreases somewhat as the transient dies out. However, the fractional sag remains constant if the pulse amplitude is measured from the base line.

PROBLEM 7.12   Choose a transistor from the Appendix to amplify a signal from a 5 kΩ source and drive a 1 kΩ load. The load must be capacitively coupled. The amplifier should provide high gain with a value of f_1 no greater than 50 Hz. Choose V_{CC}, design the amplifier, calculate values for all the components, and determine the maximum peak values of source and load voltages. Determine the percent sag for a 1 ms rectangular pulse.

7.5   HIGH-FREQUENCY CONSIDERATIONS

We learned in Section 7.2 that the low-frequency response of an *RC*-coupled amplifier is determined by the time constant of a series *RC* circuit. In this section and the following sections, we will discover that the high-frequency response is determined by the time constant of the parallel or shunt capacitance and the associated resistance. The major problem with the shunt capacitance is that it does not appear on the circuit diagram and must be either calculated from the given transistor parameters or measured directly. Therefore, we will first discuss the shunt capacitances in a transistor and then their effects on the circuit performance.

The junction capacitance associated with the depletion regions of a *p-n* junction has been previously discussed in considerable detail in connection with the junction diode. An additional capacitance, known as diffusion capacitance, which results from the stored charge caused by the injection of carriers across a junction was discussed in connection with a junction diode in Chapter 3. This diffusion capacitance was also discussed in connection with the common-base amplifier in Chapter 4, but will be discussed in some detail here because it is a major element in the determination of the high-frequency characteristics of a transistor.

The diffusion capacitance in a transistor is almost entirely the result of the injection of carriers from the emitter into the base, assuming normal operation of the transistor. These carriers diffuse across the base region and create a stored charge in the base as illustrated in Fig. 7.20, using the *p-n-p* configuration as a model. But, as the injected carriers enter the base, they produce an electric field which causes electrons to flow from the bias battery through the base terminal into the base in such a manner as to neutralize the charge. Notice that the current which flows as a result of this transfer of charge involves only the emitter and base currents and

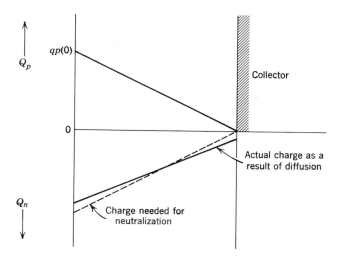

FIGURE 7.20
Stored charge in the base of *p-n-p* **transistor.**

has no effect on the collector current. Therefore, each change of base current has two components. One component contributes to the stored charge in the base region as if a capacitor were connected from base to emitter. The second component of base current causes the change of collector current desired in transistor operation. The shunting effect of the diffusion capacitance (the stored charge effect) on the base signal causes α and β to decrease with increasing frequency, as previously mentioned. The additional electrons which rush into the base to neutralize the injected charge do not appreciably increase the recombination rate in the base so long as their density is small in comparison with the doping density or concentration in the base. The electrons, like the holes, diffuse toward the collector, but the collector junction barrier potential prevents their entering the collector. Thus, a small negative charge exists at the collector junction and a net positive charge exists near the emitter junction as a result of the electron diffusion. This charge distribution creates an electric field which is just sufficient to offset the diffusion of electrons but aids the diffusion of holes and thus improves the performance of the transistor. This effect is enhanced in graded-base or drift-field transistors which have fairly heavy base doping near the emitter junction, but the doping concentration decreases to near zero at the collector junction. These transistors will be discussed later in this section.

An expression was developed in Chapter 4 for the diffusion capacitance due to the stored charge in the base of a transistor. The expression, which neglects the electric field in the base, was given as $C_D = w^2/2D_p r_e$ for a p-n-p transistor. As noted in Chapter 4, the emitter junction capacitance C_{je} adds directly to C_D. The alpha cutoff frequency, in rad/sec, was also shown to be the reciprocal of the time constant $r_e C_D$, providing C_D is large in comparison with C_{je}. Then

$$\omega_\alpha = \frac{1}{(w^2/2D_p r_e) \cdot r_e} = \frac{2D_p}{w^2} \tag{7.59}$$

where w is the effective base width and D_p is the diffusion constant of the base material. Note that ω_α is independent of I_e so long as C_D is large compared with the emitter junction capacitance C_{je}. Also observe from Eq. 7.59, as illustrated by the example in Chapter 4, that large values of ω_α, and perhaps good high-frequency performance, might be obtained by making the effective base width w very small and selecting the transistor material and doping arrangements so the diffusion constant of the charge carriers in the base region is high. A gallium arsenide n-p-n transistor holds great promise in this regard because of its high D_n.

The upper cutoff frequency of a common-base amplifier is not directly determined by the alpha cutoff frequency of the transistor. As seen in Fig. 7.21, the collector junction capacitance C_{jc} is in parallel with $(R_l + r_b)$ as well as r_c. At high frequencies, part of the current αi_e is shunted through C_{jc} and, therefore, does not pass through the load resistor R_l. Even if α is assumed to be independent of frequency, or C_1 is neglected, the load current magnitude will decrease to 0.707, $\alpha_0 i_e$, neglecting r_c, when $|I_3| = |I_c|$. Since these currents are equal when $X_{cjc} = R_i + r_b$, the cutoff frequency due to C_{jc} is

$$\omega_c = \frac{1}{(R_l + r_b)C_{jc}} \tag{7.60}$$

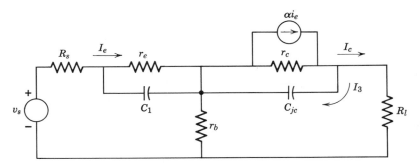

FIGURE 7.21
An equivalent circuit showing the effect of the capacitor C_{jc} on the load current at high frequencies.

For example, if $C_{jc} = 5$ pF, $r_b = 200\ \Omega$ and $R_L = 1$ kΩ, $\omega_c = 1.67 \times 10^8$ rad/sec and $f_c = 2.65 \times 10^7$ Hz. The upper cutoff frequency f_2 of the amplifier is lower than either f_α or f_c. In fact, the effects of C_1 and C_{jc} in decreasing the collector current are additive, so the upper cutoff frequency ω_2 is approximately the reciprocal of the sum of the time constants.

$$\omega_2 = \frac{1}{R_s \| r_e\, C_1 + (R_l + r_b)C_{jc}} \tag{7.61}$$

where $R_s \| r_e$ means the parallel combination of R_s and r_e which is the resistance facing C_1.

PROBLEM 7.13 A given transistor has $r_b = 200\ \Omega$, $C_1 = 200$ pF, and $C_{jc} = 5$ pF at $I_E = 1$ mA and $V_{CB} = 10$ V. Determine f_α and f_2 for a common-base amplifier with this q point if $R_s = 100\ \Omega$ and $R_l = 5$ k. *Answer:* $f_\alpha = 31.8$ MHz, $f_2 = 5.13$ MHz.

The development of the *graded-base* transistor caused a major advance in the high-frequency capabilities of the transistor. The graded base provides a built-in electric field in the base region, as previously mentioned, and a small collector junction capacitance C_{jc}. The electric field accelerates the carriers through the base and thus reduces the stored charge, or diffusion capacitance, for a given collector current.

The term *graded base* means that the doping density in the base region is not uniform but decreases as one traverses the region from the emitter junction to the collector junction. This doping distribution, or profile, which is shown in Fig. 7.22a, is obtained by diffusing the doping atoms into a heated semiconductor crystal. For example, let us assume

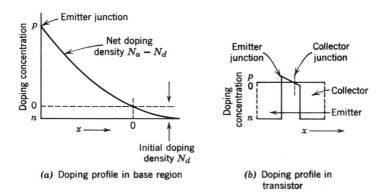

(a) Doping profile in base region

(b) Doping profile in transistor

FIGURE 7.22
Doping characteristics of a graded-base transistor.

that a graded-base n-p-n transistor is to be constructed. Then a bar of lightly doped n-type crystal is heated in an oven and the trivalent, or p, dopant is applied, probably in liquid form, to one surface of the bar. Then the acceptor atoms diffuse into the semiconductor in a manner very similar to the diffusion of carriers from the emitter into the base when forward bias is applied to the emitter junction. However, the temperature must be high (near the melting point of the crystal) to permit the doping atoms to diffuse, and then the diffusion rate of these atoms is very slow in comparison with the diffusion of charge carriers. Since the effective doping density is the difference between the donor impurity concentration N_d and the acceptor impurity concentration N_a, the end of the bar exposed to the trivalent impurity soon becomes p doped and the doping density decreases exponentially as the distance increases into the bar, as shown in Fig. 7.22a. At some distance d, the acceptor atom density is equal to the initial donor atom density and the crystal appears intrinsic. This point, which moves more deeply into the material as the diffusion process continues, becomes the collector-base junction when the diffusion is stopped by cooling the material, because the net doping changes from p-type to n-type at that point. The transistor is completed by replacing the trivalent doping material on the surface of the semiconductor with a heavily n-doped semiconductor which serves as the emitter. A process known as *alloying* is usually used to attach the emitter. Alloying is similar to welding except the continuous crystal structure is carefully maintained and the junction between the n-type and p-type materials is a plane surface. Also the lightly n-doped collector material is usually replaced by heavily n-doped material, as shown in Fig. 7.22b, from a short distance to the right of the collector junction. This heavily doped material improves the conductivity of the collector region and decreases the saturation voltage for a given collector current. Either alloying or diffusion techniques can be used to increase the doping concentration in the collector. Of course, a p-n-p transistor can be produced by using the opposite type dopants. Finally, the leads are attached and the transistor is heated in an atmosphere of either oxygen or nitrogen to form a thin layer of silicon dioxide or silicon nitride (for a silicon semiconductor) on the surface of the semiconductor. This layer is a good insulator which protects the surface from contamination and reduces surface leakage. This process is known as passivation.

As previously mentioned, the graded base produces a built-in electric field in the base because the majority carriers in the base diffuse toward the collector but are not able to enter the collector because of the potential barrier. Therefore, they accumulate and an electric field builds up until the drift current, due to the electric field, is equal to the diffusion current. This electric field therefore aids the flow of minority carriers in the base which are injected from the emitter into the base. It therefore has the same effect as increasing the diffusion constant which reduces the diffusion capacitance and increases f_α. The electric field also increases α and β because the minority carriers spend less time in the base and are less likely to recombine. The graded-base transistor also has better linearity because the electric field in a uniform-base transistor is proportional to the collector current, so β is strongly dependent upon collector current.

The graded collector junction in the graded-base transistor has very light doping on both sides of the junction in the vicinity of the junction. Therefore, the depletion region is wide on both sides of the junction, for·a given collector voltage, as compared with the uniformly-doped transistor. Therefore, the collector voltage breakdown is high and the junction capacitance comparatively low, which is an important contribution to the high-frequency performance. The lightly doped layer on the collector side of the collector junction is known as an *epitaxial* layer.

The only disadvantage of the graded-base transistor is the low emitter-breakdown voltage which results from the comparatively heavy base doping at the emitter junction and the very heavy emitter doping required to give high emitter efficiency. This low breakdown voltage is a handicap only when the transistor is used for switching and the emitter junction needs to be reverse biased. However, in these applications a diode can be connected in series with the input lead, and the diode will withstand the inverse voltage and protect the emitter-base junction in the transistor.

The physical layout of a typical graded-base, epitaxial, passivated transistor is shown in Fig. 7.23. The geometry of the transistor illustrates that the transistor is designed for high β and good collector power dissipation when the transistor is operated in the normal mode, in contrast with the *inverse mode* in which the collector is used as the emitter and vice versa.

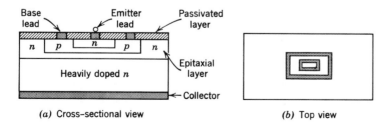

(a) Cross–sectional view   (b) Top view

FIGURE 7.23
Construction details of a typical transistor.

7.6   THE HYBRID-Π CIRCUIT AND BETA CUTOFF FREQUENCY

Although the h-parameter equivalent circuit is very useful at low and middle frequencies, it is not easily adapted to the high-frequency range where diffusion and junction capacitances must be included. The chief problem is that the h parameters vary with frequency and cannot be determined from the static characteristics. One equivalent circuit which gives good visualization of the high-frequency performance of a common-emitter transistor is the hybrid-π circuit shown in Fig. 7.24. This circuit is somewhat similar to the equivalent T circuit of Fig. 5.7a. The following differences should be noted.

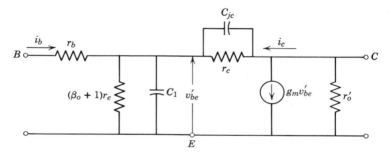

FIGURE 7.24
A hybrid-π equivalent circuit for the common-emitter transistor configuration.

1.   The diffusion and junction capacitances are included in the hybrid-π circuit.

2.   The current generator which appeared across the collector junction in the equivalent T circuit appears between the collector and emitter terminals in the hybrid-π circuit. In addition, this current is expressed in terms of the voltage across the emitter junction instead of the input current.

3.   The change in position of the dependent current generator eliminates the collector current through r_e, and only the input current i_b flows through the resistor $(\beta_o + 1)r_e$. Therefore, the factor $(\beta_o + 1)$ is required to give the proper voltage across the emitter junction and, hence, the proper input impedance for the transistor. β_o is the low-frequency value of β.

4.   The resistance r_b in the hybrid π is not quite the same as the resistance r_B in the equivalent T circuit. The hybrid-π resistance r_b must include the feedback effect of i_c through r_e in the equivalent T model.

5. The resistance r_o' has been added to the output to account for β changing with base width modulation.

The current generator $g_m v_{be}'$ must provide the same current as the equivalent T or h-parameter current generators. Thus,

$$g_m v_{be}' = \beta_o i_b \qquad (7.62)$$

But, at low frequencies where the reactance of C_1 and C_{jc} can be neglected (Fig. 7.24), $v_{be}' \simeq (\beta_o + 1)r_e i_b$. Then

$$g_m = \frac{\beta_o i_b}{v_{be}'} = \frac{\beta_o i_b}{(\beta_o + 1)r_e i_b} = \frac{\beta_o}{(\beta_o + 1)r_e} \qquad (7.63)$$

Since $\beta_o/(\beta_o + 1) = \alpha_o$ and $g_e = 1/r_e = (q/kT)I_E$,

$$g_m = \frac{\alpha_o q I_E}{kT} = \frac{q}{kT} I_C \qquad (7.64)$$

At normal room temperature, q/kT is approximately equal to 40, as previously noted, thus g_m can be easily determined as 40 I_C mhos, where I_C is the q-point value of collector current. The resistance $(h_{fe} + 1)r_e$ can also be expressed in terms of g_m, since $g_m = \alpha_o g_e$,

$$(\beta_o + 1)r_e = \frac{(\beta_o + 1)}{g_e} = \frac{\alpha_o(\beta_o + 1)}{g_m} = \frac{\beta_o}{g_m} \qquad (7.65)$$

We will designate this resistance β_o/g_m as r_π.

The hybrid-π circuit can be modified as shown in Fig. 7.25. The output resistance r_o has replaced both the resistance r_o' and the effect of resistance r_c which were shown in Fig. 7.24. The resistance r_o is a function of the driving-source resistance, but is approximately equal to $1/h_{oe}$. The effect of r_c on the output resistance is significant because, as seen from Fig. 7.24, a voltage applied at the output terminals causes a current to

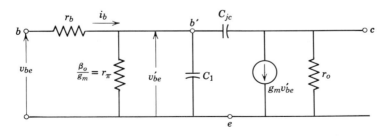

FIGURE 7.25
A modified hybrid-π circuit.

flow through r_c. This current causes a voltage v_{be}' which activates the current generator $g_m v_{be}'$. If a voltage v_o is applied to the output terminals, the current which flows through r_c is approximately v_c/r_c. Then, if the input is open, the current flows through the resistance β_o/g_m and $v_{be}' = (v_c/r_c)(\beta_o/g_m)$. Then

$$g_m v_{be}' = \frac{\beta_o v_o}{r_c} \tag{7.66}$$

Since the output resistance is the ratio of the voltage v_c to the current i_c which flows as a result of v_c, and $i_c = v_c/r_c + \beta_o v_c/r_c$

$$r_d = \frac{v_c}{(v_c/r_c) + (\beta_o v_c/r_c)} = \frac{r_c}{\beta_o + 1} \tag{7.67}$$

You may recall from chapter 5 that $r_d = r_c/(\beta_o + 1)$. The resistance r_c also has a slight loading effect on the input, but this can usually be neglected. The contribution of r_d to the output resistance r_o results from the increased diffusion current through the base region as the collector voltage increases and the base region narrows. This was the only effect considered in Chapter 5. However, the β or h_{fe} of the transistor also increases as the base width narrows because there are then fewer recombinations for a given current density. This varying β over the output voltage cycle would cause $r_\pi = \beta_o/g_m$ to vary and thus cause a greater change in collector current than the change due to increased diffusion current alone. This latter effect is accounted for by the resistor r_o' in Fig. 7.24 and was ignored in the equivalent T circuit. The resistance r_c accounted for only the change in diffusion current. Both effects are included in r_o, which is the parallel combination of r_o' and r_d when the input is open. These effects are essentially of equal magnitude when the input is open, so $r_o' \simeq r_d$. Thus, since their parallel combination is $1/h_{oe}$, either one is approximately equal to $1/2h_{oe}$. When finite values of driving-source resistance are used, r_o increases. However, r_o changes by less than a factor of two when the driving-source resistance is reduced from infinity to zero. Therefore, adequate accuracy is usually obtained if r_o is assumed to be $1/h_{oe}$.

   Figure 7.25 shows that the voltage v_{be}' decreases as the frequency increases, assuming I_b constant, because of the decreasing reactance of C_1. Also, observe that C_{jc} is in parallel with C_1 when the input is shorted (zero load resistance). Then

$$V_{be}' = \frac{I_b}{(g_m/\beta_o) + j\omega(C_1 + C_{jc})} \tag{7.68}$$

Beta cutoff frequency f_β is defined as the frequency at which the short-circuit current magnitude drops to $.707 \, \beta_o$. This frequency occurs when the j

term in the denominator of Eq. 7.68 is equal to the real term g_m/β_o, or $\omega_\beta(C_1 + C_{jc}) = g_m/\beta_o$. Then

$$\omega_\beta = \frac{g_m}{\beta_o(C_1 + C_{jc})} \tag{7.69}$$

PROBLEM 7.14 A given transistor has $\beta_o = 100$, $C_1 = 190$ pF and $C_{jc} = 10$ pF at $I_C = 5.0$ mA. Determine ω_β and f_β for this transistor.
Answer: $\omega_\beta = 10^7$ rad/sec, $f_\beta = 1.6 \times 10^6$ Hz.

A sketch of the magnitude of β as a function of frequency is given in Fig. 7.26. The frequency at which $|\beta| = 1$ is defined as f_τ. Since the reactance of $(C_1 + C_{jc})$ is inversely proportional to frequency, $|\beta|$ is also

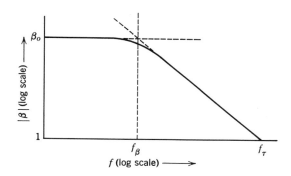

FIGURE 7.26
A sketch of β as a function of frequency.

inversely proportional to frequency for frequencies well above f_β where this reactance is small in comparison with r_π. Therefore, the high-frequency part of the curve is a straight line if logarithmic scales are used. If this straight-line portion of the curve is extended upward until it intersects the β_o line, intersection occurs at f_β. Therefore, while β decreases by the factor β_o, the frequency must increase by the factor β_o and

$$f_\tau = \beta_o f_\beta \tag{7.70}$$

And from Eq. 7.69 we can see that

$$\omega_\tau = \frac{g_m}{C_1 + C_{jc}} \tag{7.71}$$

Thus, the transistor with $\beta_o = 100$ which was found to have $f_\beta = 1.6$ MHz has a value of $f_\tau = 160$ MHz.

The magnitudes of f_τ and the alpha cutoff frequency f_α may be compared if we recall that $\omega_\alpha = 1/r_e C_1 = g_e/C_1$. Since $g_m = \alpha_o g_e$ and C_{jc} is usually small in comparison with C_1, $\omega_\tau \simeq \omega_\alpha$. Some manufacturers specify f_τ for their transistors and others specify f_α. A sufficiently accurate value of f_β may be obtained by dividing either f_α or f_τ by β_o.

PROBLEM 7.15 A given transistor has $\beta_o = 100$, $C_{jc} = 3.6$ pF and $f_\tau = 200$ MHz at $I_C = 2$ mA and $V_{CE} = 10$ V. Find f_β and C_1.

Answer: $f_\beta = 2$ MHz, $C_1 = 60$ pF.

The hybrid-π parameters can all be determined for a given transistor if the h parameters, f_τ, and C_{jc} are given, as illustrated by the following example.

Example 7.5 Let us determine the hybrid-π parameters for a transistor which has $h_{ie} = 1.5$ kΩ, $h_{fe} = 100$, $h_{oe} = 2.5 \times 10^{-5}$ mho, $f_\tau = 3 \times 10^8$ Hz and $C_{jc} = 3$ pF at $I_C = 2$ mA, $V_{CE} = 5$ V. Using these parameters,

$$g_m = 40\,I_C = 40 \times 2 \times 10^{-3} = 0.08$$
$$r_\pi = \beta_o/g_m = 100/.08 = 1.25 \text{ k}\Omega$$
$$r_b = h_{ie} - r_\pi = 1.5 \text{ k}\Omega - 1.25 \text{ k}\Omega = 250 \text{ }\Omega$$
$$r_o = 1/h_{oe} = 1/2.5 \times 10^{-5} = 40 \text{ k}\Omega$$
$$C_1 = (g_m/\omega_\tau) - C_{jc} = .08/(6.28 \times 3 \times 10^8) - 3 \text{ pF} = 37.7 \text{ pF} \quad \text{(from Eq. 7.71)}$$

PROBLEM 7.16 A given transistor has $h_{ie} = 1.2$ kΩ, $h_{fe} = 120$, $h_{oe} = 4 \times 10^{-5}$, $f_\tau = 350$ MHz, and $C_{jc} = 2$ pF at $I_C = 3$ mA, $V_{CE} = 5.0$ V. Determine the hybrid-π parameters of Fig. 2.23 for this transistor at the specified q point.

7.7 THE UPPER CUTOFF FREQUENCY OF A COMMON-EMITTER AMPLIFIER

The beta cutoff frequency f_β is the upper cutoff frequency of the common-emitter amplifier when the load resistance is very near zero and the driving-source resistance is infinite, because these are the conditions assumed when the output is shorted and the input current is constant or independent of frequency. We will now investigate the effects of the load resistance and the source resistance on the upper cutoff frequency f_2 or ω_2.

The hybrid-π equivalent circuit with load and driving source included is shown in Fig. 7.27. The collector junction capacitance C_{jc} is usually small in comparison with C_1 as previously mentioned, but the effect of C_{jc} may be surprisingly large because of the comparatively high signal voltage across it. This voltage is the difference between the voltage v_{be}' and the

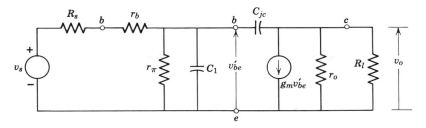

FIGURE 7.27
A hybrid-π circuit including source and load.

output voltage v_o. Since v_o is usually large compared with v_{be}', the signal current through C_{jc} may be comparable with, or even larger than, the current through C_1. The current magnitude through C_{jc} is

$$I_{jc} = (V_{be}' - V_o)\omega C_{jc} \qquad (7.72)$$

But, as seen in Fig. 7.27, $V_o = -K_v V_{be}' = -g_m V_{be}' R_{sh}$, where R_{sh} is the parallel combination of R_l and r_o. The negative sign occurs because of the polarity reversal of V_o as compared with V_{be}'. Then

$$I_{jc} = V_{be}'(1 + g_m R_{sh})\omega C_{jc} \qquad (7.73)$$

The susceptance of C_{jc} as viewed from the $b'e$ terminals is

$$B_{jc} = \frac{I_{jc}}{V_{be}'} = \omega(1 + g_m R_{sh})C_{jc} \qquad (7.74)$$

The effective capacitance of C_{jc}, as viewed from the terminals $b'e$ is therefore,

$$C_{eff} = \frac{B_{jc}}{\omega} = (1 + g_m R_{sh})C_{jc} \qquad (7.75)$$

This capacitance magnification which results when a capacitance exists between the input and the output of an amplifier is often known as the *Miller effect*. The development of the pentode tube was a result of the effort to eliminate this type of capacitance in a triode.

The equivalent circuit of Fig. 7.28 shows the effect of the junction capacitance C_{jc}. The capacitance $(1 + g_m R_{sh})C_{jc}$ has been added to the input circuit to account for the effect of C_{jc} at the terminals $b'e$. Transistor manufacturers usually give a capacitance C_{ob} (output capacitance with the base grounded) which differs from C_{jc} only by the header capacitance. Therefore, this C_{ob} can be used as an approximate value for C_{jc}. With C_{jc} known, Eq. 7.71 can be used to calculate C_1 at the desired q point.

Example 7.6   Let us consider the transistor of Problem 7.14 which has $C_1 = 190$ pF, $C_{jc} = 10$ pF, $\beta_o = 100$ and $r_\pi = \beta_o/g_m = 500$ Ω at $I_C = 5$ mA. If this transistor has $R_{sh} \simeq R_l = 1$ kΩ, $C_{eff} = (1 + 0.2 \times 10^3)10$ pF $= 2010$ pF, and the total capacitance in parallel with r_π is 2200 pF. We will now assume that the source resistance R_s is so large that the input current i_b does not vary with frequency even though the input impedance decreases as the frequency increases. Then the upper cutoff frequency occurs when the reactance of the total shunt capacitance is equal to r_π. In this example, $\omega_2 = 10^{12}/500 \times 2200 = 9 \times 10^5$ rad/sec or $f_2 = 1.44 \times 10^5$ Hz.

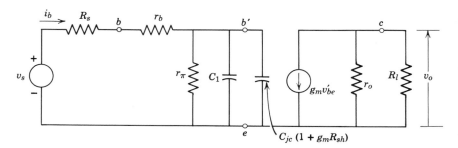

FIGURE 7.28
An equivalent circuit which shows the effect of the collector junction capacitance.

Observe that this upper cutoff frequency is less than one-tenth as high as the 1.6 MHz beta cutoff frequency which we previously calculated for this transistor. The Miller effect caused the reduction in upper cutoff frequency.

In Example 7.6 we assumed that R_s was very large and could be neglected in respect to r_π. However, from Fig. 7.28, we note that the resistance facing C_1 and C_{eff} is the parallel combination of r_π, and r_b in series with R_s. Thus, the general equation for the cutoff frequency of a bipolar transistor amplifier is

$$\omega_2 \simeq \frac{1}{[(R_s + r_b)\|r_\pi][C_1 + C_{ob}(1 + g_m R_{sh})]} \tag{7.76}$$

PROBLEM 7.17   Repeat Example 7.6 if $R_s = 400$ Ω and $r_b = 100$ Ω. (Hint: Convert the circuit to the left of r_π in Fig. 7.27 into an Equivalent Norton's circuit.)
Answer: $\omega_2 = 1.81 \times 10^6$ rad/sec or $f_2 = 2.8 \times 10^5$ Hz.

7.8   THE HIGH-FREQUENCY REGION FOR FETS, MOSFETS AND TUBES

The FET, MOSFET, and tube amplifiers also contain interelectrode capacitances. Consequently, a high-frequency equivalent circuit for these devices has the form shown in Fig. 7.29a. For most applications, it is desirable to arrange the circuit as shown in Fig. 7.29b. In this configuration, the input capacitance C_{it} is equal to C_{gs} (or C_{gk}) in parallel

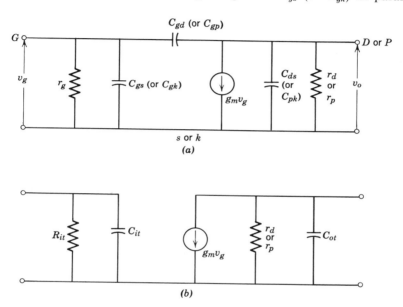

FIGURE 7.29
Equivalent circuits for FET, MOSFET, or tube devices.

with the effective value of C_{gd} (or C_{gp}). As a result of the Miller effect, the effective value of C_{gd} with a resistive load is $C_{gd}(1 + g_m R_{sh})$ or $C_{gd}(1 + K_v)$. Thus, the value of C_{it} for a FET or MOSFET is

$$C_{it} = C_{gs} + C_{gd}(1 + K_v) \tag{7.77}$$

A similar relationship can be found for a tube amplifier by simply changing the subscripts of the capacitors. In fact, the entire development in this section can be used for a vacuum tube amplifier by simply changing the capacitor subscripts.

Unfortunately, the high-frequency load is usually *not* purely resistive. Then, the term R_{sh} must be replaced by the more general term Z_{sh} where

Z_{sh} may be a complex term. Now, the input admittance to the transistor due to the capacitive elements is

$$Y_{ic} = j\omega C_{gs} + j\omega C_{gd}(1 + g_m Z_{sh}) \tag{7.78}$$

Notice that if $g_m Z_{sh}$ is complex, this input admittance now contains both a capacitive component and a real component. The effective input capacitance to the transistor is

$$C_{it} = C_{gs} + C_{gd}[1 + Re(g_m Z_{sh})] \tag{7.79}$$

where $Re(g_m Z_{sh})$ is the *real* part of $g_m Z_{sh}$. The total input resistance to the transistor is r_g in parallel with the inverse of the real component of Eq. 7.78. If we let $g_g = 1/r_g$ and $G_{it} = 1/R_{it}$, the total input conductance to the transistor is

$$G_{it} = g_g - \omega[Im(g_m Z_{sh})] \tag{7.80}$$

where $Im(g_m Z_{sh})$ is the imaginary part of $g_m Z_{sh}$. The negative sign in Eq. 7.80 results from the two j's in the equation. Notice that if we have a capacitive load, the imaginary part of $g_m Z_{sh}$ is negative, and our total input conductance will always be positive. However, if we have an *inductive* load, the imaginary part of $g_m Z_{sh}$ is positive and the *total input conductance may be negative*. In this case the amplifier may become unstable and oscillate.

In order to determine the output capacitance of the FET-type devices, notice that the current through C_{gd} in Fig. 7.29a is

$$I_{gd} = (V_g - V_o)j\omega C_{gd} \tag{7.81}$$

If we have a purely resistive load on the transistor, $V_g = -1/K_V V_o = -1/g_m R_{sh} V_o$. Then,

$$I_{gd} = -V_o\left(1 + \frac{1}{g_m R_{sh}}\right)j\omega C_{gd} \tag{7.82}$$

The negative sign appears because Eq. 7.81 was written for a current flowing from gate to drain. Therefore this sign will become positive for a current flowing from drain to gate. Then

$$B_{dg} = \frac{I_{dg}}{V_o} = \omega\left(1 + \frac{1}{g_m R_{sh}}\right)C_{gd} \tag{7.83}$$

Then, the total output capacitance of the device is equal to C_{ds} plus the effective capacitance of C_{gd} or

$$C_{ot} = C_{ds} + \left(1 + \frac{1}{g_m R_{sh}}\right)C_{gd} \tag{7.84}$$

If the gain of the transistor is very high, the term $1/g_m R_{sh}$ becomes very small and the output capacitance of the device is approximated as

$$C_{ot} \simeq C_{ds} + C_{gd} \qquad (7.85)$$

If the transistor load is not purely resistive, then the term $1/g_m R_{sh}$ becomes $1/g_m Z_{sh}$ and will be complex. However, if the gain is reasonably high, this term is still very small so Eq. 7.85 is usually sufficiently accurate. In addition, a real conductance results from $g_m Z_{sh}$ being complex. However, this real conductance is negligibly small compared to g_d and consequently can also be ignored. Thus the equivalent circuit for the FET-type devices can be reduced to that shown in Fig. 7.29b.

The FET and vacuum tube circuits in Fig. 7.4 can be reduced to the equivalent circuit shown in Fig. 7.30a. The resistances R_i include the gate (or grid resistors) in parallel with R_{it} and the resistances R_o include R_D (or R_P) in parallel with r_d (or r_p). The capacitor C_w represents the capacitance from the wires and components to ground. The central section of this equivalent circuit is the most general, so let us examine this

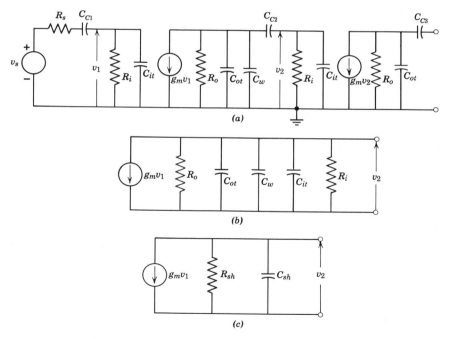

(a)

(b)

(c)

FIGURE 7.30
The high-frequency equivalent circuit for the amplifiers in Fig. 7.4: (a) detailed circuit; (b) partially simplified central section; (c) simplified central section.

section more closely. We are interested in the behavior of this circuit at high frequencies so C_{C2} is a short circuit. The central section then has the form shown in Fig. 7.30b. The capacitors C_{ot}, C_w, and C_{it} are all in parallel. These capacitors can be combined to form a single equivalent capacitor C_{sh} where

$$C_{sh} = C_{ot} + C_w + C_{it} \tag{7.86}$$

In addition, R_o and R_i are in parallel and can be combined to produce a single equivalent resistor. We have already called this resistance R_{sh}. Thus, the circuit in Fig. 7.30b can be reduced to that shown in Fig. 7.30c.

The nodal equation for Fig. 7.30c is

$$-g_m V_1 = V_2 \left(\frac{1}{R_{sh}} + sC_{sh} \right) \tag{7.87}$$

Then,

$$\frac{V_2}{V_1} = \frac{-g_m}{\dfrac{1}{R_{sh}} + sC_{sh}} \tag{7.88}$$

or

$$G_v = \frac{-g_m R_{sh}}{sC_{sh} R_{sh} + 1} \tag{7.89}$$

This relationship can be arranged to have the form

$$G_v = -g_m R_{sh} \frac{1/(C_{sh} R_{sh})}{s + [1/(C_{sh} R_{sh})]} \tag{7.90}$$

The term $g_m R_{sh}$ is equal to K_v. Let us define ω_2 as

$$\omega_2 = \frac{1}{R_{sh} C_{sh}} \tag{7.91}$$

Then, Eq. 7.90 can be written as

$$G_v = -K_v \frac{\omega_2}{s + \omega_2} \tag{7.92}$$

This general gain equation is also valid for bipolar transistors where ω_2 is found by Eq. 7.76.

The pole-zero plot for 7.92 is given in Fig. 7.31. The equation for the steady-state ac response can be found by replacing s in Eq. 7.92 with $j\omega$. Then, this equation can be written in the form

$$G_v = -K_v \frac{1}{1 + j(\omega/\omega_2)} \tag{7.93}$$

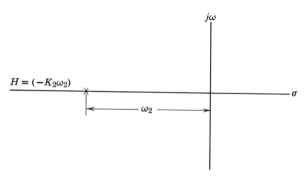

FIGURE 7.31
The pole-zero plot for Eq. 7.92.

The high-frequency gain and phase response of the amplifier can be found from either Fig. 7.31 or Eq. 7.93. Both of these plots are given in Fig. 7.32. Note that the high-frequency region begins at about $0.1\omega_2$ and the gain drops to 0.707 of the mid-band gain at ω_2. Thus, ω_2 is known as the high cutoff frequency for the amplifier. An example will be used to further clarify these ideas.

Example 7.7   A circuit is connected as shown in Fig. 7.33. The parameters of the transistors at their q points are as follows.

FET	NPN
$r_d = 100\ \text{k}\Omega$	$h_{ie} = 5.2\ \text{k}\Omega$
$g_m = 2.5 \times 10^{-3}\ \text{mhos}$	$g_m = 0.04\ \text{mho}$
$C_{dg} = 1\ \text{pF}$	$h_{fe} = 200$
$C_{gs} = 4\ \text{pF}$	$h_{oe} = 10^{-5}\ \text{mhos}$
$C_{ds} = 1\ \text{pF}$	$C_{ob} = 3\ \text{pF}$
	$f_t = 300\ \text{MHz}$

Let us determine the approximate upper cutoff frequency for the two-stage amplifier. The most accurate solution is obtained by drawing a complete equivalent circuit, or model, of the amplifier and then using a computer program such as ECAP to obtain the solution. However, an approximate pencil-and-paper solution may be obtained by drawing a simplified equivalent circuit with the effective Miller capacitances lumped with the input, or diffusion, capacitances as we have done in the preceding work. This type of equivalent circuit is drawn in Fig. 7.34. The coupling and

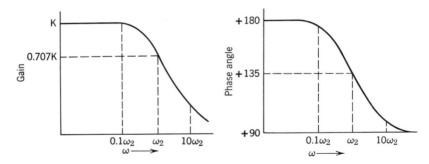

FIGURE 7.32
Frequency and phase response at high frequencies.

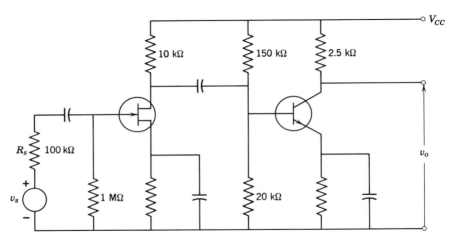

FIGURE 7.33
The amplifier circuit configuration for Example 7.7.

bypass capacitors are omitted because we are concerned only with the high-frequency response in this example.

The solution of even the simplified circuit of Fig. 7.34 appears foreboding, however, because the amplifier loads are not purely resistive and the capacitors are interacting. Nevertheless, Gray, Searle, and others[2] have shown that a good approximation of the upper cutoff frequency may be obtained by calculating the *open-circuit* time constant associated with each

---

[2] This concept is developed in *Multistage Transistor Circuits*, SEEC, Vol. 5, by R. D. Thornton, C. L. Searle, D. O. Pederson, R. B. Adler, and E. J. Angelo, Jr., John Wiley and Sons, Inc., New York, 1965, pp. 13–26.

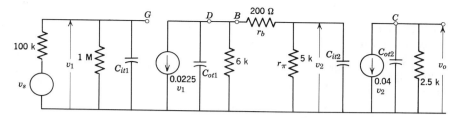

FIGURE 7.34
A simplified equivalent circuit for the amplifier of Fig. 7.33.

capacitor and then using the sum of these time constants to determine the upper cutoff frequency of the entire amplifier. An *open-circuit time constant* is defined as the time constant of a capacitor and the resistor facing it with *all other* capacitors removed from the circuit. We will use this technique in this example.

First we will find the value of C_{it1} and its open-circuit time constant. From Eq. 7.77, $C_{it1} = C_{gs} + C_{gd}(1 + K_v)$. With C_{ot1} and C_{it2} removed from the circuit, the load on the FET is purely resistive with 6 kΩ (r_d, R_D, and R_b in parallel) in parallel with 5.2 kΩ (h_{ie}) = 2.78 kΩ. Then $K_v = g_m R_{sh} = 2.5 \times 10^{-3} \times 2.78 \times 10^3 \simeq 7$ and $C_{it1} = 4$ pF $+ 1(1 + 7)$ pF $= 12$ pF. The open-circuit time constant associated with C_{it1} is $12 \times 10^{12} \times 0.9 \times 10^5 = 1.08 \times 10^{-6}$ sec.

The open-circuit time constant associated with C_{ot1} is obtained by first determining the value of C_{ot1} from Eq. 7.84 where $C_{ot} = C_{ds} + (1 + 1/K_v)C_{gd}$ $= 1$ pF $+ (1.14)1$ pF $= 2.14$ pF. The resistance facing C_{ot1} is the parallel combination of 6 kΩ and 5.2 kΩ previously calculated as 2.78 kΩ. Therefore the desired open-circuit time constant $2.14 \times 10^{-12} \times 2.78 \times 10^3 \simeq 6 \times 10^{-9}$ sec.

The effective value of C_{it2} with C_{ot2} removed from the circuit is obtained by inspection of Fig. 7.28 as $C_{it2} = C_1 + (1 + g_m R_{sh})C_{jc}$, where C_{jc} is essentially equal to C_{ob}. However, we must calculate C_1 using Eq. 7.71 where

$$C_1 = \frac{g_m}{\omega_\tau} - C_{jc} \tag{7.94}$$

Thus $C_1 = .04/6.28 \times 3 \times 10^8 - 3$ pF $\simeq 21$ pF $- 3$ pF $= 18$ pF, and $C_{it2} = 18$ pF $+ 3(1 + 0.04 \times 2.5 \times 10^3) = 18$ pF $+ 303$ pF $= 321$ pF. The resistance facing C_{it2} with C_{ot1} removed is 5 kΩ in parallel with 6.2 k$\Omega \simeq 2.76$ kΩ. Thus the open-circuit time constant associated with C_{it2} is $321 \times 10^{-12} \times 2.76 \times 10^3 \simeq 8.9 \times 10^{-7}$ sec.

Since C_{ot2} is essentially C_{ob}, and the resistance facing C_{ot2} is the 2.5 kΩ load resistor, the time constant associated with this capacitor is $3 \times 10^{-12} \times 2.5 \times 10^{-3} = 7.5 \times 10^{-9}$.

We now need to sum the four open-circuit time constants. However, the time constants associated with the output capacitances C_{ot1} and C_{ot2} are at least two orders of magnitude smaller than the time constants associated with C_{it1} and C_{it2}. Thus we may neglect these small time constants and add only the two associated with C_{it1} and C_{it2}. The sum of these two time constants is $\Sigma\tau_o = 1.08 \times 10^{-6} + 8.9 \times 10^{-7} = 1.97 \times 10^{-6}$ sec. Thus $\omega_2 = 1/1.97 \times 10^{-6} = 5.07 \times 10^5$ rad/sec $= 81$ kHz.

Observe that the time constants associated with the Miller capacitances are particularly long. Therefore, if a broadband amplifier is needed where the required f_2 is in the MHz region, circuits employing dual-gate MOSFETs, pentodes, or special circuit configurations which essentially eliminate the Miller capacitance, and which will be discussed in Chapter 10, should be used. Then the junction and wiring capacitances are of prime importance and the bandwidth calculation problem is simplified because of the lack of coupling between input and output circuits. Additional techniques for determining the bandwidth of multistage amplifiers will be discussed in Chapter 12.

PROBLEM 7.18   Repeat Example 7.7 if the FET is replaced by a 3N200 dual-gate MOSFET (Fig. 6.32) which is biased at $V_{G2S} = 4$ V, $V_{G1S} = 0$ V, and $V_{DS} = 8$ V. The input signal is applied to gate No. 1, and gate No. 2 is maintained constant. In addition, reduce R_C from 2.5 kΩ to 1 kΩ.

The value of ω_2 for most amplifiers is much beyond the audio range (up to about 15 kHz). Consequently, audio amplifiers are usually designed with little consideration for the upper cutoff frequencies. (An exception occurs if the impedance levels are exceptionally high.) However, *video* amplifiers which handle signals which will ultimately be viewed on a cathode-ray tube, or other display devices, and some high-speed pulse amplifiers must amplify signals up into the megahertz range. Then, power or voltage gain must be sacrificed to obtain additional bandwidth. Of course, the careful selection of the transistor or tube is very important in the design of a video amplifier. A suitable design procedure follows.

1.   Choose a transistor with high f_τ to minimize C_1 and low C_{ob} to minimize the Miller effect capacitance.
2.   Use the q point suggested by the transistor manufacturer, at which he specifies f_τ and C_{ob}. Smaller values of I_C will decrease f_τ and smaller values of V_{CE} will increase C_{ob}.
3.   Solve for the bias circuit components to provide adequate q point stability.
4.   Rearrange Eq. 7.91 to solve for R_{sh} which will produce the specified value of ω_2. In this step, remember C_{sh} depends on K_v, which in turn

is a function of R_{sh}. Since R_{sh} is the term we are seeking, C_{sh} should be written as a function of R_{sh} (Eqs. 7.7, 7.10 or 7.12).

5. Determine the value of R_C needed to provide the required value of R_{sh}. In broadband amplifiers, R_C is usually almost equal to the value of R_{sh}.

6. The values of C_C and C_E (or C_S or C_K) are determined by the specified value of ω_1.

PROBLEM 7.19 Determine the approximate upper cutoff frequency of the amplifier you designed for Problem 7.12. Draw the circuit diagram for the amplifier, if you have not already done so, and sketch the expected frequency response.

7.9 RISE TIME

In Section 7.4 we found the capacitors C_C or C_E had a pronounced effect on the response of the amplifier to a step function. Now, let us determine the effect the shunt capacitors have on the time response of the amplifier. Again, the input signal will be assumed to be a step function with an amplitude of $-V$ volts. Equation 7.92 states

$$G_v = \frac{V_o}{V_o} = -K_v \frac{\omega_2}{s + \omega_2} \tag{7.92}$$

Then,

$$V_o = -K_v V_i \frac{\omega_2}{s + \omega_2} \tag{7.95}$$

The Laplace transform for the step function of amplitude $-V$ is $-V/s$. When V_i is replaced by this Laplace transform, Eq. 7.95 is

$$V_o = \frac{K_v V \omega_2}{s(s + \omega_2)} \tag{7.96}$$

Equation 7.96 can be expanded by partial fractions to the following form.

$$V_o = V K_v \left(\frac{1}{s} - \frac{1}{s + \omega_2} \right) \tag{7.97}$$

The inverse Laplace transform of this function (Eq. 7.97) is

$$v_o = V K_v (1 - e^{-\omega_2 t}) \tag{7.98}$$

A sketch of the output voltage v_o as a function of time is given in Fig. 7.35. The effect of C_{sh} is to produce a finite *rise time* for the output signal. By definition, the rise time of a step function (or pulse) is the time

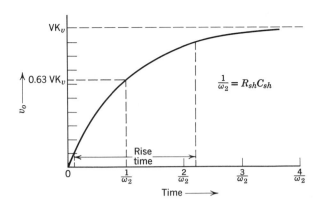

FIGURE 7.35
The effect of shunt capacitance on the time response.

required to rise from 10 percent to 90 percent of its final value. From Fig. 7.35 ((or from Eq. 7.98) the rise time, t_r, is about 2.2 time constants, where the time constant (Eq. 7.98) is $1/\omega_2$.

$$t_r = \frac{2.2}{\omega_2} = 2.2 R_{sh} C_{sh} \qquad (7.99)$$

Of course the output voltage of an RC-coupled amplifier will eventually sag or decay exponentially toward zero when a step input voltage is appled, as discussed in Section 7.4. However, the rise time is normally so short compared with the time required for significant sag that both events cannot usually be viewed simultaneously on an oscilloscope. A fast sweep is required to view the rise time, while a slow sweep is required to view the sag or decay.

In summary, the problem of solving for the frequency and phase response, or the time response, of an RC-coupled amplifier was greatly simplified by resolving the problem into three parts or frequency ranges, and from those three simplified problems the mid-frequency gain and the extremities of the pass band were obtained. This technique works well if $\omega_2 \gg \omega_1$ so there *is* a mid-frequency range. Practically all RC-coupled amplifiers meet this requirement. However, the need sometimes arises for a single equation which will characterize a single stage of an amplifier throughout its entire frequency range rather than a restricted portion of its spectrum. This equation can be synthesized by combining the low-frequency equation (Eq. 7.28) and the high-frequency equation (Eq. 7.93) to obtain

$$G_v = -K_v \frac{s\omega_2}{(s + \omega_1)(s + \omega_2)} \qquad (7.100)$$

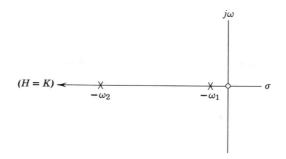

FIGURE 7.36.
The pole-zero plot for an *RC*-coupled amplifier.

Of course Eq. 7.100 can also be written for current gain by substituting *i* subscripts for *v* subscripts. Also the gains may represent either source-to-load, or device input-to-load gains.

Figure 7.36 shows the pole-zero plot for G_v of an *RC*-coupled amplifier. The scale factor has a value of *K*. A plot of gain as a function of frequency and a plot of phase angle as a function of frequency for an *RC*-coupled amplifier is given in Fig. 7.37. Since ω_2 is normally so much greater than ω_1, it is impossible to show both of these values on a linear plot. Consequently, the magnitude of ω_1 is exaggerated in Fig. 7.36. Fig. 7.37 is drawn with a log scale for ω.

PROBLEM 7.20 Prove that Eq. 7.100 is equal to Eq. 7.28 if $s = j\omega$ and $\omega \ll \omega_2$. Prove that G_v of Eq. 7.100 is equal to $-K_v$ if $s = j\omega$ and

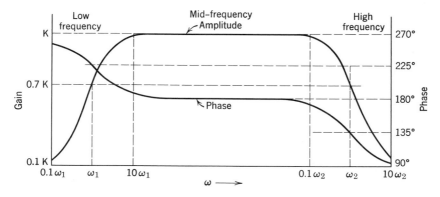

FIGURE 7.37
The gain and phase plot for an *RC*-coupled amplifier.

$\omega_1 \ll \omega \ll \omega_2$. Prove that Eq. 7.100 is equal to Eq. 7.92 if $s = j\omega$ and $\omega_1 \ll \omega$.

PROBLEM 7.21 A standard TV receiver sweeps 15,750 horizontal lines across the picture tube in one second. The resolution of the picture will be as good in the horizontal direction as in the vertical direction if the picture can change from white to black, or vice versa, in 1/700 of the length of a horizontal line. What must be the rise time and the bandwidth of the video amplifier in order to provide this resolution if 10 percent of the horizontal sweep time is used for retrace?

PROBLEM 7.22 The desired bandwidth of a video amplifier in a TV set is about 5 MHz. Assume that you are assigned the task of designing a video amplifier for a given TV receiver. The detector which drives the video amplifier has a source resistance of 1 kΩ. The load on the amplifier is the picture tube intensity grid and the grid resistor is 100 kΩ. Choose a bipolar transistor, perhaps from Appendix I, and determine the required value of resistance R_{sh}, the collector circuit resistor R_C and the voltage gain of your amplifier.

PROBLEM 7.23 Sketch the frequency and phase response of the amplifier of Fig. 7.38. You may neglect C_o.

PROBLEM 7.24 Use a computer analysis program such as ECAP to precisely obtain the frequency and phase response of the amplifier of Fig. 7.38, first neglecting C_o, and then including C_o. Compare all the results of Problem 7.23 and Problem 7.24.

PROBLEM 7.25 A circuit is connected as shown in Fig. 7.39.

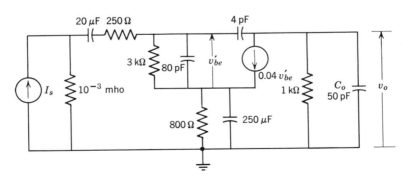

FIGURE 7.38
Amplifier circuit used in Problem 7.23 and Problem 7.24.

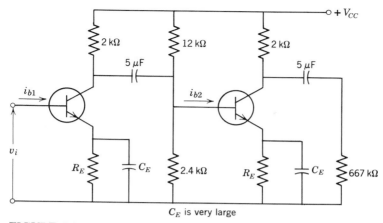

FIGURE 7.39
The circuit for Problem 7.25.

a. Determine $K_i(i_{b2}/i_{b1})$, ω_1 (low cutoff frequency) and ω_2 (high cutoff frequency) if the transistors have the following parameters.

$$h_{ie} = 1,000 \ \Omega \qquad\qquad h_{oe} = 10^{-5} \ \text{mhos}$$
$$g_m = .0625 \ \text{mhos} \qquad\qquad C_{ob} = 3 \ \text{pF}$$
$$h_{fe} = 50 \qquad\qquad f_\tau = 200 \ \text{MHz}$$

b. Sketch i_{B2} if i_{B1} is a 0.01 mA step function at time $= 0$.

PROBLEM 7.26 A triode amplifier (with a 6SN7 GT tube) is connected as shown in Fig. 7.40. The capacitances are $C_{gp} = 4$ pF, $C_{gk} = 3$ pF, and $C_{pk} = 1.2$ pF.

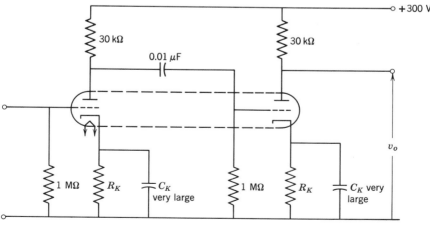

FIGURE 7.40
The circuit for Problem 7.26.

 a. The q point occurs where $v_{GK} = .-4$ V and $v_P = 140$ V. Find the values of r_p, μ, and g_m for this tube at this q point.
 b. Determine the values of R_o, R_i, and R_{sh} for this amplifier.
 c. Determine the low cutoff frequency ω_1, the voltage gain, and the high cutoff frequency ω_2 for this amplifier.

PROBLEM 7.27 Two FETs are connected as shown in Fig. 7.41. The two FETs are identical with the following parameters.

$$g_m = 2,000 \; \mu\text{mhos} \qquad r_g = 10^8 \; \Omega$$
$$r_d = 100,000 \; \text{ohms} \qquad C_{gd} = 5 \; \text{pF}$$
$$C_{gs} = 5 \; \text{pF} \qquad C_S \quad \text{is very large}$$
$$C_{ds} = 2 \; \text{pF}$$

 a. Determine the values of R_o, R_i, and R_{sh} for this configuration.
 b. Find the mid-band voltage gain $K_v = v_{g2}/v_{g1}$ for this amplifier.
 c. Find the low and high cutoff frequencies (ω_1 and ω_2, respectively) for this amplifier. Assume the wiring capacitance $C_w = 10$ pF.

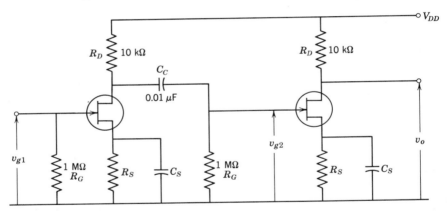

FIGURE 7.41
The circuit for Problem 7.27.

PROBLEM 7.28 Solve Problem 7.22 using a dual-gate MOSFET with $g_m = 3 \times 10^{-3}$ mho, $C_{gs} = C_{ds} = C_w = 1$ pF and $C_{gd} = 0.02$ pF.

8

Transformer Coupled Amplifiers

A transformer may be used to couple the output of an amplifier to its load. When used in this fashion, the transformer has some inherent advantages. For example, the ohmic resistance of the windings is much smaller than the ac impedance. As a result, the efficiency may be much higher than in the RC-coupled amplifier. In addition, the turns ratio may be chosen so that impedance matching may be utilized to obtain maximum power gain, and consequently, maximum voltage or current gain. In general, each of the active devices (transistors or tubes) has a poor impedance match when connected in the RC-coupled configuration. For example, if two bipolar transistors are cascaded, the transistor output resistance, $1/h_{oe}$, may be 100 kΩ or so, but the typical ac load, R_l, has been found to be about 1 kΩ or so. In contrast, if two FETs are cascaded, the output impedance of the first FET may be 10 kΩ or so, while the input impedance to the second FET may be several MΩ. Obviously, greater gain can be achieved by obtaining a better impedance match between an amplifier and its load.

8.1 THE MID-FREQUENCY RANGE

Two typical transformer-coupled amplifiers are shown in Fig. 8.1. While stabilized bias is used in Fig. 8.1a, the signal current does not flow through R_1 and R_2 but is bypassed to ground through C_B. Thus, all of the input signal is delivered to transistor T_2 regardless of the desired temperature stability. If the gate of a transistor (FET or MOSFET) or the grid of a tube is to be returned to ground, the circuit is connected as shown in Fig. 8.1b.

265

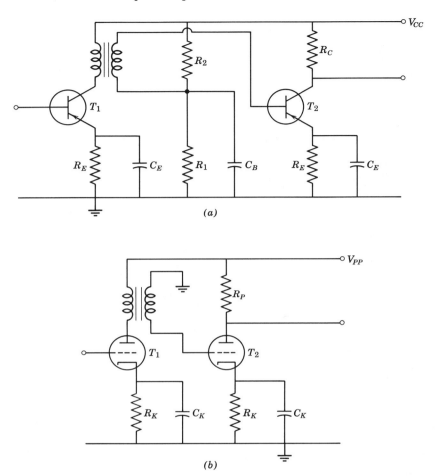

FIGURE 8.1
Typical transformer-coupled stages: (a) transistor; (b) vacuum tube.

The graphical method can be used to solve transformer circuits. However, in this configuration, the dc load, R_L, is very small (just the ohmic resistance of the transformer winding plus the emitter resistor). In contrast, the ac load, R_l, is much greater. Consequently, the dc load line is much steeper than the ac load line. An example may help clarify this concept.

Example 8.1   A transistor is connected as shown in Fig. 8.1a. The collector characteristics of this transistor (T_1 in Fig. 8.1a) are given in Fig. 8.2. The voltage V_{CC} is -8 V and $R_E = 150\ \Omega$. The dc resistance of the primary

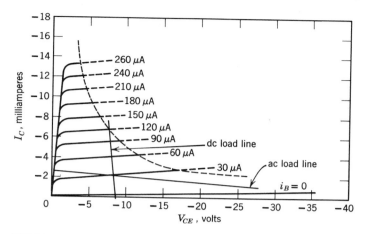

FIGURE 8.2
Characteristic curves for the transistor in Example 8.1.

winding (R_1) in the transformer is 20 Ω. We will determine the behavior of T_1 if the transformer is designed for optimum power gain.

First, the dc load line is drawn for $R_L = R_E + R_1 = 150 + 20 = 170\ \Omega$ and $V_{CC} = -8$ V. To obtain maximum power gain and hence maximum current gain, the ac load line is drawn with a slope approximately equal to that of the collector characteristic at the operating point.[1] This small slope permits the quiescent point to be chosen at a low value of collector current. For convenience, let the q point be the intersection of the dc load line and the $i_B = -30\ \mu A$ curve. Now, $1/h_{oe}$ for this transistor (at this q point) is about 13 kΩ. Then, the ac load line passes through the q point with a slope drawn for 13 kΩ.

After the quiescent point has been selected, the bias circuit components may be determined in accordance with the temperature stability requirements as discussed in Chapter 5.

The procedure now follows that given in Chapters 4 and 5. Thus, if i_B varies from $-20\ \mu A$ to $-40\ \mu A$, v_{CE} varies from -15 V to -3 V, and i_C varies from -1.8 mA to -2.6 mA. The current gain $K_i = \Delta i_C / \Delta i_B = .8/.02 = 40$. In order to find the value of base current in T_2, the turns ratio of the transformer must be known. We shall study this relationship in the remainder of this section.

---

[1] Although the output admittance of the transistor may differ appreciably from h_{oe}, the impedance match will be reasonably good if the output resistance is assumed to be $1/h_{oe}$. A more precise matching procedure is presented in Chapter 12.

Example 8.1 illustrates a very interesting behavior for the transformer-coupled amplifier. The voltage from collector to emitter v_{CE} may be much higher than the supply voltage V_{CC}. In fact, if the signal is symmetrical and drives the transistor almost into saturation, the peak value of v_{CE} is about twice as large as V_{CC}. This behavior is due to the inductive effect $(v_L = L di/dt)$ of the primary winding. As the collector current increases or decreases, a voltage opposing or aiding V_{CC} is produced in the transformer primary winding. As a result of this behavior, the transistor to be used in a transformer-coupled amplifier must be chosen with a maximum v_{CE} rating sufficiently above V_{CC} to protect the transistor.

The equivalent circuit for the transformer-coupled amplifier is given in Fig. 8.3. The current source is equal to either $g_m v_g$ or $h_{fe} i_b$. The resistance

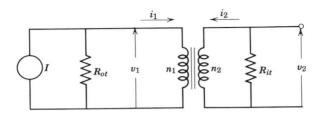

FIGURE 8.3
An equivalent circuit for a transformer-coupled amplifier.

R_{ot} is the output resistance of the tube or transistor $(1/h_{oe}, r_d, \text{ or } r_p)$ and R_{it} is the input resistance of the following tube or transistor. The transformer contains n_1 turns in the primary winding and n_2 turns in the secondary winding. These coils are wound around a magnetic core which forms a closed magnetic path for the magnetic flux. As a result of this type of construction, most of the flux linkages enclose all of the primary turns and also all of the secondary turns. Consequently, the coefficient of coupling (k) for the transformer[2] is almost unity.

If the coefficient of coupling is essentially equal to one, the voltages are related by the equation

$$v_1 = n v_2 \qquad (8.1)$$

---

[2] In the interest of brevity, transformer theory will not be covered in this text. If the reader is not familiar with this theory, he should review it in a good circuit book such as *Electrical Engineering Circuits* by Hugh Hildredth Skilling, Second Edition, John Wiley and Sons, New York, 1965.

where n is the ratio of the number of primary turns to the number of secondary turns.

$$n = \frac{n_1}{n_2} \tag{8.2}$$

Also, the currents in the transformer are related as

$$i_1 = \frac{i_2}{n} \tag{8.3}$$

Then, if Eqs. 8.1 and 8.3 are combined, we have

$$Z_1 = \frac{v_1}{i_1} = \frac{nv_2}{i_2/n} = n^2 \frac{v_2}{i_2} \tag{8.4}$$

where Z_1 is the input impedance to the primary winding. Since v_2/i_2 is equal to the impedance connected to the secondary, Z_2, we can write Eq. 8.4 as

$$Z_1 = n^2 Z_2 \tag{8.5}$$

In order to couple maximum power from a generator to a load, the impedance of the load should be equal to the conjugate of the generator output impedance. Then, in order to couple maximum power out of the current generator in Fig. 8.3, the input impedance to the transformer, Z_1, should be equal to R_{ot}. Since the load on the secondary, Z_2, is equal to R_{it}, Eq. 8.5 can be used to obtain the relationship

$$R_{ot} = n^2 R_{it} \tag{8.6}$$

If $R_{ot} = Z_1$, half of I (Fig. 8.3) will flow through R_{ot} and half will flow through Z_1. Then, $i_1 = I/2$ and $i_2 = nI/2$. If $I = h_{fe} i_b$, the value of i_2 is

$$i_2 = \frac{h_{fe} i_b n}{2} \tag{8.7}$$

But, since i_2 is actually the signal base current to transistor T_2, the current gain for the amplifier stage is

$$K_i = \pm \frac{h_{fe} n}{2} \tag{8.8}$$

The positive or negative sign in Eq. 8.8 occurs because i_2 can be taken from either side of the secondary so the output current may or may not be inverted with respect to the input current. (This versatility is another advantage of the transformer-coupled amplifier.)

If the current generator is $I = g_m v_g$ and if $R_{ot} = Z_1$, the voltage v_1 is $IR_{ot}/2$ or

$$v_1 = \frac{g_m v_g R_{ot}}{2} \tag{8.9}$$

Since $v_2 = v_1/n$ (Eq. 8.1), v_2 is

$$v_2 = \frac{g_m v_g R_{ot}}{2n} \tag{8.10}$$

The voltage v_2 is the input signal voltage to transistor (or tube) T_2, so

$$K_v = \pm \frac{g_m R_{ot}}{2n} \tag{8.11}$$

From Eq. 8.6, the value of n for an amplifier with optimum load is

$$n = \left(\frac{R_{ot}}{R_{it}}\right)^{1/2} \tag{8.12}$$

When this value is inserted into Eq. 8.8, K_i is

$$K_i = \pm \frac{h_{fe}}{2}\left(\frac{R_{ot}}{R_{it}}\right)^{1/2} \tag{8.13}$$

and Eq. 8.10 becomes

$$K_v = \pm \frac{g_m}{2}(R_{it} R_{ot})^{1/2} \tag{8.14}$$

Notice that n is greater than one if $R_{ot} > R_{it}$ (typical bipolar transistor amplifier) and n is less than one if $R_{ot} < R_{it}$ (typical coupling between FETs, MOSFETs, or tubes). Of course, the load may be a loudspeaker or some other device rather than another transistor or tube. In fact, transformers are used more often to connect loudspeakers to tubes and transistors than for any other application.

PROBLEM 8.1 If the second transistor in Example 8.1 has $h_{ie} = 1$ kΩ find the value of turns ratio n which will provide optimum power transfer. With this transformer, what is the current gain $K_i = i_{b2}/i_{b1}$?

Answer: $n = 3.6$, $K_i = 144$.

The approximation that $R_{ot} = 1/h_{oe}$ and $R_{it} = h_{ie}$ will *not* be valid under many conditions. Accurate values of R_{it} may be obtained from the equivalent circuit of a particular transistor configuration if the load impedance is known. Therefore, the design of an amplifier normally proceeds from the load toward the input. Since the bipolar transistor does not provide complete isolation of its input circuit from its output circuit, a simultaneous solution of the entire amplifier circuit would be required in order to obtain ideal impedance matching for a specified set of conditions. With the computer programs mentioned earlier, a solution of this type is quite straightforward. Approximate values of R_{ot} and R_{it} are usually sufficiently accurate for most work if a computer is not readily available. Fortunately, the input and output circuits are isolated in the FET, MOSFET, and tube amplifiers so R_{ot} and R_{it} are easy to determine.

PROBLEM 8.2   An amplifier is connected as shown in Fig. 8.1a. Calculate the current gain, $K_i = i_{b2}/i_{b1}$, for this amplifier if $R_{ot} = 30$ kΩ and $R_{it} = 1.5$ kΩ, $h_{fe} = 50$. Specify the turns ratio, n, of the transformer. Compare this gain with that of an RC-coupled amplifier using the same transistors. Assume $R_o \simeq R_{ot}$ and $R_i \simeq R_{it}$ ($R_C \gg 1/h_{oe}$ and $R_b \gg h_{ie}$).
Answer: $n = 4.47$, $K_i = 112$, $K_i = 45.4$ for the RC-coupled amplifier.

8.2 THE LOW-FREQUENCY RANGE

The transformer may be represented by an equivalent T as illustrated in Fig. 8.4a.[3] R_1 and L_1 are the resistance and leakage inductance of the primary. Also, the resistance and leakage inductance of the secondary, R_2 and L_2, have been referred to the primary by the multiplication factor n^2, where n is the turns ratio of the primary to secondary. In addition, the mutual inductance M and load resistance R_{it} have been referred to the primary. The distributed winding capacitance and circuit wiring capacitance have been neglected.

Basic textbooks which treat inductively-coupled circuits show that the mutual inductance is

$$M = k(L_{11}L_{22})^{1/2} \tag{8.15}$$

where k is the coupling coefficient, L_{11} is the self-inductance of the primary and L_{22} is the self-inductance of the secondary. Then, since inductance is proportional to the turns squared, $L_{11}/L_{22} = n^2$ or $L_{22} = L_{11}/n^2$. When this equivalent value for L_{22} is substituted into Eq. 8.15, M is

$$M = k\left(\frac{L_{11}L_{11}}{n^2}\right)^{1/2} \tag{8.16}$$

or

$$nM = kL_{11} \tag{8.17}$$

---

3 Hugh Hildredth Skilling, *Electrical Engineering Circuits*, Second Edition, John Wiley and Sons, New York, 1965, p. 341.

272 Transformer Coupled Amplifiers

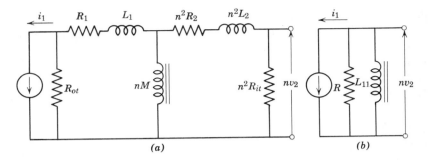

FIGURE 8.4
(a) The equivalent T circuit is used to represent the transformer in a transformer-coupled amplifier, (b) the low-frequency equivalent of (a) for a tightly coupled transformer.

Figure 8.4 shows that the total self-inductance of the primary is

$$L_{11} = L_1 + nM = L_1 + kL_{11} \tag{8.18}$$

or

$$L_1 = (1 - k)L_{11} \tag{8.19}$$

The term $(1 - k)$ is the fraction of primary flux which does not couple the secondary winding. Then L_1 is the leakage inductance of the primary. A typical coupling transformer has a coefficient of coupling k of the order of 0.99 or higher. Therefore, at low and middle frequencies, the leakage inductances may be neglected because their reactances are small in comparison with the reactance of the primary and the associated amplifier resistances R_{ot} and R_{it}. Also the resistance of the transformer windings, R_1 and R_2 are negligible in comparison with R_{ot} and R_{it}. Consequently, a simplified low-frequency circuit may be drawn as shown in Fig. 8.4b, where R is the parallel combination of R_{ot} and $n^2 R_{it}$. The output voltage is

$$nV_2 = -I_1 \frac{RsL_{11}}{R + sL_{11}} \tag{8.20}$$

$$nV_2 = -I_1 R \frac{s}{s + R/L_{11}} \tag{8.21}$$

When maximum power gain is achieved, $n^2 R_{it} = R_{ot}$ and $R = R_{ot}/2$. Let ω_1 be the frequency at which the reactance of the primary is equal to the resistance R. Then

$$\omega_1 = \frac{R}{L_{11}} \tag{8.22}$$

and

$$nV_2 = -I_1 R \frac{s}{s + \omega_1} \tag{8.23}$$

As for the RC-coupled amplifier, Eq. 8.23 can be reduced to

$$G = \pm K \frac{s}{s + \omega_1} \tag{8.24}$$

where K is the reference gain of the transformer-coupled amplifier (Eq. 8.13 or Eq. 8.14). The negative sign may or may not precede K, depending on the method of connecting the transformer. The gain equation has the same form as the RC-coupled amplifier in the low-frequency range. The frequency, phase, and time response would be identical in the two types of amplifiers in this range, providing that ω_1 is the same in each case.

The chief disadvantage of the transformer as a coupling device may be the large inductance required for satisfactory amplification of long pulses or low frequencies. Under these conditions, the transformer would be costly and bulky. As a result, the cost or weight may be a major concern and may outweigh the several advantages of a transformer. Miniature transformers are available for coupling transistor amplifiers. However, those designed for audio-frequency amplification generally have poor low-frequency response. In contrast, rectangular pulses of rather short duration (a few microseconds) may be amplified with little distortion when a miniature transformer is employed.

PROBLEM 8.3 A common-emitter transistor amplifier has $R_{ot} = 20 \text{ k}\Omega$. A coupling transformer provides an impedance match to the following transistor. What primary inductance will be required if $f_1 = 30$ Hz? *Answer:* $L = 53$ H.

PROBLEM 8.4 Approximately what primary inductance will be required for the transformer of Problem 8.3 if a 10 percent sag is permitted for a rectangular pulse of 1 msec?

8.3 THE HIGH-FREQUENCY RANGE

The leakage inductance cannot be neglected at high frequencies or large values of s. The shunt capacitance of the transistor or tube, as well as distributed capacitance of the transformer, needs to be considered. An equivalent circuit which includes the circuit capacitance is given in Fig. 8.5. In this circuit, the capacitance C_o includes the distributed capacitance of the primary, and C_i includes the distributed capacitance of the secondary. As a word of caution, considerable inaccuracy results

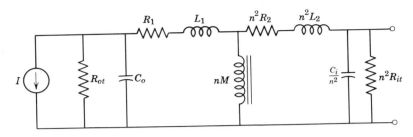

FIGURE 8.5
High-frequency equivalent circuit for a transformer-coupled amplifier.

from representing distributed capacitance by a lumped capacitor. However, fairly good qualitative ideas may be gained concerning the transformer performance. If a highly accurate equivalent circuit were devised, the circuit complexity would present a formidable solution. The reactance of the magnetizing inductance nM is so large at high frequencies that it can be removed from the equivalent circuit. Consequently, the leakage inductance n^2L_2 can be combined with L_1 to produce a total leakage inductance $2L_1$. This follows because

$$L_2 = (1 - k)L_{22} \tag{8.25}$$

or

$$n^2L_2 = n^2(1 - k)L_{22} \tag{8.26}$$

but since $n^2L_{22} = L_{11}$,

$$n^2L_2 = (1 - k)L_{11} = L_1 \tag{8.27}$$

The simplified equivalent circuit is presented in Fig. 8.6a. A further simplification is shown in Fig. 8.6b. In this circuit, the secondary resistance referred

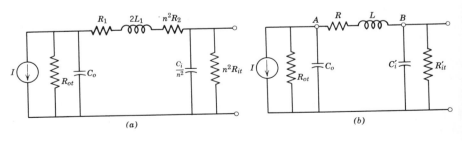

(a)   (b)

FIGURE 8.6
Simplified high-frequency equivalent circuits of the transformer-coupled amplifier.

to the primary $n^2 R_2$ has been lumped with the primary ohmic resistance to form R. To simplify the symbolism, the symbol C_i/n^2 has been changed to C_i' and $n^2 R_{it}$ has been changed to R_{it}'. The nodal equations for the circuit of Fig. 8.6b are

$$\left(G_{ot} + sC_o + \frac{1}{R + sL} \right) V_A - \frac{1}{R + sL} V_B = -I$$

$$-\frac{1}{R + sL} V_A + \left(G_{it}' + sC_i' + \frac{1}{R + sL} \right) V_B = 0 \tag{8.28}$$

where

$$G_{ot} = 1/R_{ot} \text{ and } G_{it}' = 1/R_{it}'$$

Solving for V_B, we find that

$$V_B = \frac{-I/(R + sL)}{\left(G_{ot} + sC_o + \dfrac{1}{R + sL} \right)\left(G_{it}' + sC_i' + \dfrac{1}{R + sL} \right) - \dfrac{1}{(R + sL)^2}} \tag{8.29}$$

Expanding the denominator, we have

$$V_B = \frac{-I/(R + sL)}{(G_{ot} + sC_o)(G_{it}' + sC_i') + \dfrac{G_{ot} + sC_o + G_{it}' + sC_i'}{R + sL}} \tag{8.30}$$

The transfer function of the transformer is

$$G = \frac{V_B}{I} = \frac{-1}{(G_{ot} + sC_o)(G_{it}' + sC_i')(R + sL) + (C_o + C_i')s + G_{ot} + G_{it}'} \tag{8.31}$$

$$G = \frac{-\dfrac{1}{C_o C_i' L}}{\left(s + \dfrac{G_{ot}}{C_o} \right)\left(s + \dfrac{G_{it}'}{C_i'} \right)\left(s + \dfrac{R}{L} \right) + \dfrac{C_o + C_i'}{C_o C_i' L}\left(s + \dfrac{G_{ot} + G_{it}'}{C_o + C_i'} \right)} \tag{8.32}$$

Equation 8.32 is a third-order equation.

This third-order equation can be solved by conventional means if the coefficients are in numerical form. However, if a general solution is required, all possible numerical answers must be found. The root-locus technique allows us to make a plot of all possible roots in the denominator. This root-locus method is briefly presented in Appendix II.

Equation 8.32 may be arranged in proper form for root-locus solution as follows.

$$G = \cfrac{-\cfrac{1}{C_o C_i' L \left(s + \cfrac{G_{ot}}{C_o}\right)\left(s + \cfrac{G_{it}'}{C_i'}\right)\left(s + \cfrac{R}{L}\right)}}{1 + \cfrac{C_o + C_i'}{C_o C_i' L} \cdot \cfrac{s + G_{ot} + G_{it}'/(C_o + C_i')}{\left(s + \cfrac{G_{ot}}{C_o}\right)\left(s + \cfrac{G_{it}'}{C_i'}\right)\left(s + \cfrac{R}{L}\right)}} \tag{8.33}$$

Equation 8.33 is of the form

$$G = \frac{N}{1 + HF} \tag{8.34}$$

where N is the numerator, $H = (C_o + C_i')/(C_o C_i' L)$ and F is the fraction in the denominator which involves s.

In applying the root-locus technique, the poles and zeros of F are first plotted on the s plane as shown in Fig. 8.7. Then the locus of all values of s which cause the angle of F to be $180°$ is drawn as indicated in Fig. 8.7. Again, the object is to factor the denominator of Eq. 8.32. The roots of the denominator occur at values of s which make $HF = 1/180°$. Consequently, a value of s must be found for which $H = 1/|F|$ on each

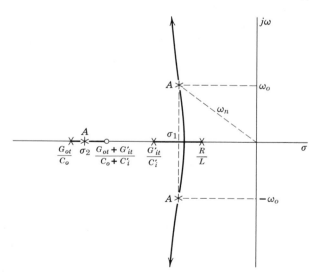

FIGURE 8.7
The root-locus plot of the denominator of Eq. 8.33.

branch of the locus. The spirule, which is a device especially designed for the solution of root-locus problems, may be used to locate the point for which $H = (C_o + C_i')/(C_o C_i' L)$ on each branch of the locus.

Each root-locus branch begins on a pole of F and terminates on a zero of F, either finite or at infinity. Thus, there are as many root-locus branches as there are poles of F. In this figure, there are three branches. Since the value of H must be zero at a pole of F and infinite at a zero of F, the values of s which are the roots of the denominator of Eq. 8.32 move away from the poles of F as H increases. Therefore, when H is small, all the roots lie on the negative real axis. The time response, then, which is obtained by taking the inverse transform of Eq. 8.32, would consist of exponential functions. These functions are of the same form as those obtained for the RC-coupled amplifier and result in the same general type of response as the RC-coupled amplifier. In order for H to be small, however, the leakage inductance L and the effective winding capacitances C_o and C_i' must be large. But C_o, C_i', and L, in conjunction with their associated resistances (or conductances), determine the location of the poles of F. To have good high-frequency response or short rise times, the poles of the transfer function and hence the poles of F must occur at large values of s. Then C_o, C_i', and L should each be small to provide good frequency and time response. This response criterion usually leads to values of H which place the poles of the transfer function (Eq. 8.33) off the real axis on the two branches which depart from the real axis and seek zeros at infinity. A typical location of the poles of the transfer function might be at the points labeled A in Fig. 8.7. The response here is said to be underdamped in contrast to the overdamped case in which all the poles are located on the real axis. Critical damping occurs when a double pole is located at the point where the root-locus branches depart from the real axis.

In the usual underdamped case, the transfer function of the transformer-coupled amplifier may be written as

$$G = \pm K \frac{1/C_o C_i' L}{(s + \sigma_1 + j\omega_o)(s + \sigma_1 - j\omega_o)(s + \sigma_2)} \tag{8.35}$$

where σ_1, ω_o, and σ_2 are obtained from Fig. 8.7

$$G = \pm K \frac{1/C_o C_i' L}{(s + \sigma_2)(s^2 + 2\sigma_1 s + \sigma_1^2 + \omega_o^2)} \tag{8.36}$$

Equation 8.36 may be written in the form

$$G = \pm K \frac{1/C_o C_i' L}{(s + \sigma_2)(s^2 + 2\zeta\omega_n s + \omega_n^2)} \tag{8.37}$$

where ω_n is the natural resonant frequency $(\omega_o^2 + \sigma_1^2)^{1/2}$ and ζ (zeta) is the damping ratio $= \sigma_1/\omega_n$.

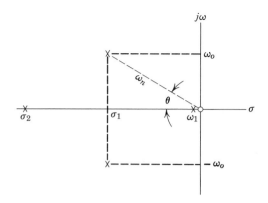

FIGURE 8.8
**The complete pole-zero plot of a typical trans-
former coupled amplifier.**

The pole-zero plot of a typical transformer-coupled amplifier, including the low-frequency pole and zero, is shown in Fig. 8.8. Since the damping ratio $\zeta = \sigma_1/\omega_n$, the cosine of the angle θ is also equal to ζ. The complete transfer function of the transformer-coupled amplifier can be written from inspection of Eqs. 8.24, 8.37, and Fig. 8.8.

$$G = \pm K \frac{s/C_o C_i' L}{(s + \omega_1)(s + \sigma_2)(s^2 + 2\zeta\omega_n s + \omega_n{}^2)} \tag{8.38}$$

The factors ζ and ω_n primarily determine the high-frequency or rise-time response characteristics of the transformer. The front edge or rise response to a step input voltage is plotted in Fig. 8.9a for various values of ζ, with ω_n held constant. The response is plotted for various values of ω_n in Fig. 8.9b with ζ held constant. The frequency response of the amplifier is presented in Fig. 8.10 for various values of ζ. It is evident from the

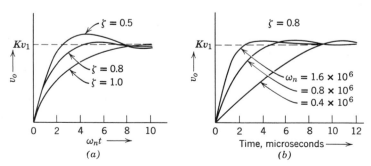

FIGURE 8.9
Front edge response of a transformer-coupled amplifier.

figures that a high resonant frequency, ω_n, will provide a short rise time and good high-frequency response. The high value of ω_n can be accomplished by providing small values of C_o, C_i', and L. For a specific value of coupling, the leakage inductance is proportional to the primary inductance which is set by the value of R_{ot} and the desired ω_1, as previously discussed. For a specific value of primary inductance, the leakage inductance can be reduced by increasing the coefficient of coupling k. High-permeability cores and special winding techniques are frequently employed in high-quality transformers to provide very tight coupling.

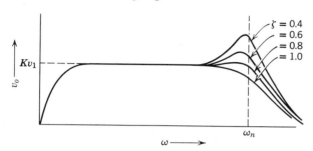

FIGURE 8.10
Frequency response of a transformer-coupled amplifier.

The distributed capacitance of the windings increases with the number of turns. Consequently, a step-up transformer would have a comparatively large secondary capacitance which is divided by n^2 in referring it to the primary. Thus C_i' may become very large if a large step-up turns ratio is used. As previously discussed, ω_n decreases as C_i' increases. As a consequence, the step-up ratio of the transformer is restricted by the acceptable value of resonant frequency. Usually, a step-up ratio of 3 is about the maximum value for a good quality transformer. In contrast, C_i' may be very small when a step-down transformer is used. Reducing the reluctance of the magnetic path is helpful in reducing winding capacitance, because the number of turns would be reduced for a specific value of inductance. In addition, special-winding techniques may also be used to minimize the distributed capacitance. The transformer connections recommended by the manufacturer must be used to realize minimum winding capacitance.

From Figs. 8.9 and 8.10 it is evident that a value of ζ much lower than 0.8 would cause an appreciable overshoot when the input is a step function, or an appreciable peak in the steady-state frequency response. Of course, small values of leakage inductance are helpful in producing large values of ζ. In addition, the resistance of the windings may be used as a control parameter. However, an increased winding resistance may result in a reduced efficiency.

PROBLEM 8.5   The gain of a transformer-coupled amplifier is

$$G_v = \frac{10^{18}}{(s + 400)(s + 10^6)(s^2 + 1.6 \times 10^5 s + 10^{10})}$$

Make a sketch of gain as a function of frequency for this amplifier.

PROBLEM 8.6   A 2N3703 transistor is connected as shown in Fig. 8.11a. The collector characteristics are given in Fig. 8.11b. The desired q point occurs when $i_C = 5$ mA.

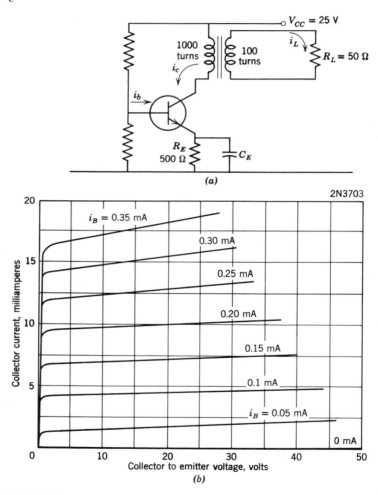

(a)

(b)

FIGURE 8.11
The circuit and characteristic curves to be used in Problem 8.6.

a. Plot the dc load line and the *q* point on the characteristic curves. Assume R_1 of the transformer is much less than 500 Ω.

b. Determine the effective ac load impedance connected to the collector circuit. Draw the ac load line on the characteristic curves.

c. Determine the current gain $K_{i1} = i_L/i_b$.

d. Determine the magnitude of primary inductance, L_1, if the desired value of ω_1 is 400 rad/sec.

PROBLEM 8.7   Design a transformer-coupled amplifier stage which uses a 2N3903 transistor. Use a 12 V supply for V_{CC}, and bias your transistors at a *q* point of $I_C = 5$ mA. The desired stability factor is $S_I = 50$ for this amplifier. The transformer has 20,000 Ω impedance on the primary side if a resistor of 1,200 Ω is connected across the secondary. Design your amplifier so maximum power will be delivered to a resistive load connected to the secondary winding of the transformer.

a. Draw the circuit diagram of your amplifier.

b. Find the values of all resistors and capacitors in your circuit if the low cutoff frequency ω_1 is assumed to be 400 rad/sec.

c. Determine the mid-band current gain and voltage gain of your amplifier.

d. Verify the mid-band current and voltage gain of your amplifier in the laboratory.

e. Also, experimentally determine ω_1, ω_2, and ζ (the damping factor) for this transformer, if possible.

PROBLEM 8.8   A 2N3903 transistor is connected as shown in Fig. 8.12. The value of i_C at the *q* point is 5 mA.

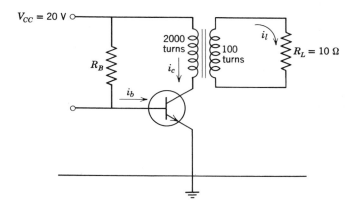

FIGURE 8.12
The configuration for Problem 8.8.

a.   Draw the dc load line on the characteristic curve and indicate the q point. Assume R_1 is very small.

b.   Determine the ac load impedance connected to the collector of the transistor and draw the ac load line on the characteristic curves.

c.   Determine the value of R_B.

d.   What is the current stability factor S_I for this amplifier?

e.   Determine the current gain (i_L/i_b) for this amplifier.

9

Small-Signal Tuned Amplifiers

The need frequently arises for an amplifier that will amplify only those frequencies which lie within a given frequency range or band. This type of amplifier is known as a tuned amplifier or band-pass amplifier. Radio and television receivers, for example, use tuned amplifiers to select one radio signal from the many which are being broadcast. Several types of tuned amplifiers are discussed in this chapter. The gain and bandwidth of each type will be of interest. It will be assumed, in the discussion of tuned amplifiers, that the input signal is a modulated signal which has a basic frequency ω_o and that the amplifier is tuned to this frequency. The required bandwidth will depend upon the characteristics of the modulation. As we will show in Chapter 20, the required bandwidth for an amplitude-modulated signal is two times the highest modulating frequency of interest. Pulse-modulated signals require a bandwidth, in Hertz, approximately twice the reciprocal of the pulse duration, or $b = 2/t_d$. Bandwidths for frequency or phase-modulated signals will be discussed in Chapter 21.

9.1 SINGLE-TUNED, CAPACITIVELY-COUPLED AMPLIFIERS

A tuned FET amplifier is shown in Fig. 9.1. Observe that the basic difference between this amplifier and the RC-coupled amplifier is the tuned circuit which replaces the resistor in the drain circuit. In order to determine the voltage gain and the bandwidth of the amplifier, the equivalent circuit of Fig. 9.2 is given. The capacitor C includes the shunt capacitance of the amplifiers and the distributed wiring capacitance. This absorption of the

283

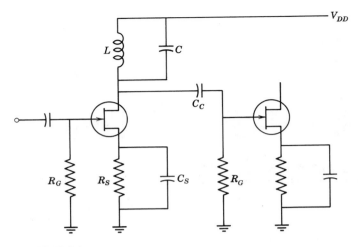

FIGURE 9.1
A tuned FET amplifier.

shunt capacitance into the tuned circuit makes possible the amplification of very high frequencies.

The equivalent circuit of Fig. 9.2 would have only parallel elements and, hence, the analysis would be comparatively easy except for the series resistance of the coil R_{ser}. Therefore, we will determine a parallel combination of resistance and inductance which will have the same impedance over the pass band as the series combination. This will be accomplished by finding the admittance of the series combination.

$$Y = \frac{1}{Z} = \frac{1}{R_{ser} + j\omega L} \qquad (9.1)$$

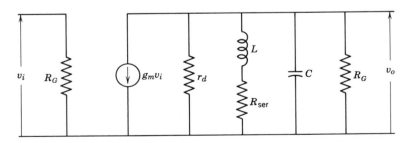

FIGURE 9.2
An equivalent circuit for the tuned amplifier of Fig. 9.1.

Rationalizing,

$$Y = \frac{R_{ser} - j\omega L}{(R_{ser} + j\omega L)(R_{ser} - j\omega L)} = \frac{R_{ser} - j\omega L}{R_{ser}^2 + (\omega L)^2} \tag{9.2}$$

But the ratio of ωL to R_{ser} is the Q of the coil, known as Q_o. Therefore, if Q_o is ten or higher, which is usually the case, $(\omega L)^2$ is at least 100 times R_{ser}^2, so this latter term can be neglected in the denominator of Eq. 9.2. Using this simplification,

$$Y \simeq \frac{R_{ser} - j\omega L}{(\omega L)^2} = \frac{R_{ser}}{(\omega L)^2} - j\frac{1}{\omega L} \tag{9.3}$$

This admittance is of the form $G + jB$ and represents a conductance $G = R_{ser}/(\omega L)^2$ in parallel with an inductive susceptance $B = 1/(\omega L)$, as shown in the dashed enclosure in the equivalent circuit of Fig. 9.3. Since the Q of a coil can be written as $Q_o = \omega L/R_{ser}$, the conductance of the parallel combination representing the coil can be written

$$G_p = \frac{R_{ser}}{(\omega L)^2} = \frac{1}{Q_o \omega L} \tag{9.4}$$

This conductance may also be expressed as a resistance

$$R_{par} = \frac{1}{G_p} = \frac{(\omega L)^2}{R_{ser}} = Q_o \omega L \tag{9.5}$$

Observe that this equivalent parallel resistance is a function of frequency. However, the tuned amplifier which uses a high Q circuit $(Q \geq 10)$ amplifies only a narrow band of frequencies near the resonant frequency ω_o. Therefore, the effective parallel resistance of the coil can be assumed constant over the pass band with the value

$$R_{par} = Q_o \omega_o L \tag{9.6}$$

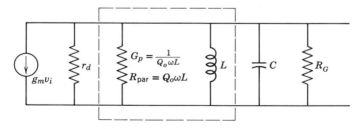

FIGURE 9.3
An equivalent circuit containing only parallel elements.

Note that the Q_o of the coil is the ratio $R_{par}/\omega_o L$ and that small values of series resistance R_{ser} which provide high Q_o give large values of effective parallel resistance R_{par} because R_{par} is proportional to the Q_o of the coil. In fact, rearranging Eq. 9.6 yields $Q_o = R_{par}/\omega_o L$. In order to emphasize this behavior, let us consider a numerical example.

Example 9.1   A given coil has an inductance of 0.1 H and a series resistance of 20 Ω. This coil is connected in parallel with a 0.1 μF capacitor. Determine the parameters of an equivalent parallel circuit near resonance.

We will assume that we have a high Q circuit ($Q = 10$ or more). Then, $\omega_o = 1/(LC)^{1/2} = 1/(10^{-1} \times 10^{-7})^{1/2} = 10^4$ rad/sec. The value of $\omega_o L$ is $10^4 \times 10^{-1} = 1000\ \Omega$. The Q_o of the coil is $Q_o = \omega_o L/R_{ser} = 1000/20 = 50$. From Eq. 9.3, the size of the inductor in the parallel circuit is the same as the inductor in the series circuit. However, from Eq. 9.6, $R_{par} = Q_o \omega_o L = 50 \times 1000 = 50{,}000\ \Omega$. Thus, the equivalent parallel circuit is a 0.1 μF capacitor in parallel with a 0.1 H coil which is also in parallel with a 50,000 Ω resistor.

Note that if the series resistance of the coil is reduced to 5 Ω, the Q_o becomes 200 and the equivalent parallel resistance becomes 200,000 Ω.

PROBLEM 9.1   A 1 mH coil with $Q_o = 100$ is connected in parallel with a 1000 pF capacitor. Determine ω_o and R_{par}.

Answer: $\omega_o = 10^6$ rad/sec, $R_{par} = 10^5\ \Omega$.

The effective parallel resistance of the coil can be combined with the other parallel resistance elements in the equivalent circuit of Fig. 9.3 to produce the simplified equivalent circuit of Fig. 9.4 where R represents the parallel combination of r_d, R_{par} and R_G of Fig. 9.3. In calculating Q_o (the coil Q) we have used $R_{par}/\omega_o L$ where R_{par} accounts for the energy loss in the coil. However, the characteristics of a tuned circuit depend upon the energy loss of the *entire* circuit. Thus, a *circuit Q* will be defined as

$$Q = \frac{R}{\omega_o L} \tag{9.7}$$

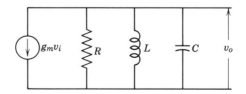

FIGURE 9.4
A simplified equivalent circuit for a tuned amplifier.

where R is the effective total shunt resistance in parallel with L and C as shown in Fig. 9.4.

The symbol Q, given without subscripts, will always represent the *circuit Q* in this discussion. The circuit Q is a very important parameter in a tuned circuit because it determines the bandwidth and affects the amplifier gain. Since the impedance of a lossless parallel tuned circuit is infinite at the resonant frequency, Fig. 9.4 shows that at resonance the total load impedance is R and the output voltage v_o of the FET amplifier is

$$v_o = -g_m v_i R \qquad (9.8)$$

Then, the voltage gain at resonance is

$$G_v = \frac{v_o}{v_i} = -g_m R \qquad (9.9)$$

Observe that the voltage gain is proportional to R which is equal to $Q\omega_o L$ at the resonant frequency. Thus, the gain is proportional to both the circuit Q and the inductive reactance of the coil. Let us consider an example.

Example 9.2 A circuit is connected as shown in Fig. 9.1. The inductance L has a value of 1 mH and a Q_o of 100. The capacitance is 1000 pF and $R_G = 1$ MΩ. The FET has $r_d = 2.5 \times 10^5$ Ω and $g_m = 2 \times 10^{-3}$ mhos. Determine the voltage gain of this circuit.

As noted in Prob. 9.1, the value of ω_o is 10^6 rad/sec and $R_{par} = 100,000$ Ω. The parallel combination of R_{par} (100,000 Ω), r_d (2.5 $\times$ 10^5 Ω), and R_G (10^6 Ω) is $R = 67$ kΩ. Then, the circuit $Q = R/\omega_o L = 6.7 \times 10^4/10^6 \times 10^{-3} = 67$. Finally, the voltage gain at resonance (Eq. 9.9) is

$$G_v = -g_m R = -2 \times 10^{-3} \times 6.7 \times 10^4 = -134.$$

PROBLEM 9.2 In the tuned FET amplifier of Fig. 9.1, $L = 1$ mH, $Q_o = 100$, $C = 500$ pF and $R_G = 1$ MΩ. The FET has $g_m = 2 \times 10^{-3}$ and $r_d = 2.5 \times 10^5$. Determine the resonant frequency and the voltage gain. *Answer:* $f_o = 225$ kHz, $G_v = -167$.

The desired circuit Q is determined by the bandwidth requirement of the amplifier, however, and not by the desired voltage gain. The relationship between the bandwidth and the circuit Q can be obtained by writing a general expression for the output voltage v_o. From Fig. 9.4,

$$v_o = g_m v_i Z = \frac{-g_m v_i}{G + j\omega C + \dfrac{1}{j\omega L}} \qquad (9.10)$$

where $G = 1/R$. If we now multiply both numerator and denominator of the right-hand side of Eq. 9.10 by R and then divide both sides of the equation by v_i,

$$G_v = \frac{-g_m R}{1 + j\omega CR + \dfrac{R}{j\omega L}} \qquad (9.11)$$

But $R/\omega_o L = Q$ and similarly $R\omega_o C = Q$, since $\omega_o L = 1/\omega_o C$ at the resonant frequency ω_o. Making these substitutions into Eq. 9.11,

$$G_v = \frac{-g_m R}{1 + jQ\left(\dfrac{\omega}{\omega_o} - \dfrac{\omega_o}{\omega}\right)} \qquad (9.12)$$

Since the numerator in Eq. 9.12 is equal to the voltage gain at resonance, when $\omega = \omega_o$, the gain decreases to the half-power value when the magnitude of the j part of the denominator is equal to unity. But there are two frequencies at which half-power gain occurs, as shown in Fig. 9.5. These frequencies are designated as ω_L and ω_H. Thus, when ω is equal to ω_L (or ω_H), the magnitude of the j part of the denominator of Eq. 9.12 is equal to one.

$$Q\left(\frac{\omega_H}{\omega_o} - \frac{\omega_o}{\omega_H}\right) = 1 \qquad (9.13a)$$

$$Q\left(\frac{\omega_L}{\omega_o} - \frac{\omega_o}{\omega_L}\right) = -1 \qquad (9.13b)$$

and

$$\frac{\omega_H}{\omega_o} - \frac{\omega_o}{\omega_H} = \frac{\omega_o}{\omega_L} - \frac{\omega_L}{\omega_o} \qquad (9.14)$$

Rearranging terms, we have

$$\frac{\omega_H}{\omega_o} + \frac{\omega_L}{\omega_o} = \frac{\omega_o}{\omega_L} + \frac{\omega_o}{\omega_H} \qquad (9.15)$$

$$\frac{\omega_H + \omega_L}{\omega_o} = \omega_o \frac{\omega_H + \omega_L}{\omega_H \omega_L} \qquad (9.16)$$

Therefore,

$$\omega_o^2 = \omega_H \omega_L \qquad (9.17)$$

Again, using Eq. 9.13a, we find

$$Q\left(\frac{\omega_H^2 - \omega_o^2}{\omega_o \omega_H}\right) = 1 \qquad (9.18)$$

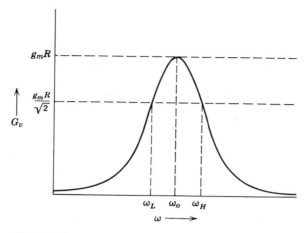

FIGURE 9.5
Frequency response of the single-tuned amplifier.

and substituting the value of $\omega_o{}^2 = \omega_H \omega_L$ from Eq. 9.17,

$$Q\left(\frac{\omega_H{}^2 - \omega_L \omega_H}{\omega_o \omega_H}\right) = Q\left(\frac{\omega_H - \omega_L}{\omega_o}\right) = 1 \qquad (9.19)$$

But $\omega_H - \omega_L$ is the bandwidth B, in radians per second, as seen from Fig. 9.5. Then,

$$B = \frac{\omega_o}{Q} \qquad (9.20)$$

Note that both sides of Eq. 9.20 may be divided by 2π to obtain the bandwidth in Hertz.

$$B \,(\text{Hertz}) = \frac{f_o}{Q} \qquad (9.21)$$

The circuit designer usually knows the desired bandwidth and his assignment is to produce an amplifier to meet the specification. Therefore, we will use this philosophy in the example to follow.

Example 9.3 Let us assume that we need to build an intermediate frequency amplifier with a resonant, or center, frequency $f_o = 455$ kHz. This amplifier is for use in a standard broadcast receiver so the desired bandwidth is 10 kHz. Therefore, the required circuit Q is 45.5. Field effect transistors with $r_d = 200$ kΩ at the desired q point are chosen for the amplifier. We will select $R_G = 10^6$ Ω and the shunt resistance, exclusive of the coil loss, is then 167 kΩ. We will call this part of the shunt resistance R_k (for R *known*). The effective parallel resistance of the coil is not yet known and cannot yet be determined from the product $Q_o \omega_o L$ because the required inductance L is not yet known, but needs to be determined. However, we may know from Q-meter measurements, or from experience, the typical value of coil Q_o at the desired resonant frequency. Let us assume $Q_o = 100$. We also have two ways of solving for the total shunt resistance R in terms of the inductance L. One relationship is, from Eq. 9.7,

$$R = Q\omega_o L \qquad (9.22)$$

Since R is the parallel combination of the known resistance R_k and $Q_o \omega_o L$, the other relationship is

$$R = \frac{R_k Q_o \omega_o L}{R_k + Q_o \omega_o L} \qquad (9.23)$$

Equations 9.22 and 9.23 can be solved simultaneously to obtain L as follows

$$Q\omega_o L = \frac{R_k Q_o \omega_o L}{R_k + Q_o \omega_o L} \qquad (9.24)$$

$$Q\omega_o L(R_k + Q_o \omega_o L) = R_k Q_o \omega_o L \qquad (9.25)$$

$$Q(R_k + Q_o \omega_o L) = R_k Q_o \qquad (9.26)$$

$$L = \frac{R_k(Q_o - Q)}{\omega_o Q_o Q} = \frac{R_k}{\omega_o}\left(\frac{1}{Q} - \frac{1}{Q_o}\right) \qquad (9.27)$$

For this design example with $Q_o = 100$, $Q = 45.5$, $R_k = 167$ kΩ and $\omega_o = 6.28 \times 4.55 \times 10^5$ rad/sec,

$$L = 1.67 \times 10^5(.022 - .01)/6.28 \times 4.55 \times 10^5 = 7 \times 10^{-4} \text{ H}.$$

The tuning capacitance can be determined from the relationship $C = 1/\omega_o^2 L = 1.75 \times 10^{-10}$ F. This tuning capacitance includes the output capacitance, C_o, of the driving transistor and the effective input capacitance, C_i, of the following transistor which, due to the Miller effect, may be a major part of the tuning capacitance. Methods of eliminating or greatly reducing this Miller capacitance, which varies with the voltage gain of the amplifier, will be discussed later. Either the inductance may be variable by the use of a slug-tuned core or the capacitance may be variable in order to tune the circuit to 455 kHz.

The voltage gain of the amplifier may be easily determined if the transconductance of the FET is known at the desired q point. Let us assume that $g_m = 10^{-3}$ mhos. Then the gain magnitude of our 455 kHz amplifier at resonance is $K_v = g_m R$. Since $R = Q\omega_o L = 45.5 \times 6.28 \times 4.55 \times 10^5 \times 7 \times 10^{-4} = 91$ kΩ, $K_v = 10^{-3} \times 91 \times 10^3 = 91$.

The design of a single-tuned capacitively coupled amplifier is summarized below

1.   The desired resonant frequency and bandwidth are assumed known, or have been determined from the signal frequency and modulation characteristics.
2.   The circuit Q is determined from the relationship $Q = f_o/B$ (Hz).
3.   The resistance R_k is determined as the parallel combination of the output resistance of the driving transistor and the input resistance of the following stage.
4.   The required tuning inductance L is determined from Eq. 9.27.
5.   The required tuning capacitance, including the output and input capacitances, is $C = 1/\omega_o{}^2 L$.
6.   The voltage gain $K_v = g_m R$ for a FET (or MOSFET or vacuum tube). The value of R is determined from Eq. 9.22 or Eq. 9.23.

This type of circuit is not used with bipolar transistors. The coupling circuits discussed in the following paragraphs will be more generally applicable to the bipolar types. Instability problems are being ignored here but will be discussed later.

PROBLEM 9.3   An *RF* amplifier is needed for a standard broadcast receiver. It is to have 20 kHz bandwidth at 1.0 MHz center frequency. Design a single-tuned capacitively-coupled amplifier for this frequency and bandwidth, using a FET with $g_m = 2 \times 10^{-3}$ mho, $R_o = 250$ kΩ, R_G (of the following transistor) $= 1.0$ MΩ, $C_o = 5$ pF and $C_i = 25$ pF for the transistors used, at the recommended q points. Determine the values for L, C and the voltage gain, if the Q_o of the coil is assumed to be 150.
Answer: $L = 422$ μH, $C = 59$ pF, $K_v = 264$.

9.2   TIME (OR TRANSIENT) RESPONSE OF TUNED AMPLIFIERS

The preceding work was based on the assumption that the amplifier and signal source had been turned on for a long time and the transient resulting from these events had decayed to zero. We will now consider the response of the tuned amplifier, as a function of time, when a step voltage is applied, which occurs when the amplifier is turned on, or when the RF input signal is switched on or off. These responses can be obtained, as discussed in Chapter 7, by writing the amplifier transfer function, or gain, as a function of s instead of $j\omega$ and then obtaining the output voltage (or current) as the product of the input voltage (or current) and the transfer function, all being in the s domain. The inverse transform then yields the output as a function of time.

The general transfer function can be written by replacing $j\omega$ in Eq. 9.11 with s. Then,

$$G_v(s) = \frac{-K_v}{1 + sCR + \dfrac{R}{sL}} \tag{9.28}$$

where $K_v = g_m R_L$, the reference gain. Equation 9.28 can be written in better form as follows.

$$G_v(s) = -\frac{sK_v}{s + s^2 CR + \dfrac{R}{L}} = \frac{-K_v}{RC}\frac{s}{s^2 + \dfrac{s}{RC} + \dfrac{1}{LC}} \tag{9.29}$$

Equation 9.29 may be written in the standard form

$$G_v(s) = \frac{-K_v}{RC}\frac{s}{s^2 + 2\zeta\omega_n s + \omega_n^{\,2}} \tag{9.30}$$

where $\omega_n = 1/\sqrt{LC}$ is the undamped resonant frequency (rad/sec) and ζ is the damping ratio. But, since $2\zeta\omega_n = 1/RC$,

$$\zeta = \frac{1}{2R\omega_n C} = \frac{\omega_n L}{2R} = \frac{1}{2Q} \tag{9.31}$$

The output voltage in the s domain is, using Eq. 9.30,

$$V_o(s) = \frac{-K_v}{RC}\frac{sV_i(s)}{s^2 + 2\zeta\omega_n s + \omega_n^{\,2}} \tag{9.32}$$

We will now assume that a negative step voltage of magnitude V is appplied to the input at time $t = 0$, so $V_i(s) = -V/s$. Then

$$V_o(s) = \frac{K_v}{RC}\frac{V}{s^2 + 2\zeta\omega_n s + \omega_n^{\,2}} \tag{9.33}$$

The denominator of Eq. 9.33 could be factored and partial fraction expansion used to obtain the inverse transform. However, the solution may be more easily obtained by arranging Eq. 9.33 into the following form.

$$V_o(s) = \frac{K_v}{RC}\frac{V}{(s + \zeta\omega_n)^2 + \omega_n^{\,2}(1 - \zeta^2)} \tag{9.34}$$

or

$$V_o(s) = \frac{K_v V}{\omega_o RC}\frac{\omega_o}{(s + \zeta\omega_n)^2 + \omega_o^{\,2}} \tag{9.35}$$

where $\omega_o = \omega_n\sqrt{1 - \zeta^2}$ is the *damped* resonant frequency and is the frequency at which maximum output voltage is obtained in a high Q circuit.

We find from a table of Laplace transform pairs that the inverse transform of Eq. 9.35 is

$$v_o(t) = \frac{K_v V}{\omega_o RC} e^{-\zeta\omega_n t} \sin \omega_o t \qquad (9.36)$$

Observe that the output voltage is an exponentially decaying sinusoid, as shown in Fig. 9.6. The time constant of the decay is $2RC$ in contrast with the usual time constant RC obtained when a circuit contains resistance and capacitance only. The oscillatory sinusoidal transient is known as *ringing* and is sometimes used to provide frequency multiplication, time markers for oscilloscope tracings, and so on.

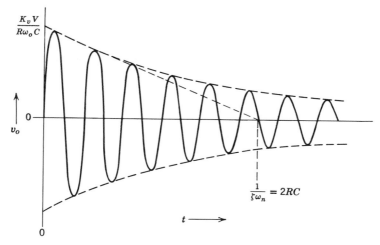

FIGURE 9.6
The time response of a typical tuned amplifier to a negative step-voltage input of magnitude V.

The pole-zero plot of the transfer function given in Fig. 9.7 gives some insight into the transient characteristics of the tuned circuit. The distance from the origin to either of the complex poles is ω_n. The maximum steady-state response occurs at ω_o which is the point on the $j\omega$ axis nearest the complex pole. The decay rate of the transient is dependent upon $\zeta\omega_n$, which is the real part of s, or σ. The damping ratio $\zeta = \cos \theta$, as seen from Fig. 9.7.

We will now suddenly apply a sinusoidal voltage $v_i = V \sin \omega_o t$ at $t = 0$ to the input of the amplifier which is tuned to ω_o rad/sec and determine

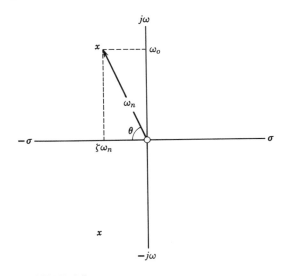

FIGURE 9.7
The pole-zero plot of the transfer function of a tuned amplifier.

the time response of the output voltage. The s-domain input voltage, from a table of transform pairs, is $V_i(s) = V\omega_o/(s^2 + \omega_o^2)$, so the s-domain output voltage is, using Eq. 9.32,

$$V_o(s) = \frac{-VK_v}{RC} \frac{\omega_o s}{(s^2 + \omega_o^2)(s^2 + 2\zeta\omega_n s + \omega_n^2)} \tag{9.37}$$

Partial fraction expansion can be used to reduce Eq. 9.37 to a sum, instead of a product, so a transform table can be used to obtain the inverse transform. Using this technique and the form of Eq. 9.35,

$$V_o(s) = \frac{-VK_v}{RC}\left[\frac{As + B}{s^2 + \omega_o^2} + \frac{Cs + D}{(s + \zeta\omega_n)^2 + \omega_o^2}\right] \tag{9.38}$$

The arbitrary constants A, B, C, and D may be evaluated by equating the numerators of Eqs. 9.37 and 9.38 (after the equations have been reduced to a common denominator).

$$(As + B)[(s + \zeta\omega_n)^2 + \omega_o^2] + (Cs + D)(s^2 + \omega_o^2) = \omega_o s \tag{9.39}$$

Expanding, we have

$$As^3 + 2A\zeta\omega_n s^2 + A\zeta^2\omega_n^2 s + A\omega_o^2 s + Bs^2 + 2B\zeta\omega_n s + B\zeta^2\omega_n^2$$
$$+ B\omega_o^2 + Cs^3 + C\omega_o^2 s + Ds^2 + D\omega_o^2 = \omega_o s \tag{9.40}$$

Equating the coefficients of equal power of s gives

$$A + C = 0 \tag{9.41}$$

$$2A\zeta\omega_n + B + D = 0 \tag{9.42}$$

$$A\zeta^2\omega_n^2 + A\omega_o^2 + 2B\zeta\omega_n + C\omega_o^2 = \omega_o \tag{9.43}$$

$$B\zeta^2\omega_n^2 + B\omega_o^2 + D\omega_o^2 = 0 \tag{9.44}$$

Some simplifications can be made by recognizing that $\zeta^2\omega_n^2$ is very small in comparison with ω_o^2 when the circuit Q is 10 or higher as previously assumed. In this case, from Eq. 9.44, $D \simeq -B$. Then from Eq. 9.42, $A \simeq 0$ and from Eq. 9.41, $C \simeq 0$. Substituting these values into Eq. 9.43, we have

$$2B\zeta\omega_n \simeq \omega_o \tag{9.45}$$

$$B \simeq \frac{\omega_o}{2\zeta\omega_n} \quad \text{and} \quad D \simeq -B \simeq -\frac{\omega_o}{2\zeta\omega_n} \tag{9.46}$$

Equation 9.38 then becomes

$$V_o = -\frac{VK_v}{2\zeta\omega_n RC}\left(\frac{\omega_o}{s^2 + \omega_o^2} - \frac{\omega_o}{s + \zeta\omega_n^2 + \omega_o^2}\right) \tag{9.47}$$

The inverse Laplace transform of Eq. 9.47 yields v_o in the time domain.

$$v_o = -\frac{K_v V}{2\zeta\omega_n RC}(\sin \omega_o t - e^{-\zeta\omega_n t}\sin \omega_o t) \tag{9.48}$$

From the substitution of $2\zeta\omega_n = 1/RC$,

$$v_o = -K_v V(1 - e^{-t/2RC})\sin \omega_o t \tag{9.49}$$

A sketch of the response which is expressed by Eq. 9.49 is given in Fig. 9.8a along with a sketch of the sinusoidal excitation. In addition, the response to an interrupted or pulse-modulated carrier is given in Fig. 9.8b. In this case, the exponential decay of the output following the cessation of the input was deduced from the known transient response of the RLC circuit. The form of this transient response was obtained from the first example in this section in which the excitation was a step voltage (see Fig. 9.6).

The desirability of a tuned amplifier, like other amplifiers, is measured by both the amplification and the preservation of the waveform of the excitation. As shown in Fig. 9.8, the tuned amplifier cannot faithfully follow instantaneous changes in the excitation amplitude. In other words, the tuned amplifier has a rise and decay time which is very similar to the RC-coupled amplifier counterpart. Two essential differences exist between

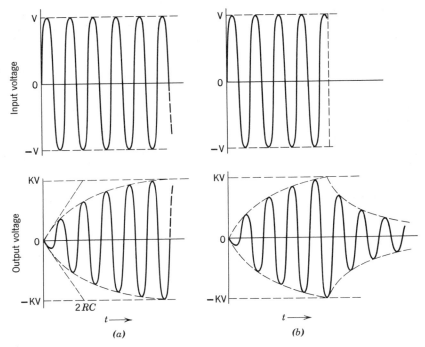

FIGURE 9.8
A sketch of the response of a tuned amplifier to (a) sinusoidal excitation of frequency ω_o, switched on at time = 0; (b) interrupted or pulse-modulated carrier excitation of frequency ω_o.

the tuned amplifier and the untuned or video amplifier. First, in the case of the tuned amplifier, it is the *envelope* of the output signal and not the individual cycle which rises and decays exponentially when the sinusoidal excitation is instantly started or stopped. Second, the time constant of the envelope rise or decay is $2RC$ in contrast to the time constant RC of the untuned or video amplifier.

Again, as in the untuned amplifier, the tuned amplifier would ideally have a very high gain and a very small rise time. But the reference gain and the envelope rise time are both proportional to the shunt resistance R. Therefore, these two criteria are in conflict. A figure of merit F_a for the amplifier might be

$$F_a = \frac{\text{reference gain}}{\text{envelope rise time}} \tag{9.50}$$

In the RC-coupled amplifier, the rise time (10–90 percent) was found to be $2.2\,RC$. This rise time was discussed in Chapter 7. Then, since the time constant of the envelope rise is $2RC$, or double that of the RC amplifier, the 10–90 percent envelope rise time is 4.4 RC. Using this rise time, we can find that the figure of merit of the tuned amplifier is

$$F_a = \frac{g_m R}{4.4RC} = \frac{g_m}{4.4C} \qquad (9.51)$$

Equation 9.51 shows that the ratio of gain to rise time improves as the shunt capacitance C decreases. Therefore, the maximum gain to rise-time ratio would be obtained when only the stray circuit capacitance is used to tune the circuit. However, this stray capacitance usually varies appreciably with temperature, amplifier gain, or supply voltage (for a transistor). Thus the resonant frequency of the amplifier may drift appreciably unless some fixed, stable capacitance is included in the tuning capacitance.

PROBLEM 9.4 The tuned amplifier of Problem 9.3 with $f_o = 1$ MHz, $B = 20$ kHz, $g_m = 2 \times 10^{-3}$ mho, $R = 132$ kΩ, $C = 59$ pF, and $L = 422$ μH has a 1 mV, 1 MHz sinusoidal signal switched into the input at time $t = 0$ and then turned *off* again at $t = 1$ ms. Determine

a. The envelope rise and fall time (10–90 percent).
b. The number of cycles produced by the tuned output circuit during the rise time.
c. The damping ratio ζ.
d. The frequency difference between ω_o and ω_n.

9.3 *Y*-PARAMETER CIRCUITS FOR RF AMPLIFIERS

In the preceding chapters of this text, *h*-parameter models have been used primarily to analyze low-frequency or AF circuits, and the hybrid-π model has been used to advantage in the design and analysis of broadband or video amplifiers. The *y*-parameter model in a special, simplified form has been used for FETs and vacuum tubes. The *y*-parameter circuit is also a convenient model for analyzing and designing tuned RF amplifiers using bipolar transistors for the following reasons.

1. The stability of a tuned amplifier can be easily predicted and controlled with the aid of the *y*-parameter model.
2. The comparatively small load and driving source resistances commonly used in RF amplifiers, for stability purposes, cause the actual input and output admittances of the transistor to closely approach its *y*-parameter value. Thus good approximate designs can be accomplished without actually solving the *y*-parameter circuit.
3. Transistor manufacturers frequently provide the *y* parameters for their transistors which are designed for RF use.

Therefore, y parameters will be used to analyze and design RF amplifiers in this chapter.

Figure 9.9 shows a y-parameter circuit and a hybrid-π circuit for comparison purposes. As a brief review of the y parameters in Chapter 4, the definitions of the y parameters are stated below.

y_{ie} is the input admittance with the output shorted. Thus y_{ie}, for a bipolar transistor, is the reciprocal of h_{ie} and h_{ie} is r_b plus the impedance of r_π, C_1, and C_{cb} in parallel. Both h_{ie} and y_{ie} are complex, of course.

y_{fe} is the ratio of short-circuit output current to input voltage. From the hybrid-π circuits, the shorted output current is $g_m v_{be}' - j\omega C_{cb} v_{be}'$, thus $y_{fe} = (g_m - j\omega C_{cb})v_{be}'/v_i$, with $R_L = 0$.

y_e is the output admittance with the input shorted. The real part of y_{oe} is somewhat smaller than h_{oe} for a bipolar transistor. y_{oe} may be found by

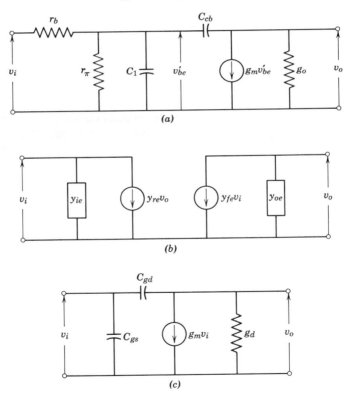

(a)

(b)

(c)

FIGURE 9.9

(a) A hybrid-π model, (b) a y-parameter model, (c) a high-frequency FET model.

applying a voltage v_o to the output, with the input shorted, and finding the total current i_o flowing into the output terminals. This current is the sum of that through g_o, C_{cb} and $g_m v_{be}'$. Then $y_{oe} = i_o/v_o$.

y_{re} is the ratio of the current flowing in the short-circuited input terminals to the voltage applied to the output terminals. Thus, $y_{re} \simeq -j\omega C_{cb}$ except for very high frequencies (approaching f_t) because $x_c \gg r_b$. The sign is negative because the current flows out of the shorted base terminal.

The y parameters of a series of RF transistors (2N4957–2N4959) are given in Fig. 9.10. The real and j parts are given separately as functions of I_C for $f = 450$ MHz. These parameters will be used in examples and problems in the remaining paragraphs in this chapter. Of course y parameters can be calculated for any specific transistor at a given q point and frequency providing enough data are given to construct a hybrid-π model. The y parameters for a FET or vacuum tube are easily obtained by inspection of the high-frequency model (Fig. 9.9c).

PROBLEM 9.5 A 2N4959 has $h_{fe} = 40$, $r_b = 50\ \Omega$, $C_{cb} = 0.4$ pF, and $f_t = 1500$ MHz at $I_C = -2.0$ mA, $V_{CE} = 10$ V. Determine the hybrid-π parameters and then calculate the common-emitter y parameters, approximately, for this transistor at the given q point and $f = 200$ MHz. Compare your parameters with those published by the manufacturer at the given frequency and q point (Fig. 9.10).

PROBLEM 9.6 Write the y parameters, in terms of the circuit elements, for a FET. The model is given in Fig. 9.9c.

9.4 INDUCTIVELY-COUPLED TUNED AMPLIFIERS

The capacitively-coupled tuned circuit of Section 9.1 did not permit impedance transformation. Therefore, field-effect transistors were used because they have high input impedance as well as high output impedance and do not need an impedance-transform type of coupling circuit. However, bipolar transistors can give considerably higher gain if the coupling circuit can provide impedance transformation as discussed in Chapter 8. One such type of coupling circuit is the inductively-coupled circuit of Fig. 9.11. Observe that this amplifier is almost identical to the transformer-coupled amplifier discussed in Chapter 8. In fact, it is a transformer-coupled amplifier in which the primary of the transformer is tuned by capacitor C and the secondary not so tightly coupled because of the different type core which is perhaps air or powdered iron.

A simplified y-parameter circuit, in which y_{re} is neglected, is given in Fig. 9.12. The output and input conductances are assumed to be

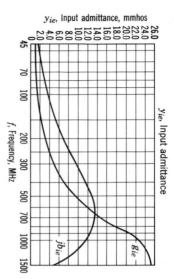

y_{ie}, Input admittance, mmhos

y_{ie}, Input admittance

f, Frequency, MHz

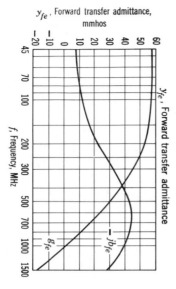

y_{fe}, Forward transfer admittance, mmhos

y_{fe}, Forward transfer admittance

f, Frequency, MHz

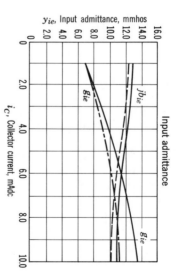

y_{ie}, Input admittance, mmhos

Input admittance

i_C, Collector current, mAdc

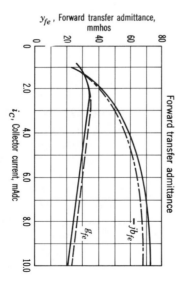

y_{fe}, Forward transfer admittance, mmhos

Forward transfer admittance

i_C, Collector current, mAdc

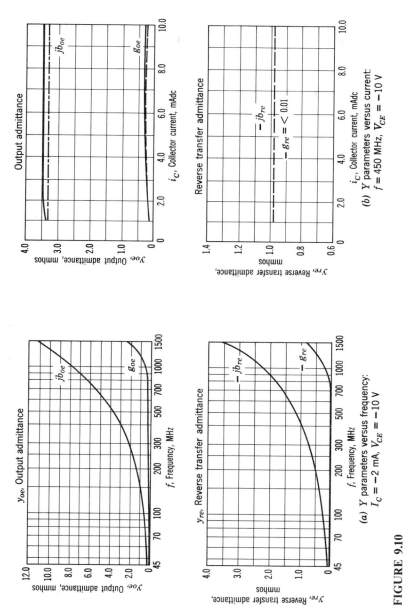

FIGURE 9.10

Common-emitter y parameters for 2N4957, 2N4958, 2N4959 transistors. (Courtesy of Motorola Semiconductor Products, Inc.)

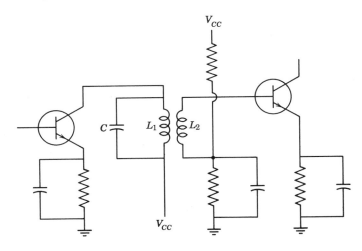

FIGURE 9.11
An inductively-coupled tuned transistor amplifier.

approximately g_{oe} and g_{ie}, respectively. The $jb_{oe} = j\omega_o C_{oe}$ and the transformed jb_{ie} are absorbed in the tuning capacitance C. Since the transformer does not meet the tight-coupling, large-inductance requirements of the untuned transformer discussed in Chapter 8, we must use a different approach to determine the effective impedance of the transformer primary. The mutual impedance of the transformer $j\omega M$, where M is the mutual inductance, is defined as the ratio of the voltage induced in one winding to the current flowing in the other winding. Therefore, referring to Fig. 9.12,

$$V_1 = (j\omega L_1 + R_1)I_1 - j\omega M I_2 \qquad (9.52)$$

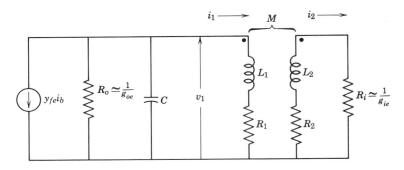

FIGURE 9.12
A simplified y-parameter equivalent circuit.

where the currents and voltages are assumed to be sinusoidal. Since the voltage induced in the secondary is $j\omega M I_1$,

$$I_2 = \frac{j\omega M I_1}{j\omega L_2 + R_i + R_2} \qquad (9.53)$$

where $R_i \simeq 1/g_{ie}$. Substituting this value of I_2 into Eq. 9.52,

$$V_1 = (j\omega L_1 + R_1)I_1 + \frac{(\omega M)^2 I_1}{j\omega L_2 + R_i + R_2} \qquad (9.54)$$

Therefore, the impedance seen at the primary terminals is

$$Z_1 = \frac{V_1}{I_1} = j\omega L_1 + R_1 + \frac{(\omega M)^2}{j\omega L_2 + R_i + R_2} \qquad (9.55)$$

The third term on the right-hand side of Eq. 9.55 is the impedance coupled into the primary (in series) as a result of the secondary current I_2. Note that this term is complex and, thus, reactance as well as resistance will be coupled into the primary. This reactance, which will affect the tuning of the primary, is a function of g_{ie}. Thus, the coupling circuit will be detuned by any change which affects g_{ie}. This undesirable detuning effect can be essentially eliminated if R_i is large in comparison with ωL_2. This condition can be met in most circuits. In addition, ωL_2 is usually much larger than R_2. Then the impedance seen at the primary terminals is

$$Z_1 \simeq j\omega L_1 + R_1 + \frac{(\omega M)^2}{R_i} \qquad (9.56)$$

This transformer primary impedance can replace the transformer in the equivalent circuit of Fig. 9.12 to produce the simplified primary circuit shown in Fig. 9.13.

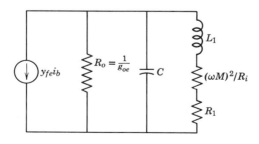

FIGURE 9.13
A simplified equivalent circuit with $R_i \gg \omega L_2$.

We now need a basis for choosing the primary inductance L_1. One sensible basis might be the obtaining of maximum power transfer and thus maximum gain as discussed in Chapter 8. However, maximum power into the primary may not yield maximum power in the load because part of the primary power is lost in the primary resistance R_1. This problem can be eliminated if we transform this loss resistance to its effective parallel value and combine it with R_o to obtain a modified output resistance R_o' as shown in Fig. 9.14. Using the relationship $R_{par} = Q_o \omega_o L_1$ at the resonant frequency (Eq. 9.6) and $R_o' = R_{par} R_o/(R_{par} + R_o)$,

$$R_o' = \frac{R_o Q_o \omega_o L_1}{R_o + Q_o \omega_o L_1} \tag{9.57}$$

If we now match the impedance of the tuned primary, Fig. 9.14, to R_o', maximum power will be transferred to the load, since the power loss in

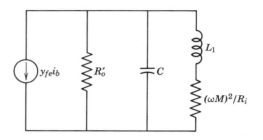

FIGURE 9.14
An equivalent circuit showing the modified output resistance R_o' which includes the primary winding loss.

the transformer has been removed. Therefore, we will transform the series resistance $(\omega M)^2/R_i$ to its equivalent parallel value and equate it to R_o'.

$$R_o' = \frac{(\omega L_1)^2}{(\omega M)^2/R_i} = \frac{L_1{}^2}{M^2} R_i \tag{9.58}$$

But from basic coupled circuit theory we know that the mutual inductance is

$$M = k\sqrt{L_1 L_2} \tag{9.59}$$

where k is the coefficient of coupling. Substituting this value for M in Eq. 9.58, we have

$$R_o' = \frac{L_1{}^2 R_i}{k^2 L_1 L_2} = \frac{L_1 R_i}{k^2 L_2} \tag{9.60}$$

We will now determine L_1 from the bandwidth requirement. Since the circuit Q is ω_o/B, which is known, and the total shunt resistance R is $R_o'/2$ in the matched amplifier, we can write

$$\frac{R_o'}{2} = Q\omega_o L_1 \tag{9.61}$$

and

$$R_o' = 2Q\omega_o L_1 \tag{9.62}$$

Substituting the value of R_o' given in Eq. 9.57 into Eq. 9.62, we have

$$\frac{R_o Q_o \omega_o L_1}{R_0 + Q_o \omega_o L_1} = 2Q\omega_o L_1 \tag{9.63}$$

and

$$\frac{R_o Q_o}{R_o + Q_o \omega_o L_1} = 2Q \tag{9.64}$$

Then

$$R_o Q_o = 2QR_o + 2QQ_o \omega_o L_1 \tag{9.65}$$

Solving for L_1, Eq. 9.65 becomes

$$L_1 = \frac{R_o(Q_o - 2Q)}{2QQ_o \omega_o} = \frac{R_o}{\omega_o}\left(\frac{1}{2Q} - \frac{1}{Q_o}\right) \tag{9.66}$$

Note that an impedance match cannot be obtained unless the primary Q_o is greater than two times the required circuit Q.

Let us now consider the design of a transistor amplifier using these ideas.

Example 9.4   Design a transistor amplifier with $f_o = 455$ kHz and bandwidth $B = 15$ kHz using transistors with $g_{oe} = 5 \times 10^{-5}$ mho and $g_{ie} = 1$ mmho, so $R_i \simeq 1$ kΩ. The Q_o of the primary of the coupling transformer is assumed to be 100. The circuit $Q = 455/15 = 30.3$ and $R_o \simeq 1/g_{oe} = 20$ kΩ. Then, using Eq. 9.66, the primary inductance $L_1 = 4.55 \times 10^{-5}$ H; $C = 1/\omega_1{}^2 L = 2.68 \times 10^{-9}$ F and $R_o' = 2Q\omega_o L_1 = 7.9$ kΩ. The secondary inductance L_2 and the coefficient of coupling k remain to be determined. Both of these parameters appear in Eq. 9.60. We have the requirement that $\omega_o L_2$ be small in comparison with R_i and Eq. 9.60 shows that minimum L_2 will occur when k is maximum. Therefore, the secondary should be tightly coupled to the primary. Values of $k \simeq 0.7$ can be obtained in air core coils if the secondary is wound on top of the primary or vice versa. Somewhat higher values of k can be obtained with ferrite cores. We will assume $k \simeq 0.7$. Then L_2 can be determined from Eq. 9.60.

$$L_2 = \frac{L_1 R_i}{k^2 R_o'} \tag{9.67}$$

The required secondary inductance is $L_2 = 11.8 \ \mu H$. We now need to verify the assumption that $\omega_o L_2$ is much smaller than R_i. Since $\omega_o L_2 = 33.7 \ \Omega$, its magnitude is only 3.4 percent of R_i, thus justifying the initial assumption. The engineer sometimes finds it necessary to design the coils for a given application. The number of turns of wire for a cylindrical coil configuration can be determined from the empirical formula

$$L = \frac{n^2 r^2}{9r + 10l} \times 10^{-6} \qquad (9.68)$$

where r is the mean radius of the coil in inches, l is the length of the coil in inches, and L is the inductance in henries. If the coil has a ferrite core and is slug tuned, the maximum value of inductance will be about twice the value obtained from the formula.

We will select a coil form with $\frac{1}{2}$ inch diameter and $\frac{3}{4}$ inch winding space for the coupling coil of this example. Then the number of primary turns $n_1 = 84$ and the number of secondary turns $n_2 = 43$. A wire table can be used to determine the size wire which will fill the winding space. More than one layer may be wound for each winding, but the Q_o will not be as high for a multilayer winding because of the increased distributed capacitance. Therefore, the winding should be confined to a single layer, if practical. If multilayer windings must be used, they can be wound with a coil winding machine to achieve high Q.

The procedure for designing an inductively-coupled tuned circuit is summarized below.

1.   The center fequency f_o or ω_o and the bandwidth requirement is either specified by the user or determined from the known signal characteristics.
2.   The circuit Q is determined from the relationship $Q = f_o/B$ (Hz).
3.   The inductance of the primary coil is determined from Eq. 9.66.
4.   The tuning capacitance is $C = 1/\omega_o^2 L_1$, including C_{oe} and transformed C_{ie}.
5.   The required secondary inductance is $L_2 = L_1 R_i / k^2 R_o'$.
6.   If you are winding your own coils, transpose Eq. 9.68 to obtain $n = L(9r + 10l)^{1/2}/r$, where L is the inductance in μH.

From the y-parameter circuit (Fig. 9.9), the signal voltage at the collector terminal is

$$v_o = \frac{-y_{fe} v_i}{y_{oe} + Y_L} \qquad (9.69)$$

However, at the resonant frequency ω_o, the total susceptance $j(b_{oe} + B_L)$ is zero, leaving only $g_{oe} + G_L$ as the total admittance through which the current $y_{fe} v_i$ flows. Then the magnitude of the voltage gain from the base to the collector at resonance is

$$K_{vc} = \frac{v_o}{v_i} = \frac{y_{fe}}{g_{oe} + G_L} \tag{9.70}$$

If maximum power transfer is achieved, $g_{oe} + G_L = 2g_{oe}'$ where g_{oe}' includes the parallel resistance $Q_o \omega_o L$ of the tuned primary.

The voltage gain of the entire stage, from base to base, is usually desired. In a lossless coupling circuit, the voltage ratio must be equal to the square root of the impedance transformation ratio which is, from the rearrangement of Eq. 9.67,

$$\frac{R_i}{R_o'} = \frac{k^2 L_2}{L_1} = \frac{k^2 n_2^2}{n_1^2} \tag{9.71}$$

Therefore, the stage gain is

$$K_v = \frac{y_{fe} k n_2}{(g_{oe} + G_L) n_1} \tag{9.72}$$

The input power to the base of the first amplifier is $V_{i1}^2 g_{ie1}$ and the power into the base of the second amplifier is $V_{i2}^2 g_{ie2}$. Then if we neglect y_{re}, the power gain of the amplifier is

$$G_p = \frac{V_{i2}^2 g_{ie2}}{V_{i1}^2 g_{ie1}} = \frac{K_v^2 g_{ie2}}{g_{ie1}} \tag{9.73}$$

PROBLEM 9.7 A 45 MHz IF amplifier for a TV receiver requires a 5 MHz bandwidth per stage. Three stages are needed and the 2N4959 transistor (Fig. 9.10) is chosen with $I_C = -2$ mA, $V_{CE} = -10$ V for each stage. Design the middle stage, using an inductively-coupled tuned circuit, which will provide 5 MHz bandwidth. Assume $Q_o = 100$ and $k = 0.7$. Calculate L_1, L_2 and C. Determine the voltage gain and the power gain at the resonant frequency. Draw a circuit diagram.

9.5 TAPPED-TUNED CIRCUITS

Impedance transformation can be accomplished in a tuned circuit by tapping a single coil, as an auto transformer, as shown in Fig. 9.15. In the equivalent circuit, Fig. 9.15b, R_o is the output resistance of the amplifying device ($1/g_{oe}$, approximately, for a transistor, r_d for a MOSFET or FET and r_p for a vacuum tube). The series resistance in the coil is R_1

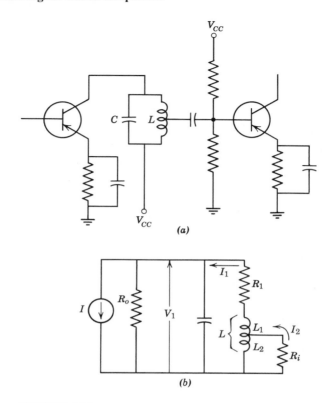

FIGURE 9.15
A tapped-tuned circuit which will provide impedance transformation.

and R_i is the input resistance of the following stage. There are n_2 turns between the tap and the RF ground terminal of the coil.

The equivalent circuit of Fig. 9.15b can be simplified somewhat by transforming the coil resistance R_1 into its equivalent parallel value $R_{par} = Q_o \omega_o L$ and then combining this resistance with the output resistance R_o to provide a modified output resistance R_o', as illustrated in Fig. 9.16.

Since there is no energy loss in the coupling circuit of Fig. 9.16b, the power delivered to the tuned circuit is equal to the power consumed by the load. Let us assume that we want the tuned coupling circuit to provide a resistance R_o' as a load for the driving amplifier so maximum power transfer will be achieved. Then the power delivered to the coupling circuit is V_1^2/R_o' where V_1 is the rms voltage applied to the tuned coupling circuit at resonance. Also, the power delivered to the actual load, which we

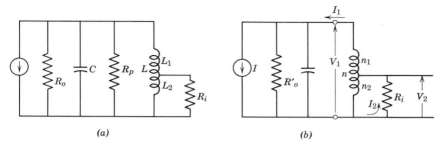

(a)   (b)

FIGURE 9.16
(a) **The equivalent parallel resistance R_p, which represents the power loss in coil L is combined with the transistor output resistance R_o to produce the modified output resistance R_o' in (b).**

have assumed is the input resistance R_i of the following amplifier, is V_2^2/R_i where V_2 is the rms voltage across R_i. Then

$$\frac{V_1^2}{R_o'} = \frac{V_2^2}{R_i} \tag{9.74}$$

We will assume that the voltage $d\phi/dt$ generated in each turn is equal to the voltage generated in every other turn. This is rigorously true if equal flux cuts every turn as it does in either tightly coupled coils or long slender coils with small mutual coupling and is a good approximation for the coil configurations normally used. Then the volts per turn in the coil are V_1/n and

$$V_2 = \frac{V_1 n_2}{n} \tag{9.75}$$

Substituting Eq. 9.75 into Eq. 9.74,

$$\frac{V_1^2}{R_o'} = \frac{n_2^2 V_1^2}{n^2 R_i} \tag{9.76}$$

Solving for n_2 explicitly,

$$n_2 = n \left(\frac{R_i}{R_o'} \right)^{1/2} \tag{9.77}$$

In the event a driving point impedance R_L other than R_o' is desired, that value may be obtained by merely replacing R_o' by R_L in Eq. 9.77, of course.

The required value of inductance L for the total coil can be determined from the bandwidth requirement and the total shunt resistance across the tuned circuit in precisely the same manner as for the inductively-coupled circuit. Thus, if maximum power transfer is desired, Eq. 9.66 is applicable. If maximum power transfer is not desired, then L can be determined from the relationship $\omega_o L = R/Q$ where R is the parallel combination of R_o' and R_L and $Q = \omega_o/B$.

The design procedure for a tapped-tuned circuit may proceed as follows.

1. Determine the circuit Q from the relationship $Q = f_o/B$ (Hz).
2. Determine L from the relationship $\omega_o L = R/Q$, where R is the total shunt resistance. For maximum power transfer, $R = R_o'/2$ and L is given by Eq. 9.66.
3. Calculate the total turns n from Eq. 9.68.
4. Calculate n_2 from Eq. 9.77.
5. Determine the tuning capacitance from $C = 1/\omega_o^2 L$.

PROBLEM 9.8 A transistor with $R_o = 10\ \text{k}\Omega$ is to be coupled to a transistor with $R_i = 1\ \text{k}\Omega$. The resonant frequency is 5 MHz and the desired bandwidth is 200 kHz. Using a tapped-tuned circuit to provide maximum power transfer, determine the total number of turns and the location of the tap. The coil is to have a $Q_o = 100$ and it is to be wound on a 0.2 inch diameter form and fill a 0.25 inch winding length. Assume $k = 0.8$.

Answer: $n = 32.8$, $n_2 = 16$.

The impedance transformation could have been accomplished by "tapping the capacitor" instead of the coil, as shown in Fig. 9.17. The relationship between C_1 and C_2 required to provide the desired impedance transformation can be determined from the energy relationship given by Eq. 9.74 for the

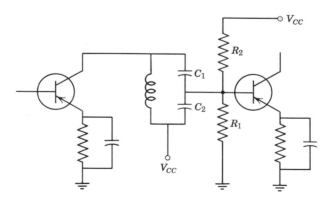

FIGURE 9.17
An alternate impedance matching system.

tapped coil. Let us assume that the current circulating around the tuned circuit is large compared with the external currents. Actually, the circulating current is Q_o times the terminal current where Q_o is determined at the terminal concerned. Then the current i through C_2 is approximately equal to the current through C_1. Since $V_2 = -jX_{C2}i$ and $V_1 \simeq -jX_C i$, where C is the capacitance of C_1 and C_2 in series and is the total capacitance that tunes L. These expressions for V_1 and V_2 may be substituted into Eq. 9.74 to yield

$$\frac{X_C{}^2}{R_o'} = \frac{X_{C2}{}^2}{R_i} \tag{9.78}$$

Replacing X_C with $1/\omega_o C$ and X_{C2} with $1/\omega_o C_2$, Eq. 9.78 becomes

$$\frac{1}{C^2 R_o'} = \frac{1}{C_2{}^2 R_i} \tag{9.79}$$

When Eq. 9.79 is solved for C_2, we have

$$C_2 = C\left(\frac{R_o'}{R_i}\right)^{1/2} \tag{9.80}$$

The tuning capacitance $C = 1/\omega_o{}^2 L$ and since capacitors in series add like resistors in parallel, $C = C_1 C_2/(C_1 + C_2)$. Therefore

$$C_1 = \frac{C_2 C}{C_2 - C} \tag{9.81}$$

The tapped-tuned circuit has an advantage over the inductively-coupled circuit because of the simplicity and availability of the single coil. However, part of the signal is shunted through the bias resistors R_1 and R_2 (Fig. 9.17).

PROBLEM 9.9 Determine the capacitor values that will replace the coil tap in Prob. 9.8.

The preceding work shows that the primary inductance and tuning capacitance of a tuned circuit are determined by the shunt resistance, the resonant frequency, the required bandwidth and the Q_o of the coil. These values of L and C may not be desirable for the following reasons.

1. The inductance may be too small to provide high, or even adequate, Q_o at the resonant frequency. Any coil has high Q_o over a limited frequency range. The smaller the coil, the higher the frequency range over which it has high Q_o.

2. A variable air capacitor may be desired, but the required capacitance may be so large that it cannot be obtained in a variable air type.

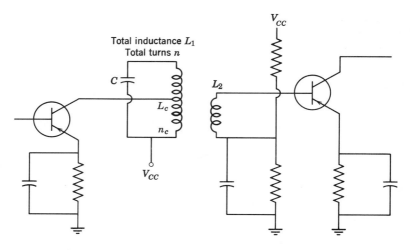

FIGURE 9.18
An inductively-coupled circuit with tapped primary.

The inductance L may be increased and the tuning capacitance C reduced to any desired value by connecting the collector to a tap on the primary coil as shown in Fig. 9.18. The required inductance between the collector tap and ground is determined in the manner previously discussed. Let us call this inductance L_c. Then, if the total desired primary inductance is L_1, the required turns ratio n_c/n can be determined from Eq. 9.77. Adapting Eq. 9.77 to the problem at hand, we have

$$n = n_c \left(\frac{L_1}{L_c}\right)^{1/2} \qquad (9.82)$$

where n is the total number of primary turns and n_c is the number of turns between V_{CC} and the collector tap.

PROBLEM 9.10   The calculated primary inductance for an inductively-coupled standard AM broadcast band amplifier is 50 μH. The amplifier is to be tuned with a variable capacitor having 365 pF maximum capacitance and the minimum frequency to be selected is 550 kHz. Determine the ratio n/n_c for a circuit like the one shown in Fig. 9.18 which will provide the desired inductance L.

FETs and vacuum tubes have higher input resistance than output resistance. Therefore, the tuned circuit is usually located in the gate or grid circuit if inductive coupling is used. Then the ohmic resistance of the tuned secondary coil provides the load resistance of the secondary and the secondary (tuning) coil provides the dc path to ground. The drain or plate resistance is often so high that impedance match cannot be attained, but adequate voltage gain is easily achieved.

9.6   DOUBLE-TUNED CIRCUITS

In the treatment of the inductively-coupled circuit in the preceding section, the requirement was made that the reactance of the untuned secondary coil be small in comparison with the resistance in series with it. An alternate method would allow the inductive reactance to be comparatively large but to cancel this reactance out at the operating frequency with a capacitance reactance, as shown in Fig. 9.19a. This circuit, known as a double-tuned circuit, has been simplified in Fig. 9.19b by the output resistance

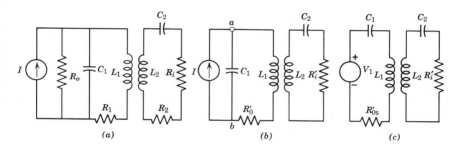

FIGURE 9.19
Equivalent circuits for the double-tuned amplifier.

R_o being transformed into its equivalent series value which may be added to the ohmic resistance R_1 to produce R_o'. R_i has also been added to the secondary resistance R_2 to produce R_i'. To further simplify the circuit, the current source to the left of terminals a and b may be replaced by a voltage source as shown in Fig. 9.19c. The driving point impedance as seen by the voltage source, using Eq. 9.54, is

$$Z_{in} = R_o' + j\left(\omega L_1 - \frac{1}{\omega C_1}\right) + \frac{(\omega M)^2}{R_i' + j[\omega L_2 - (1/\omega C_2)]} \qquad (9.83)$$

At the resonant frequency ω_o

$$Z_{in} = R_o' + \frac{(\omega_o M)^2}{R_i'} \qquad (9.84)$$

To obtain maximum power transfer, the resistance coupled into the primary must be equal to the source resistance R_o'. Then

$$R_o' = \frac{(\omega_o M_c)^2}{R_i'} \tag{9.85}$$

$$\omega_o M_c = (R_o'R_i')^{1/2} = \omega_o k_c (L_1 L_2)^{1/2} \tag{9.86}$$

where M_c and k_c are the critical values of mutual inductance and coefficient of coupling respectively which yield maximum power transfer. Using Eqs. 9.85 and 9.86, we have

$$k_c = \frac{(R_o'R_i')^{1/2}}{\omega_o(L_1 L_2)^{1/2}} = \left(\frac{R_o'}{\omega_o L_1} \cdot \frac{R_i'}{\omega_o L_2}\right)^{1/2} \tag{9.87}$$

Thus,

$$k_c = \frac{1}{(Q_1 Q_2)^{1/2}} \tag{9.88}$$

Any coupling coefficient may be expressed in terms of the critical value, k_c.

$$k = bk_c \tag{9.89}$$

where b is the ratio of actual coupling to the critical value.

To determine the bandwidth, loop equations are written for Fig. 9.19c to obtain an expression for the secondary current.

$$\left. \begin{array}{l} V_1 = Z_{11}I_1 + Z_{12}I_2 \\ 0 = Z_{12}I_1 + Z_{22}I_2 \end{array} \right\} \tag{9.90}$$

$$I_2 = -\frac{V_1 Z_{12}}{Z_{11}Z_{22} - Z_{12}{}^2} \tag{9.91}$$

At the resonant frequency ω_o, with maximum power transfer, $Z_{11} = R_o'$, $Z_{22} = R_i'$, and $Z_{12} = j\omega_o M_c$. Then, using Eq. 9.86 and Eq. 9.91, we have

$$I_{2\,\text{max}} = -j\frac{V_1(R_o'R_i')^{1/2}}{R_o'R_i' + R_o'R_i'} \tag{9.92}$$

The magnitude of the maximum secondary current

$$|I_{2\,\text{max}}| = \frac{V_1}{2(R_o'R_i')^{1/2}} \tag{9.93}$$

There may be frequencies other than ω_o at which maximum power transfer can occur. This is plausible because reactance as well as resistance is coupled into each tuned circuit and additional resonant frequencies are possible. To investigate this possibility, an effort will be made to find additional frequency and coupling combinations which will produce maximum power transfer. If these combinations exist, they will produce the same maximum secondary current as would occur at ω_o with critical coupling. Then using Eq. 9.91 and Eq. 9.92, we find that

$$\left|\frac{-jV_1}{2(R_o'R_i')^{1/2}}\right| = \left|\frac{-jV_1 b(R_o'R_i')^{1/2}}{[j\omega L_1 + (1/j\omega C_1) + R_o'][j\omega L_2 + (1/j\omega C_2)] + b^2 R_o'R_i'}\right| \tag{9.94}$$

To simplify the mathematics, let

$$L_1 = L_2 = L, \qquad C_1 = C_2 = C, \qquad R_o' = R_i' = R$$

Let $j\omega L + \dfrac{1}{j\omega C} = jX$. Then Eq. 9.94 reduces to

$$\left|\frac{V_1}{2R}\right| = \left|\frac{V_1 bR}{(jX + R)^2 + b^2 R^2}\right| \tag{9.95}$$

From cross-multiplication,

$$|2bR^2| = |-X^2 + j2XR + R^2 + b^2 R^2| \tag{9.96}$$

since $|R + jX|^2 = R^2 + X^2$

$$4b^2 R^4 = [-X^2 + (1 + b^2)R^2]^2 + 4X^2 R^2 \tag{9.97}$$

$$4b^2 R^4 = X^4 - 2X^2(1 + b^2)R^2 + (1 + b^2)^2 R^4 + 4X^2 R^2 \tag{9.98}$$

Collecting terms, we have

$$X^4 + 2(1 - b^2)X^2 R^2 + (1 - b^2)^2 R^4 = 0 \tag{9.99}$$

$$[X^2 + (1 - b^2)R^2]^2 = 0 \tag{9.100}$$

Then, $X^2 = (b^2 - 1)R^2$ or

$$X = \pm(b^2 - 1)^{1/2} R \tag{9.101}$$

Since $X = \omega L - 1/\omega C$,

$$\omega L - \frac{1}{\omega C} = \pm(b^2 - 1)^{1/2} R \tag{9.102}$$

$$\omega^2 LC - 1 = \pm(b^2 - 1)^{1/2} R\omega C \tag{9.103}$$

For the high Q case, throughout the pass band

$$R\omega C \simeq R\omega_o C = \frac{1}{Q} \tag{9.104}$$

Then, since $\omega_o{}^2 = 1/LC$,

$$\frac{\omega^2}{\omega_o{}^2} \simeq 1 \pm \frac{(b^2 - 1)^{1/2}}{Q} \tag{9.105}$$

$$\omega \simeq \pm \left[1 \pm \frac{(b^2 - 1)^{1/2}}{Q}\right]^{1/2} \omega_o \tag{9.106}$$

For real positive frequencies, the $+$ sign preceding the parenthesis in Eq. 9.106 must be used. Since $(1 \pm \chi)^{1/2} \simeq 1 \pm \chi/2$ when $\chi \ll 1$, Eq. 9.106 can be simplified to

$$\omega = \left[1 \pm \frac{(b^2 - 1)^{1/2}}{2Q}\right] \omega_o \tag{9.107}$$

Equation 9.107 indicates that when b exceeds unity there will be two frequencies which yield maximum power transfer as indicated in Fig. 9.20. Note that there is only one frequency, (ω_o), which will yield maximum power transfer when $b = 1$, and when b is less than unity there is no real frequency which will produce maximum power transfer.

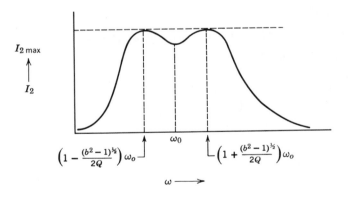

FIGURE 9.20
Response curve for a double-tuned circuit with $b > 1$.

PROBLEM 9.11   A double-tuned circuit has $Q_1 = Q_2 = 100$ and the coefficient of coupling $k = 0.02$. The resonant frequency of each tuned circuit is 1 MHz. Determine the frequencies at which maximum power transfer occurs.

The bandwidth may be obtained by finding the frequencies which yield

$I_{2\,max}/1.414$. Using Eq. 9.94 and the techniques of the preceding section, we find that

$$\left|\frac{V_1}{2(2)^{1/2}R}\right| = \left|\frac{V_1 bR}{(jX + R)^2 + b^2 R^2}\right| \tag{9.108}$$

$$\left|2(2)^{1/2}bR^2\right| = \left|-X^2 + j2XR + R^2 + b^2 R^2\right| \tag{9.109}$$

$$8b^2 R^4 = [-X^2 + (1 + b^2)R^2]^2 + 4X^2 R^2 \tag{9.110}$$

$$8b^2 R^4 = X^4 - 2(1 + b^2)X^2 R^2 + (1 + b^2)^2 R^4 + 4X^2 R^2 \tag{9.111}$$

Collecting terms, we have

$$X^4 + 2(1 - b^2)X^2 R^2 + (1 - 6b^2 + b^4)R^4 = 0 \tag{9.112}$$

The quadratic formula can be used to solve for X^2. Then

$$X^2 = -(1 - b^2)R^2 \pm [(1 - b^2)^2 R^4 - (1 - 6b^2 + b^4)R^4]^{1/2} \tag{9.113}$$

$$X^2 = (b^2 - 1)R^2 \pm (4b^2)^{1/2} R^2 \tag{9.114}$$

$$X = \pm(b^2 - 1 \pm 2b)^{1/2} R \tag{9.115}$$

$$\omega_1 L - \frac{1}{\omega_1 C} = \pm(b^2 - 1 \pm 2b)^{1/2} R \tag{9.116}$$

Multiplying through by $\omega_o C$, we have

$$\omega_1 \omega_o LC - \frac{\omega_o}{\omega_1} = \pm(b^2 - 1 \pm 2b)^{1/2} R\omega_o C \tag{9.117}$$

Since $R\omega_o C = 1/Q$ and $LC = 1/\omega_o^2$,

$$\frac{\omega_1}{\omega_o} - \frac{\omega_o}{\omega_1} = \pm\frac{(b^2 - 1 \pm 2b)^{1/2}}{Q} \tag{9.118}$$

Since this equation (9.118) must hold for both solutions ω_1 and ω_2 (or ω_L and ω_H), the relationship $\omega_o = (\omega_1\omega_2)^{1/2}$ (Eq. 9.17) must hold. Then

$$\frac{\omega_1}{\omega_o} - \frac{\omega_o^2}{\omega_1\omega_o} = \frac{\omega_1 - \omega_2}{\omega_o} = \pm\frac{(b^2 - 1 \pm 2b)^{1/2}}{Q} \tag{9.119}$$

$$\omega_2 - \omega_1 = B = (b^2 - 1 \pm 2b)^{1/2}\frac{\omega_o}{Q} \tag{9.120}$$

Since the bandwidth of an amplifier with a single-tuned circuit is ω_o/Q, the value of the radical is the ratio of the bandwidth of a double-tuned circuit to that of a single-tuned circuit with the same Q. For the case $b = 1$, the ratio is $(2)^{1/2}$.

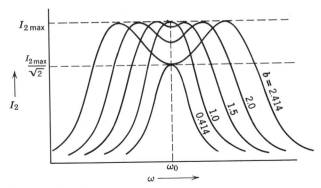

FIGURE 9.21
Frequency response as a function of coupling for a double-tuned circuit.

A plot of secondary current as a function of frequency for various values of b is given in Fig. 9.21. The values of b which give half power output at ω_o may be computed from Eq. 9.120 by letting $\omega_1 = \omega_2 = \omega_o$. Then

$$0 = (b^2 - 1 \pm 2b)^{1/2} \frac{\omega_o}{Q} \tag{9.121}$$

$$b^2 \pm 2b - 1 = 0 \tag{9.122}$$

When the quadratic formula is used to find b, we have

$$b = \pm 1 \pm (2)^{1/2} \tag{9.123}$$

Since b must be positive in realizable circuits,

$$b = (2)^{1/2} \pm 1 = 0.414 \text{ or } 2.414 \tag{9.124}$$

For values of b between those just given, the positive sign must be used inside the radical in Eq. 9.120 to obtain real values of bandwidth. If $b < 0.414$, no real values of bandwidth are possible. Actually, the circuit has a real bandwidth, but Eq. 9.120 is based on the assumption that the maximum secondary current is that which is available with critical coupling. Therefore, true values of bandwidth are obtained from Eq. 9.120 only for $b \geq 1$. When $b > 2.414$, the use of both signs inside the radical gives four frequencies at which half power is obtained. Under these conditions, the power at ω_o is less than half of the maximum value. The band has actually been split in two. This condition would be useful only under very special conditions. Under normal conditions, the bandwidth is

$$B = (b^2 + 2b - 1)^{1/2} \frac{\omega_o}{Q} \tag{9.125}$$

for $1 \le b \le 2.414$. It should be observed that the Q used in Eq. 9.125 is the loaded Q of each tuned circuit, *not* considering the coupled resistance. In a single-tuned circuit, the Q must include all loading, including the coupled resistance.

It is convenient, but not necessary, to make the parameters of the primary and secondary tuned circuits equal. The nonsymmetrical circuit is difficult to analyze and offers no special advantages. It is usually convenient to make the inductance of the coils equal. The capacitors would then be approximately equal. Good design practice would indicate that tapped coil arrangements should be used unless the shunt resistances are very high and approximately equal. The effective series resistance of each tuned circuit could then be made approximately equal. A transistor circuit employing this technique is shown in Fig. 9.22.

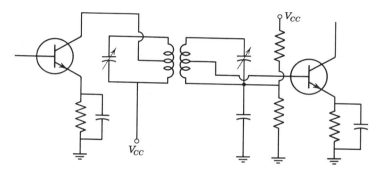

FIGURE 9.22
A double-tuned transistor circuit.

When the coil Q is high in comparison with the circuit Q, the large majority of source energy is transferred to the load. This follows from the fact that the effective series resistance caused by source and load resistances is large in comparison with the series ohmic resistance. When an impedance match is obtained, the power delivered to a specific load impedance is essentially independent of the type of coupling circuit, provided the coupling efficiency is high. The various circuits described in this chapter will each produce about the same reference gain under matched conditions. The double-tuned circuit will have the largest gain bandwidth product because of the bandwidth ratio expressed in Eq. 9.125. In addition, the frequency response characteristics of the double-tuned circuit have steeper sides and a flatter top than the single-tuned circuit. The designer has some control over the shape of the response curve by his choice of b. The best choice would depend on the application. Values for b of the order of 1.0 to 1.7 look promising for cases where uniform frequency response is desired over most of the pass band.

The development of the transient or time-domain response of a double-tuned amplifier is beyond the scope of this book. However, Martin[1] has shown that the double-tuned amplifier (or actually its staggered pair equivalent) has an overshoot in the envelope response which is comparable with the overshoot of a shunt-compensated video amplifier. Thus, the improved gain-bandwidth product or gain-rise time ratio of the double-tuned amplifier is obtained at the expense of overshoot in the envelope response. This overshoot is illustrated in Fig. 9.23 where it is assumed

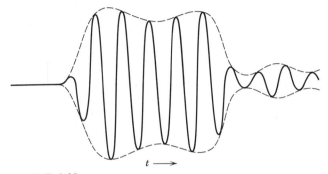

FIGURE 9.23
Typical time-domain response of a double-tuned amplifier driven by a rectangular pulse of frequency ω_o.

that the excitation is a pulse-modulated sinusoid. Martin showed that the critically or transitionally-coupled amplifier has a 4.3 percent overshoot. It has also been shown that the overshoot increases as the degree of coupling increases. Therefore, the degree of coupling b which provides the optimum frequency response characteristic does not, in general, provide optimum time-domain response.

The procedure for designing a double-tuned circuit might be as follows.

1. Choose b for the best response characteristics.
2. With ω_o and the desired bandwidth known, calculate the circuit Q from Eq. 9.125.
3. Calculate the coefficient of coupling

$$k = bk_c = \frac{b}{Q}$$

[1] Thomas L. Martin, Jr., *Electronic Circuits*, Prentice Hall, Englewood Cliffs, New Jersey, 1955, Sec. 6.6.

4. Determine the primary inductance from the relationship $Q = R_o'/\omega_o L$, where the output resistance R_o' includes the effective parallel resistance of the primary coil. However, as for the tapped-tuned circuit and the inductively-coupled tuned circuit, the effective parallel resistance of the primary coil cannot be determined until the primary inductance is known. Therefore, the relationship stated previously $(Q = R_o'/\omega_o L)$ must be used in addition to the relationship $R_o' = R_p R_o/(R_p + R_o)$ to calculate the primary inductance. Using these two relationships, the value of R_o' becomes

$$R_o' = Q\omega_o L = \frac{R_p R_o}{R_p + R_o} \tag{9.126}$$

But $R_p = Q_o \omega_o L$. Then

$$Q\omega_o L = \frac{Q_o \omega_o L R_o}{Q_o \omega_o L + R_o} \tag{9.127}$$

Simplifying,

$$Q(Q_o \omega_o L + R_o) = Q_o R_o \tag{9.128}$$

$$QQ_o \omega_o L = R_o(Q_o - Q) \tag{9.129}$$

$$L = \frac{R_o}{\omega_o}\left(\frac{Q_o - Q}{QQ_o}\right) = \frac{R_o}{\omega_o}\left(\frac{1}{Q} - \frac{1}{Q_o}\right) \tag{9.130}$$

The secondary inductance would, of course, be the same as the primary inductance if this suggested design technique is used.

5. Calculate the value of capacitance required to resonate with the inductance calculated in step 4.

6. Determine the necessary tap points on the coils to provide proper loading of the tuned circuits.

An example of a double-tuned transistor amplifier will be given to illustrate this outlined procedure.

Example 9.5 A double-tuned circuit is to be used to couple a transistor which has $R_o = 40$ k to a transistor which has $R_i = 1$ k. The center frequency is 500 kHz and the desired bandwidth is 12 kHz. The foregoing suggested procedure will be used in designing the coupling circuit.

1. The coupling coefficient 1.5 k_c will be considered optimum ($b = 1.5$).

2. $B = 12 \text{ kHz} = (1.5^2 + 3 - 1)^{1/2} \dfrac{500 \text{ kHz}}{Q}$

3. $Q = \dfrac{2.06 \times 500 \text{ kHz}}{12 \text{ kHz}} = 86$

$k = \dfrac{b}{Q} = \dfrac{1.5}{86} = 0.0175$

4. Assuming the coil $Q_o = 150$,

$$L_1 = L_2 = \frac{40 \times 10^2}{3.14 \times 10^6}\left(\frac{1}{86} - \frac{1}{150}\right) = 63 \ \mu\text{H}$$

5. $C = \dfrac{1}{(3.14 \times 10^6)^2(60 \times 10^{-6})} = 1610 \text{ pF}$

6. The tap on the secondary coil is

$$n_2 \simeq n\left(\frac{1}{40}\right)^{1/2} = 0.158n$$

The inductance could be increased and the tuning capacitance reduced from the values just calculated by tapping the primary as well as the secondary. The problem would then proceed as before but with a transformed value of R_o.

PROBLEM 9.12 A transistor having $R_o = 20 \text{ k}$ is to be coupled to a transistor having $R_i = 2 \text{ k}$. The desired bandwidth is 20 kHz and $f_o = 460 \text{ kHz}$. Design a double-tuned circuit using $b = 1.2$ and $Q_o = 100$.

PROBLEM 9.13 Use pentode tubes with $r_p = 1 \text{ M}\Omega$ for the amplifier of Problem 9.12. In case the tuning capacitors are too small for good frequency stability (several times the tube capacitance), use loading resistors to reduce the load resistance of both primary and secondary.

In addition to the circuits considered, the double-tuned circuit can be arranged in either a tee or pi network, as shown in Fig. 9.24. Many

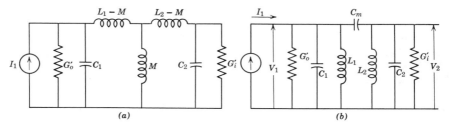

(a)   (b)

FIGURE 9.24
Tee- and pi-type double-tuned coupling networks.

variations of these circuits are possible and may be found in a radio handbook.

The circuit of Fig. 9.24b is convenient because the degree of coupling may be easily controlled. This circuit may be analyzed using nodal equations.

$$Y_{11}V_1 + Y_{12}V_2 = I_1$$

$$Y_{12}V_1 + Y_{22}V_2 = 0$$

(9.131)

where $Y_{11} = G_o' + j\left(\omega C_1 - \dfrac{1}{\omega L_1}\right)$

$Y_{22} = G_i' + j\left(\omega C_2 - \dfrac{1}{\omega L_2}\right)$

$Y_{12} = j\omega C_m$

G_o' and G_i' include the effective shunt resistance of L_1 and L_2, respectively. Solving for the output voltage, we have

$$V_2 = \frac{I_1 Y_{12}}{Y_{11} Y_{22} + Y_{12}{}^2}$$

(9.132)

The value of Y_{12} which yields maximum output voltage V_2 at the resonant frequency may be obtained by differentiating Eq. 9.132 with respect to Y_{12} and by equating the derivative to zero. Equation 9.133 results.

$$|Y_{12}| = (G_o' G_i')^{1/2}$$

(9.133)

This value of Y_{12} could be called the critical or transitional value. Using the relationship $R_o' = Q_1 \omega_o L$, we have

$$G_o' = \frac{1}{Q_1 \omega_o L_1} = \frac{\omega_o C_1}{Q_1}$$

(9.134)

Also, .

$$G_i' = \frac{1}{Q_2 \omega_o L_2} = \frac{\omega_o C_2}{Q_2}$$

(9.135)

Making these substitutions into Eq. 9.133, we see that

$$|Y_{12}|(\text{critical}) = \omega C_{mc} = \omega_o \frac{(C_1 C_2)^{1/2}}{(Q_1 Q_2)^{1/2}}$$

(9.136)

where C_{mc} is the critical value of coupling capacitance. Using the critical coupling coefficient $k_c = 1/(Q_1 Q_2)^{1/2}$ previously defined in Eq. 9.88, we see that Eq. 9.136 becomes

$$\omega C_{mc} = \omega_o k_c (C_1 C_2)^{1/2} \tag{9.137}$$

Compare Eq. 9.137 with Eq. 9.88. Since ω is very nearly ω_0 within the pass band,

$$C_{mc} \simeq k_c (C_1 C_2)^{1/2} \tag{9.138}$$

Also, a general value of coupling capacitance C_m can be defined in terms of the critical or transitional value C_{mc}.

$$C_m = b C_{mc} = b k_c (C_1 C_2)^{1/2} \tag{9.139}$$

It should be evident that the nodal equations of the capacitively-coupled parallel circuit are identical to the loop equations of the inductively-coupled series circuit. Therefore, the equations for bandwidth, Eq. 9.125, and frequencies at which maximum power transfer occurs, Eq. 9.107, which were developed for the series circuit, apply equally well to the parallel, capacitively-coupled circuit.

PROBLEM 9.14   The circuit of Fig. 9.24b is used with $f_o = 460$ kHz and bandwidth $= 20$ kHz. Using $b = 1.5$ and $G_o = G_i = 50$ μmhos determine $Q_1 = Q_2$ and C_m for the circuit. The Q_o of each coil is 100.

9.7   TUNED-AMPLIFIER STABILITY

Amplifier stability is a term used to indicate the freedom from oscillation in an amplifier. A tuned amplifier is particularly susceptible to oscillation because of its normally high gain and its LC coupling circuits. These coupling circuits can cause a relative phase shift of 180° at frequencies below resonance and thus cause regenerative (or in-phase) feedback. You will recall that the common-emitter type amplifier causes a voltage polarity reversal as the signal goes through the amplifier. Therefore, if a feedback path which provides another polarity reversal exists between the output and the input, the feedback signal causes regeneration and perhaps oscillation in the amplifier. The regeneration narrows the bandwidth of the amplifier even though the feedback may not be sufficient to cause oscillation.

The undesired feedback in a tuned amplifier may come from three causes. The first is improper circuit layout or inadequate shielding which is the responsibility of the engineer who designs the circuit. Unwanted coupling can exist between coupling coils which are quite widely separated on a chassis unless the coils are enclosed in metal cans which confine the coils' magnetic and electric fields. The magnetic fields are confined at high frequencies because of eddy currents induced in the shield can by the magnetic fields from the coil. These eddy currents cause opposing magnetic fields which reduce the external magnetic fields to almost zero. Capacitive and inductive coupling can also exist between the signal-carrying output leads and input leads of an amplifier unless these leads are very short and well separated. Therefore, careful chassis, or circuit board, layout is essential for a stable amplifier.

The second cause of instability is coupling between amplifier stages through the power supply leads. This coupling results from either long power supply leads or high power supply impedance. Decoupling filters between stages and bypass capacitors at the amplifier terminals are used to reduce this type of coupling. Decoupling is discussed in Chapter 11.

The third cause of instability is the capacitive coupling due to collector junction capacitance in transistors, drain-gate capacitance in FETs or MOSFETs and plate-grid capacitance in vacuum tubes. These capacitances couple energy from the output circuit of the amplifier back into the input circuit and thus can cause instability, or oscillation, in the amplifier under certain conditions. The conditions which cause this instability and the techniques for preventing it will be the major topic of this section.

The y-parameter circuit of Fig. 9.25 may represent any type of amplifier.

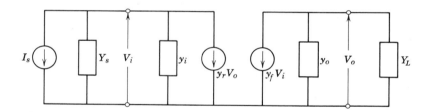

FIGURE 9.25
A y-parameter amplifier circuit.

We will assume that the driving-source current I_s is sinusoidal. Then, the voltage at the amplifier input terminals is

$$V_i = \frac{I_s}{Y_s + y_i + Y_1} \tag{9.140}$$

where Y_1 is the equivalent admittance of the dependent generator $y_r V_o$. This admittance is the ratio of the current $y_r V_o$ to the voltage V_i across the generator. Therefore,

$$Y_1 = \frac{y_r V_o}{V_i} \tag{9.141}$$

Thus Y_1 is $y_r G_V'$ where G_V' is the voltage gain from the input terminals to the collector terminals. Then, since $V_o = -y_f V_i/(y_o + Y_L)$,

$$Y_1 = -\frac{y_r y_f}{y_o + Y_L} \tag{9.142}$$

Substituting this expression for Y_1 into Eq. 9.140

$$V_i = \frac{I_s}{Y_s + y_i - [(y_r y_f)/(y_o + Y_L)]} \tag{9.143}$$

This equation reveals the instability problem. The denominator of Eq. 9.143 may become zero if both the real parts and the j parts become zero as a result of the negative sign. If this should happen, the input voltage V_i and hence the output voltage would become infinite, theoretically, for any finite source current I_s. In fact, input voltage V_i could occur when $I_s = 0$. This is the condition for instability. We will investigate the conditions under which the denominator of Eq. 9.143 can become zero, or vanish. Of course the voltages V_i and V_o do not become infinite because they will limit at approximately the power supply voltage.

The conditions under which the denominator of Eq. 9.143 vanishes can be found quite easily by graphical techniques. First a polar plot of $y_o + Y_L = (g_o + G_L) + j(b_o + B_L)$ is made in Fig. 9.26a. The conductance

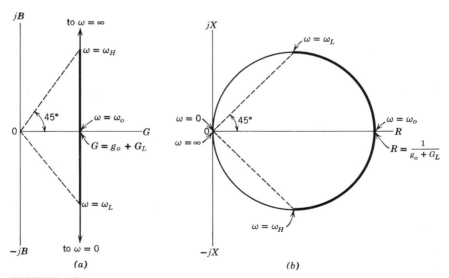

FIGURE 9.26
(a) **Polar plot of** $y_o + Y_L$, (b) **polar plot of** $Z = 1/(y_o + Y_L)$.

$(g_o + G_L)$ is assumed to remain constant, which is a good approximation only in the vicinity of the pass band from ω_L to ω_H. However, our interest lies primarily in this frequency range. The load is assumed to be a tuned circuit so B_L can vary from $-\infty$ to ∞. A capacitive load produces only positive values of jB.

Since $y_o + Y_L$ appears in the denominator of the expression for Y_1, we will need to take its reciprocal before multiplying by $-y_r y_f$. This reciprocal is easily obtained from the polar plot of $y_o + Y_L$ as shown in Fig. 9.26b. For any given ω, the magnitude of Z is the reciprocal of the magnitude of Y and the angle of Z is the same magnitude as the angle of Y but has the opposite sign. Thus the vertical line representing Y in Fig. 9.26a plots

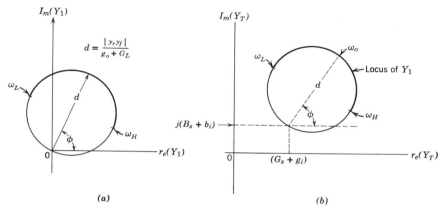

(a)   (b)

FIGURE 9.27
(a) **Polar plot of** Y_1, (b) **polar plot of** $Y_s + y_i + Y_1$.

into a circle with diameter $1/(g_o + G_L)$ on the Z plane (Fig. 9.26b). The pass band occupies a half circle as shown. The top half circle occurs below resonance when the total load is inductive.

The plot of Y_1 can now be made by multiplying Z by the polar form of $-y_r y_f = |y_r y_f|/\phi$ where $\phi = 180° + Arg(y_r) + Arg(y_f)$. The magnitudes and angles of y_r and y_f are nearly constant through the pass band and therefore will be assumed constant. Then the radial lines representing Z in Fig. 9.26b will each be multiplied by $|y_r y_f|$, and their angles will all be shifted by the angle ϕ. Thus the circular shape is preserved and the diameter is shifted by the angle ϕ as shown in Fig. 9.27a. The final step in obtaining the plot of $Y_s + y_i + Y_1$ (Eq. 9.140) is the addition of $Y_s + y_i$ to the plot of Y_1, of course. This addition is easily accomplished by placing the origin of the Y_1 circle at the coordinate $(G_s + g_i) + j(B_s + b_i)$ on the plot of $Y_T = Y_s + y_i + Y_1$, as illustrated in Fig. 9.27b.

The goal of the preceding graphical analysis is to find the conditions under which Y_T, the denominator of Eq. 9.140, becomes zero. The amplifier design will then preclude the possibility of these conditions occurring. We can observe from Fig. 9.27b that Y_T cannot be zero with the set of parameters assumed because the circle which represents the locus of all possible values of Y_T, measured from the origin, does not intersect either the real or j axis and the zero values lie on these axes. The circle will intersect the real Y_T axis if the input circuit, or driving source, is tuned below resonance so $j(B_s + b_{ie})$ is negative. However, the real part of Y_T cannot vanish if the circle does not intersect the j or $I_m(Y_T)$ axis and the amplifier is stable, provided it is properly constructed. We will now consider an example of the design of an amplifier.

Example 9.6   Assume that we are assigned the task of designing a 200 MHz, band pass amplifier. The 2N4959 transistor in the common-emitter configuration is chosen as the amplifying device and the y parameters for $I_E = -2.0$ mA and $V_{CE} = -10$ V, which we will use, are given in Fig. 9.10. We read these parameters as $y_{fe} = (54 - j22)$ mmho $= 58\underline{/-22°}$, $y_{re} = -j0.5$ mmho $= 0.5\underline{/-90°}$, $y_{ie} = (2.8 + j6.8)$ mmho, and $y_{oe} = (0.2 + j1.5)$ mmho. Let us attempt to obtain maximum power gain by matching impedances. Then $G_s = g_{ie}$ and $G_L = g_{oe}$, $d = |y_{re}\,y_{fe}|/2g_{oe} = 58 \times 0.5 \times 10^{-6}/2 \times 0.2 \times 10^{-3} = 65$ mmho and $\phi = 180° - 22° - 90° = 68°$. Also $G_s + g_{ie} = 5.6$ mmho. We will assume that the input circuit is at resonance, so $j(B_s + b_{ie}) = 0$. The locus of Y_T, using these values, is drawn in Fig. 9.28a. Observe that

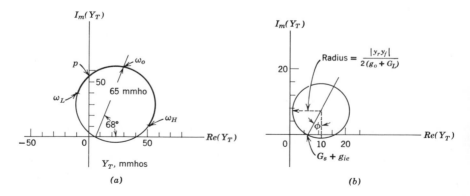

FIGURE 9.28
(a) Locus of Y_T for the amplifier of Example 9.6, (b) amplifier stabilized by increasing G_L.

the real part of Y_m is zero at point p which lies at a frequency between ω_o and ω_L. The amplifier will oscillate at approximately ω_p, the frequency at point p. At this frequency, below resonance, the tuned input circuit will have a negative susceptance $(-jB_s)$ which will move the circle downward until the point p lies at the origin of the coordinates. If we have need for an oscillator instead of an amplifier, the design above will suffice. However, the circuit could be simplified by replacing the tuned input circuit with an inductive susceptance having about 57 mmhos magnitude, as seen in Fig. 9.28a.

There are several measures which can be employed to stabilize the amplifier of Example 9.6 by confining the Y_T locus (circle) to the right half plane. There are three ways to reduce the diameter of the circle so that the circumference will not cross the jB axis, as shown in Fig. 9.28b.

1. The forward transadmittance Y_f can be reduced by decreasing the q-point collector or drain current. However, the other y parameters will also change as a result of the q-point change.

2. The load conductance can be increased, thus increasing the magnitude of the denominator of the term $|y_r y_f|/(G_o + G_L)$. Once the graphical construction of Fig. 9.28b is understood, the locus of Y_T need not be drawn. Figure 9.28b shows that the required condition for stability is: *The radius of the circle must be less than the real* (or *x*-axis) *coordinate of the center of the circle.* Then

$$\frac{|y_r y_f|}{2(g_o + G_L)} < G_s + g_i + \frac{|y_r y_f|}{2(g_o + G_L)} \cos \phi \qquad (9.144)$$

or

$$\frac{|y_r y_f|}{2(g_o + G_L)} < \frac{G_s + g_i}{1 - \cos \phi} \qquad (9.145)$$

3. The reverse transadmittance can be reduced, theoretically to zero, by a technique known as *neutralization* or *unilateralization*. This neutralization is accomplished by providing a feedback current, equal in magnitude but opposite in phase, to cancel the feedback current through y_r, as shown in Fig. 9.29. The voltage $(-aV_o)$ which causes the neutralizing current I_n to flow can be obtained by applying V_{CC} (or V_{DD}, etc.) to a tap on the output coil, as illustrated in Fig. 9.29b, or from the secondary of an inductively-coupled circuit, as shown in Fig. 9.29c. Of course, the secondary connection must provide a polarity reversal as indicated by the polarity dots.

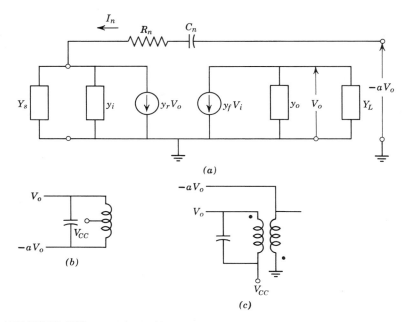

FIGURE 9.29
(a) **A generalized neutralizing circuit,** (b) **a tapped circuit for obtaining** $-aV_o$, (c) $-aV_o$ **obtained from the secondary of the coupling transformer.**

The amplifier is properly neutralized when $I_n = -y_r V_o$. Then

$$\frac{-aV_o}{R_n + \dfrac{1}{j\omega_o C_n}} = -y_r V_o \qquad (9.146)$$

and

$$R_n - j\frac{1}{\omega_o C_n} = \frac{a}{y_r} \qquad (9.147)$$

The values of R_n and C_n can be easily determined by putting a/y_r in rectangular form and then equating R_n to the real part and $1/\omega_o C_n$ to the j part. Very frequently the real part of y_r is negligible, as in Example 9.6. Then R_n may be omitted and $\omega_o C_n = B_r/a$. When R_n is negligible and $a = 1$, the neutralizing capacitance is equal to C_{ob}, C_{dg} or C_{pg}, depending upon the type of amplifying device, which causes the internal feedback.

Although neutralization permits maximum power gain and will theoretically reduce the Y_1 locus (circle) to a point, there are disadvantages to this type of stabilization. The first problem is that y_r varies quite widely among devices of the same type and therefore C_n must be adjusted for each individual circuit. This adjustment is expensive in a mass-produced amplifier. Second, y_r varies with q point. Thus a properly neutralized amplifier at one q point may not be properly neutralized at a different q point. Therefore, mismatching is a more popular stabilizing technique than neutralization. Sometimes a combination of the two is used. A special circuit arrangement known as the *cascode* connection that almost eliminates y_r will be discussed in Chapter 10.

In addition to reducing the diameter of the Y_1 locus, stabilization can be achieved by increasing G_s, as seen in Fig. 9.28 or Eq. 9.145. Often G_s and G_L are increased simultaneously by connecting a resistor in parallel with the transistor output to reduce R_o' but using the impedance transformation ratio determined for the original (higher) R_o'. This will increase G_s for the following transistor, and in a cascade of several stages G_s and G_L are both increased by the same ratio.

Up to this point, we have assumed that the performance of the amplifier will be satisfactory if the amplifier is stable, or free from oscillation. However, the bandwidth is inadequate and the maximum response does not occur at resonance if the diameter of the Y_1 circle is not small compared with $(g_i + G_s)$. The reason for this behavior is illustrated in Fig. 9.30. The input voltage $V_i = I_s/Y_T$ (Eq. 9.140) and the magnitude of Y_T is the length of the vector from the origin to the Y_T locus (Fig. 9.30a). Therefore V_i peaks near ω_L because of the close proximity of this part of the locus to the origin. Since the output voltage $V_o = V_i G_V$, the output peaks at some frequency between ω_o and ω_L, and the bandwidth of the amplifier is reduced considerably from the design value as seen in Fig. 9.30b. The

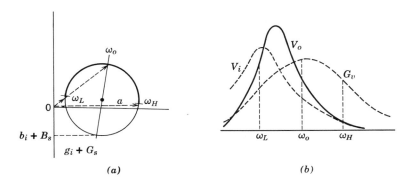

(a)   (b)

FIGURE 9.30
(a) Y_T **locus**, (b) V_i **and** V_o **as functions of frequency.**

design value is the bandwidth of G_s, and will be achieved only when V_i is independent of frequency, which will occur when the Y_1 locus shrinks to the point a and the input is untuned. The bandwidth of amplifiers containing two or more tuned circuits will be discussed in Chapter 12. Therefore, in order to realize the design bandwidth as well as to provide a margin of stability safety factor, the radius of the locus of Y_1 should be much smaller than the distance from the $I_m(Y_T)$ axis to the center of the Y_1 circle. We will let this ratio be k and incorporate this ratio in Eq. 9.144.

$$\frac{|y_r y_f|}{2(g_o + G_L)} = k \left[G_s + g_i + \frac{|y_r y_f|}{2(g_o + G_L)} \cos \phi \right] \tag{9.148}$$

Then

$$\frac{|y_r y_f|}{(g_o + G_L)} = \frac{2k(G_s + g_i)}{1 - k \cos \phi} \tag{9.149}$$

We need to reconsider the procedure for designing a tuned RF amplifier in the light of the stability requirement. The following procedure may be used when $G_L > g_o$.

When the required load conductance G_L is larger than the transistor output conductance plus the coil loss (our former $1/R_o'$), either neutralization or mismatching must be used to stabilize the amplifier. Mismatching may be accomplished by either of two methods. The first is to make the impedance transformation ratio of the coupled circuit smaller than that required for an impedance match. Then G_L of the transistor under consideration and G_S of the following transistor are increased by the same ratio as the decrease in the transformation ratio, as previously mentioned. The second method is to connect a resistor across the collector (or drain) tuned circuit, treat this resistor as part of r_o' (or $1/g_o'$) and then design the coupling circuit for maximum power transfer. This latter technique is illustrated in Fig. 9.31, where the total load conductance G_L and output conductance g_o

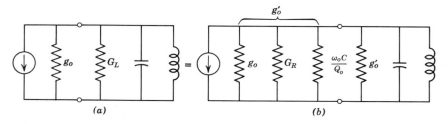

(a)   (b)

FIGURE 9.31
Equivalent circuits for a mismatched tuned amplifier where a resistor with conductance G_R is connected across the tuned circuit to provide the mismatch.

of Eq. 9.149 are shown in Fig. 9.31a. The circuit of Fig. 9.31b shows that the total load conductance G_L consists of the sum of the conductance of the loading resistor G_R, the coil loss conductance $1/Q_o \omega_o L$, which is the same as $\omega_o C/Q_o$, and the conductance g_o' coupled from the input of the next stage or actual load. The conductance g_o' is the actual output conductance g_o plus the conductances G_R and $\omega_o C/Q_o$ which represent the losses on the collector side of the tuned coupling circuit. Thus, maximum power transfer is achieved when the coupled conductance is also g_o'. Then Fig. 9.31 shows that

$$g_o + G_L = 2g_o' = 2\left(g_o + G_R + \frac{\omega_o C}{Q_o}\right) \tag{9.150}$$

The circuit Q is the resonant frequency divided by the bandwidth and is also the ratio of the total shunt resistance to $\omega_o L$, or alternatively, $\omega_o C$ to the total conductance. Thus,

$$Q = \frac{f_o}{B(Hz)} = \frac{\omega_o C}{2g_o'} = \frac{\omega_o C}{g_o + G_L} \tag{9.151}$$

Equation 9.151 may be solved explicitly for C.

$$C = \frac{Q(g_o + G_L)}{\omega_o} \tag{9.152}$$

This expression for C may be substituted into Eq. 9.150 to obtain

$$g_o + G_L = 2\left[g_o + G_R + \frac{Q(g_o + G_L)}{Q_o}\right] \tag{9.153}$$

and

$$(g_o + G_L)\left(1 - \frac{2Q}{Q_o}\right) = 2(g_o + G_R) \tag{9.154}$$

The loading conductance G_R may be obtained explicitly from Eq. 9.154,

$$G_R = \frac{g_o + G_L}{2}\left(\frac{Q_o - 2Q}{Q_o}\right) - g_o \tag{9.155}$$

If G_R is negative, the value of G_L is adequate to provide the desired stability without the additional loading resistor.

PROBLEM 9.15   Design the 200 MHz amplifier of Example 9.6 for 10 MHz bandwidth and stability ratio $k = 5$. Use single-tuned inductive coupling and design your circuit to couple two identical amplifiers. Assume $Q_o = 100$.

PROBLEM 9.16   A 2N4959 transistor has the following common-base y

parameters at $I_C = -2$ mA, $V_{CE} = -10$ V, $f = 200$ MHz; $y_{ib} = (55 - j15)$ mmho, $y_{ob} = (0.1 + j1.5)$ mmho, $y_{fb} = (-53 + j17)$ mmho, and $y_{rb} = (0 - j0.22)$ mmho.

   a.   Sketch the Y_T locus and check the validity of Eq. 9.144.

   b.   Design a common-base amplifier, using the 2N4959, with stability ratio $k = 5$ and compare the stage voltage gain of the common base with the common emitter.

PROBLEM 9.17   Design the 200 MHz amplifier of Problem 9.15 using a perfectly neutralized, maximum gain amplifier. Compare the stage voltage gain of this amplifier with the mismatched amplifier of Problem 9.15. Assume $Q_o = 100$.

PROBLEM 9.18   A FET with $g_{mo} = 5$ mmho, $I_{DSS} = 2.0$ mA, $r_d = 100$ kΩ, $C_{gd} = 1.0$ pF and $G_{gs} = 2.0$ pF is used as a 45 MHz amplifier with $b = 5.0$ MHz. Design a double-tuned circuit to couple identical stages with stability ratio $k = 5$. Determine the reference voltage gain. Assume $Q_o = 100$.

10

Direct-Coupled Amplifiers

The need frequently arises for an amplifier which will faithfully reproduce very slowly varying signals. The very low cutoff frequency required for such an amplifier may eliminate capacitive or transformer coupling from practical consideration and leave only direct coupling as a feasible solution. The main disadvantage of direct coupling is that thermal currents generated in the amplifier are amplified along with the signal currents. Thus, thermal stability problems are increased and thermal currents may mask signal currents. Therefore, particular attention must be paid to thermal stability in a direct-coupled or dc amplifier. Four amplifier types will be studied in this chapter. They are the Darlington connection, *npn-pnp* arrangements, differential amplifiers, and the cascode configuration.

10.1 THE DARLINGTON CONNECTION

One method of direct coupling bipolar transistors, known as the Darlington connection, is shown in Fig. 10.1. In this arrangement the emitter current of transistor T_1 is the base current of transistor T_2. If R_L is small, $i_{E1} = (\beta_1 + 1)i_{B1}$ and $i_{C2} = \beta_2 i_{B2}$. Then the ratio $i_{C2}/i_{B1} = (\beta_1 + 1)\beta_2$. The current i_{C1} adds to i_{C2} in the load resistor, but if β_2 is large, i_{C1} is negligible and the total amplification factor is approximately the product $\beta_1\beta_2$. Three transistors are sometimes used in the Darlington connection to produce a current gain approximately equal to $\beta_1\beta_2\beta_3$.

The thermal currents are also amplified in the Darlington connection as stated previously. The discussion in Chapter 5 showed that the thermal current amplification is equal to the current stability factor in any given

335

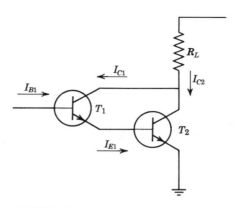

FIGURE 10.1
The Darlington connection.

stage. Resistors may be used in the Darlington circuit to reduce the termal currents as shown in Fig. 10.2. The thermal current $S_{I1}I_{CO1}$ from transistor T_1 is divided between resistor R_{B2} and the input of transistor T_2. Also part of the thermal current I_{CO2} of transistor T_2 may flow through R_{B2} to reduce the stability factor S_{I2}. Note that R_E is not bypassed because the expected signal frequencies are too low for effective bypassing. Therefore, for good voltage gain, R_E must be small in comparison with R_L. The signal gain is also reduced because part of the signal current is shunted

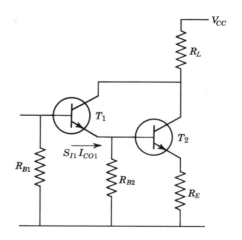

FIGURE 10.2
A Darlington-connected amplifier with linear thermal stabilization.

through the stabilizing resistors. In fact, a change in thermal current is indistinguishable from a signal current. Therefore, with this linear-type stabilization, high gain with adequate thermal stability can be achieved only by the use of silicon transistors which have very small values of I_{CO}.

We will now consider the design of a Darlington amplifier.

Example 10.1   A circuit is connected as shown in Fig. 10.2. The load resistor R_L is determined by the amplifier application. Let us assume that $R_L = 100$ Ω and $V_{CC} = 20$ V. We will next choose $R_E = 10$ Ω. The transistor selected for T_2 has $h_{FE2} = 100$, and $r_{b2} = 12$ Ω. Since the maximum value of $i_C \simeq 20$ V/$(R_L + R_E) \simeq 180$ mA, the average value of collector current i_{C2} should be about 90 mA. Then, $h_{ie2} = r_{b2} + (h_{FE2} + 1)r_{e2}$, but $(h_{FE2} + 1)r_{e2} = h_{FE2}/g_{m2}$. The value of g_{m2} is $I_{E2}(\text{mA})/25 \simeq 90/25 = 3.6$, and $h_{FE2}/g_{m2} = 100/3.6 = 28$ Ω. Therefore, $h_{ie2} = 40$ Ω at this average collector current. Now, transistor T_2 has an unbypassed resistor in the emitter circuit so the input resistance of transistor T_2 is $h_{ie2} + (h_{FE2} + 1)R_E = 40 + (101)10 = 1050$ Ω. A sensible choice of value for R_{B2} might be about the same as the input resistance of T_2. Then half of the signal current will be shunted through R_{B2}. We will choose $R_{B2} = 1$ kΩ. Then the total resistance in the emitter circuit of T_1 (Fig. 10.2) is approximately $R_{eq} = 1.05k \times 1k/2.05k = 512$ Ω. We will now select transistor T_1 with $h_{FE1} = 120$ and $r_{b1} = 100$ Ω. The average base current of transistor T_2 is $i_{C2}/\beta_2 = 90$ mA$/100 = 0.9$ mA. Assuming v_{BE} of transistor T_2 to be 0.6 V and recognizing that the voltage drop across R_E at the average value of emitter current is $I_{E2} R_E \simeq .09$ A$(10$ $\Omega) = 0.9$ V, the voltage across R_{B2} is 1.5 V and the current through R_{B2} is 1.5 V/$1k = 1.5$ mA at this average value of current. Then the average emitter current of transistor T_1 is $I_{RB2} + I_{B2} = 1.5 + 0.9 = 2.4$ mA and the average base current of T_1 is $i_{E1}/h_{FE1} = 2.4$ mA$/120 = 20$ μA.

The average input impedance of transistor T_1 is $(h_{FE1} + 1)(R_{eq}) + h_{ie1} \simeq h_{FE1} R_{eq} + h_{FE1}/g_{m1} = 120(512) + 120/(2.4/25) = 61.5k + 1.25k \simeq 63$ kΩ. Note that this high input impedance is due to the impedance in the emitter circuit, R_{eq}. Let us again sacrifice about one-half of the signal current and select $R_{B1} = 68$ kΩ. Then the total current gain for the Darlington amplifier is $K_i \simeq \beta_1/2$ times $B_2/2$ or $\beta_1\beta_2/4 = 3000$ and the voltage gain $K_v = v_0/v_I = K_i R_L/R_i = 3000 \times 100/63k = 4.77$. Note that transistor T_1 acts as an emitter follower and provides current gain but not voltage gain.

Diodes can be used to stabilize the Darlington amplifier as shown in Fig. 10.3. If the reverse saturation current I_{S2} of diode D_2 is equal to the thermal current I_{CO2} of transistor T_2, the current stability factor of transistor T_2 is unity. Similarly, if the saturation current I_{S1} is equal to the

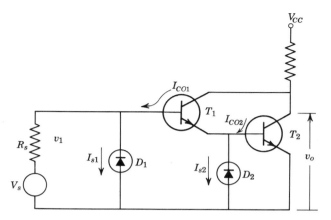

FIGURE 10.3
A diode-stabilized Darlington amplifier.

thermal current I_{CO1}, the current stability factor of transistor T_1 is one. The only problem arises in finding diodes which match the transistors in thermal currents. Since the resistance of a reverse-biased diode is very high, the approximate current gain of the amplifier of Fig. 10.3, in terms of the transistor betas, is $\beta_1\beta_2$.

Example 10.2 Let us assume that the Darlington amplifier of the preceding example has the stabilizing resistors replaced by diodes. Then at the average currents previously determined, the total current gain is approximately $K_i = h_{FE2} \times h_{FE1} = 120 \times 100 = 12{,}000$, the approximate input resistance of transistor T_1 is $\beta_1 h_{ie2} + h_{ie1} = 120 \times 40 + 1.25 \text{ k} = 6.05 \text{ k}\Omega$ and the approximate voltage gain $K_v = v_o/v_i = i_o R_o/i_i R_i = K_i R_o/R_i = 1.2 \times 10^4 \times 100/6.05 \text{ k} = 198$.

We have assumed that the signal source provided forward bias for the Darlington amplifier. If the signal source does not have a dc component which will provide this bias, a resistor must be connected between the base of transistor T_1 and V_{CC} to provide the required bias. Some modern devices contain a Darlington configuration in a single container.

PROBLEM 10.1 A given germanium power transistor, which has $h_{ie} = 10$ Ω, $\beta = 120$, and $I_{CO} = -80 \ \mu A$ at the desired q point, is driven by a Darlington-connected germanium transistor with $h_{ie} = 1500 \ \Omega$, $\beta = 150$, and $I_{CO} = -5 \ \mu A$ at its desired q point. Assume h_{re} and h_{oe} are negligible. The amplifier is diode stabilized. If the load resistance in the collector circuit of the power transistor is 30 Ω and $V_{CC} = -20$ V, determine the low-frequency input resistance and the voltage gain of the amplifier.

Answer: $R_i = 2700$ Ω, $K_v \simeq 200$.

PROBLEM 10.2   If the Darlington-connected amplifier of Problem 10.1 is capacitively coupled to its driving source, draw a circuit diagram of the amplifier and determine the value of bias resistance if $I_{C2} = 300$ mA. What should be the values of I_S for the diodes? Assume that all values are given at $T = 25°C$.

Answer: $R \simeq 1.2$ MΩ, $I_{S1} = 5$ μA, $I_{S2} = 80$ μA.

10.2   *npn-pnp* COMBINATIONS

A dc amplifier can be constructed by alternating *npn* and *pnp* types as shown in Fig. 10.4. This amplifier is diode stabilized and the input voltage is assumed to provide forward bias for the transistors. Observe that the collector current of transistor T_1 is the base current of transistor T_2.

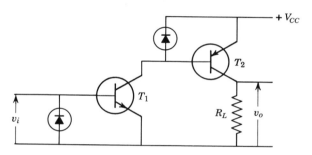

FIGURE 10.4
An *npn-pnp* **dc amplifier.**

Therefore, the input impedance of transistor T_1 is much lower and the voltage gain much higher for this amplifier as compared with the Darlington amplifier. The analysis of this type of amplifier is straightforward, as shown by the following example.

Example 10.3   Let us use the same transistor T_1 as in the Darlington amplifiers of Examples 10.1 and 10.2 and a *pnp* transistor with similar characteristics to T_2 and compare the current gain, input resistance, and voltage gain to the Darlington amplifier values at the same average current values. In Example 10.1 we found that $h_{ie2} = 40$ Ω, $h_{FE2} = 100$, $R_L = 100$ Ω, $h_{ie1} = 1.25$ kΩ, and $h_{FE1} = 120$. Then the current gain is $K_i \simeq \beta_1\beta_2 = 120 \times 100 = 1.2 \times 10^4$, $R_{in} \simeq h_{ie1} = 1.25$ kΩ, and $K_v = K_i R_o/R_i = 1.2 \times 10^4 \times 100/1.25$ k $= 960$. Note that the voltage gain is increased by the same ratio as the reduction of input resistance. Observe also that the dc potential in the output is zero when the dc potential of the input is zero. Sometimes this is a distinct advantage.

The alternating *npn-pnp* arrangement may be extended to include any desired number of transistors. For example, Fig. 10.5 uses three transistors. Notice that the zero signal output and input potential can be the same only if an even number of transistor stages is used in the amplifier.

As previously mentioned, the biggest problem in building a diode-stabilized amplifier is finding diodes with the required reverse saturation currents. Sometimes a transistor is used in place of a diode as shown in Fig. 10.6. Since the thermal current in the collector circuit of a transistor is $S_I I_{CO}$, and S_I is approximately equal to R_B/R_E as discussed in Chapter

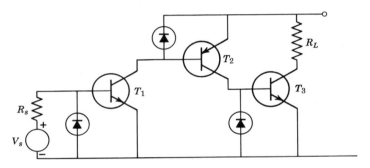

FIGURE 10.5
A three-stage dc amplifier.

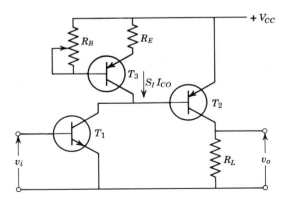

FIGURE 10.6
Thermal stabilization by the use of a transistor.

5, the thermal current in transistor T_3 can be controlled by adjusting R_B. In fact, if I_{CO} in transistor T_3 is large enough, this transistor can compensate for the thermal currents in both T_1 and T_2, and no diodes are required in the two-stage amplifier.

PROBLEM 10.3 Assume that the amplifier of Fig. 10.6 has $R_L = 30\ \Omega$ and $V_{CC} = 20$ V. The characteristics of T_2 are: $h_{ie} = 10\ \Omega$, $\beta = 120$, h_{re} and h_{oe} are negligible. The characteristics of T_1 are: $h_{ie} = 1500\ \Omega$, $\beta = 150$, h_{re} and h_{oe} are negligible at the respective q points. What is the voltage gain and what is the input resistance of the amplifier?

Answer: $K_v = 360$, $R_i = 1500\ \Omega$.

PROBLEM 10.4 Referring to the amplifier of Problem 10.3, transistor T_2 has $I_{CO} = 80\ \mu A$; transistor T_1 has $I_{CO} = 1\ \mu A$ and $S_I = 50$; transistor T_3 has $I_{CO} = 10\ \mu A$. What must be the approximate ratio R_B/R_E to thermally stabilize the amplifier?

Answer: 13.

Vacuum tubes are difficult to use in dc amplifiers. The reason for this difficulty is due to the plate requiring a positive potential and the grid requiring a negative potential with respect to the cathode. Thus, if an amplifier is connected as shown in Fig. 10.7, the cathode potential of

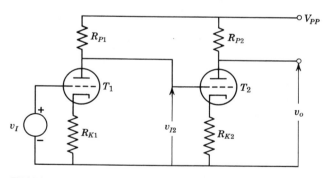

FIGURE 10.7
A vacuum tube dc amplifier.

tube T_2 must be more positive than the plate of T_1 for proper bias. Under these conditions, R_{K2} must be quite large. Then, since R_{K2} cannot be bypassed with a capacitor for a dc amplifier, the voltage gain of tube T_2 will be approximately equal to R_{P2}/R_{K2}. However, if R_{P2} is made large, the current through the tube T_2 decreases and R_{K2} must be increased to obtain the proper bias. Consequently, the voltage gain of the second tube is quite small. Other vacuum tube configurations are possible, but, in general, dc tube amplifiers are difficult to design.

The FET has the same problems as the vacuum tube unless both p-channel and n-channel devices are intermixed. Then, configurations similar to the p-n-p and n-p-n combinations are possible.

As mentioned previously, the MOSFET is easy to use in dc amplifiers. Since both the gate and the drain require the same polarity of bias voltage, the circuit would appear as shown in Fig. 10.8. The value of R_{D1} must be chosen so the quiescent value of drain voltage for T_1 is equal to the desired quiescent value of gate voltage for T_2. Resistors could be added between the sources and ground to provide more design flexibility.

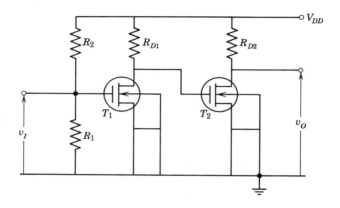

FIGURE 10.8
A MOSFET dc amplifier.

PROBLEM 10.5   Two 6J5 triodes are connected as shown in Fig. 10.7. The desired q point for both tubes is at $v_{GK} = -6$ V and $i_p = 3.5$ mA. Determine the values of R_{K1}, R_{P1}, R_{K2}, and R_{P2} for this amplifier if $V_{PP} = 250$ V. What is the voltage gain of tube T_1 and of tube T_2?

PROBLEM 10.6   Two p-channel MOSFETs are connected as a dc amplifier. Draw the circuit diagram of this amplifier. The characteristics of these MOSFETs are given in Fig. 6.19. If V_{DD} is -20 V and the desired quiescent values of $v_G = -10$ V, find the values of all resistors in the circuit. What is the voltage gain of each MOSFET?

10.3 DIFFERENTIAL AMPLIFIERS

The next type of dc amplifier to be considered is the differential amplifier. There are two basic types, balanced and unbalanced. The circuit diagram of a balanced differential amplifier is given in Fig. 10.9. In this amplifier the input signal v_i is balanced with respect to ground. With this type of signal, the forward bias of transistor T_1 is increased while the forward bias of transistor T_2 is decreased. If the transistors are matched and linear, the emitter current of one transistor increases by the same amount the emitter current of the other transistor decreases and the current through the common-emitter resistor R_E remains constant. Therefore, the voltage across R_E remains constant and no degeneration is caused by this resistor. If the transistors are not well matched, a small voltage will appear across R_E, but this voltage will tend to degenerate the higher gain transistor and regenerate the lower one, and thus improve the balance. Since the impedance between the base and the emitter of each transistor is h_{ie}, the input impedance from base to base is $2h_{ie}$.

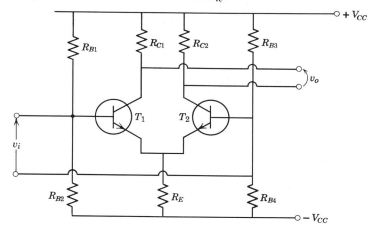

FIGURE 10.9
A balanced differential amplifier.

The collector potential of one transistor (Fig. 10.9) increases while the collector potential of the other transistor decreases. The output voltage, which is the difference between the collector potentials may be fed to a balanced load or to another balanced amplifier. The balanced amplifier which follows may conveniently use the opposite type transistors (p-n-p to follow n-p-n) as shown in Fig. 10.10. However, the transistors may all be of the same type. Note that the two transistors in a balanced amplifier

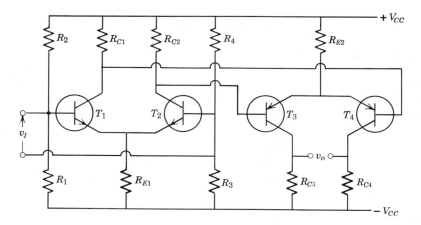

FIGURE 10.10
A two-stage balanced amplifier.

appear to be in series when viewed from the output terminals as they do when viewed from the input terminals. Therefore, the output resistance of the balanced amplifier is twice that of a single transistor amplifier at the same q point. However, the current gain and voltage gain of the blanced amplifier are the same as a single transistor similarly biased and with the emitter resistor perfectly bypassed.

The main advantage of the balanced amplifier is that in-phase input signals which are applied to the two bases do not produce an output signal, which is the difference between the two collector potentials. These in-phase signals are called *common mode* signals and include thermally generated currents. Changes in v_{BE} due to temperature changes, and extraneous signals are not transferred from one stage to the next providing the two transistors are well matched and maintained at the same temperature. Also, the individual transistors may have excellent q-point stability because of the high permissible value of emitter circuit resistance R_E.

The effective value of R_E can be increased greatly, and yet allow the desired value of emitter current to flow, if a transistor is used to replace the resistor R_E as shown in Fig. 10.11. Since the collector current of the emitter-circuit transistor T_3 is determined almost entirely by the stabilized bias circuit D_1, R_1, R_2 and R_E', the sum of the emitter currents $2I_E$ of the differential amplifier is held constant and, therefore, the q-point collector currents and voltages of the differential amplifier are held constant. The diode D_1 in the stabilized-bias circuit compensates for the temperature variation of v_{BE} in transistor T_3.

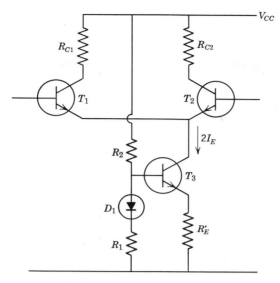

FIGURE 10.11
A differential amplifier with improved q-point stabilization.

When an input signal is applied to a differential amplifier, each collector voltage varies with respect to V_{CC} or ground, as previously noted. Therefore, an output signal which is referenced to V_{CC} or ground can be obtained from the difference amplifier to drive a single-ended amplifier or load. The amplifier then has a balanced input and an unbalanced output, and the voltage gain is decreased by a factor of two.

The differential amplifier can be used with both the input and the output unbalanced, as shown in Fig. 10.12. In this circuit, the base of one transistor is used as the ground reference. At first glance, it may appear that the unbypassed emitter resistor R_E will cause serious degeneration and low gain in the amplifier. However, the common-base input impedance h_{ib} of transistor T_2 is in parallel with R_E. Since h_{ib} is normally very small in comparison with R_E, the input impedance of transistor T_1 is

$$R_{it} \simeq h_{ie1} + (h_{fe1} + 1)h_{ib2} \qquad (10.1)$$

But if the two transistors are alike, $(h_{fe1} + 1)h_{ib2} = h_{ie1}$. Therefore,

$$R_{it} \simeq 2h_{ie1} \qquad (10.2)$$

When the output of a differential amplifier is unbalanced, the common-mode signals are not cancelled in the output because the output voltage

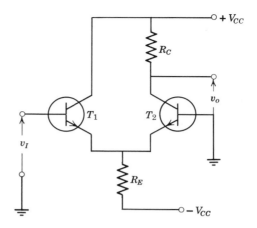

FIGURE 10.12
A differential amplifier with unbalanced input
and output.

is referenced to a fixed point, V_{CC} or ground. However, the amplifier gain for common-mode signals may be very small because h_{ib2} is not in parallel with R_E when in-phase signals are applied to both bases. These in-phase signals could be extraneous signals in a balanced input or thermally induced signals ΔI_{CO} and ΔV_{BE} in either a balanced or unbalanced input. Thus, an index of the goodness of a differential amplifier is the *common-mode rejection ratio (CMRR)* defined as follows

$$CMRR = \frac{\text{Voltage gain for difference signals}}{\text{Voltage gain for common-mode signals}} \quad (10.3)$$

Since the voltage gain for difference signals is

$$K_v = \frac{K_i R_L}{R_i} \simeq \frac{\beta R_L}{2h_{ie}} \quad (10.4)$$

The voltage gain for common-mode signals is approximately $R_L/2R_E$, since both emitter currents flow through R_E (which doubles the effectiveness of R_E). Then

$$CMRR \simeq \frac{(\beta R_L/2h_{ie})}{(R_L/2R_E)} = \frac{\beta R_E}{h_{ie}} \quad (10.5)$$

Therefore, a high common-mode rejection ratio is obtained when R_E is very large in comparison with $h_{ie}/\beta = h_{ib}$. Notice that the unbalanced difference amplifier of Fig. 10.12 is actually a common-collector amplifier directly coupled to a common-base amplifier. As noted previously, the voltage gain of a common-collector amplifier is about one. However, for the unbalanced difference amplifier, the load on T_1 (Fig. 10.12) is the input impedance of a common-base amplifier which is very low. In fact, the input signal is applied across two forward-biased junctions (the base-emitter junctions of T_1 and T_2) in series. As a result, the voltage on the emitters is about one-half of the input voltage. Thus the voltage gain of the common-collector stage is about 0.5 if two identical transistors are used.

The differential amplifier with unbalanced input and output (Fig. 10.12) is often called an *emitter-coupled amplifier*. This configuration has two distinct advantages over the single common-emitter amplifier in addition to the good thermal stability previously discussed. First, the emitter-coupled amplifier has a much wider bandwidth because the voltage gain of the common-collector stage is about 0.5 so the Miller effect is essentially eliminated in the first stage. In addition, the common-base amplifier (T_2 of Fig. 10.12) has a very low input impedance and is driven from a low source impedance. Thus $R_{sh} C_{sh}$ will be small and ω_2 will be large. Second, the output signal is well isolated from the input by the grounded base of T_2. Consequently, the emitter-coupled amplifier is popular as an *RF* amplifier.

Example 10.4 Two identical transistors are connected as shown in Fig. 10.12. The resistor $R_C = 1 \text{ k}\Omega$ and $R_E = 500 \Omega$. The transistor parameters are:

$$h_{ie} = 1 \text{ k}\Omega \qquad h_{FE} = 100$$
$$h_{re} \simeq 0 \qquad h_{oe} = 10^{-5} \text{ mhos}$$

Let us determine the characteristics of this amplifier.

First, the input impedance is $2 h_{ie} = 2 \text{ k}\Omega$. Since T_2 is a common-base amplifier, the output impedance of transistor T_2 is very high (about $h_{FE}/h_{oe} \simeq 10 \text{ M}\Omega$). Then, the output impedance of the amplifier is R_C in parallel with the output impedance of transistor T_2 or $R_o \simeq 1 \text{ k}\Omega$.

The voltage gain of T_1 is $\simeq 0.5$. The input impedance to transistor T_2 is $h_{ib} = h_{ie}/(h_{FE} + 1) = 1000/101 \simeq 10 \Omega$. The voltage gain of T_2 is $\simeq R_C/h_{ib} = 1000/10 = 100$. Then the total voltage gain of the amplifier is $0.5 \times 100 = 50$.

The current gain of $T_1 \simeq h_{FE} = 100$ since $1/h_{oe} \ll$ the load on T_1 which is approximately the input impedance to T_2 or h_{ib}. In addition, since $R_C \ll$ the output resistance of transistor T_2, the current gain of $T_2 \simeq \alpha = 0.99 \simeq 1$. Thus, the total current gain of the amplifier is about $100 \times 1 = 100$. As a check, note that $K_v = K_i R_o/R_i = 100 \times 1000/2000 = 50$.

The *CMRR* is $h_{FE} R_E/h_{ie} = 100 \times 500/1000 = 50$. Thus, the input signal v_I will be amplified 50 times as much as any temperature-induced voltage at the emitter-base junction of T_1 if T_2 is maintained at the same temperature.

PROBLEM 10.7   Repeat Example 10.4 if $R_C = 2$ kΩ and $R_E = 1$ kΩ.

Answer: $R_i = 2$ kΩ, $R_o = 2$ kΩ, $K_v = 100$, $K_i = 100$, $CMRR = 100$.

Transistor manufacturers provide two transistors made on the same silicon or germanium chip. These transistors are closely matched and maintain essentially the same temperature. Therefore, they provide very good thermal characteristics in differential amplifier applications. Balanced amplifiers are very commonly used in integrated circuits because of their universal application and excellent thermal characteristics. Integrated circuits will be discussed in more detail in Chapter 15.

PROBLEM 10.8   Two 2N3903 transistors are connected as shown in Fig. 10.12. $R_C = 2$ kΩ and $R_E = 1$ kΩ. The quiescent collector current of T_1 is 10 mA, and the quiescent collector current of T_2 is 5 mA. Determine R_i, R_o, K_v, K_i and $CMRR$ for this configuration.

10.4   THE CASCODE AMPLIFIER

An amplifier configuration somewhat related to the unbalanced differential amplifier is shown in Fig. 10.13a. This configuration is known as a *cascode* amplifier. Usually, operation from a single power supply is desirable so the configuration shown in Fig. 10.13b is used. In this form (Fig. 10.13b), the resistors R_3 and R_4 maintain the base of transistor T_2 at a dc potential above 0 V. The capacitor C is normally connected from the base of transistor T_2 to ground to remove any signals which may be capacitively coupled into this base circuit.

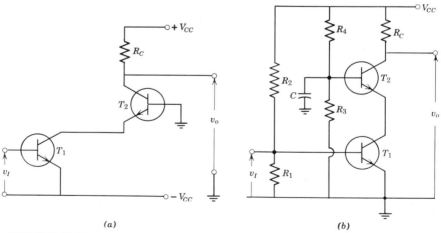

(a)   (b)

FIGURE 10.13

A cascode amplifier: (a) two-power supply configuration; (b) one-power supply configuration.

Electrically, the cascode amplifier is essentially an unbalanced differential amplifier with the common-collector stage replaced by a common-emitter stage. Of course the load on the common-emitter stage is very low (the input impedance of the emitter of T_2) so the voltage gain of this stage is very low. The analysis is similar to that followed in Example 10.4.

Both the cascode and the unbalanced differential amplifiers have such low voltage gains in the first stage that the Miller effect (on input capacitance) is very small. From Chapter 7, we note the shunt capacitive effect becomes more troublesome for high impedance circuits. Consequently, let us examine the behavior of a FET cascode amplifier.

Example 10.5 Two identical FETs are connected as shown in Fig. 10.14. The driving source resistance $R_s = 10^6\ \Omega$, $R_D = 5\ \text{k}\Omega$, $C_{gs} = 5\ \text{pF}$, and $C_{gd} = 4\ \text{pF}$. The pinch-off voltage of each transistor is two volts and $V_{DD} = 30\ \text{V}$. The g_m of each transistor is 2×10^{-3} mhos. Let us determine the characteristics of this amplifier.

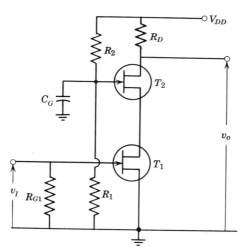

FIGURE 10.14
A FET cascode amplifier.

The load impedance of transistor T_1 is the impedance looking into the source of transistor T_2. This impedance is $\Delta v_s / \Delta i_D$, but since v_G is held constant by C_G, Δv_s is equal to Δv_{SG} or $-\Delta v_{GS}$. However, $\Delta i_D / \Delta v_{GS} = g_{m2}$. Thus, as previously noted, the magnitude of the impedance looking into the source of T_2 is $1/g_{m2}$. Then, the voltage gain of transistor T_1 is $g_{m1}\,(1/g_{m2})$. Since both transistors are identical, the voltage gain of transistor T_1 is *one*.

The effective input resistance to T_1 is r_{gs} (very large) and the effective input capacitance to T_1 is $C_{gs} + (K_v + 1)C_{gd} = C_{gs} + 2C_{gd} = 5 + 2 \times 4 = 13$ pF.

Since the input resistance to T_2 is low $(1/g_m = 500\ \Omega)$, the input capacitance of this transistor (C_{gs}) does not limit the upper cut-off frequency of the amplifier. The gate grounding capacitor C must be very large in comparison with C_{gs} or C_{gd} in order to maintain the gate at signal ground.

The resistors R_1 and R_2 establish the q-point drain voltages V_{DS} across both transistors. In Fig. 10.14, transistor T_1 has zero gate to drain bias so transistor T_2 must also have zero gate to drain voltage so the drain currents of the two transistors will be identical. Of course, both transistors should have drain voltages V_{DS} higher than the pinch-off voltage V_P. In fact, this requirement can provide the basis for determining R_1 and R_2. With zero gate bias, the voltage drop across R_1 is equal to the q-point drain voltage V_{DS} of transistor T_1. Then,

$$V_{DS1} = \frac{V_{DD}R_1}{R_1 + R_2} \tag{10.6}$$

or

$$R_1 = \frac{V_{DS1}R_2}{V_{DD} - V_{DS1}} \tag{10.7}$$

The pinch-off voltage of our transistors is two volts, so let us choose V_{DS1} at least one or two volts greater than V_P. Let us choose $V_{DS1} = 5$ V. We must also choose a value for R_2. Let $R_2 = 220$ kΩ so it will not load the power supply. Then $R_1 = 5 \times 220/(30 - 5) = 44$ kΩ. The drain supply voltage for transistor T_2 is $30 - 5 = 25$ V. The design of the cascode amplifier can now proceed as though it were a single FET with $V_{DD} = 25$ V.

The voltage gain of transistor T_2 is $g_m R_D = 2 \times 10^{-3} \times 5 \times 10^3 = 10$. The voltage gain of the entire amplifier is $1 \times 10 = 10$.

The output impedance of this amplifier is R_D in parallel with the output impedance of transistor T_2. In this example, $R_o \simeq R_D = 5$ kΩ.

Reverse bias will sometimes be desired, depending upon the temperature stability requirements, the magnitude of input signal or the impedance of the load. A bias resistor and bypass capacitor must then be included in the source circuit of transistor T_1, as shown in Fig. 10.15, to provide the proper bias. Transistor T_2 must have the same drain current and will automatically adjust to the same bias if both transistors have V_{DS} greater than V_P. However, for the n-channel transistors shown, the gate potential of transistor T_2 is more negative than the drain potential of transistor T_1 by the amount of the bias voltage V_{GS}. But this same amount of voltage

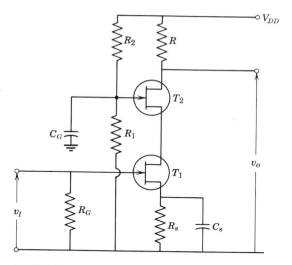

FIGURE 10.15
A FET cascode amplifier for ac signals.

is dropped across the bias resistor of transistor T_1, so the voltage drop across R_1 is still equal to the drain-to-source voltage V_{DS} of transistor T_1. Therefore, the design of the biased cascode amplifier can proceed in the same manner as for a single FET with source bias, but the effective drain supply voltage is V_{DD} minus the V_{DS} of transistor T_1.

The absence of coupling between the output and the input circuits of the cascode amplifier is a great advantage in some applications, particularly in tuned radio-frequency amplifiers (Chapter 9). Also we should note that stray wiring capacitance should be kept to an absolute minimum since it adds directly to the FET capacitance which is very small in the cascode configuration.

The transistor output impedance of the cascode configuration is very high, as shown below. The equivalent circuit of the cascode configuration is as shown in Fig. 10.16a. However, the voltage v_o applied to the output terminals does *not* cause current to flow in the gate circuit of transistor T_1 at low and moderate frequencies so $v_{gs1} \simeq 0$. On the other hand, the source-to-ground voltage of T_2 is caused by the $i_d r_d$ drop of T_1 in the source circuit of T_2. Therefore, since the gate of T_2 is at ground potential,

$$v_{gs2} = -v_{sg2} = -i_d r_d \tag{10.8}$$

The circuit of Fig. 10.16a is simplified to that shown in Fig. 10.16b by transforming the current source $g_m v_{gs2}$ to a voltage source $g_m r_d \, v_{gs2}$ and

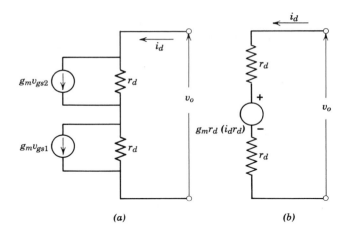

(a)   (b)

FIGURE 10.16
The equivalent circuit for a cascode FET configuration:
(a) **current source model;** (b) **modified voltage source model.**

then replacing v_{gs2} by $-i_d r_d$. The polarity of the voltage generator is reversed to eliminate the negative sign. Observe from Fig. 10.16b that

$$v_o = i_d(2r_d) + g_m r_d^2 i_d \tag{10.9}$$

and

$$r_{ot} = \frac{v_o}{i_d} = 2r_d + g_m r_d^2 \tag{10.10}$$

Simplifying,

$$r_{ot} = r_d(2 + g_m r_d) \tag{10.11}$$

Since $g_m r_d$ is much greater than one, the output resistance r_{ot} is extremely high. For example, a typical FET with $g_m = 2 \times 10^{-3}$ mho and $r_d = 10^5 \, \Omega$ would have $r_{ot} = 10^5(2 + 200) \simeq 20$ MΩ when placed in the cascode configuration. The output resistance of bipolar transistors or tubes in cascode may be found approximately by Eq. 10.10 by replacing r_d with the appropriate output resistance of the single device.

PROBLEM 10.9   A given FET has $g_m = 2 \times 10^{-3}$ mho, $C_{gd} = 2$ pF, and $C_{gs} = 3$ pF at the q point $I_D = 1.5$ mA, $V_{DS} = 5$ V, $V_{GS} = -0.1$ V. If $g_m = 3 \times 10^{-3}$ mho, $R_l = 6$ kΩ, and the driving-source resistance is 50 kΩ, determine the upper cutoff frequency f_2 for an amplifier using this single FET.
Answer: $f_2 = 77.7$ kHz.

PROBLEM 10.10 Calculate values of R_S and R_1 (Fig. 10.15) for a cascode amplifier to replace the amplifier of Prob. 10.9 using two of the FETs of Problem 10.9 at the q points given, with $V_{DD} = 25$ V and $R_3 = 100$ kΩ. Determine the upper cutoff frequency f_2.
Answer: $R_S = 67$ Ω, $f_2 = 455$ kHz.

PROBLEM 10.11 A given transistor amplifier has a current-stabilizing transistor in the emitter circuit as shown in Fig. 10.11. What is the *CMMR* if $h_{ie} = 1$ kΩ and $\beta = 100$ for transistors T_1 and T_2 and the output resistance of the stabilizing transistor T_3 is 200 kΩ?
Answer: 2×10^4.

PROBLEM 10.12 What is the approximate voltage gain of the amplifier of Problem 10.11 if the output is unbalanced and the load resistance is 2 kΩ?
Answer: 100.

PROBLEM 10.13 Two 2N3903 FETs are used in the cascode configuration for the *RF* amplifier in an FM tuner. The center frequency $f_o = 100$ MHz and the desired bandwidth is 1 MHz. The driving source is a twin-lead transmission line with $R_S = 300$ Ω and the load is another cascode FET amplifier. Assume $Q_o = 150$. Design the amplifier and calculate the voltage gain v_o/v_s.

11

Power Amplifiers

A typical amplifier consists of several stages of amplification. Most of these stages are small-signal, low-power devices. For these stages, efficiency is usually unimportant, distortion is negligible, and the equivalent circuits accurately predict the amplifier behavior. In contrast, the final stage of an amplifier (and in some cases an additional driver stage) is usually required to furnish appreciable signal power to its load. Typical loads include loud-speakers, antennas, positioning devices, and so on. These amplifiers are commonly called power amplifiers. Because of this relatively high power level, the efficiency of the power amplifier is important. Also distortion becomes a problem because the amplifier parameters vary appreciably over the signal cycle. Therefore, the equivalent circuits are only rough approximations and graphical methods assume increased importance. Heat dissipation also becomes a problem. This chapter will discuss these problems which are peculiar to power amplifiers, namely: distortion, efficiency, push-pull configurations, and thermal conduction.

11.1 AMPLIFIER DISTORTION

The power amplifier is usually expected to deliver the maximum power, for which it is capable, to the load. Therefore, it may be driven from approximately cutoff on one half cycle to saturation on the next half cycle. Also, the load line will probably pass near the maximum dissipation curve as shown in Fig. 11.1a. Therefore, h_{fe} and h_{ie} may vary widely over the signal cycle. Fortunately, these two nonlinearities tend to cancel as shown in Fig. 11.2. The upper part of this figure was constructed by plotting i_C as a function of i_B from values taken along the load line in Fig. 11.1a. The lower curve in Fig. 11.2 is the inverted input characteristic obtained directly from Fig. 11.1b.

354

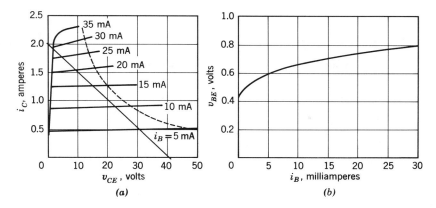

FIGURE 11.1
Characteristics of a typical silicon power amplifier.

If v_{BE} is known, the value of i_B can be determined directly from the lower curve in Fig. 11.2. However, one usually knows the waveform of the driving source voltage v_S. In this case, we note that $v_S = i_B R_s + v_{BE}$ where R_s is the internal resistance of the driving source. As we have already noted, equations of this type represent straight lines on the transistor characteristic curves. In fact, note that a value v_S may be obtained by drawing a line with slope R_s from a value of v_{BE} to the $i_B = 0$ axis. To help clarify this concept, let us consider an example.

Example 11.1 The transistor whose characteristics are given in Fig. 11.2 is connected to a voltage source whose open-circuit output voltage is $v_S = 0.72 + 0.28 \sin \omega t$. This voltage source has an internal impedance equal to 6.7 Ω. Let us determine the waveform of i_C for this input signal.

When $t = 0$, the signal component (the sinusoid) is zero. Thus, $v_S = 0.72$ V which is the q point for the transistor in this example. This voltage appears on the curve in Fig. 11.2 at point C_1. Now, v_{BE} will be less than v_S due to the drop across R_s which is equal to $i_B R_s$. Fortunately, we have already graphically solved problems of this type when we drew load lines on the characteristic curves of diodes or transistors. In this example, the intercept with the v_{BE} axis occurs at the voltage v_S. The slope of the dashed line from point C_1 toward the input characteristic i_B versus v_{BE} curve has a value of $R_s = \Delta v_{BE}/\Delta i_B$ as shown in Fig. 11.2. In this example, $R_s = 6.7\ \Omega$ so if $\Delta i_B = 15$ mA, the value of $\Delta v_{BE} = .015 \times 6.7 = .07$ V. A dashed line with this slope is drawn from point C_1 until this line intercepts the input characteristic curve at point C_2. This point C_2 represents the actual voltage v_{BE} when $v_S = 0.72$ V. In this example, the q-point value of v_{BE} is $\simeq 0.65$ V.

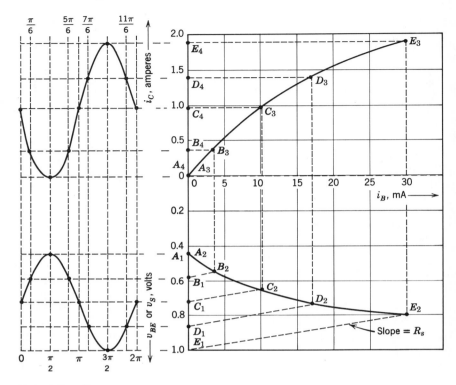

FIGURE 11.2
Construction used to determine the distortion in a transistor amplifier.

The value of i_B is found by projecting point C_2 up to the i_B axis. In this case, the q-point value of i_B is 10.5 mA. This same value of i_B can be projected up to the dynamic output curve (point C_3). This point C_3 determines the value of output collector current i_C when projected over to the i_C axis (point C_4). Thus, the quiescent collector current is approximately 0.98 A. Additional values of v_S are projected in this fashion until i_C can be drawn. Since R_s is a constant, all of the lines from v_S to the input characteristic curve will be parallel. In this example, we have projected values of v_S for $t = 0$, $t = \pi/6$, $t = 5\pi/6$, $t = \pi$, $t = 7\pi/6$, etc., so we could construct the i_C versus time curve.

PROBLEM 11.1   Repeat Example 11.1 if R_S is changed to 20 Ω.

Note from the construction in Fig. 11.2 that the output signal is not sinusoidal because the positive half cycle of collector current is smaller than the negative half cycle. When the output signal does not have the same

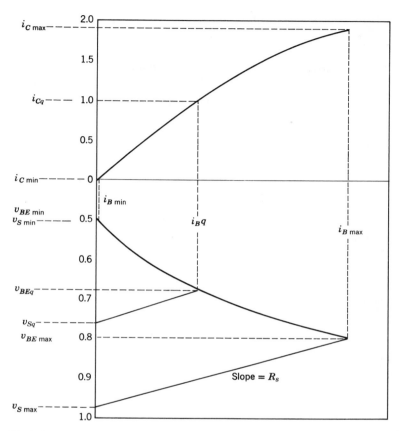

FIGURE 11.3
Arrangement for determining the optimum R_s.

shape as the input signal, the output signal is said to be *distorted*. If the usable portion of the input characteristic has more curvature than the transfer $(i_C$ versus $i_B)$ characteristic, there exists a value of source resistance R_s which will make the positive half cycles of output current equal to the negative half cycles. This equalization and, hence, minimum (or near-minimum) distortion can be obtained because the source resistance dilutes the effect of the nonlinearity of the input characteristic.

The value of driving source resistance R_s which gives minimum distortion can be determined from the relationships shown in Fig. 11.3. First, the value of i_{Cq} is located midway between $i_{C\,max}$ and $I_{C\,min}$ to give equal successive half cycles of output current. Then the corresponding values of $i_{B\,max}$, i_{Bq}, and $i_{B\,min}$ are determined from the transfer curve. Then the corresponding values of $v_{BE\,max}$, v_{BEq}, and $v_{BE\,min}$ are obtained from the input characteristics. Also, the corresponding source potentials may be expressed as

$$v_{S\,max} = v_{BE\,max} + i_{B\,max}\,R_s \tag{11.1}$$

$$v_{Sq} = v_{BEq} + i_{Bq}\,R_s \tag{11.2}$$

$$v_{S\,min} = v_{BE\,min} + i_{B\,min}\,R_s \tag{11.3}$$

But, v_S is assumed to be sinusoidal or symmetrical. Therefore,

$$v_{S\,max} - v_{Sq} = v_{Sq} - v_{S\,min} \tag{11.4}$$

Rearranging Eq. 11.4

$$v_{S\,max} + v_{S\,min} = 2\,v_{Sq} \tag{11.5}$$

Substituting the values of v_S given by Eqs. 11.1, 11.2, and 11.3 into Eq. 11.5, we obtain

$$v_{BE\,max} + v_{BE\,min} + (i_{B\,max} + i_{B\,min})R_s = 2(v_{BEq} + i_{Bq}\,R_s) \tag{11.6}$$

Solving for R_s

$$R_s = -\frac{v_{BE\,max} + v_{BE\,min} - 2v_{BEq}}{i_{B\,max} + i_{B\,min} - 2i_{Bq}} \tag{11.7}$$

PROBLEM 11.2   Determine the value of source resistance which will give minimum (or near minimum) distortion for the amplifier with characteristics given in Figs. 11.1 or 11.2.

Answer: $R_s = 2.5\ \Omega$.

The value of R_s which provides low distortion for the preceding problem is unreasonably small for most sources, such as a driver transistor. Therefore, the distortion will probably not be minimized. Consequently, we need to be able to determine the percent distortion of the amplifier when the source resistance is not optimized so we can make a judgment as to the acceptability of the distortion level.

Harmonic (or waveform) distortion is analyzed by assuming the driving source to be sinusoidal and then finding the amplitude coefficients of the first few terms of the Fourier series which represents the distorted output. For example, the output current can be represented by the series

$$i_C = I_0 + I_1 \cos \omega t + I_2 \cos 2\omega t + I_3 \cos 3\omega t - \tag{11.8}$$

The series may contain only cosine terms, as given, when the $t = 0$ axis is chosen at a point of even symmetry as shown later. The problem can be simplified further if the output distortion can be identified as being either

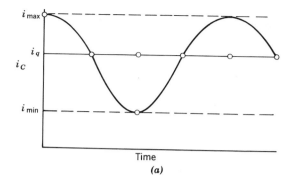

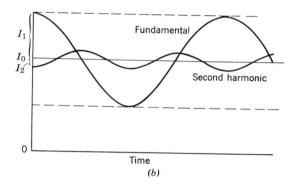

FIGURE 11.4
Second harmonic distortion: (*a*) **total current;**
(*b*) **frequency components.**

primarily composed of even harmonics or primarily composed of odd
harmonics. For example, if successive half cycles are dissimilar as shown in
Fig. 11.4*a*, even harmonics must be present and they usually predominate
over odd harmonics which may also be present. Note that the total current
in Fig. 11.4*a* resembles very closely the collector current waveform of
Fig. 11.2 which resulted from nonoptimum source resistance. However, this
waveform can be obtained by adding the dc component I_0, the fundamental
component $I_1 \cos \omega t$ and the second harmonic component $I_2 \cos 2\omega t$. Thus,
for this special case where successive half cycles are dissimilar, we can
assume the distribution to be primarily second harmonic and write the
following relationship.

$$i_C = I_0 + I_1 \cos \omega t + I_2 \cos 2\omega t \qquad (11.9)$$

But when $t = 0$, $i_C = i_{C\,max}$. In addition, $\cos \omega t = 1$ and $\cos 2\omega t = 1$. Therefore, using Eq. 11.9,

$$i_{C\,max} = I_0 + I_1 + I_2 \tag{11.10}$$

Also, when $\omega t = \pi/2$ or $90°$, $i_C = i_{C_q}$, $\cos \omega t = 0$, and $\cos 2\omega t = -1$. Then

$$i_{C_q} = I_0 - I_2 \tag{11.11}$$

And when $\omega t = \pi$ or $180°$, $i_C = i_{C\,min}$, $\cos \omega t = -1$, and $\cos 2\omega t = 1$. Thus

$$i_{C\,min} = I_0 - I_1 + I_2 \tag{11.12}$$

These equations can be solved simultaneously to yield I_0, I_1, and I_2 in terms of $i_{C\,max}$, $i_{C\,min}$, and i_{C_q}, which we know. For example, if Eq. 11.12 is subtracted from Eq. 11.10, I_1 is obtained.

$$I_1 = \frac{i_{C\,max} - i_{C\,min}}{2} \tag{11.13}$$

Similarly, if Eq. 11.11 is added to Eq. 11.10 and Eq. 11.13 is substituted for I_1 in the result, I_0 is obtained

$$I_0 = \frac{i_{C\,max} + i_{C\,min} + 2I_{C_q}}{4} \tag{11.14}$$

Then I_2 may be found by substituting Eq. 11.14 into Eq. 11.11

$$I_2 = \frac{i_{C\,max} + i_{C\,min} - 2i_{C_q}}{4} \tag{11.15}$$

The percentage of second harmonic distortion in the amplifier is

$$\% \text{ second harmonic} = \frac{I_2}{I_1} \times 100 \tag{11.16}$$

PROBLEM 11.3 Determine the peak amplitude of the fundamental, the peak value of the second harmonic and percent second harmonic distortion in the amplifier represented by Fig. 11.2 with the maximum level input signal given.

Answer: $I_1 = 0.97$ A, $I_2 = .04$ A, second harmonic = 2.6%.

The 2.6 percent value of distortion determined for the amplifier of Fig. 11.2 is not objectionable for many applications but is unacceptable in a high-fidelity amplifier where one percent distortion is usually considered to be the upper acceptable limit.

Sometimes successive half cycles of output current or voltage are similar, as they may be when optimum source resistance is used or when balanced push-pull amplifiers which are discussed in Section 11.3 are employed. However, these half cycles are not usually sinusoidal because the peaks of

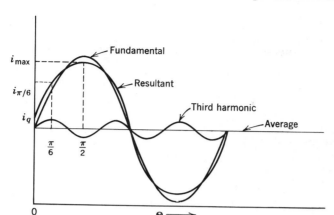

FIGURE 11.5
Typical third harmonic distortion.

the output signal tend to be flattened because of the decreased h_{fe} at the extremes of the current excursion. This type of distortion does not contain even harmonics which cause successive half cycles to be dissimilar. Figure 11.5 shows that third harmonics can cause this type of distortion. Usually, the third harmonic is much larger than the higher order odd harmonics so the collector current can be expressed in terms of the dc component, the fundamental, and the third harmonic as

$$i_C = I_0 + I_1 \sin \omega t + I_3 \sin 3\omega t \tag{11.17}$$

But when $\omega t = 0$, Fig. 11.5 and Eq. 11.17 show that

$$i_q = I_0 \tag{11.18}$$

Also, when $\omega t = \pi/6$, the driving source voltage is one-half the maximum value, as shown in Fig. 11.5, and with $\sin \pi/6 = 0.5$ and $\sin 3\pi/6 = 1$ substituted into Eq. 11.17, this equation becomes

$$i_{\pi/6} = I_0 + 0.5\,I_1 + I_3 \tag{11.19}$$

When $\omega t = \pi/2$, Fig. 11.5 and Eq. 11.17 show that

$$i_{max} = I_0 + I_1 - I_3 \tag{11.20}$$

These equations can be solved simultaneously to obtain I_1 and I_3 in terms of i_q, $i_{\pi/6}$ and i_{max}. The results are

$$I_1 = \frac{2(i_{\pi/6} + i_{max} - 2i_q)}{3} \tag{11.21}$$

$$I_3 = \frac{2i_{\pi/6} - i_{max} - i_q}{3} \tag{11.22}$$

Example 11.2   We will determine the value of fundamental component and third harmonic component for a current which has the waveform shown in Fig. 11.3 with $i_q = 0.97$ A, $i_{max} = 1.94$ A and $i_{\pi/6} = 1.5$ A.

The fundamental component is found from Eq. 11.21. $I_1 = 2(1.5 + 1.94 - 2 \times 0.97)/3 = 3/3 = 1$ A. Similarly, From Eq. 11.22, we find $I_3 = (2 \times 1.5 - 1.94 - 0.97)/3 = 0.03$ A. The percent third harmonic distortion is $I_3/I_1 \times 100 = (.03/1) \times 100 = 3\%$.

PROBLEM 11.4   Determine the value of the fundamental component, the third harmonic component, the dc component, and the percent third harmonic distortion for a current with $i_q = 500$ mA, $i_{max} = 975$ mA, and $i_{\pi/6} = 775$ mA.
Answer: $I_0 = 500$ mA, $I_1 = 500$ mA, $I_3 = 25$ mA, percent third harmonic $= 5\%$.

11.2   POWER OUTPUT AND EFFICIENCY

We will now discuss a method of determining the power output and the efficiency of an amplifier. The voltage drop and power loss in emitter-circuit stabilizing resistors will be neglected. These losses may be included by increasing V_{CC} to compensate for the IR drop across R_E and then using this higher V_{CC} to calculate the power delivered by the power supply.

Figure 11.6 shows how the power output and efficiency of an amplifier can be determined from the collector characteristics. The maximum power available is of major interest and that power is obtained when the transistor is biased half way between saturation and collector-current cutoff and is driven from saturation to cutoff. The maximum power output for a sinusoidal driving signal is then $V_c I_c$ where V_c and I_c are the rms values of collector voltage and current. But these rms values can be obtained by dividing the peak-to-peak values by two, to obtain peak values, and then dividing by $\sqrt{2}$ to obtain rms values. Therefore,

$$P_o = \frac{(v_{CE\,max} - v_{CE\,min})(i_{C\,max} - i_{C\,min})}{8} \tag{11.23}$$

The amplifier of Fig. 11.6 can deliver the same signal power to the load with approximately half as much input power from the power supply if the load is transformer coupled, as shown in Fig. 11.7. The improved efficiency occurs because the dc power loss in the load is eliminated. If the ohmic resistance of the transformer primary is negligible, the dc load line is nearly vertical. The ac signal will swing as high above V_{CC} as it swings below V_{CC}. Thus, V_{CC} is now the average value of v_{CE}. An example will be used to

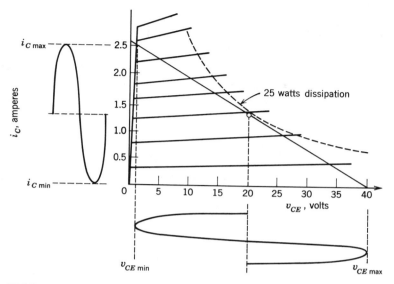

FIGURE 11.6
Current, voltage, and power relationships in an amplifier.

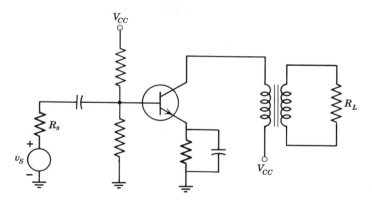

FIGURE 11.7
A transformer-coupled power amplifier.

illustrate the contrast between the amplifier with R_L in the collector circuit and a transformer-coupled amplifier.

Example 11.3   The collector characteristics of an amplifier with a 16 Ω load is shown in Fig. 11.6. We will determine the collector circuit efficiency if R_L is connected between the collector and V_{CC} and then compare this efficiency to that of a transformer-coupled amplifier (Fig. 11.7) which operates along the same load line.

From Eq. 11.23, the ac signal power is $P_o = (40 - 1)(2.45 - 0)/8 = 11.9$ W. The collector power input for the amplifier with the resistance in the collector circuit is $P_i = V_{CC} I_{C\,\text{ave}} = 40 \times 1.225 = 49$ W. The collector circuit efficiency is $\eta_c = (P_o/P_i)100 = (11.9/49)100 = 24\%$.

When the transformer-coupled amplifier (Fig. 11.7) is considered, V_{CE} will swing as high above V_{CC} as it swings below V_{CC}. Then, $V_{CC} = (V_{CE\,\text{max}} + V_{CE\,\text{min}})/2 = (40 + 1)/2 = 20.5$ V. The ac power output is the same as above. However, the power input is now $P_i = V_{CC} I_{C\,\text{ave}} = 20.5 \times 1.225 = 25.1$ W. The collector-circuit efficiency is now $\eta_c = (11.9/25.1) = 47.4\%$, which is essentially twice as great.

The power output would be increased and, hence, the efficiency increased if the collector voltage v_{CE} could be driven to zero on one half cycle and the collector current i_C could be driven to zero on the next half cycle. Then, using Fig. 11.6 and Eq. 11.23,

$$P_{o\,\text{max}} = \frac{2V_{CC}\,2I_{Bq}}{8} = \frac{V_{CC}\,I_{Cq}}{2} \tag{11.24}$$

The maximum efficiency is then

$$\text{max eff} = \frac{(V_{CC}\,I_{Cq})/2}{V_{CC}\,I_{Cq}} \times 100 = 50\% \tag{11.25}$$

This is the maximum theoretical efficiency of a class A amplifier with sinusoidal input. The efficiency can approach 100 percent if the input signal is a square wave. A class A amplifier is defined as an amplifier in which collector current flows during the entire cycle. We have considered only class A amplifiers to this point in our work.

The class A amplifier is inefficient because of the relatively large q-point current which flows through the amplifier all the time. This q-point current must be equal to the peak signal current. Otherwise the collector current is cut off during part of the signal cycle and serious distortion results. Note that the average collector power input to a class A amplifier is essentially constant. However, the signal output power increases with the signal level. The collector dissipation is the collector power input minus the signal power output. Consequently, the collector power dissipation decreases and the collector efficiency increases as the signal level increases.

PROBLEM 11.5   The transistor whose characteristics are given in Fig. 11.6 is to be operated with $V_{CC} = 40$ V and R_L (connected from collector to V_{CC}) = 20 Ω. Determine the maximum value of ac power output, collector power input, and collector efficiency.

Answer: $P_o \simeq 9.7$ W, $P_i \simeq 39.2$ W, $\eta_c = 24.7\%$.

PROBLEM 11.6   A circuit is connected as shown in Fig. 11.7. The characteristics of the transistor are given in Fig. 11.6. The maximum collector voltage is 40 V and the ac impedance of the transformer primary is 20 Ω. Determine the maximum value of ac power output, collector power input, and collector efficiency, neglecting the ohmic resistance of the transformer.

Answer: $P_o \simeq 9.7$ W, $P_i \simeq 20$ W, $\eta_c \simeq 48.5\%$.

11.3 PUSH-PULL AMPLIFIERS

A high efficiency amplifier can be built using two transistors in an arrangement known as push-pull, shown in Fig. 11.8. This amplifier is similar to the balanced amplifier discussed in Chapter 10 except that transformer coupling is used to provide balanced signals of opposite polarity to the two bases. Transformer coupling is also used to add the balanced

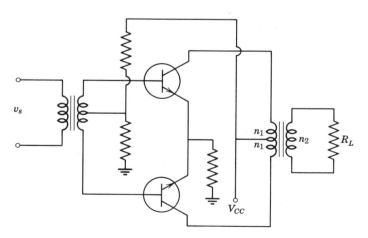

FIGURE 11.8
A transformer-coupled push-pull amplifier.

outputs of the two transistors in a single load R_L. The push-pull amplifier can be operated class A and will provide about twice the power output
obtainable from a single transistor under similar circumstances. Of course,
the total collector input power also doubles (two stages) so the collector
efficiency remains the same. The distortion of the push-pull connection
is lower, however, because the even harmonic distortion is cancelled as
a result of the balanced arrangement. The positive half cycle of one
transistor adds to the negative half cycle of the other transistor to provide
an output with similar successive half cycles as shown in Fig. 11.9.

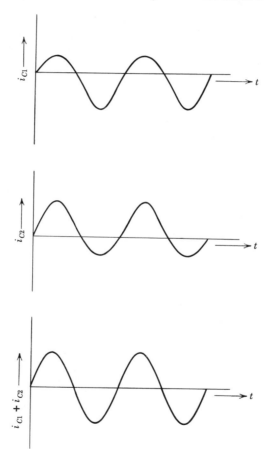

FIGURE 11.9
**Illustration of the cancellation of even harmonics
in the output of a balanced-push-pull amplifier.**

The main advantage of a push-pull amplifier results from the fact that the q-point currents of the individual transistors can be drastically reduced and the efficiency, therefore, markedly increased. In fact, each transistor can be biased at very nearly cutoff, so each transistor delivers power to the load only during one-half of the signal cycle. Collector current flows approximately half the time in each transistor. This type of operation is known as *class B*.

When each transistor is biased precisely at cutoff, distortion occurs in the output because h_{fe} decreases rapidly with collector current at very small values of collector current. This type of distortion is known as *crossover distortion* and is illustrated in Fig. 11.10a. In this figure, the

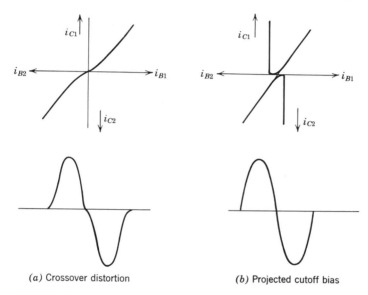

(a) Crossover distortion   (b) Projected cutoff bias

FIGURE 11.10
Illustration of the crossover distortion which results from cutoff bias and the reduction of distortion by the use of projected cutoff bias.

transfer characteristic of one transistor is inverted and reversed as compared to the other transistor so the sum of the two curves is the total transfer curve of the push-pull amplifier. Most of the nonlinearity of the total transfer curve occurs at small values of i_C as shown.

The crossover distortion can be eliminated by providing a small forward-bias current as shown in Fig. 11.10b. The minimum value of bias is determined by extending the relatively straight portion of the transfer curve of one transistor until the extension crosses the $i_C = 0$ axis. The value of i_B found at this intersection is known as *projected-cutoff bias.* If the transfer curve for the other transistor is inverted and reversed and the projected transfer curves are made to coincide at the projected-cutoff bias value of i_B, the total transfer curve is essentially straight and little distortion occurs. Since the magnetic flux in the output transformer, which results from the dc components of collector currents, tends to cancel, the net flux which is effective in producing voltage in the secondary is proportional to the difference between the individual collector currents. Therefore, the total or *composite* transfer characteristic is the difference between the individual transfer characteristics at any point.

PROBLEM 11.7   The transfer characteristic for a 2N2147 power transistor is given in Fig. 11.11. Determine the projected-cutoff bias for this transistor.
Answer: $I_B \simeq 2$ mA.

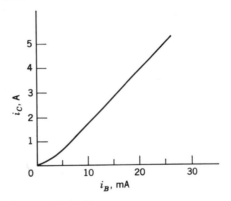

FIGURE 11.11
**The transfer characteristic of a typical
power transistor.**

A set of composite (or total) collector characteristics can be drawn by inverting one set and reversing it as compared with the other, and vertically aligning the individual q points. However, the bottom half of such a composite set contains the same information as the top half, so only the top half, as shown in Fig. 11.12, is needed. The composite (total) load line passes through the composite q point which is at $i_C = 0$ and $v_{CE} = V_{CC}$. The operation of the individual transistor follows the composite load line except during the time when both transistors are conducting. Then the total collector current is the difference between the individual collector currents. Since each transistor works into one-half of the output transformer primary, the composite load line represents the impedance of only one-half of the primary, which is one-fourth of the total primary impedance.

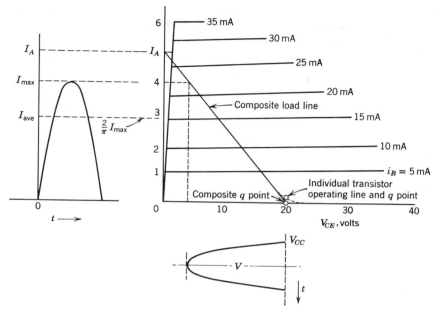

FIGURE 11.12
Basic relationships in a class B push-pull amplifier.

The maximum permissible collector current and the maximum power output for the class B, push-pull amplifier can be determined as functions of the permissible transistor dissipation with the aid of Fig. 11.12. The signal is assumed to be sinusoidal. The total transistor power dissipation is

$$P_d = P_i - P_o \qquad (11.26)$$

where P_i is the collector power input from the power supply and P_o is the signal power output, as previously discussed. But, assuming cutoff bias,

$$P_i = I_{ave} V_{CC} = \frac{2}{\pi} I_{max} V_{CC} \qquad (11.27)$$

and

$$P_o = \frac{I_{max}^2}{2} R_L \qquad (11.28)$$

Then, substituting Eq. 11.28 and Eq. 11.27 into Eq. 11.26,

$$P_d = \frac{2}{\pi} I_{max} V_{CC} - \frac{I_{max}^2}{2} R_L \qquad (11.29)$$

We need to find the value of I_{max} which will give maximum power dissipation $P_{d\,max}$. This can be done by differentiating P_d with respect to I_{max} and equating this derivative to zero.

$$\frac{dP_d}{dI_{max}} = \frac{2}{\pi} V_{CC} - I_{max} R_L = 0 \tag{11.30}$$

and, solving for the I_{max} which gives maximum power dissipation,

$$I_{max} = \frac{2}{\pi} \frac{V_{CC}}{R_L} = \frac{2}{\pi} I_A \tag{11.31}$$

where I_A is the current axis intercept (V_{CC}/R_L) of the load line as shown in Fig. 11.12. Substituting this value of I_{max} (Eq. 11.31) into Eq. 11.29, the maximum power dissipation can be found in terms of I_A and V_{CC}.

$$P_{d\,max} = \frac{4}{\pi^2} I_A V_{CC} - \frac{2}{\pi^2} I_A^2 R_L \tag{11.32}$$

But $I_A R_L = V_{CC}$, so

$$P_{d\,max} = \frac{4}{\pi^2} I_A V_{CC} - \frac{2}{\pi^2} I_A V_{CC} = \frac{2}{\pi^2} I_A V_{CC} \tag{11.33}$$

However, the power dissipation capabilities of the two transistors P_d is usually known or obtainable, and the maximum value of I_A can be found from Eq. 11.33

$$I_A(max) = \frac{\pi^2}{2} \frac{P_d}{V_{CC}} \simeq \frac{5P_d}{V_{CC}} \tag{11.34}$$

A method of determining P_d will be discussed in Section 11.5. The intercept current I_A must be reduced as the q-point collector current I_{Cq} is increased above cutoff. An empirical alteration of Eq. 11.34 takes the q-point current of the individual transistor into account

$$I_A(max) \simeq \frac{5P_d}{V_{CC}} - 3I_{Cq} \tag{11.35}$$

The minimum load resistance, maximum power output and maximum efficiency can now be determined if V_{CC} and P_d are known. We will illustrate this idea by an example.

Example 11.4   Let us consider the amplifier with characteristics given in Fig. 11.12 and $V_{CC} = 20$ V. Let us assume that $P_d = 20$ W total for the two transistors and $I_{Cq} = 100$ mA. Then, using Eq. 11.35, $I_A = 5 - 0.3 = 4.7$ A. The minimum value of load resistance is $R_L = V_{CC}/I_A = 20/4.7 = 4.25\ \Omega$. The maximum power output, assuming the saturation voltage is negligible is $P_o = V_{CC} I_A/2 = 47$ W. At this power output, the collector power input is $P_i = I_{ave} V_{CC} = [I_A(2/\pi) + I_{Cq}]V_{CC} = (4.7 \times .636 + .1)20 = 62$ W. The maximum collector circuit efficiency is $(47/62)100 = 75\%$.

Figure 11.13a shows the power output, power dissipation, and percent efficiency of a class B amplifier as a function of peak output signal amplitude I_{max}. The class B amplifier is assumed to be biased at cutoff. The power output, power dissipation, and efficiency of a class A amplifier, either push-pull or single-ended, is given as a function of I_{max} in Fig. 11.13b.

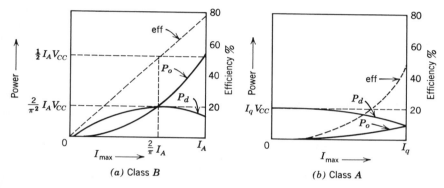

(a) Class B (b) Class A

FIGURE 11.13
Power output, power dissipation, and efficiency characteristics of class A and class B amplifiers.

Both amplifiers have the same power dissipation capabilities $P_{d\,max}$. Note the increased power output capability of the class B amplifier as compared with the class A. For example, if the power dissipation capability is 10 W in either amplifier, the maximum power output from the class A amplifier is 5 W, and the maximum power output from the class B amplifier is 25 W, neglecting the saturation voltage.

The balanced source voltage can be obtained from a phase *inverter* or *splitter* instead of a transformer with some saving in cost and weight. A differential amplifier can be used as a phase inverter as shown in Fig. 11.14. The collector voltages of the differential pair T_1 and T_2 provide the opposite polarity voltages required for the push-pull amplifier. The diodes in the base circuits of the push-pull amplifier provide projected-cutoff bias and clamp the bases at this bias voltage. A small resistance may be needed in series with the diodes to adjust the bias. Without the diodes, the base-emitter junctions (diodes) would clamp the bias voltage to the peak of the base driving voltage.

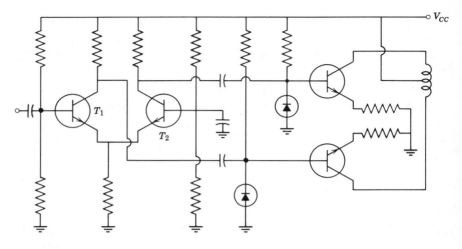

FIGURE 11.14
A differential amplifier used as a phase splitter.

PROBLEM 11.8   Two transistors with a total power dissipation capability of 15 W are used in a class B push-pull amplifier with $I_{C_q} = 50$ mA and $V_{CC} = 25$ V. Determine the minimum permissible value of ac load resistance and the maximum power output, assuming sinusoidal signals, for this amplifier. Neglect the transistor saturation voltage.
Answer: $R_L = 8.4$ Ω, $P_o = 35.6$ W.

11.4   COMPLEMENTARY-SYMMETRY AMPLIFIERS

Both the input and output transformers can be eliminated in a push-pull amplifier type know as a complementary-symmetry amplifier. This amplifier uses one *n-p-n* and one *p-n-p* transistor with similar characteristics as shown in Fig. 11.15. The bias resistors R_1, R_2, R_3, and R_4 provide the desired forward bias, usually projected-cutoff bias, for the transistors. When the input signal v_i goes through a positive half cycle, the emitter junction of the *n-p-n* transistor T_1 is forward biased and the emitter junction of the *p-n-p* transistor is reverse biased, causing the collector current of T_1 to pass through the load resistor R_L from left to right. During the next half cycle when the input voltage is negative, the *p-n-p* transistor is forward biased and the *n-p-n* transistor is cut off so the collector current of T_2 flows through the load resistor R_L from right to left. When the signal current is zero, the two transistor currents are nearly equal, so essentially no direct current flows through the load resistor. Note that this amplifier, like any push-pull amplifier can be operated class A or at any q point between

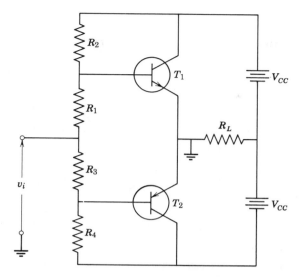

FIGURE 11.15
A complementary-symmetry, push-pull amplifier.

class A and class B. Operation at q points between class A and class B is known as class AB. However, high-power amplifiers are usually operated at projected-cutoff class B because of the high efficiency of this mode.

The circuit of Fig. 11.15 has a serious limitation. That is, the power supply which is represented by two batteries does not have a ground point. Therefore, the output signal voltage appears between the power supply and ground, and this power supply would not be usable for other amplifiers in the system. The shunt capacitance between the power supply and ground or chassis would also limit the high-frequency response of the amplifier.

The emitter-follower version of the complementary-symmetry amplifier permits the grounding of the power supply as shown in Fig. 11.16. However, as compared with the common-emitter configuration, the input impedance is high and the required input voltage v_i is large. The forward bias for the transistors in Fig. 11.16 is provided by the two forward-biased diodes. The low dynamic resistance of these diodes as well as their negative temperature coefficients provide improved thermal stability of the transistors as compared with the circuit of Fig. 11.15. A small resistance can be connected in either series or parallel with the diodes to provide precise adjustment of the q point.

The driver transistor for the complementary-symmetry amplifier can be

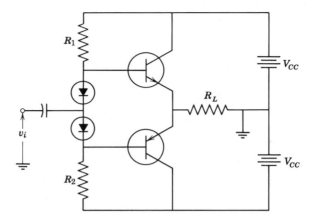

FIGURE 11.16
A common-collector configuration of the complementary-symmetry amplifier.

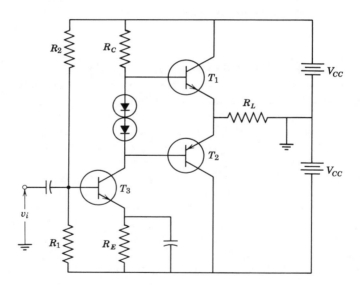

FIGURE 11.17
A complementary-symmetry amplifier with a direct-coupled driver.

directly coupled as shown in Fig. 11.17. Let us consider the design of a typical amplifier of this type to improve our understanding of its operation.

Example 11.5 We will assume that V_{CC} (each supply in Fig. 11.17) = 15 V and $R_L = 100~\Omega$. Then the current axis intercept I_A for either T_1 or T_2 is $I_A = 15~\text{V}/100~\Omega = 150$ mA. From Eq. 11.34 using projected-cutoff bias, the power dissipation requirement for the two transistors T_1 and T_2 is $I_A V_{CC}/5 = 450$ mW. The silicon complementary pair 2N3703 and 2N3705 have adequate current capability and power dissipation and voltage ratings for this application and, therefore, will be used. These transistors have $\beta_o \simeq 100$ over the collector current range 10 mA to 150 mA. Projected-cutoff bias is approximately $I_C = 1$ mA and $I_B = 15~\mu$A. Therefore, the forward-biasing diodes, with perhaps some additional resistance, should be chosen so these q point values are obtained. We should expect that the diodes will be low-current silicon diodes to match the characteristics of the emitter junctions of the transistors.

When the input signal v_i is positive, the collector of the driver T_3 goes negative so the transistor T_1 is cut off and transistor T_2 conducts. Then the emitter end of the load resistor has approximately the same potential as the collector of the driver. When the driver transistor T_3 is driven to saturation, the potential of the emitter end of the load resistor R_L differs from the potential of the negative end of the lower V_{CC} by the drop across R_E plus the saturation voltage of transistor T_3 plus the voltage V_{BE} of transistor T_2. Therefore, the voltage $I_E R_E$ of the driver transistor should be fairly small; we will use one volt.

When the input signal goes negative, the collector of the driver transistor T_3 goes positive, so transistor T_1 conducts and transistor T_2 is cut off. Then the emitter end of the load resistor approaches the potential of the positive end of the upper V_{CC}. However, both the collector current of T_3 and the base current of T_1 must flow through the collector load resistor R_C. Therefore, at the peak negative input voltage v_i, the transistor T_3 is cut off and the maximum-signal base current of transistor T_1 must flow through R_C. Thus, the difference between the maximum positive load potential and the positive power supply potential is the voltage $i_{B\,\text{max}} R_C$ plus the voltage V_{BE} of transistor T_1. So the voltage $i_{B\,\text{max}} R_C$ should not exceed a volt or so. We will allow 1.5 V. Then with i_C (of T_1) max = 150 mA and $\beta_o = 100$, $i_{B\,\text{max}} = 1.5$ mA, and $R_C = 1.5~\text{V}/1.5~\text{mA} = 1$ kΩ.

We can now design the driver stage with $R_C = 1$ kΩ. At quiescent conditions, the voltage drop across the driver load R_C is approximately $V_{CC} = 15$ V. Therefore, the q point collector current is 15 mA, and the value of R_E required for one volt drop is $1/.015 = 67~\Omega$. The driver operates as a class A amplifier with $V_{CC} = 30$ V and maximum dissipation $P_d = 15$ V

$(15 \text{ mA}) = 225 \text{ mW}$. The 2N3704 transistor which is similar to the 2N3705 except $\beta_o \simeq 200$ has adequate voltage and dissipation ratings for this application. The bias resistors R_1 and R_2 can be determined by the techniques discussed in Chapter 5. We will assume that a current stability factor of 25 is adequate for this silicon transistor. Then $R_B = 25 \times 67 = 1.7 \text{ k}\Omega$, $V_B = I_B R_B + V_{BE} - I_E R_E = 1.63 \text{ V}$, $R_1 = (30/28.37)1.7k = 1.8 \text{ k}\Omega$ and $R_2 = (30/1.63)1.7k = 32 \text{ k}\Omega$. The coupling capacitor C and the emitter bypass capacitor C_E can be determined when the driving-source resistance is known. The bypass capacitor C_E is frequently omitted to provide higher input resistance and better linearity (less distortion) at the expense of lower power gain.

You may have observed that the power dissipation rating of the class A driver must be about the same as either of the class B pair. Therefore, a high-power amplifier will require a high-power driver unless a class B intermediate amplifier is used as shown in Fig. 11.18. This figure also

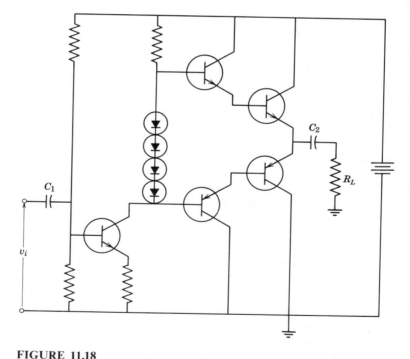

FIGURE 11.18
A high-power complementary-symmetry amplifier using Darlington intermediate amplifiers.

shows that a single power supply can be used if the load resistor R_L is capacitively coupled to the amplifier. The amplifier design can proceed in the same manner as previously discussed, except the effective β of the Darlington pair is treated as $\beta_1\beta_2$, as previously discussed. The capacitor C_2 is calculated in the usual manner. Since the output resistance of the (emitter follower) amplifier is very low, the coupling capacitance can be determined from the values of f_1 and R_L. For example, if $R_L = 10\ \Omega$ and $f_1 = 32$ Hz, determined by C, $C = 1/\omega_1 R_L = 500\ \mu F$.

The availability of complementary-symmetry pairs of transistors is limited, especially in the high-power types. This problem is resolved by using the *quasicomplementary-symmetry* circuit of Fig. 11.19. Note that this amplifier has two transistors of the same type in the output. However, the lower transistor is driven by the collector current of the driver amplifier instead of its emitter current. In other words, the lower amplifier consists of a

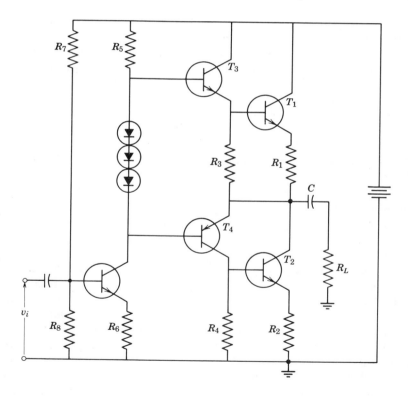

FIGURE 11.19
A quasicomplementary-symmetry amplifier.

pnp-npn combination rather than a Darlington connection. The polarity reversal in this arrangement permits the transistor T_2 to be an *n-p-n* instead of a *p-n-p* type. At first glance one might suspect that the input resistance of the driver T_4 is much lower than the input resistance of T_3. This is not true because the output voltage $\Delta i_{C2} R_L$ is between the emitter of T_4 and ground, and the base-to-ground signal voltage v_{B4} of transistor T_4 is essentially equal to the output voltage. Therefore,

$$R_{it} = \frac{\Delta v_{B4}}{\Delta i_{B4}} \simeq \frac{\Delta i_{C2} R_L}{i_{B4}} = \beta_2 \beta_4 R_L \qquad (11.36)$$

Thus, the input resistance of transistor T_4, with $R_L = 10\ \Omega$, $\beta_2 = 100$, and $\beta_4 = 200$ is $R_{it} = 200{,}000\ \Omega$. The input resistance of transistor T_3 is also approximately $200{,}000\ \Omega$.

The resistors R_1, R_2, R_3, and R_4 (Fig. 11.19) are needed to provide adequate thermal stability when germanium transistors are used. The resistors R_1 and R_2 should be small in comparison with R_L to avoid serious loss of output power. For example, if $R_L = 8\ \Omega$, R_1 and R_2 should not exceed about $1\ \Omega$. The resistors R_3 and R_4 should be of the same order of magnitude as the input resistance of the power transistors T_1 and T_2, as discussed in the Darlington amplifier section of Chapter 10. Thus, these resistors should be of the order of $100\ \Omega$ in a high-power amplifier. Note that the input resistance of T_2, as seen from the viewpoint of R_4 is $h_{ie2} + (\beta_2 + 1)R_2$. Although h_{ie2} varies widely over the signal cycle, it is usually small in comparison with $(\beta + 1)R_2$. The same relationship holds for the input resistance of transistor T_1 as seen from the viewpoint of R_3. Note that the output resistances of transistors T_3 and T_4 are in parallel with R_3 and R_4, respectively. The diodes between the bases of T_3 and T_4 cause this output resistance to be quite low, so long as emitter current flows in transistors T_3 and T_4. Thus the current stability factors of transistors T_1 and T_2 are much lower than the ratio R_3/R_1 or R_4/R_2 when current flows in transistors T_3 and T_4.

The complementary-symmetry amplifier can be arranged in a common-emitter configuration by using two *pnp-npn* combinations as shown in Fig. 11.20. The capacitor C_E maintains the emitters of transistors T_3 and T_4 at ac ground potential and the resistors R_3 and R_4 compensate for differences between transistors T_3 and T_4 and transistors T_1 and T_2. Either R_3 or R_4 can be adjusted until the V_{CE} of transistor T_1 is equal to the V_{CE} of transistor T_2. Lack of this balance in the output transistors decreases the available power output. The available power output from this configuration is somewhat higher than the common-collector configurations previously considered because the peak-to-peak output voltage is equal to the total power supply voltage minus the saturation voltages of transistors

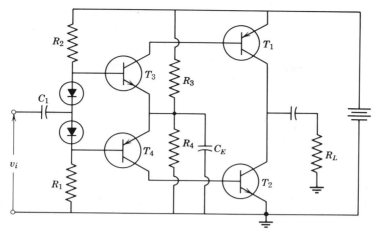

FIGURE 11.20
A common-emitter complementary-symmetry amplifier.

T_1 and T_2. These saturation voltages may be only a few tenths of a volt. The power gain of the common-emitter configuration is also higher, and therefore the required voltage and power gain can be obtained with fewer transistors than the common-collector configurations. The current through resistors R_1 and R_2 should be several times the maximum base current of transistors T_3 and T_4. Projected cutoff biase is desirable.

PROBLEM 11.9   If the β_o of all transistors in Fig. 11.20 are approximately 100, $R_L = 8$ Ω and the power supply voltage $= 40$ V, dtermine R_1 and R_2 values so that the current through them is 3 times i_{b3} max, and determine R_3 and R_4 so that their current is $0.3i_{E3}$ max.
Answer: $R_1 = R_2 = 27$ kΩ, $R_3 = R_4 = 2.7$ kΩ.

A variation of the complementary-symmetry circuit of Fig. 11.20, shown in Fig. 11.21, is recommended by Haas.[1] This circuit has improved q point stability because the collector currents of transistors T_1 and T_2 flow through R_1 and R_2, and the voltages across these resistors stabilize the q point currents as follows.

$$V_{D1} + V_{D2} \simeq V_{BE3} + V_{EB4} + I_C(R_1 + R_2) \tag{11.37}$$

$$I_C \simeq \frac{V_{D1} + V_{D2} - (V_{BE3} + V_{EB4})}{R_1 + R_2} \tag{11.38}$$

---

[1] G. C. Haas, "Design Factors and Considerations in Full Complementary Symmetry Audio Power Amplifiers," *Journal of the Audio Engineering Society*, Vol. 16, No. 8, July 1968.

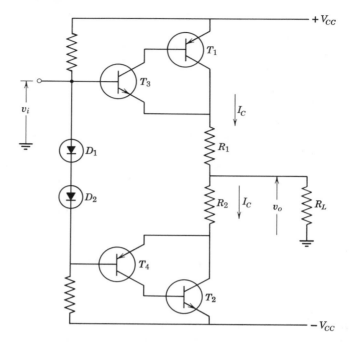

FIGURE 11.21
A full complementary-symmetry amplifier with stabilized q point.

where I_C is the q point collector current of the output transistors. The emitter currents of transistors T_3 and T_4 also flow through R_1 and R_2 but are very small compared with I_C and are therefore neglected. Since the thermal coefficients of the diodes can match those of the transistors T_3 and T_4 quite well, I_C is thermally stable providing the diodes are maintained at about the same temperature as the driving transistors. This thermal stability is achieved at the cost of reduced gain, however. The circuit of Fig. 11.21 behaves as an emitter-follower configuration because the input signal voltage v_i is the sum of the output signal voltage v_o and $(i_C R_1 + v_{be3})$. Therefore, the voltage gain is slightly less than one.

The biasing diodes are not needed, and consequently omitted, in the novel complementary-symmetry circuit of Fig. 11.22. In this circuit, the base-emitter voltages of the driving transistors T_3 and T_4 provide the projected-cutoff bias for the output transistors T_1 and T_2. The current through R_1 and R_2 is the base current for T_1 plus the emitter current for T_4. Similarly, the current through R_3 and R_4 equals $i_{B2} + i_{E3}$. Thus the driving transistors are in competition with the output transistors for the

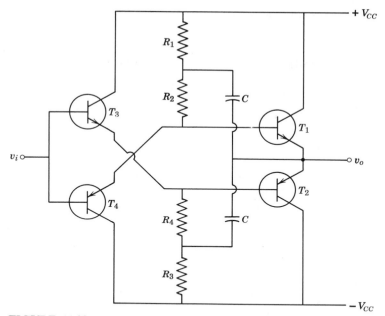

FIGURE 11.22
A complementary-symmetry circuit in which the V_{BE} voltages of the driving transistors provide bias for the output transistors.

currents through the resistors. When v_i goes positive, i_{E4} decreases so i_{B1} increases, and if transistor T_4 is driven to cutoff, the current through R_1 and R_2 must saturate T_1 in order to obtain full output. The positive input voltage tends to increase the emitter current of T_3 which will drive T_2 into cutoff. Thus T_3 appears as an emitter follower with $R_3 + R_4$ as a load. A negative input voltage v_i reverses the roles of the transistors. The capacitor C provides bootstrapping in the output circuit which holds the currents through the resistors essentially constant, thus providing the following advantages.

1. The output transistors can be driven into saturation so the peak output voltage is approximately equal to V_{CC}.
2. The input impedance of the circuit is increased because the ac impedance of the bootstrapped resistors is very high.

The circuit is an emitter-follower configuration characterized by low distortion, high input impedance, and essentially unity voltage gain. The input impedance, with bootstrapping, is the product of the driver β, the output transistor β_1 and R_L.

The main disadvantage of the circuit of Fig. 11.22 is that the q-point emitter current of the drivers must at least equal the maximum base current requirement of the output transistor. Thus, the dissipation requirement of the drivers may be high.

The bootstrapping technique shown in Fig. 11.22 can be used to advantage in the circuits of Figs. 11.19 and 11.21, or any emitter-follower type complementary-symmetry circuit in which maximum power output is desired.

PROBLEM 11.10 Add a voltage amplifier to the circuit of Fig. 11.21. Draw a circuit diagram, choose suitable transistor types, and determine suitable component values assuming $R_L = 8 \, \Omega$ and $-V_{CC}$ to $+V_{CC}$ potential difference is 34 V. Determine the approximate peak input voltage and maximum sinusoidal power output for your amplifier.

PROBLEM 11.11 Repeat Problem 11.10, except use the circuit of Fig. 11.22. Determine the power dissipation requirement for each transistor.

11.5 THERMAL CONDUCTION AND THERMAL RUNAWAY

As noted in Chapter 5, small-signal transistors rely on air convection and connecting lead conduction to remove heat from the transistor to the surrounding atmosphere. Therefore, the heat dissipation rating is based on the ambient temperature, usually 25°C, and this rating must be reduced as the ambient temperature is increased above the reference temperature. You may recall that the reciprocal of the derating factor is known as *thermal resistance*.

Power transistors (above about 1 watt) are usually fastened securely to a large metal plate or chassis which serves as a heat sink. The collector of the power transistor is usually connected electrically and mechanically to the transistor case. Maximum thermal conductivity is obtained when the transistor case is electrically and mechanically fastened to the heat sink. However, an insulator is usually required to electrically isolate the collector from the heat sink as shown in Fig. 11.23.

Every transistor has a maximum permissible junction temperature $T_{j\,\text{max}}$ which ranges from about 85°C to 110°C for germanium and about 130°C to 175°C for silicon. Thermal resistance θ_T is defined as the ratio of temperature rise in degrees centrigrade (or Kelvin) to the power conducted in watts. Therefore, the junction temperature can be related to the power being dissipated P_d, the thermal resistance θ_T and the ambient temperature T_a by the following equation, previously given in Chapter 5.

$$T_j = \theta_T P_d + T_a \qquad (11.39)$$

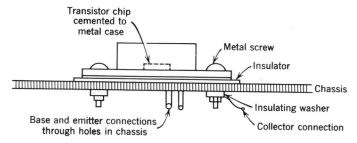

FIGURE 11.23
Heat sink arrangement for a typical power transistor.

The thermal resistance θ_T consists of three parts.

1. The thermal resistance between the collector junction and the transistor case θ_{jc}.
2. The thermal resistance between the transistor case and the heat sink θ_{cs}.
3. The thermal resistance between the heat sink and the ambient surroundings θ_{sa}.

Heat conduction is comparable to electrical conduction. Thus, the equivalent circuit of Fig. 11.24 represents the *thermal relationships* of the transistor and its heat sink. The capacitors shown represent the thermal capacitance of the three parts of the circuit. The thermal capacitance C_j of the collector junction is very small because the mass of the transistor chip,

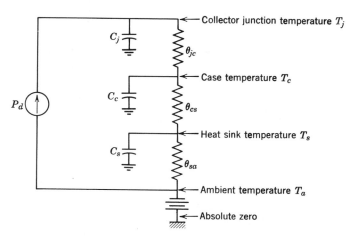

FIGURE 11.24
An equivalent thermal circuit.

and hence its heat capacity, is very small. Therefore, the thermal time constant $\theta_{jc} C_j$ is of the order of milliseconds. The significance of this time constant is that the load line can pass through the area above the maximum dissipation curve *providing the excessive dissipation does not continue for more than a few milliseconds.* This situation may occur in a class *B* or *AB* amplifier. The thermal capacitance C_c of the case is much greater than that of the junction. Therefore, a transistor amplifier which is designed to operate with a heat sink can operate at least several seconds without the heat sink before the transistor is damaged. Similarly, the thermal capacity of the heat sink is usually much greater than that of the transistor case, so the amplifier may operate for several minutes with an inadequate heat sink.

The thermal resistance θ_{jc} from the collector junction to the case is usually given by the manufacturer for each type of power transistor. Also, thermal resistance θ_{cs} data are usually available for the various transistor mounting systems.[2] The thermal resistance θ_{sa} of a 1/8-inch thick sheet of bright aluminum is given as a function of area (both sides) in Fig. 11.25. When the heat sink is horizontal, the thermal resistance increases by about

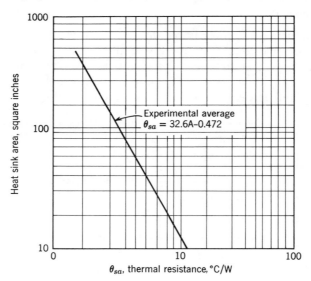

FIGURE 11.25
Heat sink versus thermal resistance. (Courtesy of of Motorola Semiconductor Products Division.)

[2] See *Motorola Power Transistor Handbook*, First Edition, Motorola Semiconductor Products Division, Phoenix, Arizona, p. 23.

10 percent because of the reduced convection. On the other hand, black painting or anodizing of the heat sink will lower its thermal resistance because of increased heat radiation.

We can now determine the dissipation capability of a power transistor with a specific heat sink system by applying "Ohms law" for thermal circuits wherein power dissipation P_d is comparable to current, temperature difference is analogous to potential difference, or voltage, and thermal resistance compares with electrical resistance, of course. The following example illustrates this procedure.

Example 11.6 Let us consider a transistor with $\theta_{jc} = 1°C/W$ mounted on a 1/8-inch aluminum 10 inch × 10 inch sheet. A mica washer coated with silicone grease is used to insulate the transistor from the chassis. The thermal resistance of this washer is 0.5°C/W. The thermal resistance of the aluminum heat sink is $\theta_{sa} = 2.5°C/W$, as seen in Fig. 11.25, if mounted vertically, and 2.75°C/W if mounted horizontally. The total thermal resistance from junction to ambient is $\theta_T = 1.0 + 0.5 + 2.5 = 4°C/W$ with vertical heat sink. If the maximum junction temperature of the transistor is 175°C and the ambient temperature is 25°C, the maximum transistor dissipation is $P_{d\,max} = (175 - 25)/4 = 37.5$ W. However, if the maximum expected ambient temperature is 75°C, the maximum safe transistor dissipation is $P_{d\,max} = (175 - 75)/4 = 25$ W. The maximum power output from this transistor in a class A amplifier with $T_a = 75°C$ is $P_o = 25/2 = 12.5$ W. However, if two of these transistors are mounted on the same heat sink with the same type washers, the total thermal resistance is $\theta_T = (1.5/2) + 2.5 = 3.25°C/watt$. The maximum dissipation of both transistors at $T_a = 75°C$ is $P_{d\,max} = 100/3.25 \simeq 30$ W. The maximum sinusoidal power output from these two transistors when operated class B at 75°C ambient temperature is $P_o = 2.5 \times 30 = 75$ W, as can be noted in Fig. 11.13a.

A power transistor is often given a power dissipation rating with the case temperature held at 25°C. For example, the transistor in the example above can dissipate $(175 - 25)/1 = 150$ watts at 25°C case temperature. This case temperature can be maintained by cooling the case with circulating water. Normally, this rating is intended to indicate only the thermal resistance between junction and case. For example, if a transistor is rated as having $T_{j\,max} = 100°C$ and $P_{d\,max} = 150$ W at 25°C case temperature, we may calculate that $\theta_{jc} = (100 - 25)/150 = 0.5°C/W$.

The transistor is not completely protected when the heat sink is adequate for the required dissipation; the transistor can be destroyed by *thermal runaway*. This thermal runaway results from the increasing collector current caused by the increasing I_{co}. You may recall from Chapter 5 that $\Delta I_c = S_I \Delta I_{co}$, where S_I is the current stability factor. We were not concerned about thermal runaway in the RC-coupled amplifier because the dc load line did not cross the maximum dissipation curve so the transistor could not be destroyed by heat. In fact, the power dissipation decreases with increasing collector current when the q point is to the left of the load line center. However, in a transformer-coupled or complementary-symmetry amplifier the dc load line is almost vertical as shown in Fig. 11.26. Therefore, the dc collector voltage V_{CE} is almost independent of the

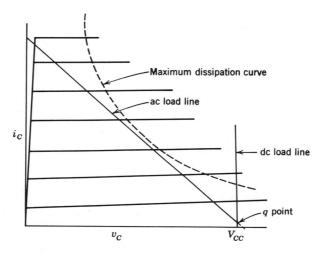

FIGURE 11.26
Illustration of the susceptibility of a transformer-coupled amplifier or complementary-symmetry amplifier to thermal runaway.

collector current, at the value V_{CC}, and the power dissipation is essentially proportional to the average collector current.

Thermal runaway occurs when the rate of increase of collector junction temperature exceeds the ability of the heat sink to remove the heat. The portion of heat dissipation which results from I_{CO} is

$$P_d = S_I I_{CO} V_{CC} \qquad (11.40)$$

Then the rate of increase of this power dissipation with time is

$$\frac{dP_d}{dt} = S_I V_{CC} \frac{dI_{CO}}{dt} \qquad (11.41)$$

The rate of heat conduction away from the collector junction to the ambient surroundings is the reciprocal of the thermal resistance, or $1/\theta_T$.

Thus, in order to prevent a continual build-up of heat, the following inequality must hold,

$$\frac{1}{\theta_T} > S_I V_{CC} \frac{dI_{CO}}{dt} \tag{11.42}$$

If I_{CO} doubles for each 10°C temperature increment, the I_{CO} increases by about 7 percent per °C. Then

$$\frac{dI_{CO}}{dt} = 0.07 I_{CO} \tag{11.43}$$

and

$$\frac{1}{\theta_T} > S_I V_{CC}(0.07 I_{CO}) \tag{11.44}$$

Note that the current stability factor S_I must be controlled in order to prevent thermal runaway. Using Eq. 11.44 to obtain S_I explicitly,

$$S_I < \frac{14.3}{\theta_T V_{CC} I_{CO}} \tag{11.45}$$

The thermal current I_{CO} must be determined at the maximum expected junction temperature.

Example 11.7 Let us consider a class B amplifier with $V_{CC} = 16$ V, $R_L = 8\ \Omega$ and $I_{CO} = 0.1$ mA at 25°C and doubles for each 10°C increase. Then $I_A = 16/8 = 2.0$ A and the maximum dissipation occurs when $I_{C\,max} = 0.636 I_A = 1.27$ A and $I_{ave} = 0.636(1.27) = 0.81$ A. Therefore, the maximum power dissipation is $V_{CC} I_{ave} = 16 \times 0.81 = 13$ W. We will assume that θ_T is 4.0°C/W as in the preceding example and the maximum ambient temperature is 40°C. Then, from Eq. 11.39, $T_{j\,max} = 13 \times 4 + 40 = 92$°C and I_{CO} at this temperature is

$$I_{CO} = 0.1 \text{ mA}(2)^{(92-25)/10} = 0.1 \text{ mA}(2)^{6.7} \simeq 10 \text{ mA}$$

The thermal-stability factor must be less than $S_I = 14.3/(4 \times 16 \times .01) = 22.3$ to insure that thermal runaway will not occur.

PROBLEM 11.12 If the supply voltage V_{CC} is raised to 20 V and the thermal resistance θ_T is reduced to 3°C/W, all other conditions remaining the same as in Example 11.7, determine the maximum value of the stability factor S_I to insure thermal stability.

Answer: $S_I < 12.5$.

PROBLEM 11.13 Design a transistor power amplifier using an emitter-follower type complementary-symmetry arrangement which will provide full power output to an 8 Ω speaker. The driving source is a radio tuner with 1.0 V rms output voltage, open circuit, and 1.0 kΩ internal resistance R_s. The total power supply voltage is 40 V. Use transistors of your choice, including those with characteristics given in Appendix I. Draw a circuit diagram. Calculate the maximum sinusoidal power output, the maximum permissible stability factor using the heat sink of your choice, and determine the voltage gain of your amplifier. You should use a gain, or volume, control something like the one shown in Fig. 11.27 to control the output power of the amplifier.

Answer: $P_o \simeq 20$ W, $K_v = v_o/v_s = 14$.

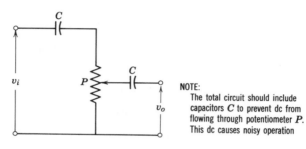

FIGURE 11.27
A typical gain, or volume, control circuit.

PROBLEM 11.14 Design the amplifier of Problem 11.13 using a quasi-complementary-symmetry arrangement for the power amplifier instead of the complementary-symmetry configuration.

PROBLEM 11.15 Use the common-emitter, complementary-symmetry arrangement for the output stage instead of the emitter-follower arrangement in the design of the amplifier of Problem 11.13.

PROBLEM 11.16 Assuming that a power amplifier with voltage gain = 15 and input resistance = 1 MΩ or higher is needed, replace the input transistor in the amplifier of Fig. 11.19 with an FET and determine the g_m required to provide the required voltage gain if $R_5 = 4.7$ kΩ and the other transistors are as given in the example associated with Fig. 11.19. The power supply voltage is 30 V. Determine the required value of I_D.
Answer: $g_m = 3.3$ mmho, $I_D \simeq 3$ mA.

PROBLEM 11.17 Modify the amplifier of Problem 11.16 by adding bootstrapping to increase the output power and improve the linearity of the amplifier.

12

Multistage Amplifiers

Most practical amplifiers require more gain than can be obtained from a single stage. Consequently, it is common practice to feed the output of one amplifier stage into the input of the next stage (Fig. 12.1). When amplifiers are connected in this fashion, they are called *cascaded amplifiers* or *multistage amplifiers*. There are a few concepts, unique to cascaded amplifiers, which will be considered in this chapter.

12.1 GAIN AND BANDWIDTH CONSIDERATIONS IN CASCADED AMPLIFIERS

Most of the cascaded amplifier stages are used to obtain either a voltage gain or a current gain. However, in most cascaded amplifiers, it is ultimately the power gain that is important. When the proper level of signal has been obtained, a power amplifier stage is used to produce sufficient power to activate the required load device (loudspeaker—servo motor—antenna, etc.). If a voltage gain is required, we can calculate the total gain by using the equation for voltage gain of one stage. Thus, from Fig. 12.1, the voltage gain for stage 1 is

$$G_1 = \frac{V_2}{V_1} \tag{12.1}$$

In addition, the voltage gain for stage 2 is

$$G_2 = \frac{V_3}{V_2} \tag{12.2}$$

The gain for additional stages can be written in a similar manner. Then, the total amplifier voltage gain G_A for n cascaded stages is

$$\frac{V_2}{V_1} \times \frac{V_3}{V_2} \times \frac{V_4}{V_3} \times \cdots \frac{V_{n+1}}{V_n} = \frac{V_{n+1}}{V_1} \tag{12.3}$$

389

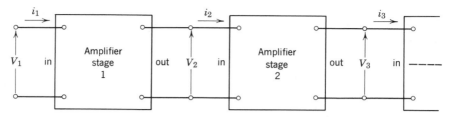

FIGURE 12.1
Cascaded amplifier stages.

or

$$G_A = G_1 \times G_2 \times G_3 \times \cdots G_n \qquad (12.4)$$

Obviously, a similar derivation could have been achieved for current gains or for power gains. In either of these cases, the total amplifier gain is equal to the product of the individual stage gains as indicated by Eq. 12.4. In general, the individual stage gains are functions of s and, consequently, the amplifier gain G_A is also a function of s. For steady-state sinusoidal signals, s becomes $j\omega$ and G_1, G_2, etc., become magnitudes at given phase angles. Then, G_A will be equal in magnitude to the products of all the magnitudes with a phase angle equal to the sum of the individual stage phase shifts.

A reference gain K has been defined for the individual stages. An amplifier reference gain K_A can be found from the individual stage gains.

$$\pm K_A = (\pm K_1) \times (\pm K_2) \times (\pm K_3) \times \cdots (\pm K_n) \qquad (12.5)$$

The K_A term will be positive if the total number of phase inversions is even and negative if the total number of inversions is odd. Since K is a magnitude only, this relationship does not involve s.

We have previously defined ω_1 and ω_2 as the lower and upper cutoff frequencies, respectively. These are the frequencies at which the voltage or current gain of one stage has been reduced to 0.707 of its reference value. (Power gain is reduced to 0.5 of its reference value for resistive loads.) Now, if we have an amplifier with two identical stages of amplification, the voltage gain at ω_1 will be reduced by a factor of 0.707 in each stage. Thus, the amplifier gain at ω_1 (and also ω_2) will be

$$0.707(-K_{v1}) \times 0.707(-K_{v2}) = 0.5K_A \qquad (12.6)$$

In fact, for n identical cascaded stages of amplification, the gain at ω_1 and ω_2 will be $(0.707)^n K_A$.

To reestablish a meaningful amplifier bandwidth, let us define ω_L as the lower cutoff frequency of the cascaded amplifier and ω_H as the upper cutoff frequency of the cascaded amplifier. At these frequencies, the gain of the amplifier will be $0.707K_A$. In order to arrive at some relationship between ω_1 and ω_L and also between ω_2 and ω_H, consider an RC-coupled amplifier containing n identical cascaded stages. The voltage gain per stage is given for the low frequencies by Eq. 7.24 as

$$G = -K \frac{s}{s + \omega_1} \tag{7.24}$$

For sinusoidal steady state, $s = j\omega$. Then, Eq. 7.24 can be written as

$$G = -K \frac{j\omega}{j\omega + \omega_1} \tag{12.7}$$

or

$$G = -K \frac{1}{1 - j(\omega_1/\omega)} \tag{12.8}$$

The magnitude of this function can be written as

$$|G| = K \frac{1}{[1^2 + (\omega_1/\omega)^2]^{1/2}} \tag{12.9}$$

If there are n cascaded stages, the magnitude of the amplifier gain from Eq. 12.4 is

$$|G_A| = |G|^n = K^n \left(\frac{1}{[1 + (\omega_1/\omega)^2]^{1/2}} \right)^n \tag{12.10}$$

Since $K^n = K_A$, we can write

$$|G_A| = K_A \frac{1}{[1 + (\omega_1/\omega)^2]^{n/2}} \tag{12.11}$$

Now, if ω is to be equal to ω_L, the term multiplying K_A must be equal to 0.707 or $1/(2)^{1/2}$. Then

$$2^{1/2} = \left[1 + \left(\frac{\omega_1}{\omega_L} \right)^2 \right]^{n/2} \tag{12.12}$$

or

$$2^{1/n} = 1 + \left(\frac{\omega_1}{\omega_L} \right)^2 \tag{12.13}$$

This equation is solved for $\omega_L{}^2$ to yield

$$\omega_L{}^2 = \frac{\omega_1{}^2}{2^{1/n} - 1} \tag{12.14}$$

or

$$\omega_L = \frac{\omega_1}{[2^{1/n} - 1]^{1/2}} \tag{12.15}$$

A similar solution of Eq. 7.92 yields

$$\omega_H = \omega_2 [2^{1/n} - 1]^{1/2} \tag{12.16}$$

PROBLEM 12.1   Derive Eq. 12.16 from Eq. 7.92.

PROBLEM 12.2   A three-stage amplifier is constructed of identical stages. Each individual stage has a pass band from 30 Hz to 20,000 Hz. Determine low cutoff and high cutoff frequency of the three-stage amplifier.

Answer: $\omega_L \simeq 59$ Hz, $\omega_H \simeq 10,200$ Hz.

PROBLEM 12.3   We desire to make a four-stage amplifier. Determine the low and high cutoff frequencies for each stage if the amplifier must have a bandwidth of 60 Hz to 15,000 Hz.

Answer: $\omega_1 \simeq 26$ Hz, $\omega_2 = 34,500$ Hz.

Table 12.1 gives the relationships between the lower and upper cutoff frequencies of the stage and the lower and upper cutoff frequencies of the amplifier for n identical stages.

TABLE 12.1

n	1	2	3	4	5	6	7	8
ω_1/ω_L	1	0.644	0.510	0.435	0.387	0.348	0.333	0.30
ω_2/ω_H	1	1.55	1.96	2.3	2.58	2.88	3.00	3.33

From Eqs. 12.15 and 12.16, or from Table 12.1, ω_L will be greater than ω_1 if n is greater than one and ω_H will be less than ω_2 if n is greater than one. Thus, the bandwidth of the amplifier decreases as the number of cascaded stages increases. Or, if the amplifier bandwidth is to remain constant, the stage bandwidth must increase as the number of cascaded stages increases. This last statement leads to an interesting dilemma. If a high-gain, very wide band amplifier is desired, stages must be cascaded to obtain the higher gain. However, as more stages are cascaded, the bandwidth of each stage must be increased. Unfortunately, as noted in Section 7.7, the gain-bandwidth product *may be* a constant. Under these conditions, the increased bandwidth results in reduced gain per stage. Thus, in order to compensate for the reduced gain per stage, more stages of greater bandwidth are required. This process can be carried so far that the total *amplifier gain* (for a given bandwidth) *may actually decrease as additional cascaded stages are added.*

PROBLEM 12.4   If gain × bandwidth = 409 × 10^6 rad/sec, and $f_H = 12$ MHz, calculate K_A for 3, 7, and 8 cascaded stages. Determine the bandwidth per stage (f_2) and gain per stage for each amplifier.

Answer: For $n = 7$, $K_A = 136$, $f_2 = 32$ MHz, $K = 2.03$.

When n identical, single-tuned amplifiers are cascaded, it can be shown[1] that

$$B_n = B(2^{1/n} - 1)^{1/2} \tag{12.17}$$

where B_n is the bandwidth of the total amplifier and B is the bandwidth of each stage in the amplifier. The behavior of these tuned circuits is much the same as the behavior of the RC-coupled amplifier just considered.

Tuned circuits do have a rather unique advantage if the cascaded stages are *not* tuned to the same resonant frequencies. Tuned amplifiers of this type are known as stagger-tuned amplifiers. Again, the treatment of these amplifiers is deleted in the interest of compactness. However, excellent and detailed treatments exist in the literature.[2]

So far, the work on cascaded amplifiers has been primarily concerned with gain and bandwidth of the amplifier. If a step function is applied to a single-stage amplifier, the rise time was given in Chapter 7 as being approximately equal to $2.2/\omega_2$. If this same concept is to be applied to a cascaded amplifier, the expected rise time will be about $2.2/\omega_H$. In addition, Martin[3] has used a work by Elmore[4] to show that the overall rise time of the cascaded amplifier T_{AR} is

$$T_{AR} = (T_{R1}^2 + T_{R2}^2 + T_{R3}^2 + \cdots)^{1/2} \tag{12.18}$$

where

T_{R1} is the rise time of stage one
T_{R2} is the rise time of stage two, etc.

Thus, if n identical stages with a rise time of T_{RS} are cascaded, the rise time of the total amplifier T_{AR} will be

$$T_{AR} = T_{RS}(n)^{1/2} \tag{12.19}$$

---

[1] For derivations of this equation see *Electronic Circuits* by T. L. Martin, Prentice-Hall, Englewood Cliffs, New Jersey, pp. 171–174, 1955.

[2] The design of stagger-tuned circuits will not be included here. However, excellent derivations exist in *Vacuum Tube Amplifiers* by G. E. Valley and Wallman, Vol. 18 of the MIT Radiation Lab series, McGraw-Hill Book Co., New York, pp. 176–200, 1948, or *Electronic Circuits* by T. L. Martin, Prentice-Hall, Inc., Englewood Cliffs, New Jersey, pp. 186–205, 1955.

[3] *Op. cit.*, Martin, pp. 222–224.

[4] W. C. Elmore, "Transient Response of Damped Linear Networks with Particular Regard to Wideband Amplifiers," *J. Appl. Phys.*, Vol. 19, pp. 55–62, January 1948.

These two equations assume no overshoot in the rise waveform and are accurate to within 10 percent when as few as two stages are involved.

PROBLEM 12.5 Design a video amplifier with a bandwidth of 30 Hz to 4 MHz. The total voltage gain must be at least 1000. Use cascaded 2N5457 FETs and assume that the following data apply.

$$C_{iss} = 4.5 \text{ pF} = C_{gs} + C_{gd}$$

$$C_{gd} = 1.5 \text{ pF}$$

$$g_m = 3000 \ \mu\text{mhos}$$

$$\max R_G = 10 \text{ M}\Omega$$

$$r_d = 0.1 \text{ M}\Omega$$

Assuming wiring capacitance per stage is 1 pF.

- a. Determine the gain-bandwidth product of one stage.
- b. Determine how many stages are required.
- c. Determine ω_1 and ω_2.
- d. Determine R_L and C_c.
- e. Determine R_s and C_s.
- f. Draw the diagram for the total amplifier. List all values.

PROBLEM 12.6 Design a video amplifier with a bandwidth of 30 Hz to 4 MHz. The total current gain must be at least 1000. Choose your transistors and circuit configuration and assume that the interstage wiring capacitance is 1 pF.

12.2 dB GAIN

Power gain in bel units is defined as

$$B = \log \frac{P_2}{P_1} \tag{12.20}$$

where "log" is the logarithm to the base 10 and B is in *bels*. P_2/P_1 is the power ratio between the points in question.

The bel unit is convenient because it reduces a multiplication problem in the case of the gain of a cascaded amplifier to an addition problem. Nevertheless, the bel is an inconveniently large unit because a power gain of 10 is only 1 bel. Therefore, the decibel (dB) has been accepted as the practical unit. The dB unit has the additional advantage that a power change of 1 dB in an audio system is barely discernible to the ear, which has a logarithmic response to intensity changes.

$$dB = 10 \log \frac{P_2}{P_1} \tag{12.21}$$

Also

$$dB = 10 \log \frac{V_2{}^2/R_2}{V_1{}^2/R_1} \qquad (12.22)$$

If the resistance is the same at the two points of reference,

$$dB = 10 \log\left(\frac{V_2}{V_1}\right)^2 = 20 \log \frac{V_2}{V_1} \qquad (12.23)$$

Similarly,

$$dB = 20 \log \frac{I_2}{I_1} \qquad (12.24)$$

The impedance levels are frequently of secondary importance in a voltage of current amplifier. Therefore, the Eqs. 12.23 and 12.24 are sometimes loosely used without regard to the relative resistance levels.

PROBLEM 12.7   An amplifier consists of four stages, each of which has a voltage gain of 20. What is the dB gain of each stage? What is the total gain in dB? Assume that the resistance levels of each stage are the same.

PROBLEM 12.8   What is the dB level at the half-power frequencies, f_1 and f_2, compared with the mid-frequency or reference level?

It is often convenient to express a power level in dB with regard to a given reference level. One commonly used reference level is 6 mW. Another commonly used reference level is 1 mW. When this (1 mW) reference level is used, the dB units are usually called *volume units* (vu) or dBm. On the other hand, the open-circuit output voltage of a microphone is usually rated in dB with reference to 1 V when the standard excess acoustical pressure is one microbar, or one-millionth of standard barometric pressure.

PROBLEM 12.9   What is the power level of 30 dB? 40 vu?
Answer: 60 W, 10 W.

PROBLEM 12.10   What is the open-circuit output volage of a microphone which has -56 dB level? Assume the excess acoustical pressure to be 1 microbar.

12.3   STRAIGHT-LINE APPROXIMATIONS OF GAIN AND PHASE CHARACTERISTICS (BODE PLOTS)

The analysis given in Section 12.1 is sufficient for the analysis and design of identical cascaded stages. This section will present a method of analysis which can be used on any amplifier. However, as an introduction, we will apply this method to a single RC-coupled stage. The transfer function of a single RC stage was given by Eq. 7.100 as

$$G = -K \frac{s\omega_2}{(s + \omega_1)(s + \omega_2)} \tag{7.100}$$

For ac steady state, $s = j\omega$. Now, if ω is assumed to be small in comparison with ω_1, the term $(s + \omega_1)$ becomes essentially ω_1. In addition, since $\omega_1 \ll \omega_2$ in a typical amplifier, the term $(s + \omega_2)$ becomes essentially ω_2. Then, Eq. 7.98 becomes

$$G \simeq -K \frac{j\omega}{\omega_1} \tag{12.25}$$

Thus, the gain of the amplifier is proportional to frequency and the phase angle of G is nearly 270°. An interesting method of expressing the magnitude relationship exists. In musical terms, the frequency doubles every octave. Thus, from Eq. 12.25, the gain doubles for every octave increase in frequency. When Eq. 12.23 or 12.24 is used, 20 log 2 is approximately 6. Thus, we can also state *the gain increases* 6 dB *per octave frequency increase or* 20 dB *per decade*. A plot of dB gain versus frequency over the range where Eq. 12.25 applies will be a straight line if frequency is plotted on a logarithmic scale. As noted, Eq. 12.25 is valid for $\omega \ll \omega_1$. However, as ω approaches ω_1, the accuracy of this approximation decreases. Nevertheless, as an approximation let us assume that Eq. 12.25 is valid for $\omega \le \omega_1$. Then, the plot of G versus ω and phase angle versus ω will be as shown for $\omega \le \omega_1$ in Fig. 12.2.

Now, if $s = j\omega$ and $\omega_1 \ll \omega \ll \omega_2$, then $(s + \omega_1) \approx j\omega$ and $(s + \omega_2) \approx \omega_2$. Then, Eq. 7.98 becomes

$$G = -K \tag{12.26}$$

In this range, the gain is a constant and is independent of frequency while the phase angle remains constant at about 180°. If this condition is assumed to exist (again, this is a rough approximation) for $\omega_1 \ge \omega \ge \omega_2$, the plots will have the form given in Fig. 12.2.

Finally, if $s = j\omega$ and $\omega \gg \omega_2$, then, $(s + \omega_1) \approx j\omega$ and $(s + \omega_2) \approx j\omega$. Under these conditions, Eq. 7.98 becomes

$$G = -K \frac{\omega_2}{j\omega} \tag{12.27}$$

In this range, G is reduced by one-half for each octave frequency increase or G decreases by 6 dB per octave or 20 dB per decade frequency increase. In addition, the phase angle becomes nearly 90°. If these approximations are assumed to be valid for $\omega \ge \omega_2$, the plots will be as shown (for $\omega \ge \omega_2$) in Fig. 12.2.

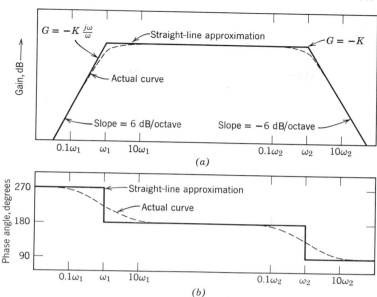

FIGURE 12.2
Straight line approximations for $G = -Ks\omega_2/(s + \omega_1)(S + \omega_2)$; (a) **gain**;
(b) **phase shift**.

The actual plots of gain and frequency (Fig. 7.98) are shown as dashed lines in Fig. 12.2. The straight-line approximations are seen to be fairly good approximations. In fact, with a correction factor which can be applied near the singularities, or corners, accurate plots can be obtained. With the insight gained from the straight-line analysis of the RC-coupled stage, we are now ready to derive some general rules which can be applied in plotting gain and phase curves.

The total amplifier gain G_A was given in Eq. 12.4 as

$$G_A = G_1 \times G_2 \times G_3 \times \cdots G_n \qquad (12.4)$$

In general, the stage gains (G_1, G_2, etc.) are functions of s. Consequently, G_A is also a function of s and will have the form

$$G_A = H \frac{s^n + a_1 s^{n-1} + a_2 s^{n-2} + \cdots a_n}{s^m + b_1 s^{m-1} + b_2 s^{m-1} + \cdots b_m} \qquad (12.28)$$

In actual circuits, any number of the coefficients a_1, b_1, a_2, b_2, etc., may be zero. The polynomial in the numerator and the polynomial in the denominator can be factored. When these polynomials are factored, the terms will have the following three forms.

$$\text{Form } 1 = s^k \tag{12.29}$$

$$\text{Form } 2 = (s + a)^k \tag{12.30}$$

$$\text{Form } 3 = (s^2 + 2\zeta\omega_n s + \omega_n{}^2)^k \tag{12.31}$$

where k is an integer 1, 2, 3, etc., which differs from one if repeated roots are present. Thus, Eq. 12.28 can be factored (in fact, since G_A is usually written as the product of the individual stage gains, the terms are usually already factored) and written as

$$G_A = Hs^h \frac{(s + Z_1)(s + Z_2) \cdots}{(s + P_1)(s + P_2) \cdots} \tag{12.32}$$

where h can have a value of zero, a positive integer, or a negative integer, and Z_1, Z_2, P_1, P_2, etc., may be complex numbers if they are derived from terms such as those of Eq. 12.31. Now, if we take the logarithm[5] of Eq. 12.32 we have

$$\log G_A = \log H + h \log(s) + \log(s + Z_1) + \log(s + Z_2) + \cdots$$
$$- \log(s + P_1) - \log(s + P_2) - \cdots \tag{12.33}$$

Thus, if we work with logarithms of the different factors in Eq. 12.32, the response of the different factors can be added or subtracted to obtain the response of the total amplifier.

The log H is a constant and is independent of frequency. However, for the steady-state solution, s becomes $j\omega$ and

$$h \log(s) = h \left[\log \omega + j \frac{\pi}{2} \right] \tag{12.34}$$

As already noted, the term log ω is equal to a gain increase of 6 dB/octave or 20 dB/decade. The effect of $h \log(s)$ on G_A is plotted in Fig. 12.3 for several values of h.

When considering factors of the form given by Eq. 12.30, it is convenient to draw, first of all, a straight-line approximation for the gain and phase characteristics and then correct this approximation in order to arrive at the actual gain and phase curves. Thus, we note that when $s = j\omega$, Eq. 12.30 can be written as

$$\log[(j\omega + a)^k] = k \log(j\omega + a) \tag{12.35}$$

If $\omega \ll a$, Eq. 12.35 reduces to $(k \log a)$ which is a constant. Thus, for a straight-line approximation, the gain and phase curves will be as shown in

---

[5] In taking the logarithm of a complex number, the complex number should be reduced to a magnitude and a phase angle such as M^θ. Then, log $M^\theta = \log M + j\theta$ where θ is in radians.

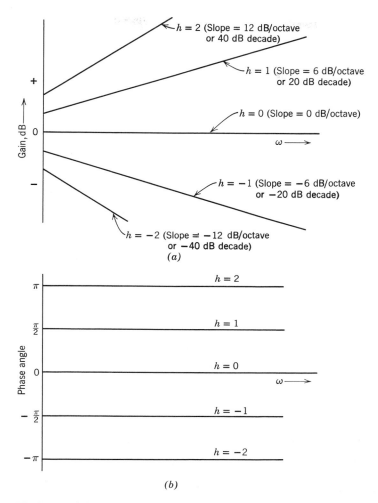

FIGURE 12.3
A plot of $h \log(s)$ for several values of h: (a) gain plot; (b) phase angle plot.

Fig. 12.4 (by the dashed lines) for values below $\omega = a$. However, if $\omega \gg a$, Eq. 12.35 becomes

$$k\left[\log \omega + j\frac{\pi}{2}\right]$$

In this case, the slope is ($k \times 6$ dB/octave) and the phase angle is $k\pi/2$. For the straight-line approximation, these curves hold for $\omega > a$ (Fig. 12.4).

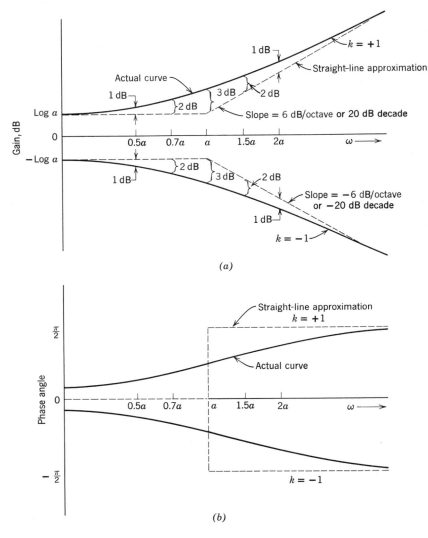

FIGURE 12.4
A plot of $k \log(j\omega + a)$ for $k = +1$ and -1: (a) gain; (b) phase angle.

The two straight lines for the gain curve must intercept at $\omega = a$. When actual values near $\omega = a$ are calculated, the curves are as shown by the solid lines. Table 12.2 lists the actual error between the straight-line

TABLE 12.2

ω or f	0.3a	0.5a	0.7a	a	1.4a	2a	3a
Ratio	0.96	0.9	0.82	0.7	0.82	0.9	0.96
Departure	0.5 dB	1 dB	2 dB	3 dB	2 dB	1 dB	0.5 dB

approximation and the actual curves for a few significant values of ω. Note that k will be positive if the term $(s + a)^k$ is in the numerator of Eq. 12.32 and will be negative if the term is in the denominator.

The straight-line approximation for the phase angle (Fig. 12.4b) is a poor approximation to the actual phase plot. Fortunately, a very good straight-line approximation can be drawn as shown in Fig. 12.5. In this figure, the

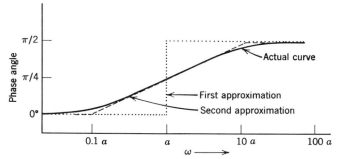

FIGURE 12.5
A plot of $\log(j\omega + a)$ **showing the first and second straight-line approximations.**

first straight-line approximation is drawn as a dotted curve with a value of 0 for $\omega \leq a$ and a value of $\pi/2$ for $\omega \geq a$. The actual phase angle departures from this first straight-line approximation are given in Table 12.3. The second straight-line approximation is given in Fig. 12.5 as a dashed

TABLE 12.3

Frequency	0.1a	0.3a	0.5a	0.7a	a	1.4a	2a	3.3a	10a
Departure	5.7°	17.5°	26.6°	35.3°	+45°	−35.3°	−26.6°	−17.5°	−5.7°

line. This second approximation has a value of zero for $\omega \leq 0.1a$. Then, from $\omega = 0.1a$ and phase $= 0$, a straight line is drawn to the point where $\omega = 10a$ and the phase angle $= \pi/2$. (Of course, the ω axis must be plotted on a logarithmic scale for this straight-line approximation or for any other Bode plot.) This second straight-line approximation is very close to the actual phase plot (the solid line in Fig. 12.5). The largest error in this approximation is $5.7°$ and occurs at $\omega = 0.1a$ and $10a$. A slight fillet at these locations produces a very good approximation to the actual curves.

When conjugate poles or zeros exist (as in tuned amplifiers), the factors have the form given by Eq. 12.31. Again, a straight-line approximation can be used to simplify the plotting procedure. When $s = j\omega$, Eq. 12.31 can be written as

$$\log(-\omega^2 + 2j\zeta\omega_n\omega + \omega_n^2)^k = k \log(-\omega^2 + 2j\zeta\omega_n\omega + \omega_n^2) \quad (12.36)$$

Now, if $\omega \ll \omega_n$, Eq. 12.36 becomes $2k \log \omega_n$ which is a constant. This straight-line approximation is used for all values of $\omega < \omega_n$ (Fig. 12.6). In contrast, if $\omega \gg \omega_n$, Eq. 12.36 becomes $2k \log \omega + kj\pi$. Note that if k is one, the slope of the straight-line approximation will be 12 dB/octave or 40 dB/decade. The interception of the two straight-line gain curves occurs at $\omega = \omega_n$. The straight-line approximations and the actual curve for a typical case are shown in Fig. 12.6. However, since Eq. 12.36 contains ζ as well as ω_n, the exact shape of the curve near ω_n is a function of ζ. Curves which can be used to determine the proper correction for different values of ζ are given in Fig. 12.7. The corrections listed by these curves should be multiplied by k to obtain the exact corrections to be used. The curves given in Fig. 12.7 represent the response of a double pole in the denominator of the gain equation.

The concepts just developed will now be used in an example to clarify their use.

Example 12.1 The voltage equation for an amplifier is given by the following equation.

$$G_A = 2 \times 10^{12} \frac{s^2}{(s + 100)(s + 200)(s + 100{,}000)^2} \quad (12.37)$$

Plot the amplitude and phase of G_A as a function of frequency.

As a first step, let us find the mid-band gain of this amplifier. Since we have two low-frequency poles ($\omega = 100$ and $\omega = 200$), two zeros at $\omega = 0$, and two high-frequency poles (both at $\omega = 100{,}000$), Eq. 12.37 has the form of two cascaded RC amplifiers. Then, Eq. 12.37 can also be written as

$$G_A = \frac{Ks^2\omega_2^2}{(s - \omega_{11})(s + \omega_{12})(s + \omega_2)^2} \quad (12.38)$$

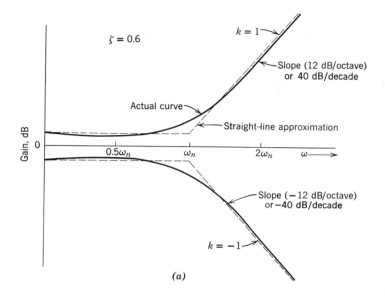

(a)

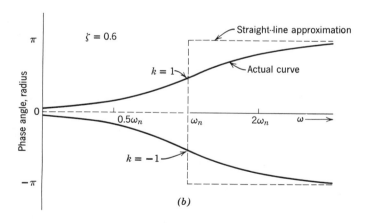

(b)

FIGURE 12.6
A plot of $k \log(-\omega^2 + 2j\omega_n \omega + \omega n^2)$ **for** $k = +1$ **and** $k = -1$;
(a) **magnitude**; (b) **phase angle**.

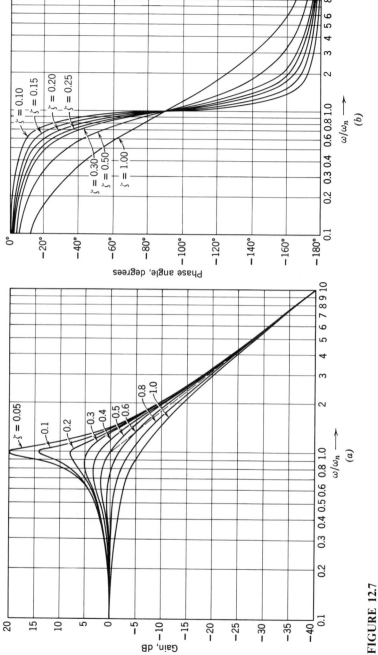

FIGURE 12.7
Magnitude and phase of $\omega_n^2/(s^2 + 2\zeta\omega_n s + \omega_n^2)$: (a) magnitude; (b) phase.

Now, ω_2 is 10^5 and $K\omega_2{}^2$ is 2×10^{12}. Consequently, the mid-band gain K of the amplifier is 200 or 46 dB. In the mid band, $\omega_{11} < \omega_{12} \ll \omega \ll \omega_2$ and Eq. 12.38 reduces to $G_A = K$. Thus, the gain is constant at 46 dB.

The straight-line approximation for this amplifier can be drawn as shown in Fig. 12.8. The gain curve drops off at 12 dB/octave or 40 dB/decade for frequencies above ω_2 (10^5 rad/sec) due to the double poles at this

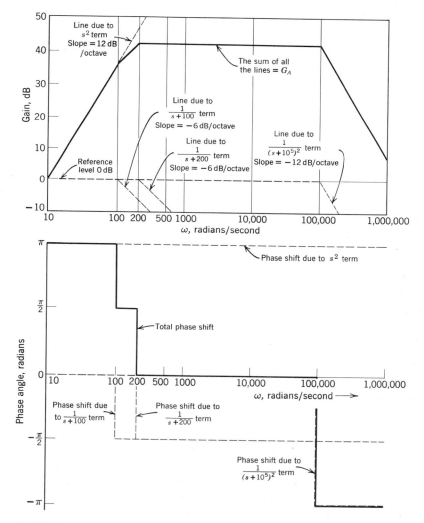

FIGURE 12.8
The straight-line approximations for Eq. 12.37.

frequency. In addition, at ω_{12} (200 rad/sec), the curve drops off at 6 dB/octave or 20 dB/decade. This curve steepens to 12 dB/octave or 40 dB/decade as the frequency drops below $\omega = \omega_{11}$ (100 rad/sec).

This total straight-line approximation can also be considered to be the sum of a series of straight-line terms due to each factor in Eq. 12.37 as shown in Fig. 12.8. In general, as we traverse the gain curve in the direction of increasing ω, the slope increases 6 dB/octave or 20 dB/decade as we pass each zero, and decreases 6 dB/octave or 20 dB/decade as we pass each pole.

In the final step, a correction table is tabulated as shown in Table 12.4.

TABLE 12.4
Correction Values to be Used on Fig. 12.8

Frequency	Correction for $\dfrac{1}{s+100}$ term		Correction for $\dfrac{1}{s+200}$ term		Correction for $\dfrac{1}{(s+100,000)^2}$ term		Total Correction	
	dB	Phase	dB	Phase	dB	Phase	dB	Phase
50	−1	−26.6°					−1	−26.6°
70	−2	−35.3°					−2	−35.3°
100	−3	45.0°	−1	−26.6°			−4	+18.4°
140	−2	+35.3°	−2	−35.3°			−4	0°
200	−1	+26.6°	−3	45.0°			−4	+71.6°
280			−2	35.3°			−2	+35.3°
400			−1	+26.6°			−1	+26.6°
50,000					−2	−53.2°	−2	−53.2°
70,700					−4	−70.6°	−4	−70.6°
100,000					−6	−90.0°	−6	−90.0°
141,400					−4	+70.6°	−4	+70.6°
200,000					−2	+53.2°	−2	+53.2°

When these total corrections are applied to the straight-line approximation (shown dashed in Fig. 12.9), the actual gain and phase curves shown as solid lines in Fig. 12.9 result. These are the required curves.

The foregoing example indicates the process to be used in plotting gain and phase characteristics for any amplifier whose characteristic equation is known. In fact, it is often possible to start from a known gain curve and synthesize an amplifier which will have this gain curve.

Gain and phase plots are usually placed in line vertically so that they can be easily correlated. This arrangement also eliminates some effort in making the phase plot because the break frequencies in the gain plot are

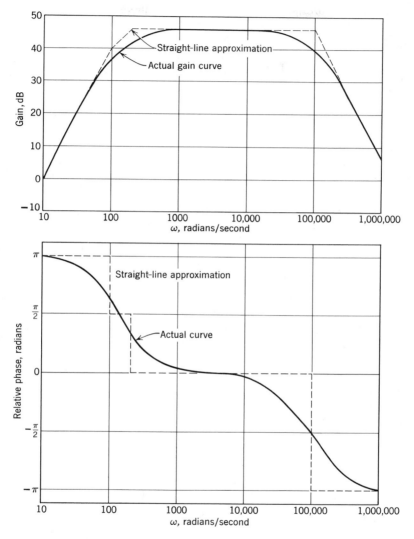

FIGURE 12.9
The actual gain and phase curves for Eq. 12.37.

obvious and may also be used for the phase plot. This arrangement will be used for the study of feedback amplifier stability in Chapter 13.

PROBLEM 12.11   A two-stage RC-coupled amplifier has $f_1 = 10$ Hz and 20 Hz. The values of f_2 are 10^5 Hz and 5×10^5 Hz. Draw the straight-line gain and linear phase approximations, then sketch the actual gain and phase curves.

12.4 DECOUPLING FILTERS

You learned in Chapter 9 that instability or oscillation may occur in a tuned amplifier unless signal feedback from the amplifier output to the input is minimized by keeping the signal leads short and well separated. Also, shielding and neutralization may be required to insure stability.

Although instability is not a problem in a single RC-coupled stage because the feedback is usually not in proper phase to reinforce the input signal, instability can readily occur in a multistage amplifier if the opportunity exists for signal feedback from the output of one amplifier to the input of a preceding amplifier. One common feedback path is through a common power supply lead (V_{CC}) as shown in Fig. 12.10. In this figure, two stages are shown in block diagram form. Each stage has a current gain of 100, so the total current gain $K_i = i_{c2}/i_i = 10,000$. The power supply always has some internal impedance which is shown as R_p, so the output current i_{c2} causes a signal voltage v_p to appear on the power supply lead. This signal component $v_p = i_{c2} R_p$ causes a feedback current to flow through R_C and R_2 into the input of the first amplifier. Since the resistance of R_C is usually

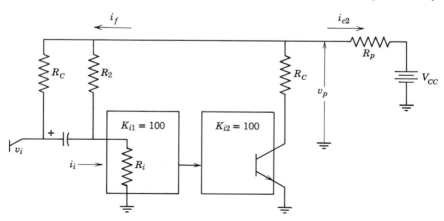

FIGURE 12.10
A semiblock diagram showing how feedback can occur through a power supply lead.

much lower than the resistance of R_2, let us assume that the parallel combination of R_C and R_2 is essentially equal to the resistance R_C. Then the feedback current is

$$i_f \simeq \frac{v_p}{R_C + R_i} \qquad (12.39)$$

Since $v_p = i_{c2} R_p$, this feedback current can be expressed in terms of the output current i_{c2}.

$$i_f = \frac{i_{c2} R_p}{R_C + R_i} \qquad (12.40)$$

Each of the two common-emitter amplifiers produces a polarity reversal, so the feedback current is in phase with the assumed input current and increases the total current of the amplifier. In fact, if the feedback current is equal to the assumed input current without feedback, the feedback current will provide the required input current and the output current will be maintained with no external signal. This is the condition for oscillation or instability. Therefore, if the ratio of i_f/i_{c2} is equal to or greater than the reciprocal of the current gain $1/K_i$, the amplifier will oscillate. The power supply resistance R_p which will cause oscillation can be thus determined from Eq. 12.40 by letting $i_f/i_{c2} = 1/K_i$ and solving for R_p.

$$R_p = \frac{i_f}{i_{c2}}(R_C + R_i) = \frac{R_C + R_i}{K_i} \qquad (12.41)$$

Of course, R_p must be less than this value for stable operation. For example, let us assume that $R_C = 10$ kΩ and $R_i = 2$ kΩ for the amplifier of Fig. 12.10. The value of R_p which will cause oscillation is 12 k$\Omega/10^4 = 1.2\,\Omega$. Therefore, R_p should not be larger than about 1 Ω. This very small permissible value of R_p can be realized only by using a well-regulated power supply and keeping the power supply leads short. (Regulated power supplies will be discussed in Chapter 17.) Although a battery is shown as the power supply in Fig. 12.10, the battery only represents the various types of power supplies which might be used. The rectifier-filter type power supplies frequently used in equipment intended for ac input power usually have a capacitor in the output of the power supply. This capacitor has low impedance at the higher signal frequencies, but since its capacitive reactance is inversely proportional to frequency, there will be some low frequency where the feedback is adequate for oscillation unless the low-frequency gain of the RC-coupled amplifier is restricted by the judicious choice of the lower cutoff frequency of the amplifier.

Instability due to signal feedback through the power supply leads is usually prevented in a multistage amplifier by using decoupling filters.

Figure 12.11a shows a decoupling filter consisting of R_D and C_D. For the feedback currents, the circuit can be simplified to that shown in Fig. 12.11b. The decoupling filter acts as a voltage divider as shown in Fig. 12.11c to reduce the signal component v_p across the power supply to the value v_f. Since v_f is now the signal voltage which forces the feedback current i_f through R_C and R_i, the decoupling filter reduces the feedback current by the ratio v_f/v_p. The voltage divider will be effective if X_{cd} is small in comparison with R_D at the low-frequency cutoff f_1 of the amplifier. But the dc component of collector current for the lower level stage (or stages) must pass through R_D, so the value of R_D is determined by the permissible dc voltage drop across R_D. Sometimes R_D is used to reduce V_{CC} by a specified amount.

Example 12.2   Let us assume that the desired V_{CC} for the higher level stage is 25 V and the desired V_{CC} for the lower level stages is 15 V. Then,

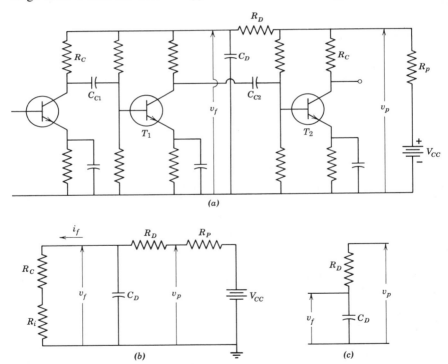

(a)

(b)

(c)

FIGURE 12.11
A decoupling filter used to reduce the feedback current through the power supply leads: (a) **actual circuit;** (b) **simplified circuit for feedback currents;** (c) **further simplified circuit.**

if the q-point collector current of each lower level stage is 1 mA, the desired value of R_D is 5 kΩ. Let us assume that the amplifier is RC coupled with $f_1 = 16$ Hz and that we wish to reduce the signal feedback by a factor of .01. Then $X_{cd} = .01 \, R_D$ at 16 Hz and the required value of $C_D = 1/2\pi(16)(50) = 200 \, \mu F$.

The impressive reduction of feedback realized in the preceding decoupling filter was obtained at the expense of a 200 μF capacitor. Although this value of capacitor is not prohibitively large or expensive, a saving in cost and space may be realized by using a transistor in the decoupling filter circuit as shown in Fig. 12.12. This filter is known as an *active* filter because the

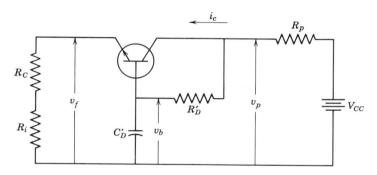

FIGURE 12.12
An *active* decoupling filter.

transistor is known as an *active* element in contrast with a passive R, L, or C element. The transistor amplifier appears to be in the common-collector (or emitter-follower) configuration because R_C and R_i are in the emitter circuit. Therefore, the feedback voltage v_f is essentially equal to the signal voltage v_b between the base and ground. But v_b is the voltage across the filter capacitor C_D'. Thus, R_D' and C_D' serve as the decoupling filter in the active circuit. But R_D' may be much larger than the R_D in the passive circuit because i_b instead of i_c flows through R_D'. Therefore, C_D' may be much smaller than C_D.

Example 12.3 Let us consider an active filter as a replacement for the passive filter of the preceding example. Also, let us assume that the transistor has $\beta_0 = 100$ so $I_B = 20 \, \mu A$ and V_{BE} (dc value) = 0.5 V. Then $R_B' = 9.5$ V/ 20 $\mu A = 4.75 \times 10^5 \, \Omega$, and the .01 reduction in feedback current can be obtained at 16 Hz with $D_D = 1/2\pi(16) \, (4.75 \times 10^3) = 2.1 \, \mu F$.

Observe that the required value of filter capacitance for the active filter is reduced by a factor almost equal to β_o. For example 12.3, the value of

C_D' might be increased to 10 μF with little increase in cost or size to provide improved decoupling. A Darlington-connected pair of transistors is often used in high-current applications where I_C may be hundreds of milliamperes or perhaps amperes.

PROBLEM 12.12 A three-stage amplifier is connected as shown in Fig. 12.10. If the current gain of each stage is 150, $R_C = 8$ kΩ, $R_2 = 200$ kΩ and $R_i = 3$ kΩ, determine the maximum value of power supply resistance R_p for stable operation.

Answer: $R_p < 0.47$ Ω.

PROBLEM 12.13 The lower cutoff frequency of the amplifier of Problem 12.12 is 10 Hz and $R_p = 10$ Ω. Design an active decoupling filter which will reduce the feedback current to 0.2 of the value which will cause oscillation. Use a silicon transistor with $\beta_0 = 100$ and allow $C_{CE} = 4.0$ V for the filter transistor. The current through the filter is 1.0 mA.

Answer: $R_D' = 350$ kΩ, $C_D' = 5$ μF.

PROBLEM 12.14 *a.* Draw the straight-line approximation curves (gain and phase) for an amplifier whose gain is given by the equation

$$G_A = \frac{10^{19}s^3}{(s + 100)(s + 1000)^2(s^2 + 6 \times 10^4 s + 10^{10})(s + 10^5)}$$

b. Draw the actual gain phase curves for this amplifier.

PROBLEM 12.15 Repeat Problem 12.14 if the gain equation is

$$G_A = \frac{10^{24}s(s + 200)^2}{(s + 100)(s + 400)(s + 800)(s + 10^6)^3}$$

PROBLEM 12.16 Three identical stages are cascaded. The gain equation for one stage is

$$G_s = \frac{200s}{(s + 100)(s + 10^6)}$$

a. Plot gain versus frequency curves for the total amplifier. From this curve, find ω_L and ω_H for the total amplifier.

b. Use the method outlined in Section 12.1 to determine ω_L and ω_H. How do these values compare with those in part *a*?

PROBLEM 12.17 It is desirable to construct an amplifier which has a gain-versus-frequency curve with the same shape as the one given in Fig. 12.13.

a. Find a gain function G_A which would give this type of response.

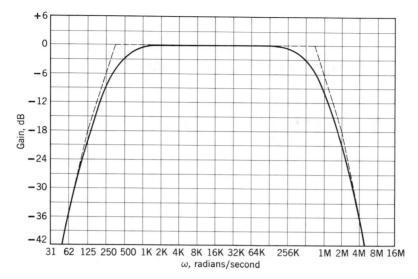

FIGURE 12.13
Characteristics for Problem 12.17.

b. Break this function up into functions which represent *RC*-coupled stages.

c. Design an amplifier which will provide the desired response, using transistors of your choice from either the Appendix or the manufacturer's data. Assume that the wiring capacitance is negligible.

Answer:
(b) $G_A = K_1[s/(s + 125)(s + 2 \times 10^6)] \times K_2[s/(s + 350)(s + 7 \times 10^5)]$
$$\times K_3[s/(s + 7 \times 10^5)(s + 350)]$$

13

Amplifier Noise

The maximum usable gain of a cascade amplifier is usually determined by the noise generated in the amplifier. High gain is required when the available input signal is very weak. However, if the input signal is not stronger than the noise generated in the first stage of a cascaded amplifier, the noise may make the signal unusable. The signal source always generates some noise and therefore has a finite ratio of signal to noise. Ideally, the amplifier would not generate noise, and therefore the signal-to-noise ratio in the output of the amplifier would be the same as the signal-to-noise ratio out of the source. However, the amplifier always adds noise of its own and thus the signal-to-noise ratio in the output of the amplifier is always lower than the signal-to-noise ratio of the source.

Three fundamental sources of noise in a transistor amplifier are:

1. Diode noise which results from the random injection of charge carriers across the depletion region.

2. Resistor noise which results from the random motion of electrons in a resistance at temperatures above $0°K$.

3. A third type of noise is known as $1/f$ noise because its magnitude is approximately inversely proportional to frequency. This type of noise is effective only at lower audio frequencies, usually 1 kHz or less. This $1/f$ noise is caused by a phenomenon known as *cathode flicker* in vacuum tubes and results primarily from surface leakage in transistors and semiconductor diodes. This type of noise varies quite widely among units of the same type device and has not been theoretically characterized. Improvements in transistor surface treatment have greatly reduced the $1/f$ noise in comparison with earlier models. In applications which require a low-noise amplifier the transistor or tube should be hand picked for low $1/f$ noise.

In resistors and diodes, the average currents are very predictable when a

414

known constant voltage is applied to the device, but small, random fluctuations about the average values result from the random motion of electrons or charge carriers. This random motion is due to the kinetic energy of the carriers which is proportional to the absolute temperature of the material. These random variations are called noise because they are amplified along with the signal and cause background noise radiation from a loud-speaker or headset in an audio system. These noise signals also cause fuzzy oscillograph tracings or "snow" on a TV picture tube. The noise will not be audible or the snow noticeable if the signals are very large in comparison with the noise. Therefore, the *signal-to-noise ratio* determines whether or not the noise will be disturbing. The tolerance level of the user will depend upon the program material, of course. For example, a noisy but intelligible long-distance voice communication may be satisfactory, but a noticeable background noise accompanying a musical concert may be annoying.

Since the first stage of an amplifier amplifies its own noise as well as the input signal, the combined noise and signal power available as an input signal to the second stage is usually large in comparison with the noise power generated by the second stage. Therefore, the first stage is the crucial one in considering noise contribution in an amplifier. Inductive reasoning shows that if the stages of a cascade amplifier each generate equal noise power and have equal power gains, the contribution of a given stage to the total noise power in the output of the amplifier is $N_c = N_o/G_p^{(n-1)}$, where N_o is the noise generated in the amplifier stage, G_p is the power gain of the amplifier stage, and n is the numbered position of the stage ($n = 1$ for the first stage, and so on).

13.1   RESISTOR NOISE

First, we will determine the noise voltage of a resistor. This noise voltage fluctuates randomly and, therefore, does not have discrete frequency components such as one might obtain from the Fourier analysis of a periodic fluctuation. Instead, the noise voltage or current has a continuous noise spectrum over a very broad band of frequencies extending well beyond the upper frequency limits of transistor amplifiers. Therefore, the noise power in the output of an amplifier is proportional to the bandwidth of the amplifier in Hz. Since Boltzmann's constant k relates temperature to energy, a reasonable relationship between noise power p_n, temperature T and bandwidth Δf is

$$p_n = kT\Delta f \tag{13.1}$$

The Δf is used instead of B for bandwidth because the effective noise power bandwidth is different than the half-power bandwidth we have previously

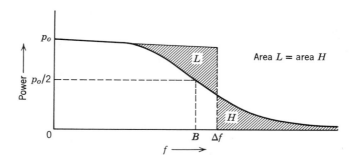

FIGURE 13.1
Power-versus-frequency curve showing the difference between half-power bandwidth B and noise bandwidth Δf.

defined. The difference between the noise bandwidth and the half-power bandwidth is illustrated in Fig. 13.1. The half-power bandwidth, as used heretofore, is the frequency at which the power is one-half the low-frequency value in a low-pass amplifier. However, Δf is the bandwidth which makes the rectangular area $p_o \Delta f$ (Fig. 13.1) equal to the total area under the total power response curve, or in mathematical terms,

$$p_o \Delta f = \int_0^\infty p \, df \qquad (13.2)$$

Thus the area labeled L in Fig. 13.1 is equal to the area H. For a single RC-coupled amplifier or a single tuned stage Δf is approximately $1.5B$. However, Δf approaches B as the number of stages, or frequency-discriminating circuits, are added because the slopes of the response curves become steeper. Therefore, little error is introduced if Δf is assumed equal to B in most applications.

Nyquist[1] postulated that the noise power given by Eq. 13.1 is the maximum power that can be transferred from a noisy (normal) resistor to a noiseless resistor. Since maximum power is transferred when the load resistance is equal to the source resistance, a Thevenin's equivalent circuit can be drawn for the noisy resistor as shown in Fig. 13.2 A simple calculation will show that this equivalent circuit will transfer the maximum power $p_n = kT\Delta f$. Of course, if two normal resistors are connected together, they both generate noise power so there is no net transfer of energy if they are at the same temperature. The voltage $\bar{v}_n$ is not an rms voltage in the usual sense because it cannot represent a frequency component.

[1] H. Nyquist, "Thermal Agitation of Electric Charge in Conductors," *Physical Review*, Vol. 32, 1928, p' 110.

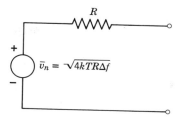

FIGURE 13.2
A Thevenin's equivalent circuit
for the noisy resistor.

Therefore, the bar is placed over v_n to indicate that it is a fictitious voltage which will yield the proper noise power transfer. Therefore, $\bar{v}_n$ can be treated as an rms voltage.

PROBLEM 13.1   Verify that the noise generator of Fig. 13.2 transfers maximum power $kT\Delta f$.

PROBLEM 13.2   Determine the effective noise voltage of a 100 kΩ resistor at 300°K, as measured with a noiseless voltmeter with a 1 MHz band-pass ($k = 1.38 \times 10^{-23}$ J/°K).

Answer: 91 μV.

The Thevenin's equivalent noise source can be transformed to a Norton's or current source as shown in Fig. 13.3. A simple calculation will show that the short-circuit current of the Thevenin's circuit (Fig. 13.2) is

$$i_n = \frac{\sqrt{4kTR\Delta f}}{R} = \sqrt{4kTG\Delta f} \tag{13.3}$$

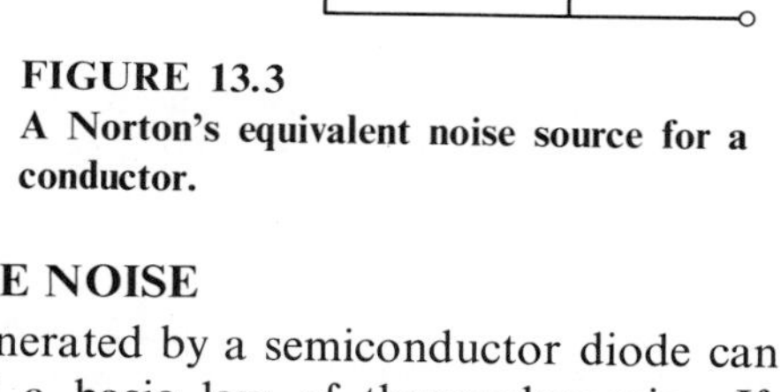

$$\bar{i}_n = \sqrt{4kTG\Delta f} \quad (\uparrow) \qquad \lessgtr G = \frac{1}{R}$$

FIGURE 13.3
A Norton's equivalent noise source for a
conductor.

13.2   DIODE NOISE

The noise generated by a semiconductor diode can be deduced from resistor noise and a basic law of thermodynamics. If the diode and a resistance, at the same temperature, are connected together with no external voltage applied, no net noise power will flow between them. Otherwise one of them must get hotter and the other cooler. Therefore, the noise current flow resulting from noise generated in the diode must be equal to the noise current flow resulting from noise generated in the resistor. But the noise in a diode, exclusive of the noise generated in the ohmic resistance of the doped semiconductor, is due to the random variations in current across the junction. However, in a diode with no bias voltage, the saturation current flows in one direction and an injection current equal to the saturation current flows in the opposite direction, as discussed in Chapter 3. Although the average current is zero, the noise components of the two oppositely directed currents are additive, because the random variations depend only on the magnitude of current and not on the direction. Thus the magnitude of *noise-producing* current in the unbiased diode is $2I_S$.

As discussed in Chapter 3, the dynamic conductance of a diode with zero bias is

$$G_o = \frac{q}{kT} I_S \tag{13.4}$$

If this value of conductance is substituted into Eq. 13.3, the noise component of an unbiased diode becomes

$$i_{nd} = \sqrt{4qI_S \Delta f} \tag{13.5}$$

But the noise-producing current in an unbiased diode is $I = 2I_S$, as discussed above. Therefore, the noise component of diode current can be expressed in terms of the noise-producing diode current by substituting $I_S = I/2$ in Eq. 13.5 to obtain

$$i_{nd} = \sqrt{2qI\Delta f} \tag{13.6}$$

Schottky and others have shown that this (Eq. 13.6) relationship holds for any value of noise-producing diode current I. When reverse bias is applied, $I = I_S$. When forward bias is applied, $I = I_S + I_I$, where I_I is the total injection current. But I_I is normally very large compared with I_S, so I is essentially the q point or average value of diode current when the diode is forward biased.

The noise bandwidth of a diode usually exceeds the useful frequency range of the diode, so the noise bandwidth Δf is usually the noise bandwidth of the amplifier or measuring instrument which follows the diode.

PROBLEM 13.3   Determine the noise current of a diode at 300°K with 1 mA average current, as measured with a meter with 1 MHz bandwidth. *Answer:* $i_n = 1.8 \times 10^{-8}$ A.

13.3 TRANSISTOR NOISE

We are now prepared to identify the noise sources in a transistor and to make some noise voltage (or equivalent resistance) calculations. Thermal currents (I_{CO}, I_{EO}) have noise components and, therefore, increase the noise of a transistor. Thus, silicon transistors have less noise than germanium transistors if other characteristics are similar. Consequently, we will assume that the transistor which is chosen for low noise is silicon and the thermal currents may be neglected. The main noise sources in a silicon transistor are then the forward-biased emitter junction (diode noise) and the ohmic resistance in the base r_b. A hybrid-π equivalent circuit with these noise sources shown is given in Fig. 13.4. The emitter

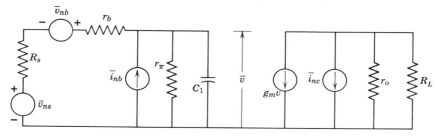

FIGURE 13.4
A noise equivalent circuit for the transistor.

junction diode noise has two effects. First, it produces a noise voltage across the junction. The equivalent noise current $\bar{i}_{nb}$ is placed in parallel with r_π to produce the proper contribution to the noise voltage $\bar{v}$ across the junction. The second effect of the emitter junction noise current is the direct transmittal of this current, reduced by the ratio α, to the collector circuit. This component, which is not amplified by the transistor, is represented by the noise current source $\bar{i}_{nc}$ in the collector circuit. The resistors r_o and r_π are noiseless resistors because they are fictitious resistors which account for the transistor characteristics but do not represent ohmic resistance.

The following conclusions can be drawn from the foregoing discussions.

1. The collector circuit noise current $\bar{i}_{nc}$ can be small if the q-point collector current is small.

2. The noise current $\bar{i}_{bn}$ can be minimized for a given q-point collector current if h_{FE} is large.

3. The contribution of the base resistance noise is small if r_b is small in comparison with the source resistance R_s.

4. The noise in the output of the amplifier can be minimized by restricting the bandwidth of the amplifier to that required by the signal.

Thus, a silicon transistor with high h_{FE} at low values of I_C will have a low noise figure when driven by a source with $R_s \gg r_b$. The transistor should be selected for low $1/f$ noise if very low noise is required.

A noise figure of merit for an amplifier is the *spot noise figure, F.* This figure is defined as

$$F = \frac{\text{Noise power delivered by an amplifier to a load}}{\text{Noise power delivered if } R_s \text{ (at 290°K) is the only noise source}}$$

This ratio is expressed for a narrow-frequency band at a specific frequency because the amplifier noise is a function of frequency. The noise figure is usually given in dB at 1 kHz.

The noise contributed by both the amplifier and the driving source can be represented by a noise voltage $\bar{v}_n = (\bar{v}_{ns}^2 + \bar{v}_{na}^2)^{1/2}$ at the input of the amplifier, as shown in Fig. 13.5, where $\bar{v}_{na}$ is the equivalent noise voltage

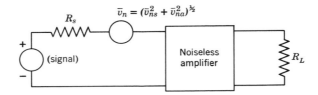

FIGURE 13.5
The representation of amplifier and driving-source noise by an equivalent noise voltage $\bar{v}_n$.

of the amplifier referred to the amplifier input. The equivalent noise voltage $\bar{v}_n$ can be represented as the noise voltage which will be produced by a resistance R_n at 290°K (the standard reference temperature). Then the noise figure F is equal to R_n/R_s, as shown by the following relationships. The equivalent input noise current to the amplifier is

$$\bar{i}_n = \frac{\bar{v}_n}{R_s + Z_i} = \frac{(4kTR_n \Delta f)^{1/2}}{R_s + Z_i} \tag{13.7}$$

where Z_i is the input impedance of the transistor at the specified fequency. The noise output voltage is then

$$\bar{v}_{no} = \bar{i}_n G_i R_L = \frac{(4kTR_n \Delta f)^{1/2} G_i R_L}{R_s + Z_i} \tag{13.8}$$

Similarly, the noise voltage in the output which results only from the noise voltage of the source is

$$\bar{v}_{nos} = \frac{(4kTR_s \Delta f)^{1/2} G_i R_L}{R_s + Z_i} \tag{13.9}$$

The noise power in the load is $\bar{v}_{no}{}^2/R_L$, whereas if the amplifier were noiseless, the noise power in the load would be $\bar{v}_{nos}^2/R_L$. By definition, the spot noise figure is the ratio of these two powers. Then, using Eqs. 13.8 and 13.9, the spot noise figure is

$$F = \frac{4kTR_n\Delta f G_i{}^2 R_L{}^2}{4kTR_s\Delta f G_i{}^2 R_L{}^2} = \frac{R_n}{R_s} \qquad (13.10)$$

Observe that the noise figure F is independent of the load resistance. Also, the input impedance Z_i does not appear explicitly in Eq. 13.10. However, the equivalent noise resistance R_n is a function of the transistor input impedance.

We will now find the equivalent noise resistance R_n in terms of the transistor parameters given in Fig. 13.4. Since R_n represents the total noise resistance including the source resistance, let us define a noise resistance R_a which, at 290°K, will produce noise equivalent to the noise generated by the amplifier. Then $R_n = R_s + R_a$, and in terms of R_a

$$F = \frac{R_s + R_a}{R_s} \qquad (13.11)$$

We will first find the noise voltage $\bar{v}_{na}$ which, if placed in the input of the transistor in series with the driving source, will produce the same output noise current as the transistor. Figure 13.4 shows that there are three noise sources in the transistor—one voltage source $\bar{v}_{nb}$ and two current source i_{nb} and i_{nc}. The voltage $\bar{v}_{nb}$ is already in series with the driving source, and we need to convert the two current sources to equivalent voltage sources in series with $\bar{v}_{nb}$ as shown in Fig. 13.6.

The equivalent voltage $\bar{v}_{nb}'$ results from the current source i_{nb}. This current source can be easily transformed to the equivalent voltage source $\bar{v}_{nb}'$ with the aid of the equivalent circuit given in Fig. 13.7, wherein all noise sources except i_{nb} have been turned off. The voltage $\bar{v}_{nb}'$ in the

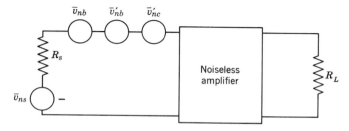

FIGURE 13.6
An equivalent noise circuit for the transistor.

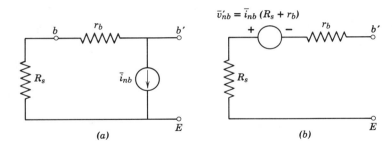

FIGURE 13.7
Equivalent circuits (*a*) **Norton and** (*b*) **Thevenin for the base current noise component** $\bar{i}_{nb}$.

Thevenin's circuit (Fig. 13.7*b*) is the open-circuit voltage at terminals $b' - E$ of the Norton's circuit (Fig. 13.7*a*). This voltage is

$$\bar{v}_{nb}' = \bar{i}_{nb}(R_s + r_b) \tag{13.12}$$

The equivalent noise voltage $\bar{v}_{nc}'$ can be found from the equivalent circuit given in Fig. 13.8. Since $\bar{i}_{nc} = g_m \bar{v}$, $\bar{i}_{nc}$ can be expressed in terms of $\bar{v}_{nc}'$ since $\bar{v} = \bar{v}_{nc}' Z_\pi / (R_s + r_b + Z_\pi)$.

$$\bar{i}_{nc} = \frac{g_m \bar{v}_{nc}' Z_\pi}{R_s + r_b + Z_\pi} \tag{13.13}$$

Note that Z_π is the impedance of r_π and $j\omega C_1$ in parallel. Using Eq. 13.13 to obtain $\bar{v}_{nc}'$ explicitly,

$$\bar{v}_{nc}' = \bar{i}_{nc} \frac{(R_s + r_b + Z_\pi)}{g_m Z_\pi} \tag{13.14}$$

Now the effective noise voltage $\bar{v}_{na}$ which we are seeking is the effective sum of the three noise components we have found. But these component noise voltages are uncorrelated; that is, their phase or frequency components

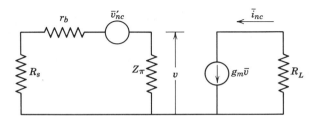

FIGURE 13.8
An equivalent circuit used to determine $\bar{v}_{nc}'$ **in terms of** $\bar{i}_{nc}$.

are unrelated. However, the noise power of each component is proportional to the square of the component voltage, and these noise powers can be added directly. Therefore,

$$\bar{v}_{na}{}^2 = \bar{v}_{nb}{}^2 + (\bar{v}_{nb}')^2 + (\bar{v}_{nc}')^2 \tag{13.15}$$

or, using Eqs. 13.12 and 13.14,

$$\bar{v}_{na}{}^2 = \bar{v}_{nb}{}^2 + i_{nb}{}^2(R_s + r_b)^2 + \frac{i_{cn}{}^2 |R_s + r_b + Z_\pi|^2}{g_m{}^2 |Z_\pi|^2} \tag{13.16}$$

We now need to express these voltage components in terms of the base resistance r_b and the q-point diode current components I_B and I_C which cause them. Letting $\bar{v}_{na}{}^2 = 4kTR_a \Delta f$, as previously suggested,

$$4kTR_a \Delta f = 4kTr_b \Delta f + 2qI_B \Delta f(R_s + r_b)^2 + \frac{2qI_C \Delta f |R_s + r_b + Z_\pi|^2}{g_m{}^2 |Z_\pi|^2} \tag{13.17}$$

Dividing Eq. 13.17 through by $4kT\Delta f$,

$$R_a = r_b + \frac{q}{2kT} I_B(R_s + r_b)^2 + \frac{qI_C}{2kT} \frac{|R_s + r_b + Z_\pi|^2}{g_m{}^2 |Z_\pi|^2} \tag{13.18}$$

This expression for R_a can be simplified if the relationships $I_B = I_C/h_{FE}$ and $qI_C/kT = g_m$ are used. Then

$$R_a = r_b + \frac{g_m(R_s + r_b)^2}{2h_{FE}} + \frac{g_m|R_s + r_b + Z_\pi|^2}{2(g_m Z_\pi)^2} \tag{13.19}$$

Equation 13.19 shows that R_a, and therefore the noise figure F, is a function of frequency because Z_π decreases as the frequency increases. We have found that $\beta_o = g_m r_\pi$, and in general $\beta = g_m Z_\pi$; therefore, Z_π is equal to β/g_m and $h_{ie} = r_b + Z_\pi$, so Eq. 11.19 can be written

$$R_a = r_b + \frac{g_m(R_s + r_b)^2}{2h_{FE}} + \frac{g_m(R_s + h_{ie})^2}{2\beta^2} \tag{13.20}$$

This expression points up the importance of small I_C and, hence, small g_m with large β and large h_{FE} (dc beta) for low noise at low frequencies.

PROBLEM 13.4 A transistor with $r_b = 200\ \Omega$, $\beta = 100$, and $h_{FE} = 80$ at $I_C = 0.1$ mA is used in an amplifier with $R_s = 10$ kΩ. Determine the equivalent noise resistance R_a and the noise figure F.
Answer: $R_a = 3{,}050\ \Omega$, $F_a = 1.30$.

Sometimes the amplifier noise is expressed as a *noise temperature* T_a instead of a noise figure. The noise temperature is the temperature which the source resistance R_s must have in order to produce the same noise as the amplifier. Then

$$(4kT_a R_s \Delta f)^{1/2} = [4k(290°K)R_a \Delta f]^{1/2} \qquad (13.21)$$

and

$$T_a = 290 \frac{R_a}{R_s} °K \qquad (13.22)$$

PROBLEM 13.5   Determine the noise temperature of the amplifier in Problem 13.4.

Answer: $T_a = 88.5°K$.

The noise figure increases at frequencies above f_β because the signal gain decreases while the transistor noise remains constant. This effect shows up in the last term of Eq. 13.20 because β and h_{ie} are both inversely proportional to frequency for frequencies above f_β. However, as Eq. 13.20 shows, R_a does not increase appreciably with frequency until the magnitude of h_{ie} becomes smaller than R_s.

PROBLEM 13.6   Determine the noise figure of the amplifier in Problems 13.4 and 13.5 at $f = 20f_\beta$.

Answer: $F = 2.08$.

13.4   NOISE OPTIMIZATION FOR LOW-FREQUENCY TRANSISTOR AMPLIFIERS

We have considered some general principles which can be used to design a low-noise amplifier, but do not yet know how to determine an optimum driving-source resistance for a given q point (I_C) or vice versa. Let us investigate this optimization by writing an expression for the noise temperature, using Eq. 13.22 and Eq. 13.20

$$T_a = 290 \left[\frac{r_b}{R_s} + \frac{g_m(R_s + r_b)^2}{2R_s h_{FE}} + \frac{g_m(R_s + r_b + r_\pi)^2}{2R_s \beta_o^2} \right] \qquad (13.23)$$

The minimum value of T_a can be obtained by differentiating T_a with respect to R_s and equating this derivative to zero to find the optimum R_s, assuming I_C and hence g_m and β_o to be constant. However, the derivative can be taken more readily if the squared terms in Eq. 13.23 are expanded and the approximation $r_b \ll (R_s + r_\pi)$ is made, as follows.

$$T_a = 290 \left[\frac{r_b}{R_s} + \frac{g_m}{h_{FE}} \left(\frac{R_s}{2} + r_b + \frac{r_b^2}{2R_s} \right) + \frac{g_m}{\beta_o^2} \left(\frac{R_s}{2} + r_\pi + \frac{r_\pi^2}{2R_s} \right) \right]$$

$$(13.24)$$

Collecting terms,

$$T_a = 290 \left[\frac{1}{R_s} \left(r_b + \frac{g_m {r_b}^2}{2h_{FE}} + \frac{g_m {r_\pi}^2}{2{\beta_o}^2} \right) + R_s \left(\frac{g_m}{2h_{FE}} + \frac{g_m}{2{\beta_o}^2} \right) + \frac{g_m r_b}{h_{FE}} + \frac{g_m r_\pi}{{\beta_o}^2} \right]$$

(13.25)

Differentiating,

$$\frac{dT_a}{dR_s} = 290 \left[-\frac{1}{{R_s}^2} \left(r_b + \frac{g_m {r_b}^2}{2h_{FE}} + \frac{g_m {r_\pi}^2}{2{\beta_o}^2} \right) + \left(\frac{g_m}{2h_{FE}} + \frac{g_m}{2{\beta_o}^2} \right) \right] = 0$$

(13.26)

Using the relationships $r_\pi = \beta_o/g_m$ and ${\beta_o}^2 \gg h_{FE}$,

$$-\frac{1}{{R_s}^2} \left(r_b + \frac{g_m {r_b}^2}{2h_{FE}} + \frac{1}{2g_m} \right) + \frac{g_m}{2h_{FE}} = 0 \qquad (13.27)$$

Solving for ${R_s}^2$,

$$R_s^2 = \frac{2h_{FE}}{g_m} r_b + r_b^2 + \frac{h_{FE}}{g_m^2} \qquad (13.28)$$

and

$$R_{s\,(opt)} = \left[\frac{h_{FE}}{g_m} \left(2r_b + \frac{1}{g_m} \right) + r_b^2 \right]^{1/2} \qquad (13.29)$$

In low-noise amplifiers with small g_m, $r_b \ll h_{FE}/g_m$, so the r_b^2 term in Eq. 13.29 can usually be neglected. Then

$$R_{s(opt)} \simeq \frac{\sqrt{h_{FE}(2r_b g_m + 1)}}{g_m} \qquad (13.30)$$

Very frequently, the driving-source resistance is given and the value of q-point collector current is desired. The optimum g_m and hence I_C can be determined in terms of R_s from Eq. 13.30. Thus

$$R_s^2 g_m^2 = 2h_{FE} r_b g_m + h_{FE} \qquad (13.31)$$

and

$$g_m^2 - \frac{2h_{FE} r_b g_m}{R_s^2} - \frac{h_{FE}}{R_s^2} = 0 \qquad (13.32)$$

Using the quadratic equation,

$$g_m = \frac{h_{FE} r_b}{R_s^2} \pm \left[\left(\frac{h_{FE} r_b}{R_s^2} \right)^2 + \frac{h_{FE}}{R_s^2} \right]^{1/2} \qquad (13.33)$$

The positive sign is needed in Eq. 13.33 to give positive values of g_m. Then

$$g_{m(\text{opt})} \simeq \frac{h_{FE}\,r_b + \sqrt{(h_{FE}\,r_b)^2 + h_{FE}\,R_s^{\,2}}}{R_s^{\,2}} \qquad (13.34)$$

PROBLEM 13.7 A transistor with $h_{FE} = 80$ and $r_b = 400\ \Omega$ at $I_C = 100\ \mu A$ is used as a low-noise amplifier at the given I_C. Determine the optimum driving-source resistance.

PROBLEM 13.8 A low-noise amplifier is needed for a microphone with $10\ k\Omega$ internal resistance. Use the transistor of Problem 13.7 and determine optimum I_C, assuming h_{FE} remains 80.

Manufacturers frequently give noise-figure contours similar to the one shown in Fig. 13.9 for their low-noise transistors. These contours greatly simplify the problem of selecting the best value of I_C for a fixed source resistance. For example, if $R_s = 2\ k\Omega$, a good value of I_C is 50 μA, which will yield a noise figure less than 3 dB at $f = 1$ kHz.

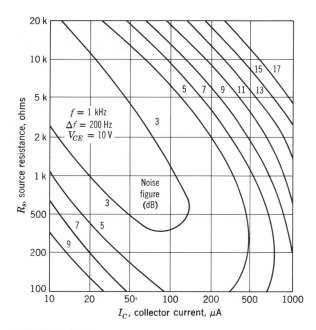

FIGURE 13.9
Constant noise-figure contours for a typical low-noise transistor (2N2443).

13.5 NOISE OPTIMIZATION IN RF AMPLIFIERS

Although Eq. 13.19 can be used to calculate the noise resistance and consequently the noise figure of an amplifier, as previously discussed, this equation gives no clues toward the optimization of noise in a high frequency or *RF* amplifier. However, as previously noted, the amplifier noise resistance does not increase appreciably until the operating frequency f is well above frequency f_β. Then $Z_\pi \simeq 1/j\omega C_1$ and Eq. 13.19 can be simplified using this approximation.

$$R_a = r_b + \frac{g_m(R_s + r_b)^2}{2h_{FE}} + \frac{g_m|R_s + r_b + 1/j\omega C_1|^2}{2(g_m/\omega C_1)^2} \qquad (13.35)$$

But the square of a complex impedance is the square of the resistance plus the square of the reactance. Using this relationship for the numerator of the last term in Eq. 13.35, this equation becomes

$$R_a = r_b + \frac{g_m(R_s + r_b)^2}{2h_{FE}} + \frac{g_m[(R_s + r_b)^2 + 1/(\omega C_1)^2]}{2(g_m/\omega C_1)^2} \qquad (13.36)$$

You may recall that $\omega_\tau = g_m/C_1$, as given in Eq. 7.71. Making this substitution in Eq. 13.36 and separating the last term into its two components,

$$R_a = r_b + \frac{g_m(R_s + r_b)^2}{2h_{FE}} + \frac{g_m(R_s + r_b)^2\omega^2}{2\omega_\tau^2} + \frac{1}{2g_m} \qquad (13.37)$$

Equation 13.37 shows that g_m should be small to minimize the second term on the right, and in addition, ω should be small in comparison with ω_τ to minimize the third term on the right. However, when I_C is small, both g_m and ω_τ are small. Therefore, small g_m and large ω_τ are incompatible. As g_m is decreased to make the second term smaller, the third term increases because $(\omega/\omega_\tau)^2$ increases more rapidly than g_m decreases. Experience teaches that near optimum conditions are usually obtained by making the magnitudes of the increasing and decreasing terms equal. If these were the only terms in the equation, complete optimization would be obtained in this manner. Following this procedure,

$$\frac{g_m(R_s + r_b)^2}{2h_{FE}} = \frac{g_m(R_s + r_b)^2\omega^2}{2\omega_\tau^2} \qquad (13.38)$$

Simplifying,

$$\frac{1}{h_{FE}} = \frac{\omega^2}{\omega_\tau^2} = \frac{f^2}{f_\tau^2} \qquad (13.39)$$

In a high Q band-pass amplifier, the frequency f can be considered constant

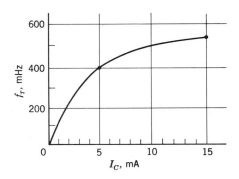

FIGURE 13.10.
f_τ **as a function of** I_C **for a 2N3298 transistor.**

at f_o so far as noise is concerned. Then the first step in the noise optimization is to obtain a plot of f_τ as a function of I_C, as shown in Fig. 13.10, and choose the value of I_C such that, using Eq. 13.39,

$$f_\tau = \sqrt{h_{FE}}\, f_o \qquad (13.40)$$

where f_o is the desired center, or resonant, frequency. This value of I_C then establishes $g_m \simeq 40\, I_C$. If we use this optimized value of I_C, Eq. 13.38 may be substituted into Eq. 13.37 to obtain an expression for R_a that is not a function of either ω or ω_τ. Then the noise figure $F = 1 + (R_a/R_s)$ for this partially optimized condition can be written as

$$F = 1 + \frac{r_b}{R_s} + \frac{g_m(R_s + r_b)^2}{h_{FE}\, R_s} + \frac{1}{2g_m R_s} \qquad (13.41)$$

The driving-source resistance R_s can now be optimized for this q point using the technique $dF/dR_s = 0$.

$$\frac{dF}{dR_s} = \frac{g_m}{h_{FE}} - \frac{1}{R_s^2}\left(r_b + \frac{g_m r_b^2}{h_{FE}} + \frac{1}{2g_m}\right) = 0 \qquad (13.42)$$

Then

$$R^2_{s(\mathrm{opt})} = \frac{h_{FE}\, r_b}{g_m} + r_b^2 + \frac{h_{FE}}{2g_m^2} \qquad (13.43)$$

Figure 13.11 illustrates the improvement in noise temperature at high frequencies that results from increasing the q-point collector current from 30 μA to 1 mA. The improvement results from increased f_τ.

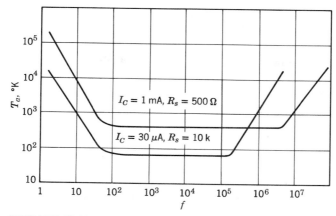

FIGURE 13.11
Noise temperature as a function of frequency for a typical common-emitter amplifier at $T = 295°\text{K}$.

PROBLEM 13.9 A 2N3298 transistor is to be used as a 20 MHz amplifier with a 2 MHz bandwidth. If this transistor has $h_{FE} = 90$ and $r_b = 100\ \Omega$, determine the value of I_C which satisfies Eq. 13.39 and calculate $R_{s(\text{opt})}$ and noise figure F for this value of I_C.

When the required center frequency f_o of an RF amplifier approaches the maximum obtainable f_τ, it is impossible to make $f_\tau = \sqrt{h_{FE}\,f_o}$, of course. The value of I_C must then be chosen to provide near maximum f_τ. The optimum driving-source resistance can then be determined by differentiating Eq. 13.37 with respect to R_s and equating the derivative to zero. This technique yields

$$R_{s(\text{opt})}^2 = \frac{r_b + (1/2g_m) + (g_m r_b^2/2h_{FE}) + (g_m r_b^2 f_o^2/2f_\tau^2)}{(g_m/2h_{FE}) + (g_m f_o^2/2 f_\tau^2)} \qquad (13.44)$$

The values of $R_{s(\text{opt})}$ obtained from Eq. 13.44 can be used in Eq. 13.37 to calculate the noise figure F.

Transistor manufacturers frequently recommend values of I_C and R_s for typical values of f_o and list typical values of noise figure F for these suggested values.

13.6 NOISE IN FET AND TUBE AMPLIFIERS

The field-effect transistor has only the channel resistance and surface leakage as principal sources of noise, since the only diode current is the thermal saturation current of the reverse-biased gate diode. Therefore, the noise figure of a silicon FET may be very low. Since the diode current is very small, the medium-frequency (1 kHz) noise figure is determined primarily by the ratio r_{ch}/R_s where r_{ch} is the channel resistance. A typical value for F is 1.1 for high values of source resistance (of the order of 1 MΩ). The $1/f$ noise is present, however, and the noise figure rises at low frequencies. Also, there is some diode current, so the noise figure increases as the power gain decreases at high frequencies. The noise figure for a typical FET is given as a function of frequency and driving-source resistance in Fig. 13.12.

The triode vacuum tube has primarily diode noise which results from the random electron emission from the cathode. However, the space charge surrounding the cathode tends to suppress the randomness of the emission current and reduces the tube noise well below the value which would occur in an emission-limited tube. This smoothing effect complicates the derivation of a theoretical noise figure determination. However, an expression for the equivalent grid circuit noise resistance has been developed[2] for the triode and experimentally verified. This expression is

$$R_a = \frac{2.5}{g_m} \qquad (13.45)$$

This noise resistance does not include the source resistance. Therefore, the noise figure F would be determined from the relationship $F = 1 + (R_a/R_s)$, and high values of source resistance give low-noise figures.

Like the transistor, the vacuum tube has a $1/f$ noise known as cathode flicker which increases the noise figure at low frequencies. Also, the noise figure increases as the frequency becomes high and the electron transit time reduces the gain of the amplifier. The additional high-frequency noise is known as induced grid noise. Low-noise triode tubes and low-noise transistors have noise figures of approximately equal magnitude.

The tetrode or pentode tube has a source of noise in addition to those found in triodes. This noise, known as *partition noise*, results from the random division of the space current between the screen grid and the plate. The noise of a tetrode or pentode can be expressed as an equivalent grid circuit resistance by the following equation.

$$R_a \simeq \frac{2.5}{g_m} + \frac{20I_{G2}}{I_K g_m} \qquad (13.46)$$

where I_K is the average cathode current. The second term on the right-hand side of Eq. 13.46 accounts for the partition noise.

[2] K. R. Spangenberg, *Vacuum Tubes*, McGraw-Hill Book Company, Inc., 1948, pp. 310–312.

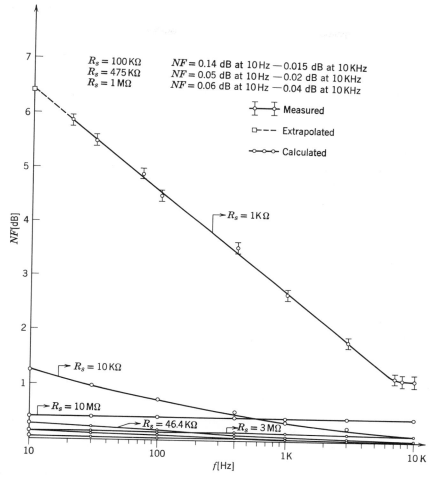

FIGURE 13.12
The noise figure of a typical FET amplifier as a function of driving-source resistance and frequency.

The noise generated by a vacuum tube may be much higher than the values obtained by the foregoing equations because of defects in an individual tube or circuit. For example, gas in a tube causes random collisions which increase the noise. Also, faulty electrode contacts may generate noise. Heater-cathode interaction may cause noise and hum unless the bias between the heater and the cathode prevents an exchange of charge carriers between the two. Usually, low-noise tubes are hand picked after noise tests have been made.

13.7 MEASUREMENTS OF THE NOISE FIGURE OF AN AMPLIFIER

A noise test may be easily conducted on an amplifier if a noise diode is used as a signal source. The noise diode is connected in parallel with the desired source resistance as shown in Fig. 13.13a. The noise diode must have a high internal impedance compared with R_s in order to act as a true current source. Therefore, a temperature-limited vacuum diode is usually used. First the noise output voltage V_{no} is measured with an rms-activated meter while the noise diode is turned off. Then the output noise power results only from the equivalent amplifier noise source $\bar{v}_n$ (Fig. 13.13b).

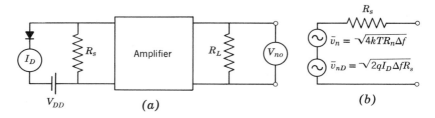

FIGURE 13.13
A circuit used to determine the noise figure of an amplifier : (a) block diagram; (b) equivalent input circuit.

Next, the diode current is increased by adjusting the supply voltage V_{DD} until the output voltage increases by a factor of $\sqrt{2}$, which indicates that the noise power in the output has doubled and the Thevenin's equivalent voltage of the diode noise source $\bar{v}_{nd}$ is equal to equivalent noise voltage $\bar{v}_n$. Then, from Fig. 13.13b,

$$2qI_D\,\Delta f R_s^2 = 4kTR_n\,\Delta f \tag{13.47}$$

and

$$R_n = \frac{qI_D R_s^2}{2kT} \tag{13.48}$$

But the noise figure F is

$$F = \frac{R_n}{R_s} = \frac{qI_D R_s}{2kT} \tag{13.49}$$

At the standard reference temperature $T = 295°\text{K}$, $q/2kT = 20$, and

$$F = 20I_D R_s \tag{13.50}$$

The noise figure obtained by the procedure described will give an integrated noise figure over the pass band of the amplifier or meter, whichever is less. A filter can be placed in series with the output meter to determine the spot noise figure in the pass band of the filter.

13.8   SIGNAL-TO-NOISE-RATIO-CALCULATIONS

An example will illustrate the calculation of signal-to-noise ratio (*SNR*) at both the input and output of an amplifier. The *SNR* is a power ratio.

Example 13.1   A microphone is rated as having -60 dB (1 mV) open-circuit output voltage, 25 kΩ impedance (resistance), and frequency response from 50 to 15,000 Hz (assumed noise bandwidth). Determine its signal-to-noise ratio (*SNR*). (Assume $T = 290°$K.) $SNR = \bar{v}_o{}^2/4kTR\Delta F = 10^{-6}/4 \times 1.38 \times 10^{-23}$ $(290)(2.5 \times 10^4)(1.5 \times 10^4) = 1.65 \times 10^5$. This ratio is usually expressed in dB. Thus $SNR = 10 \log (1.65 \times 10^5) = 52$ dB.

The *SNR* at the output of the amplifier is reduced by the noise figure F of the amplifier, since the output noise, referred to the input, is represented by $(R_s + R_a)$. Then, if $1/f$ noise is neglected, the signal-to-noise ratio of an amplifier with $F = 2$, or 3 dB, when used with this microphone is 8.3×10^4 (or 49 dB). Note that the noise figure in dB is subtracted from the input *SNR* in dB to obtain the output *SNR* in dB.

PROBLEM 13.10   A given strain gauge has 10 kΩ internal resistance and produces a 100 microvolt rms open-circuit signal. This signal is amplified by a bipolar silicon transistor which has $r_b = 200$ Ω, $\beta_o = 100$, $h_{FE} = 80$, and $f_\tau = 5$ MHz at $I_C = 100$ μA $V_{CE} = 10$ V. Determine the theoretical signal-to-noise ratio both into the amplifier and out of the amplifier, neglecting $1/f$ noise, if the noise band-width of both the source and the amplifier is 10 kHz. Assume $T = 290°$K.
Answer: 6250 (38 dB), 4800 (36.8 dB).

PROBLEM 13.11   A three-stage RF amplifier must have a total voltage gain of 6400 and a 200 kHz bandwidth at a center frequency of 5.0 MHz. If the transistor of Fig. 13.9 is used for each stage and all stages are identical, determine the gain and bandwidth of each stage. Which values of I_C and R_s listed in Fig. 13.9 give the best noise figure, and what is the value of this noise figure?
Answer: $K_v = 40$, $B = 392$ kHz, $F = 2.38$ (3.8 dB), approximately.

PROBLEM 13.12   A given dynamic microphone is rated -57 dB output (open circuit), 20 kΩ resistance, and 40 Hz to 15 kHz frequency response. Choose a suitable transistor, or transistors, and design a low-noise amplifier which will provide approximately 1 V rms output. Determine the approximate SNR in the output of your amplifier.

PROBLEM 13.13   A given capacitor microphone is rated -74 dB and has $C = 10^{-9}$ F as an internal capacitance. This microphone will provide uniform frequency response from 20 to 20,000 Hz if it feeds into an amplifier having a high input resistance. Design an amplifier which will provide approximately 1 V rms output over the 20–20,000 Hz range when used with this microphone. You choose the transistors. Determine the approximate SNR in the output of your amplifier.

PROBLEM 13.14   Derive an expression for the optimum driving-source resistance for a FET amplifier. The variables might be the channel resistance and the thermal saturation current I_D .

14

Negative Feedback

The performance of an amplifier can be altered by the use of feedback; that is, by adding part or all of the output signal to the input signal. If there are an even number of polarity reversals (or no polarity reversals) between the input and the output of the amplifier, the feedback is said to be positive. This type of feedback is used in oscillator circuits. On the other hand, if there is an odd number of polarity reversals in the amplifier so the feedback signal tends to cancel the input signal in the mid-frequency range, the feedback is said to be negative. This negative feedback, which is the subject of this chapter, can reduce distortion, increase the bandwidth, change the output impedance and the input impedance, and stabilize the gain of an amplifier. All of these improvements are obtained at the expense of reduced mid-frequency gain.

14.1 THE EFFECT OF FEEDBACK ON GAIN, DISTORTION, AND BANDWIDTH

The block diagram of an amplifier with feedback is given in Fig. 14.1. An expression for the gain of the amplifier with feedback will be developed without regard to whether the feedback is negative or positive. The voltage v_a which actually drives the amplifier is the algebraic sum of the input voltage v_i and the feedback voltage v_f, or

$$v_a = v_i + v_f \tag{14.1}$$

The feedback factor $^1\beta_v$ is the ratio of the feedback voltage v_f to the

[1] In this chapter β will be used as the feedback factor, not as the transistor current amplification factor which will be designated as h_{fe}. This β is commonly used in the literature to represent the feedback factor, or ratio.

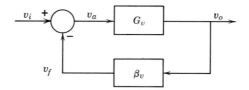

FIGURE 14.1
The block diagram of an amplifier with negative feedback.

output voltage v_o. This ratio is usually obtained by a resistive voltage divider as shown later. Thus,

$$v_f = \beta_v v_o \tag{14.2}$$

We will let G_v be the voltage gain or amplification of the amplifier without feedback. Thus,

$$v_o = G_v v_a \tag{14.3}$$

The voltage gain G_{vf} of the amplifier with feedback is

$$G_{vf} = \frac{v_o}{v_i} \tag{14.4}$$

But, using Eq. 14.1, $v_i = v_a - v_f$ and substituting the value of v_o from Eq. 14.3,

$$G_{vf} = \frac{G_v v_a}{v_a - v_f} \tag{14.5}$$

Substituting $\beta_v v_o$ for v_f (Eq. 14.2) and then dividing both the numerator and denominator of Eq. 14.5 by v_a, we have

$$G_{vf} = \frac{G_v v_a}{v_a - \beta_v v_o} = \frac{G_v}{1 - \beta(v_o/v_a)} \tag{14.6}$$

Finally, substituting G_v for v_o/v_a in Eq. 14.6,

$$G_{vf} = \frac{G_v}{1 - G_v \beta_v} \tag{14.7}$$

In the preceding work, the symbol G_v was used to represent the voltage gain of the amplifier. Thus G_v must be a function of frequency $j\omega$, or s, in agreement with the gain expressions in earlier chapters. Also, G_v must include a negative sign if the amplifier has an odd number of common-emitter stages (the negative feedback configuration) so a polarity reversal occurs in the amplifier. Then $G_v = -K_v F(s)$, where K_v is the reference or mid-frequency gain and $F(s)$ is the part of the gain expression which varies with frequency. Using this expression, Eq. 14.7 becomes

$$G_{vf} = \frac{-K_v F(s)}{1 + K_v \beta_v F(s)} \qquad (14.8)$$

This is the basic equation for *negative* feedback. Because of the polarity reversal, the input signal to the amplifier is the arithmetic *difference* between the input signal and the feedback signal at mid-frequencies. The positive sign in the denominator identifies the negative feedback. In most applications we are interested in the characteristics of the amplifier in the mid-frequency range. Thus the mid-frequency voltage gain with feedback is

$$K_{vf} = \frac{K_v}{1 + K_v \beta_v} \qquad (14.9)$$

As in Chapter 7, K_{vf} is the magnitude of the mid-frequency gain with feedback. This gain is accompanied by a negative sign (as in Chapter 7) when we wish to consider the signal inversion due to an odd number of polarity inversions between the input and output signals. Equation 14.9 is used for most applications because it yields a simple number for the voltage gain. However, when considering frequency response (or similar problems), the general relationship of Eq. 14.8 must be used.

Equation 14.9 could be written in terms of current gains K_{if}, K_i and a current feedback ratio β_i instead of their voltage counterparts. Making these substitutions into Eq. 14.9,

$$K_{if} = \frac{K_i}{1 + K_i \beta_i} \qquad (14.10)$$

In a given amplifier, the feedback reduces both the current gain and the voltage gain by the same factor, since the relationship $K_{vf} = K_{if} Z_L/Z_i$ holds for any amplifier, with or without feedback. Therefore, $1 + K_v \beta_v = 1 + K_i \beta_i$ and $K_v \beta_v = K_i \beta_i$. Thus the term $K\beta$ can mean either $k_v \beta_v$ or $K_i \beta_i$. This reduction in gain is the disadvantage of negative feedback, and additional gain must be provided in the amplifier to compensate for the feedback. Several advantages accrue, however, which make negative feedback very attractive.

One desirable characteristic of negative feedback is improved gain stability. You may observe from Eq. 14.9 or Eq. 14.10 that the product $K\beta$ (meaning either $K_v \beta_v$ or $K_i \beta_i$) may be large in comparison with one.

Then

$$K_f = \frac{K}{1 + K\beta} \simeq \frac{K}{K\beta} = \frac{1}{\beta}\bigg|_{K\beta \gg 1} \tag{14.11}$$

Thus, the gain becomes almost independent of the amplifier characteristics and depends primarily on the resistance ratio of a voltage divider.

Another desirable characteristic of negative feedback is the reduction of harmonic or nonlinear distortion. The reason for this reduction may be seen from Fig. 14.2, where the amplifier is assumed to distort the sinusoidal

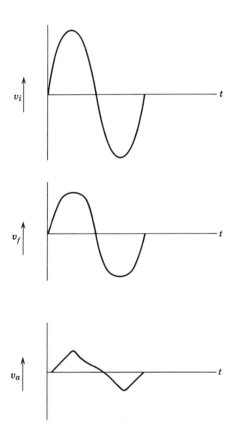

FIGURE 14.2
A sketch showing how the feedback voltage
v_f **predistorts the amplifier input voltage**
v_a **to partially compensate for the amplifier**
distortion.

input voltage by flattening the peaks. The feedback voltage v_f has the same waveform as the output voltage. Therefore, the flattened peaks of the feedback voltage subtract less from the input voltage, thus accentuating the peaks of the amplifier input voltage v_a and predistorting v_a in a manner which will partially compensate for the flattening caused by the amplifier.

The amount of distortion reduction caused by negative feedback can be determined with the aid of Fig. 14.3. The amplifier with gain K has

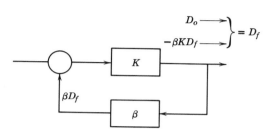

FIGURE 14.3
A block diagram illustrating the reduction of distortion in the output of an amplifier.

distortion D_o appearing in its output without feedback. After feedback is applied, distortion D_f appears in the output. But distortion D_f is first reduced by the factor β and applied to the input of the amplifier which in turn amplifies βD_f by the gain and adds it, in opposite polarity, to the original distortion D_o to obtain the total distortion D_f. Thus,

$$D_f = D_o - K\beta D_f \tag{14.12}$$

Solving explicitly for D_f in terms of D_o,

$$D_f = \frac{D_o}{1 + K\beta} \tag{14.13}$$

Note that the distortion is reduced by the same factor as the gain. Frequently, the feedback factor β is chosen to reduce the distortion a given amount.

The bandwidth of an amplifier is also increased by the use of negative feedback. We will use the expression of voltage gain for an RC-coupled stage to show this. You may recall that the expression for gain (voltage or current) for the low and middle-frequency range is

$$G = \frac{-K}{1 - jf_1/f} \tag{14.14}$$

where K is the reference, or mid-frequency, gain and f_1 is the low-frequency half-power point. Using this expression for G in the general feedback formula, Eq. 14.8,

$$G_f = \frac{-K/(1 - jf_1/f)}{1 + [K\beta/(1 - jf_1/f)]} = \frac{-K}{1 - j(f_1/f) + K\beta} = \frac{-K}{1 + K\beta - jf_1/f}$$

(14.15)

Then dividing both numerator and denominator of Eq. 14.15 by $1 + K\beta$,

$$G_f = \frac{-K/(1 + K\beta)}{1 - (jf_1/[(1 + K\beta)f])}$$

(14.16)

You may observe from Eq. 14.16 that the low-frequency cutoff for the amplifier with feedback is $f_{1f} = f_1/(1 + K\beta)$.

Similarly, we can determine the effect of feedback on the upper cutoff frequency of an RC-coupled amplifier, since we know the gain expression for middle and high frequencies is

$$G = \frac{-K}{1 + jf/f_2}$$

(14.17)

Using this value of G in the feedback Eq. 14.8,

$$G_f = \frac{[-K/(1 + jf/f_2)]}{[1 + K\beta/(1 + jf/f_2)]}$$

(14.18)

Equation 14.18 can be rearranged in the same manner as Eq. 14.13 to give

$$G_f = \frac{K/(1 + K\beta)}{1 + jf/(1 + K\beta)f_2}$$

(14.19)

Observe that the upper half-power frequency of the amplifier with feedback is $f_{2f} = (1 + K\beta)f_2$. Sometimes negative feedback is used for the specific purpose of increasing the bandwidth of an amplifier. However, when several stages are included in the feedback loop, or transformer coupling is used, the improvement in bandwidth is not so simply related to the feedback factor as in the preceding example.

An example of a feedback amplifier may help clarify some of the foregoing concepts.

Example 14.1   An amplifier which is assumed to consist of a single RC-coupled stage is connected as shown in Fig. 14.1. The characteristics of the amplifier without feedback are: voltage gain $K_v = 1000$, distortion = 6%, $f_1 = 20$ Hz, and $f_2 = 200 \, kHz$. Determine the characteristics of the amplifier with negative feedback when the voltage feedback ratio $\beta_v = 0.01$.

The value of $1 + K_v\beta_v$ or $(1 + K\beta)$ appears in each equation, so let us evaluate this term. Then $1 + K\beta = 1 + 1000 \times 0.01 = 1 + 10 = 11$. From Eq. 14.9, $K_{vf} = K_v/(1 + K\beta) = 1000/11 = 90.9$. (If we use the approximation given by Eq. 14.11, $K_{vf} \simeq 1/.01 = 100$.) The distortion with feedback (Eq. 14.13) is $D_f = D_o/(1 + K\beta) = 6\%/11 = 0.54\%$. The low-frequency half-power frequency with feedback f_{1f} is $f_1/(1 + K\beta) = 20/11 = 1.8$ Hz. The upper half-power frequency is $f_{2f} = (1 + K\beta)f_2 = 11 \times 200$ kHz $= 2.2$ MHz with negative feedback.

PROBLEM 14.1   Repeat example 14.1 if $\beta_v = 0.1$.

Answer: $(1 + K\beta) = 101$, $K_{vf} = 9.9$, $D_f = 0.059$, $f_{1f} = 0.198$ Hz, $f_{2f} = 20.2$ MHz.

PROBLEM 14.2   Assume that the acceptable distortion level in the amplifier of Example 14.1 is 1 percent. Determine the feedback factor β_v which will reduce the distortion from 6 percent to 1 percent. What is the voltage gain of the amplifier with this amount of feedback?

Answer: $\beta_v = 0.005$, $K_{vf} = 167$.

14.2   THE EFFECT OF FEEDBACK ON INPUT IMPEDANCE AND OUTPUT IMPEDANCE

Negative feedback will either increase or decrease the input impedance of an amplifier depending upon whether the feedback signal is added in series or parallel with the input signal. Let us first consider the series connection shown in Fig. 14.4. The input impedance Z_{if} is the ratio of input

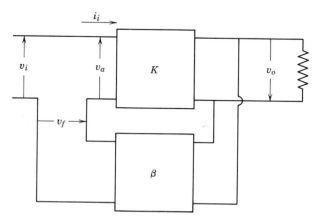

FIGURE 14.4
Series input connection for a feedback circuit.

voltage v_i to the input current i_i. Thus

$$Z_{if} = \frac{v_i}{i_i} = \frac{v_a + v_f}{i_i} \qquad (14.20)$$

but $v_f = K\beta v_a$, so

$$Z_{if} = \frac{v_a + K\beta v_a}{i_i} = \frac{v_a}{i_i}(1 + K\beta) \qquad (14.21)$$

However, v_a/i_i is the impedance Z_i of the amplifier without feedback. Therefore, in terms of Z_i,

$$Z_{if} = Z_i(1 + K\beta) \qquad (14.22)$$

Example 14.2   An amplifier with $Z_i = 1$ kΩ has a voltage gain $K_v = 1000$ and $\beta_v = .01$. With the feedback connection shown in Fig. 14.4, the input impedance with feedback is $Z_{if} = 1\ k(11) = 11\ k\Omega$.

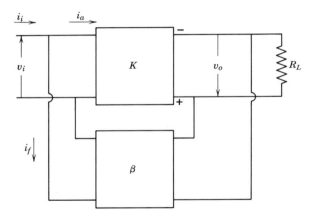

FIGURE 14.5
Parallel input connection for a feedback circuit.

We will now consider the parallel arrangement of input and feedback circuits shown in Fig. 14.5. In the parallel circuit, the input current i_i is the sum of amplifier input current i_a and the feedback current i_f. Thus, the input admittance with *feedback* is

$$Y_{if} = \frac{i_i}{v_i} = \frac{i_a + i_f}{v_i} \qquad (14.23)$$

But, using the amplifier current gain K_i and the current feedback factor $\beta_i = i_f/i_o$, the feedback current $i_f = i_a K_i \beta_i$. Making this substitution for i_f in Eq. 13.23,

$$Y_{if} = \frac{i_a + i_a K_i \beta_i}{v_i} = \frac{i_a}{v_i}(1 + K_i \beta_i) \qquad (14.24)$$

Since the input admittance of the amplifier without feedback Y_i is i_a/v_i, Y_{if} can be written in terms of Y_i.

$$Y_{if} = Y_i(1 + K\beta) \qquad (14.25)$$

As mentioned previously, $K_i \beta_i = K_v \beta_v$. We can also show this by recalling that $K_v = K_i R_L/R_i$. Also, $\beta_v = v_f/v_o = i_f R_i/i_o R_L = \beta_i R_i/R_L$. Thus, the product $K_v \beta_v = K_i \beta_i$, providing the input resistance R_i as seen from the feedback network is the same as the input resistance as seen by the driving source. This equality can exist only if the input current is the driving-source current and the source resistance R_s is included as part of the input resistance. This subject will be discussed in greater detail and examples will be given in Section 14.3. At the moment let us accept the fact that $K_i \beta_i$ may equal $K_v \beta_v$.

Example 14.3   An amplifier has $K_v = 1000$, $\beta_v = .01$ and $Z_i = 1000\ \Omega$ and is connected with parallel feedback as shown in Fig. 14.5. Then the input admittance without feedback, $Y_i = 1/Z_i = 10^{-3}$ mho and the input admittance with feedback $Y_{if} = 10^{-3}(11) = 1.1 \times 10^{-2}$ mho. The input impedance with feedback is $Z_{if} = 1/Y_{if} = 90.9\ \Omega$.

Up to this point, the feedback network has been shown as connected in parallel with the output terminals. This parallel connection is commonly known as *voltage* feedback because the feedback quantity (either current or voltage) is proportional to the output voltage. This voltage feedback is illustrated in Fig. 14.6. The output impedance with feedback is desired and can be determined by turning the input current source off and applying a voltage source v_2 to the output terminals. The current i_2 which flows is the sum of the current i_o through the output resistance, which may include the load resistance, and the current i_a which flows into the amplifier as a result of the feedback. The current flow into the feedback network should be, and usually is, negligible. Then

$$Y_{of} = \frac{i_2}{v_2} = \frac{i_o + i_a}{v_2} \qquad (14.26)$$

But with the driving-current source off, the feedback current i_f is also

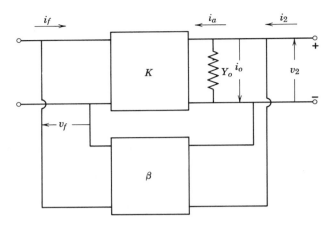

FIGURE 14.6
The parallel output connection, or voltage feedback.

the amplifier input current, so $i_a = i_f K_i = i_o \beta_i K_i$. Making this substitution in Eq. 14.26, Y_{of} is

$$Y_{of} = \frac{i_o + i_o \beta_i K_i}{v_2} = \frac{i_o}{v_2}(1 + K_i \beta_i) \qquad (14.27)$$

But i_o/v_2 is the output admittance Y_o without feedback. Therefore, Y_{of} can be written in terms of Y_o

$$Y_{of} = Y_o(1 + K\beta) \qquad (14.28)$$

Example 14.4 An amplifier is connected as shown in Fig. 14.6 and has $K_v = 1000$ and $\beta = .01$. The output resistance $R_o = 500\ \Omega$ before feedback is applied. Then, with feedback, $Y_{of} = 2 \times 10^{-3}(11) = 2.2 \times 10^{-2}$ mho and $R_{of} = 45\ \Omega$.

You may recall that we assumed the loading of the feedback circuit on the output of the amplifier was neglected. If this loading is not negligible, it can be included as part of Z_o.

You may have observed the regularity with which the factor $(1 + K\beta)$ appears as a modifier in negative feedback circuits. Let us now investigate the final feedback connection—the series arrangement of feedback and load in the amplifier output. This connection, which is known as current feedback because the feedback quantity (either current or voltage) is proportional to the output current, is shown in Fig. 14.7. Again, voltage v_2 is applied to the output terminals and current i_2 flows as a result. But i_2 is the sum of the current i_o through the output resistance

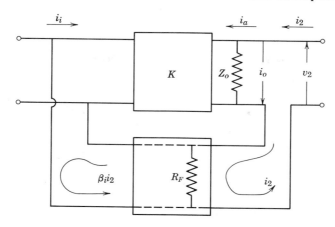

FIGURE 14.7
The series output connection, or current feedback.

and the current i_a which flows into the amplifier as a result of the feed-back. Note that the feedback current $\beta_i i_2$ is opposite in direction to the assumed amplifier input current i_i. But with the driving-source current turned off, the feedback current *is* the amplifier input current, so $i_i = -\beta_i i_2$. Then

$$i_2 = i_o + i_a = i_o - K_i \beta_i i_2 \tag{14.29}$$

or

$$i_2(1 + K\beta) = i_o \tag{14.30}$$

If the voltage drop $i_2 R_F$ across the feedback resistor is small in comparison with v_2, then

$$i_o \simeq \frac{v_2}{Z_o} \tag{14.31}$$

and

$$i_2(1 + K\beta) \simeq \frac{v_2}{Z_o} \tag{14.32}$$

Thus, since the output resistance with feedback Z_{of} is v_2/i_2,

$$Z_{of} = \frac{v_2}{i_2} = Z_o(1 + K\beta) \tag{14.33}$$

Example 14.5 The amplifier of the preceding example with $K_v = 1000$, $\beta_v = .01$ and $R_o = 500\ \Omega$ has the series output connection, or current feedback. Then, the output resistance with feedback, $R_{of} = 500\,(11) = 5500\ \Omega$.

PROBLEM 14.3 A given amplifier has $K_v = 10^4$, $R_i = 3\ k\Omega$, and $R_o = 3\ k\Omega$. This amplifier is to be inserted in a transmission line with characteristic impedance $Z_o = 150\ \Omega$. If feedback is used to make the impedance of the amplifier match the line in both directions, determine the feedback factor, the type connections (series or parallel), and the voltage gain with feedback.
Answer: $\beta_v = 1.9 \times 10^{-3}$, $K_{vf} = 500$.

14.3 FEEDBACK CIRCUITS

Many types of feedback circuits can be devised but only a few typical circuits will be discussed here. We will first consider the circuit of Fig. 14.8 which has the parallel output connection, or voltage feedback, and the series input connection. The feedback is negative because one polarity reversal occurs between the emitter of transistor T_1 where the feedback is applied and the collector of transistor T_2 where the

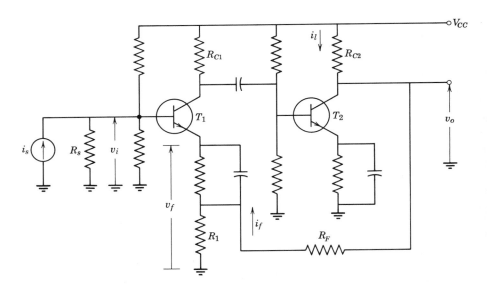

FIGURE 14.8
A feedback circuit with parallel output connection, or voltage feedback, and series input connection.

output voltage is obtained. Since $i_f = v_o/(R_F + R_1)$ and $v_f = i_f R_1$, the voltage feedback factor is approximately

$$\beta_v = \frac{v_f}{v_o} = \frac{R_1}{R_F + R_1} \qquad (14.34)$$

The impedance looking into the emitter circuit of T_1 is in parallel with R_1 so far as the feedback signal is concerned, and this may cast serious doubts upon the accuracy of Eq. 14.34. However, the feedback voltage v_f is nearly equal to the input voltage v_i when normal amounts of feedback are used, so the signal voltage between the emitter and base is small in comparison with v_f. Therefore, most of the feedback current i_f flows through R_1 and Eq. 14.34 is fairly accurate. The resistor R_1 should be small enough to permit good voltage gain in the first stage and R_F should be large in comparison with the load impedance Z_L so R_F will not seriously load the output.

Example 14.6 Let us use the foregoing assumptions and analyze the circuit shown in Fig. 14.8. Both transistors have $h_{fe} = 150$, $h_{ie} = 2$ kΩ, $h_{re} \simeq 0$, $h_{oe} \simeq 10^{-5}$ mho. The value of R_{C2} is 1 kΩ and is essentially the total load impedance of transistor T_2. The value of $R_s = 10$ kΩ and $R_{C1} = 2$ kΩ. The bias circuit resistance $R_b = 12$ kΩ.

We will choose $R_1 = 100$ Ω so the input impedance of transistor T_1 is $Z_{i1} = h_{ie} + (h_{fe} + 1)R_1 = 17$ kΩ. The total shunt resistance at the input of stage 1 is the parallel combination of R_s, R_b, and Z_{i1}, so $R_{sh} = 1/(1/10 \text{ k} + 1/12 \text{ k} + 1/17 \text{ k}) = 4.14$ kΩ. Then the current gain for stage 1 is $K_{i1} = i_{c1}/i_s = h_{fe}R_{sh}/Z_{i1} = 150 \times 4.14 \text{ k}/17 \text{ k} = 36.4$. The input impedance to transistor T_2 is $Z_{i2} = h_{ie2} = 2$ kΩ. The total shunt load on transistor T_1 is the parallel combination of R_{C1}, R_b and h_{ie2} (the value of $1/h_{oe}$ is much greater than this parallel combination so it can safely be ignored). This shunt resistance is about 925 Ω. The current gain for stage 2 is $K_{i2} = h_{fe}R_{sh}/R_i = 150 \times 925/2,000 \simeq 69$. Then the total current gain of the amplifier $= K_i = K_{i1}K_{i2} = 36.4 \times 69 = 2510$, and the total voltage gain is $K_v = v_o/v_i = K_i R_{L2}/R_{sh} = 2510 \times 10^3/4.14 \times 10^3 = 606$.

We will assume that the feedback is to be used to reduce the distortion of this amplifier by a factor of 6. Then the value of $1 + K_v\beta_v = 6$ so $K_v\beta_v = 5$. Thus, $\beta_v = 5/K_v = 5/606 = 0.00825$. Then from Eq. 14.34, we have $R_F = (R_1/\beta_v) - R_1 = (100/.00825) - 100 \simeq 12$ kΩ. This value of R_F is more than ten times the load resistance R_{C2} so we will assume it is large enough to have negligible loading effect on R_{C2}. The Z_o without feedback is $\simeq R_{C2}$ (since $1/h_{oe} \gg R_{C2}) = 1,000$ Ω. The output resistance with feedback is $Z_{of} \simeq 1 \text{ k}/(1 + K_v\beta_v) = 1 \text{ k}/6 = 167$ Ω. Also the input impedance to transistor T_1 is $Z_{if} = Z_i (1 + K\beta) = 17$ k $\times 6 = 102$ kΩ. The current gain K_{if} is $2510/6 = 420$ and the voltage gain K_{vf} is $606/6 = 101$.

PROBLEM 14.4   Repeat Example 14.6 if the h_{fe} of both transistors is 200. Assume that we wish to reduce the distortion to 10 percent of the value without feedback.

Answer: $K_i \simeq 367$, $K_v \simeq 84$, $\beta_v = 0.0119$, $R_F \simeq 8.3$ kΩ, $Z_{of} \simeq 100$ Ω, $Z_{if} = 220$ kΩ.

A typical circuit which uses both parallel input connection and parallel output connection is shown in Fig. 14.9. Observe that an odd number of common-emitter stages must be included in the feedback loop to obtain negative feedback. We will analyze this circuit in the following example.

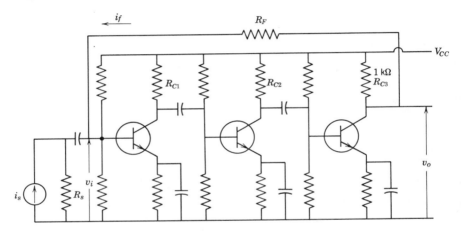

FIGURE 14.9
A feedback circuit with parallel input and output connections.

Example 14.7   A circuit is connected as shown in Fig. 14.9. All three transistors have $h_{fe} = 100$, $h_{ie} = 1$ kΩ, $h_{re} \simeq 0$, $h_{oe} = 10^{-5}$ mho. The bias circuit resistance of each stage is $R_b = 10$ kΩ. The source resistance $R_s = 2$ kΩ, $R_{C1} = R_{C2} = 2$ kΩ, and $R_{C3} = 1,000$ Ω. We wish to reduce the distortion of this circuit to one-tenth of its open-circuit value and then determine the characteristics of the circuit with feedback.

First, we must determine the amplifier characteristics with R_F open. We note that the values of R_C for each stage are much less than $1/h_{oe}$ so we can neglect h_{oe}. The input impedance to each transistor is h_{ie} which is equal to 1 kΩ. The shunt resistance on the input of the first stage is the parallel combination of R_s, R_b, and h_{ie}. Thus $R_{sh} = 1/(1/2$ k $+ 1/10$ k $+ 1/1$ k$)$

$= 625$ Ω. The current gain of the first stage is $K_{i1} = i_{c1}/i_s = h_{fe} R_{sh}/h_{ie} = 100 \times 625/1{,}000 = 62.5$. Note that if the first transistor is regarded as a current source and R_{C1} is used for R_s, the circuit for the second stage is identical to the circuit for the first stage. Therefore, $K_{i2} = K_{i1} = 62.5$. A similar observation for the circuit driving the third stage indicates $K_{i3} = K_{i1}$ also. Thus, the total open-loop current gain is $K_i = K_{i1} \times K_{i2} \times K_{i3} = 62.5^3 = 2.42 \times 10^5$. Since $R_{C3} \ll 1/h_{oe}$, the output impedance, $Z_o = 1{,}000$Ω. Then the total open-loop voltage gain is $K_v = K_i Z_o/R_{sh} = 2.42 \times 10^5 \times 10^3/625 = 3.87 \times 10^5$.

Since we desire a reduction of distortion by a factor of 10, the term $(1 + K\beta) = 10$. Then $K\beta = 9$. Now, since the output voltage v_o is very large (3.87×10^5 times as large) in comparison with the input voltage v_i, the feedback current i_f is very nearly equal to v_o/R_F. Also the output signal current $i_o = v_o/R_{C3}$. Therefore, the feedback current factor, or ratio, is

$$\beta_i = \frac{i_f}{i_o} \simeq \frac{v_o/R_F}{v_o/R_C} = \frac{R_C}{R_F} \qquad (14.35)$$

Now, $K_i = 2.42 \times 10^5$ and $K_i \beta_i = 9$. Consequently, $\beta_i = 9/2.42 \times 10^5 = 3.71 \times 10^{-5}$. From Eq. 14.35, we note that $R_F = R_C/\beta_i = 10^3/3.71 \times 10^{-5} = 2.69 \times 10^7$ Ω. The input impedance of the amplifier, including R_s, can be found from Eq. 14.25 as $Z_{if} = Z_i/(1 + K\beta) = Z_i/10 = 625/10 = 62.5$ Ω. The output impedance (from Eq. 14.28) is 100 Ω.

Let us now find out if $K_i \beta_i$ does equal $K_v \beta_v$ for this circuit. For this amplifier, $\beta_v = Z_i/(R_F + Z_i) = 625/2.69 \times 10^7 = 2.32 \times 10^{-5}$. Then, $K_v \beta_v = 2.32 \times 10^{-5} \times 3.87 \times 10^5 = 9$.

PROBLEM 14.5   Find a value of R_F which will reduce the distortion of the amplifier in Example 14.7 by a factor of 50. Find the value of K_{if}, K_{vf}, Z_{if} and Z_{of} with this value of R_F connected in the circuit.
Answer: $R_F = 4.95$ MΩ, $K_{if} = 4.84 \times 10^3$, $Z_o = 20$ Ω.

Figure 14.10 shows a feedback system with series output connection, or current feedback, and parallel input connection. Since the output current flows through the resistor R_1 in the emitter circuit of the output stage, the voltage across this resistance is proportional to the output current. Therefore, the feedback current is proportional to the output current. If the signal voltage across $R_1(i_o R_1)$ is large in comparison with v_i, the current feedback factor is approximately

$$\beta_i = \frac{i_f}{i_o} = \frac{R_1}{R_F} \qquad (14.36)$$

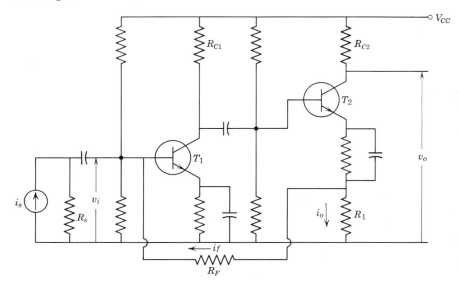

FIGURE 14.10
A feedback circuit with series output connection (current feedback) and parallel input connection.

PROBLEM 14.6 Assume that the amplifier of Fig. 14.10 uses the same components as the amplifier of Example 14.6. The unbypassed resistor $R_1 = 100\ \Omega$ has been changed from the input stage to the output stage, however, so the input impedance as seen by the current source i_s is the parallel combination of R_s, R_b and h_{ie} or 1.33 kΩ for the values given previously. We will assume that the total current gain (i_o/i_s) is the same as before, or 2,590. Then if the desired value of $K_i\beta_i = 9$, determine the required value of R_F. Since the output impedance without feedback was assumed to be approximately 1 kΩ, determine the output impedance and the input impedance, as seen by the current-driving source I_s, with the feedback specified.
Answer: $R_F = 2.88 \times 10^4\ \Omega$, $R_{of} = 10$ kΩ, $R_{if} = 133\ \Omega$.

Negative feedback can be used to provide q-point stability in a dc amplifier, as shown in Fig. 14.11. This amplifier is the quasi-complementary-symmetry amplifier given in Fig. 11.19 with an additional input stage T_1 to provide the extra gain needed for adequate feedback. The series input connection and parallel output connection for the feedback network was chosen to provide high input impedance and low output impedance for the amplifier, which is often desirable.

Example 14.8 A circuit is connected as shown in Fig. 14.11. The q-point collector current of transistor T_1 is about 72 μA, which is the q-point base current of the following transistor. We will hold the emitter potential of transistor T_1 at 16 V with respect to ground, with no direct current through the feedback resistor R_F. Then the dc feedback will stabilize the potential at point O in the output at approximately 16 V, which will permit maximum output signal amplitude. Therefore, the total emitter circuit resistance for T_1 needs to be 16 V/72 μA = 220 kΩ. This high resistance is bypassed by capacitor C and 220 Ω additional resistance is added in series to provide ac or signal feedback as well as dc feedback. We will choose the feedback resistor R_F to provide the value of β_v which will give 15 V peak output signal when 0.5 V peak signal is applied at the input. This will require $K_{vf} = 30$. Since the voltage gain K_v without feedback is large in comparison with 30, we can easily approximate $\beta_r = 1/30 = .033$ and $R_F \simeq 220 \times 30 = 6.6$ kΩ. The dc feedback factor

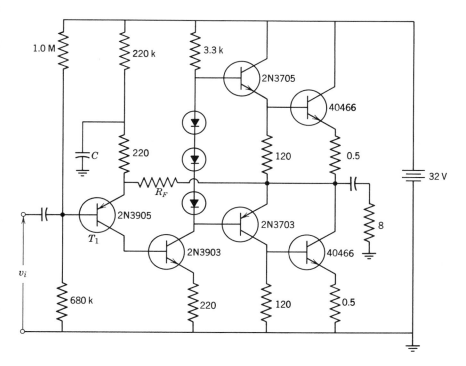

FIGURE 14.11
A quasicomplementary-symmetry amplifier with feedback.

$\beta_v = 220 \text{ k}/(220 \text{ k} + 6.6 \text{ k}) \simeq 1$. Thus, the dc voltage at point O must be approximately the same as the dc voltage at the emitter of T_1. The voltage gain of the amplifier without feedback is approximately 600 and h_{fe} of T_1 is approximately 70 at $I_C = 72 \ \mu A$.

PROBLEM 14.7 Determine the ac input resistance of transistor T_1, Fig. 14.11, using the transistor parameters and the feedback circuit given in Example 14.8.

Answer: $R_{if} = 840 \text{ k}\Omega$.

14.4   STABILITY OF FEEDBACK CIRCUITS

The general equation for gain with negative feedback was developed in Section 14.1 and is repeated here for convenience.

$$G_f = \frac{-KF(s)}{1 + K\beta F(s)} \qquad (14.37)$$

where G_f is the gain with feedback, or closed-loop gain, $-KF(s)$ is the gain without feedback, or forward open-loop gain, and β is the feedback factor or ratio. Observe that the denominator of Eq. 14.37 will be equal to zero if the term $K\beta \, F(s) = -1$. Then the gain G_f will be infinite, or ouput will occur with no input and the amplifier will oscillate or be unstable. This situation can arise at either a high frequency or a low frequency where the relative phase shift of the amplifier is 180° with respect to the mid-frequency phase, or the feedback voltage is actually in phase with the signal voltage. Thus, if the magnitude of total loop gain $K\beta F(s)$ is equal to at least *one* at a frequency for which the relative phase shift is 180°, oscillation will occur. Therefore, as previously stated, the condition for oscillation is

$$K\beta F(s) = -1 = 1 \underline{|\pm 180} \qquad (14.38)$$

Root locus is one technique used to determine the stability of an amplifier, or system. As the name implies, root locus is the locus of all possible roots, in the s domain, of the denominator of Eq. 14.37. Therefore, the root locus is an s-plane plot of all points where $KF(s)\beta(s) = -1 = 1 \underline{|\pm 180°}$. The locus has as many branches as $KF(s)\beta(s)$ has poles. Each branch begins on a pole and terminates on a zero, either finite or at infinity, of $KF(s)\beta(s)$. The root locus progresses along its path toward the zero as the reference gain K increases from zero toward infinity. The amplifier with feedback is unstable for values of $K\beta$ which cause the locus to cross the $j\omega$ axis because sustained oscillation will then occur since poles will occur in the feedback amplifier at real frequencies. The development of the rules for plotting a root locus are beyond the scope of this book. However, the rules for making root locus plots are given in the Appendix. The root-locus plots will be given as visual aids and for the benefit of those who have studied the root-locus technique. The Bode plots, discussed below, are used as the basic analysis and design tool.

The gain-phase plots, or Bode plots, discussed in Chapter 12, can be used to predict the stability of an amplifier with feedback. In fact, with the aid of a Bode plot, the feedback network can be designed so that the amplifier will not only be stable, but will provide near-optimum frequency and transient response. This design will be the subject of this section.

Example 14.9 We will now consider the stability of the three-stage RC-coupled amplifier of Fig. 14.12. Both the low and high half-power

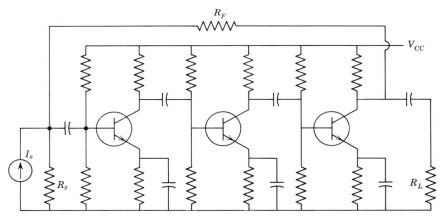

FIGURE 14.12
A three-stage amplifier with feedback.

frequencies are determined for each stage, either by calculation or by measurement, as discussed in Chapter 7 and Chapter 12. Let us assume that all three stages have a low half-power frequency at 100 Hz and have three different upper half-power frequencies at 10^5, 3×10^5 and 10^6 Hz. The gain magnitude and phase plots are sketched in Fig. 14.13 by using the asymtote technique discussed in Chapter 12. The mid-frequency current gain of the amplifier is assumed to be 10^5 and the total load resistance in the output of the amplifier is assumed to be 1 kΩ. The current-gain axis is labeled in dB as well as numerical gain. The relative phase is plotted on the same frequency scale as the gain magnitude. Observe that the relative phase shift is 180° at two frequencies, one low and the other high. These

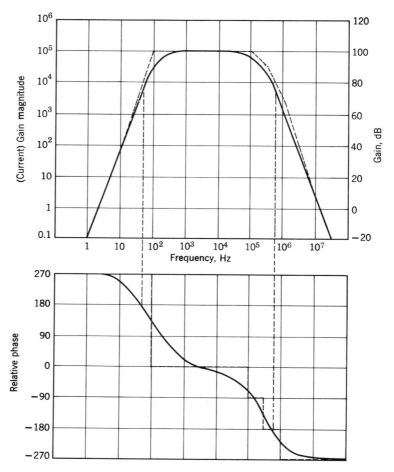

FIGURE 14.13
A Bode plot for the amplifier of Fig. 14.12.

frequencies are approximately 40 Hz and 6×10^5 Hz. At the lower frequency, the forward current gain $|G_i|$ of the amplifier is 7×10^3 or 77 dB and at the higher frequency $|G_i| = 5 \times 10^3$ or 74 dB at 180° relative phase. But the total loop gain $G\beta$ must be less than one at these 180° relative phase frequencies or the amplifier will oscillate when feedback is applied. Therefore, the magnitude of the feedback factor β_i which will cause marginal oscillation, considering each frequency individually, is $\beta_i = 1/7 \times 10^3 = 1.4 \times 10^{-4}$ or -77 dB at 40 Hz and $\beta_i = 2 \times 10^{-4}$ or -74 dB at 6×10^5 Hz. Note that the smaller amount of feedback is usable at the lower frequency because the half-power frequencies of all the stages are the same. Therefore, if this smaller feedback is used, the amplifier will be stable at the higher frequency, but will marginally oscillate at the lower frequency. Of course, oscillation is intolerable at any frequency, so the feedback factor must be reduced not only to provide stability but also to provide satisfactory frequency response and transient response. The root-locus sketch of Fig. 14.14 provides some insight into the reduction of β,

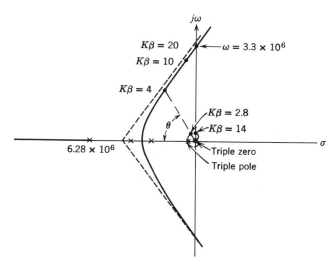

FIGURE 14.14
A root-locus sketch for the amplifier of Fig. 14.12 (not to scale).

or rather $K\beta$, required to produce satisfactory response. Those familiar with root locus will recognize that the damping ratio $\zeta = \cos \theta$ where θ is the angle between a line drawn from the origin to a given point on the locus and the negative real axis. But $Q = 1/2\zeta$, as shown in Eq. 9.31 (Chapter 9) and Q should not exceed about 1.0 for good transient response. Therefore, θ should not exceed about 60°, and typical values of loop gain $K\beta$ found by a spirule, or algebraic calculation, at the intersection of the 60° radial and the root locus are about 20 percent of the value found at the intersection with the $j\omega$ axis. This gain reduction depends upon the amount of separation between the amplifier poles, but the factor of 5 is a commonly accepted value. Thus the total loop gain magnitude should not exceed about 0.2 or -14 dB at the frequencies which give 180° phase shift. Then, there are no peaks in the frequency response and no more than 5 percent overshoot in the transient response. This reduction of gain (a factor of 5, or 14 dB) is known as *gain margin*. Smaller values of gain margin may be used if peaks in the response or ringing in the transient response can be tolerated. A sketch of the time and frequency response of a feedback amplifier with 3 dB gain margin in contrast with the same amplifier with 14 dB gain margin is given in Fig. 14.15. A step current input is used to excite the transient and only the low-frequency transient is shown. Gain margin values between 6 dB and 14 dB are frequently used.

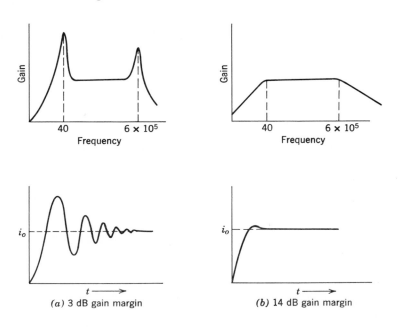

(a) 3 dB gain margin   (b) 14 dB gain margin

FIGURE 14.15
Frequency and time response for two values of gain margin.

We will complete the feedback design for the amplifier of Fig. 14.12 using a 14 dB or factor of 5 gain margin. The feedback factor required to reduce the loop gain magnitude to 0.2 or -14 dB at 40 Hz is $\beta_i = 0.2/7 \times 10^3 = 2.8 \times 10^{-5}$ or -77 dB $- 14$ dB $= -91$ dB. The total mid-frequency loop gain with this magnitude of feedback is $K_i \beta_i = 10^5 \times 2.8 \times 10^{-5} = 2.8$, or 100 dB $- 91$ dB $= 9$ dB.

In Example 14.9 the value of permissible loop gain is disappointingly small because of the modest improvement in the amplifier characteristics. A smaller gain margin would permit somewhat higher loop gain at the expense of peaks in the frequency response or ringing in the transient or time response. A capacitor can be used in the feedback circuit to improve the characteristics of this amplifier. This technique is known as phase-lead compensation and will be discussed after we have considered a two-stage amplifier.

Example 14.10   The stability of the two-stage RC-coupled amplifier shown in Fig. 14.8 will now be considered. We will assume that there are two low half-power frequencies at 10 and 100 Hz and two high half-power frequencies at 10^5 and 10^6 Hz. The mid-frequency voltage gain was previously calculated to be 833. A Bode plot for this amplifier is given in Fig. 14.16 and a root-locus sketch is given in Fig. 14.17. Observe that the

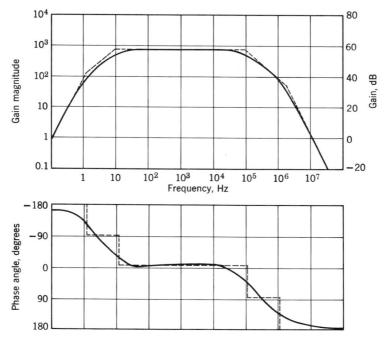

FIGURE 14.16
A Bode plot for the amplifier of Fig. 14.8.

phase plot does not cross the $\pm 180°$ values and the root-locus branches do not cross the $j\omega$ axis. Therefore, the amplifier will be stable for any finite value of loop gain. Thus, the amplifier is said to be *unconditionally stable*. However, the transient response may be unsatisfactory or severe peaks may occur in the frequency response if the loop gain is not limited.

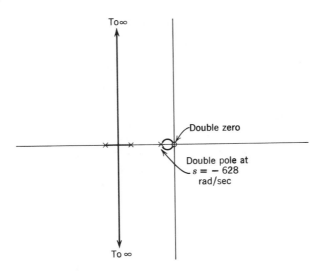

FIGURE 14.17
A sketch of the root locus of the two-stage amplifier of Example 14.10 (not to scale).

The *gain margin* specification given for the three-stage amplifier is not appropriate because there is no frequency at which the phase is 180°, except zero or infinity, where the gain is zero. Therefore, another margin known as *phase margin* is used to obtain the desired transient characteristics. For example, a 45° phase margin means that the amplitude of the total loop gain is one when the phase shift is 45° less than 180°, or ±135°. For the Bode plot of Fig. 14.16, the forward voltage gain of the amplifier is approximately 60 or 36 dB at both the low- and high-frequency 135° phase points. This phase margin compares roughly with the 14 dB gain margin as can be seen in Fig. 14.13. Thus, the value of feedback factor β_v required to reduce the loop gain to one at this phase margin is $\beta_v = 1/60 = .0167$ or -36 dB. The mid-frequency value of loop gain is therefore $K_v \beta_v = 833 \times .0167 = 13.9$ or 59 dB $- 36$ dB $= 23$ dB for this 45° phase margin. This is a satisfactory value of loop gain for many applications. Of course, higher loop gain can be realized with a smaller phase margin.

PROBLEM 14.8 The phase margin of the amplifier in Example 14.10 is reduced to 30°. Determine the mid-frequency loop gain $K_v \beta_v$.

Answer: 27.8 or 29 dB.

14.5 PHASE-LEAD COMPENSATION

The relatively large amount of feedback which can be applied to the two-stage amplifier in comparison with the three-stage amplifier may lead us to believe that no more than two stages should be included in the feedback loop. However, the technique known as phase-lead compensation, mentioned previously, can make the phase characteristics of the three-stage amplifier comparable with the uncompensated two-stage amplifier, and thus permit a similar closed-loop gain. You may have observed that the relative phase shift in the amplifier causes the instability and that each term of the form $K/(s + a)$ contributes relative phase shift up to $90°$. The two-stage amplifier was unconditionally stable because only two such terms appear at high frequencies because of the shunt capacitance and two appear at low frequencies because of the coupling and bypass capacitance. The phase shift at low frequencies due to the coupling capacitors can be eliminated by using direct coupling. Therefore, the low-frequency phase characteristics of the three-stage amplifier will be similar to the two-stage amplifier if direct coupling is used to eliminate the coupling and emitter bypass capacitors in one stage of the three-stage amplifier. Phase-lead compensation will be used to effectively eliminate one of the high-frequency terms in the amplifier gain expression.

The phase-lead network consists simply of a small capacitance C_F in parallel with the feedback resistor R_F as shown in Fig. 14.18. Since the impedance (Z_F) of R_F and C_F in parallel is usually very much larger than either R_L or R_i, particularly in a three-stage amplifier, the output current is very nearly $I_l = V_o/R_L$ and the feedback current I_f is very nearly V_o/Z_F. Therefore, the current feedback factor is

$$\beta_i' \simeq \frac{I_f}{I_l} = \frac{R_L}{Z_F} \qquad (14.39)$$

But, in the s domain

$$Z_F = \frac{R_F(1/sC_F)}{R_F + 1/sC_F} = \frac{R_F}{1 + sR_F C_F} \qquad (14.40)$$

Since the time constant $R_F C_F = 1/\omega_f$, where ω_f is the break-frequency or crossover frequency of the network, $R_F = 1/\omega_f C_F$, then

$$Z_f = \frac{R_F}{1 + s/\omega_f} = \frac{1}{C_F} \frac{1}{(s + \omega_f)} \qquad (14.41)$$

and, using Eq. 14.39,

$$\beta_i' = R_L C_F(s + \omega_f) = \frac{R_L}{R_F} R_F C_F(s + \omega_f) = \beta_i \frac{s + \omega_f}{\omega_f} \qquad (14.42)$$

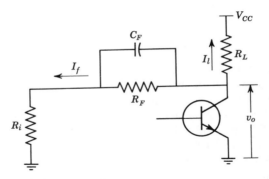

FIGURE 14.18
A phase-compensated feedback network.

where β_i is the mid-frequency current feedback factor. We will now write
the complete equation for the high-frequency gain of the three-stage
RC-coupled amplifier with phase-compensated feedback.

$$G_{if} = \frac{K_1 K_2 K_3 \omega_{21} \omega_{22} \omega_{23}/(s + \omega_{21})(s + \omega_{22})(s + \omega_{23})}{1 + K_1 K_2 K_3 \omega_{21} \omega_{22} \omega_{23} \beta_i (s + \omega_f)/\omega_f (s + \omega_{21})(s + \omega_{22})(s + \omega_{23})}$$

$$(14.43)$$

Observe that the term $(s + \omega_f)$ in the numerator of the term $G_i \beta_i'$ will
cancel one of terms of the form $(s + \omega_2)$ in the denominator if ω_f is equal
to one of the values of ω_2. Let us make $\omega_f = \omega_{22}$, then the $G_i \beta_i'$ term is

$$G_i \beta_i' = \frac{K_1 K_2 K_3 \beta_i \omega_{21} \omega_{23}}{(s + \omega_{21})(s + \omega_{23})} \qquad (14.44)$$

This term has the gain magnitude characteristics of a three-stage amplifier
but the relative high-frequency phase characteristics (for the total loop gain)
of a two-stage amplifier. However, the forward open-loop gain magnitude
can be plotted as though the $(s + \omega_f)$ term is in the amplifier gain G_i instead
of the feedback path β_i'. Then the gain magnitude can also be plotted
as though the three-stage amplifier has only two high-frequency break points
or poles. This plot will not give the correct high-frequency forward gain
but will give the correct feedback design.

Example 14.11   We will now apply phase-lead compensation to the three-
stage amplifier with feedback (Example 14.9) shown in Fig. 14.12, letting
$f_f = f_{22} = 3 \times 10^5$ Hz. Also, direct coupling will be used to couple two of
the stages. The modified forward gain magnitude is sketched in Fig. 14.19.
Also, the relative phase is sketched as though the phase compensation is
in the amplifier. We will allow 45° phase margin which occurs at

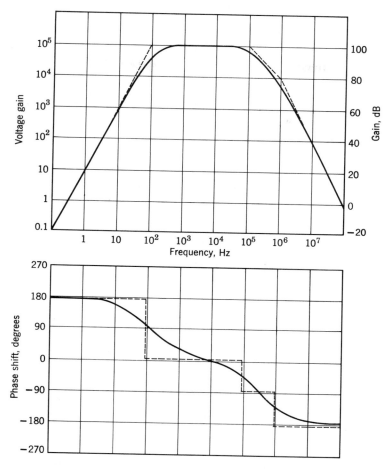

FIGURE 14.19
A Bode plot for the three-stage compensated-feedback amplifier.

approximately $f = 30$ Hz and $f = 10^6$ Hz. Note that the low-frequency $45°$ phase margin point has higher forward gain than the comparable high-frequency point. As previously mentioned, this situation results from the two break points or poles occurring at the same frequency (100 Hz). Therefore, the gain-phase characteristics would be improved and a higher closed-loop gain permitted if one of the coupling circuits had a considerably different value of f_1, for example, 10 Hz. Of course, additional direct coupling could eliminate the pole completely. Therefore, we will be primarily concerned with the high-frequency gain at the $45°$ phase margin. This current gain magnitude is 6×10^3 or 77 dB. Therefore, the value of β_i which will give $45°$ phase margin is $\beta_i = 1/6 \times 10^3 = 1.67 \times 10^{-4}$ or -77 dB. The mid-frequency closed-loop gain is, therefore, $K_i \beta_i = (10^5)(1.67 \times 10^{-4}) = 16.7$ or 100 dB $- 77$ dB $= 23$ dB, and the amplifier distortion is reduced by the factor $(1 + K\beta) = 17.7$. This is a very satisfactory value of loop gain for most applications. The value of R_F (Fig. 14.12) required to give this value of feedback is (with $R_L = 1$ kΩ) $R_F = R_L/\beta_i = 6 \times 10^5$ Ω. The value of C_F required to cancel the pole or break-frequency at 3×10^5 Hz is $C_F = 1/R_F \omega_f = 0.8$ pF. This value of capacitance is near the minimum value obtainable from a parts store. The value of R_F could be decreased and the value of C_F increased proportionally if the feedback is taken from a tap in the load circuit as shown in Fig. 14.20. In this

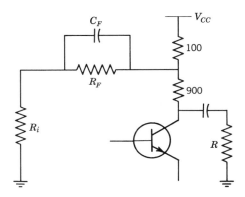

FIGURE 14.20
Illustration of a tapped output to reduce R_F and increase C_F.

circuit R_F is decreased by a factor of 10 and C_F is increased by a factor of 10 compared with the values calculated when the feedback is obtained directly across the output.

The value of $K\beta$ determined at a specified phase margin may be larger than the designer desires to use. Smaller values of $K\beta$ can be used, of course, with improved stability (larger gain or phase margins).

PROBLEM 14.9   A given three-stage transistor amplifier has stage upper cutoff frequencies of 10^5, 5×10^5, and 3×10^6 Hz. The amplifier input resistance $R_i = 10$ kΩ, the load resistance $R_L = 1$ kΩ, and the driving-source resistance $R_s = 5$ kΩ. The current gain $I_l/I_s = 10^4$. Direct coupling is used so stability at low frequencies is not a problem, assuming adequate power supply decoupling is employed. A feedback circuit similar to the one given in Fig. 14.12 is used except a capacitor C_F is placed in parallel with R_F and C_F will be chosen so the zero in the feedback network will cancel the amplifier pole at 5×10^5 Hz. Make a Bode plot similar to Fig. 14.19 and determine the values of R_F, C_F, and $K\beta$ for a 45° phase margin.

Answer: $R_F = 200$ kΩ, $C_F = 1.6$ pF, $K\beta = 50$.

The capacitor C_F in the compensating circuit actually creates a pole as well as a zero in the feedback factor β_i'. The pole did not appear because we neglected R_i. When $R_F \gg R_i$, the pole can usually be neglected because the frequency of the pole is high compared with the frequency of the zero. Then, the pole has little influence on the Bode plot at frequencies where $|K\beta| \gg 1$.

PROBLEM 14.10   Derive an expression for β_i that includes R_i instead of neglecting it. What is the ratio of the frequency of the pole to the frequency of the zero?

The gain phase plots of an amplifier which has been constructed can be readily obtained by the use of a good signal generator and a dual trace oscilloscope, so the output and input voltages can be compared as to magnitude and phase. The gain and phase plot for an amplifier design can be easily obtained by using a computer program known as ECAP (Electronic Circuit Analysis Program) if a computer with this program is available.

14.6   LOG GAIN PLOTS

The Bode plots discussed in the preceding section permit the design of stable feedback circuits which provide satisfactory frequency and transient response, but they do not show the frequency characteristics or bandwidth of the amplifier after the feedback has been applied. Therefore, we will develop in this section a technique which will show the closed-loop, and open-loop gain characteristics and, as an added bonus, make the feedback design process easier. This proposed technique is just a slight extension and simplification of the Bode plot method previously discussed.

Let us consider the amplifier of Problem 14.9 without the compensating capacitor C_F. The Bode plot for this amplifier is given in Fig. 14.21. Observe that the amplifier gain without feedback is about 2000 at a phase margin of 45°, as shown by the dashed lines. Then the feedback factor which will provide this 45° phase margin is $1/2000 = 5 \times 10^{-4}$, as previously discussed. Also, as previously discussed, the gain of an amplifier *with feedback* is approximately $1/\beta$ so long as $K\beta$ is large in comparison with 1. Therefore, within the limits of this approximation, the gain of the amplifier *with*

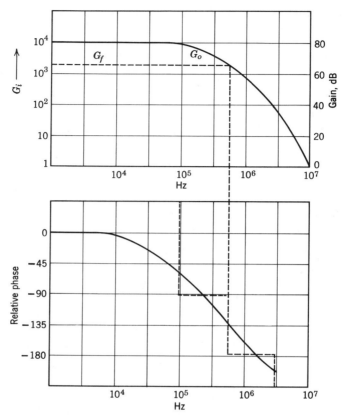

FIGURE 14.21
A Bode plot for the uncompensated amplifier of Problem 14.7.

feedback is 2,000 which is the *same* as the gain without feedback at the frequency which produces $-135°$ relative phase shift. In other words, the straight-line approximation of the closed-loop gain curve (with feedback) intersects the open-loop forward gain curve at the frequency at which $G\beta = 1$. This relationship should be expected if one reviews the rules for making straight-line approximations. At frequencies above the point of intersection $G\beta$ is less than one, so the closed-loop and open-loop gains are equal, within the limits of the straight-line approximation. The *phase margin* is the difference between the relative phase of the amplifier at the frequency of the intersection of the straight-line approximations of the open-loop and closed-loop gains and 180°. The upper cutoff frequency, or bandwidth, of the amplifier with feedback is approximately at the first break-frequency in the straight-line approximation of the *closed-loop* gain curve. This break-frequency is at the intersection of the closed- and open-loop gain curves, as shown in Fig. 14.21 at $f = 5 \times 10^5$ Hz, when the feedback does not include compensation.

A close relationship exists between the relative phase of the open-loop amplifier and the break-frequencies of the straight-line gain plot of the amplifier. For example, the $-135°$ relative phase occurs at approximately the second break-frequency of the open-loop amplifier, as shown in Fig. 14.21. Therefore, the two gain curves must intersect at a frequency no greater than the second break-frequency if no compensation is used and the desired phase margin is at least 45°. Thus the phase plot is not required for the feedback design, provided the break-frequencies are reasonably well separated.

The effect which compensation in the feedback circuit has on the bandwidth of the closed-loop amplifier can now be shown. Since the gain with feedback $G_f \simeq 1/\beta'$, and $\beta' = \beta(s + \omega_f)/\omega_f$ as given in Eq. 14.42, G_f has a break-frequency or pole at f_f, as shown in Fig. 14.22. The value of f_f was chosen to be 5×10^5 Hz, which is the same as the value used in Problem 14.9. Note that the bandwidth of the amplifier with feedback is the same as f_f because of the pole at f_f. The zero in the feedback circuit corrects the phase by 90°, approximately, so the intersection of the G and G_f curves can occur at the third break-frequency of G instead of the second break-frequency. Observe that the difference in the slopes of the two curves at their intersection is the same for both the compensated and uncompensated feedback systems, Figs. 14.21 and 14.22. In other words,

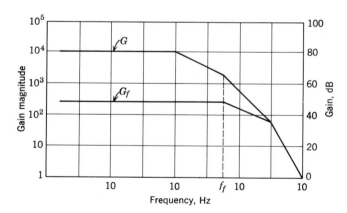

FIGURE 14.22
Straight-line gain approximations for G and G_f for the amplifier of Fig. 14.21 with compensated feedback.

the difference in the slopes of G and G_f (straight-line approximations) should not exceed 20 dB per decade, or 6 dB per octave, if the phase margin is to be at least 45°. Note that $G_f = 200$ and β is about ten times as high as the uncompensated case with similar phase margin. Also observe that the bandwidth of the feedback amplifier could be increased if the frequency f_f were increased to the neighborhood of 10^6 Hz, or perhaps 2×10^6 Hz. However, the phase margin is maximum when f_f is the same as the center break-frequency.

The actual closed-loop gain G_f is within a few dB of the straight-line approximation if the phase margin is at least 45°. However, small gain or phase margins cause severe peaking of the actual frequency response where $G\beta'$ approaches -1, as discussed in Section 13.4. This peaking is illustrated in Fig. 14.23 for several values of phase margin.

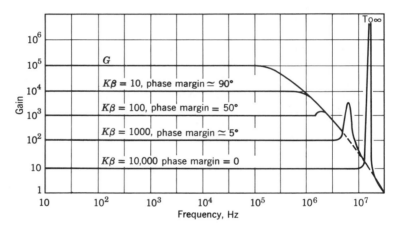

FIGURE 14.23
Frequency response for several values of phase margin.

The preceding feedback examples have shown that the stability problems decrease and more feedback can be applied to the amplifier when the half-power frequencies of the amplifier are widely separated, in contrast with their occurrence in a narrow frequency range. Therefore, stabilization can be obtained by adding shunt capacitance in one of the stages, so its upper cutoff frequency is low in comparison with the other stages. The benefit of this technique can be seen from the log gain plot of Fig. 14.24. As illustrated, the open-loop gain of the amplifier may decrease greatly before the next higher stage-cutoff frequency is reached. As noted, the requirement for stability is that the closed-loop gain G_f intersect the open-loop gain curve G at a higher gain, or lower frequency, than the second break-

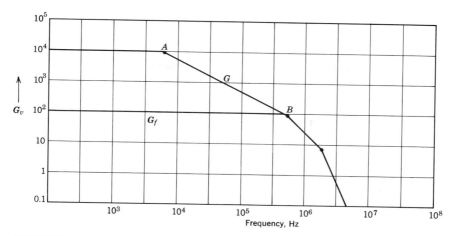

FIGURE 14.24
Illustration of stabilization by lowering the upper cutoff frequency of one stage.

frequency when no compensation is employed in the feedback network. Then, the closed-loop gain G_f can be small in comparison with the open-loop gain G over the mid-frequency range, or in other words, $K\beta$ can be large. In the amplifier of Fig. 14.24, $K\beta \simeq 100$. One convenient design technique is to determine the desired value of feedback factor β from some criterion such as distortion reduction or the fixing of the closed-loop gain. The closed-loop gain line is then drawn at the value $1/\beta$. The second break point B is then placed slightly below the closed-loop gain line G_f to allow adequate phase margin. This second break-frequency is determined from either theoretical calculations or lab measurements, as previously discussed. It was given as 5×10^5 Hz for the amplifier of Fig. 14.24. The desired open-loop gain curve G is then constructed by drawing a straight line from this desired second break point B, with a slope -20 dB per decade, until it intersects the horizontal part of the open-loop gain curve. This intersection is shown as point A in Fig. 14.24, and the horizontal part of the open-loop gain curve represents the mid-frequency gain of the open-loop amplifier. The upper cutoff frequency of that stage which had the lowest frequency break point must then be lowered by adding capacitance to coincide with the frequency indicated at point A (5 kHz in Fig. 14.24).

The upper cutoff frequency of an amplifier can be easily lowered by either increasing the resistance or shunt capacitance in the amplifier. One easy method is to add a capacitor in parallel with the load resistor as shown in Fig. 14.25a. You may recall that the upper cutoff frequency of a stage is determined by the total shunt resistance and the effective shunt

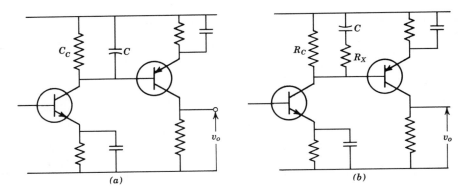

FIGURE 14.25
Typical methods of reducing the upper cutoff frequency of an amplifier.

capacitance. The total shunt resistance R_{sh} in parallel with the capacitor C is the load resistor R_C in parallel with both the output admittance (approximately h_{oe}) of the preceding transistor and the input resistance (approximately h_{ie}) of the following transistor. The capacitor C is usually much larger than the other capacitances in the amplifier. Therefore, since $\omega_2 = 1/R_{sh} C$,

$$C = \frac{1}{R_{sh} \omega_2} \qquad (14.45)$$

where ω_2 is the upper cutoff frequency of the amplifier, determined as point A by the compensation technique illustrated in Fig. 14.24. For example, the lowest break-frequency, point A, in Fig. 14.24 is 5×10^3 Hz or 3.14×10^4 rad/sec. Thus, if $R_C = 10$ kΩ, $h_{oe1} = 20$ μmhos, and $h_{ie2} = 4$ kΩ, $R_{sh} \simeq 2.7$ kΩ, and $C = 1/2.7 \times 3.14 \times 10^7 = .012$ μF.

If a small resistance R_X is placed in series with the added capacitance C, as shown in Fig. 14.25b, the relative phase shift can be decreased at higher frequencies in the neighborhood of the second break frequency at point B, with a consequent increase of phase margin. The greatest phase correction occurs at the radian frequency $\omega_x = 1/R_X C$. Therefore, ω_x should be approximately equal to $\omega_B = 2\pi f_B$, where f_B is the second break-frequency at point B, Fig. 14.24. If this relationship is used in the example of Fig. 14.24 where $\omega_B = 100\, \omega_A$, then $R_X = R_{sh}/100 = 27$ Ω. The resistance R_X creates a pole in β' which produces a zero in the total open-loop gain.

PROBLEM 14.11 A given amplifier has an open-loop voltage gain $K_v = 5{,}000$ in the midfrequency range. Its three stages have upper cutoff frequencies of 10^5, 3×10^5 and 10^6 Hz.

a. What maximum value of feedback factor β_v can be used, assuming a satisfactory phase margin is maintained and no compensation is used? What is the approximate upper cutoff frequency of the amplifier with this feedback?

Answer: $\beta_v \simeq 6 \times 10^{-4}, f_2 \simeq 3 \times 10^5$ Hz.

b. What maximum value of feedback factor β_v can be used (with satisfactory phase margin) if phase-lead compensation is used in the feedback network? What is a good, or optimum, value of the $R_F C_F$ time constant? What voltage gain and bandwidth will the amplifier have with this feedback?

Answer: $\beta_v \simeq 1.8 \times 10^{-3}$, $K_{vf} \simeq 500$, $\tau_f \simeq 1.8 \times 10^{-7}$, $f_2 \simeq 9 \times 10^5$ Hz.

c. It is desired that the voltage gain of the amplifier with feed-back be 100. Assume that you will decrease the bandwidth of one stage sufficiently to obtain good stability with an uncompensated feedback network. What should be the upper cutoff frequency of the modified stage? What value of shunt capacitance is needed if $h_{oe1} = 40$ μmho, $R_C = 5.0$ kΩ, and $h_{ie2} = 2.0$ kΩ? (R_b, if any, can be neglected.)

Answer: $f_2 = 6$ kHz, $C \simeq .02$ μF.

d. What will be the upper cutoff frequency of the amplifier of part *c*? What value of resistance R_X (Fig. 14.25*b*) will give optimum phase margin, approximately?

Answer: $f_2 \simeq 3 \times 10^5$ Hz, $R_x \simeq 27$ Ω.

e. What measures must be taken to provide stability and good transient response at low frequencies for all values of feedback?

PROBLEM 14.12 The amplifier of Fig. 14.26 has $V_{CC} = 40$ V and R_L is an 8 Ω loudspeaker. The amplifier should provide its full output power with 1 V peak input signal V_i. Choose the transistor types and all components for the amplifier. Assume that the maximum distortion of the amplifier without feedback is 5 percent, determine the maximum distortion with feedback. Also determine the approximate maximum power output, with sinusoidal input signal, the approximate output impedance and input impedance of the amplifier. Determine the dc voltage gain of the amplifier. What does the dc voltage gain have to do with the thermal stability? The capacitor C_1 is a decoupling and filter capacitor. C_2 is a feedback-compensating capacitor.

PROBLEM 14.13 Add an input stage, if necessary, and design a feedback circuit for the amplifier of Fig. 11.20. The amplifier should produce full output power with $V_i = 1$ V peak. Choose your own V_{CC} and R_L values and design the circuit.

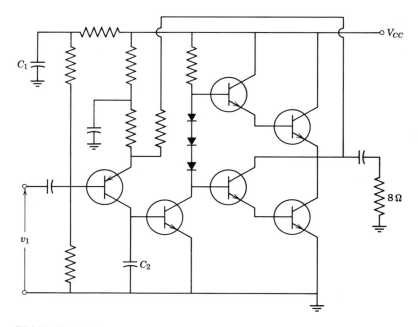

FIGURE 14.26
A quasicomplementary-symmetry af amplifier with negative feedback.

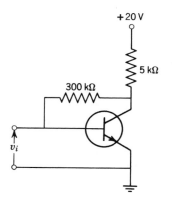

FIGURE 14.27
An amplifier with self bias.

PROBLEM 14.14   An emitter follower can be viewed as an amplifier with feedback, where $\beta = 1$. Using this approach, derive expressions for the reference voltage gain and input impedance of an emitter follower.

PROBLEM 14.15   An amplifier with self bias as shown in Fig. 14.27 can be viewed as an amplifier with feedback. Determine the current gain and the input impedance (for small signals) if $\beta_o = 100$ and $r_b = 200 \ \Omega$, using the feedback approach.

15

Linear Integrated Circuits

One advantage of a transistor over a vacuum tube is the great size reduction possible for a given power output. Thus, it is quite surprising to open a conventional transistor container and discover that the actual portion of germanium or silicon is so much smaller than the total container. Most of the volume is used for mounting the leads and protecting the silicon or germanium chip from an unfavorable environment. To utilize some of this waste space, some semiconductor manufacturers began mounting several diodes in one container. Then, two transistors in one case began to appear on the market. By making two internal connections, two transistors can be mounted in a Darlington configuration as shown in Fig. 15.1. As noted in Chapter 10, the current gain of a

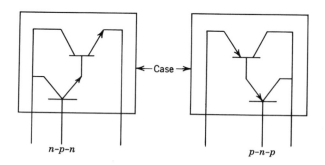

FIGURE 15.1
Two transistors internally connected to form a Darlington configuration.

472

Darlington configuration is approximately the product of the current gains of the individual transistors. Thus, very high current gains from a single, three-lead "transistor" case are possible.[1]

As manufacturing processes were refined, the complexity of circuitry on a single ship increased. Now, complex switching circuits (for use in digital computing circuits) and entire amplifiers are packed as single units. Basically, two different approaches are used in constructing these *integrated circuits*. One procedure produces *thin-film circuits* while another procedure produces *semiconductor monolithic circuits*.

15.1   THIN-FILM CIRCUITS

The thin-film circuits are constructed on a flat plate of insulating material (glass, ceramic, etc.) which is known as the substrate. Then, by silk screening, sputtering, vacuum evaporation, or other processes, thin layers of conducting, insulating, or resistive material are deposited to form the wiring and passive elements (resistors and capacitors) in the circuit. At the present time, it is not feasible to construct useful inductors in the limited space of an integrated circuit. Therefore, if inductors *must be used*, leads are provided so that external inductors may be connected to the integrated circuit. Figure 15.2 shows how the different elements may be constructed. Of course, the film thickness has been exaggerated in this diagram. The active elements (transistors, diodes, etc.) may be added as separate chips with connecting leads welded to the thin film. Some manufacturers are perfecting techniques which will permit them to deposit the active elements

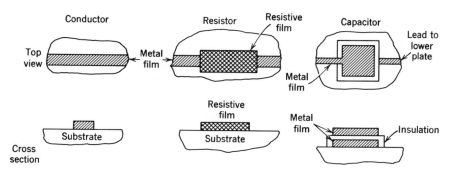

FIGURE 15.2
Typical thin-film construction.

---

[1] For example, Motorola advertises Uniblock Darlingtons with h_{FE} values from 5,000 to 75,000.

on the   substrate much as the passive elements are deposited. In fact, Westinghouse[2] reports constructing thin-film circuits, including transistors, on a flexible substrate! A typical thin-film circuit is shown in Fig. 15.3.

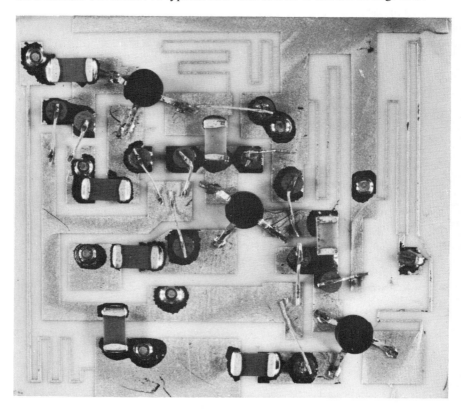

FIGURE 15.3
A typical thin-film circuit. Courtesy of Motorola Semiconductor Products, Phoenix, Ariz.

15.2   MONOLITHIC CIRCUITS

Most of the integrated circuits on the market at the present time are semiconductor monolithic circuits. The substrate of these circuits is a silicon (at least at the present time) chip. The n and p regions are then diffused down into this chip. The procedure used in constructing a transistor from a chip will help clarify this process. As shown in Fig. 15.4a, a

[2] "Flexible Thin-Film Transistors Stretch Performance, Shrink Cost," by Peter Brody and Derrick Page, *Electronics*, Vol. 41, No. 17, August 19, 1968, pp. 100–103.

substrate of p-doped silicon is subjected to an oxidizing atmosphere (a small amount of water vapor speeds up the process considerably) at an elevated temperature. As a result of this step, a thin layer of silicon oxide is formed over the substrate (Fig. 15.4a). By the use of a photo-engraving process, the layer of silicon oxide is removed over the required collector area as shown in Fig. 15.4b. The silicon chip is then placed in an

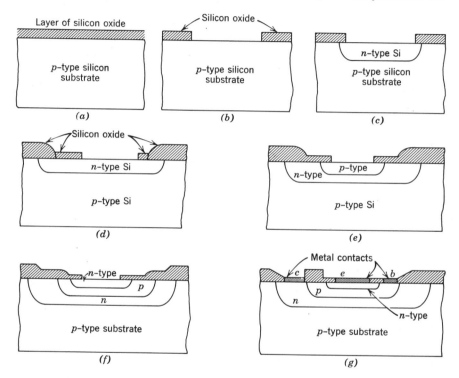

FIGURE 15.4
The process used in constructing an n-p-n transistor in a silicon substrate.

n-type atmosphere at an elevated temperature. The n-type atoms diffuse into the silicon (but do not diffuse into the silicon oxide) creating a layer of n-type silicon as shown in Fig. 15.4c. Again, a layer of silicon oxide is formed over the entire chip. (The old silicon-oxide layer becomes a little thicker.) The photo-engraving process is repeated to remove the silicon oxide over the required base region as shown in Fig. 15.4d. The chip is then placed in a p-type atmosphere at an elevated temperature and a layer of p-type silicon is formed in the n-type collector region as shown in Fig. 15.4e. Again, a layer of silicon oxide is formed over the entire chip. The photo-engraving process is repeated to remove the silicon oxide over the required emitter area. Then the chip is subjected to an n-type atmosphere until a layer of n-silicon is formed in the base region p-silicon (Fig. 15.4f). A final silicon-oxide layer is formed over the chip and the silicon oxide is again removed over the points where ohmic contact is to be made with the silicon (emitter, base, and collector lead junctions). By vacuum evaporation or sputtering, a layer of metal is deposited onto the chip as shown in Fig. 15.4g. This metal is also deposited over the insulating silicon oxide to form connecting leads to the other circuit elements. If the junction between the substrate and the collector is maintained in a reverse-bias condition, the leakage current will be small and the transistor is effectively insulated from the substrate. In Fig. 15.4, the substrate should be connected directly to the negative lead of the power supply to ensure proper bias of the substrate. The configuration shown in Fig. 15.4e could be used (with proper ohmic contacts) for a diode and the configuration shown in Fig. 15.4c could be used for a resistor. In the latter case, the n material is the resistor and an increase of length or decrease of width of this region would increase the resistance.

Of course, the techniques described above could be used to produce field-effect transistors (FET) or MOSFET transistors. Actually, fewer steps are required to construct MOSFET circuits than for bipolar circuits. Both types of circuits are presently being manufactured on a commercial basis.

A typical monolithic integrated circuit is shown in Fig. 15.5. In some instances it is desirable to combine thin-film and monolithic techniques to produce *hybrid integrated circuits.*

The design of integrated circuits is beyond the scope of this book. However, some interesting considerations which are used in designing these integrated circuits will be noted. In the past, the active circuit elements (tubes, transistors, etc.) have been relatively expensive when compared to the passive circuit elements (resistors, capacitors, and inductors). Consequently, conventional designs usually incorporate as few active elements as possible. This design criterion is still used with thin-film circuits which have discrete active elements. In contrast, a whole new concept is used in designing monolithic circuits or thin-film circuits with deposited active elements. All of the transistor collectors in the circuit can be formed at the same time and with the same process. Similarly, all the bases are formed simultaneously and all the emitters are formed in a given time period. Consequently, an integrated circuit with ten or fifteen transistors can be produced at a cost essentially the same as an integrated circuit with four or five transistors. In addition, the cost of the chip with the circuit in place is very small (10 percent or so) in comparison to the cost of testing and packaging the

FIGURE 15.5
A monothithic integrated circuit. Courtesy of Fairchild Semiconductors.

circuit. Thus, when designing monolithic circuits, *the cost of additional active elements is almost negligible.* In fact, since a transistor uses less space on the silicon chip than a resistor, it is desirable to replace resistors with transistors where possible! Also, it is possible to use the junction capacitance of a reverse-biased diode instead of constructing a capacitor in the circuit.

As shown in Fig. 15.4, a *p*-type substrate is used for *n-p-n* type monolithic circuits. Conversely, an *n*-type substrate is required for *p-n-p* type monolithic transistors. Therefore, it seems impossible to construct both *n-p-n* and *p-n-p* transistors on the same monolithic substrate, and thus take advantage of complementary symmetry and other convenient *n-p-n–p-n-p* arrangements. However, a *lateral p-n-p* transistor which uses a *p*-type substrate has been developed, as shown in Fig. 15.6, thus making possible the use of both *n-p-n* and *p-n-p* transistors on the same monolithic chip. The lateral *p-n-p* transistor has very low β (of the order of one), however, so a high gain *n-p-n* driver is always used with it to provide the needed current gain.

Thermally-stable, high-resistance resistors were difficult to produce and space consuming until recently. Therefore, a transistor, acting as a current source, is often used as a high load resistance for another transistor. However, high-resistance, low-volume, thermally stable resistors can now be created by diffusing chromium into silicon.

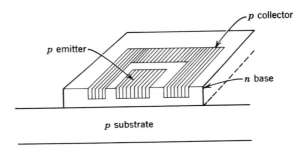

FIGURE 15.6
A lateral *p-n-p* transistor.

15.3 BASIC LINEAR INTEGRATED CIRCUITS

The integrated-circuit manufacturer attempts to fill many of the needs of the user with a single type of integrated circuit, thereby creating large sales volume for each type. Since the differential amplifier, discussed in Chapter 10, can have excellent thermal stability and is useful for either dc or ac amplifiers in either balanced or single-ended operation, the differential amplifier is used in almost all linear integrated circuits.

A versatile amplifier should have high input impedance so it will not load the commonly used driving sources, low output impedance so it can drive a wide variety of load resistance, high gain (which can be controlled over wide limits with feedback), and broad bandwidth which can also be controlled by feedback and compensation. The techniques used for accomplishing or controlling these characteristics will be discussed in the remainder of this chapter.

A simple, basic, linear integrated circuit (*IC*) is shown in Fig. 15.7. This circuit consists of a differential amplifier (T_1 and T_2) with a current source T_5 acting as a high emitter impedance, as discussed in Chapter 10, and a pair of emitter followers (T_3 and T_4) to provide low output

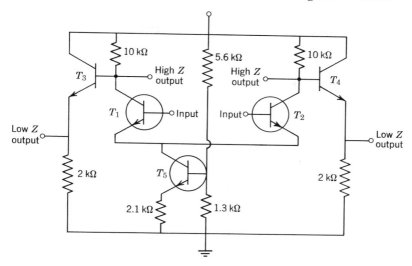

FIGURE 15.7
Schematic diagram of Fairchild μA 730*IC*.

impedance. This circuit may be ideally suited for some applications, but has the following disadvantages for other applications.

1.   Insufficient voltage or current gain.
2.   Only moderate input impedance.
3.   Large (volts) offset voltage between both the input and output terminals and ground, probably requiring the use of blocking capacitors.
4.   Bias currents for transistors T_1 and T_2 must be provided from the external circuit.

The integrated circuit shown in Fig. 15.8 eliminates most of the short-comings listed above. In this circuit the input differential amplifier (T_1 and T_2) is followed by a second differential amplifier (T_3 and T_4) to provide additional gain, and two emitter followers (T_5 and T_6) are connected in cascade to provide low impedance output. The resistor R_5 is placed in the emitter circuit of transistor T_5 to provide a dc voltage drop so the output terminal will be at approximately dc ground potential. The input terminals are at approximately dc ground potential because of the balanced supply voltage $V+$ and $V-$. The transistors T_7, T_8, and T_9 are used to control the bias current in the other transistors, as discussed later. Transistor T_9 also provides a *bootstrapping* function which increases the voltage swing available at the output terminal and actually provides voltage gain between the base of transistor T_4 and the output of the amplifier. We will illustrate this bootstrapping technique by assuming that the base of transistor T_5 is driven

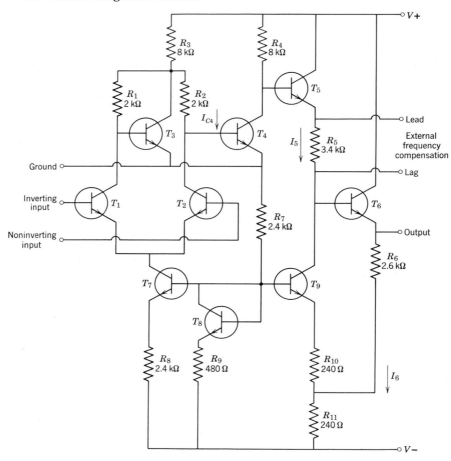

FIGURE 15.8
Schematic diagram of Fairchild μA 702 IC.

positive, therefore causing the emitter of T_5 and the base of T_6 to become more positive. If the bootstrapping circuit were missing, the potential at the base of T_6, and hence the output potential, would not rise as much as the emitter potential of T_5 because of the increased IR drop across R_5 as the current through it would increase. However, the increased emitter current of the output transistor T_6 flows through resistor R_{11} in the emitter circuit of T_9 and thus tends to reverse bias T_9. But the reduced forward bias of T_9 reduces its collector current which flows through R_5. Therefore, the *total* current through R_5 actually *decreases* and the base potential (and hence the emitter potential) of T_6 experiences a *greater* rise than either the base or emitter potential of T_5. Another way to look at the output circuit is to recognize that positive feedback is applied from the output through T_9 (as a common-base amplifier) to the input of T_6. The loop gain must be less than one so oscillation will not occur.

Transistors T_7, T_8, and T_9 provide stabilized bias to all the amplifiers in the IC, as illustrated in Fig. 15.9. This bias circuit is identical to its counterpart in Fig. 15.8 but is rearranged slightly for convenience. Transistor T_8 is connected as a diode so the voltage drop across $R_7 + R_9$ is $V-$ minus V_{EB}. Thus the collector and emitter currents of T_8 are fixed almost completely by the supply voltage $V-$ if the transistor current gains are high so the base currents of T_9 and T_7 can be neglected. Also, since the transistors are essentially identical, their values of V_{EB} are very nearly equal so their emitter potentials are almost identical. Therefore, the emitter, and hence collector, currents of the biasing transistors are controlled almost completely by their emitter circuit resistors.

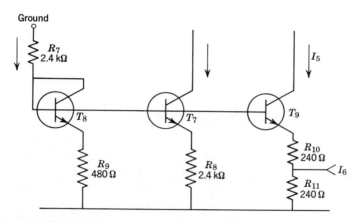

FIGURE 15.9
The stabilized-bias circuit of Fig. 15.8.

Example 15.1 The bias currents for the circuit of Figs. 15.8 and 15.9 will be determined as an example of the bias circuit design and operation. We will assume that $V- = -8$ V and $V_{BE} = 0.5$ V for all transistors. Then for transistor T_8, $I_C \simeq |I_E| = (8 - 0.5)$ V$/(2.4$ k $+ 480)\Omega = 2.6$ mA. The voltage drop across the emitter resistor of T_8 is therefore approximately $(2.6$ mA$)$ $(480\ \Omega) = 1.25$ V. The voltage across the 2.4 kΩ must also be 1.25 V, so the emitter current of transistor T_7 is 1.25 V$/2.4$ k$\Omega = 0.52$ mA. The collector current of T_7 is essentially 0.52 mA, and this current divides equally between T_1 and T_2 whose collector currents must therefore each be about

0.26 mA and the voltage drop across R_1 and R_2 is $(2\ k\Omega)(0.26\ mA) =$ 0.52 V. The voltage drop across R_3 must therefore be $V+$ (8 V) minus $(0.52\ V + V_{BE}) = 6.98\ V$. The current through R_3 is then $6.98\ V/8\ k\Omega = 0.87\ mA$ and the collector current of transistor $T_3 = (0.87 - 0.52)\ mA = 0.35\ mA$. The collector current of transistor T_4 must also be 0.35 mA since T_3 and T_4 form a balanced differential amplifier. The emitter current of transistor T_9 is not so easy to calculate because only current I_5 (approximately) flows through R_{10} but both I_5 and I_6 flow through R_{11}. We *do* know that the sum of the voltage drops in the emitter circuit is 1.25 V. Therefore,

$$I_5 R_{10} + (I_5 + I_6)R_{11} = 1.25 \tag{15.1}$$

we can find the value of I_5 required to provide zero dc output voltage, which is desired. If we neglect the base currents, Fig. 15.8 shows that

$$I_5 R_5 + I_{C4} R_4 = V_+ - 2V_{BE} \tag{15.2}$$

We have already found I_{C4} to be approximately 0.35 mA.

PROBLEM 15.1 Continue this example and determine both I_5 and I_6 assuming $V_{BE} = 0.5$ V and $V_+ = 8$ V.

Answer: $I_5 \simeq 1.23$ mA, $I_6 = 2.7$ mA.

15.4 OPERATIONAL AMPLIFIERS

The term *operational amplifier* was coined by people in the analog computer field and was used to designate an amplifier which has very high open-loop gain, high input impedance, and low output impedance. This type of amplifier is very versatile because feedback can be used to control its characteristics, thus making it useful for a wide variety of applications. Operational amplifiers usually have values of open-loop voltage gain between 10,000 and 100,000. A triangular symbol is used to represent the entire amplifier as shown in Fig. 15.10. The amplifier is normally an integrated circuit with differential input, as previously discussed. The feedback circuit must always be connected to the *inverting* input for negative feedback, but the signal may be applied to either terminal, depending upon the required input impedance of the amplifier. When the signal is applied to the noninverting input, as shown in Fig. 15.10, the amplifier input impedance is very high at both input terminals because the feedback voltage is nearly equal to the input voltage and the two voltages appear as a common-mode input signal which sees a very high input impedance because of the very high common-mode impedance in the emitter circuit. The difference between the signal source and feedback signals is the

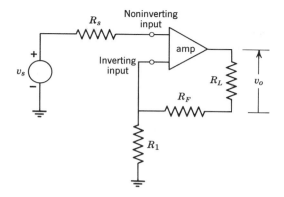

FIGURE 15.10
An operation amplifier used as a linear amplifier
with stable gain and very high input impedance.

differential signal which is effective in driving the amplifier. In other words,
the feedback arrangement of Fig. 15.10 is series input and the open-loop
differential input impedance of the amplifier is multiplied by the factor
$(1 + K\beta)$ as a result of the feedback, as discussed in Chapter 14. Since
the effective amplifier impedance is very high, the feedback factor is

$$\beta = \frac{R_1}{R_1 + R_F} \qquad (15.3)$$

and the reference gain of the amplifier is

$$K_{vf} = \frac{1}{\beta} = \frac{R_1 + R_F}{R_1} \qquad (15.4)$$

assuming the open-loop gain is much higher than the closed-loop gain and
therefore $K\beta \gg 1$, as discussed in Chapter 14. The amplifier may require
compensation if small values of closed-loop gain are used. However,
some integrated circuits have built-in compensation which provides stable
operation for all values of closed-loop gain down to unity. Compensation
techniques will be discussed in section 15.5.

Sometimes it is desirable to have very low impedance and very small
voltage at the amplifier input terminals. These characteristics are obtained
when the input and feedback signals are both applied to the inverting
input terminal of the amplifier as shown in Fig. 15.11. This is the
parallel input connection discussed in Chapter 14. The noninverting input
terminal is maintained at signal ground potential by capacitor C. The
amplifier input current i_a is very small, usually much smaller than one

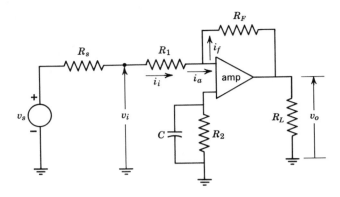

FIGURE 15.11
An operational amplifier with the input signal applied to the inverting input terminal.

microampere, so both i_i and i_f are very large in comparison with i_a. Then $i_i \simeq i_f$. Also, the input voltage of the amplifier is very small. Being the differential input voltage, it is equal to the output voltage divided by the voltage gain of the amplifier. Therefore, the voltage at the inverting input terminal is of the order of microvolts and can be assumed to be essentially zero in comparison with either the output voltage or the driving-source voltage. Then,

$$i_i \simeq \frac{v_s}{R_s + R_1} \tag{15.5}$$

and

$$|i_f| \simeq \frac{v_o}{R_F} \tag{15.6}$$

But $i_i \simeq i_f$, so using Eqs. 15.5 and 15.6,

$$\frac{v_s}{R_s + R_1} = \frac{v_o}{R_F} \tag{15.7}$$

Therefore, the closed-loop voltage gain (v_o/v_s) of the amplifier is

$$K_{vf}' = \frac{v_o}{v_s} = \frac{R_F}{R_s + R_1} \tag{15.8}$$

Usually, the voltage gain from the output to the input terminals labeled v_i is wanted rather than the ratio v_o/v_s. This gain is

$$K_{vf} = \frac{v_o}{v_i} = \frac{R_F}{R_1} \tag{15.9}$$

The resistor R_1 can be made large in comparison with R_s. Then $K_{vf}' \simeq K_{vf}$.

Example 15.2   An example will illustrate the use of the relationships above and demonstrate the small error introduced by the approximations which were used. We will assume that an amplifier is needed which will provide an input impedance of 10 kΩ and a voltage gain $v_o/v_i = 100$. The driving-source resistance is 1 kΩ. An operational amplifier with open-loop voltage gain $= 5 \times 10^4$ and input resistance $= 100$ kΩ will be used. The resistor $R_1 = 10$ kΩ (Fig. 15.11) is used to provide the desired input resistance and $R_F = 100 \ R_1 = 1$ MΩ will provide the desired gain. The voltage gain from voltage source to output is $v_o/v_s = 10^6/(1.1 \times 10^4) = 91$. We will now determine the errors resulting from the assumption that $i_a \simeq 0$ and $v_{ia} \simeq 0$, assuming that the output signal is 5.0 V. The amplifier input voltage is then $v_{ia} = 5/(5 \times 10^4) = 10^{-4}$ V, or 100 microvolts and the signal input current to the amplifier is 10^{-4} V$/10^5 \ \Omega = 10^{-9}$ A or 1 nA. The feedback current $i_f \simeq$ 5 V$/10^6 \ \Omega = 5\mu$A which is 5000 times as large as the amplifier input current i_a. Since the current from the driving source flowing through R_1 can differ from i_f only by the amount i_a, this difference can be only one part in 5000 or 0.02%. Similarly, the driving-source voltage is 5 V$/91 = .055$ V which is 550 times as large as the $v_{ia} = 10^{-4}$ V which we neglected. Therefore, the approximations are extremely good and the gain of the amplifier is essentially as stable as the resistance value of R_1 and R_F.

The bias currents for the differential input stage of the operational amplifier normally flow through the external resistors connected to the input terminals. Therefore, the dc resistance from each of the two input terminals and ground should be approximately the same. Otherwise, the $I_B R$ drop across the two resistors will be different and a dc *offset* voltage will appear between the two input terminals. This offset voltage is multiplied by the closed-loop gain of the amplifier and appears in the output. Of course, a small offset voltage appears at the input due to imperfect balance in the input differential amplifier, but this offset can be minimized by proper balancing of the input circuit resistance. For example, the resistor R_2 in Fig. 15.11 should be equal to the parallel combination of R_F and $R_1 + R_s$, assuming there is no blocking capacitor in the driving source. If there is *no* dc path through the driving source, then $R_2 = R_F$.

The operational amplifier is commonly used as a *mixer* or *summer* as shown in Fig. 15.12. Using the reasoning above $i_1 + i_2 + i_3 + i_4 = i_f$. But $i_1 = v_1/R_1$, $i_2 = v_2/R_2$, $i_3 = v_3/R_3$, $i_4 = v_4/R_4$, and $i_f = v_o/R_F$. Then,

$$\frac{v_o}{R_F} = \frac{v_1}{R_1} + \frac{v_2}{R_2} + \frac{v_3}{R_3} + \frac{v_4}{R_4} \tag{15.10}$$

and

$$v_o = \frac{R_F}{R_1} v_1 + \frac{R_F}{R_2} v_2 + \frac{R_F}{R_3} v_3 + \frac{R_F}{R_4} v_4 \tag{15.11}$$

Of course, any desired number of inputs can be used. If a simple summing or adding operation is desired, then $R_1 = R_2 = R_3 = R_4 = R_F$. However, the voltage at any or all inputs can be amplified as indicated by Eq. 15.11. The resistance of R_5 should be equal to the total dc (external) resistance from the inverting input terminal to ground, of course, for lower offset voltage.

PROBLEM 15.2 Use the circuit of Fig. 15.12 to add the signals from three microphones which have output voltages of 2 mV, 4 mV, and 10 mV, respectively, at normal voice levels of 1 microbar acoustic pressure. It is desired that each microphone produce 0.1 V at the output of the amplifier and the load resistance for any of the microphones should be 5 kΩ or higher. Design the mixer-circuit using an operational amplifier with an open-loop gain of 5×10^4 with $R_L \geq 2$ kΩ. Assume $R_s \ll 5$ kΩ.

Answer: $R_1 = 5$ kΩ, $R_2 = 10$ kΩ, $R_3 = 25$ kΩ, $R_F = 250$ kΩ, R, (noninverting input) = 3.1 kΩ.

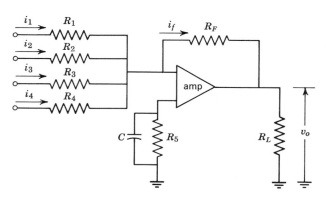

FIGURE 15.12
A summing amplifier.

15.5 AMPLIFIER COMPENSATION

Some compensation techniques which will provide stable operation and acceptable transient response of an amplifier with feedback were discussed in Chapter 14. The ideas developed there are very briefly reviewed here with the aid of Fig. 15.13 where curve *a* is the straight-line approximation of the open-loop frequency response curve of a representative operational amplifier with 30,000 or 89 dB reference gain and break frequencies at 150 kHz, 1.5 MHz, and 15 MHz. You may recall from Chapter 14 that approximately 45° or more phase margin will be maintained

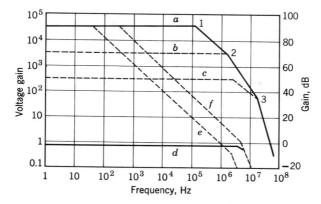

FIGURE 15.13
Frequency response curves.

if the straight-line approximation of the closed-loop response curve intersects the open-loop response curve at a point where the difference in their slopes is only 6 dB per octave or 20 dB per decade. Therefore, no compensation is required in either the amplifier or the feedback network if the closed-loop reference gain is 3,000 or higher, as shown by curve b. This gain corresponds to 20 dB of feedback $(1 + K\beta = 10)$. Curve c shows that the closed-loop gain can be reduced to approximately 300 if phase-lead compensation is used in the feedback network with $R_F C_F \simeq 1/6\pi \times 10^6$. $(1 + K\beta = 100$, or 40 dB of feedback.) Sometimes a lower closed-loop gain such as 10 is desired, or even a gain of 1, perhaps, which is common for a summing amplifier. Then the *amplifier* must be compensated to achieve adequate stability. A simple type of compensation is achieved by adding capacitance in parallel with the output of one of the amplifier stages, as discussed in Chapter 14. If wide bandwidth is not important, no compensation is required in the feedback network if the compensating capacitor produces a sufficiently low break-frequency so the 6 dB/octave slope of the open-loop gain curve of the compensated amplifier intersects the closed-loop gain curve before the second break-frequency is encountered. For example, in Fig. 15.13 the desired closed-loop gain curve d is drawn at unity gain. The compensating capacitor is added in the output of the amplifier stage which had the break-frequency at 150 kHz, which is reduced to about 40 Hz by the compensation. The open-loop gain curve e then decreases at 20 dB per decade from 30,000 gain at 40 Hz to unity gain at 1.2 MHz which is less than the second break-frequency at 1.5 MHz. Thus, the amplifier is stable with an uncompensated feedback network. Some operational amplifiers have *built-in* compensation of this type and are therefore stable for any amount of uncompensated feedback to unity gain. Bandwidth is sacrificed in this type of amplifier, however, because additional bandwidth could be obtained by using phase-lead compensation in the feedback network, as shown by curve *f*, where the point of intersection of *f* with the closed-loop gain curve can occur at a frequency higher than the second break-frequency. You may recall that the upper cutoff frequency of the amplifier occurs approximately at the first break-frequency in the closed-loop response. This break-frequency is located at the intersection with the open-loop response curve when no phase-lead compensation is used in the feedback network, or it is the break frequency produced by the compensating network when phase-lead compensation is used.

Observe from Fig. 15.13 the large amount of bandwidth that is sacrificed when an operation amplifier with built-in compensation is used with small amounts of feedback which produce high closed-loop gains.

PROBLEM 15.3 Determine the approximate bandwidth of the amplifier of Fig. 15.13 if the closed-loop gain is 3000 and the amplifier is (*a*) internally compensated or (*b*) uncompensated.

Answer: (*a*) 400 Hz, (*b*) 1 MHz.

PROBLEM 15.4 How much capacitance is needed to compensate the amplifier of Fig. 15.13 if curve *e* is the desired open-loop response and the total effective resistance between the two collector terminals to which the compensating capacitor is attached is 8 kΩ.

Answer: 0.5μF.

15.6 COMPENSATION FOR OPTIMUM RISE TIME AND SLEW RATE

The *slew rate* of an amplifier is the rate at which the output voltage of the amplifier rises toward the supply voltage when some or all of the stages of the amplifier are either cut off or in saturation. In other words, the amplifier is not operating in its normal linear mode. Slew rate is usually expressed in volts/μsec. This slew rate becomes important when fast-rising rectangular pulses are applied to the input of an amplifier which has high open-loop gain, such as an operational amplifier, but has a large feedback factor which reduces the gain to a small value. Although the input pulse may seemingly be too small to overload the amplifier, the delay time and rise time of the amplifier prevent the feedback pulse from providing adequate cancellation of the leading edge of the input pulse. Thus, a large, sharp, disabling spike is applied as a differential input signal to the amplifier, as shown in Fig. 15.14. The output voltage does not then rise at the rate predicted on the basis of linear amplifier theory because the amplifier is not entirely operative, but rises at the slower slew rate instead.

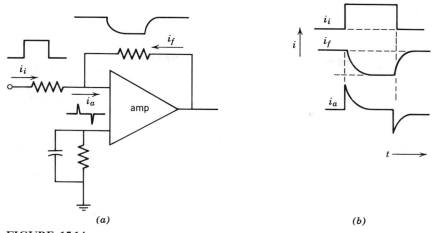

(a)   (b)

FIGURE 15.14
Illustration of the large input spikes which result from fast-rising pulses, large amounts of feedback, and amplifier delay and rise time.

Thus, slew rate becomes an important parameter in amplifier design. Compensating capacitors can greatly decrease or slow the slew rate when they are placed near the output of an amplifier because the lower level stages are then cut off and the output rises slowly due to the large time constants caused by the large compensating capacitance. On the other hand, a compensating capacitance near the input of the amplifier may reduce the magnitude of the input spike and help maintain linear operation of the amplifier as illustrated in Fig. 15.15 where the compensating capacitor C

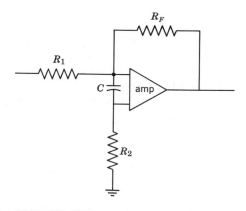

FIGURE 15.15
Input compensation which increases rise time of the circuit.

is placed at the input of the amplifier. This capacitor slows the rate of rise of the input pulse and thus greatly reduces the size of the input spike to the amplifier, thereby permitting linear amplifier operation over a wide range of magnitude of input signals. This single compensating capacitor may not provide optimum rise time for the amplifier, however, because a compensating capacitance usually needs to either absorb or remove a low-frequency break point in the open-loop gain curve in order to obtain wide bandwidth and the input capacitor C may not accomplish this function. However, input compensation may be used in conjunction with another compensating scheme to provide maximum bandwidth and hence minimum rise time as well as immunity from slow slew rates. A compensating scheme which has these characteristics will be illustrated by an example which follows.

Example 15.3   We will assume that a high-speed summing amplifier is needed which will provide less than 100 ns (nanosecond) rise time and less than 5 percent overshoot at unity gain. The desired input resistance is 12 kΩ. A μA715C integrated circuit is chosen as the operational amplifier because of its published frequency characteristics. An equivalent circuit of the μA715C is given in Fig. 15.16. The input stage of this amplifier is a Darlington-connected cascode differential amplifier utilizing transistors Q_1, Q_3, and Q_{16} on one side and Q_2, Q_4, and Q_{17} on the other side of the differential pair. The second stage is a differential amplifier consisting of Q_{18} and Q_{19}. Compensation terminals are brought out at both the input and the output of this stage. A Darlington-connected emitter-follower (Q_{21} and Q_{22}) which prevents loading and provides dc level adjustment in the output of the second stage, is followed by a Darlington-connected common-emitter stage (Q_{10} and Q_{11}) which drives the complementary-symmetry emitter-follower output Q_{14} and Q_{15}. All the other transistors in the circuit are used to provide the proper bias currents for the amplifying circuits, as previously discussed.

The straight-line approximation of the frequency-response curve of the μA715C is given in Fig. 15.17. The open-loop response curve (without compensation) shows that there are three break-frequencies at about 110 kHz, 1 MHz, and 20 MHz, respectively. The desired closed-loop gain (with feedback) is unity. Therefore, the desired open-loop response with compensation intersects the closed-loop response at a frequency somewhat lower than the third break-frequency. This point of intersection was chosen at 10 MHz, which provides a theoretical rise time of $2.2/(2\pi \times 10^7) = 35$ ns. A compensation scheme can hopefully be found which will cause the open-loop voltage gain to rise uniformly at the rate of 6 dB/octave or 20 dB/decade until the low-frequency open-loop gain of 30,000 is reached.

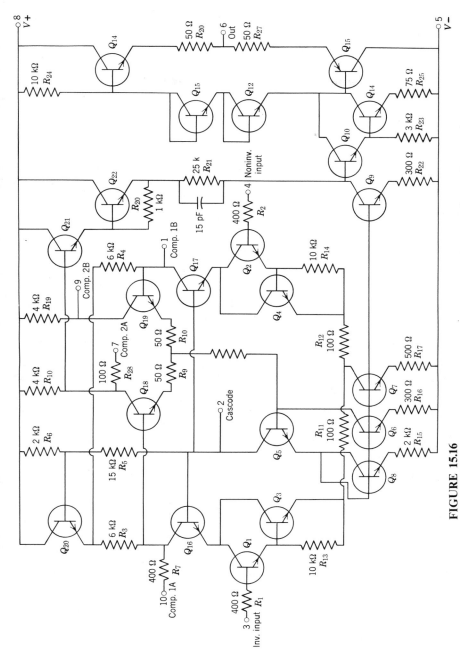

FIGURE 15.16
Equivalent circuit of the μA715C. (Courtesy of Fairchild Semiconductor Company.)

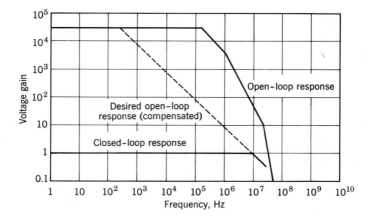

FIGURE 15.17
Frequency response curves for the μA715C.

The single break-frequency to the left of the intersection of the open-loop and closed-loop gain curves then occurs at $10^7/(3 \times 10^4) = 330$ Hz.

Part of the compensation is accomplished at the amplifier input to relieve the slew-rate problem, as shown in Fig. 15.18. The 1.3 kΩ resistor is connected in series with the compensating capacitor C to eliminate the phase shift of this capacitor at higher frequencies, thus permitting the use of additional compensation at higher frequencies. The equivalent circuit of the input configuration in Fig. 15.18 is given in Fig. 15.19a. The voltage

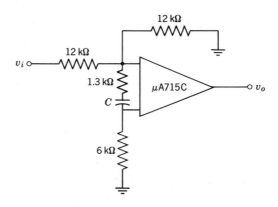

FIGURE 15.18
Input compensation circuit.

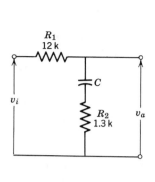

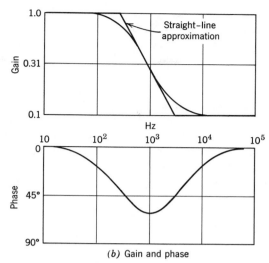

(a) Equivalent circuit   (b) Gain and phase

FIGURE 15.19
The equivalent compensated input circuit of Fig. 15.16 and its relative gain and phase.

v_a is the voltage at the input terminals of the amplifier. The transfer function of this circuit is found as

$$v_a = v_i \frac{R_2 + \dfrac{1}{sC}}{R_1 + R_2 + \dfrac{1}{sC}} \qquad (15.12)$$

After simple algebraic manipulation, Eq. 15.12 can be arranged as

$$\frac{v_a}{v_i} = \frac{R_2}{R_1 + R_2} \frac{s + \dfrac{1}{R_2 C}}{s + \dfrac{1}{(R_1 + R_2)C}} \qquad (15.13)$$

The frequency and phase response of the equivalent open-loop input circuit (with the feedback circuit grounded at the output end as shown in Fig. 15.18) is given in Fig. 15.19b. Notice that the straight-line approximation for the gain curve has a value of one from zero to $\omega = 1/(R_1 + R_2)C$. Then, the gain decreases 6 dB per octave (or 20 dB/decade) until $\omega = 1/R_2 C$. The gain then remains constant at a value of $R_2/(R_1 + R_2)$.

We will use the circuit shown in Fig. 15.18 to reduce the gain to 0.1 of the dc gain. The resistor R_1 has a value of 12 kΩ since the desired input impedance is 12 kΩ. Then, the value of R_2 is found from the relationship $0.1 = R_2/(R_1 + R_2)$. For our case, $R_2 = 1.3$ kΩ. From Fig. 15.17, the open-loop response of the amplifier should begin dropping off at 6 dB/octave when $f = 330$ Hz. The input circuit shown in Fig. 15.18 can be used to begin this decrease in gain if the pole in Eq. 15.13 occurs at 330 Hz. Then, $C = 1/(2\pi \times 330 \times [12\text{ k} + 1.3\text{ k}]) = 0.036$ μF. The zero of Eq. 15.13 occurs at $f = 1/(2\pi \times 3.6 \times 10^{-8} \times 1.3 \times 10^6) = 3.3$ kHz. The curves in Fig. 15.19b are drawn using these break-frequencies.

The next part of the compensation strategy is to compensate the second differential amplifier stage in the μA715C so its lower break-frequency, or pole, will occur at 3.3 kHz and thus continue the 6 dB/octave roll-off initiated by the input compensation. This compensation can be accomplished with minimum capacitance and, hence, maximum slew rate if the compensating capacitance is included in a feedback circuit around the second stage as shown in Fig. 15.20. The 500 Ω resistor is included internally in

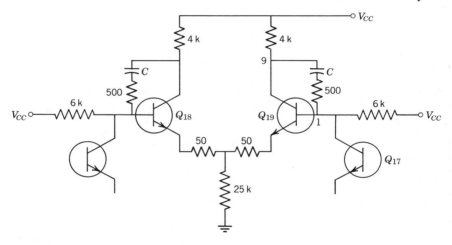

FIGURE 15.20
Circuit showing the compensating capacitor in a feedback loop around the second differential amplifier.

series with compensation terminals 7 and 10. A 500 Ω resistance must be added externally in series with C between 1 and 9 to keep the differential amplifier balanced. Although balance is not necessary for satisfactory operation, it does simplify the theoretical compensation calculations. With the amplifier balanced, we can look at only one side of it, as shown in Fig. 15.21a. The open-loop gain characteristics were deduced by observing that the emitter circuit of this stage (Fig. 15.16) has a 25 kΩ

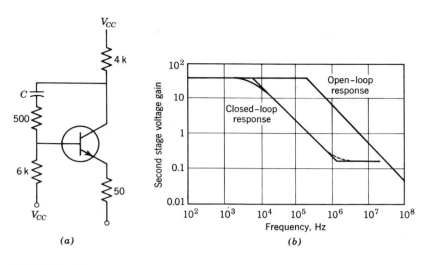

FIGURE 15.21
Equivalent circuit and frequency response of the second differential stage.

resistor in it, across which the dc voltage drop may not exceed about 25
volts if $+15$ and -15 V power supplies are used, as suggested by the
manufacturer. Thus, the q point collector currents of each of the two
transistors are about 0.5 mA each. Then the internal emitter resistance is,
approximately, $r_e = 25/0.5 = 50$ Ω. If the transistor current gain is high,
the input voltage is about the same as the voltage drop across the total
emitter circuit resistance so the approximate voltage gain K_v is $R_C/(r_e + R_E)$,
as discussed in Chapter 7. Therefore, $K_v \simeq 4k/100 = 40$. The break frequency,
or pole, at 110 kHz was assumed to be the upper cutoff frequency of this
stage, since the input stage is cascode and the other common-emitter stage
which drives the output is preceded by an emitter follower which increases
its bandwidth. With these two assumptions, the open-loop frequency
response curve for this stage, given in Fig. 15.21b, was drawn. However, the
desired compensated response has a break-frequency, or pole, at 3.3 kHz to
continue the 6 dB/octave roll-off initiated by the input compensation.

Since the compensating capacitor C (Fig. 15.21a) is in a feedback loop,
the feedback equation must be used to calculate the required capacitance
to produce a break-frequency, or pole, at 3.3 kHz. This equation, developed
in Chapter 14 is

$$G_f = \frac{K}{1 + K\beta(\omega)} \tag{15.14}$$

where G_f is the gain with feedback, K is the forward open-loop gain and $\beta(\omega)$ is the feedback factor which is a function of frequency because of the capacitor C. The pole, or break-frequency, in the closed-loop gain G_f occurs when $K\beta(\omega)$ has a magnitude of one and a phase angle of $90°$. At the break-frequency, X_c is much larger than the resistances in the circuit, as seen later, so the phase angle is near $90°$. Then, at the break-frequency f_b

$$|K\beta(\omega_b)| = 1 \tag{15.15}$$

But K has already been estimated at 40, and inspection of Fig. 15.21 shows that

$$\beta(\omega) = \frac{6k\|r_{in}}{(6k\|r_{in}) + 500 + (1/j\omega C)} \tag{15.16}$$

Since $r_{in} = r_b + (r_e + R_E)h_{fe} \simeq 100 h_{fe}$, and since a typical value for h_{fe} is 100, r_{in} is of the order of 10 kΩ, and 6 kΩ in parallel with 10 kΩ is about 4 kΩ. Then

$$\beta(\omega) = \frac{4 \times 10^3}{4.5 \times 10^3 + (1/j\omega C)} \tag{15.17}$$

Rationalizing Eq. 15.17,

$$\beta(\omega) = \frac{4 \times 10^3 (j\omega C)}{1 + (4.5 \times 10^3)(j\omega C)} \tag{15.18}$$

Substituting this expression for $\beta(\omega)$ and $K = 40$ into Eq. 15.15

$$\left| \frac{40 \times 4 \times 10^3 (j\omega_b C)}{1 + (4.5 \times 10^3)(j\omega_b C)} \right| = 1 \tag{15.19}$$

Since the magnitude of the denominator of the fraction on the left of Eq. 15.19 must be equal to the magnitude of the numerator, but the j term in the numerator is nearly 40 times as large as the j term in the denominator, this denominator j term must be negligible in comparison with *one*. Then, the magnitude of the numerator of Eq. 15.19 must be one.

$$|j1.6 \times 10^5 \omega_b C| = 1 \tag{15.20}$$

The j in the left side of Eq. 15.20 can be dropped because of the magnitude sign. Its presence indicates the needed $90°$ relationship previously discussed. Then,

$$C = \frac{1}{(1.6 \times 10^5)\omega_b} = \frac{1}{(1.6 \times 10^5)(2.1 \times 10^4)} = 280 \text{ pF} \tag{15.21}$$

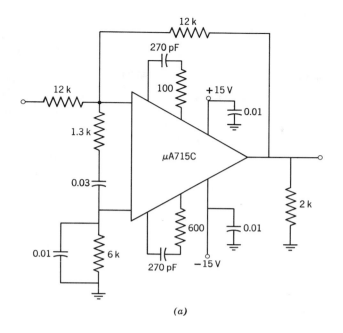

(a)

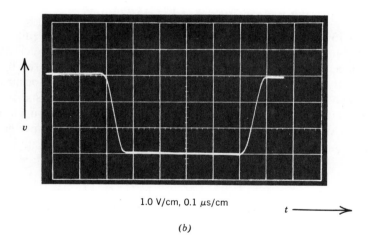

1.0 V/cm, 0.1 μs/cm

(b)

FIGURE 15.22
(a) **Circuit diagram, and** (b) **oscillogram of the time response of the**
unity-gain amplifier.

The nearest stock size is 270 pF. The 500 Ω resistor in series with C causes a zero in the response of the second stage at $\omega = 1/(500 \times 2.7 \times 10^{-10}) = 7.4 \times 10^6$ or $f = 1.2 \times 10^6$. Therefore, the 6 dB/octave roll-off in response stops at this frequency as shown by the closed-loop response curve of Fig. 15.21b. However, one of the other stages in the μA715C has a break-frequency or pole which occurs at about 1.0 MHz, as seen in Fig. 15.17. The resistance in series with C in the feedback network of the second stage can be increased slightly to move the zero in the second stage back to 1.0 MHz and then the pole at 1.0 MHz will neutralize the effect of the zero and the 6 dB/octave roll-off will continue until the compensated open-loop gain of the entire μA715C crosses the closed-loop gain curve at unity gain as desired. The resistance required to place the zero in the second stage at 1.0 MHz is $R = 1/\omega C = 1/(6.28 \times 10^6 \times 2.7 \times 10^{-10}) \simeq 600\ \Omega$.

The phase margin of the operational amplifier can be increased by adding a phase-lead capacitor in parallel with the 12 kΩ feedback resistor. The break-frequency of the phase-lead network should be near the intersection of the open-loop and closed-loop gain curves at 10 MHz. This criterion gives $C_f = 1/(6.28 \times 10^7 \times 1.2 \times 10^4) = 1.3$ pF. However, the amplifier *without* the phase-lead capacitor, as shown in Fig. 15.22a, gave ideal time response to a rectangular input pulse, as shown in Fig. 15.22b.

PROBLEM 15.5   A given *IC* operational amplifier has the same equivalent circuit as the μA715C except the 25 kΩ resistance in the second differential stage is reduced to 12 kΩ, the open-loop voltage gain is 5×10^4, the three break-frequencies in the open-loop frequency-response curve occur at 150 kHz, 2.0 MHz, and 20 MHz, and there is no built-in resistance in series with the compensation terminals. Design a compensation circuit similar to the one in Example 15.3 using $R_F = R_1 = 12$ kΩ for unity gain. *Typical answer:*   Same input compensation. Second stage $C = 200$ pF, $R = 400\ \Omega$.

15.7   ADDITIONAL TYPES OF INTEGRATED CIRCUITS

Many types of linear integrated circuits are available to meet special needs. Their circuits, performance characteristics, and applications are available from the various manufacturers. Also, selection guides are occasionally published in the electronics periodicals. The electronics engineer should become familiar with these sources of information. Only a few types of special linear *IC*s can be presented here.

One special type of linear *IC* is the wideband video amplifier. The circuit diagram of a typical video amplifier is given in Fig. 15.23a and

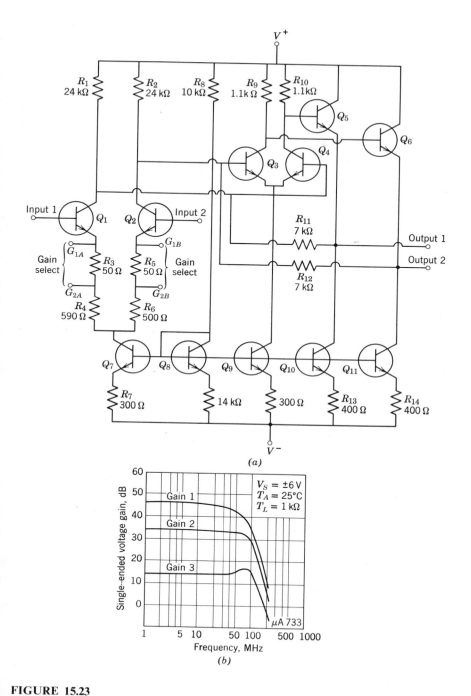

FIGURE 15.23

(a) Circuit diagram, and (b) frequency response curves of the μA733 video amplifier. (Courtesy of Fairchild Semiconductor Company.)

its frequency response curves for $R_s = 50 \, \Omega$ are given in 15.23b. The amplifier consists of two differential stages with negative feedback around the second stage, and a balanced emitter-follower output. The gain of the amplifier is controlled by the degenerative resistance in the emitter circuit of the first differential amplifier. The maximum voltage gain of 400 is obtained by connecting terminal G_{1A} to G_{1B}. The intermediate gain of 100 is obtained by connecting terminal G_{2A} to G_{2B}. The low gain (10) is obtained without any connections of the gain-select terminals. Continuously variable gain from about 10 to 400 may be obtained by connecting a 10 kΩ potentiometer between terminals G_{1A} and G_{1B}. No external feedback or compensation is needed for any of these gain values. The amplifier may be used either single-ended input or single-ended output or both. In fact, the voltage gains given in Fig. 15.23b are for single-ended output and are therefore 1/2 of, or 6 dB lower than, the gain values given above.

Another special type of integrated circuit is the RF/IF amplifier. These amplifiers differ from the operational or video amplifiers in that the amplifier output impedance needs to be high, rather than low, to permit proper bandwidth in the tuned coupling circuit, as discussed in Chapter 9. Some amplifiers bring out both high-impedance and low-impedance terminals to accommodate tuned as well as untuned loads. Provision for automatic gain control is also convenient in an RF/IF amplifier because they are often used in radio receivers where automatic gain control is needed to maintain relatively uniform output voltage in spite of the wide variety of input signal levels.

The major part of a radio receiver is available in a single IC, as shown in Fig. 15.24. This μA719 amplifier has three differential, or emitter-coupled, stages between the input and output of section 1. The input, usually from a tuned circuit, is applied between the high input 1 and the low input 1 terminals. The output of section 1 is obtained at high impedance between the collector of Q_8, which is emitter coupled to Q_6, and $V+$. A tuned circuit would be used in the output if section 1 is used only as an RF or IF amplifier. However, either AF or video can be obtained across a resistance (about 25 kΩ with RF bypass) if a parallel-tuned circuit is connected between the QUAD terminal and the low INPUT 1 terminal, so the transistors Q_6, Q_7, and Q_8 become a quadrature FM detector.[3] Section 2 has about 50 MHz bandwidth and will serve as an RF, video, or AF amplifier with about 35 mmhos transconductance if the DECOUPLE terminal is bypassed to ground or about 8 mmho transconductance if the DECOUPLE terminal remains free.

The final, special-type integrated circuit to be considered is the voltage

---

[3] Quadrature detectors are discussed in Chapter 21.

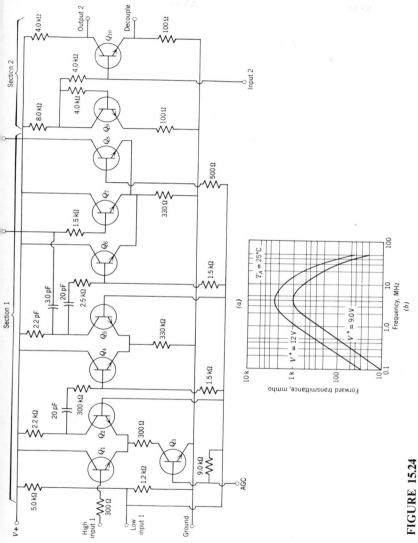

FIGURE 15.24

A high-gain RF/IF amplifier, FM detector and AF/video amplifier, μA719. (Courtesy of Fairchild Semiconductor Company.)

comparator. The comparator is normally used to indicate whether or not the noninverting input is positive with respect to the inverting input. The indication, of course, is the output voltage which changes from negative to positive very quickly and distinctly as the relative polarity of the input changes. Therefore a high-gain, nonsaturating amplifier is needed. The non-saturating feature is needed because of the time required to pull a transistor out of saturation. This time is known as *storage time* and is discussed in Chapter 22.

The schematic diagram and the transfer characteristics of a typical comparator are given in Fig. 15.25. The differential input stage Q_1 and Q_2 is followed by a second differential stage Q_3 and Q_4, which in turn drives a single emitter follower Q_7. As the noninverting input goes positive with respect to the inverting input, the base and emitter of Q_7 go positive until the base of the diode-connected transistor Q_6 becomes positive with respect to its emitter. Then resistors R_4 and R_5 are clamped at almost the same potential at their lower ends, and the collector currents of both Q_3 and Q_4 flow through these two resistors. But since Q_3 and Q_4 are the two halves of a differential amplifier, the sum of their collector currents is constant and the drops across R_4 and R_5 are constant. Thus, the output voltage cannot rise above about 3 V, as shown by the transfer characteristics of Fig. 15.25, which is $V+$ minus the sum of the drop across R_5 plus V_{BE7} plus the 6.2 V reference voltage. When the noninverting input goes negative with respect to the inverting input, the transistor Q_4 is driven to saturation, which also places Q_8 approximately in saturation because the output voltage must be $(6.2 \text{ V} + V_{BE7})$ negative with respect to the collector of Q_4, which is only $(6.2 + V_{CE \text{ sat}})$ positive with respect to ground. Therefore, the output voltage is clamped at approximately -0.5 V which is $(V_{BE7} - V_{CE \text{ sat}})$ while the input is negative, as shown by the voltage transfer characteristic in Fig. 15.25. Note that approximately 1 mV input signal is required to drive the output to its clamped voltage in either direction. The biasing circuits control the currents so that the transistors are not driven appreciably into saturation during normal operation.

The voltage comparator is used in digital voltmeters or analog-to-digital converters where a stable, precise reference voltage is applied to the inverting input, and the output voltage operates some type of read-out when the reference voltage is exceeded. The comparator is also convenient for changing sine waves to square waves, cleaning up a noisy binary signal, or performing many other functions described in the literature.

A careful study of the manufacturers data and literature is almost essential prior to the choosing and application of a linear *IC* to a specific circuit problem. The study of this chapter will hopefully help you to understand the literature and data sheets.

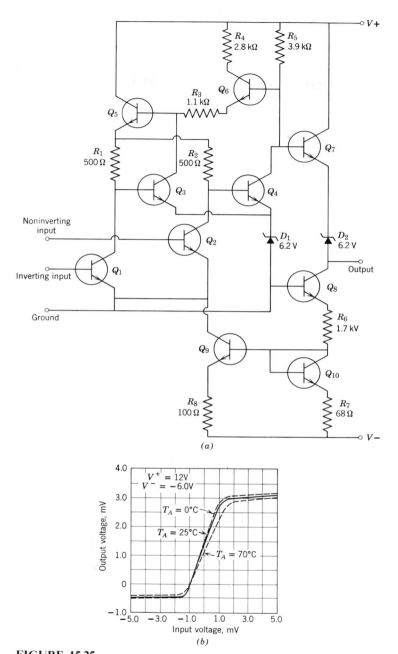

FIGURE 15.25
The schematic diagram and voltage transfer characteristics of a typical
voltage comparator, μA710. (Courtesy of Fairchild Semiconductor
Company.)

PROBLEM 15.6 The comparator of Fig. 15.25 is used to produce a square wave from a 1.0 V peak, 60 Hz sine wave. What will be the approximate peak-to-peak magnitude and rise time of the square wave? Hint: Use $\sin \theta \simeq \theta$ (in radians) for small angles.

Answer: 3.5 V, 5 μs.

PROBLEM 15.7 For the amplifier of Fig. 15.24, determine the voltage gain of section 1 when $V_+ = 12$ volts, the signal is a 10.7 MHz IF and the impedance of the tuned circuit is 5 kΩ.

Answer: 6,000.

PROBLEM 15.8 What is the gain-bandwidth product and the rise time of the amplifier of Fig. 15.23 with single-ended (*a*) gain 1, (*b*) gain 2?

Answer: (*a*) 8,000 MHz, 9 ns, (*b*) 5,000 MHz, 3.5 ns.

PROBLEM 15.9 A μA715C amplifier is to be used to provide voltage gain = 5,000. The driving-source resistance is 10 kΩ and the amplifier should present a very high impedance to the source. Devise a feedback network and compensation, if necessary, to give the desired characteristics. What is the expected bandwidth and rise time of your amplifier?

Answer: $R_1 = 10$ kΩ, $R_F = 5$ MΩ with 10 : 1 voltage divider in the output. $B = 1$ MHz, $t_r = 0.35$ μs.

PROBLEM 15.10 Repeat Problem 15.9 for voltage gain = 1000.

Answer: $R_F = 90$ kΩ, $C_F = 1$ pF with 100 : 1 voltage divider in the output. $B \simeq 2$ MHz, $t_r \simeq 170$ ns.

PROBLEM 15.11 A μA715C is to be used as a summing amplifier with 10 kΩ input resistance, voltage gain = 10 and minimum rise time. Design a feedback and compensation system for the amplier. What is the expected rise time of your amplifier?

16

Active Filters

Electrical filters are circuits which pass certain bands of frequencies and reject other frequencies. The tuned amplifiers of Chapter 9 are examples of band pass filters. In the radio frequency spectrum, the inductors and capacitors are reasonably small, but in the audio frequencies, the inductors in particular are very large. Several techniques such as gyrators, negative impedance converters, positive impedance inverters, and operational amplifier circuits have been developed to eliminate the use of inductors in filters. These circuits are known as active filters since they require the use of active elements as well as RC circuit components. Of all the techniques developed, the operational amplifier circuits are the most promising. The operational amplifiers (op amps) are used for so many applications that their cost is much below that of the active elements in the other configurations. Consequently, this chapter will be restricted to the use of op amps in active filters.

16.1 THE OP AMP CONFIGURATION

A typical configuration for an active filter is shown in Fig. 16.1. The blocks Y_a, Y_b, and Y_c represent passive networks and are normally composed of RC elements. By using the standard y notation, the following six equations can be written.

$$I_1 = y_{11a} V_1 + y_{12a} V_2 \qquad (16.1)$$

$$I_2 = y_{21a} V_1 + y_{22a} V_2 \qquad (16.2)$$

$$I_3 = y_{11b} V_3 + y_{12b} V_o \qquad (16.3)$$

$$I_4 = y_{21b} V_3 + y_{22b} V_o \qquad (16.4)$$

$$I_5 = y_{11c} V_5 + y_{12c} V_6 \qquad (16.5)$$

$$I_6 = y_{21c} V_5 + y_{22c} V_6 \qquad (16.6)$$

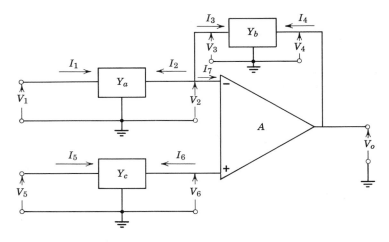

FIGURE 16.1
An op amp active filter configuration.

The parameter y_{11a} is the input admittance of the network Y_a with the output terminals shorted ($V_2 = 0$). Also, y_{12a} is the transfer admittance (I_1/V_2) with the input terminals shorted ($V_1 = 0$). Since Y_a, Y_b and Y_c are passive (RC) circuits, y_{21a} is equal to y_{12a}. Finally, y_{22a} is the input admittance of the output terminals (I_2/V_2) with the input terminals shorted ($V_1 = 0$).

Now, if the input impedance of the op amp is very high (a good assumption), I_7 and I_6 are essentially zero. If I_7 is zero, $-I_3 = I_2$ and Eq. 16.2 and Eq. 16.3 can be combined as

$$y_{21a} V_1 + y_{22a} V_2 = -y_{11b} V_3 - y_{12b} V_o \qquad (16.7)$$

If I_6 is zero, Eq. 16.6 becomes

$$0 = y_{21c} V_5 + y_{22c} V_6 \qquad (16.8)$$

or

$$V_6 = -V_5 \frac{y_{21c}}{y_{22c}} \qquad (16.9)$$

If the voltage gain of the op amp is very high (another good assumption), the input voltage of the op amp ($V_6 - V_2$) will be very small for a reasonable output voltage. Then, $V_6 \simeq V_2 = V_3$ and the value of V_6 found by Eq. 16.9 can be substituted back into Eq. 16.7 for V_2 and V_3.

$$y_{21a} V_1 - y_{22a} \frac{y_{21c}}{y_{22c}} V_5 = y_{11b} \frac{y_{21c}}{y_{22c}} V_5 - y_{12b} V_o \qquad (16.10)$$

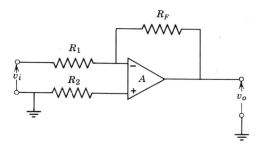

FIGURE 16.2
An op amp with negative feedback.

Solving this relationship for V_o , we have

$$V_o = -\frac{y_{21a}}{y_{12b}} V_1 + \frac{y_{21c}(y_{22a} + y_{11b})}{y_{22c} y_{12b}} V_5 \qquad (16.11)$$

Now, V_o is the output voltage for our filter and V_1 and V_5 are the input voltages, therefore Eq. 16.11 is the desired relationship for the output voltage.

We can verify the validity of Eq. 16.11 by the following example.

Example 16.1   An inverting op amp circuit is connected as shown in Fig. 16.2. Let us use Eq. 16.11 to determine the voltage gain of the circuit.

First, we must determine the y parameters of the networks (resistors in this case). The circuit for network Y_a is shown in Fig. 16.3a. The configuration for finding y_{11a} is shown in Fig. 16.3b.

$$y_{11a} = \left.\frac{I_1}{V_1}\right|_{V_2=0} = \frac{1}{R_1} \qquad (16.12)$$

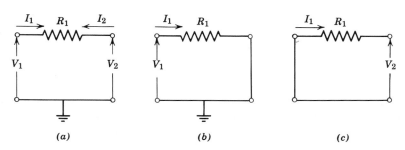

(a) $\qquad\qquad$ (b) $\qquad\qquad$ (c)

FIGURE 16.3
The network for Y_a in Fig. 16.2 : (a) actual network; (b) configuration for y_{11};
(c) configuration for y_{12}.

Notice that y_{22a} will also be equal to $1/R_1$. The configuration for finding y_{12} is given in Fig. 16.3c.

$$y_{12a} = \frac{I_1}{V_2}\bigg|_{V_1=0} = -\frac{1}{R_1} \tag{16.13}$$

The negative sign exists because a voltage with the polarity of V_2 would cause I_1 to be a negative current (opposite to the direction shown). Of course, $y_{21} = y_{12}$ for this network. The y parameters for all three networks are tabulated in Table 16.1.

TABLE 16.1
The y Parameters for the Networks in Fig. 16.2.

Parameter	Value
y_{11a}, y_{22a}	$1/R_1$
y_{12a}, y_{21a}	$-1/R_1$
y_{11b}, y_{22b}	$1/R_F$
y_{12b}, y_{21b}	$-1/R_F$
y_{11c}, y_{22c}	$1/R_2$
y_{12c}, y_{21c}	$-1/R_2$

In Fig. 16.2, the value of V_5 is zero. Then from Eq. 16.11, the value of V_o is

$$V_o = -\frac{1/R_1}{1/R_F} V_1 = -\frac{R_F}{R_1} V_1 \tag{16.14}$$

This value agrees with that found in Chapter 15 (Eq. 15.9) for the same configuration.

PROBLEM 16.1   An amplifier is connected in the noninverting configuration shown in Fig. 15.10. Use Eq. 16.11 to determine if Eq. 15.4 is valid.

Equation 16.11 is a very general relationship but is a bit awkward to use. Normally, the noninverting input is returned through a resistance to ground as in Example 16.1. Then, V_5 is zero and Eq. 16.11 is reduced to the following simple configuration.

$$V_o = -\frac{y_{21a}}{y_{12b}} V_1 \tag{16.15}$$

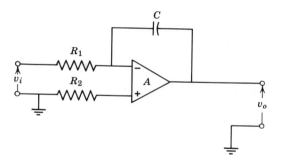

FIGURE 16.4
An op amp integrator circuit.

This relationship is normally used to design active filters. However, before we consider a typical filter, let us consider another op amp configuration.

Example 16.2 A circuit is connected as shown in Fig. 16.4. We will determine the transfer function for this configuration.

As found in Example 16.1, the value of y_{21a} is $-1/R_1$. The transfer function y_{12b} is found from the configuration in Fig. 16.5 to be $y_{12b} = -sC$. Then, from Eq. 16.15, the value of V_o is

$$V_o = -\frac{1}{sR_1C} V_1 = -\frac{1}{R_1C} \frac{V_1}{s} \tag{16.16}$$

The output signal, V_o, is the *integral* of the input signal, V_1. The circuit shown in Fig. 16.4 is commonly used as an integrator in analog computer circuits.

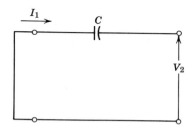

FIGURE 16.5
The circuit to find y_{12b} for Fig. 16.4.

PROBLEM 16.2 Develop a configuration which can be used as a differentiator[1] circuit. Write the equation which describes the transfer function of this circuit.

16.2 TYPICAL *RC* NETWORK CONFIGURATIONS

In order to design typical active filters, the engineer must be acquainted with the y parameters for many circuit configurations. The y parameters for several different circuit configurations are tabulated in Table 16.2. Other configurations are, of course, possible, but this selection provides a fair degree of circuit design flexibility.

PROBLEM 16.3 Verify that the y parameters for the first two circuit configurations of Table 16.2 are correct.

PROBLEM 16.4 The sixth circuit in Table 16.2 is known as a bridged T network. Verify that the given y parameters for this configuration is correct.

16.3 OP AMP FILTER DESIGN

Active filters are designed by using Eq. 16.15 and the circuits listed in Table 16.2 (or similar circuits). The procedure can be best explained by considering an example.

Example 16.3 A student desires to construct a *color organ*. He wishes to have a red light with the light intensity proportional to the amplitude of the low-frequency component of an audio signal. The intensity of a yellow light is to be controlled by the amplitude of the mid-frequency component of the audio signal. Finally, a blue light is to be controlled by the amplitude of the high-frequency component of the audio signal. In order to separate the three audio components, the student decides he must have a low-pass filter which will pass frequencies from zero to 1000 rad/sec (about 160 Hz) to activate the red light. He also needs a band-pass filter from 1000 rad/sec to 10,000 rad/sec (about 1,600 Hz) to activate the yellow light. Finally, he needs a high-pass filter from 10,000 rad/sec to the upper end of the audio spectrum to activate the blue light.

The desired characteristics for the three filters are given in Fig. 16.6.

---

[1] Integration of a voltage waveform produces an output waveform that is smoother than the input signal. In contrast, differentiation produces the opposite effect. Consequently, the analog computer circuits are normally designed around integrator circuits. (Said in another way, differentiator circuits degrade the signal-to-noise ratio.)

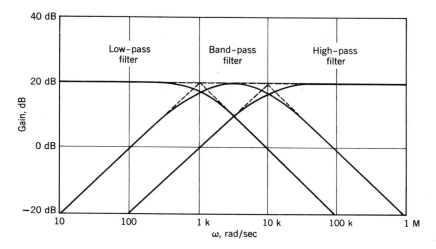

FIGURE 16.6
The response curves for Example 16.3.

We will design the band-pass filter. From Fig. 16.6 we note that the transfer function of the filter is

$$\frac{V_o}{V_i} = \frac{As}{(s + 10^3)(s + 10^4)} \tag{16.17}$$

Notice that Eq. 16.17 is identical to that of an RC-coupled amplifier with $\omega_1 = 10^3$ rad/sec and $\omega_2 = 10^4$ rad/sec. Then, the constant A must be equal to $K\omega_2$ where K is the mid-band gain of the amplifier. The desired mid-band gain of our filter is $+20\,\text{dB}$ or a voltage gain of 10. Then $A = 10 \times 10^4 = 10^5$. From Eq. 16.15 we have

$$\frac{V_o}{V_i} = -\frac{y_{21a}}{y_{12b}} = \frac{10^5 s}{(s + 10^3)(s + 10^4)} \tag{16.18}$$

We note that the first circuit in Table 16.2 will provide a zero and one of the poles. Let us use this circuit for our Y_a. Then,

$$y_{21a} = -\frac{1}{R_1} \frac{s}{[s + (1/R_1 C_1)]} \tag{16.19}$$

At higher frequencies the impedance of C_1 is very small so the input impedance to the amplifier is essentially equal to R_1. In order to maintain a high impedance to the filter (so we will not load down the signal

TABLE 16.2
Typical RC Circuits and Their y-Parameters.

Circuit configuration	Y-parameters
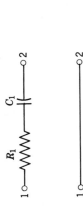	$$y_{11} = y_{22} = \frac{1}{R_1} \frac{s}{\left(s + \dfrac{1}{R_1 C_1}\right)}$$ $$y_{12} = y_{21} = -\frac{1}{R_1} \frac{s}{\left(s + \dfrac{1}{R_1 C_1}\right)}$$
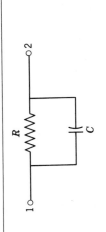	$$y_{11} = y_{22} = C\left(s + \frac{1}{RC}\right)$$ $$y_{12} = y_{21} = -C\left(s + \frac{1}{RC}\right)$$
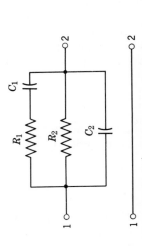	$$y_{11} = y_{22} = C_2 \frac{s^2 + s\left(\dfrac{1}{R_1 C_1} + \dfrac{1}{R_2 C_2} + \dfrac{1}{R_1 C_2}\right) + \dfrac{1}{C_1 C_2 R_1 R_2}}{\left(s + \dfrac{1}{R_1 C_1}\right)}$$ $$y_{12} = y_{21} = -C_2 \frac{s^2 + s\left(\dfrac{1}{R_1 C_1} + \dfrac{1}{R_2 C_2} + \dfrac{1}{R_1 C_2}\right) + \dfrac{1}{C_1 C_2 R_1 R_2}}{\left(s + \dfrac{1}{R_1 C_1}\right)}$$

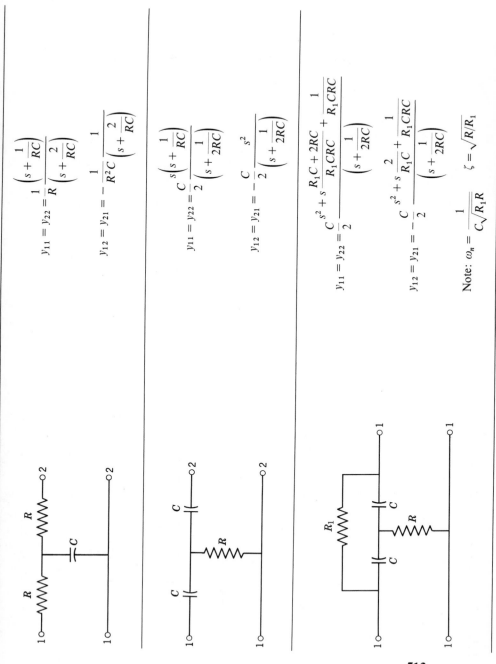

$$y_{11} = y_{22} = \frac{1}{R}\,\frac{\left(s + \dfrac{1}{RC}\right)}{\left(s + \dfrac{2}{RC}\right)}$$

$$y_{12} = y_{21} = -\frac{1}{R^2 C}\,\frac{1}{\left(s + \dfrac{2}{RC}\right)}$$

$$y_{11} = y_{22} = \frac{C}{2}\,\frac{s\left(s + \dfrac{1}{RC}\right)}{\left(s + \dfrac{1}{2RC}\right)}$$

$$y_{12} = y_{21} = -\frac{C}{2}\,\frac{s^2}{\left(s + \dfrac{1}{2RC}\right)}$$

$$y_{11} = y_{22} = \frac{C}{2}\,\frac{s^2 + s\dfrac{R_1 C + 2RC}{R_1 CRC} + \dfrac{1}{R_1 CRC}}{\left(s + \dfrac{1}{2RC}\right)}$$

$$y_{12} = y_{21} = -\frac{C}{2}\,\frac{s^2 + s\dfrac{2}{R_1 C} + \dfrac{1}{R_1 CRC}}{\left(s + \dfrac{1}{2RC}\right)}$$

Note: $\omega_n = \dfrac{1}{C\sqrt{R_1 R}}$ $\zeta = \sqrt{R/R_1}$

TABLE 16.2—Continued

Circuit configuration	Y-parameters
	$$y_{11} = y_{22} = C_1\,\dfrac{s^2 + s\,\dfrac{C_2 R + 2C_1 R}{C_1 RC_2 R} + \dfrac{1}{RC_1 RC_2}}{\left(s + \dfrac{2}{RC_2}\right)}$$ $$y_{12} = y_{21} = -C_1\,\dfrac{s^2 + s\,\dfrac{2}{RC_2} + \dfrac{1}{RC_1 RC_2}}{\left(s + \dfrac{2}{RC_2}\right)}$$
	$$y_{11} = y_{22} = C_2\,\dfrac{s^2 + s\,\dfrac{2R_1 C_1 + R_2 C_1 + R_2 C_2}{2R_1 R_2 C_1 C_2} + \dfrac{R_1 + R_2}{2R_1{}^2 R_2 C_1 C_2}}{s + \dfrac{2}{R_1 C_1}}$$ $$y_{12} = y_{21} = -C_2\,\dfrac{s^2 + s\,\dfrac{2R_2 C_2 + R_1 C_1}{R_1 R_2 C_1 C_2} + \dfrac{2R_1 + R_2}{R_1{}^2 R_2 C_1 C_2}}{s + \dfrac{2}{R_1 C_1}}$$

$$y_{11} = \frac{C_1 C_2}{C_1 + C_2} \frac{\left(s + \dfrac{1}{R_1 C_1}\right)\left(s + \dfrac{1}{R_2 C_2}\right)}{\left(s + \dfrac{R_1 + R_2}{R_1 R_2 (C_1 + C_2)}\right)}$$

$$y_{22} = \frac{1}{R_2} \frac{\left(s + \dfrac{1}{R_1(C_1 + C_2)}\right)}{\left(s + \dfrac{R_1 + R_2}{R_1 R_2(C_1 + C_2)}\right)}$$

$$y_{12} = y_{21} = -\frac{C_1}{R_2(C_1 + C_2)} \frac{\left(s + \dfrac{1}{R_1 C_1}\right)}{\left(s + \dfrac{R_1 + R_2}{R_1 R_2(C_1 + C_2)}\right)}$$

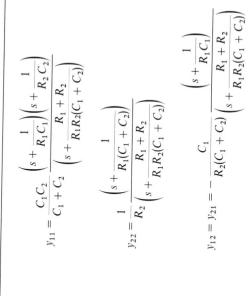

$$y_{11} = y_{22} = \frac{C}{2} \frac{\left(s + \dfrac{1}{RC}\right)\left(s + \dfrac{R_A + R}{R R_A C}\right)}{\left(s + \dfrac{2R_A + R}{2R_A RC}\right)}$$

$$y_{12} = y_{21} = -\frac{C}{2} \frac{\left(s + \dfrac{1}{RC}\right)^2}{\left(s + \dfrac{2R_A + R}{2R R_A C}\right)}$$

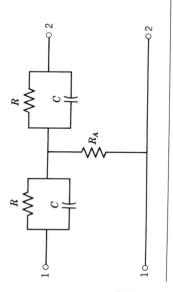

TABLE 16.2—*Continued*

Circuit configuration	Y-parameters

$$y_{11} = y_{22} = \frac{C(C+C_A)}{2C+C_A}\,\frac{\left(s+\dfrac{1}{RC}\right)\left(s+\dfrac{1}{R(C+C_A)}\right)}{\left[s+\dfrac{2}{R(2C+C_A)}\right]}$$

$$y_{12} = y_{21} = -\,\frac{C^2}{C_A+2C}\,\frac{\left(s+\dfrac{1}{RC}\right)^2}{\left[s+\dfrac{2}{R(C_A+C)}\right]}$$

$$y_{11} = \frac{1}{R_1}\,\frac{s\left(s+\dfrac{C_1+C_2}{R_2C_1C_2}\right)}{s^2 + s\,\dfrac{R_1C_1+R_1C_2+R_2C_2}{R_1R_2C_1C_2} + \dfrac{1}{R_1R_2C_1C_2}}$$

$$y_{22} = \frac{R_1+R_2}{R_1R_2C_1}\,\frac{s\left(s+\dfrac{1}{R_1C_2+R_2C_2}\right)}{s^2 + s\,\dfrac{R_1C_1+R_1C_2+R_2C_2}{R_1R_2C_1C_2} + \dfrac{1}{R_1R_2C_1C_2}}$$

$$y_{12} = y_{21} = -\,\frac{C_2}{R_1C_1}\,\frac{1}{s^2 + s\,\dfrac{R_1C_1+R_1C_2+R_2C_2}{R_1R_2C_1C_2} + \dfrac{1}{R_1R_2C_1C_2}}$$

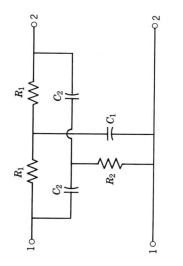

$$y_{11} = y_{22} = \frac{C_2}{2} \frac{s^3 + s^2\left(\dfrac{1}{R_2 C_2} + \dfrac{2}{R_1 C_1} + \dfrac{2}{R_1 C_2}\right) + s\left(\dfrac{2}{R_1 R_2 C_1 C_2} + \dfrac{2}{R_1^2 C_1 C_2} + \dfrac{1}{R_1 R_2 C_2^2}\right) + \dfrac{1}{R_1^2 R_2 C_1 C_2^2}}{\left(s + \dfrac{2}{R_1 C_1}\right)\left(s + \dfrac{1}{2R_2 C_2}\right)}$$

$$y_{12} = y_{21} = -\frac{C_2}{2} \frac{s^2 + s^2\,\dfrac{2}{R_1 C_1} + s\,\dfrac{2}{R_1^2 C_1 C_2} + \dfrac{1}{R_1^2 R_2 C_1 C_2^2}}{\left(s + \dfrac{2}{R_1 C_1}\right)\left(s + \dfrac{1}{2R_2 C_2}\right)}$$

source), let us choose $R_1 = 10^5 \, \Omega$. If y_{21a} is to provide the pole at $\omega = -10^3$ rad/sec, then $1/R_1 C_1 = 10^3$ or $C_1 = 1/(10^3 \times R_1) = 1/10^8 = 10^{-8}$ F or $C_1 = 0.01 \, \mu F$. When these values are substituted back into Eq. 16.19, y_{21a} is

$$y_{21a} = -10^{-5} \frac{s}{(s + 10^3)} \tag{16.20}$$

Now,

$$\frac{y_{21a}}{y_{12b}} = -10^{-5} \frac{s}{(s + 10^3) y_{12b}} \tag{16.21}$$

However, an expression for $-y_{21a}/y_{12b}$ is given in Eq. 16.18. Consequently,

$$\frac{10^{-5} s}{(s + 10^3) y_{12b}} = \frac{10^5 s}{(s + 10^3)(s + 10^4)} \tag{16.22}$$

or

$$y_{12b} = 10^{-10}(s + 10^4) \tag{16.23}$$

The second circuit in Table 16.2 has y_{12} of the form given by Eq. 16.23, so this configuration will be used for Y_b. The actual y_{12} has a negative sign so V_o of Eq. 16.18 will be inverted with respect to V_1. Since we are applying the signal to the inverting input, this behavior would be expected. Now, from Table 16.2,

$$y_{12} = -C\left(s + \frac{1}{RC}\right) \tag{16.24}$$

Comparing Eq. 16.23 and Eq. 16.24, $C = 10^{-10}$ F or 100 pF. (Notice that C could be made larger if R_1 of circuit Y_a is made smaller.) Then $1/RC = 10^4$ so $R = 1/(10^4 \times C) = 1/10^{-6} = 1 \, M\Omega$. Our band-pass filter would have the form shown in Fig. 16.7. There is no dc path to the input terminals, but there is a path through the 1 MΩ resistor to the output terminals. Therefore, if the offset voltage is to be maintained near zero, R_2 of Fig. 16.7 should be 1 MΩ also.

PROBLEM 16.5   Design a low-pass filter with the characteristics given in Fig. 16.6.

PROBLEM 16.6   Design a high-pass filter with the characteristics given in Fig. 16.6.

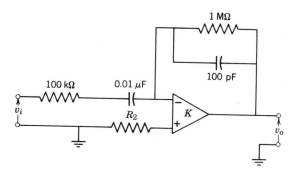

FIGURE 16.7
The band-pass filter of Example 16.3.

16.4 STABILITY OF ACTIVE FILTERS

Of course, each active filter should be checked for stability. The method outlined in Chapter 15 can be used to determine the circuit stability.

First, the op amp is stabilized by creating a dominant low-frequency pole so the response drops off at 6 dB/octave (or 20 dB/decade) until the gain is reduced to one. If the feedback were purely resistive, this step would be sufficient to insure stable operation for any amount of feedback. Unfortunately, since the feedback networks of filters may be quite complex, the foregoing procedure *does not* insure freedom from oscillation for the filter circuit. However, if the op amp is compensated so the response drops off at 6 dB/octave, the gain equation of the op amp (over its useful frequency range) is

$$G_a = \frac{-K\omega_h}{s + \omega_h} \qquad (16.25)$$

where K is the dc open-loop gain of the amplifier; ω_h is the open-loop -3 dB break-frequency for the compensated amplifier.

The closed-loop gain of the amplifier is

$$G_f = \frac{G_a}{1 + \beta G_a} \qquad (16.26)$$

where β is the transfer characteristic of the feedback path. This transfer function β is equal to

$$\beta = \frac{V_2}{V_o} \tag{16.27}$$

where V_2 and V_o have the values given in Fig. 16.1. Equation 16.2 is repeated here for convenience.

$$I_2 = y_{21a}V_1 + y_{22a}V_2 \tag{16.2}$$

If there is no input signal, $V_1 = 0$ and

$$I_2 = y_{22a}V_2 \tag{16.28}$$

Now, if the input impedance of the amplifier (Fig. 16.1) is very high, $I_7 \simeq 0$ and $I_2 \simeq -I_3$. Then, Eq. 16.28 can be written as

$$I_3 \simeq -I_2 = -y_{22a}V_2 \tag{16.29}$$

Equation 16.3 states that

$$I_3 = y_{11b}V_3 + y_{12b}V_o \tag{16.3}$$

Note that $V_2 = V_3$ in Fig. 16.1. Then, Eq. 16.29 and Eq. 16.3 can be combined to yield the following relationship

$$-y_{22a}V_2 = y_{11b}V_2 + y_{12b}V_o \tag{16.30}$$

This relationship can be rearranged to produce the following equation

$$\beta = \frac{V_2}{V_o} = \frac{-y_{12b}}{y_{11b} + y_{22a}} \tag{16.31}$$

The term βG_a in Eq. 16.26 is found by combining Eq. 16.25 and Eq. 16.31.

$$\beta G_a = \frac{K\omega_h y_{12b}}{(s + \omega_h)(y_{11b} + y_{22a})} \tag{16.32}$$

Since y_{12b}, y_{11b}, and y_{22b} may each be expressed as a function of s, Eq. 16.32 can be written in the form

$$\beta G_a = \frac{a_n s^n + a_{n-1}s^{n-1} + \cdots a_1 s + a_o}{b_m s^m + b_{m-1}s^{m-1} + \cdots b_1 s + b_o} \tag{16.33}$$

Note that there will be n finite zeros in Eq. 16.33 and m finite poles.

The term βG_a expresses the open-loop gain of the amplifier and the feedback path. If the magnitude of βG_a is less than one where the phase shift of βG_a is $\pm 180°$, the circuit will be stable. A root-locus plot of

Eq. 16.26, or a Bode plot of Eq. 16.33, can be used to determine the stability of the active filter. In order to illustrate this procedure, let us consider an example.

Example 16.4 Let us determine if the configuration of Fig. 16.7 is stable. The value of β from Eq. 16.31 is

$$\beta = \frac{-y_{12b}}{y_{11b} + y_{22a}} = \frac{-\{-C[s + (1/RC]\})}{C[s + (1/RC)] + (1/R_1)\{s/[s + (1/R_1C_1)]\}} \quad (16.34)$$

When the actual circuit values are substituted into this equation, we have

$$\beta = \frac{10^{-10}(s + 10^4)}{10^{-10}(s + 10^4) + [(10^{-5}s)/(s + 10^3)]} \quad (16.35)$$

When both numerator and denominator of Eq. 16.35 are multiplied by $(s + 10^3)$, the value of β is

$$\beta = \frac{10^{-10}(s + 10^4)(s + 10^3)}{10^{-10}(s + 10^4)(s + 10^3) + 10^{-5}s} \quad (16.36)$$

or

$$\beta = \frac{(s + 10^4)(s + 10^3)}{s^2 + 1.11 \times 10^5 s + 10^7} \quad (16.37)$$

The denominator of Eq. 16.37 can be factored. Then β is

$$\beta = \frac{(s + 10^4)(s + 10^3)}{(s + 1.11 \times 10^5)(s + 90)} \quad (16.38)$$

Using the expression for G_a in Eq. 16.25,

$$\beta G_a = \frac{-K\omega_h(s + 10^4)(s + 10^3)}{(s + \omega_h)(s + 1.11 \times 10^5)(s + 90)} \quad (16.39)$$

If we assume $\omega_h \simeq 2000$, the root-locus plot for G_f will be as shown in Fig. 16.8. Notice that this plot does indicate that the filter will be stable for all values of amplifier gain.

PROBLEM 16.7 Use a Bode plot to prove that Eq. 16.39 represents a stable configuration. Assume $\omega_h = 2000$ rad/sec.

PROBLEM 16.8 Determine if the circuits which you designed in Problem 16.5 and Problem 16.6 are stable.

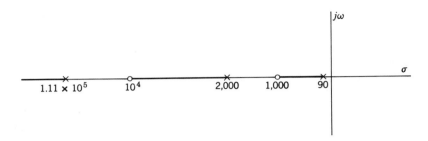

FIGURE 16.8
The root-locus plot for Eq. 16.39 with $\omega_h = 2{,}000$.

16.5   COMPLEX ACTIVE FILTERS

The foregoing filters have been quite simple. As the filter requirements become more demanding, the circuit complexity increases. One approach is to use more complex RC circuits to obtain the additional poles and zeros. In another approach, the transfer function $-y_{21a}/y_{12b}$ is divided into two or more simpler functions. A separate op amp is then used to produce each separate transfer function. Because of the op amp characteristics (low output impedance and high input impedance), the separate filter sections do not interact appreciably. Accordingly, the individual filter sections can be cascaded to provide the total filter configuration. An example will be used to illustrate this concept.

Example 16.5   The filter of Example 16.3 with a pass band from 1000 rad/sec to 10,000 rad/sec does not have sufficient rejection of a signal near the pass band. Consequently we desire the response to drop off at 12 dB/octave (or 40 dB/decade) outside of the pass band.

The equation for the transfer function of the filter must have two zeros at (or very near) zero and two poles to produce the low-frequency break-point. Then two additional poles are required to produce the high-frequency break-point. The transfer function will have the following form

$$\frac{V_o}{V_i} = \frac{As^2}{(s + \omega_1)^2(s + \omega_2)^2} \tag{16.40}$$

We are tempted to let ω_1 be 10^3 rad/sec, but then our gain will be down 6 dB at 10^3 rad/sec. Consequently, we must use the concepts of Chapter 12. Then, from Eq. 12.15, or from Table 12.1, we find ω_1 must be 643 rad/sec. Similarly, ω_2 must be 1.556×10^4 rad/sec. If we desire the mid-band gain to be 10, $A = K\omega_2{}^2 = 10 \times (1.556 \times 10^4)^2 = 2.42 \times 10^9$. Thus the transfer function is

$$G_f = \frac{2.42 \times 10^9}{(s + 643)^2(s + 1.556 \times 10^4)^2} \tag{16.41}$$

A study of Table 16.2 reveals that this function cannot be achieved by using any combination of the given circuits and a single op amp. In fact, the high- and low-frequency drop-off of 12 dB/decade indicate that the circuit would almost certainly oscillate if the proper circuit configuration *could* be found. Therefore let us divide the transfer function into two identical parts

$$G_f = \frac{4.92 \times 10^4}{(s + 643)(s + 1.556 \times 10^4)} \cdot \frac{4.92 \times 10^4}{(s + 643)(s + 1.556 \times 10^4)} \tag{16.42}$$

Each part of this function now has the form of Eq. 16.18. Thus, the required filter can be constructed from two filters with the form shown in Fig. 16.7 connected in cascade. The total filter configuration would be as shown in Fig. 16.9.

In Example 16.4, we found that a circuit similar to each unit in Fig. 16.9 was stable. Either a root-locus plot or a Bode plot indicates that each unit in Fig. 16.9 is indeed stable. Thus, since there is negligible interaction between units, the total configuration is also stable.

It is possible to purchase two op amps in a single flat-pack case, so the cost and size of this filter need not be twice as much as the filter in Example 16.4.

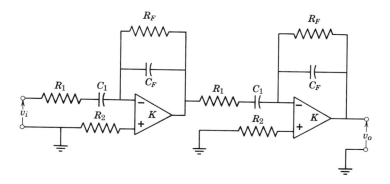

FIGURE 16.9
The band-pass filter of Example 16.5.

PROBLEM 16.9   Determine the values of each of the circuit elements in Fig. 16.9 if $R_1 = 100,000 \ \Omega$.

PROBLEM 16.10   Design a high-pass filter with a cutoff frequency of $\omega = 10^3$ rad/sec that has a response which drops off at 12 dB/octave or 40 dB/decade. Use one op amp if possible. Determine if your filter is stable.

PROBLEM 16.11   Design a low-pass filter with a cutoff frequency of $\omega = 10^3$ rad/sec. The response should drop off at 12 dB/octave or 40 dB/decade in the stop band. Use one op amp if possible. Determine if your filter is stable.

PROBLEM 16.12   Design a filter that has the following transfer function

$$G_f = \frac{-10(s + 100)(s + 100,000)}{(s + 1,000)(s + 10,000)}$$

Make a gain plot of this response. Determine if your filter is stable.

PROBLEM 16.13   Design a band-rejection filter which has the following transfer function

$$G = \frac{10(s + 8,000)(s + 10,000)}{(s + 2,000)(s + 40,000)}$$

Make a gain plot of this response. Determine if your filter is stable.

17

Power Supplies

Essentially all of the electronic devices discussed in the preceding chapters of this book require a dc power source for their operation. Up to this point, little attention has been given to the problem of obtaining this dc power. In fact, the inference may have been made that one or more batteries will supply adequate power for any device. This is possible, but in most instances the battery is neither the most convenient nor the most economical means of obtaining dc power. It is usually much more convenient and economical to obtain, by the use of rectifiers and appropriate filters, dc power from the ac power line. Some of the commonly used rectifying and filtering systems are discussed in this chapter.

The requirements of a power supply differ widely among the various electronic devices. The primary characteristics which need to be considered in the design of a power supply follow.

1.   The dc voltage or voltages required by the device, which is known as the load on the supply, is of primary importance.

2.   The power supply must be able to furnish the maximum current requirement of the load.

3.   The variation of the dc output voltage with change in load current may be important. The voltage regulation of a power supply is defined as the change in output voltage, when the current is changed from no load to full load, divided·by the full load voltage. The voltage regulation is expressed as a percent, as shown in the following equation.

$$\% \text{ regulation} = \frac{\text{no load voltage} - \text{full load voltage}}{\text{full load voltage}} \times 100 \quad (17.1)$$

4.   The rapid variations of the output voltage which result from imperfect filtering must be considered. These voltage variations have a fundamental frequency which is related to the ac power line frequency

and are called ripple voltage or simply *ripple*. The ripple is defined as the ratio of the rms value of the ripple voltage to full load dc voltage. This ripple is usually expressed in percent as indicated below.

$$\% \text{ ripple} = \frac{\text{rms ripple voltage}}{\text{full load dc voltage}} \times 100 \qquad (17.2)$$

A commonly encountered term in Power supply design is the *Peak-to-peak ripple ratio* (*pprr*) which is defined as $(V_{o\,\text{max}} - V_{o\,\text{min}})/V_{o\,\text{ave}}$. Power supply characteristics in addition to those listed are also considered in power supply design. These characteristics will be considered as the need arises

17.1 CAPACITOR INPUT FILTERS

Some basic rectifier circuits were considered in Chapter. 3. These circuits included both half-wave and full-wave rectifiers using either solid-state or electron-tube diodes. Also, the capacitor was considered as a filtering element in Chapter 3. Whenever the filter capacitor immediately follows the rectifier, the filter is known as a capacitor input filter. Three basic rectifier circuits are shown in Fig. 17.1. These circuits were introduced in Chapter 3.

Equations that express the approximate relationships between peak input voltage, average output voltage, peak-to-peak ripple ratio, load resistance, and filter capacitance were developed in Chapter 3. However, those equations were based on the following assumptions.

1.   The peak-to-peak ripple ratio (*pprr*) is small (about 10 percent or less).
2.   The resistance of the transformer windings is negligible.
3.   The ohmic, or bulk, resistance of the rectifier diodes is negligible.

Unfortunately, these constraints are not always met in actual practice. In fact, the *iR* drops in the transformer and the rectifiers may be significant, even though the resistances of these elements are small, because the peak currents may be very high in comparison with the average current in the load. Therefore, J. Phillip Stringham, a systems design engineer of Ball Brothers Research Corporation, developed a computer program and obtained a solution, which is *not* based on the above assumptions, for the full-wave rectifier with a capacitive filter. Stringham prepared a nomograph from his solution, which provides the needed relationships among the parameters with accuracy limited only by the tolerances of the components and the preciseness of reading the nomograph. The nomograph parameters

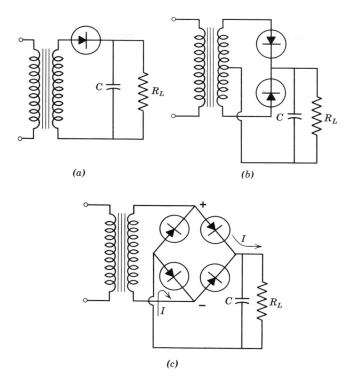

(a)

(b)

(c)

FIGURE 17.1
Typical rectifier circuits incorporating a single capacitor filter :
(a) half-wave rectifier; (b) full-wave rectifier using a tapped
transformer; (c) a bridge rectifier circuit.

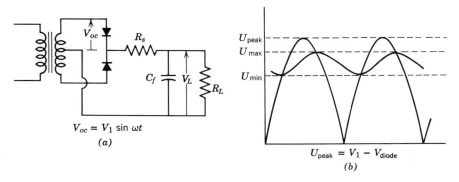

$V_{oc} = V_1 \sin \omega t$

(a)

$U_{peak} = V_1 - V_{diode}$
(b)

FIGURE 17.2
A full-wave rectifier circuit : (a) circuit diagram; (b) voltages defined.

are illustrated in Fig. 17.2 and the nomograph is given in Fig. 17.3. A second nomograph prepared by Stringham, which can be used to determine the peak diode currents, is given in Fig. 17.4.

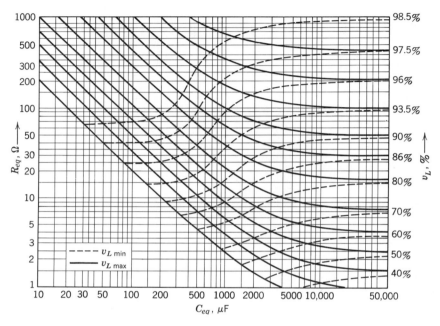

FIGURE 17.3
Nomograph to determine the circuit parameters of Fig. 17.2.

17.2   CIRCUIT AND VOLTAGES FOR THE FULL-WAVE RECTIFIER

Figure 17.2a shows the circuit diagram of the full-wave rectifier with the open circuit (no load) voltage V_{oc} across one-half of the secondary. The resistance R_s includes the dynamic resistance of the diode plus the ac resistance looking into one-half of the secondary of the transformer. The value of R_s could be experimentally determined by disconnecting the filter capacitor and plotting the voltage V_L across R_L as a function of $I_L = V_L/R_L$. Then R_s is the slope of the curve, $\Delta V_L/\Delta I_L$. Typical values of R_s for some representative power supply ratings are given in Table 17.1. This source resistance is included in the nomograph of Fig. 17.3 by defining the parameter R_{eq} as

$$R_{eq} = R_L/R_s \tag{17.3}$$

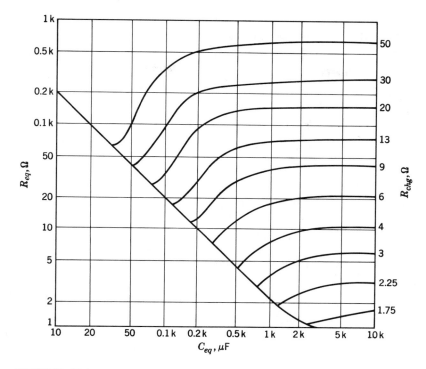

FIGURE 17.4
Nomograph to determine the peak diode currents.

Also, this nomograph can be used for any primary power frequency f by defining C_{eq} as

$$C_{eq} = R_s C_f(f/60) \qquad (17.4)$$

The nomograph gives $v_{L\,\text{max}}$ and $v_{L\,\text{min}}$ as percentages of $V_{L\,\text{peak}}$, where $V_{L\,\text{peak}}$ is the peak voltage across the load when the load resistance is very high, or when R_L approaches infinity. Thus $V_{L\,\text{peak}}$ can be obtained from the relationship

$$V_{L\,\text{peak}} = V_1 - V_{\text{diode}} \qquad (17.5)$$

where V_{diode} is the forward voltage drop across the diode when the diode current is small (perhaps 10 percent of the average value).

Finally, after the parameters R_{eq} and C_{eq} have been determined (Eqs. 17.3 and 17.4), a parameter known as R_{chg} can be found from the

TABLE 17.1
Typical R_s as a Function of Secondary Voltage and Current Ratings.

	10 V	25 V	50 V	100 V	200 V
0.02 A	70 Ω	120 Ω	220 Ω	420 Ω	820 Ω
0.2 A	7.0 Ω	12.0 Ω	22.0 Ω	42.0 Ω	82.0 Ω
1 A	1.4 Ω	2.4 Ω	4.4 Ω	8.4 Ω	16.4 Ω
2 A	0.7 Ω	1.2 Ω	2.2 Ω	4.2 Ω	8.2 Ω
5 A	0.3 Ω	0.5 Ω	0.9 Ω	1.7 Ω	3.3 Ω

nomograph given in Fig. 17.4. Then the peak diode current I_{peak} can be determined from the following equation.

$$I_{peak} = \frac{V_{L\,peak}}{R_s R_{chg}} \tag{17.6}$$

The proper use of the nomographs is illustrated by the following examples.

Example 17.1   We want to determine the minimum output voltage v_L, the peak-to-peak ripple voltage, and the peak diode current of a full-wave circuit such as Fig. 17.2a when V_{oc} (1/2 secondary) $= 35$ V, $R_L = 10$ Ω, $R_s = 0.35$ Ω, $C_f = 5000$ μF, $f = 60$ Hz, and $V_{diode} = 0.5$ V.

1. $R_{eq} = R_L/R_s = 10/0.35 = 28.6$ (from Eq. 17.3). $C_{eq} = R_s C_f f/60 = 0.35 \times 5000 \times 60/60 = 1750$ μF (Eq. 17.4).

2.   Using these values of R_{eq} and C_{eq}, read $v_{L\,max}$ (%) and $v_{L\,min}$ (%) from Fig. 17.3, then compute the V_{peak} and the minimum and maximum load voltages.

$$v_{L\,max}(\%) \simeq 90\% \text{ (dashed line)}$$
$$v_{L\,min(\%)} \simeq 80.5\% \text{ (solid line)}$$
$$V_{peak} = (1.414 \times 35 - 0.5) = 49 \text{ V}$$
$$v_{L\,max} = 49 \times 0.90 = 44.1 \text{ V}$$
$$v_{L\,min} = 49 \times 0.805 = 39.4 \text{ V}$$

Peak-to-peak ripple voltage $= 44.1 - 39.4 = 4.7$ V

3.   The parameters R_{chg} can be obtained from Fig. 17.4.

$$R_{chg} \simeq 8.5 \text{ Ω}$$
$$I_{peak} = V_{peak}/R_s R_{chg} = 49/0.35 \times 8.5 = 16.5 \text{ A}$$

Example 17.2   The transformer and diodes of Example 17.1 are used in the circuit of Fig. 17.2a, but the frequency is increased to 400 Hz and the load resistance is increased to 12 Ω. A voltage regulator will follow the rectifier and filter. The voltage input to the regulator must not fall below 41 V. We need to determine the required filter capacitance C_f and the peak diode currents I_{peak}.

1. $R_{eq} = R_L/R_s = 12/0.35 = 34 \ \Omega$

    $v_{L \ min}(\%) = 41 \times 100/49 = 83.6\%$

    $C_{eq} \simeq 2500 \ \mu F$ (Fig. 17.3)

    $C_f = C_{eq} \times 60/R_s f = 2500 \times 60/0.35 \times 400 = 1070 \ \mu F$ (Eq. 17.4)

2. Using Fig. 17.4,

    $R_{chg} \simeq 8 \ \Omega$

    $I_{peak} = V_{L \ peak}/R_s R_{chg} = 49/0.35 \times 8 = 17.5 \ A$ (Eq. 17.6)

The nomographs can be used for the bridge rectifier but R_s includes the dynamic resistance of two diodes in series and $V_{peak} = V_1 - 2V_{diode}$.

The power supply discussed above has rather poor regulation unless $R_L C \gg T$ because the output voltage is a strong function of the rate of discharge of the filter capacitor through the load. This drop adds to the diode drop and the voltage drop in the transformer windings. A sketch of V_O as a function of I_L is given in Fig. 17.5 for two different values of capacitance in a typical rectifier circuit.

Power supply *output resistance* is a common and useful term to express the relationship between output voltage and load current. This dynamic output resistance is defined as

$$r_o = \frac{\Delta V_O}{\Delta I_L} \qquad (17.7)$$

For example, the power supply that has the output characteristics given in Fig. 17.5 has an average value of output resistance $r_o = 9 \ V/1 \ A = 9 \ \Omega$ when the filter capacitance is 1000 μF.

A full-wave *bridge* rectifier was discussed in Chapter 3 and is shown in Fig. 17.6 for convenience. Observe that this rectifier does not require a center-tapped transformer. Therefore the transformer is required only for voltage transformation and may be eliminated if this function is not required. Note that the capacitor C charges to the maximum value of the *full secondary* voltage minus the drop across two forward-biased diodes. Therefore, the bridge rectifier provides twice as much dc output voltage, for the same full secondary voltage, as the circuit of Fig. 17.2. The peak inverse voltage

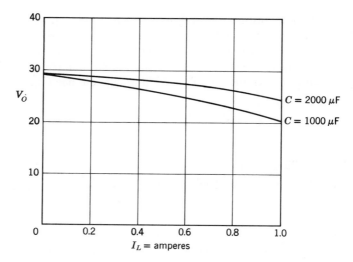

FIGURE 17.5
Output voltages as a function of output current for two different values of filter capacitance.

across the diodes is the same in both circuits, however, so the ratio of output voltage to peak inverse voltage is twice as high in the bridge circuit. The bridge rectifier has the disadvantage of requiring four diodes, and in very low voltage circuits the rectifier efficiency may be seriously reduced because two diodes are in series with the load.

PROBLEM 17.1   Determine the average value of output resistance for the power supply with characteristics shown in Fig. 17.5 with $C = 2000 \ \mu\text{F}$. *Answer:*   6 Ω.

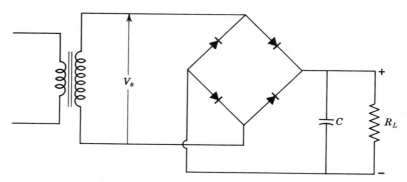

FIGURE 17.6
A full-wave bridge rectifier circuit.

PROBLEM 17.2 Design a power supply to provide 25 V and 1.5 A dc output with a 3.75 V peak-to-peak ripple voltage. Use a center-tapped transformer with 120 V, 60 Hz primary power. List component ratings.

Answer: $C = 2200 \ \mu\text{F}$, $V_s = 49$ V, peak inverse $= 60$ V.

PROBLEM 17.3 Design a power supply that will provide 40 V at 2 A dc output, using a bridge rectifier and 120 V, 60 Hz primary power. Use *pprr* = 0.1.

Answer: $C = 3300 \ \mu\text{F}$, peak inverse voltage $= 56$ V.

PROBLEM 17.4 A power supply for a vacuum tube amplifier must produce 200 V at 200 mA. We desire the peak-to-peak ripple to be no greater than 2 V. Design a power supply for this amplifier and note the ratings of each component used in this power supply.

17.3 EMITTER-FOLLOWER REGULATORS

Reference diodes (Zener diodes) were discussed in Chapter 3 and a simple circuit, that will maintain a fairly constant voltage across a load resistor, was analyzed. A circuit of this type can be used to improve the voltage regulation and reduce the ripple in a power supply. However, the load currents usually encountered require high-power dissipation capability in the reference diode and the power supply efficiency is low because of the power loss in this diode. One or more transistors can be used in conjunction with the reference diode, however, to greatly increase the efficiency of the regulator by reducing the current through the reference diode. A typical circuit of this type is shown in Fig. 17.7. This circuit is known as an *emitter-follower regulator* because the output voltage follows

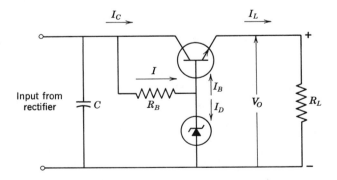

FIGURE 17.7
An emitter-follower regulator.

(is nearly equal to) the reference voltage; the difference being the base-emitter voltage V_{BE}.

The emitter-follower regulator operates in the same manner as the simple Zener regulator except that the currents through the reference diode are reduced by the factor $(\beta + 1)$ because of the transistor. To illustrate this behavior, let us assume both the input voltage and the reference-diode voltage to be constant, then the current I through R_B is constant. A reduction in load current I_L will reduce the base current I_B since $I_B = I_L/(h_{FE} + 1)$, but the reference diode current I_D will increase because the base voltage will tend to rise. Thus the sum of the currents $(I_B + I_D)$ is essentially constant. The diode voltage, and hence the base voltage, does change by the amount $\Delta V_B = \Delta I_D r_d$ where r_d is the dynamic resistance of the diode, discussed in Chapter 3. The load voltage V_O increases more than the base voltage because of the reduced V_{BE}. If the change in load current is small so that r_d and r_e are essentially constant over the current range, the change in output voltage is

$$\Delta V_O = \Delta I_L r_e + \Delta I_D r_d \tag{17.8}$$

But $\Delta I_D \simeq \Delta I_B = \Delta I_L/(h_{FE} + 1)$, so

$$\Delta V_O \simeq \Delta I_L \left(r_e + \frac{r_d}{h_{FE} + 1} \right) \tag{17.9}$$

and the output resistance of the regulator is

$$r_o = \frac{\Delta V_O}{\Delta I_L} \simeq r_e + \frac{r_d}{h_{FE} + 1} \tag{17.10}$$

Of course, if the input voltage does not remain constant, the current I through R_B changes with input voltage and this current change must be absorbed by the reference diode.

Example 17.3 An example will illustrate the design of an emitter-follower regulator. Let us assume that the required output voltage is 20 V and the maximum load current is 1.0 A. Let us also assume that the chosen transistor is silicon with $h_{FE} = 50$ and $V_{BE} = 0.6$ V at $I_E = 1.0$ A. The reference diode current is minimum when the load current is maximum and we shall choose this minimum diode current to be 5 mA. Then, the reference diode voltage should be $(V_O + V_{BE}) = 20.6$ V at $I_D = 5$ mA. We must choose the minimum voltage that can be tolerated across R_B, keeping in mind that the higher this voltage is, the higher the required input voltage and hence the lower the efficiency of the power supply. However, we shall see shortly that low values of R_B, which result from choosing a low minimum voltage, result in high diode dissipation and inferior regulator characteristics.

In this example, we will choose 2 V as the minimum voltage, $V_{R\,min}$, across R_B. Now the value of R_B can be calculated from the following relationship.

$$R_B = \frac{V_{R\,min}}{I_{D\,min} + I_{L\,max}/(h_{FE} + 1)} \qquad (17.11)$$

For this example, $R_B = 2/(5 \text{ mA} + 20 \text{ mA}) = 80 \ \Omega$.

The minimum voltage input to the regulator is $20.6 + 2 = 22.6$ V. This is the voltage $v_{O\,min}$ shown in Fig. 17.3 as $v_{L\,min}$.

The required value of filter capacitance and the voltage rating of the transformer can be determined from Fig. 17.3 after we have chosen the current rating of the transformer secondary so we can determine R_{eq}. If the load on the power supply is expected to draw near 1.0 A dc for long periods of time, then the transformer secondary must have a current rating greater than 1.0 A because the high peak currents that flow in the transformer windings cause higher power dissipation than sinusoidal currents of the same average value. Therefore, a generous safety factor (perhaps 30 percent) must be allowed for the current rating of the transformer. On the other hand, a 1.0 A rating is adequate if the load draws 1.0 A only occasionally. Audio-frequency amplifiers have this latter characteristic and we shall assume our load to be of that type. Then, from Table 17.1, $R_s \simeq 2.4 \ \Omega$, and from Eq. 17.3, $R_{eq} = 20 \ \Omega/2.4 \ \Omega \simeq 8 \ \Omega$. We can now move to the right along the $R_{eq} = 8 \ \Omega$ line on Fig. 17.3. Observe that as we move to the right, the ripple decreases and the capacitance C_{eq} increases. Let us assume that a 10 percent difference between $v_{L\,max}$ and $v_{L\,min}$ is satisfactory. This difference is obtained approximately at $C_{eq} = 5,000 \ \mu F$ where $v_{L\,min} = 67\%$ and $v_{L\,max} = 76\%$, and the actual difference is 9%. Thus, using Eq. 17.4, the actual filter capacitance $C_f = C_{eq}/R_s = 5000/2.4 \simeq 2,000 \ \mu F$. The maximum *open-circuit* voltage V_{peak} out of the rectifier is therefore $v_{L\,min}/0.67 = 22.6/0.67 = 33.7$ V, and we must add approximately 1.0 V for the drop across the two rectifiers, giving 34.7 as the peak open-circuit voltage across the transformer secondary. However, transformers are normally rated at full-load voltage, which is about 10 percent less than the open-circuit voltage, but the open-circuit voltage we have determined is at *minimum* line voltage, which is about 10 percent lower than the nominal line voltage. Therefore 34.7 V is also the peak secondary voltage at full load and nominal input voltage. The secondary rms voltage rating is, therefore, $34.7 \times 0.707 = 24.5$ V. Either a 24 V or a 25 V transformer would be suitable.

The maximum power dissipation of the reference diode must yet be determined. This dissipation will occur when the primary voltage is maximum and the load current is minimum. The maximum full load rms secondary voltage of 27 V occurs when the primary voltage is 130 V. The corresponding no-load voltage is 10 percent above this value or about 30 V. The peak voltage into the rectifiers is therefore $(2)^{1/2}(30) = 42$ V and the maximum voltage applied to the regulator is approximately $42 - 1 = 41$ V, allowing 0.5 V for each rectifier at minimum current. Let us assume that the minimum load current is zero. Then the ripple is approximately zero and the dc voltage applied to the regulator is nearly equal to the 41 V maximum value. The current through R_B is $I = (41 - 20.6)\text{V}/80\ \Omega = 255$ mA, which all flows through the reference diode since $I_B = 0$ when $I_L = 0$. Thus, the maximum power dissipation of the reference diode is approximately 20.6 V (0.255 A) = 5.25 W. A 10 W diode would probably be purchased.

Notice that the high dissipation requirement of the reference diode in the preceding circuit resulted from the low value of R_B, which in turn resulted from the low value of minimum voltage (2 V) that we allowed across R_B. The large change in reference-diode current also causes a wider voltage variation than may be desired. On the other hand, an increase of the minimum voltage across R_B requires higher input voltages and lower efficiency. Of course, R_B is larger and ΔI_D is smaller if the maximum load current is smaller.

The characteristics of the emitter-follower regulator can be improved and the reference diode dissipation can be reduced greatly in high-current regulators if a Darlington-connected amplifier is used in the circuit (Fig. 17.8). The maximum base current I_{B1} is then $I_L/(h_{FE2} + 1)(h_{FE1} + 1)$, which is usually less than 1 mA. Then a low-power reference diode with minimum currents of the order of 1 mA can be used. In this circuit, R_B can be much larger than in Example 17.3.

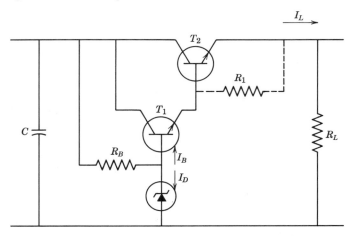

FIGURE 17.8
An emitter-follower regulator using a Darlington-connected amplifier.

Example 17.4   Let us add a silicon transistor T_1 to the circuit of Example 17.3 so it has the form given in Fig. 17.8. Assume that T_1 has $h_{FE1} = 100$ and $V_{BE} = 0.6$ V at $I_E = 20$ mA. Then $I_{B1\ max} \simeq 20/100 = 0.2$ mA. Then, if $I_{D\ min} = 1.0$ mA and we allow a 2.0 V minimum across R_B, the value of R_B is 2.0 V/1.2 mA = 1.67 kΩ. Any standard value between 1.5 kΩ and 1.8 kΩ would be suitable. The design would then proceed as before, except the additional 0.6 V must be added to 20.6 V to yield a 21.2 reference-diode voltage. The voltages all along the line back to the primary voltage must therefore be appropriately increased. The important result is that $I_{D\ max} = 19.8$ V/1.67 kΩ = 12 mA and the maximum power dissipation of the reference diode is (21.2 V)(12 mA) = 254 mW, approximately. In addition, the change of I_D is $\Delta I_D = 11$ mA. The maximum dissipation of the power transistor (T_2) must also be considered. This dissipation occurs at maximum line voltage and maximum load current. The minimum input voltage to the Darlington regulator is 22.6 + 0.6 = 23.2 V at full load and 105 V input to the transfer primary. Then, if the input to the transformer rises to 130 V, $v_{min} = 23.2 \times 130/105 = 28.8$ V. Since the ripple is about 10 percent, the average input voltage to the regulator is about 5 percent above 28.8 V or 30.3 V. The average voltage across the regulating transistor under these conditions is 30.3 − 20 = 10.3 V and the maximum power dissipation is 10.3 V × 1 A = 10.3 W. If germanium transistors are used the dashed resistor R_1 (Fig. 17.8) may be needed to reduce the thermal currents, as was discussed in Chapter 10.

The reduction of the ripple voltage by the emitter-follower regulator can be determined from the equivalent circuit of Fig. 17.9. This equivalent circuit was initially discussed in Chapter 3. The ripple voltage is the ac component in either the input or output of the regulator. The ripple

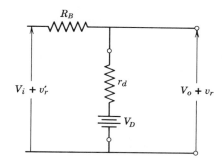

FIGURE 17.9
An equivalent circuit used to determine
the ripple voltage at the regulator output.

voltage in the output is essentially the same as the ripple voltage, v_r, across r_d. Thus,

$$v_r = \frac{v_r'r_d}{R_B + r_d} \tag{17.12}$$

where v_r' is the ripple voltage across the filter capacitor. The ripple voltage is maximum at full load and should be calculated for this condition. For example, if $r_d = 30\ \Omega$ (a typical value) at $I_D = 1$ mA and the peak-to-peak ripple voltage across the capacitor is about 3.5 V, as considered in the preceding examples, the peak-to-peak ripple voltage in the output of the Darlington-connected circuit example is $v_r = 3.5 \times 30/1.7$ k$\Omega = 0.06$ V.

Sometimes the good regulation of a regulated supply is not needed but low ripple voltage is required. Then the reference diode in the circuit of Fig. 17.8 may be replaced by a capacitor as shown in Fig. 17.10.

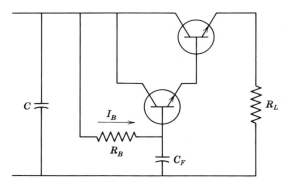

FIGURE 17.10
An active filter for reducing ripple.

This circuit can be designed in the same manner as the emitter-follower regulator except there is no reference diode current and jX_C replaces r_d in determining the output ripple from Eq. 17.12.

Example 17.5   Let us assume that the active filter circuit of Fig. 17.10 will be used to replace the Darlington-connected emitter-follower regulator of Fig. 17.8, the power supply requirements and transistors being the same as in Example 17.3. Thus, $V_{O\ min} = 20$ V, $I_{L\ max} = 1$ A, $h_{FE2} = 50$, $h_{FE1} = 100$, $C = 2000$ μF. Now $I_{B\ max}$ is $\simeq 1$ A/(50)(100) or 0.2 mA. If we allow a 2 V maximum drop across R_B, the value of R_B is 2 V/0.2 mA $= 10$ kΩ. The value of v_r can be calculated from Eq. 17.10 for a given value of C_F. However, a preferable procedure is to specify the desired value of v_r and calculate the required value of C_F. If jX_C is used to replace r_d in Eq. 17.12, this equation becomes $v_r = jv_r'X_C/(R_B + jX_C)$. However, if $R_B \gg X_C$, the denominator of this relationship is $\simeq R_B$ and the magnitude of v_r is

$$|v_r| = \frac{v_r'X_{CF}}{R_B} \tag{17.13}$$

Solving for X_{CF},

$$X_{CF} = \frac{1}{2\omega C_F} = \frac{|v_r|R_B}{v_r'} \tag{17.14}$$

Solving for C_F,

$$C_F = \frac{v_r'}{2\omega R_B|v_r|} \tag{17.15}$$

where ω is the angular frequency $(2\pi f)$ of the primary power frequency. The factor 2 appears because of the full-wave rectification. In this example, let us specify the peak-to-peak ripple voltage in the output to be 0.01 V. Then with $v_r' = 3.5$ V and $f = 60$ Hz, $C_F = 3.5/(4\pi \times 60 \times 0.01 \times 10^4) = 47\ \mu\text{F}$.

The filter capacitor C_F can also be used in parallel with a reference diode to further reduce the ripple voltage in the output of an emitter-follower regulator.

PROBLEM 17.5 Design an emitter-follower regulator that will provide 25 V at 2 A maximum to a load. Use $h_{FE2} = 50$ and $h_{FE1} = 100$, and use the circuit of Fig. 17.8. Allow for a primary voltage range of 105 V to 130 V (60 Hz). Specify component ratings and determine the ripple voltage in the output, assuming $r_d = 25\ \Omega$ and $I_{D\ min} = 1$ mA. Use $pprr = 0.15$ approximately.

Answer: For $V_{R\ min} = 2$ V, $R_B = 1.4$ kΩ, Zener = 26.2 V at 1 mA, 460 mW actual max diss; $C = 3000\ \mu\text{F}$, Trans. sec = 32.9 V.

PROBLEM 17.6 Use an active filter instead of the regulator in the power supply of Problem 17.4 and determine the value of capacitance C_F that will provide a 5 mV peak-to-peak maximum ripple in the output.
Answer: $C_F = 220\ \mu\text{F}$.

17.4 CLOSED-LOOP REGULATORS

Although the emitter-follower regulators provide satisfactory performance for many applications, their output resistance cannot be reduced below the value given by Eq. 17.10. Also, large values of C_F (Fig. 17.10) are required to provide very low values of ripple. On the other hand, regulators that employ the principle of negative feedback can provide almost any desired value of output resistance and ripple quite easily. The basic philosophy of the closed-loop regulator is illustrated by the block diagram of Fig. 17.11. A fraction of the output voltage ηV_O is compared with a reference voltage V_{REF}, and their difference is amplified and used to control the series regulator, which in turn controls the output voltage. A typical

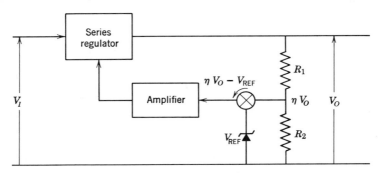

FIGURE 17.11
The block diagram of a closed-loop regulator.

circuit diagram that will perform this basic function is shown in Fig. 17.12. In low-current regulators, the series regulator may be a single transistor, but a Darlington-connected amplifier is usually used as shown in Fig. 17.12 and as previously used in the emitter-follower regulator (Fig. 17.8). The differential amplifier consisting of transistors T_1 and T_2 provides both the voltage comparison and the amplification functions. The resistors R_1 and R_2 are chosen so that their ratio provides the desired ratio between V_{REF} and V_O, while their sum provides a bleeder current through the resistors that is large compared to the base current of transistor T_1. If these conditions are met,

$$V_{REF} \simeq \frac{R_2}{R_1 + R_2} V_O \tag{17.16}$$

and the current I_R through R_1 and R_2 is

$$I_R = \frac{V_O}{R_1 + R_2} \tag{17.17}$$

These equations can be solved simultaneously to find R_1 and R_2. By substituting Eq. 17.17 into Eq. 17.16

$$V_{REF} = R_2 I_R \tag{17.18}$$

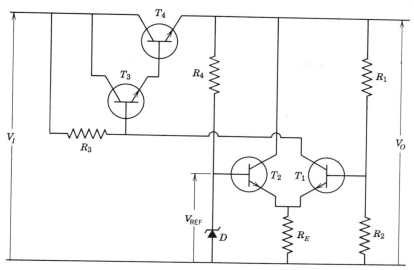

FIGURE 17.12
A typical closed-loop regulator circuit.

and

$$R_2 = \frac{V_{REF}}{I_R} \tag{17.19}$$

Substituting this expression for R_2 into Eq. 17.17 and solving for R_1

$$R_1 = \frac{V_O}{I_R} - R_2 \tag{17.20}$$

The resistor R_4 is chosen so that the current through the reference diode D is large in comparison with the base current of transistor T_2. Then, the current through D is essentially constant and hence the reference voltage V_{REF} is very constant. The diode D should have essentially zero temperature coefficient if V_{REF} and hence V_O are to be independent of temperature.

The operating principles of the closed-loop regulator will be illustrated by assuming that the output voltage V_O becomes more positive because of either a reduction in load current or an increase of input voltage or both. Then, the base of transistor T_1 becomes more positive than V_{REF} and the current through T_1 increases. The increased current through T_1 causes the drop across resistor R_3 to increase. Then, the forward bias of the series regulator transistors T_3 and T_4 is also decreased. This decreased forward bias increases V_{CE} and therefore reduces the output voltage, or in other words, tends to cancel the assumed rise in output voltage. Note that a polarity reversal must be provided in the amplifier so that negative feedback is obtained by the closed loop.

The effectiveness of the amplifier and feedback system in improving the characteristics of the power supply will now be investigated. A semiblock-diagram similar to the one in Fig. 17.11 is drawn in Fig. 17.13, as an aid in the investigation. Observe that the output voltage v_{CE} of the series

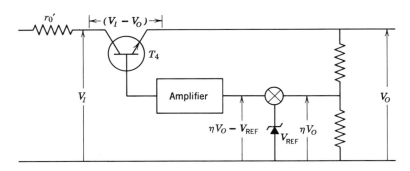

FIGURE 17.13
A block diagram of a closed-loop regulator.

regulator transistor T_4 is the difference between the output voltage V_O and the input voltage V_I or $(V_I - V_O)$. Let K_v be the voltage amplification between the input of the amplifier and the output (V_{CE}) of the series regulator (Transistor T_4). Then

$$V_I - V_O = (\eta V_O - V_{REF})K_v \qquad (17.21)$$

But we are primarily interested in the ratio of the change in output voltage to the change in the input voltage in order to determine the reduction in ripple or input voltage variations. Therefore, we write Eq. 17.21 in terms of voltage variations instead of total voltages.

$$\Delta(V_I - V_O) = \Delta(\eta V_O - V_{REF})K_v \qquad (17.22)$$

But $\Delta V_{REF} = 0$ if the reference voltage is constant. Then

$$\Delta V_I - \Delta V_O = \Delta \eta V_O K_v \qquad (17.23)$$

Solving for ΔV_O, we have

$$\Delta V_O(1 + \eta K_v) = \Delta V_I \qquad (17.24)$$

or

$$\Delta V_O = \frac{\Delta V_I}{1 + \eta K_v} \qquad (17.25)$$

Observe from Eq. 17.25 that variations in the output voltage are reduced by the factor $(1 + \eta K_v)$ compared with variations of V_I that are caused by ripple, primary voltage variations, and so forth. Note Eq. 17.25 is a standard feedback equation with β replaced by η.

If both sides of Eq. 17.25 are divided by the change in load current ΔI_L, the effectiveness of the regulator in reducing the power supply output resistance may be obtained. Thus

$$\frac{\Delta V_O}{\Delta I_L} = \frac{\Delta V_I / \Delta I_L}{1 + \eta K_v} \qquad (17.26)$$

But $\Delta V_O / \Delta I_L$ is the output resistance of the power supply after regulation and $\Delta V_I / \Delta I_L$ is the output resistance r_o' of the filter before regulation, where we should think of the change in voltage ΔV_I as being caused by the change of current ΔI_L. Then

$$r_{of} = \frac{r_o'}{1 + \eta K_v} \qquad (17.27)$$

where r_{of} is the output resistance of the regulated supply. Observe that, as in other feedback circuits, the output resistance is reduced by the factor $(1 + \eta K_v)$. As noted in Chapter 11, low-power supply impedance is helpful in maintaining stability in a multistage amplifier.

Example 17.6 Let us consider the design of the type regulator shown in Fig. 17.12 with the following data and specifications given: $V_O = 20$ V, $I_{L\,max} = 1.0$ A, $h_{FE4} = 80$, $h_{FE3} = h_{FE2} = h_{FE1} = 100$, $pprr = 0.15$. Notice in Example 17.3 that $0.10\ V_{peak}$ is about equal to $0.15\ V_{AVE}$ since $v_{AVE} = 0.7\ V_{peak}$. All transistors are silicon. R_3 (which was R_B in the preceding examples) can be determined as before. At maximum load $I_{B3} = 1$ A/$(80 \times 100) = 0.12$ mA. If we assume that $V_{I\,min}$ occurs at $I_{L\,max}$, this is the worst case. Under these conditions, the minimum collector current flows in T_1. We shall let this minimum current be 0.2 mA. Then, if we allow 2.0 V minimum across R_3, $R_3 = 2/0.32$ mA $= 6$ kΩ. The rectifier circuit and filter capacitor could now be determined as in Example 17.3. We shall select $V_{REF} = 6.8$ V since diodes in this voltage area have minimum dynamic resistance and minimum thermal coefficient. It is preferable to calculate the remaining resistance values in the circuit using nominal input voltage, rather than minimum. This minimum V_I is approximately $20 + (2 \times 0.6) + 2 = 23.2$ V. The peak-to-peak ripple at full load is about

$24 \times 0.15 = 3.6$ V so that the average V_I at 105 primary volts (assuming 117 V, 60 Hz nominal input power) is $23.2 + 3.6/2 = 25$ V. (We guessed 24 V in calculating this average above.) Then the *nominal* average $V_I = 25 \times 117/105 = 27.8$ V. Under this condition, the average current through $R_3 = (27.8 - 21.2)/6$ k$\Omega = 1.1$ mA. Since $I_{B3 \text{ max}} = 0.12$ mA, the nominal collector current in transistor T_1 is $1.1 - 0.12 \simeq 1.0$ mA. Thus $I_{B1} = 1.0$ mA$/100 = 10$ μA. Since the current through the voltage divider circuit should be very large in comparison with I_{B1}, we choose this bleeder current $I_R = 10$ mA, which is only 0.1 percent of the full load current. Then, by using Eqs. 17.23 and 17.24, $R_2 = 6.8$ V$/10$ mA $= 680$ Ω, $R_1 = 2$k$\Omega - 680 = 1.32$ kΩ. An adjustable resistance that will span the 1.32 kΩ value is usually used for R_2 so that the ouput voltage can be adjusted.

At nominal conditions the base current $I_{B2} = I_{B1} = 10$ μA. The current through R_4 should be large compared with this value so that the current through D varies only a few percent. We choose this current to be approximately 2 mA. Then $R_4 = (20 - 6.8)$ V$/2$ mA $= 6.6$ kΩ. A standard 6.8 kΩ resistor will suffice.

The amplifier voltage gain K_v must be calculated to determine the characteristics of the regulated supply. The easiest method of calculating voltage gain is probably using the relationship $K_v = K_i R_L/R_{\text{in}}$. The following approximations will be used: $K_i = h_{FE}$, $R_{\text{in}} = 2h_{ie}$ for the differential amplifier, $R_{\text{in}} = (h_{ie3} + h_{FE3} h_{ie4})$ for the Darlington amplifier, and $h_{ie} \simeq \beta_o/g_m$. Then $h_{ie} \simeq 100/0.04 = 2.5$ kΩ and $R_{\text{in}} \simeq 5.0$ kΩ for the differential amplifier. In addition, $h_{ie} \simeq 80/40 = 2.0$ Ω for transistor T_4 at $I_L = 1$ A and $h_{ie} = 100/0.48 = 220$ Ω for transistor T_3. Then, $R_{\text{in}} = 220 + 200 = 420$ Ω for the Darlington connection. This resistance, in parallel with R_3, is the load resistance for the differential amplifier. This fact points up the need for the Darlington amplifier to provide a suitable (400 Ω) load resistance for the differential amplifier. The load resistance for the series regulator is the output resistance r_o' as seen in Fig. 17.13. Then, for the differential amplifier, $K_v \simeq 100(400)/5$ k$\Omega = 8$. The value of r_o' can be accurately obtained from a plot of V_I as a function if I_L, as given in Fig. 17.5. The value of r_o' can also be obtained by calculating the increase of the average value of V_I as I_L decreases from its full load value to essentially zero and then taking $\Delta V_I/\Delta I_L$. This voltage rise is approximately the peak ripple voltage plus about a 2 V reduced drop in the rectifier diodes and the transformer windings. The total voltage rise in this example is, therefore, about $(0.15 \times 25/2) + 2 \simeq 4$ V and since ΔI_L is 1 A, $r_o' \simeq 4$ Ω. Then the voltage gain of the series regulator is $K_i r_o'/R_{\text{in}} = 100 \times 80 \times 4/420 = 76$. Thus the total gain $K_v = 8 \times 76 = 608$ and $\eta = V_{\text{REF}}/V_O = 6.8/20 = 0.34$, so $\eta K_v = 207$ and $(1 + \eta K_v) = 208$. Therefore, the peak-to-peak ripple voltage is $3.7/208 = 0.018$ V and $r_{of} = 4/208 = 0.02$ Ω for the regulated power supply.

Integrated-circuit operational amplifiers can be conveniently used to provide excellent characteristics in a regulated power supply. These amplifiers usually provide open-loop gain values of several thousand. The operational amplifier can replace the differential amplifier directly in the circuit of Fig. 17.12 providing the output voltage V_O and V_{REF} do not exceed the permissible output voltage range of the operational amplifier. Many different regulator configurations can be devised, but the circuit of Fig. 17.14 is one

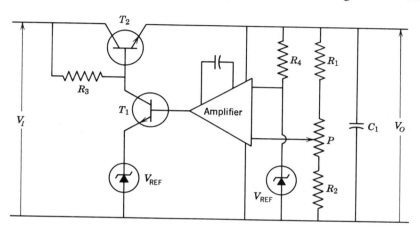

FIGURE 17.14
A regulator that incorporates an operational amplifier.

arrangement that allows the regulated output voltage to be used as the power source for the operational amplifier. The reference voltage V_{REF} should be near $V_O/2$. The transistor T_1 is used as a common-emitter driver, instead of being Darlington connected, to provide a base potential that is the desired output potential for the operational amplifier. The voltage gain of T_1 is low because its collector load resistance is R_3 in parallel with h_{ie2}, both of which are small. The potentiometer P is included to provide adjustable output voltage. The capacitor C_1 maintains low output impedance at high frequencies where the gain of the regulating amplifier is low and therefore cannot provide very low output impedance. The capacitor C_2 and the load resistance R_L may provide the low-frequency pole required to compensate the amplifier. Also, feed back may be required around the op amp to reduce its gain and broaden its bandwidth to provide stability.

PROBLEM 17.7 Design a closed-loop regulator similar to the circuit of Fig. 17.12 that will deliver a 0.5 A maximum at 25 V to a load. Assume that the rectifier and filter provide voltage V_I with 0.1 *pprr*. All transistors have $h_{FE} \simeq \beta = 100$. Using $V_{REF} = 7$ V, calculate suitable component values and estimate the peak-to-peak ripple voltage and the output resistance for the regulated supply.

Typical answer: $R_1 = 7$ kΩ, $R_2 = 18$ kΩ, $R_3 = 3.9$ kΩ, $R_4 = 18$ kΩ, $pprr = 3.25/470 = 7$ mV, $r_{of} = 10.5/470 = 0.022$ Ω.

PROBLEM 17.8 The circuit of Fig. 17.14 is used as a voltage regulator to provide 1.0 A at 15 V to the load. Transistor T_2 has $h_{FE} = \beta = 80$ and transistor T_1 has $\beta = 100$. The operational amplifier has open-loop voltage gain $= 4000$ for loads of 100 Ω or greater and has open-loop poles at 10^5 Hz, 3×10^5 Hz and 10^6 Hz. If $V_{REF} = 7.0$ V and $pprr = 0.15$, determine suitable values for all the resistors and the potentiometer resistance if the output voltage is to be adjustable from 14 V to 16 V. Determine the peak-to-peak ripple in the output and the output resistance r_{of} if $r_o' = 4$ Ω, and $r_{d\,ref} = 10$ Ω. Use bleeder current $= 5$ mA at $V_O = 16$ V. Use feedback and compensation on the op. amp, if necessary, for stability.

17.5 CURRENT LIMITERS

One problem with transistor power supplies is that the ordinary fuse acts too slowly to protect the semiconductors in either the power supply or the load. This problem is usually solved by some technique that limits the maximum output current from the power supply to a value that will protect the semiconductors in either the power supply or the load, or both. Numerous current-limiting circuits have been devised but only three representative circuits will be discussed here. We shall first consider the circuit shown in Fig. 17.15. This is the regulator circuit of Fig. 17.14 with the resistor R_E and the diodes D_1 and D_2 added to perform the current-limiting function. The load current flows through R_E and tends to forward bias the silicon diodes D_1 and D_2. However, this forward bias is insufficient to cause appreciable current flow through the diodes until the limiting value of current is reached. Then, the diodes current rises sharply with a small increase of emitter current through R_E. Since the base current is $I_B = I - I_D$, the increasing diode current tends to hold the base current constant, which in turn maintains the emitter and hence the load current almost constant. The regulator tries desperately to hold the output voltage V_O constant, but succeeds only in cutting off the collector current in transistor T_1, and the voltage V_O decreases sharply toward zero.

A more quantitative view of the current limiter can be gained from Fig. 17.16, which is used in conjunction with Example 17.7.

Example 17.7 The load current I_L in the circuit of Fig. 17.15 is to be limited to approximately 1 A. Determine the value of R_E if $h_{FE2} = 100$, and determine the approximate short-circuit current of the power supply.

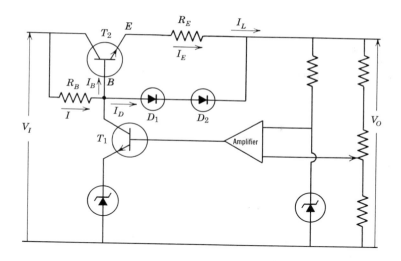

FIGURE 17.15
A current-limiting circuit.

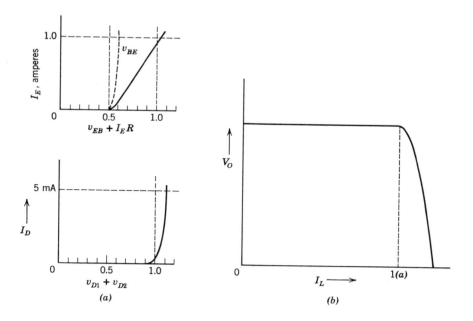

FIGURE 17.16
Current and voltage relationships in the current-limiting circuit.

The characteristics of the two diodes in series and v_{BE} versus I_E are given in Fig. 17.16a, where $v_{BE} = 0.6$ V at $I_E = 1$ A. Also, the value of collector current of transistor T_1 was assumed to be 5 mA at full load current. Therefore current limiting will begin and the output voltage will begin to drop sharply when the diode current $I_D = 5$ mA. Under these conditions, the value of I_{C1} is zero. At $I_D = 5$ mA, the voltage across the two diodes in series is (from the characteristics of Fig. 17.16) 1.1 V. Therefore, $v_{RE} = (1.1 - 0.6)$ V $= 0.5$ V and $R_E = 0.5$ V/1 A $= 0.5\ \Omega$, with power dissipation $= I^2 R_E = 0.5$ W. Let us assume that $v_O = 20$ V and $V_I = 26.1$ V at $I_L = 1$ A. Then $IR_B = 26.1 - 21.1 = 5$ V and $R_B = 5$ V/(10 mA + 5 mA) = 330 Ω. When the power supply is shorted, $v_O = 0$ and the voltage across $R_B = 25$ V, approximately, then $I = 25/330\ \Omega = 75$ mA. Of this current, approximately 65 mA flows through the diodes. We assume that this current raises the diode voltage drop about 0.1 V or 10 percent, and hence raises the voltage across R_E about 0.1 V or 20 percent, which results from a 20 percent or 200 mA increase of current through R_E. Thus the load current will increase to 1A + 200 mA + 65 mA = 1.265 A when the output is shorted.

A current limiter that uses a single transistor instead of two diodes is shown in Fig. 17.17. The principle of operation of this circuit is the same as the diode circuit of Fig. 17.15. The transistor T_3 draws essentially no current until the load current that flows through R produces around a 0.5 V drop. Then the collector current I_{C3} rises sharply with load current and both I_B and I_L are held essentially constant at the predetermined value of I_L. This circuit has a somewhat sharper cutoff than the diode-limiting circuit

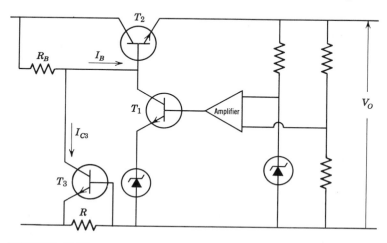

FIGURE 17.17
A transistor current limiter.

of Fig. 17.15 because I_{C3} does not flow through the load as the diode current did. In either circuit, the power dissipation in both T_2 and R_B is much higher than normal while the output is shorted, since nearly all the input voltage appears across these elements when the output is shorted. Therefore, the heat sink for T_2 and the dissipation rating of R_B must be adequate to prevent the destruction of these components. In either of the circuits above, the resistor R adds directly to the unregulated output resistance r_o' and is effectively reduced by the regulator to $R/(1 + \eta K_v)$, which produces a negligible increase in r_o.

The high power dissipation problem does not occur in the current-limiting circuit of Fig. 17.18 because the load current is reduced to a small fraction of its maximum value as soon as the maximum value is reached. This circuit is the current limiter *only* and does not include a regulator, which could follow the limiter. In this circuit I_L flows through R_1 and produces a small voltage drop (about 2 V). Resistor R_2 is less than $\beta_1 R_1$, so that transistor T_1 is normally in saturation with very small v_{CE1}. Transistor T_2 will not conduct until $v_{CE1} + I_L R_3$ rises to approximately 0.5 V (for silicon transistors). Then transistor T_2 begins to conduct and I_{C2} robs the base current of transistor T_1 and pulls it out of saturation. This action increases v_{CE1} and thus increases the forward bias of T_2. This process is cumulative, or regenerative, so that transistor T_1 is very quickly cut off and the only current that flows into the load must flow through R_2 and R_4. But these resistors normally carry only base currents and, therefore, allow only currents that are much smaller than the full load current. Transistor T_2 becomes saturated and, therefore, does not have high dissipation. However, it must safely conduct the current v_I/R_2. The circuit will not recover, even if the short is removed, until the input

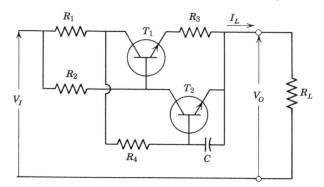

FIGURE 17.18
A current limiter that reduces load current to a small value when the current limit is exceeded.

voltage is removed, which resets the circuit. The capacitor C is included to slow the circuit response so that the circuit will not be tripped by capacitive loads.

PROBLEM 17.9   Devise a circuit that will permit the selection of 50 mA, 250 mA, or 1.0 A as the current-limiting values for the circuit of either Fig. 17.15 or 17.17. Specify resistance values.

Answer: Switch, $R = 10\ \Omega$, $R = 2\ \Omega$, $R = 0.5\ \Omega$.

PROBLEM 17.10   The transistors in Fig. 17.18 are silicon. Determine suitable values for the resistors in Fig. 17.18 if $h_{FE1} = h_{FE2} = 100$, and $I_{L\ max} = 1$ A. Determine $I_{C2\ max}$ if $V_I = 25$ V and the circuit is tripped. Assume $v_{CE\ sat} = 0.1$ V and $V_{BE} = 0.6$ V. For conduction, $V_{R2} = 2.0$ V at $I_L = 1$ A.

Answer: $R_1 = 2.5\ \Omega$, $R_2 < 200\ \Omega$, $R_3 = 0.5\ \Omega$, $R_4 < \beta R_2$, perhaps 4.7 kΩ, $I_{C2\ max} = 125$ mA.

17.6   SILICON-CONTROLLED RECTIFIERS AND TRIACS

Voltage or current regulation may also be accomplished through the use of special semiconductor devices known as thyrodes, silicon-controlled switches (SCSs), silicon-controlled rectifiers (SCRs), or triacs. We shall first consider how these devices operate and then study several circuits that employ them.

The thyrode, or SCR, is constructed as shown in Fig. 17.19a. The action of this device can be explained by the equivalent circuit, which is shown in Fig. 17.19b. This equivalent circuit shows an n-p-n transistor and a p-n-p transistor interconnected. (In fact, the SCR action can be produced by connecting two transistors as shown in Fig. 17.19b.) If the gate is negative (or near zero volts), the n-p-n transistor will be in the cutoff condition with essentially no collector current. The collector current of the n-p-n transistor furnishes the base current for the p-n-p transistor and vice versa. Thus, if one transistor is cut off, the other transistor is also cut off, and the impedance from anode to cathode is very high.

If the bias on the gate is made positive until collector current flows in the n-p-n transistor, this collector current becomes the base current for the p-n-p transistor, which begins to conduct. The collector current of each transistor becomes the base current for the other transistor. A cumulative action is therefore initiated since an increase of current in one unit causes an increase of current in the other unit. This cumulative action culminates when both transistors are driven into saturation. When both transistors are

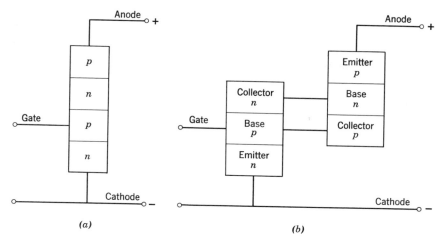

FIGURE 17.19
The silicon-controlled rectifier: (*a*) actual construction; (*b*) equivalent electrical circuit.

saturated, the impedance from anode to cathode is very low. Thus, the gate is said to "trigger" the SCR "ON."

As soon as the self-regeneration action commences, the gate loses control over the action. Generally, the collector current from the *p-n-p* unit is much larger than the external gate current. As a result, the external gate circuit can turn the SCR ON but has difficulty turning the switch OFF. (The action is much the same as the action in the gas triode or thyratron.) To turn the SCR OFF, the gate bias must be in the reverse direction and the anode voltage must be reduced essentially to zero. In this condition, no current flows through the *p-n-p* unit, and the gate regains control of the circuit.

If the gate is maintained at cutoff (negative potential) and the voltage on the anode is varied, the characteristics are as indicated in Fig. 17.20. As the negative anode voltage is increased, avalanche breakdown of the SCR occurs. On the other hand, as the anode voltage is made more positive, the center junction is reverse biased while the other two junctions are forward biased. As the center junction approaches avalanche breakdown, the avalanche current across this junction has the same polarity as a positive gate current. Thus, as avalanche breakdown approaches the SCR turns itself ON. This anode "turn-on" potential (with reverse-biased gate) is known as the *breakover voltage* of the SCR.

In normal operation, the SCR is operated with an anode potential below the breakover voltage. Then, the device is turned ON at the appropriate

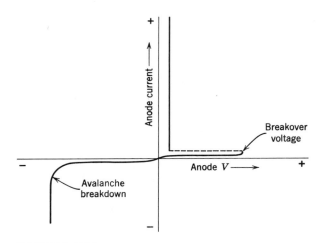

FIGURE 17.20
The current-voltage characteristics of the SCR with gate reverse biased.

time by the gate. However, remember the device can be turned ON by high anode potentials. In addition, if the anode potential *changes* at a sufficiently high rate, current from the junction capacitances in the SCR may be large enough to supply the gate with sufficient current to turn the SCR ON. The symbol for the SCR is shown in Fig. 17.21a.

Another *p-n-p-n* device is known as a *Shockley diode*. The construction of this device is similar to an SCR without the gate lead. The Shockley diode can

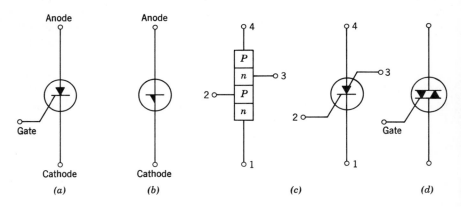

FIGURE 17.21
Symbolic representation of *p-n-p-n* devices: (*a*) SCR; (*b*) Shockley diode; (*c*) the SCS; (*d*) the triac.

be turned ON by a high potential (which exceeds the breakover voltage) and turned OFF by a zero or reverse potential. The symbol for a Shockley diode is shown in Fig. 17.21*b*.

A more versatile *p-n-p-n* arrangement has a lead brought out from each area. A device with this configuration is known as a *silicon-controlled switch* (SCS). The configuration of an SCS and its symbolic representation is shown in Fig. 17.21*c*.

Two SCRs may be mounted in the same case for full-wave applications. These devices are known as *triacs* and have the anode of each device connected internally to the cathode of the other device. A single gate lead (connected internally to the two gates) is provided. These devices can be switched from a blocking to a conducting state for either polarity of applied anode potential with positive or negative gate triggering. The symbol for a triac is given in Fig. 17.21*d*.

17.7 APPLICATIONS OF SCRs

The SCR (and its numerous variations) lends itself well to the control of power to various devices. For example, a simple form of overvoltage protection can be achieved by the circuit shown in Fig. 17.22. Under normal operation, the voltage across the SCR is below the anode breakover voltage. In addition, the voltage is also below the breakdown voltage of the Zener diode so that the Zener is nonconducting. The avalanche voltage of the Zener diode is chosen so that a surge in the supply voltage will cause the Zener to conduct. The current flowing through the Zener diode produces a positive potential on the gate of the SCR causing it to fire. Since the value of resistance in R_3 is very low, the SCR draws a heavy current from the

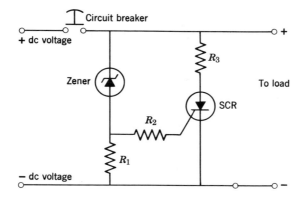

FIGURE 17.22
A "crowbar" type of overvoltage circuit protection.

source and opens the circuit breaker, thus protecting the load from the voltage surge. This action is similar to shorting out the supply with a crowbar whenever an overvoltage ocurs. Thus the name *crowbar* type of protection. To understand the design of the circuit components, let us consider an example.

Example 17.8   A circuit is to be constructed as shown in Fig. 17.22. The load draws 800 mA at 24 V. We shall assume that the power supply is capable of furnishing this amount of power and has a large capacitor in the output. A circuit breaker with a current rating of 1 A (slightly above the 800 mA required by the load) will be used.

An SCR is selected that will have a breakover voltage above 24 V and a current rating high enough to throw the circuit breaker. The 2N3559 will be used since it has a forward blocking voltage of 30 V and a continuous forward anode current rating of 1.6 A. The surge current rating is 18 A! Let us limit this surge current to 12 A. Then, R_3 would have a value of V/12A or 2 Ω. (Often the wiring will provide enough resistance to safely limit the surge current.) The 12 A will quickly open the 1 A circuit breaker.

The Zener diode must have 24 V for its avalanche value. Let us use a 1N4749 A. This silicon zener has an avalanche voltage of 24 V $\pm$ 5% and a power dissipation of 1 W. (If this breakdown tolerance is too great, a diode can be chosen from several diodes through the use of the curve tracer to give a more precise value of avalanche voltage.) This type of diode has a maximum reverse leakage current of 5 μA maximum. The gate of the 2N3559 will not trigger at a voltage less than 0.3 V for temperatures to 75°C. Therefore, let us choose R_1 to have a value of 0.3 V/$(5 \times 10^{-6}) = 60$ kΩ. At 25°C (room temperature) the different units of 2N3559 will fire for values of gate voltage between 0.35 and 0.8 V. Thus, if the Zener has an avalanche voltage of 24.0 V, the SCR will fire when the power supply voltage is increased above 28 V. (Some units will fire when the voltage is as low as 24.35 V.)

The resistor R_2 is included to limit the current flow to the gate of the SCR. The gate can stand a surge of 250 mA for up to 8 ms. However, the zener diode can only stand a surge of 190 mA. If the maximum expected voltage surge is 30 V, the current through R_1 will be $(30 - 24)$ V/60,000 Ω = 0.1 mA. Since this current is negligible compared to 190 mA, the value of R_2 is $(30 - 24)$ V/0.19 A = 31.6 Ω. Use a 33 Ω resistor. (The data for the 2N3559 came from the Texas Instrument Handbook and the data for the 1N4749 A came from the Motorola Handbook.)

PROBLEM 17.11   Design a crowbar circuit like that shown in Fig. 17.22 that will protect a load of 700 mA at a voltage of 180 V. The following devices have the ratings given:

2N3562: silicon-controlled rectifier
forward blocking voltage = 200 V
continuous forward anode current = 1.6 A
surge current = 18 A
minimum gate voltage to trigger = 0.3 V at 75°C
gate trigger voltage at 25°C = 0.35 to 0.8 V
maximum gate surge current = 250 mA for 8 ms
1M180ZS5: Zener diode
Zener voltage = 180 V
maximum reverse leakage current = 5 μA
maximum surge current $\simeq$ 25 mA

In the foregoing example, the circuit breaker in Fig. 17.22 removes all potential from the SCR. Thus, when the circuit is again activated, the gate maintains control of the SCR. In some applications, the power supply may have a current-limiting feature. (This current limiter may be as simple as a series resistor.) In circuits of this type, the circuit breaker can be eliminated. The circuit then assumes the form shown in Fig. 17.23. In this circuit, when an overvoltage occurs, the SCR fires and effectively shorts out the output circuit. The SCR continues to conduct as long as the power is left on. Naturally, an SCR is chosen with a current rating high enough to handle the short-circuit from the power supply.

In order to reapply power to the load, the reset switch in Fig. 17.23 is closed. This reset switch shorts out the SCR element and current ceases to fiow in the SCR. With no current in the SCR, the gate regains control and the SCR is turned OFF. When the reset switch (which is usually a simple

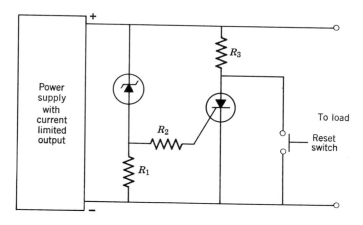

FIGURE 17.23
A modified crowbar circuit.

push-button switch) is released, power is again delivered to the load. As was mentioned previously, if the anode potential rises too fast, charge stored at the junction capacitances may retrigger the gate. To eliminate this effect, a capacitor may be connected across the SCR so that the anode potential will increase at a slower rate. Fortunately, in most circuits the wiring and inter-electrode capacitances are sufficient to prevent transient triggering of the gate.

As the name implies, SCR devices can often be used as controlled rectifiers. For example, the SCR is widely used in a light-dimmer control or for the speed control of simple series motors (such as electric saws and drills). A simple control of this type is shown in Fig. 17.24. To visualize the effect

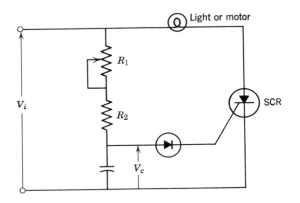

FIGURE 17.24
A light dimmer or motor speed control.

of this circuit, an analysis of the RC circuit must be considered. If the current to the gate of the SCR is assumed to be quite small, the current through the capacitor is

$$I_c = \frac{V_i}{(R_1 + R_2) + (1/j\omega C)} \tag{17.28}$$

where V_i is the input voltage. The voltage across the capacitor is $I_c Z_c$ or

$$V_c = \frac{V_i}{(R_1 + R_2) + (1/j\omega C)} \cdot \frac{1}{j\omega C} = \frac{V_i}{j\omega C(R_1 + R_2) + 1} \tag{17.29}$$

Note that if $j\omega C(R_1 + R_2)$ is much smaller than 1, V_c is essentially equal to V_i. In contrast, if $j\omega C(R_1 + R_2)$ is much larger than 1, V_c becomes approximately equal to $V_i/[j\omega C(R_1 + R_2)]$. This later value of V_c will lag V_i by approximately 90° and will have a magnitude much less than V_i. Thus, for a given value of ωC, the *magnitude* and the *phase angle* of the gate voltage (V_c in this analysis) can both be adjusted by changing $R_1 + R_2$ (or really just R_1) in Fig. 17.24.

By adjusting the magnitude and the phase angle of the gate voltage, the time when the SCR fires can be controlled. This effect is illustrated in Fig. 17.25. In Fig. 17.25a, the value of $\omega C(R_1 + R_2)$ is much smaller than 1, so the gate voltage has almost the same magnitude and the same phase as the

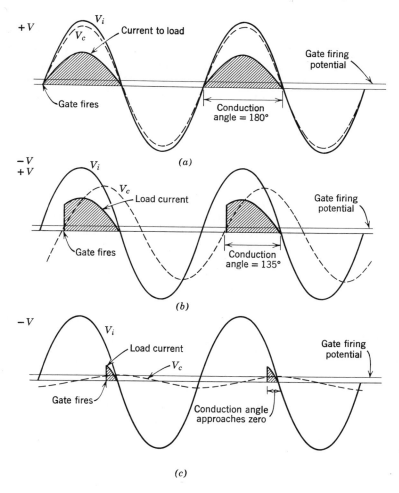

FIGURE 17.25

The effect of adjusting R_1 in Fig. 17.24 on the voltage and current wave-forms: (a) $\omega C(R_1 + R_2) \ll 1$; (b) $\omega C(R_1 + R_2) \simeq 1$; (c) $\omega C(R_1 + R_2) \gg 1$.

applied voltage. Since the gate potential exceeds the firing potential very early in the cycle, current flows as if the SCR were a conventional diode. At the end of the positive half cycle, the anode voltage and current are reduced to zero so that the gate regains control of the SCR. The voltages then reverse on the anode and gate. (The diode in Fig. 17.24 protects the gate against the large reverse-bias voltage.) With reversed potentials, almost no current flows through the SCR. As the gate and anode voltage becomes positive, the cycle repeats.

In Fig. 17.25b, the value of $\omega C(R_1 + R_2)$ is approximately equal to 1. Then, from Eq. 17.29, the gate voltage lags the applied voltage by 45°. In this case, the gate will not permit the SCR to fire until the voltage across the load and the SCR has progressed 45° through the positive half-cycle. Thus, the current waveform has the shape shown in Fig. 17.25b. Notice that the average value of this current waveform is less than the average value of the current waveform shown in Fig. 17.25a. Since the power delivered to the load is I^2R, the power to the load is reduced as the conduction angle (the period when the SCR is conducting) is reduced.

In Fig. 17.25c, the value of $\omega C(R_1 + R_2)$ is much greater than 1. The gate voltage now lags the applied voltage by almost 90°. In addition, the amplitude of the gate potential is reduced to such a small magnitude that the gate will not fire until the gate voltage is almost at its peak value. Thus, the gate fires when the applied voltage to the load and SCR has almost reached the end of its positive half cycle. A short pulse of current flows and then ceases as the applied voltage to the load and SCR drops to zero.

The design considerations for this type of circuit will be illustrated through the use of an example.

Example 17.9 Let us design a light-dimmer circuit to handle a 100 W light globe. We shall use the circuit shown in Fig. 17.26. The SCR must be able to handle about 1 A (current of the 100 W light) and block at least 175 V (the 110 V line may increase to 125 V with 175 V peak). Since the 2N3562 has a blocking voltage of 200 V and an average anode current of 1 A, it will be used. The gate will have current pulses longer than 8 ms, so the average gate power dissipation of 100 mW must be used to determine the maximum allowable gate current in this circuit. Curves of gate-cathode voltage versus gate current (present in Texas Instrument Manual) indicate typical 2N3562 devices have 1.3 V from the gate to cathode when the gate current is 150 mA. If the rms values of 1.3 V and 0.15 A are used, the average power dissipation would seem to be 195 mW. However, since this circuit is a half-wave circuit, the actual average dissipated power is really only 98 mW, which is below the allowable gate dissipation. Maximum gate current flows when R_1 is zero. Then, the value of R_2 should limit the gate

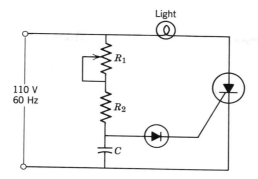

FIGURE 17.26
Light-dimmer circuit used in Example 17.9.

current to 0.15 A rms. The value R_2 is 110 V/0.15 = 730 Ω. To allow an extra safety margin, let us choose R_2 to be 1000 Ω.

From Eq. 17.29, we note that if R_1 is zero, $\omega CR_2 \ll 1$. Since ω is $2\pi \times 60 = 377$ rad/sec, we have $C \ll 1/377 \times 10^3$ or $C \ll 2.66$ μF. Let us pick a value approximately one-fifth of this value. Then $C = 0.5$ μF will be used. As a final calculation, $\omega C(R_1 + R_2) \gg 1$ when all of R_1 is in the circuit. Then, $377 \times 5.0 \times 10^{-7}(R_1 + R_2) \gg 1$ or $(R_1 + R_2) \gg 5{,}200$ Ω. Again, let us choose $(R_1 + R_2)$ about five times as large as this value or $R_1 = 25{,}000$ Ω.

The maximum reverse voltage permitted on the gate is 5 V, so the diode shown in Fig. 17.26 will be required. This diode must be able to withstand the total reverse voltage applied to the circuit as noted by Eq. 17.29. Thus, the diode must have a reverse voltage rating greater than 175 V and must be able to pass a current equal to the maximum gate current (150 mA in this example) when forward biased.

The circuit shown in Fig. 17.26 has one serious limitation. The maximum current this circuit will pass is a half-wave signal. Since a regular 117 V light globe is intended for use in a full-wave circuit, the light from the lamp in Example 17.9 will vary from essentially zero to about one-third of its brilliance in a full-wave circuit. There are two ways to achieve the full light output from the circuit shown in Fig. 17.26. These two modifications are shown in Figs. 17.27 and 17.28.

The circuit shown in Fig. 17.27 is a modification suggested by *RCA*.[1] In this configuration, the current to the light (or load) is controlled from zero to a half-wave value with the switch S_1 open. (The circuit is identical to that used in Example 17.9.) When S_1 is closed, the diode D_2 acts as a

[1] RCA Transistor Manual, Radio Corporation of America, 1964, p. 377.

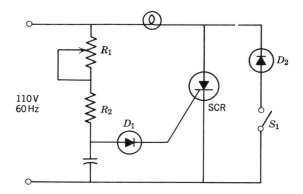

FIGURE 17.27
A light-dimmer circuit with an extended range of adjustment.

half-wave rectifier and the SCR with its associated circuitry can be adjusted to furnish as much of the other half of the wave as desired. Consequently, with switch S_1 open, the brilliance of the light can be adjusted from zero to about one-third of the lamp's full brilliance. With switch S_1 closed, the brilliance can be adjusted from one-third to full brilliance of the lamp.

The second modification of Fig. 17.26 is shown in Fig. 17.28. In this circuit, the SCR is replaced by a triac. Since the triac can switch from a blocking to a conducting state for either polarity of applied anode voltage and with either positive or negative gate triggering, the current through the load will have the waveform shown in Fig. 17.29. In this circuit, the current is controlled from zero to a full conduction value by adjusting R_1 through the proper range.

Of course, many other control circuits are possible. A large number of

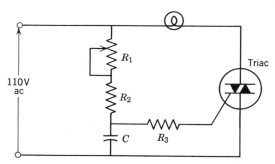

FIGURE 17.28
A full-wave control circuit using a triac.

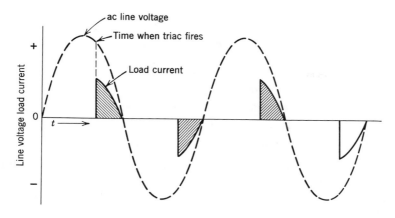

FIGURE 17.29
The voltage and current waveform for the circuit in Fig. 17.28.

these circuits require the use of a *unijunction* transistor. The operating principles of the unijunction transistor will be given in Chapter 22. Then, an adjustable voltage supply that uses the unijunction device to trigger the SCRs will be considered in detail.

PROBLEM 17.12 A circuit is connected as shown in Fig. 17.26. The circuit elements are as given in Example 17.9. Sketch the input voltage V_i, the capacitor voltage V_c (if the gate loading is assumed to be negligible), and the current through the lamp i_L if R_1 is equal to 10,000 Ω. What is the conduction angle in this case?

PROBLEM 17.13 A circuit is connected as shown in Fig. 17.28. The triac is an MAC2 and has the following specifications.

    peak blocking voltage = 200 V
    rms conduction current = 8 A
    peak gate power = 10 W
    peak gate current = 2 A
    typical ON voltage = 1 V
    typical gate trigger voltage = 0.9 V (2 V maximum)

Find the values of each circuit element in Fig. 17.28 for control over the current range from zero to essentially full wave. What maximum wattage rating may the light bulb have? What is the maximum power dissipation of the triac?

PROBLEM 17.14 A digital logic circuit (Chapter 23) requires a power supply with the following characteristics.

$V_{out} = 5$ V at 2 A

maximum peak-to-peak ripple $= 5$ mV

regulation from no load to full load $= 5$ mV

Even if circuit components in the power supply fail, the output should not rise above 6 V max.

Design a power supply which will meet these requirements.

PROBLEM 17.15 Design a laboratory power supply which has the output voltage adjustable from 0 V to 30 V. The current capability is 2.5 amperes but the supply should be constructed so current limiting can be adjusted for 30, 50, 100, 250, 500, 1000, or 2500 mA. When adjusted for a given output, the output voltage must not change more han 30 mV as the current varies from 0 to 2.5 A (current limiter set at 2.5 A). In addition, the maximum peak-to-peak ripple voltage must be no greater than 30 mV.

PROBLEM 17.16 The transistor T_3 in Fig. 17.17 can be replaced by an SCR to provide a current limiting circuit that will reduce the load current to a very small value during overload. Use this technique to design the current limiter for Problem 17.15. Explain the principles of operation of your circuit.

18

Large-Signal Tuned Amplifiers

In Chapter 11, the class B amplifier was found to have a higher theoretical efficiency than the class A amplifier. Unfortunately, the single class B amplifier produced a large amount of distortion in the signal. As we discovered, this distortion takes the form of adding harmonics to the desirable fundamental frequencies. By the use of a push-pull configuration, we were able to cancel the even harmonics in the class B amplifier and both high-efficiency and good ouput signal waveform were achieved.

If we desire to amplify a radio frequency (RF) signal, the tuned circuits introduced in Chapter 9 can be used to eliminate the undesirable harmonics from a class B amplifier. The tuned circuit (a pass-band configuration) passes the desired fundamental components but rejects the unwanted harmonic components. In fact the tuned circuit allows the use of a mode of operation known as *class C*, which has a higher efficiency than a class B amplifier. The class C amplifier passes current (collector, drain, or plate) for less than 180° of the signal cycle. In this chapter we will examine class B and class C tuned amplifiers.

18.1 HARMONIC REJECTION BY A TUNED CIRCUIT

In a class B or class C amplifier, the collector (or plate or drain) current flows for 180° or less of the input signal cycle. Consequently, the signal out of the amplifying device is a series of current pulses. In order to gain more insight into the operation of a tuned circuit, let us consider an idealized example.

Example 18.1 A typical class B or class C amplifier is connected as shown in Fig. 18.1 (for class B operation, V_{GG} and $V_{BB} = 0$ V). Let us assume the

563

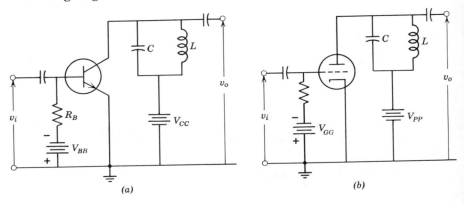

FIGURE 18.1
A class B or class C tuned amplifier : (a) transistor circuit ; (b) vacuum tube circuit.

input signal is a square wave with a period of $1/f_o$ where f_o is the resonant frequency of the tuned circuit in Fig. 18.1.

The technique of replacing the series resistance in the coil with an equivalent parallel resistance, R_{par}, (Chapter 9) can be used to reduce the circuits in Fig. 18.1 to the equivalent circuit of Fig. 18.2a. Then the output resistance of the tube or transistor, R_{ot}, can be combined in parallel with R_{par} to produce the equivalent resistance, R. The circuit then has the form shown in Fig. 18.2b.

The output voltage, V_o, of Fig. 18.2b is

$$V_o = \frac{-I}{[sC + (1/R) + (1/sL)]} \qquad (18.1)$$

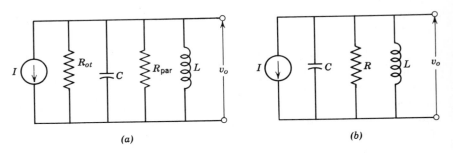

FIGURE 18.2
Equivalent circuits for the amplifiers of Fig. 18.1 : (a) first circuit ; (b) final form.

This equation can also be written as

$$V_o = \frac{-I(sLR)}{s^2LCR + SL + R}$$ (18.2)

or

$$V_o = \frac{-I}{C} \frac{s}{s^2 + (1/RC)s + (1/LC)}$$ (18.3)

Now, if I is $g_m V_i$, Eq. 18.3 can be written as

$$V_o = -V_i \frac{g_m}{C} \frac{s}{s^2 + (1/RC)s + (1/LC)}$$ (18.4)

This equation can also be written in the standard form,

$$V_o = -V_i \frac{g_m}{C} \frac{s}{s^2 + 2\zeta\omega_n s + \omega_n^2}$$ (18.5)

where $\omega_n^2 = 1/LC$ and $\zeta = 1/2Q$, as discussed in Chapter 9.

Now if the input voltage is a step function with a magnitude V, the value of V_i is V/s. Then V_o is

$$V_o = -\frac{g_m V}{C} \frac{1}{s^2 + 2\zeta\omega_n s + \omega_n^2}$$ (18.6)

This equation can also be expressed as

$$V_o = -\frac{g_m V}{C\omega_n} \frac{\omega_n}{(s + \zeta\omega_n)^2 + (1 + \zeta^2)\omega_n^2}$$ (18.7)

Equation 18.7 can be written as

$$V_o \simeq -\frac{g_m V}{C\omega_o} \frac{\omega_o}{(s + \zeta\omega_n)^2 + \omega_o^2}$$ (18.8)

where $\omega_o = \omega_n(1 - \zeta^2)^{1/2}$ is the *damped* resonant frequency. The inverse Laplace function for Eq. 18.8 is

$$v_o = -\frac{g_m V}{\omega_o C} e^{-\zeta\omega_n t} \sin \omega_o t$$ (18.9)

Notice that the input signal (Fig. 18.3a) can be produced by adding a series of positive and negative step functions (the solid, dashed, dotted, etc., lines) as shown in Fig. 18.3c. The response of the series of step functions is a series of damped sinusoids (Eq. 18.9) as shown in Fig. 18.3. (The dashed sinusoid is the response due to the dashed step function.) When all of these damped sinusoids are added together, the total output response is a sinusoid as shown in Fig. 18.3b.

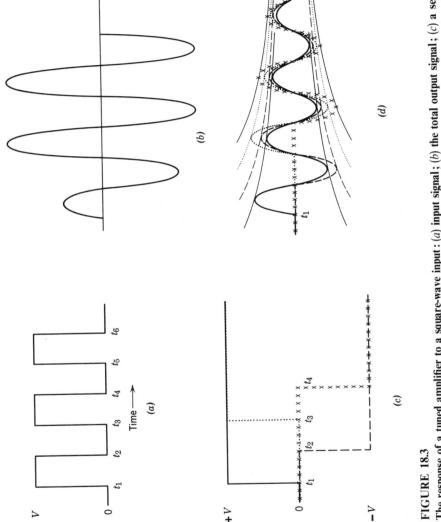

FIGURE 18.3
The response of a tuned amplifier to a square-wave input: (a) input signal; (b) the total output signal; (c) a series of step functions which can be added to produce the input signal; (d) the output signals from the step functions of part (c).

Notice that essentially all of the higher order terms in the input wave of Example 18.1 have been eliminated and only the fundamental frequency appears in the output. The filtering action of the tuned circuit is quite obvious.

Since the theoretical efficiency of a class B amplifier is higher than the theoretical efficiency of a class A amplifier, the reader may suspect that the theoretical efficiency of a class C amplifier may be even higher. Such is the case, as will be demonstrated later in this chapter. As a result of this higher efficiency, the usual large-signal tuned amplifier is a class C amplifier. Since most large-signal tuned amplifiers are class C, the remainder of this chapter will be directed to the design and analysis of class C amplifiers. Of course, the ideas pertaining to class C amplifiers can be extended to include class AB or class B amplifiers, so the methods outlined are rather general.

Higher power (hundreds or thousands of watts) and high-frequency (above about 0.5 MHz) transistors are either very expensive or nonexistant, depending on the power and frequency requirements. Consequently, transistors have not replaced vacuum tubes to any great extent in high-power, high-frequency applications such as radio broadcast transmitters.

PROBLEM 18.1   A tube is connected as shown in Fig. 18.1. Assume g_m is 2000 micromhos and r_p is 1 MΩ. The capacitor C has a value of 400 pF and L has an inductance of 10 mH. If the Q of the tuned circuit is 50:

   a.   Find the voltage gain of this amplifier for a signal with the same frequency as the ω_o of the tuned circuit.

   b.   Find the gain of this amplifier for a signal with a frequency of $2\omega_0$.

Answer: (*a*) $G_v = -400$;   (*b*) $G_v = 6.67\underline{/90°}$.

18.2   BASIC BEHAVIOR OF CLASS C TUNED CIRCUITS

The class C amplifier is biased below cutoff. Hence, if a sinusoidal input waveform is applied, the plate or collector current flows for less than 180° of a cycle. The actual angle over which current flows is called the *conduction angle*. Since the plate or collector circuit contains a fairly high Q-tuned circuit, the voltage variation in the plate or collector circuit is essentially a sinusoid. Of course, this plate or collector voltage is 180° out of phase with the input voltage when the output circuit is tuned to the input signal frequency. Accordingly, the voltages and currents in a typical class C tube amplifier are as shown in Fig. 18.4. In this amplifier, the control grid is driven positive. In fact, the control grid is usually driven positive to such an extent that the maximum control grid potential is equal to the minimum plate potential. This condition usually is used to obtain maximum power

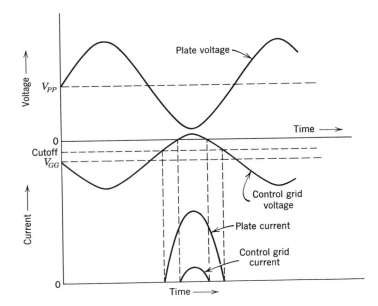

FIGURE 18.4
Voltage and current relationships in a class _C_ amplifier.

output. If the control grid becomes more positive than the plate, the control grid may draw so much current that the plate current actually _decreases._ Hence, the optimum power output usually occurs when the maximum control grid potential is equal to the minimum plate potential.

The class _C_ amplifier requires a graphical analysis that is different from the methods employed previously in this text. The method used in the preceding chapters assumed that the plate current was a fairly linear function of the input voltage. However, as shown in Fig. 18.4, the plate current (or collector current) may be a very nonlinear function of plate (or collector) voltage. Fortunately, the relationship between grid (or base) voltage and plate (or collector) voltage _is_ linear as shown in Fig. 18.4 when the input signal is sinusoidal and the load is a tuned circuit. Consequently, a very useful set of tube or transistor characteristics can be produced if grid (or base) voltage is plotted as a function of plate (or collector) voltage with the plate (or collector) current held constant. These curves are known as _constant current curves._ A set of constant current curves for a high-power tetrodes is shown in Fig. 18.5. The constant current curves for high-power tubes are normally supplied by the manufacturer. However, constant current curves can be plotted from the information contained in a set of conventional plate (or collector) characteristic curves. Notice that the tube amplification factor is the reciprocal of the slope of the constant current curves (Fig. 18.5).

PROBLEM 18.2   Draw a set of constant current curves for a 6J5 tube. The plate characteristics are given in Appendix I.

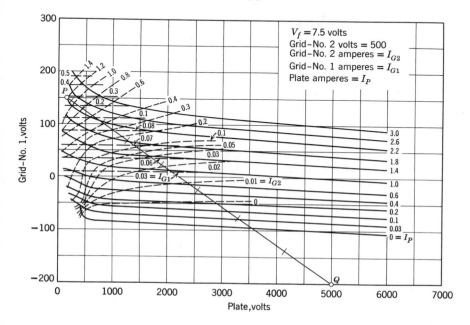

FIGURE 18.5
Constant current curves for a 4-1000 A tube. (Courtesy of Radio Corporation of America.)

18.3   GRAPHICAL CLASS *C* ANALYSIS AND DESIGN

The constant current curves can be used to solve class *C* amplifier problems. The usual procedure is as follows

1.   Plot the quiescent operating point on the constant current curves. (See point *Q* in Fig. 18.5)
2.   Choose a value for the maximum grid (or base) voltage and the minimum plate (or collector) voltage. As these voltages occur at the same

time, this point can also be located on the constant current curves. (See point P of Fig. 18.5.)

3.   Since the ac plate (or collector) voltage is 180° out of phase with the control grid (or base) excitation, the plot of instantaneous plate (or collector) potential versus the instantaneous grid (or base) potential is a straight line.[1] Thus, a straight line can be drawn on the constant current characteristics connecting the two points P and Q. (See the line connecting P and Q in Fig. 18.5.)

4.   Values of plate (or collector) current and grid (or base) current (also screen current for tetrodes or pentodes) can be found for given values of grid or base voltage. Thus, the waveforms for all voltages and currents in the class C amplifier can be plotted. (See Fig. 18.9.)

5.   The average values of these currents can be found by graphically integrating the current waveforms over one complete cycle. In addition, a Fourier analysis can be used to determine the fundamental components of these current waveforms.

6.   When the average currents are known, the power requirements from the power supplies can be determined. In addition, with the fundamental current components known, the signal power input and the signal power output can be determined.

In order to find the average (dc) currents and the fundamental components of the signal currents, integration must be performed. Usually, a graphical approach is necessary. Therefore, a review of graphical integration may be in order at this time. To integrate the area under the curve of Fig. 18.6, the base of the area is divided into a series of uniform lengths Δx as shown. Each small area in Fig. 18.6 is treated as if it were a trapezoid. Then the area is equal to 1/2 (sum of the two parallel sides) times the altitude (Δx). Now, since the integral of y from a to b is the total area under the curve or the sum of all the small areas,

$$\int_a^b y \, dx \simeq \Delta x \left[\left(\frac{y_1}{2} + \frac{y_2}{2} \right) + \left(\frac{y_2}{2} + \frac{y_3}{2} \right) + \left(\frac{y_3}{2} + \frac{y_4}{2} \right) + \cdots \right] \quad (18.10)$$

or

$$\int_a^b y \, dx \simeq \Delta x \left[\frac{y_1}{2} + y_2 + y_3 + y_4 + \cdots + y_{n-1} + \frac{y_n}{2} \right] \quad (18.11)$$

---

[1] If the phase angle between two ac voltages of the same frequency is different from 0° or 180°, a plot of one voltage versus the other voltage is an ellipse. In fact, the dimensions of the ellipse can be used as a means of determining the phase angle between the two voltages. For a more detailed analysis of this approach see F. E. Terman, *Radio Engineering Handbook*, McGraw-Hill Book Company, New York, New York, 1943, pp. 947–949.

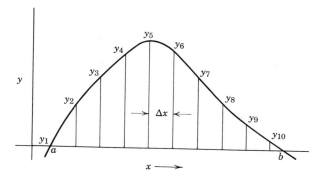

FIGURE 18.6
A method of graphical integration.

Of course, the smaller Δx becomes, the more accurate the graphical integration will be.

The method of solution for a class C amplifier can be further clarified by the use of an example.

Example 18.2 A 4-1000A tube is to be used as a class C amplifier in a circuit as shown in Fig. 18.7. The screen grid is maintained at a constant potential of 500 V. The characteristic curves of this tube are shown in Fig. 18.5. If a 5000 V plate supply (V_{PP}) and a -200 V bias supply (V_{GG}) is to be used, find the characteristics of the amplifier.

The constant current curves are reproduced in Fig. 18.8. The quiescent operating point is denoted in Fig. 18.8 as the point Q. The second point

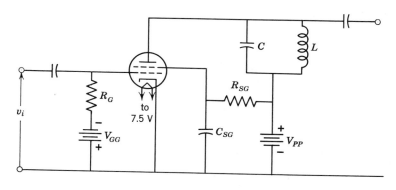

FIGURE 18.7
The class _C_ amplifier for Example 18.2.

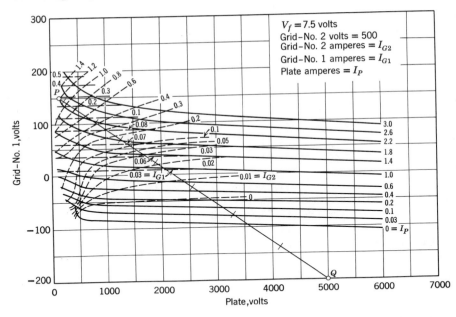

FIGURE 18.8
Graphical solution of a class C amplifier. (Curves are courtesy of Radio Corporation of America.)

is taken where the control grid potential is maximum and the plate potential is minimum. As already mentioned, the usual optimum point occurs when the maximum grid potential is approximately equal to the minimum plate potential. Let the control grid swing to $+150$ V and let the plate swing down to $+150$ V. This value of grid and plate voltage is shown in Fig. 18.8 as point P. The line from P to Q of Fig. 18.8 is drawn and is known as the *operating line* and determines not only the relationship between the control grid voltage and the plate voltage, but also the relationship between the control grid potential and the various currents in the tube. Accordingly, the line PQ is marked for grid voltage intervals corresponding to Δt intervals. In this example, Δt is chosen so that $\omega \Delta t$ is equal to 10 electrical degrees.

From the operating line in Fig. 18.8, the values of voltages and currents in Table 18.1 are found. These values can be used to plot the various voltages and currents in the circuits. This plot is given in Fig. 18.9. Also from the values in Table 18.1, the magnitudes of the dc currents and the fundamental components of the grid and plate signal currents can be found.

We can simplify the graphical Fourier analysis by choosing the proper

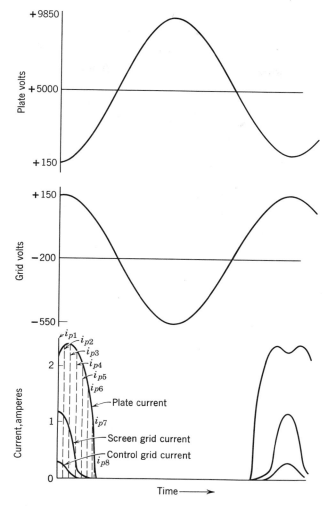

FIGURE 18.9
Voltages and currents of the class *C* amplifier in Example 18.1.

value as reference. For example, in Fig. 18.8, if $t = 0$ or $\theta = 0$ at $v_{g\,max}$, only cosine terms will be present in the Fourier series because the value of the function at $+\theta°$ will be equal to the value of the function at $-\theta°$. Because of this simplification, the $v_{g\,max}$ point of Fig. 18.9 will be used as the reference $(\theta = 0)$ in the following analysis. The 10° intervals along the

TABLE 18.1
A list of Current and Voltage Values in a Class C amplifier.

Phase Angle θ	$\cos \theta$	Grid 1 Voltage	Plate Voltage	Plate Current	Grid 2 Current	Grid 1 Current
0°	1.000	+150 V	150 V	2.2 A	1.2 A	0.3 A
10°	0.985	+144.5	250	2.4	1.1	0.26
20°	0.940	+129	475	2.4	0.8	0.18
30°	0.866	+103	800	2.2	0.45	0.093
40°	0.766	+68	1275	1.85	0.18	0.048
50°	0.643	+25	1900	1.2	0.029	0.005
60°	0.500	−25	2550	0.47	0.005	0
70°	0.342	−80.4	3300	0.03	0	0
80°	0.174	−139.1	4150	0	0	0
90°	0.000	−200	5000	0	0	0

operating line can be located by first determining the grid or plate voltage at these intervals.[2]

Now, from the Fourier analysis of Fig. 18.9, the dc value of plate current I_P is given by

$$I_P = \frac{1}{2\pi} \int_{-\pi}^{\pi} i_P \, d\theta \tag{18.12}$$

In Fig. 18.9, the value of $\Delta\theta$ is 10° or $\pi/18$ rad. Hence, the graphical integration from 0 to π becomes

$$\int_{0}^{\pi} i_P \, d\theta = \frac{\pi}{18} \left(\frac{i_{P1}}{2} + i_{P2} + i_{P3} + \cdots + i_{P7} + \frac{i_{P8}}{2} \right) \tag{18.13}$$

where i_{P1}, i_{P2}, etc., have the values indicated in Fig. 18.9. Now, since the area under the curve (Fig. 18.9) from $-\pi$ to 0 is equal to the area under the curve from 0 to π (for the given reference), the average plate current is (Eq. 18.12)

$$I_P \simeq 2 \times \frac{1}{2\pi} \times \frac{\pi}{18} \left(\frac{i_{P1}}{2} + i_{P2} + i_{P3} + \cdots i_{P7} + \frac{i_{P8}}{2} \right) \tag{18.14}$$

[2] For example, at these intervals, $v_{G1} = -200 + 350 \cos(n\Delta\theta)$ where $\Delta\theta = 10°$ and $n = 0, 1, 2, 3$, etc. The values of v_{G1} at these intervals are listed in Table 18.1.

For the values in Table 18.1, the value of I_P is

$$I_P \simeq \frac{1}{18}\left(\frac{2.2}{2} + 2.4 + 2.4 + 2.2 + 1.85 + 1.2 + 0.47 + 0.03 + \frac{0}{2}\right) \quad (18.15)$$

or

$$I_P \simeq \frac{11.65}{18} = 0.647 \text{ A} \quad (18.16)$$

By similar analysis, the direct or average current to the screen grid and to the control grid can be found. For the values in Table 18.1, the magnitude of I_{G1} (the average control grid current) is 44.5 mA and the magnitude of I_{G2} (the average screen grid current) is 175 mA.

The peak value of the fundamental component of the plate current is also needed. To find this value, the required Fourier expansion is

$$I_{p1} = \frac{1}{\pi}\int_{-\pi}^{\pi} i_P \cos\theta \, d\theta \quad (18.17)$$

Because of the symmetry involved, this can be written as

$$I_{p1} = \frac{2}{\pi}\int_{0}^{\pi} i_P \cos\theta \, d\theta \quad (18.18)$$

Accordingly,

$$I_{p1} \simeq \frac{2}{\pi} \times \frac{\pi}{18}\left[\frac{i_{P1}\cos 0°}{2} + i_{P2}\cos 10° + i_{P3}\cos 20° + i_{P4}\cos 30°\right.$$

$$\left. + \cdots + I_{P(n-1)}\cos(10n-20)° + i_{Pn}\cos(10n-10)°\right] \quad (18.19)$$

For the values in Table 18.1, the peak magnitude of the fundamental value of plate current, I_{p1}, is

$$I_{p1} \simeq \frac{10.059}{9} = 1.118 \text{ A} \quad (18.20)$$

By similar analysis, the peak magnitude of the fundamental component of control grid current I_{g1} is 81.7 mA. The magnitude of the fundamental component of screen current could also be found, but the screen has essentially no ac component of voltage. Consequently, the power at the fundamental frequency is zero, and the magnitude of fundamental current is not required.

From the values of voltages and currents just derived, the power relationships of the tube can be established. The power drawn from the power supply by the plate of the tube is

$$P_I = V_{PP} I_P \tag{18.21}$$

For this example, $P_I = 5000 \times 0.647 = 3235$ W.

The load impedance which the tuned plate circuit must offer at the fundamental frequency can now be found. Since the tuned circuit is at resonance, this impedance is purely resistive and has a value given by the relationship

$$R_L = \frac{V_p}{I_{p1}} = \frac{V_{PP} - V_{P\,min}}{I_{p1}} \tag{18.22}$$

where V_p and I_{p1} are the peak values of fundamental voltage and current. In this case, since $V_{PP} = 5000$ V and $V_{P\,min} = 150$ V,

$$R_L = \frac{4850}{1.118} = 4340 \ \Omega \tag{18.23}$$

Now the signal power output P_o is

$$P_o = \frac{V_p I_{p1}}{2} \tag{18.24}$$

For this example,

$$P_o = \frac{(4850)1.118}{2} = 2710 \text{ W} \tag{18.25}$$

The total power dissipated by the plate of the tube is

$$P_d = P_I - P_o \tag{18.26}$$

Thus, $P_d = 3235 - 2710 = 525$ W. This value of plate dissipation is well within the limits of the tube, therefore operation at this level is permissible.

The plate efficiency η_p of the tube is given by the relationship

$$\eta_p = \frac{P_o}{P_I} \times 100 \tag{18.27}$$

For this example,

$$\eta_p = \frac{2710}{3235} = 83.8\% \tag{18.28}$$

Other information which can be found is the effective input resistance. This resistance R_{in} is given by the relationship

$$R_{in} = \frac{V_{g1}}{I_{g1}} \tag{18.29}$$

Thus,

$$R_{in} = \frac{350}{0.0817} = 4280 \ \Omega \tag{18.30}$$

where V_{g1} and I_{g1} are the peak values of the control grid voltage and the fundamental component of grid signal current. The grid driving power P_g can be found by the relationship,

$$P_g = \frac{V_{g1}I_{g1}}{2} \tag{18.31}$$

In this example,

$$P_g = \frac{0.0817 \times 350}{2} = 14.3 \ \text{W} \tag{18.32}$$

Since the maximum control grid dissipation for a 4–1000 A tube is 25 W, this grid power dissipation is permissible. However, the stage which supplies the signal for this stage must be capable of furnishing the grid power required by this stage. The results of this example are tabulated below.

The peak signal voltage on the grid V_{g1} is 350 V.
The peak fundamental component of grid current I_{g1} is 81.7 mA.
The input resistance R_{in} is 4280 Ω.
The average dc voltage in the grid circuit V_{GG} is -200 V.
The average dc current in the grid circuit I_{G1} is 44.5 mA.
The input power to the grid circuit P_g is 14.3 W.
The average dc current in the screen circuit I_{G2} is 175 mA.
The voltage on the screen grid V_{G2} is $+500$ V.
The average dc current in the plate circuit I_P is 0.647 A.
The average dc voltage on the plate V_{PP} is 5000 V.
The peak signal voltage on the plate V_p is 4850 V.
The fundamental component of plate current I_{p1} is 1.118 A.
The required value of plate load impedance R_L is 4340 Ω.
The power output P_o is 2710 W.
The plate power input P_I is 3235 W.
The plate dissipation P_d is 525 W.
The plate efficiency η_p is 83.8%.
The plate conduction angle θ is 17° to $(180° - 17°)$ or 146°.

The foregoing example illustrates the method of approach to solve class *C* (or class *B* or class *AB*) problems. The list of known quantities at the end of the example indicates the effectiveness of the method. Unfortunately, if one bias or signal voltage level is changed, the entire process must be repeated for the new condition. Consequently, in order to optimize a circuit a number of solutions must be tried. Hence, a trial and error approach is indicated. On the brighter side, a little experience in the design of class C amplifiers reduces the trial and error to a minimum.

The waveform of the plate current in Fig. 18.9 deserves a word of explanation. At maximum grid voltage, the control and screen grids are *robbing* the plate current from the plate. Since the control grid is so much nearer the cathode than the plate, this action is not unexpected. In addition, since the screen grid is so much more positive than the plate, the screen has a tendency to take a larger portion of the total space current. An examination of the constant current curves (Fig. 18.8) indicates the reason for this type of behavior. Over most of the range of the constant current curves, these curves are straight. However, at the end of the curves where the plate voltage is very low, the curves rise gradually at first and then more sharply. As the control grid potential becomes more positive, the plate potential for the beginning of the rise becomes more positive. If the plate does not swing low enough to shift the operating line into the rising portion, the plate current will have a waveform indicated by Fig. 18.10, curve *a*. If the plate swings low enough, the operating line of Fig. 18.8 may become parallel to the constant current curves over the low plate voltage end of the operating line. If this condition occurs, the plate current is saturated for low plate voltages. This condition is shown in Fig. 18.10, curve *b*. If the plate swings to an even lower potential, the operating line may actually recross one or more constant current curves. In this case, the plate current has the waveform shown in Fig. 18.10, curve *c*.

The curves of Fig. 18.9 were drawn to aid the reader in visualizing the operation of the amplifier   and to illustrate the procedure of solution. However, in most solutions the sketch is not required. In fact, all the values required for the solution of the problem can be obtained from Table 18.1.[3]

PROBLEM 18.3   A 6J5 tube is to operated class C. V_{GG} is -30 V and V_{PP} is $+300$ V. Assume the peak magnitude of control-grid signal is 34 V and the peak magnitude of the plate signal is 260 V. Plot the plate current waveform. Use the curves from Problem 18.2.

PROBLEM 18.4   Repeat Example 18.2 with $V_{GG} = -200$ V, $V_{PP} = 5000$ V, and $V_{G2} = 500$ V. However, in this problem let the maximum control-grid potential be 145 V and the minimum plate potential be 250 V. Compare the results of this problem with the results of Example 18.2.

---

[3] To aid in determining the values of Table 18.1, a plastic overlay is made by the Eimac Company. This overlay is known as the "No. 5 Tube Performance Computer" and is available from the Manager of Amateur Service Department, Eimac, A Division of Varian Associates, 301 Industrial Way, San Carlos, California.

In Example 18.2 we assumed that the input voltage was sinusoidal. This assumption is valid only if the effective output resistance of the stage which drives the amplifier is small in comparison with the lowest value of the input resistance of the amplifier. Obviously, this assumption will not always be valid. In fact, it is usually *not* valid for transistor amplifiers with their low input impedance. Thus, let us consider another class *C* amplifier example where the source impedance cannot be ignored.

Example 18.3 A 2N1899 transistor is connected as shown in Fig. 18.11. The characteristic curves for the 2N1899 are given in Fig. 18.12.

To begin, let us draw the constant current curves for the transistor. First, the curves of constant base current are plotted in Fig. 18.13 from the v_{BE} versus i_B curves of Fig. 18.12. When the constant base current curves have been drawn, the curves for constant collector current can be obtained from the i_C versus v_{CE} curves. The required value of i_B and v_{CE} are obtained from Fig. 18.12 and plotted in Fig. 18.13.

The curves of Fig. 18.13 can be used if the impedance of the voltage source which excites the base of the transistor is negligible. However, if the impedance of the source is *not* negligible, a modification to the constant current curves of Fig. 18.13 must be made. This modification is achieved by observing that in Fig. 18.11 the open-circuit generator voltage v_S is

$$v_S = i_B R_s + v_{BE} \tag{18.33}$$

The curves of Fig. 18.13 are curves of v_{BE} versus v_{CE}. By adding the $i_B R_s$ component, a set of curves for v_S versus v_{CE} can be constructed as shown in Fig. 18.14. In these curves, each constant base current curve has been shifted upward by the voltage value of $i_B R_s$ where R_s is 10 Ω. Since the collector current is a function of the base current, the constant collector current curves must be shifted upward to maintain their relative positions in relation to the new constant base current curves. With the curves of Fig. 18.14 available, the solution proceeds as shown in Example 18.2. Note, however, that a new set of curves must be drawn if the generator impedance is changed.

PROBLEM 18.5 In Example 18.3, the value of V_{CC} is 40 V and the value of V_{BB} is 1 V. If v_S swings up to $+3$ V and v_{CE} swings down to 3 V, determine the values of the following parameters: average dc base current, average dc collector current, peak fundamental component of base current, peak fundamental component of collector current, base input resistance, effective output load resistance, signal power input to the base, signal power output, dc power input, and collector efficiency.

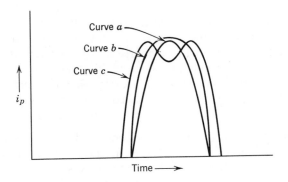

FIGURE 18.10
Typical plate current waveforms in class C operation.
In curve a, the minimum plate potential is much larger
than the maximum grid potential. In curve b, the
minimum plate potential is low enough to cause plate
current saturation. In curve c, the maximum grid
potential is high enough to rob the plate of current
while the plate potential is minimum.

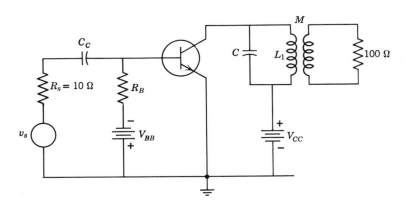

FIGURE 18.11
A transistor class C amplifier.

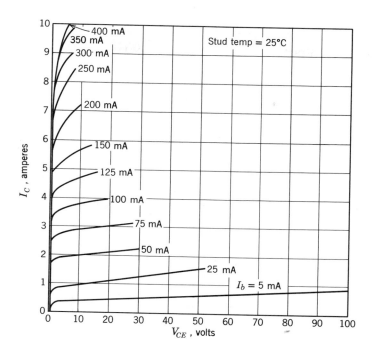

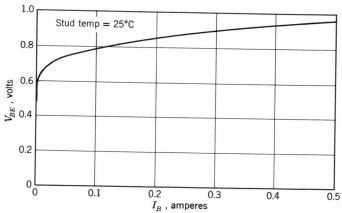

FIGURE 18.12
The characteristic curves of a 2N1899 power transistor.

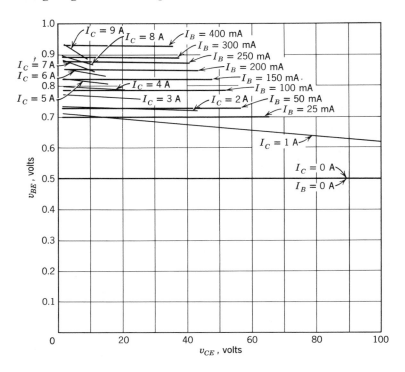

FIGURE 18.13
Constant current curves for the transistor of Fig. 18.12.

18.4   DETERMINATION OF THE TUNED-CIRCUIT PARAMETERS

In Example 18.2, many electrical quantities were found, but no attempt was made to determine the values of the tuned-circuit components except to note the magnitude of resistance which the tuned circuit should offer at resonance. Actually, the configuration of the tuned circuit determines the method of approach that must be used. In general, all the ideas developed in Chapter 9 can be used for the large-signal tuned circuits. However, there is one striking difference in the philosophy of design in large-signal as opposed to small-signal coupling circuits. In the small-signal amplifier, efficiency is not of prime concern. The power loss in the coupling circuit results in smaller stage gain than would be possible with lossless circuits, but the loss of a few decibels gain is not serious in a high-gain amplifier. As noted in Chapter 9,

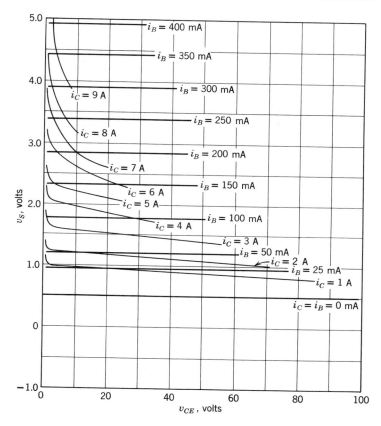

FIGURE 18.14
Modified constant current curves for the transistor of Fig. 18.12 when the signal voltage source has an internal impedance of 10 Ω.

the bandwidth of a small-signal amplifier is of major importance because this type of amplifier is used in applications which normally require the selection of a specified band of frequencies and the rejection of all other frequencies. On the other hand, the large-signal tuned amplifier is usually used in radio transmitters or other applications where power output and waveform of the signal are important but frequency selection is *not*. In the tuned power amplifier the desired output is a sinusoid or modulated sinusoid and the Q's of the coupling circuits need be only high enough to adequately eliminate the harmonics. The coupling efficiency increases as the circuit Q decreases, as will be shown. The efficiency is important in high-power amplifiers where increased efficiency results in materially reduced operating costs.

Considering a single-tuned, inductively-coupled circuit as shown in Fig. 18.15, the power loss in the tuned circuit is $I^2 R_{ser}$, where I is the current and

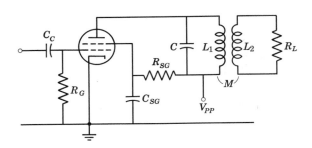

FIGURE 18.15
A typical class C coupling circuit.

R_{ser} is the resistance of the primary coil at the resonant frequency. The power transferred to the secondary circuit is $I^2 R_{Ls}'$ where R_{Ls}' is the resistance coupled into the tuned circuit as a result of the load current flowing in the secondary. As discussed in Chapter 9, the ohmic resistance of the secondary is usually very small in comparison with the load resistance. Then the efficiency of the coupled circuit becomes

$$\eta = \frac{\text{coupled power}}{\text{loss power} + \text{coupled power}} \times 100 = \frac{I^2 R_{Ls}'}{I^2 (R_{ser} + R_{Ls}')} \times 100$$

$$= \frac{R_{Ls}'}{R_{ser} + R_{Ls}'} \times 100 \qquad (18.34)$$

The efficiency may be obtained in terms of the coil Q_o and circuit Q, since $R_{ser} = \omega_0 L/Q_0$ and $(R_{Ls}' + R_{ser}) = \omega_0 L/Q$ then

$$R_{Ls}' = \frac{\omega_0 L}{Q} - R_{ser} = \frac{\omega_0 L}{Q} - \frac{\omega_0 L}{Q_o} \qquad (18.35)$$

$$\eta = \frac{\omega_o L(1/Q - 1/Q_o)}{\omega_o L(1/Q)} \times 100 = \frac{Q_o - Q}{Q_o} \times 100 \qquad (18.36)$$

It may be seen from Eq. 18.36 that Q_o must be large in comparison with the circuit Q in order for the efficiency to approach 100 percent. This relationship generally holds for coupled circuits. The maximum practical efficiency can be obtained by selecting a coil with a Q_o as high as possible and then choosing a circuit Q which is just adequate to provide good waveform. Values of circuit Q between 12 and 20 are commonly used.

Example 18.4 The tube in Example 18.2 must operate at a frequency of 10^6 rad/sec. The tuned circuit has as actual Q_o of 200 and the desired loaded Q is 20. The actual value of load resistance is 100 Ω and the coupling circuit of Fig. 18.15 is selected to provide the power transfer.

To simplify the calculations, we will assume that the resistance of coil L_2 is much less than R_L and also that the reactance of L_2 at resonance is much less than R_L.

From Example 18.2, the resistance of the loaded tuned circuit at resonance is 4340 Ω. For a parallel-tuned circuit, the Q is given by the relation

$$Q = \frac{R_p}{\omega_o L_1} = \frac{\omega_o C}{G} \qquad (18.37)$$

where R_p is the parallel resistance, G is the conductance of the resonant circuit, and C is the capacitance of the tuned circuit. Hence, for this example

$$C = \frac{QG}{\omega_o} = \frac{20}{4340 \times 10^6} = 4.6 \times 10^{-9} \text{ F} \qquad (18.38)$$

The value of L_1 can be found from the relation

$$\omega_o = \frac{1}{\sqrt{L_1 C}} \qquad (18.39)$$

In this case

$$L_1 = \frac{1}{10^{12} \times 4.6 \times 10^{-9}} = 2.18 \times 10^{-4} \text{ H} \qquad (18.40)$$

The loaded tuned-circuit parallel resistance R_p is 4340 Ω. Part of this resistance $R_{Lp}{}'$ is coupled from the load back to the tuned circuit and the remainder of this resistance R_{par} is due to the resistance in the tuned circuit. Since the unloaded Q_0 is known (200), Eq. 18.37 can be used to calculate R_{par}.

$$Q_0 = \frac{R_{par}}{\omega_0 L_1} = 200 = \frac{R_{par}}{10^6 \times 2.18 \times 10^{-4}} \qquad (18.41)$$

or

$$R_{par} = 200 \times 2.18 \times 10^{-4} \times 10^6 = 43,400 \ \Omega \qquad (18.42)$$

Now, the parallel combination of R_{par} and $R_{Lp}{}'$ must be equal to R_p.

Thus, we can write

$$R_{Lp}' = \frac{R_p R_{par}}{R_{par} - R_p} = \frac{4340 \times 43,400}{43,400 - 4340} = 4820 \ \Omega \qquad (18.43)$$

From Eq. 9.60, the magnitude of L_2 can be determined if the coefficient of coupling k is known. If k is assumed to be 0.7, L_2 is

$$L_2 = \frac{L_1 R_L}{k^2 R_{Lp}'} \qquad (18.44)$$

For this example

$$L_2 = \frac{2.18 \times 10^{-4} \times 100}{0.49 \times 4820 \ \Omega} = 9.22 \ \mu\text{H} \qquad (18.45)$$

From Eq. 18.36, the coupling efficiency is given as

$$\eta = \frac{Q_o - Q}{Q_o} \times 100 = \frac{200 - 20}{200} \times 100 = 90\% \qquad (18.46)$$

Thus, in this example, 90 percent of the power output from the tube is coupled into the load and 10 percent of this power is lost in the coupling circuit.

PROBLEM 18.6   Repeat Example 18.4 if the Q of coil L_1 is 100. How much power would be coupled to the load under these conditions?

PROBLEM 18.7   The tube in Example 18.2 is connected to a coupling circuit which is shown in Fig. 18.16. If the resonant frequency is 1 MHz and the Q of the tuned circuit with load is 20, find the values of the tuned-circuit components. The Q_o of the coil L is 200 at resonance.

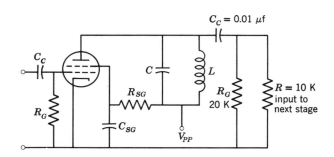

FIGURE 18.16
The circuit for Problem 18.7.

18.5   SELF-BIAS CIRCUITS FOR CLASS C AMPLIFIERS

The use of two or three batteries (Fig. 18.1) or power supplies is undesirable in class C amplifiers. Accordingly, means of self-bias for class C amplifiers was achieved at an early date. The circuit of Fig. 18.16 illustrates a typical self-biased stage. The screen circuit of Fig. 18.16 operates in the same manner as described in Chapter 6. As before, the value of R_{SG} is given by the relationship

$$R_{SG} = \frac{V_{PP} - V_{G2}}{I_{G2}} \tag{18.47}$$

Also, the value of C_{SG} is chosen so X_{csg} is small compared with R_{SG} in parallel with r_{sg} at the operating frequency.

The class C amplifier differs from the small-signal amplifiers in the method employed for control-grid or base bias. In small-signal *tube amplifiers*, the bias was obtained from a resistance in the cathode circuit. This type of bias is satisfactory if low-power stages are considered. However, the power lost in the cathode circuit becomes rather large in high-power stages. Fortunately, a different means of bias is possible if the control grid draws current; the flow of electrons from the control grid to ground through R_G (Fig. 18.16) can provide a negative potential on the control grid. Accordingly, the size of the control grid resistor R_G is found by the relationship

$$R_G = \frac{V_G}{I_G} \tag{18.48}$$

where V_G is the required grid bias potential and I_G is the dc grid current drawn by the grid. The value of C_C (Fig. 18.16) must be large enough to store the charge produced by the pulse of grid current, so an essentially constant current flows through R_G between current pulses. If the old rule of thumb that X_C be equal to approximately $0.1\ R_G$ is applied, this condition is usually fulfilled. This type of bias is known as *grid-leak* bias.

The self-bias system described previously has one bad characteristic. If the ac driving signal should be removed from the control grid, the bias on this stage will be reduced to zero. Zero bias will allow large currents to flow through the tube. These large currents usually destroy the tube in a short time. Consequently, protective circuits are usually incorporated in this type of circuit to remove the plate supply voltage if the excitation fails. Sometimes enough cathode bias or fixed bias is used in conjunction with grid-leak bias to protect the tube. Frequently an RF choke is connected in series with the grid-leak resistor so essentially no signal currents flow through the grid resistor. Otherwise, signal power is wasted.

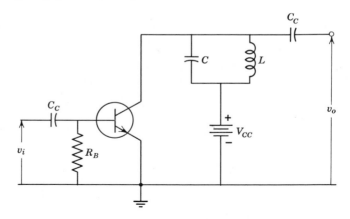

FIGURE 18.17
A self-bias class C amplifier.

The transistor amplifier can be connected as shown in Fig. 18.17. Then, the current flowing through the resistor R_B produces reverse bias in the same manner as grid-leak bias in the tube amplifier. Thus,

$$R_B = \frac{V_B}{I_B} \qquad (18.49)$$

where R_B is the magnitude of bias resistance required. The voltage V_B is the quiescent base voltage and I_B is the average value of base current. However, in the transistor, zero bias is also cutoff bias. Therefore, no protective devices are necessary, since the collector current reduces to essentially zero when the excitation is removed.

The value of base bias required for class C operation of a transistor amplifier cannot be specified in terms of the cutoff bias (as for a vacuum tube) because the "cutoff" bias voltage of a transistor is actually a forward bias. However, the transistor class C bias can be determined in terms of the collector current conduction angle (Fig. 18.18). In this figure, an operating line is shown on a modified set of constant current characteristics (similar to those of Fig. 18.14). The desired bias voltage is V_B. The base-emitter voltage at which collector current begins to flow is V_X, and the peak source voltage, which occurs at the collector saturation point P, is V_P. From Fig. 18.18 it can be seen that

$$\frac{V_X - V_B}{V_P - V_B} = \cos\frac{\theta}{2} = \frac{V_X - V_B}{(V_P - V_X) + (V_X - V_B)} \qquad (18.50)$$

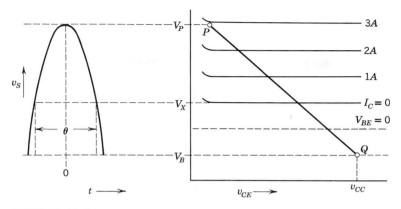

FIGURE 18.18
The relationships used to determine the transistor base bias voltage as a function of conduction angle.

where θ is the collector current conduction angle. Then, solving for $V_X - V_B$

$$V_X - V_B = \frac{(V_P - V_X)\cos \theta/2}{1 - \cos \theta/2} \tag{18.51}$$

or

$$V_X - V_B = \frac{V_P - V_X}{\sec \theta/2 - 1} \tag{18.52}$$

The potential difference $V_P - V_X$ may be determined from the input characteristics as well as from the modified constant current curves since $V_P = V_{BE\,max} + i_{B\,max} R_S$. After determining V_P, the conduction angle θ is then chosen and the potential difference between V_X and V_B is readily determined by the use of Eq. 18.52. For example, if the typical class C conduction angle 120° is chosen, $\cos 60° = 0.5$, $\sec 60° = 2$ and $V_X - V_B = V_P - V_X$.

In designing transistor class C amplifiers, the emitter-to-base reverse breakdown voltage V_{EBO} must be considered. If the combination of bias potential V_B and peak base signal voltage V_b exceeds this breakdown voltage, steps must be taken to protect the transistor. While several configurations are possible, the simplest way to protect the transistor is to insert a diode in series with the base. The polarity of this diode should permit current to flow when forward bias is applied to the diode-base combination. Then, when reverse bias is applied, the diode will essentially prohibit the flow of reverse current in the base circuit. Of course, the reverse voltage breakdown of the diode should be greater than the maximum reverse bias applied to the diode-base combination. This method of transistor base protection is also used in some of the pulse circuits (for example, Fig. 22.25) in Chapter 22.

PROBLEM 18.8   The transistor in Fig. 18.17 is the one whose characteristics are shown in Fig. 18.14. The V_{CC} supply is 50 V and it is desired to swing the collector current to 8A with I_B of 300 mA. A conduction angle of 110° is desired. The resonant frequency is one MHz and the desired circuit Q is 15. Find the values of all pertinent voltages and currents (both ac and dc), the input power, the output power, the efficiency of the circuit and all circuit components. Neglect the loss resistance in the coils and assume $k = 0.7$. The signal source resistance is 10 Ω.

Answer: $P_o = 77.5$ W, $P_I = 90.5$ W, $P_d = 13.0$ W, diode protection of the base is required.

18.6   EFFICIENCY OF CLASS C AMPLIFIERS

The high efficiency of the class C amplifier can be understood by reference to Fig. 18.19. The instantaneous power which the plate or collector must dissipate is equal to the instantaneous value of plate (or collector) voltage times the instantaneous value of current. Since current only flows while the plate voltage is low, the plate or collector dissipation is relatively low. However, the total power into the stage is equal to the V_{PP} supply (or collector supply) voltage multiplied by the average current. The difference between the input power and the plate or collector dissipation is the useful output power.

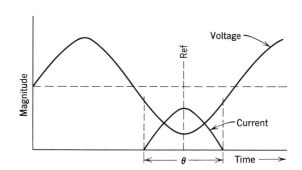

FIGURE 18.19
A plot of plate or collector current and voltage in a class C amplifier.

The *theoretical maximum collector or plate efficiency* can be calculated in the following manner. From Fig. 18.19, the average power into the circuit is

$$P_i = \frac{1}{2\pi} \int_{-\pi}^{\pi} V_{CC}\, i(t)\, d(\omega t) \tag{18.53}$$

where V_{CC} is the collector supply voltage. (This voltage would be V_{PP} for a tube circuit), and $i(t)$ is the current as a function of time. Since the waveforms are symmetrical about the reference point (Ref. of Fig. 18.19), Eq. 18.53 can be written as

$$P_i = \frac{2}{2\pi} \int_{0}^{\pi} V_{CC}\, i(t)\, d(\omega t) \tag{18.54}$$

Now, if the current $i(t)$ is assumed to be a section of a sinusoid within the conduction angle θ, $i(t)$ can be written as

$$i(t) = I\,(\cos \omega t - \cos \theta/2) \tag{18.55}$$

Notice that $I(1 - \cos \theta/2)$ is the peak value of current in the collector circuit and that $i(t)$ is zero at $\omega t = \theta/2$. Equation 18.54 can now be written as

$$P_i = \frac{V_{CC}I}{\pi} \int_{0}^{\theta/2} (\cos \omega t - \cos \theta/2)\, d(\omega t) \tag{18.56}$$

or

$$P_i = \frac{V_{CC}I}{\pi} [\sin (\theta/2) - (\theta/2) \cos (\theta/2)] \tag{18.57}$$

Although the actual collector (or plate) voltage cannot swing all the way to zero in an actual amplifier, this condition would still give the limiting value of power output. Accordingly, the average collector dissipation, P_d, can be written as

$$P_d = \frac{1}{2\pi} \int_{-\pi}^{\pi} (V_{CC} - V_{CC} \cos \omega t) i(t)\, d(\omega t) \tag{18.58}$$

The term $(V_{CC} - V_{CC} \cos \omega t)$ is the relationship which describes the voltage on the collector. Now, if the point Ref. on Fig. 18.19 is taken as reference and it is observed that $i(t) = 0$ for $\pi > t > \theta/2$, Eq. 18.58 can be written as

$$P_d = \frac{1}{\pi} \int_{0}^{\theta/2} V_{CC}(1 - \cos \omega t) I(\cos \omega t - \cos \theta/2)\, d(\omega t) \tag{18.59}$$

The power output P_o of the stage is given by the relationship

$$P_o = P_i - P_d = \frac{V_{CC} I}{\pi} \left[\int_0^{\theta/2} \cos^2 \omega t \, d(\omega t) - \cos \theta/2 \int_0^{\theta/2} \cos \omega t \, d(\omega t) \right]$$

(18.60)

or

$$P_o = \frac{V_{CC} I}{4\pi} (\theta - \sin \theta)$$

(18.61)

Therefore, the maximum collector efficiency η_p of a class C amplifier with a conduction angle θ is

$$\eta_p = \frac{P_o}{P_i} = \frac{V_{CC} I/4\pi(\theta - \sin \theta)}{V_{CC} I/\pi(\sin \theta/2 - \theta/2 \cos \theta/2)}$$

(18.62)

$$\eta_p = \frac{\theta - \sin \theta}{4 \sin \theta/2 - 2\theta \cos \theta/2}$$

(18.63)

When θ is equal to π, the stage would be operating as a class B amplifier. As a check on Eq. 18.63, when $\theta = \pi$, the maximum efficiency is 0.785 or 78.5 percent. This is the value of maximum efficiency for a class B amplifier as found in Chapter 11. As another point of interest, note that as $\theta \to 0$, the $\sin \theta \to \theta$ and the $\sin \theta/2 \to \theta/2$. When these values of θ are substituted into Eq. 18.63, the result is an indeterminate form (0/0). However, when L'Hospital's rule is applied, the limit of Eq. 18.63 as $\theta \to 0$ is 1. Hence, the efficiency approaches 100 percent as $\theta \to 0$. However, as $\theta \to 0$, the power output also approaches zero for a finite collector or plate current. Accordingly, a compromise must be made between efficiency and output power. A typical conduction angle is 120°, and the theoretical efficiency for this angle is about 85 percent.

In the usual design procedure, the collector or plate current is allowed to approach the permissible maximum. Then, a rough approximation of the power output can be found from Eq. 18.61.

The value of θ is made as small as possible to keep the efficiency high but still produce the required output power. The value of V_{CC} in Eq. 18.61 is made as high as practical to keep the conduction angle low and consequently the efficiency high. However, after all these preliminary calculations, solutions of the type indicated in Example 18.2 or 18.3 must be found to verify the behavior of the circuit.

In many design problems, the stage under development is required to produce a certain amount of power at a required voltage level to drive the next stage in the amplifier. In these problems, the type of coupling circuit must be considered as well as the operating voltages and currents of the transistor. As a final word of caution, Eq 18.63 gives the *maximum theoretical efficiency* of a circuit. Actual circuits will *always* have efficiencies *lower* than the values indicated by Eq. 18.63 for a given θ. (Actual amplifiers are not driven to zero collector volts, and the collector current is usually not sinusoidal.)

PROBLEM 18.9   The characteristics of a 4-65 A tube are given in Fig. 18.20. Design an amplifier using this tube to drive the tube of Examples 18.2 and 18.4. List all voltage and current requirements for this driver stage.

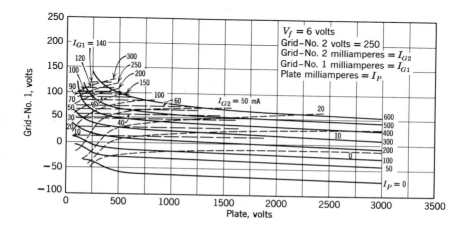

FIGURE 18.20
Constant current curves for a 4-65 A tube. (**Courtesy of Radio Corporation of America.**)

PROBLEM 18.10   Design a class C power amplifier to operate at 1 MHz, using a 2N1899 transistor with $V_{CC} = 50$ V. The driving-source resistance is 3 Ω. Use i_C max $= 8$ A, i_B max $= 0.3$ A and conduction angle $= 120°$. Determine power output, collector efficiency, and driving power. Is base circuit protection required?

PROBLEM 18.11   A 4-1000 A tube is to be used a a class C amplifier. Reference to a tube manual indicates the proper power output can be achieved if a control grid bias supply of -200 V and a plate supply of 5500 V is used. The screen is maintained at $+500$ V. If the grid signal is 325 V peak, find the values of currents, voltages, powers, and efficiency.

PROBLEM 18.12   A triode class C amplifier is analyzed and found to have the following voltages and currents.

Control grid supply $= -100$ V
Plate supply $= +10,000$ V
Peak signal plate voltage $= 9000$ V
Peak signal plate current $= 2$ A
Peak signal grid voltage $= 2000$ V
Peak signal grid current $= 0.1$ A
Average plate current $= 1.2$ A
Average grid current $= 0.06$ A

a.   What is the signal power output?
b.   What is the plate power input?
c.   What is the plate efficiency?
d.   What is the signal power required to drive the grid circuit?
e.   If self-bias is used, what size of grid resistor is required?
f.   What impedance must the load in the plate circuit have?

PROBLEM 18.13   A circuit is connected as shown in Fig. 18.11. The characteristics of this transistor are given in Fig. 18.12. The internal

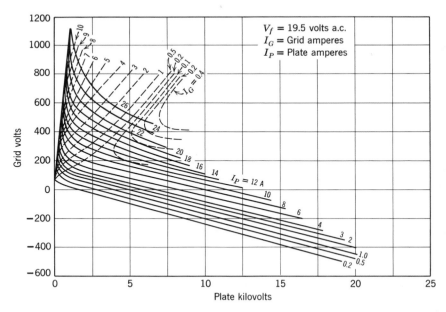

FIGURE 18.21
Constant current curves for a 9C21 tube. (Courtesy of Radio Corporation of America.)

resistance R_s of the generator is 2000 Ω and V_{CC} is -20 V. The maximum collector current should not exceed 8.0 mA. A conduction angle of 120° is desirable. The resonant frequency is to be 500 kHz and the desired circuit Q is 20. Find the values of all ac and dc voltages and currents, also the input power, the output power and the efficiency of the circuit. If the resistance of the coil is negligible and $k = 0.7$, find the value of all circuit elements. Assume $f_\beta \gg 500$ kHz.

PROBLEM 18.14 The characteristics of a 9C21 triode are given in Fig. 18.21. Design an amplifier using this tube and a V_{PP} supply of 10,000 V. The control grid is to be biased at -500 V and to have a voltage swing of 960 V peak. The power output should be approximately 27 kW. List all voltages, currents, and power as well as efficiency. Sketch the circuit and give circuit element values if the resonant frequency is 2 MHz and the circuit Q is 20.

19

Oscillator Circuits

In this chapter, an electronic oscillator will be defined as a device that generates a sinusoidal voltage or current waveform. As with most electronic devices, a source of dc power is required for operation. In general, either one-terminal pair devices such as klystrons, tunnel diodes, and so forth, or two-terminal pair devices such as conventional transistors, tubes, etc., can be used as the active elements. The active elements must work in conjunction with passive (R, L, and C) networks. In some microwave devices, the values of R, L, and C are distributed over the circuit rather than being separate lumped elements. However, even in these instances equivalent circuits of R, L, and C elements can be developed to help visualize the action of the device.

19.1 OSCILLATORS WITH TWO-TERMINAL ACTIVE ELEMENTS

If the active element in an oscillator contains only two terminals, this active element must be connected either in series or in parallel with the passive elements of the circuit. Accordingly, the circuit can be represented as shown in Fig. 19.1. If the active element behaves as a negative conductance, oscillations can occur if this negative conductance of the active elements is greater than or equal to the positive conductance of the passive circuit because there is then no damping in the circuit. To determine how oscillations can occur, and at which frequency and at what amplitude a given circuit will oscillate, the following procedure is followed. First, a polar plot is made *of the negative* of the admittance of the passive circuit as a function of frequency. Second, a polar plot is made of the admittance of the active element as some circuit parameter is varied. This second polar plot is made on the same chart (and to the same scale) as the first polar plot. Usually, the circuit parameter which varies the admittance of the

596

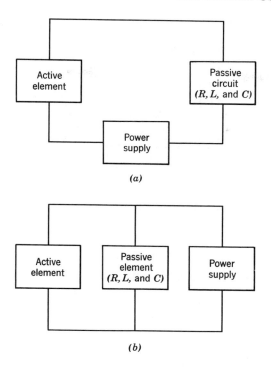

(a)

(b)

FIGURE 19.1
**A simplified block diagram of an oscillator with a
two-terminal active element: (*a*) series connections; (*b*) parallel connection.**

active element is the magnitude of the ac signal in the circuit, although other
circuit parameters can be used. Wherever the two plots intersect, the negative
conductance is equal to the positive conductance, and stable oscillations will
occur. The frequency of oscillation and usually the magnitude of oscillation
can be determined from the plot.

The reader was introduced to the tunnel diode in Chapter 3 (Section 3.11).
To illustrate the method just discussed, an example will be given for a
tunnel-diode oscillator.

Example 19.1   The characteristic curve of a tunnel diode is given in Fig. 19.2.
This tunnel diode is connected as shown in Fig. 19.3*a*. The diode specifications
list a total shunt capacitance C_d of 7 pF. (An inductance of 6×10^{-9} H
is also present but will be ignored in this example). Find the frequency of

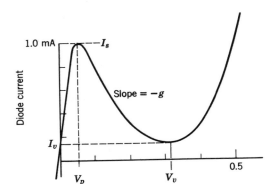

FIGURE 19.2
The characteristic curve of a tunnel diode.

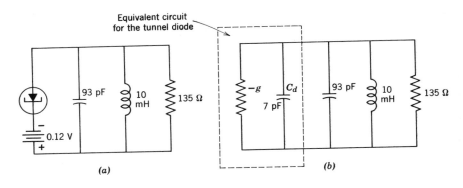

FIGURE 19.3
A tunnel diode oscillator: (*a*) actual circuit; (*b*) equivalent circuit.

oscillation and the magnitude of voltage across the tuned circuit for this configuration.[1] If the resistance of L is ignored, the tunnel diode will be biased at $+0.12$ V. Thus the tunnel diode can be replaced by an equivalent circuit consisting of a negative conductance $-g$ and capacitance C_d, as shown in Fig. 19.3*b*.

---

[1] This circuit is idealized to the extent that a battery of 0.12 V is not practical. However, this voltage can be obtained from power supplies or resistor-battery-capacitor combinations. Unfortunately, great care should be taken or the diode will oscillate (or switch) with the battery circuit instead of with the tuned circuit. Additional tunnel diode oscillator circuits are given in,"Designing Tunnel Diode Oscillators," by Wen-Hsiung Ko in *Electronics*, February 10, 1961, Vol. 34, No. 6, pp. 68–72.

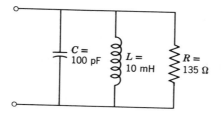

FIGURE 19.4
The equivalent ac circuit for the passive
portion of the circuit in Fig. 19.3.

The passive portion of the circuit (including the capacitance C_d) will have
the configuration shown in Fig. 19.4. The admittance of the circuit of Fig. 19.4
can be written as

$$Y(s) = sC + \frac{1}{sL} + \frac{1}{R} \tag{19.1}$$

or

$$Y(s) = C\,\frac{s^2 + s(1/RC) + (1/LC)}{s} \tag{19.2}$$

When the values of R, L, and C from Fig. 19.4 are substituted into Eq. 19.2,

$$Y(s) = 10^{-10}\,\frac{s^2 + s \times 7.4 \times 10^7 + 10^{12}}{s} \tag{19.3}$$

Now, for steady-state alternating current, the complex frequency s becomes
$j\omega$ in Eq. 19.3. Accordingly,

$$Y(j\omega) = 10^{-10}\,\frac{-\omega^2 + j(7.4 \times 10^7)\omega + 10^{12}}{j\omega} \tag{19.4}$$

Therefore, a polar plot of $Y(j\omega)$ as ω varies is as given in Fig. 19.5.
The next step requires the construction of a curve which represents the
admittance of the tunnel diode. The characteristic curve of the tunnel diode
has been reproduced in Fig. 19.6. Since the diode is based at $+0.12$ V, the
Q point would be located as shown in Fig. 19.6. The slope of the characteristic
curve at Q is indicated by the line AB. *The slope of this line AB is the
admittance of the diode with no ac signal*

$$-g = \frac{0.6 \times 10^{-3}\ \text{A}}{0.08\ \text{V}} = 7.5 \times 10^{-3}\ \text{mhos}$$

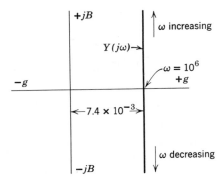

FIGURE 19.5
A plot of $Y(j\omega)$ from Eq. 19.4

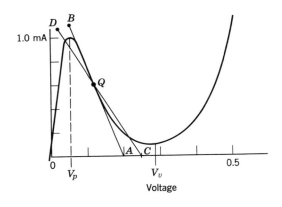

FIGURE 19.6
The effect of signal level on the admittance of a tunnel diode.

This value of g is plotted as point A in Fig. 19.7.

If an ac signal is superimposed on the voltage at Q, the voltage varies along the line AB (Fig. 19.6) for small signals. However, as the ac signal increases in magnitude, the voltage must follow the characteristic curve. Accordingly, as the signal becomes larger, the voltage digresses from the AB curve. As the ac voltage becomes larger, the *average slope of dI/dV* decreases and becomes equal to the value indicated by the line CD when the magnitude of the ac voltage is about 0.17 V peak-to-peak. The slope of this line indicates that the average admittance of the diode with an ac signal of 0.17 V peak-to-peak is about

$$-g = \frac{1.2 \times 10^{-3} \text{ A}}{0.25 \text{ V}} = 4.8 \times 10^{-3} \text{ mhos}$$

This value of g is plotted on Fig. 19.7 as point B. As the magnitude of the ac signal increases, the value of $-g$ decreases. In fact, for an ac signal of about 0.25 V peak-to-peak, the average g is about zero. If the ac signal increases beyond this value, the average g actually becomes positive. Accordingly, the plot of g for the tunnel diode is as shown in Fig. 19.7.

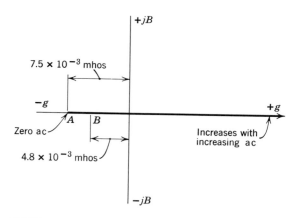

FIGURE 19.7
A plot of g versus signal amplitude for the tunnel diode of Fig. 19.3.

The plot of $Y(j\omega)$ for the passive circuit was given in Fig. 19.5. A plot of $-Y(j\omega)$ and the plot of g versus signal amplitude (Fig. 19.7) are reproduced on a single set of coordinate axes in Fig. 19.8. The intersection of the two plots, point P, is the required solution. In this example, the resonant frequency is 10^6 rad/sec and the magnitude of signal for $g = -7.4 \times 10^{-3}$ mhos is about 0.12 V peak-to-peak. This is the value of ac voltage for which the operating line just begins to leave the line AB in Fig. 19.6.

In the foregoing example, if the circuit were at quiescent conditions and the ac signal were zero, the tunnel diode conductance would have the value given by point A of Fig. 19.8 Under these conditions, the negative conductance of the tunnel diode would be greater than the positive conductance of this resonant circuit. Therefore, oscillations would build up in the circuit until the ac signal was large enough to shift the value of negative conductance in the tunnel diode to point P of Fig. 19.8. At point P, the negative conductance of the tunnel diode is equal to the positive conductance of the passive circuit and the net conductance is zero. Hence, a pole-zero plot would show the poles on the $j\omega$ axis. Therefore, the damping is zero and steady-state oscillations occur. Under these conditions, the circuit is in equilibrium and will continue operation at this signal level.

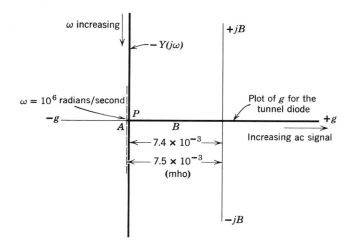

FIGURE 19.8
A plot of $-Y(j\omega)$ **from Fig. 19.5 and the plot** $f(g)$ **from Fig. 19.7.**

The admittance plots of many devices are more complicated than the straight-line plot of Fig. 19.7. In addition, the plots of $Y(j\omega)$ for many passive circuits are more involved than the straight-line plot of Fig. 19.5. Even so, the method outlined in Example 19.1 can be applied to these more complicated circuits.

PROBLEM 19.1 Determine the slope of the characteristic curve of Fig. 19.2 at diode voltages of 0.05, 0.06, 0.07... 0.18, 0.19 V. Find the average of these slopes to determine the value of $-g$ when the ac signal of Example 19.1 is 0.14 V peak-to-peak or 0.05 V rms.

19.2   OSCILLATORS WITH FOUR-TERMINAL ACTIVE ELEMENTS

Many oscillator circuits use triodes, pentodes, transistors, and so on as the active elements. In all these oscillators, energy of the proper magnitude and phase is fed from the output back to the input circuit. This type of oscillator is a type of feedback amplifier. To visualize the requirements of an oscillator, consider the typical feedback circuit shown in Fig. 19.9.

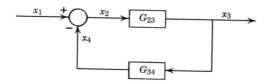

FIGURE 19.9
A typical feedback circuit.

The transfer function G_{vf} was found in Chapter 14 to be

$$G_{vf} = \frac{G_v}{1 - G_v \beta_v} \tag{14.7}$$

In this chapter we will adopt the nomenclature given in Fig. 19.9. Using this nomenclature, Eq. 14.7 becomes

$$G_{13} = \frac{x_3}{x_1} = \frac{G_{23}}{1 - G_{23} G_{34}} \tag{19.5}$$

Whereas most oscillators use positive feedback, some oscillators actually require *negative feedback* for proper operation. In fact, as already noted in Chapter 14, an ordinary amplifier with negative feedback may become unstable and oscillate.

If the circuit is an oscillator, the signal x_1 must be zero and the signal x_3 must be finite. Accordingly, from Eq. 19.5,

$$\frac{G_{23}}{1 - G_{23} G_{34}} = \infty \tag{19.6}$$

or

$$1 - G_{23} G_{34} = 0 \tag{19.7}$$

Hence,

$$G_{23} G_{34} = 1 \tag{19.8}$$

This criterion just establishes the instability of the circuit. In addition to this criterion, in order to fulfill our definition of an oscillator the output should be sinusoidal.

The root-locus plot, which was discussed in Chapter 14, is a solution of Eq. 19.7 as the reference gain K of the circuit is adjusted. Accordingly, the point of operation must lie on the root-locus plot. In addition, if the output of the circuit is to be a steady-state sinusoidal waveform, the point of operation must lie on the $j\omega$ axis of the s plane. Consequently, the root-locus plot of a circuit must cross the $j\omega$ axis in the s plane if the circuit is to be used as an oscillator. In addition to this restriction, no other poles of G_{13} can be in the right half of the s plane for the value of K which produces oscillation. This restriction must be enforced to prevent the instability of a pole in the right half plane from "swamping" out the oscillation.

PROBLEM 19.2 Prove that if $x_2 = x_1 + x_4$ (positive feedback) in Fig. 19.9, then

$$G_{13} = \frac{G_{23}}{1 - G_{23}G_{34}}$$

19.3 RC OSCILLATORS

Generally, an RC oscillator consists of an amplifier with feedback provided by an RC circuit. The most common type of RC oscillator uses a series RC circuit in series with a parallel RC circuit to provide positive

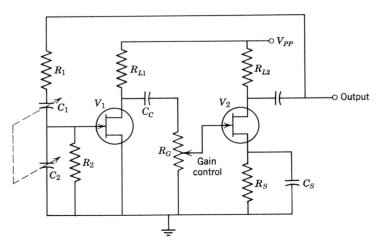

FIGURE 19.10
An RC bridge oscillator.

feedback to the amplifier, as shown in Fig. 19.10. This feedback circuit provides in-phase positive feedback at only *one* frequency since the phase shift in the parallel part of the circuit is opposite to that in the series part, and these phase shifts are equal at only *one* frequency. An equivalent circuit for Fig. 19.10 is given in Fig. 19.11. In usual practice $R_1 = R_2$ and $C_1 = C_2$. Then the phase shifts are equal when each is 45° or $R = X_c$ in each branch. Thus the frequency of oscillation is $\omega_o = 1/RC$, assuming there is no phase shift through the amplifier. At this frequency, the impedance of the series branch is twice that of the parallel branch, so the magnitude of the feedback ratio is 1/3. Thus, the forward or open-loop gain of the amplifier must be 3 in order to provide oscillation.

Although the above characteristics may not be intuitively obvious, they can be rather easily obtained by a feedback analysis, as given below. We will let the forward gain of the amplifier $G_{23} = K$, neglecting the effects of the coupling and shunt capacitances in the amplifier. Referring to Fig. 19.11, the feedback ratio G_{34} can be written as

$$G_{34} = \frac{x_4}{x_3} = \frac{R_2(1/sC_2)/[R_2 + (1/sC_2)]}{R_1 + 1/sC_1 + R_2(1/sC_2)/[R_2 + (1/sC_2)]} \qquad (19.9)$$

This equation can be simplified to

$$G_{34} = \frac{1}{R_1 C_2} \frac{s}{s^2 + s\dfrac{R_1 C_1 + R_2 C_2 + R_2 C_1}{C_1 C_2 R_1 R_2} + \dfrac{1}{C_1 C_2 R_1 R_2}} \qquad (19.10)$$

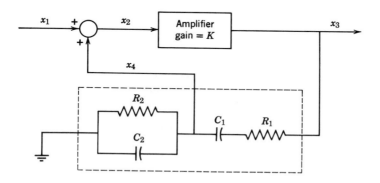

FIGURE 19.11
An equivalent circuit for Fig. 19.10.

Equation 19.10 can be further simplified by letting $R_1 = R_2 = R$ and $C_1 = C_2 = C$. Then,

$$G_{34} = \frac{1}{RC} \frac{s}{s^2 + s\dfrac{3}{RC} + \dfrac{1}{R^2 C^2}} \tag{19.11}$$

Since this circuit has positive feedback, Eq. 19.5, repeated below, applies.

$$G_{13} = \frac{G_{23}}{1 - G_{23} G_{34}} \tag{19.5}$$

When K (the amplifier gain) is substituted for G_{23} and Eq. 19.11 is substituted for G_{34}, Eq. 19.5 becomes

$$G_{13} = \frac{K}{1 - \dfrac{K}{RC} \dfrac{s}{s^2 + s(3/RC) + 1/R^2 C^2}} \tag{19.12}$$

The root-locus plot for Eq. 19.12 is given in Fig. 19.12. From Fig. 19.12, it is obvious that the circuit in Fig. 19.10 can be used as an oscillator.

The frequency of oscillation as well as the required amplifier gain for oscillation can be found either graphically or analytically. In the graphical analysis, careful construction of the root-locus plot is required. A spirule simplifies finding the required value of K. In some cases, the problem is so complicated that the analytical approach is rather difficult to evaluate. Hence, the graphical approach is the logical method of solution. However, the graphical approach can be very time consuming.

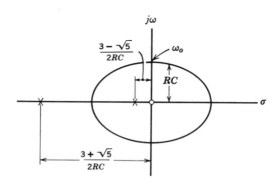

FIGURE 19.12
The root locus for the circuit of Fig. 19.10.

In this particular problem, the analytical approach is rather simple. The first step is to multiply both numerator and denominator of Eq. 19.12 by $(s^2 + s(3/RC) + 1/R^2C^2)$. Then, Eq. 19.12 becomes

$$G_{13} = K \frac{s^2 + s(3/RC) + 1/R^2C^2}{s^2 + s[(3-K)/RC] + 1/R^2C^2} \tag{19.13}$$

This equation can be written as

$$G_{13} = K \frac{s^2 + 2\zeta_1 \omega_n s + \omega_n^2}{s^2 + 2\zeta_2 \omega_n s + \omega_n^2} \tag{19.14}$$

Referring back to Fig. 19.12, we see that the two roots of the denominator must occur at $+j\omega_o$ and $-j\omega_o$ (for steady-state oscillations). Accordingly, the denominator of G_{13} must have the factors

$$(s + j\omega_o)(s - j\omega_o) = \text{denominator of } G_{13} \tag{19.15}$$

where ω_o is the radian frequency of oscillation, or

$$s^2 + \omega_o^2 = \text{denominator of } G_{13} \tag{19.16}$$

When Eq. 19.16 is equated to the denominator of Eq. 19.14, the following relationship results.

$$s^2 + \omega_o^2 = s^2 + 2\zeta_2 \omega_n s + \omega_n^2 \tag{19.17}$$

Thus, if there are to be steady-state oscillations, the damping factor ζ_2 must have a value of zero. When the values of ζ_2 and ω_n from Eq. 19.13 are substituted into Eq. 19.17, the coefficients of like powers of s can be equated to yield

$$0 = \frac{3-K}{RC} \tag{19.18}$$

$$\omega_o^2 = \frac{1}{R^2C^2} \tag{19.19}$$

From Eq. 19.18,

$$K = 3 \tag{19.20}$$

and from Eq. 19.19,

$$\omega_o = \frac{1}{RC} \tag{19.21}$$

Figure 19.12 illustrates that the gain of the amplifier in an oscillator circuit must be maintained at a constant value. Figure 19.12 shows that a decrease of gain will shift the poles from the $j\omega$ axis into the left half plane. Under these conditions, the signal output will be an exponentially decaying sinusoidal wave. In contrast, an increase of gain will shift the poles from the $j\omega$ axis into the right half plane. In this case, the output waveform is an exponentially increasing sinusoidal wave. Consequently, it is important for the gain of the amplifier to be maintained at a constant level.

Many circuits have a sort of "built-in" gain stabilization factor. We can visualize this type of gain stabilization by referring to the typical dynamic transfer characteristic in Fig. 19.13. For small ac signals, the operation is along the steepest part of the dynamic curve indicated by the line AB. However, as the ac signal increases in magnitude, operation extends into the less steep portions of the dynamic transfer characteristic and the *average slope* decreases, as indicated by the line CD. Consequently, the average gain $\Delta V_o/\Delta V_i$ decreases as the ac signal level increases.

The typical linear oscillator circuits are adjusted so that the gain is slightly higher than the gain necessary for oscillations with very small signals. Hence, the poles in the root-locus plot are slightly to the right of the $j\omega$ axis (in the right half plane) when the oscillator is first turned on. Consequently,

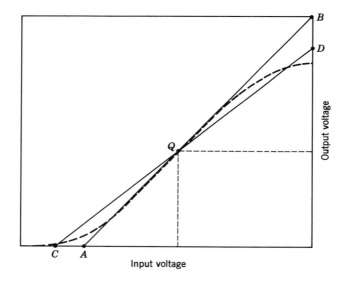

FIGURE 19.13
A typical dynamic characteristic curve.

the level of signal increases. As the signal level increases, the gain of the amplifier decreases and the poles of the root-locus plot slide back to the *jω* axis. If the signal increases or decreases beyond this value, the gain of the circuit will change in such a direction as to return the operating point back to the *jω* axis. Unfortunately, this method of gain control causes distortion in the output signal because of operation into the nonlinear portions of the characteristic curves.

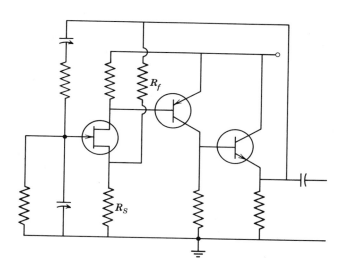

FIGURE 19.14
An *RC* oscillator with negative feedback gain control.

The circuit of Fig. 19.14 is an imporoved version of the oscillator of Fig. 19.10. In Fig. 19.14, the forward gain of the amplifier is controlled by negative feedback. The negative feedback path consists of the resistor R_f and the resistor R_S. In addition to increased gain stability, the negative feedback improves the waveform of the ouput signal. The negative feedback circuit must be designed so that the gain $K_f = 3$ as given by Eq. 19.20. The direct coupling eliminates the low-frequency phase shift and the emitter-follower output prevents excessive loading by the feedback circuits and the actual oscillator load.

The *RC* oscillator configuration of Fig. 19.14 is commonly known as a Wein–Bridge oscillator because of its bridge characteristics, which are illustrated in Fig. 19.15. Observe that the amplifier input is between nodes

A and *B* and the amplifier output is applied across the opposite nodes. Since the open-loop voltage gain of the amplifier is of the order of thousands, the bridge is very nearly balanced in normal operation. Also note that the transistor has little loading effect on either the positive or negative feedback circuits because of the small current which flows in the branch *AB* in comparison with the currents in the other branches.

The circuit of Fig. 19.14 still has a problem, inasmuch as oscillations will not build up unless the voltage gain of the amplifier with its negative feedback is greater than 3, so the *s*-plane poles are in the right half plane during the starting period. The gain must then reduce to precisely 3 during steady-state oscillation as a result of overdriving an amplifier stage, as previously discussed. However, the accompanying distortion is undesirable, and techniques have been developed to automatically control the gain without introducing distortion.

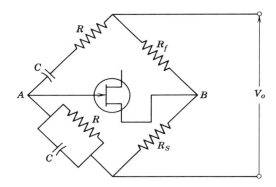

FIGURE 19.15
Illustration of the bridge circuit arrangement of the oscillator of Fig. 19.14.

One method of automatic gain control is shown in Fig. 19.16. In this circuit the FET F_1 acts as a variable resistance in the negative feedback loop to control the gain of the amplifier. You may recall that the drain resistance of a FET is essentially the ohmic channel resistance when the magnitude of the drain-to-source voltage is less than the pinch-off voltage and this channel resistance is controlled by the gate voltage. Thus, the feedback factor and hence the amplifier gain can be controlled by the output voltage magnitude. When power is first applied to the circuit, the gate of F_1 is forward biased by the positive supply, and the channel resistance is

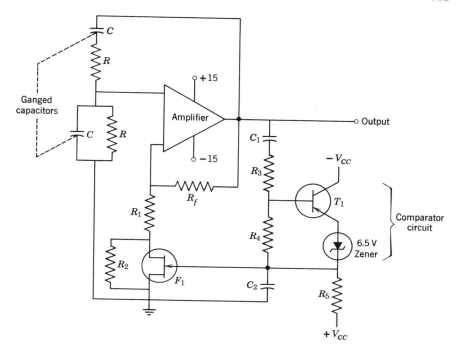

FIGURE 19.16
An automatic gain control system for a Wein-Bridge oscillator.

very low. Then the negative feedback factor is essentially $R_1/(R_f + R_1)$ which is chosen to be considerably less than 1/3. Thus the initial turn-on and noise cause oscillations to build up at the frequency ω_o until the output voltage negative peak exceeds the Zener diode voltage plus the forward V_{BE} of T_1. Then T_1 conducts and the emitter current of T_1 increases the voltage drop across R_5 and thus charges capacitor C_2 negatively. Therefore, F_1 becomes reverse biased and the channel resistance increases until the negative feedback factor becomes 1/3 and the amplifier gain becomes precisely 3. The supply voltages $+V_{CC}$ and $-V_{CC}$ must be large enough so that the peak output voltage (about 7 V if $V_E = 6.5$ V) is well within the linear range of the amplifier. The resistor R_3 linearizes the impedance in series with C_1 to prevent a dc voltage change across C_1. A power amplifier may follow the oscillator to provide buffering for the oscillator, higher output power, and lower output impedance. The resistors R are usually changed by a *band switch* to provide a wide range of frequency for the oscillator.

PROBLEM 19.3 Design a Wein–Bridge oscillator with automatic gain control. Choose the semiconductor devices, including *IC*s if desired, and calculate the circuit components assuming the desired frequency range is 10 Hz to 1 MHz. Tuning capacitors, including trimmers, usually have about a 10 to 1 capacitance ratio.

PROBLEM 19.4 An *RC* oscillator is shown in Fig. 19.17. (*a*) Write the expression for the closed-loop voltage gain or transfer function. (*b*) Sketch the root locus. (*c*) Determine the required g_m for each FET. (*d*) Determine the frequency of oscillation in terms of R_L and C. You may neglect stray capacitance and assume $R_G \gg R_L$.

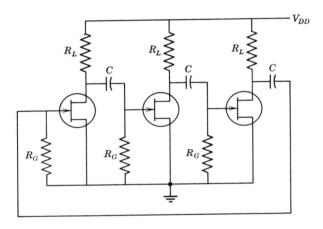

FIGURE 19.17
A three-stage *RC* oscillator.

19.4 *LC* OSCILLATORS

RC oscillators are used almost exclusively in the frequency range below about 500 kHz. However, the shunt capacitance in the *RC*-coupled stages becomes troublesome at higher frequencies, and tuned amplifiers which utilize this shunt capacitance as part of the tuning capacitance, as discussed in Chapter 9, are used as the basic amplifying device. In fact, an amplifier which has capacitive coupling between the input and output circuits may oscillate if both the input and load circuits are inductive. The conditions required for oscillation were discussed in Section 9.6. The oscillator which results when these conditions are intentionally met is known as a tuned-plate, tuned-grid (or tuned-collector, tuned-base, etc.) oscillator.

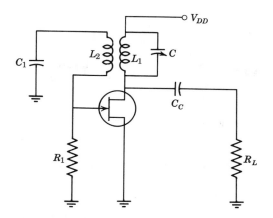

FIGURE 19.18
A tuned-drain oscillator.

The single-tuned inductively-coupled amplifier can be used as an oscillator if some of the energy in the output circuit is coupled back to the input circuit, as shown in Fig. 19.18.

The pole-zero plot for the voltage gain of the tuned-drain amplifier is given in Fig. 19.19. The root-locus plot for positive feedback is also given in this figure. Observe that the frequency of oscillation is very nearly the resonant frequency of the tuned circuit if the circuit Q is high because the

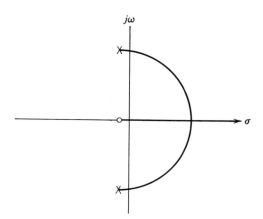

FIGURE 19.19
Pole-zero and root-locus plot for the tuned-drain
oscillator.

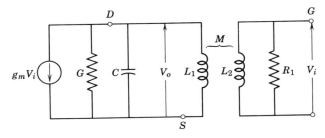

FIGURE 19.20
An equivalent circuit for the tuned-drain oscillator.

amplifier (open-loop) poles are very near the $j\omega$ axis. The actual frequency of oscillation and the mutual inductance requirements may be determined from an analysis of the equivalent circuit of Fig. 19.20. Small-signal operation is assumed initially because the equivalent circuit is valid only for small signals. Since the transformer primary impedance is $j\omega L_1 + (\omega M)^2/R_1$ (Eq. 9.56), assuming $R_1 \gg \omega L_2$, the nodal equation for the drain node may be written

$$g_m V_i = \left[G + j\omega C + \frac{1}{j\omega L_1 + (\omega M)^2/R_1} \right] V_o \qquad (19.22)$$

where G includes the conductance of the load resistance R_L. If $R_1 \gg \omega L_2$, as previously assumed, the gate voltage v_i is essentially $j\omega M I_d$, where I_d is the signal current through L_1. Then, since $I_d = V_o/(j\omega L_1 + \omega^2 M^2/R_1)$,

$$V_i = \frac{j\omega M V_o}{j\omega L_1 + [(\omega M)^2/R_1]} \qquad (19.23)$$

The voltage gain with positive feedback may now be written using $G_{23} = V_o/V_i$, which is obtainable from Eq. 19.22, and $G_{34} = V_i/V_o$ from Eq. 19.23.

$$G_{13} = \frac{G_{23}}{1 - G_{23} G_{34}} = \frac{g_m/[G + j\omega C + R_1/(j\omega L_1 R_1 + \omega^2 M^2)]}{1 - \dfrac{j\omega M g_m}{(j\omega L_1 + \omega^2 M^2/R_1)(G + j\omega C) + 1}} \qquad (19.24)$$

Oscillation occurs when the denominator of Eq. 19.24 is zero. Then

$$j\omega L_1 G - \omega^2 L_1 C + \frac{\omega^2 M^2 G}{R_1} + \frac{j\omega^3 M^2 C}{R_1} + 1 - j\omega M g_m = 0 \qquad (19.25)$$

Equating the real terms to zero

$$\omega^2 L_1 C = \frac{\omega^2 M^2 G}{R_1} + 1 \tag{19.26}$$

or

$$\omega^2 = \frac{1}{L_1 C - (M^2 G/R_1)} \simeq \frac{1}{L_1 C} + \frac{GM^2}{L_1{}^2 C^2 R_1} \tag{19.27}$$

Since $1/L_1 C = \omega_n{}^2$, where ω_n is the undamped resonant frequency of the tuned circuit,

$$\omega = \omega_n \sqrt{1 + \frac{(\omega_n M)^2 G}{R_1}} \tag{19.28}$$

Equation 19.28 shows that the frequency of oscillation depends on the load conductance G, the input resistance R_1, and the coupling impedance $\omega_n M$. The oscillation frequency will be very nearly ω_n if $(\omega_n M)^2 G/R_1$ is small in comparison with unity, which is true if the circuit Q is high so G is small, or R_1 is large, as it may be when the amplifier is an FET.

The required value of mutual inductance may be obtained by equating the imaginary terms of Eq. 19.25 to zero. Then

$$\frac{\omega^3 M^2 C}{R_1} - \omega M g_m + \omega L_1 G = 0 \tag{19.29}$$

or

$$M^2 - \frac{g_m R_1 M}{\omega^2 C} + \frac{L_1 G R_1}{\omega^2 C} = 0 \tag{19.30}$$

Using the quadratic formula,

$$M = \frac{g_m R_1}{2\omega^2 C} \pm \sqrt{\left(\frac{g_m R_1}{2\omega^2 C}\right)^2 - \frac{L_1 G R_1}{\omega^2 C}} \tag{19.31}$$

$$M = \frac{g_m R_1}{2\omega^2 C}\left(1 \pm \sqrt{1 - \frac{4\omega^2 L_1 C G}{g_m{}^2 R_1}}\right) \tag{19.32}$$

If the circuit Q is high, $\omega^2 \simeq 1/L_1 C$. Then

$$M \simeq \frac{g_m R_1 L_1}{2}\left(1 \pm \sqrt{1 - \frac{4G}{g_m{}^2 R_1}}\right) \tag{19.33}$$

The solution which results from the positive sign preceding the radical in Eq. 19.33 may be discarded because it yields unrealizable values of M.

Normally, $4G/g_m{}^2 R_1 \ll 1$; therefore, the approximation $(1 - x)^{1/2} \simeq 1 - x/2$ can be used to simplify Eq. 19.33. Then,

$$M \simeq \frac{L_1 G}{g_m} \qquad (19.34)$$

Note that the required value of M is independent of frequency, provided that a variable capacitor is used and L_1 remains fixed. The value of M obtained from Eq. 19.34 results in class A operation. Larger values of M cause increased base drive and result in class B or class C operation, depending on M. The resistor R_1 and capacitor C_1 (Fig. 19.18) automatically provide the increased reverse bias required for class B or class C operation as the mutual inductance M is increased and the gate junction draws current on the positive peaks of the input voltage.

A bipolar transistor can be used instead of the FET in Fig. 19.18. Then g_m is replaced by the y_{fe} of the transistor and y_{fe} becomes complex at frequencies above f_β. Thus the analysis becomes more complicated at these higher frequencies. Also, the lower input resistance R_1 of the bipolar transistor causes the frequency to be more dependent upon the transistor parameters.

The analysis for the FET is directly applicable to a tube oscillator. However, the frequency of the tube oscillator drifts considerably during its warm-up period because of the changing interelectrode capacitance. This frequency drift can be minimized by the use of negative temperature coefficient capacitors in the tuning circuit.

The Hartley oscillator circuit shown in Fig. 19.21 is similar to the tuned-collector circuit except a single, tapped coil is used and the tuning capacitor tunes the entire coil. In this arrangement, the coupling between the collector and base circuits does not depend on the mutual inductance between L_1 and L_2 because an ac voltage across L_1 is also applied across

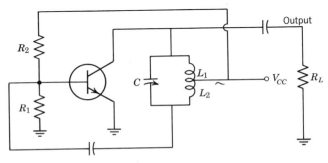

FIGURE 19.21
A Hartley oscillator circuit.

the series combination of C and L_2. Therefore, signal current will flow through L_2, and the voltage across L_2 is the feedback voltage to the base-emitter junction. The signal currents through L_1 and L_2 are essentially equal (because the tuned circuit current is large compared with the collector current), so that the ratio of collector voltage to base voltage is essentially L_1/L_2. This ratio is the voltage gain, and the reciprocal L_2/L_1 is the feedback ratio. Thus $K\beta = 1$, provided that L_2/L_1 is large enough to cause oscillation. Note that one end of the tuned circuit is at the same signal potential as the collector, and the other end of the tuned circuit is the same signal potential as the base. The coil tap is at the same signal potential as the emitter. Since the base and collector are at opposite potentials with respect to the emitter, the feedback is positive. This signal arrangement always holds for the Hartly oscillator. Since the input signal is not referenced to ground, any one of the three coil or electrode terminals may be at signal-ground potential. The oscillator operation is unaltered by the choice of ground point except, of course, that the output terminal must *not* be at the signal-ground point. An example will be used to illustrate one method of designing a Hartley oscillator.

Example 19.2   A 2N4957 transistor with y parameters given in Fig. 9.10 is to be used in the Hartley oscillator circuit of Fig. 19.21. Let us assume the load resistance R_L to be 10 kΩ and the Q_o of the coil to be 100. The oscillator frequency is to be 1.0 MHz. The circuit Q should be high to insure good frequency stability. In this example, we will design for $Q = 50$. The base driving power is very small in comparison with the power furnished to the load or dissipated in the tuned circuit and, therefore, its effect on the circuit Q will be neglected. Figure 9.10 shows that $g_{oe} \simeq 0.1$ mmho at $I_C = 2$ mA, so the parallel combination of R_L and R_o is $R_X = 5$ kΩ.

The inductance L_1 can be determined by the following method. The total shunt collector circuit resistance is (using Eq. 9.7)

$$R_{sh} = Q\omega_o L_1 \qquad (19.35)$$

The portion of this shunt resistance contributed by the coil resistance in the tuned circuit is

$$R_{par} = Q_o \omega_o L_1 \qquad (9.6)$$

But R_{sh} is the parallel combination of R_{par} and R_X. Then, using Eqs. 9.6 and 19.35

$$R_X = \frac{R_{par} R_{sh}}{R_{par} - R_{sh}} = \frac{Q_o Q}{Q_o - Q} \, \omega_o L_1 \qquad (19.36)$$

and

$$L_1 = \frac{R_x(Q_o - Q)}{\omega_o(Q_o Q)} \qquad (19.37)$$

In this example

$$L_1 = \frac{5 \times 10^3(50)}{6.28 \times 10^6(100)(50)} = 8.0 \ \mu\text{H}$$

The class A voltage gain of the amplifier is $y_{fe}R_{sh}$. Also R_{sh} is 5 kΩ in parallel with $Q_o \omega_o L_1 = 5$ kΩ, or 2.5 kΩ. The Q-point collector current will be chosen as 2.0 mA. Then y_{fe} (Fig. 9.10) is 58 mmhos and

$$G_v = y_{fe}R_{sh} = 58 \times 10^{-3}(2.5 \times 10^3) = 145$$

The oscillator will operate class A if

$$L_2 = L_1/G_v = 8/145 = 0.055 \ \mu\text{H}$$

However, a change in parameters or loading might stop the oscillation in this class A mode. The oscillator will be much more dependable if the inductance L_2 is increased by a factor of at least 4 or 5. The oscillator will then have much better amplitude stability and greater power output. Then

$$L_2 \simeq 5(0.055) \ \mu\text{H} = 0.27 \ \mu\text{H}$$

The tuning capacitance can be found from the relationship

$$C = \frac{1}{\omega_o^2 L} = \frac{1}{(6.28 \times 10^6)^2(8.27 \times 10^{-6})} = 3023 \ \text{pF}$$

The bias components are chosen to provide about 2.0 mA quiescent collector current and the blocking capacitor C should have reactance equal to approximately one-tenth of the bias resistance.

The Colpitts oscillator shown in Fig. 19.22 is almost identical to the Hartley except that the tuned circuit capacitance, instead of the inductance, is tapped. Also an RF choke has been added to permit the application of direct current to the collector and to present a very high impedance at the oscillation frequency. The design of a Colpitts oscillator may follow the pattern given for the Hartley oscillator in Example 19.2 but with $j\omega L_1$ and $j\omega L_2$ replaced by $1/j\omega C_1$ and $1/j\omega C_2$, respectively.

PROBLEM 19.5 Design a Colpitts oscillator which uses a 2N4957 transistor and has the same specifications and load resistance as the Hartley oscillator of Example 19.2.

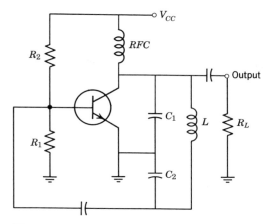

FIGURE 19.22
A Colpitts oscillator.

PROBLEM 19.6 Design a Hartley oscillator using a 2N2844 field-effect transistor. Draw a circuit diagram and determine suitable circuit components if $f_o = 1$ MHz, $V_{DD} = -15$ V, and the external load resistance $R_L = 10$ kΩ. The coil $Q_o = 150$ and the desired circuit $Q = 100$. Design for class C operation. Specify the coil tap point n_2/n.
Answer: $L_1 = 5.3$ μH, $n_2/n \simeq 1/6$.

19.5 CRYSTAL-CONTROLLED OSCILLATORS

A general class of oscillators which achieve very good frequency stability because of the exploitation of a high Q circuit is the *crystal-controlled* oscillator. In the "crystal" oscillator, the conventional LC circuit is replaced by a quartz crystal. The crystal has the property of producing a potential difference between its parallel faces when the crystal is strained or deformed. Conversely, when a potential difference is applied across the faces of a crystal, it will deform or change shape. This property, which is known as the *piezoelectric effect* after its discoverer, causes the crystal to behave as a very high Q resonant circuit. The crystal will vibrate readily at its *mechanical resonant frequency*, but because of its associated electrical properties the crystal behaves as though it were an LC circuit with extremely high Q (of the order of thousands). The crystal is cut into very thin slices and then carefully ground to the desired resonant frequency. The orientation of the slice, with reference to the crystal axes, determines the properties of the crystal, such as vigor of oscillation and variation of frequency with temperature.

The equivalent electrical circuit of a crystal is given in Fig. 19.23. The crystal itself behaves as a series RLC circuit. However, the electrical connections must be made to the crystal faces by conducting electrodes or plates, known as a crystal holder. The crystal holder provides a capacitance, shown as C_h in Fig. 19.23, which is in parallel with the crystal circuit. Thus the crystal behaves as a series resonant circuit at its natural resonant frequency, but at a slightly higher frequency the net inductive reactance of the crystal resonates with the crystal holder capacitance to produce parallel resonance. The parallel resonant frequency is only slightly higher than the series resonant frequency because the equivalent inductance of the crystal may be of the order of henries. This extremely high equivalent inductance accounts for the extremely high Q of the crystal and provides a very impressive rate of change of reactance with frequency.

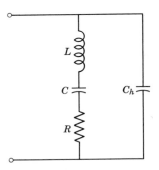

FIGURE 19.23
The equivalent circuit of a
crystal mounted in a holder.

The reactance of a typical crystal in a holder is sketched as a function of frequency in Fig. 19.24. Note that the reactance is inductive only between the series resonant frequency ω_s and the parallel resonant frequency ω_p. These frequencies differ by a very small percentage (a few hundred Hz per MHz); therefore, the effective inductance changes very rapidly with frequency in this region.

The crystal can replace the tuning inductor in a conventional circuit as illustrated by the Colpitts-type circuit of Fig. 19.25. The oscillator may be designed as a conventional oscillator and the crystal will provide the proper inductance for operation very near its natural resonant frequency. A change of tuning capacitance changes the impedance of the tuned circuit but has little

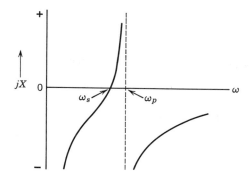

FIGURE 19.24
A sketch of reactance as a function of frequency for a crystal in a holder.

influence on the frequency of oscillation because of the compensating change of effective inductance.

In the circuit of Fig. 19.26a, the crystal operates in its series mode. At the resonant frequency of the crystal, the oscillator operates as a Hartley circuit. The circuit of Fig. 19.26b uses the crystal in its series mode to couple two transistors, one of which is operating in the common-base configuration and the other in the common-collector configuration. Many other circuits may be devised or found in the literature.

Crystals can be used in conjunction with integrated circuits to produce stable, fixed-frequency oscillations. A basic IC crystal oscillator circuit is shown in Fig. 19.27. Observe that this circuit is quite similar to the Wein-Bridge oscillator. The esssential differences are:

1. The frequency-selective network in the positive feedback network is a crystal instead of a series-parallel RC network.
2. A band-pass, or low-pass filter as shown, is used in the output to attenuate harmonics and thus provide good output waveform without the use of automatic gain control.

The forward gain of the amplifier is controlled by the negative feedback circuit so that vigorous oscillation, but essentially class A operation, is obtained. The bandwidth of the amplifier, with feedback, must be adequate for the desired frequency of oscillation, of course. Video amplifiers usually provide wider bandwidth than operational amplifiers. Also IC video amplifiers often provide convenient gain control and are therefore convenient to use as the IC amplifier. If an operational amplifier is used, it will normally

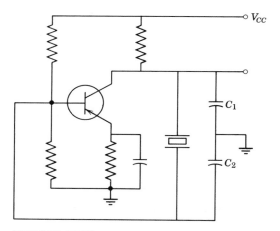

FIGURE 19.25
A Colpitts oscillator with a crystal for a tuned circuit.

require compensation to insure stability. Otherwise, spurious oscillations may occur.

The chief disadvantages of crystals are as follows.

1.   They are fragile, especially the high-frequency crystals, and consequently can be used only in low-power circuits.

2.   The oscillator frequency cannot be adjusted appreciably. However, the parallel or holder capacitance has some effect on the frequency in the parallel mode.

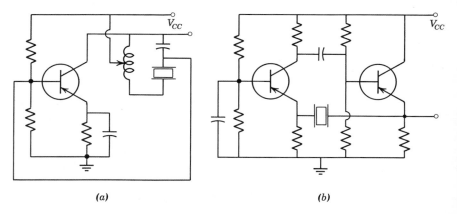

(a)   (b)

FIGURE 19.26
Some typical crystal oscillator circuits.

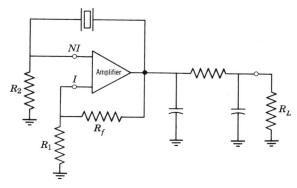

FIGURE 19.27
An *IC* crystal oscillator.

PROBLEM 19.7 A quartz crystal has the following electrical character-istics.

$L = 3.2$ henries
$C = 0.05$ pF
$R = 4000\ \Omega$
$C_h = 6$ pF

(*a*) Determine the value of f_s and f_p for this crystal. (*b*) What is the Q of this crystal? (*c*) Design a Colpitts oscillator which uses this crystal and a 2N4957 transistor. The value of R_L is 10 kΩ and the transistor output resistance is 40 kΩ.

Answer: (*a*) $f_s = 398$ kHz, $f_p = 401$ kHz; (*b*) $Q = 2000$.

PROBLEM 19.8 Design a 10 MHz crystal oscillator using an *IC* of your choice. Assume that the crystal has the following electrical parameters, approximately.

$L = .025$ H
$C = .01$ pF
$R = 5$ kΩ
$C_h = 1$ pF

Determine the values of suitable external components for the circuit.

PROBLEM 19.9 The phase-shift oscillator shown in Fig. 19.28 is a rather common type of *RC* oscillator. The tube and load resistor form an amplifier, and the three capacitors C in conjunction with the three resistors R form the feedback path. Make a root-locus plot for this circuit, and determine the required transconductance and the oscillation frequency.

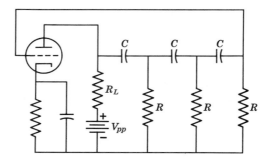

FIGURE 19.28
A phase-shift oscillator.

PROBLEM 19.10   A reflex klystron is used for generating microwave frequencies. A typical reflex klystron contains a built-in resonant cavity, which can be represented as a parallel-tuned circuit. (*a*) Determine the value of L, C, and R for this tuned circuit if the resonant frequency is 10^{10} Hz and the Q of the circuit is 1000. The conductance of the tuned circuit at resonance is 20 micromhos. (*b*) Make a plot of $-Y(j\omega)$ for this tuned circuit.

An electron beam passes through a gap in the resonant cavity. The electrons in this beam are stopped by the electric field of a negative "repeller" electrode and repelled back through the gap in the resonant cavity. The electron beam interacts with the electric field of the gap to produce a conductance g in parallel with the capacitor of the tuned circuit. A plot of the value of g as a function of N is shown in Fig. 19.29. The parameter N is the number of cycles which the electric field across the gap has completed between the time a given reference electron first passed through the gap to the time when this same electron returns to the gap. (*c*) At what value of N will oscillations first begin? (*d*) What is the frequency of these oscillations? The radius of the admittance spiral decreases as the ac signal increases. (*e*) At what value of N (for $0 < N < 2$) will the ac signal be maximum? (*f*) As N is increased above this value for maximum signal, the oscillations will cease. At what value of N will oscillations cease? (*g*) What frequency corresponds to the value of N in part *f*?

PROBLEM 19.11   Convert the amplifier of Problem 18.9 into an oscillator. This oscillator will provide the excitation for the tube of Examples 18.2 and 18.4.

PROBLEM 19.12   Design an oscillator that will operate over the frequency range 25 MHz to 30 MHz, using an *IC* of your choice. Determine suitable values for the external components.

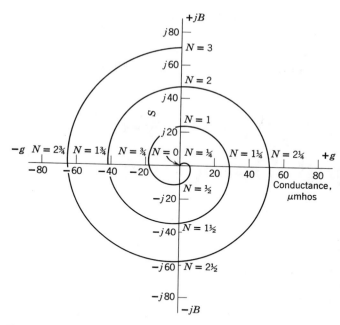

FIGURE 19.29
A plot of electron beam conductance in a reflex klystron.

20

Amplitude Modulation and Detection

With the exception of the pulse circuits, the goal in most previous amplifiers has been to preserve the waveform of the input signal. However, in some applications, the form of the signal is *intentionally* changed. Generally, these devices are known as *modulators*, and the process is known as *modulation* which means "to change."

Modulators are needed in a great number of electronic systems. For example, in a radio transmitter an oscillator generates the basic radio frequency signal which is commonly known as the *carrier*. If this carrier were merely amplified and broadcast, there would be no intelligence transmitted, and the system would be useless. Thus, somewhere in the transmitter the carrier must be changed or modulated by the intelligence which is to be transmitted. The intelligence can then be recovered at radio receivers by a device known as a *detector*. The carrier may be changed in any of several ways such as *in amplitude* or *in frequency*. In this chapter only *amplitude modulation* will be considered.

20.1 AMPLITUDE MODULATION

In order to gain some understanding of the fundamental principles of amplitude modulation, a simple special case will be considered in which a carrier with maximum amplitude A_c and natural frequency ω_c will be modulated by a sinusoidal (single-frequency) signal which has a natural frequency ω_m. A sketch of the modulated carrier voltage as a function of time, along with the carrier and the modulating signal, is given in Fig. 20.1. As shown in this figure, the maximum amplitude variation from the unmodulated value is MA_c, where M is known as the *modulation index*.

626

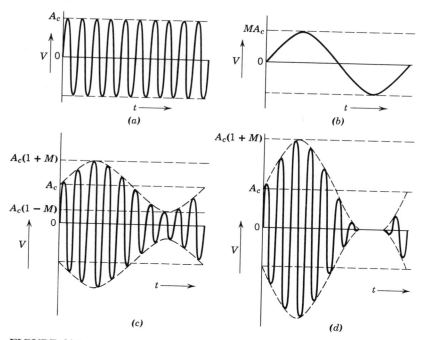

FIGURE 20.1
Modulation: (a) the carrier, (b) the modulating signal, (c) an amplitude-modulated carrier, (d) an overmodulated carrier.

When M has a value of one, the amplitude of the modulated wave varies between $2A_c$ and zero. In this case, the carrier is said to be 100 percent modulated. When M has the value 0.5, the amplitude of the modulated carrier varies between 1.5 A_c and 0.5 A_c. The carrier is then said to be 50 percent modulated, and so forth. Observe that if M exceeds unity, the carrier is completely interrupted for a time, the envelope (see Fig. 20.1) of the carrier no longer has the same form as the modulating signal, and the carrier is said to be *overmodulated*. Overmodulation naturally causes distortion in the system. It may be seen from Fig. 20.1 that the voltage of the modulated wave may be expressed by Eq. 20.1

$$v = A_c[1 + M \cos \omega_m t] \cos \omega_c t \qquad (20.1)$$

This expression could have just as well been in terms of current instead of voltage. Also Eq. 20.1 would be more general if arbitrary phase angles were included in the expression. However, the results would not be altered by the increased generality. Expanding Eq. 20.1, we have

$$v = A_c \cos \omega_c t + M A_c \cos \omega_c t \cos \omega_m t \qquad (20.2)$$

Substituting the trigonometric identity $\cos a \cos b = 1/2[\cos(a + b) + \cos(a - b)]$ into Eq. 20.2, we see that

$$v = A_c \cos \omega_c t + \frac{MA_c}{2} \cos(\omega_c + \omega_m)t + \frac{MA_c}{2} \cos(\omega_c - \omega_m)t \quad (20.3)$$

It may be seen from Eq. 20.3 that the effect of the modulation is to *produce two new frequencies*, which are called side frequencies. The upper side frequency is the *sum* of the carrier frequency and the modulating frequency whereas the lower side frequency is the *difference* between the carrier and modulating frequencies. Therefore, a tuned amplifier which is called upon to amplify a modulated carrier must have sufficient bandwidth to include the side frequencies. Notice that the modulating frequency is *not* included in the modulated wave. In case the modulating signal were derived from a symphony orchestra, each frequency component would produce a pair of side frequencies. Consequently, the highest frequency components present in the modulating signal would determine the required bandwidth of the tuned circuits in the radio transmitting and receiving equipment. If these tuned circuits have insufficient bandwidth, the highest modulating frequencies will not be reproduced by the receiver. Collectively, the upper side frequencies are known as the upper sideband and the lower side frequencies are known as the lower sideband.

Any modulating waveform which is represented as a function of time may be resolved into frequency components by Fourier analysis. Thus, the bandwidth requirements may be determined whenever the waveform of the modulating signal is known as a function of time.

Example 20.1   Consider the rectangular pulse shown in Fig. 20.2*a* to be the modulating signal. The pulse duration is t_d and the period is T. Using Fourier analysis, the frequency components of the modulating signal are found as follows. The time reference is taken so that only cosine terms will appear in the Fourier series. Then

$$A_n = \frac{2}{T} \int_{-T/2}^{T/2} V \cos\left(\frac{2\pi n}{T} t\right) dt = \frac{4}{T} \int_0^{T/2} V \cos\left(\frac{2\pi n}{T} t\right) dt \quad (20.4)$$

Since the function V is zero from $t_d/2 < t < T/2$, Eq. 20.4 becomes

$$A_n = \frac{4}{T} \int_0^{t_d/2} V \cos\left(\frac{2\pi n}{T} t\right) dt = \frac{T4V}{2\pi Tn} \left[\sin\left(\frac{2\pi n}{T} t\right)\right]_0^{t_d/2} \quad (20.5)$$

then

$$A_n = \frac{2V}{\pi n} \sin\left(\frac{\pi n t_d}{T}\right) \quad (20.6)$$

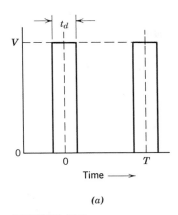
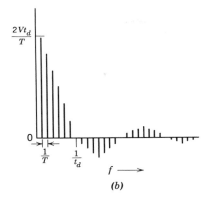

FIGURE 20.2
(a) **A rectangular pulse-modulating signal,** (b) **the frequency spectrum of a rectangular pulse.**

Multiplying both the numerator and denominator of Eq. 20.6 by t_d and re-arranging, we have

$$A_n = \frac{2Vt_d}{T} \frac{\sin(\pi n t_d/T)}{n\pi t_d/T} \tag{20.7}$$

Thus, the frequency components of the modulating wave are an infinite series as follows.

$$V_{(\omega)} = \frac{2Vt_d}{T} \left[\frac{\sin[(\pi t_d)/T]}{(\pi t_d)/T} \cos\left(\frac{2\pi}{T} t\right) + \frac{\sin[(2\pi t_d)/T]}{(2\pi t_d)/T} \cos \frac{4\pi}{T} t \cdots \right] \tag{20.8}$$

The coefficients of these harmonically related frequency components are of the form $(\sin x)/x$. These frequency components, which are spaced at the interval $f = 1/T$, are shown graphically in Fig. 20.2b. It was assumed in Fig. 20.2 that the period T is large in comparison with the pulse duration t_d. Of course, all the components cannot be shown because they form an infinite series. The frequency components of the amplitude-modulated carrier are shown in Fig. 20.3 along with a sketch of the modulated wave. To include all the side frequencies, the bandwidth of the amplifiers which amplify this signal must be infinite. Of course, an infinite bandwidth is practically impossible to attain, so a compromise must be reached. Frequently, the accepted compromise is that all the side frequencies up to the first zero amplitude component be included, since the frequency components beyond this point are of rather small amplitude. This choice would result in a modulated signal which departs considerably from the ideal rectangular shape, but nevertheless it provides a convenient reference point because the

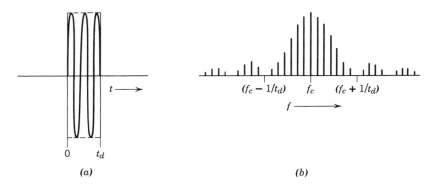

FIGURE 20.3
(a) **A pulse-modulated wave,** (b) **the frequency components of a pulse-modulated wave.**

first zero amplitude modulating component occurs (from Eq. 20.7) at the lowest value of n for which

$$A_n = \frac{2Vt_d}{T} \frac{\sin(n\pi t_d/T)}{n\pi t_d/T} = 0 \qquad (20.9)$$

The lowest values of n at which Eq. 20.9 will hold is

$$\frac{n\pi t_d}{T} = \pi \qquad (20.10)$$

or

$$n = \frac{T}{t_d} \qquad (20.11)$$

Since the frequency components are separated by the frequency $1/T$, the width of a single sideband would be (from Eq. 20.11)

$$\frac{B}{2} = n\left(\frac{1}{T}\right) = \frac{1}{t_d} \qquad (20.12)$$

The bandwidth requirement would then be $2/t_d$. Wider bandwidth would naturally provide better waveform of the modulated signal.

The bandwidth requirement for a pulse-modulated wave can also be determined from the required rise time of the modulation envelope. The analysis of the time response of a tuned amplifier, given in Chapter 9, showed that the rise time of the modulation envelope is approximately $4.4/B$, where B is the bandwidth of the tuned amplifier in radians per second. Therefore, the bandwidth may be determined from the desired rise time.

PROBLEM 20.1 A 100 MHz carrier is modulated by a rectangular pulse which has a duration of 1 μsec and a repetition rate of 1000 pulses per second (pps). If the side frequencies up to the first zero magnitude component on each side of the carrier are to be included, what bandwidth will be required of an amplifier for this modulated wave? How many side frequencies would be included in this case, and what circuit Q would be required of an amplifier which incorporates a single-tuned circuit? What will be the envelope rise time of this amplifier?

PROBLEM 20.2 Repeat Problem 20.1 but include all side frequencies up to the second zero component.

Returning to the case of a single modulating frequency, we can see from Eq. 20.3 that when the modulation index M is unity (100 percent modulation) the amplitude of either side frequency is one-half that of the carrier. Since power is proportional to the square of the amplitude, each side frequency will have one-fourth as much power as the carrier, and the total side frequency power will be one-half as great as the carrier power. Again, this relationship holds only for 100 percent sinusoidal modulation.

20.2 MODULATING CIRCUITS

A large variety of circuits may be used to provide amplitude modulation. Only a few typical circuits will be included in this work for the purpose of illustrating the basic principles. One common method of accomplishing modulation is by using the modulating signal to vary the plate or collector voltage of a class B or class C amplifier as shown in Fig. 20.4. Since in class B or class C operation, the peak amplitude of the signal voltage in the output is very nearly equal to the supply voltage, the amplitude of the output voltage follows very closely the variation in voltage supplied from the modulator. The modulation is said to be *linear* when the envelope of the modulated wave has the same waveform as the modulating signal. Observe that the modulation may be linear even though the modulated amplifier may be very nonlinear so far as the waveform of each RF cycle is concerned. The tuned coupling circuit essentially eliminates the harmonics of the carrier frequency as well as the modulating frequency, so the output is a modulated wave of the form of Fig. 20.1c. The capacitor C provides a low impedance path to the carrier currents so these currents do not flow through the modulation transformer. On the other hand, the capacitor C must not bypass the modulating frequencies.

The power requirement of the modulator as well as the modulator load resistance may be determined from the basic current and voltage relationships of the modulated amplifier. To obtain 100 percent modulation, the

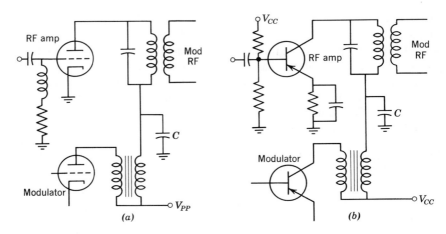

FIGURE 20.4
Typical circuits for (a) plate and (b) collector modulation.

maximum value of the modulating voltage V_m must be equal to the power supply voltage (V_{PP} or V_{CC}). Under these conditions, the RF output of the modulated amplifier is equal to zero at the negative peak of the modulating signal. Then, using the tube circuit as an example, we have

$$V_{m\,max} = V_{PP} \tag{20.13}$$

Also, since the average plate current of the modulated amplifier is reduced from I_P to zero during this negative half-cycle of the modulating signal, the maximum value of the modulating signal current is

$$I_{m\,max} = I_P \tag{20.14}$$

The modulator power output is

$$P_{mod} = \frac{V_{m\,max}I_{m\,max}}{2} = \frac{V_{PP}I_P}{2} \tag{20.15}$$

Thus, as shown by Eq. 20.15, the power output from the modulator must be equal to one-half the power supplied to the RF amplifier by the power supply. Therefore, the power from the modulator provides the power for the sideband frequencies. It was previously shown that the power in the side frequencies is one-half the carrier power for this case.

The effective load on the modulator may be easily determined since the maximum amplitudes of both the modulator voltage and current are known.

$$R_L = \frac{V_{m\,max}}{I_{m\,max}} = \frac{V_{PP}}{I_P} \tag{20.16}$$

The value of supply voltage is known, and the value of I_P may be determined from the graphical analysis of the class B or class C amplifier, as discussed in Chapter 18. Also, I_P is easily measured by a dc meter in the plate circuit of the RF amplifier.

The turns ratio of the modulation transformer should be chosen so that the load resistance determined by Eq. 20.16 will present the desired load resistance for the modulating tube or transistor.

The modulator may typically be any of the power amplifier circuits discussed in Chapter 11. Push-pull circuits are usually used to modulate high-power amplifiers. For pentode and tetrode RF amplifiers, it is helpful to modulate the screen grid voltage as well as the plate voltage, since the plate potential has a very small effect on the magnitude of plate current except at low values of plate voltage. Also, the screen current becomes high when the plate voltage is low, unless the screen voltage is reduced at the same time.

PROBLEM 20.3 A class C amplifier which has 1 kW output and is 80 percent efficient is to be plate modulated. The plate supply voltage $V_{PP} = 2000$ V. Determine the required power output and load resistance for the modulator. The modulator is expected to provide 100 percent modulation. *Answer:* $P_{\text{mod}} = 625$ W, $R_L = 3200 \ \Omega$.

PROBLEM 20.4 Draw the circuit diagram for a modulated pentode amplifier in which both plate and screen grid are modulated.

The linearity of plate or collector modulation is usually good. In the tube circuit, better linearity is attained when the modulated amplifier has grid-leak bias rather than fixed bias. The improved linearity results from the variation of the bias over the modulation cycle. During the negative half of the modulation cycle, the grid draws more current because the plate potential is reduced. Thus, the bias is increased and assists in the reduction of plate current. Conversely, during the positive portion of the modulation cycle, the grid current is decreased because of the increased plate voltage. Consequently, the bias is decreased, thus enhancing the plate current increase. This varying bias tends to offset the tendency toward flattening of the peaks of the modulated wave due to saturation effects.

Collector modulation presents a special problem in the transistor circuit. The base drive must be large enough to saturate the transistor at the peak of the modulation cycle in order to provide linear modulation and high efficiency. Consequently, the transistor may be highly overdriven during the modulation troughs when the collector current should be comparatively small. However, excess charge[1] is stored in the base during the time the

---

[1] This excess stored charge is discussed in more detail in Chapter 22.

transistor is in saturation and the collector current cannot decrease until this excess charge has been removed. As a result of this delay, the large values of collector current are not confined to the period during which the collector voltage is low and the collector dissipation is increased. In fact, the collector current may increase during the portion of the collector voltage cycle when the current would normally decrease. This enlarged, out-of-phase current may seriously decrease the efficiency of the amplifier. Grid-leak type, or RC bias, will curtail the excess stored charge because the base current increases rapidly when the transistor is driven into saturation and the resulting increased bias decreases the excessive base drive.

The main disadvantage of plate modulation is the large modulating power required when the modulated amplifier is of high power. Of course, a lower level stage could be modulated in a high-power transmitter, and the modulated wave could then be amplified. But class C amplifiers are not suitable for amplifying modulated waves because the waveform of the modulation is not preserved. Of course, class B amplifiers are suitable, but their reduced efficiency is a serious handicap when the power level is high.

Fortunately, grid or base modulation requires much less modulating power than does plate modulation. Typical grid and base modulation circuits are given in Fig. 20.5. The RF amplifiers could be operated either class B or class C. In these circuits, the modulating voltage varies the grid (or base) bias as shown in Fig. 20.6. The dynamic transfer characteristic for a typical

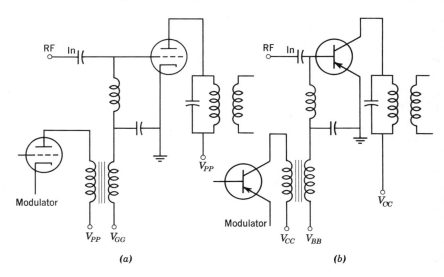

(a)   (b)

FIGURE 20.5
(a) **Grid-modulated and** (b) **base-modulated amplifiers.**

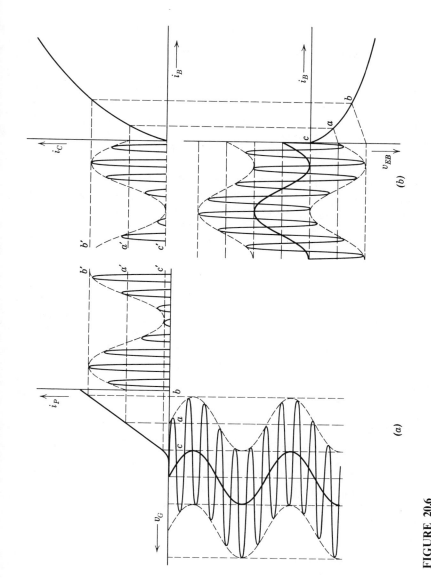

FIGURE 20.6
Voltage and current relationships in (a) grid and (b) base-modulated amplifiers.

vacuum tube is given in Fig. 20.6a, and the current transfer characteristics of a typical transistor are given in Fig. 20.6b. For the transistor, the effects of the driving-source resistance R_s, as discussed in Chapter 11, are included. In this figure, the source resistances of the carrier and the modulator are assumed to be the same, and in practice these resistances should be adjusted to approximately the same value.

In the RF amplifiers of Fig. 20.6 it is desirable to vary the output current from the maximum design value (point b') at the crest of the modulating signal to essentially zero (point c') at the trough of the modulating signal. Therefore, the RF amplifier current peaks should be adjusted to the average of these two values (point a') when the carrier is unmodulated. It may be seen from Fig. 20.6 that the maximum amplitude of the modulating voltage should be the potential difference between point a and either point b or point c on the input voltage axis. This potential difference will be called V_m. Considering the transistor, we know that the bias current varies from the value obtained when the carrier is unmodulated (i_{Ba}) to zero as the modulating voltage varies through its negative half-cycle. Therefore, the transducer[2] power required from the modulator is

$$P_m = \frac{V_m i_{Ba}}{2} \qquad (20.17)$$

As stated previously, the output resistance of the modulator and its associated circuits should be approximately the same as the output resistance of the carrier source and its associated circuits. These resistances should have the value which gives best linearity as determined by the technique discussed in Chapter 11.

In the tube amplifier, the grid draws no current when the carrier is unmodulated. Hence, the grid draws current and therefore presents a load to the modulator only on the positive peaks of the modulation cycle. This varying load will cause distortion of the modulation envelope unless one of the following conditions is met.

1.   The output impedance of the modulator is small in comparison with the minimum load resistance which occurs at the modulation peaks.

2.   A loading resistor, which is small in comparison with the minimum load resistance caused by the grid current, is placed across the output of the modulator.

Grid or base modulation has two serious disadvantages when compared with plate or collector modulation. First, the power output and efficiency of

---

[2] The transducer power is the power furnished by the current generator or voltage generator in the equivalent circuit of the source. This is the power delivered to the internal resistance of the source in addition to the load.

the grid-modulated amplifier is comparatively low. Since the unmodulated plate current peaks can be only about half as large as in the plate-modulated circuit, the power output and efficiency suffer severely. Second, the adjustment of the grid- (base) modulated amplifier is more critical and a high degree of linearity is more difficult to attain.

Any of the electrodes of a tube or transistor could be used as the modulated element. For example, emitter or cathode modulation is very similar to base or grid modulation. The main difference is that the emitter (cathode) current is much larger than the base (grid) current so the modulator power must be much greater. On the other hand, the linearity might be better, especially in a tube amplifier, because the cathode current is a fairly linear function of the modulating voltage.

With the advent of integrated circuits, additional methods of modulation present themselves. One type of integrated circuit is known as a *multiplier*, or *gated video amplifier*. (Several multiplier circuits are on the market including the MC1595 by Motorola.) If two signals, v_1 and v_2, are applied to the inputs of a multiplier circuit, the output signal is the product $v_1 v_2$ of the two input voltages. Thus, if the multiplier is connected as shown in Fig. 20.7, the output

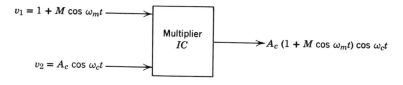

FIGURE 20.7
A multiplier used as a modulator.

signal will be an AM signal as shown. Of course, due to the permissible power dissipation of the *IC*, this circuit is a low-signal level circuit. However, for low-power applications, the compact size of *IC*s make this arrangement very attractive.

20.3 THE MODULATION PROCESS

By this time, the inquisitive reader should have posed the question, "What basic difference exists between the linear amplifier which only amplifies the applied signals and the modulator which produces the sum and difference of the applied frequencies in addition to their possible amplification?" A clue to the answer might be found in Chapter 11 where the production of new frequencies called harmonics is considered. In that case, it was the nonlinearity of the amplifier parameters that generated the new frequencies. This nonlinear relationship between the output current i_O and the input voltage v_i can be expressed by the power series

$$i_O = A_0 + A_1 v_i + A_2 v_i^2 + A_3 v_i^3 + \cdots \qquad (20.18)$$

Now, if the input signal v_i consists of two sinusoidal signals, for example, a carrier frequency and a modulating frequency, then

$$v_i = A_c \cos \omega_c t + A_m \cos \omega_m t \qquad (20.19)$$

Substituting this value of v_i into Eq. 20.18, we have

$$i_O = A_0 + A_1(A_c \cos \omega_c t + A_m \cos \omega_m t) + A_2(A_c \cos \omega_c t + A_m \cos \omega_m t)^2$$
$$+ A_3(A_c \cos \omega_c t + A_m \cos \omega_m t)^3 + \cdots \qquad (20.20)$$

It we assume that the nonlinearity is such that the fourth and all higher order terms are negligibly small, Eq. 20.20 can be expanded to

$$i_O = A_0 + A_1 A_c \cos \omega_c t + A_1 A_m \cos \omega_m t + A_2 A_c^2 \cos^2 \omega_c t$$
$$+ A_2 A_m^2 \cos^2 \omega_m t + 2A_2 A_c A_m \cos \omega_c t \cos \omega_m t$$
$$+ A_3 A_c^3 \cos^3 \omega_c t + 3A_3 A_c^2 A_m \cos^2 \omega_c t \cos \omega_m t$$
$$+ 3A_3 A_c A_m^2 \cos \omega_c t \cos^2 \omega_m t + A_3 A_m^3 \cos^3 \omega_m t \qquad (20.21)$$

Most of the terms of Eq. 20.21 have a familiar form. As previously discussed, the A_0 term is merely a bias term and the next two terms involving A_1 are the input frequency components which appear in the output current. The next three terms (4th, 5th, and 6th) result from the second-order term of the series. Two of these terms are squared and result in second harmonics of both of the input frequencies, as shown in Chapter 11. The other second-order term is the product term $2A_2 A_c A_m \cos \omega_c t \cos \omega_m t$. Using the trigonometric identity $\cos a \cos b = 1/2[\cos(a + b) + \cos(a - b)]$, we see that this term becomes $A_2 A_c A_m[\cos(\omega_c + \omega_m)t + \cos(\omega_c - \omega_m)t]$. In this case the frequencies $(\omega_c + \omega_m)$ and $(\omega_c - \omega_m)$ are the sum and difference frequencies, or the sideband frequencies, of the modulated wave. Thus, it is seen that a second-order nonlinearity of amplifier parameters causes modulation.

Continuing the investigation, we see that the remaining four terms of Eq. 20.21 result from the third-order term of the series. If the trigonometric identity $\cos^3 a = (3/4)\cos a + (1/4)\cos 3a$ is used, it is seen that the cubic cosine terms produce third harmonics of the input frequencies as well as contribute to the fundamental. The remaining two terms of the form $\cos^2 a \cos b$ could be written as $\cos a(\cos a \cos b)$ which, in turn could be written as $1/2(\cos a)[\cos(a + b) + \cos(a - b)]$. Using the identities again, we could write this term as $1/4[\cos(2a + b) + \cos b + \cos(2a - b) - \cos b]$. When $\omega_m t$ is substituted for a, and $\omega_c t$ for b, it is seen that sideband frequencies appear at twice the modulating frequency; or, in other words, there is distortion in the modulation components. On the other hand, when $\omega_c t$ is substituted for a, and $\omega_m t$ for b, in the foregoing trigonometric term, it is seen that the second harmonic of the carrier frequency $2\omega_c$ also has sidebands. Therefore, the cubic term of the series produces second harmonic distortion of the modulation envelope, third harmonic distortion of the input frequencies, and sidebands of the $2\omega_c$ term. From the preceding analysis, it should be evident that if the fourth-order term were included in the series, fourth harmonics of the input frequencies and third harmonic distortion terms of the modulation envelope would occur in the amplifier current, and so on.

The various frequency components which appear in the output current of the nonlinear amplifier are displayed graphically in Fig. 20.8. The

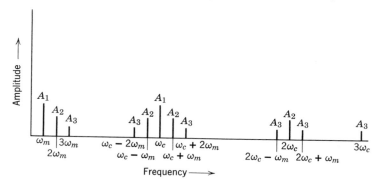

FIGURE 20.8
A display of the frequency components which appear in the output current of a nonlinear amplifier which has two frequencies, ω_c and ω_m, applied to its input. Only second- and third-order nonlinearities are considered.

relative magnitudes of these various components depend on the relative magnitudes of the first-order coefficient A_1, the second-order coefficient A_2, and the third-order coefficient A_3, as shown in Fig. 20.8.

If the amplifier has only a second-order nonlinearity and a filter were used to select only the carrier and the sideband frequencies, the output of the amplifier would be (from Eq. 20.21 and the trigonometric identities)

$$i_o = A_1 A_c \cos \omega_c t + A_2 A_c A_m[\cos(\omega_c + \omega_m)t + \cos(\omega_c - \omega_m)t] \quad (20.22)$$

If we write Eq. 20.22 in the form of Eq. 20.3,

$$i_o = A_1 A_c \left\{ \cos \omega_c t + \frac{A_2 A_m}{A_1} [\cos(\omega_c + \omega_m)t + \cos(\omega_c - \omega_m)t] \right\} \quad (20.23)$$

By comparison with Eq. 20.3, the modulation index of this second-order or "square-law" modulator is

$$M = \frac{2A_2 A_m}{A_1} \tag{20.24}$$

A good "square-law" modulator would then have a large value of A_2, or high degree of second-order nonlinearity, but no higher order of nonlinearity. Unfortunately, this type of device may be difficult to find. However, examples of this type of modulated amplifier appear later in this chapter.

A slight digression at this point may be in order. It should be recalled that in Chapter 11 the objective was to minimize the non-linearity so that the output waveform would be identical to the input waveform and hence contain the same frequency components in the same relative magnitudes. The harmonics present in the output were used as an index of the degree of nonlinearity of the amplifier. Although the harmonics are a good *index* of the nonlinearity of an audio amplifier, which is used in the reproduction of sound, they *are not* the cause of the dischordant, unpleasant sounds which come from a nonlinear amplifier. The harmonically related frequencies are harmonious, and although an increased harmonic content will change the timbre of the sound, it will not cause the sound to be unpleasant. It is the modulation terms (sum and difference frequencies) produced by the nonlinearity which cause the dischordant, unpleasant sounds. The sum and difference frequencies may not be harmonious with the original frequencies. Thus, a poor sound reproducer may sound acceptable for a solo performance but unacceptable for an ensemble.

PROBLEM 20.5   A pentode amplifier has $g_m = 4000 + 200v_G$ μmhos. The Q point is $V_G = -10$ V. Two sinusoidal signals, each having a maximum amplitude of 5 V are applied to the input. Their frequencies are 150 Hz and 400 Hz. The load is 5 kΩ resistive, and the amplifier is well designed. Assuming the load resistance to be very small in comparison with the plate resistance, determine the amplitude of the sum and difference frequencies and second harmonics in the output. What is the percent modulation, assuming the 400 Hz frequency to be the carrier?
Answer: fund. = 50 V peak, 2nd Har = 12.5 V peak, Sum = Diff = 25 V peak, % Mod = 100.

20.4   SINGLE SIDEBAND TRANSMISSION

In a modulated wave the sidebands carry the intelligence. Yet the total sideband power is usually much less than the carrier power. Also, the upper sideband carries the same information as the lower sideband. Therefore, amplitude modulation seems to be an inefficient way to transmit intelligence. To increase the efficiency of transmission, the carrier and one sideband are sometimes eliminated. This type of transmission is known as *single sideband* transmission and results in reduced power and bandwidth requirement for a specific transmission effectiveness. When single sideband transmission is used, the carrier (from a local oscillator) must be inserted in the receiver to recover the intelligence. As will be shown later, the local oscillator which provides the carrier must have very good frequency stability to faithfully reproduce the modulating signal.

One of the problems of single sideband transmission is the separation of one sideband from the other sideband and carrier. If low-frequency components are present in the modulating signal, some side frequencies are very near the carrier frequency. A commonly used circuit which eliminates the carrier and thus reduces the filtering requirement for single sideband production is the balanced modulator. In a typical balanced modulator circuit such as is shown in Fig. 20.9, the carrier is applied in phase to the

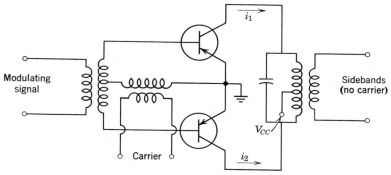

FIGURE 20.9
A balanced modulator circuit.

inputs of the two transistors while the modulating signal is applied in opposite phase to the two inputs. The amplifiers are operated class *B* so modulation takes place in each amplifier. The carrier is suppressed in the output because the carrier is applied "in phase" to the inputs. The output is tuned to the desired sideband frequency, thus essentially eliminating the modulating signal. Additional filtering is required to eliminate the undesired sideband.

A brief analysis will illustrate this modulation process. It will be assumed that the amplifier is a square-law device when biased at cutoff. This is not strictly true, but it is a reasonable approximation when the source impedance is low. FETs and MOSFETs are square-law devices as discussed in Chapter 6. Then, assuming the amplifiers to be identical, we see that

$$i_1 = A_0 + A_1 v_{b1} + A_2 v_{b1}{}^2 \Big\}$$
$$i_2 = A_0 + A_1 v_{b2} + A_2 b_{b2}{}^2 \Big\} \qquad (20.25)$$

but

and

$$v_{b1} = V_c \cos \omega_c t + V_m \cos \omega_m t \Big\}$$
$$v_{b2} = V_c \cos \omega_c t - V_m \cos \omega_m t \Big\} \qquad (20.26)$$

When these values of v_b are substituted into Eq. 20.25,

$$i_1 = A_0 + A_1(V_c \cos \omega_c t + V_m \cos \omega_m t) + A_2(V_c \cos \omega_c t + V_m \cos \omega_m t)^2$$
$$i_2 = A_0 + A_1(V_c \cos \omega_c t - V_m \cos \omega_m t) + A_2(V_c \cos \omega_c t - V_m \cos \omega_m t)^2$$

$$(20.27)$$

Because of the push-pull arrangement, only the components of collector current which are of opposite polarity will be effective in producing an output. Then the effective output current is

$$i_e = i_1 - i_2 = 2A_1 V_m \cos \omega_m t + 4A_2 V_c V_m \cos \omega_c t \cos \omega_m t \quad (20.28)$$

The tuned output will eliminate the modulating frequency component, so the effective output voltage is

$$v_o = B \cos \omega_c t \cos \omega_m t = \frac{B}{2} [\cos(\omega_c + \omega_m)t + \cos(\omega_c t - \omega_m t)] \quad (20.29)$$

where the constant B includes the tuned circuit constants in addition to $4A_2 V_c V_m$. It was assumed in Eq. 20.29 that the tuned output circuit accepted each side frequency equally well.

The waveform of the modulation envelope is altered by the removal of one of the sidebands. Figure 20.10 shows the waveform of a single side frequency and a carrier. Note that the envelope is not sinusoidal. This same waveform is produced whenever two signals having nearly the same frequency are mixed or added. The amplitude of the combination then varies as the difference between their two frequencies because of the alternate reinforcement and cancellation of their instantaneous values.

PROBLEM 20.6   Add the instantaneous values of a 1 V, 9 Hz signal to those of a 1 V, 10 Hz to verify the waveform of Fig. 20.10.

The *IC* multiplier discussed in Section 20.2 can be used as a balanced modulator. When the circuit is connected as shown in Fig. 20.11, the output signal contains only the two sidebands. A filter can then be used to remove one of these sidebands.

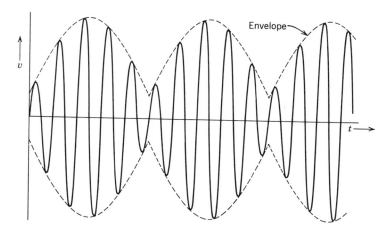

FIGURE 20.10
The waveform of a carrier and single side frequency.

PROBLEM 20.7   Determine what constitutes the difference between the configurations shown in Fig. 20.7 and Fig. 20.11. Determine the value of V_o in terms of V_m and V_c in Fig. 20.11.

Both sidebands are *not* transmitted with the carrier suppressed because the carrier which must be inserted in the receiver must then maintain the proper phase relationship with the sidebands to have the intelligence properly recovered. This preseiseness of oscillator control is practically impossible. At least the required circuit complexity would be prohibitive for the benefit gained. Sometimes the carrier and one sideband are transmitted to reduce the required bandwidth. This is done in modern television practice, where a small part of the other sideband is included to reduce the requirements of the filter which must discriminate between the carrier and the deleted sideband.

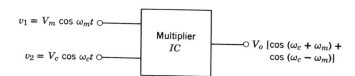

FIGURE 20.11
A suppressed carrier modulator.

20.5   DETECTION OF AMPLITUDE-MODULATED WAVES

The process by which the original modulating signal, or intelligence, is recovered in the receiving equipment is known as *detection* or *demodulation*. It was previously noted that an amplitude-modulated wave normally consists of the carrier and the sideband frequencies *only* and does not contain the modulating frequencies. Therefore, the modulating signal, or frequencies, must be reproduced in the receiver to complete the transmission of the intelligence. For example, in the broadcast of entertainment, the carrier may be modulated by an orchestra. This carrier with its sidebands are radiated from the transmitting antenna in the form of electromagnetic waves. These waves, in turn, induce small voltages into the receiving antenna. These voltages are usually amplified by tuned amplifiers with sufficient bandwidth to include the sidebands. If the receiver included only linear amplifiers, the amplified carrier and sidebands would be fed to a loudspeaker. However, this would be futile because the loudspeaker cannot respond to the carrier or sideband frequencies because they are radio frequencies. Therefore, the receiver must include a detector.

Since each modulating frequency is the difference between a sideband frequency and the carrier frequency, it seems evident that a nonlinear device is needed to recover the modulating frequencies from the modulated wave. The square-law device is one possible candidate for a detector. Again, the vacuum tube transistor and FETs may be operated as square-law devices. Thus, they may be used in this mode to provide detection as illustrated in Fig. 20.12. It may be seen from Fig. 20.12c and Fig. 20.12d that the square-law detector essentially eliminates one-half of the modulated wave. In other words, the detector acts as a rectifier.

It should be observed that the current pulses in the output of the detector vary in amplitude in accordance with the modulation envelope or modulating signal. The capacitor C (referring to Fig. 20.12) is placed in parallel with the load resistance for the purpose of bypassing the carrier and sideband frequencies so that only the recovered modulating signal will appear in the output. In a FET amplifier, the resistor R_1 permits a fairly constant current to flow through R_S and thus provides a fairly constant self-bias for the transistor. This bias will fluctuate somewhat with signal level because the transistor current is highly dependent on signal level, but if the bleeder current through R_1 is made large in comparison with the average transistor current, the bias may be maintained near cutoff for a wide range of signal levels. The source bypass capacitor C_S should be large enough to bypass the detected modulating frequencies.

We assume the detector to have only second-order nonlinearity, the output current i_o is

$$i_O = A_0 + A_1 v_i + A_2 v_i^2 \tag{20.30}$$

If the input voltage is a carrier and two side frequencies which were produced by a sinusoidal modulating signal having the natural frequency ω_m, then from Eq. 20.3,

$$v_i = V_c \cos \omega_c t + \frac{MV_c}{2} \cos(\omega_c + \omega_m)t + \frac{MV_c}{2} \cos(\omega_c - \omega_m)t \quad (20.31)$$

Substituting this value of v_i into Eq. 20.30, we have

$$i_0 = A_0 + A_1 \left[V_c \cos \omega_c t + \frac{MV_c}{2} \cos(\omega_c + \omega_m)t + \frac{MV_c}{2} \cos(\omega_c - \omega_m)t \right]$$

$$+ A_2 \left[V_c \cos \omega_c t + \frac{MV_c}{2} \cos(\omega_c + \omega_m)t + \frac{MV_c}{2} \cos(\omega_c - \omega_m)t \right]^2 \quad (20.32)$$

Since the bypass and dc blocking capacitors in the output eliminate all components of output voltage except those which may result from the squared term of Eq. 20.32, the useful value of output current is

$$i_o' = A_2 \left\{ V_c^2 \cos^2\omega_c t + MV_c^2 \cos \omega_c t[\cos(\omega_c + \omega_m)t + \cos(\omega_c - \omega_m)t] \right.$$

$$\left. + \frac{MV_c^2}{2} [\cos(\omega_c + \omega_m)t + \cos(\omega_c - \omega_m)t]^2 \right\} \quad (20.33)$$

If we expand Eq. 20.33 and retain only those products which can produce components that will be retained in the output, i_o' becomes

$$i_o'' = A_2 MV_c^2 \cos \omega_c t \cos(\omega_c + \omega_m)t + A_2 MV_c^2 \cos \omega_c t \cos(\omega_c - \omega_m)t$$

$$+ \frac{A_2 M^2 V_c^2}{2} \cos(\omega_c + \omega_m)t \cos(\omega_c - \omega_m)t \quad (20.34)$$

When we replace the product terms of Eq. 20.34 with their trigonometric identities and retain only the terms which will be retained in the output, i_o'' becomes

$$i_o''' = \frac{A_2 MV_c^2}{2} \left(\cos \omega_m t + \cos \omega_m t + \frac{M}{2} \cos 2\omega_m t \right) \quad (20.35)$$

After filtering, the output voltage is

$$v_o = A_2 MV_c^2 R_L \left(\cos \omega_m t + \frac{M}{4} \cos 2\omega_m t \right) \quad (20.36)$$

Observe from Eq. 20.36 that a second harmonic of the modulating frequency appears in the output, and that the amplitude of this second harmonic is one-fourth M as large as the amplitude of the recovered modulating frequency.

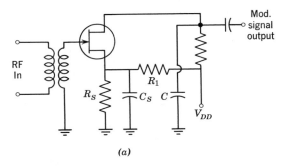

(a)

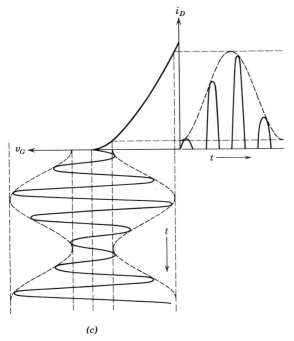

(c)

FIGURE 20.12
(a) **FET detector,** (b) **transistor detector, and** (c) **and** (d)
illustration of the detection process by the use of the transfer
characteristics.

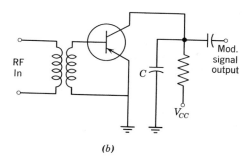

RF
In

C

Mod.
signal
output

V_{CC}

(b)

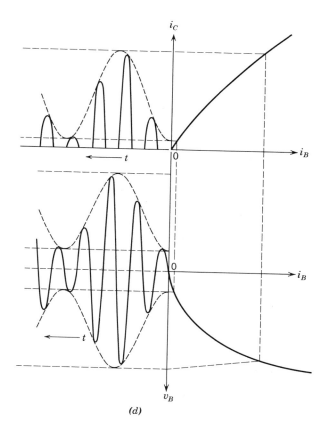

i_C

0

i_B

t

0

i_B

t

v_B

(d)

PROBLEM 20.8   Prove that the square-law detector will recover the modulating signal without distortion when single sideband transmission is used.

PROBLEM 20.9   If single sideband transmission were used and the carrier suppressed at the transmitter, what would be the effect on the detector output frequencies if the local oscillator which provides the carrier in the receiver were tuned to $\omega_c + \Delta\omega$ instead of ω_c ?

20.6   LINEAR DETECTORS

In reference to Fig. 20.12c and Fig. 20.12d, it seems that distortionless (or linear) detection could be accomplished by a linear rectifier. A linear rectifier is a device which conducts only during alternate half cycles of the input signal, and during the conducting half cycles the output current is proportional to the input voltage. The characteristics of a linear rectifier and the distortionless detection which may be obtained with this rectifier are shown in Fig. 20.13. It may be seen from the figure that the peak amplitude of each current pulse in the output is proportional to the peak amplitude of the input voltage during that particular conducting half cycle. Thus the peak, and therefore the average, values of the output current pulses follow the amplitude of the input voltage precisely during conducting half cycles and have the same waveform as the modulation envelope. Whether the output voltage would approach the peak or average

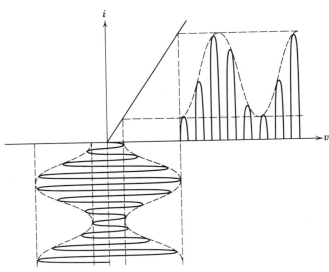

FIGURE 20.13
The detection characteristics of a linear rectifier.

value of the input voltage depends on the type of filter used in the output. If the filter is a bypass capacitor, as previously indicated, and the internal resistance of the rectifier is small in comparison with the load resistance, the output voltage will tend to follow the peaks of the input voltage as explained in Chapter 3. Thus, the capacitor charges essentially to the peak input voltage during the conducting half cycles, but there is not time for appreciable discharge through the high resistance load during the non-conducting half cycles. If the bypass capacitor is too large, the time constant of the discharge will be so large that the detector output will not be able to follow the modulation envelope when the modulation envelope decreases amplitude rapidly. This situation, which may occur when the modulating frequencies are high, is known as *negative clipping* and will be discussed in more detail later.

As discussed in Chapter 3, the vacuum diode or semiconductor diode has a linear dynamic characteristic when the load resistance is large in comparison with the internal resistance of the diode. The voltage drop across the diode then becomes insignificant in comparison with the load voltage which then closely follows the input voltage. Thus, the properly designed diode detector may be a linear detector. Typical diode detector circuits are shown in Fig. 20.14. In the semiconductor diode circuit, Fig. 20.14a, R_D is the diode load resistor, C_F is the RF bypass capacitor and C_C is the dc blocking and load-coupling capacitor. R_L represents the actual load on the detector. In this circuit, R_L is a gain control (called a volume control in audio circuits).

In addition to detection, a circuit has been added which is called the *automatic gain control* (AGC). The AGC voltage is the average value of the detector output voltage since R_1 and C_1 act as a filter to remove the modulating signal as well as the RF from the AGC system. This AGC voltage is therefore proportional to the amplitude of the carrier in a continuous wave system and may be used to automatically control the gain of one or more RF amplifier stages. For example, if *p-n-p* transistors are used as RF amplifiers ahead of the detector shown in Fig. 20.14a and the diode is connected as shown, the AGC voltage will tend to reverse bias the controlled amplifiers and hence reduce their gain. Thus, for small input signals the RF amplifier will have high gain, but as the magnitude of the input signal increases, the gain of the RF amplifier decreases. This effect tends to keep the detector output relatively constant and prevents overdriving the RF amplifiers. Overdriving an amplifier means to cause its operation to extend into the saturation and cutoff regions. This type of operation in an RF amplifier changes the wave-form of the modulation envelope, thus causing severe distortion. AGC is effectively and easily applied to a dual-gate MOSFET as discussed in Chapter 6.

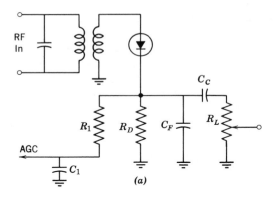

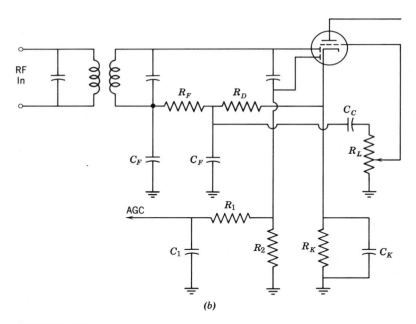

FIGURE 20.14
(a) A typical semiconductor diode detector and AGC circuit, (b) a duo-diode triode tube used as a combination detector, AGC, and audio amplifier.

The circuit of Fig. 20.14b has a few added features that are worthy of mention. Double tuning has been used in the input coupling circuit. The once popular duo-diode triode tube has been used as a combination detector, AGC and audio amplifier circuit. Since the amplifier section uses cathode bias, the detector load resistor R_D is returned to the cathode rather than to the ground in order to prevent biasing the detector plate as well as the amplifier grid. A π-type RF filter, consisting of the resistor R_F and the two capacitors C_F, has been used to reduce the RF in the output. The AGC voltage is obtained from a second diode plate which is coupled to the first by a capacitor. The AGC diode load resistor is R_2 and the AGC filter is R_1 and C_1. This AGC arrangement is known as delayed AGC because the AGC diode is reverse biased. This reverse bias is provided by the cathode bias resistor since the AGC load resistor returns to ground. Thus no AGC voltage will appear when the peak amplitude of the input signal is less than the bias on the AGC diode plate. Therefore, this type of AGC gives improved RF gain for very small signals.

Since the diode load resistor develops a dc voltage which tends to reverse bias the diode, it may seem that the diode may not conduct during the negative half cycles of the modulation envelope. To investigate this possibility, the relationship between the dc output voltage and the RF input voltage is shown in Fig. 20.15 for a typical diode detector. In this figure, it is assumed that the diode load resistor is bypassed for RF so that the dc load voltage approaches the peak values of the RF input voltage. However, the dc load voltage is somewhat less than the peak input voltage because of the internal resistance of the diode and the partial discharge of the RF bypass capacitor between the input voltage peaks as discussed in Chapter 3. Thus, the dc load voltage decreases as the load resistance decreases. Simultaneously, the diode current increases, because both the voltage drop across the diode and the discharge between the input peaks increase with decreasing load resistance, as previously discussed.

Since the axes of the plot of Fig. 20.15 are the load voltage and load current, load lines may be drawn on the rectification characteristics as shown. The use of the load line may be most easily explained by an example.

Example 20.2 Consider a 150 kΩ dc load resistance for the diode of Fig. 20.15. Assuming that the peak carrier voltage at the input of the detector is 20 V, the dc voltage V_Q across the diode load resistor is about 17.5 V as indicated. So long as the carrier is unmodulated, the dc voltage across the load resistor will remain constant and the quiescent operating point on the load line will be at point Q. On the other hand, when the carrier is modulated, the operating point must move up and down along the dc load

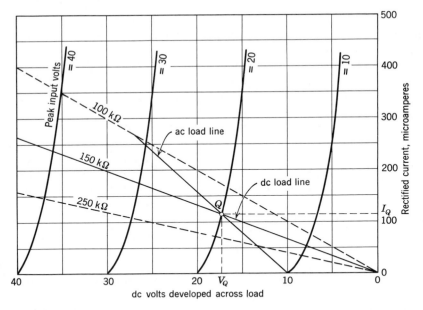

FIGURE 20.15
Rectification characteristics of a diode detector.

line, since the detector input voltage is continually changing. This condition is true if the dc load is the *only* load on the detector. When the carrier has 100 percent modulation, the input voltage must vary between 0 and 40 V peak, so the output voltage will vary between 0 and 36 V peak. Thus, the detection is essentially linear.

As indicated in Fig. 20.14, the detector usually has part of its load isolated from the dc load resistor by a blocking capacitor. Therefore, the ac load resistance for the detector is usually smaller than the dc load resistance. Consequently, an ac load line should be drawn through the point Q as shown in Fig. 20.15. In this case, the detector operating point moves up and down along the ac load line as the modulated signal is applied to the input. Thus, a modulation crest causes a greater increase of rectified current than would occur if the dc load resistor had been the total load. Similarly, the rate of decrease of rectifier current is more rapid during a trough, or negative half cycle, of the modulation than it would be if only the dc load were present. Notice from Fig. 20.15 that this effect causes the diode current to be reduced to zero before the input voltage is reduced to zero. This is known as negative clipping. Notice that negative clipping occurs only when the percentage of modulation exceeds a certain value. In this example, where the ac load resistance is one-half the value of the dc load resistance, the rectifier current is reduced to zero when the input voltage is reduced to one-half the quiescent, or carrier, value. Therefore, negative clipping would occur when the percentage of modulation exceeds 50 percent.

A simple relationship exists between the ratio of ac load resistance to dc load resistance and the maximum modulation index that can occur without clipping. Referring to Fig. 20.15, we can see that the output voltage which results from the unmodulated carrier is

$$V_Q = I_Q R_{dc} \qquad (20.37)$$

Also, the maximum change in the output voltage which can occur without negative clipping is

$$(\Delta V)_{max} = I_Q R_{ac} \qquad (20.38)$$

But the maximum modulation index which can occur without negative clipping is the rato of $(\Delta V)_{max}$ to V_Q. Then

$$M_{max} = \frac{(\Delta V)_{max}}{V_Q} = \frac{I_Q R_{ac}}{I_Q R_{dc}} = \frac{R_{ac}}{R_{dc}} \qquad (20.39)$$

At high modulating frequencies the RF bypass capacitor may significantly reduce the ac impedance and thus can cause negative clipping, as previously mentioned. Also, it is apparent that any load which is coupled to the detector through a capacitor must be large in comparison with the dc load resistance to permit a high percentage of modulation without appreciable distortion.

PROBLEM 20.10   The detector of Fig. 20.14a has $R_D = 50 \text{ k}\Omega$, $R_1 = 1 \text{ M}\Omega$, $R_L = 250 \text{ k}\Omega$, $C_F = 500 \text{ pF}$, and $C_1 = C_C = 0.1 \ \mu\text{F}$. Determine the maximum percentage of modulation which can be accepted without causing negative clipping.
Answer: $M_{max} = 80\%$.

It should be observed that a linear detector does not provide distortionless demodulation when only one sideband is present, because the modulation envelope does not have the same shape as the modulating signal when single sideband transmission is used. On the other hand, it may be proven (Prob. 20.8) that a square-law detector provides distortionless demodulation when single sideband transmission is used.

The impression may have been conveyed that tube amplifiers and transistors automatically provide square-law detection when they are biased near cutoff. This is not true. These amplifying types of detectors may provide fairly linear detection if they are properly designed for this purpose. For example, the triode tube may provide linear detection if the load resistance is large in comparison with the plate resistance. Also, the proper adjustment of driving-source resistance may provide good linearity for a transistor detector.

In the foregoing paragraphs, the term RF has sometimes been used to mean the carrier and sidebands whereas *audio* has been used to indicate the modulation. This usage is much too specialized. In some applications, such as automatic control, the carrier may be low frequency of a few hundred Hertz whereas the modulating signal may have a period of several seconds.

The effective input resistance of a detector must be obtained in some manner before the tuned coupling circuit can be properly designed. The effective input resistance may be determined if the average input power can be found in terms of the input voltage. The average input power over one cycle is

$$P = \frac{1}{2\pi} \int_{-\pi}^{\pi} vi \, d\theta \tag{20.40}$$

Assuming the input voltage to be an unmodulated carrier so that $v = V_c \cos \theta$, we see that

$$P = \frac{1}{2\pi} \int_{-\pi}^{\pi} V_c i \cos \theta \, d\theta \tag{20.41}$$

For a biased amplifier-type detector or a diode detector which does not have an RF bypass capacitor, the input current i flows only on alternate half cycles. Then, if it is assumed that during these conducting half cycles the input current is

$$i = \frac{V_c \cos \theta}{R_i} \tag{20.42}$$

where R_i is the average input resistance during the conducting half cycle, the average input power is

$$P = \frac{1}{2\pi} \int_{-\pi/2}^{\pi/2} \frac{V_c^2 \cos^2 \theta}{R_i} \tag{20.43}$$

$$P = \frac{1}{2\pi} \left(\frac{\pi V_c^2}{2R_i} \right) = \frac{V_c^2}{4R_i} \tag{20.44}$$

The effective input resistance may be defined as

$$R = \frac{(V_{rms})^2}{P} = \frac{V_c^2/2}{P} = 2R_i \tag{20.45}$$

An accurate value of R_i may be difficult to obtain because of the dependence of R_i on the magnitude of the input voltage in most cases. However, an average value of input resistance which is obtained at an average value of input voltage is sufficiently accurate for good design.

The effective input resistance of a diode detector in which the load resistance is large in comparison with the forward resistance of the diode and an adequate RF bypass capacitor is connected in parallel with the load resistance may be easily found if we assume that the total input power is dissipated in the load resistor. Since the dc voltage across the load resistor R_D is approximately equal to V_c, the power dissipated in the load is

$$P = \frac{V_c^2}{R_D} \qquad (20.46)$$

where P is the average power in the load when the carrier is unmodulated. But this power is approximately the same as the RF input power which may be defined as

$$P = \frac{V_c^2}{2R_i} = \frac{V_c^2}{R_D} \qquad (20.47)$$

where R_i, again, is the effective input resistance of the detector. Therefore,

$$R_i \simeq \frac{R_D}{2} \qquad (20.48)$$

where R_D, as before, is the dc load resistance. When the carrier is modulated it would at first seem that the ac load resistance of the diode instead of the dc resistance R_D should be used to calculate the input resistance because additional power is coupled through the dc blocking capacitors into the RC-coupled part of the load. However, this additional power comes from the power in the sidebands. Therefore, the effective input resistance remains approximately $R_D/2$ as long as negative clipping does not occur.

20.7 FREQUENCY CONVERTERS

In the preceding paragraphs, it was shown that both modulation and detection involve two basic processes. The first of these processes is the production of sum and difference frequencies by a nonlinear device; the second is the separation of the desired output components from the undesired ones by some type of a filter. In the early days of radio, the idea was conceived that these two processes could be used to change or translate any given frequency to a different frequency. The superheterodyne receiver is based on this principle. This receiver contains an oscillator which is known as a *local oscillator*. The output from this oscillator is mixed with the incoming signal and the combination is applied to a nonlinear device. The output of this nonlinear device is tuned to the difference between the oscillator frequency and the frequency of the incoming signal. The difference frequency is known as the *intermediate frequency* (IF). This system has two main advantages when compared with a conventional amplifier. First, higher gain and better stability may be obtained from an amplifier at the lower intermediate frequency. Second, the intermediate frequency may remain constant even though a wide variety of input frequencies may be selected because the local oscillator frequency may be varied in such a manner as to produce a constant difference between the oscillator frequency and the variable input frequency. Therefore, the design of the IF amplifier may be optimized whereas the design of the RF amplifier must be a compromise since it is required to operate over a wide range of frequencies.

Example 20.3   The standard broadcast superheterodyne receiver is a good example of the principle of frequency conversion. This receiver is required to tune over the 550 to 1600 kHz frequency range and it should provide essentially constant gain and constant bandwidth over this range. A commonly used intermediate frequency is 455 kHz. To provide this intermediate frequency, the local oscillator must vary from 1005 to 2055 kHz and must always remain 455 kHz above the selected input frequency. This maintenance of a constant difference frequency over the band is known as *tracking*. The local oscillator frequency could be below the input signal frequency, but then its frequency range would have to be 95 to 1145 kHz. This frequency ratio is too high to cover with a single band. When the selected input frequency is 1000 kHz and the sideband frequencies extend from 995 to 1005 kHz, the oscillator frequency is 1455 kHz. Then the frequencies apppearing in the output current of the nonlinear device are:

1. The original frequencies   995–1005 kHz
                                     1455 kHz
2. The sum frequencies   2450–2460 kHz
3. The difference frequencies   450–460 kHz
4. Harmonic frequencies   1990 kHz and upwards

When a tuned circuit which has a resonant frequency of 455 kHz and a bandwidth of 10 kHz is placed in the output, the difference frequencies, which include the new 455 kHz carrier with its sideband frequencies, are retained and all the other frequencies are rejected.

The nonlinear device which produces the sum and difference frequencies is frequently called a *mixer*. This term is not very appropriate, however, because it is also used to designate linear devices in which two or more signals are mixed, but in which additional frequencies are not produced. The nonlinear device is more appropriately called a *first detector* or *modulator*. The combination of the first detector and the local oscillator is known as a *frequency converter*.

The block diagram of a typical superheterodyne receiver is shown in Fig. 20.16. The RF amplifier shown is frequently omitted in entertainment-type receivers. The advantages of including this RF amplifier are briefly discussed in a following paragraph.

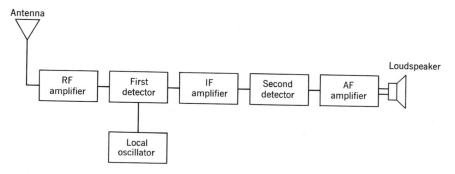

FIGURE 20.16
The block diagram of a typical superheterodyne receiver.

Some typical converter circuits are shown in Fig. 20.17. In these circuits the oscillators are the tuned gate or base type, although other types could be used. The dashed lines between the input and oscillator-tuned circuits indicate that the tuning capacitors are *ganged* so that single-knob tuning may be used. In the bipolar transistor circuit, the oscillator signal is applied to the emitter. In the MOSFET circuit, the oscillator signal is applied to gate No. 2 of the MOSFET. The RF input signal is applied to gate No. 1 of the MOSFET.

In any converter circuit the oscillator signal must cause the transconductance of forward current transfer ratio to vary over the oscillator output cycle. The input signal is usually so small that the amplifier parameters will not be affected appreciably by the input voltage variations. The ratio of the difference frequency component of current in the output to the signal voltage input is known as *conversion transconductance*. To more fully understand the conversion problem, a special case will be considered in which the input signal is so small that the amplifier parameters are essentially unaffected by the input signal. Then

$$i_o = I_0 + g_m v_i \tag{20.49}$$

where v_i is considered to be a sinusoidally modulated carrier.

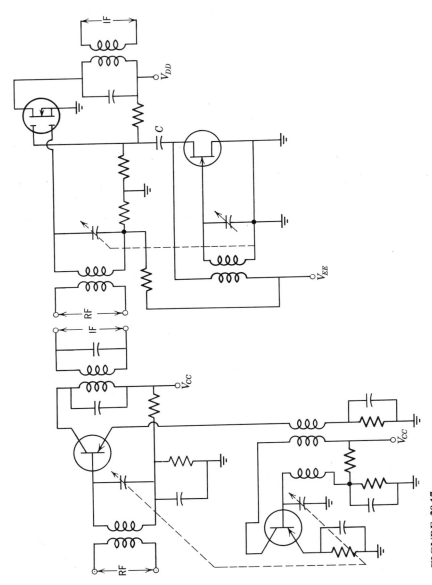

FIGURE 20.17
Typical frequency converter circuits.

$$v_i = V_c \left[\cos \omega_c t + \frac{M}{2} \cos(\omega_c + \omega_m)t + \frac{M}{2} \cos(\omega_c - \omega_m)t \right] \quad (20.50)$$

It will be assumed that the transconductance of the amplifier varies linearly with the applied oscillator voltage as it does in a square-law device. This relationship may be expressed by the following equation.

$$g_m = g_{mo}(1 + A \cos \omega_o t) \quad (20.51)$$

where g_{mo} is the quiescent value of transconductance, ω_o is the natural oscillator frequency, and A is a constant which is proportional to the amplitude of the oscillator voltage. Substituting Eq. 20.50 and Eq. 20.51 into Eq. 20.49, we have

$$i_O = I_0 + g_{mo} V_c \left[\cos \omega_c t + \frac{M}{2} \cos(\omega_c + \omega_m)t \right.$$
$$\left. + \frac{M}{2} \cos(\omega_c - \omega_m)t \right] (1 + A \cos \omega_o t) \quad (20.52)$$

or

$$i_O = I_0 + g_{mo} V_c \left[\cos \omega_c t + \frac{M}{2} \cos(\omega_c + \omega_m)t + \frac{M}{2} \cos(\omega_c - \omega_m)t \right] + g_{mo} A V_c$$
$$\times \left[\cos \omega_c t \cos \omega_o t + \frac{M}{2} \cos(\omega_c + \omega_m)t \cos \omega_o t \right.$$
$$\left. + \frac{M}{2} \cos(\omega_c - \omega_m)t \cos \omega_o t \right] \quad (20.53)$$

Observe that the first bracketed term of Eq. 20.53 contains the carrier and sideband frequencies of the input signal which will be rejected by the filter in the output. The product terms in the second bracket produce sum and difference frequencies when the trigonometric identities are used. The sum frequencies will be rejected by the output filter, assuming that this filter is tuned to the difference frequency, so the only terms of importance in the output are

$$i_o' = \frac{g_{mo} A V_c}{2} \left[\cos(\omega_c - \omega_o)t + \frac{M}{2} \cos(\omega_c - \omega_o + \omega_m)t \right.$$
$$\left. + \frac{M}{2} \cos(\omega_c - \omega_o - \omega_m)t \right] \quad (20.54)$$

The conversion transconductance is

$$g_c = \frac{|i_o'|}{|v_i|} = \frac{g_{mo} A V_c}{2V_c} = \frac{g_{mo} A}{2} \quad (20.55)$$

Observe from Eq. 20.55 that the conversion transconductance is proportional to the amplitude of the oscillator signal as well as to the quiescent transconductance when the transconductance is proportional to the oscillator signal. It may be seen from Eq. 20.51 that the transconductance would vary from zero to $2g_{m_o}$ when $A = 1$. The conversion transconductance would then be $g_{m_o}/2$. This is about the maximum attainable conversion transconductance because larger oscillator voltages cause g_m to be zero over an appreciable part of the negative half cycles of oscillator voltage, whereas a saturating or decreasing value of g_m may occur as the grid or base is driven strongly into the forward region. A sensible design procedure may be to bias the amplifier about halfway between the maximum and zero values of g_m and then provide an oscillator signal of sufficient amplitude to vary g_m between the zero and maximum value.

A single amplifier may be used as both the modulator and the oscillator as shown in the circuits of Fig. 20.18. The vacuum tube circuit (a) employs a tube known as a *pentagrid converter*. Grids 1 and 2 serve as the oscillator grid and plate, respectively. A Hartley oscillator circuit is used in which one end of the tuned circuit is grounded. Therefore, grid, 2, which acts as the oscillator anode, may be at RF ground potential and serve the additional function of a screen grid. Most of the electrons which pass through the oscillator grid also pass through the oscillator anode and become the space current in the tube. The input signal is applied to grid 3, which controls the space current which passes through the oscillator section. The remaining elements of the tube, grids 4 and 5 and the plate, have the same functions as the screen grid, suppressor grid, and plate of the pentode amplifier. In this converter, the oscillator should operate either class B or class C so that the tube space current flows through the oscillator section in pulses. The signal grid transconductance is then varied from zero to its maximum value over the oscillator cycle, and conversion takes place in the same manner as described for the separate oscillator circuit. The chief advantage of the pentagrid converter circuit lies in the reduction of the number of tubes required. An additional advantage may be the improved isolation between the signal and oscillator-tuned circuits since the mixing of the signals is accomplished in the electron stream of the tube rather than by inductive or capacitive coupling.

The transistor circuit of Fig. 20.18b operates in the same general manner as does the pentagrid converter circuit except that the entire transistor acts as both the oscillator and the detector. The oscillator shown is a tuned-collector, common-base circuit. The isolation between the oscillator and input signal circuits is not as good in the transistor circuit as it is in the pentagrid converter circuit.

The circuits shown are only representative types. There are several other

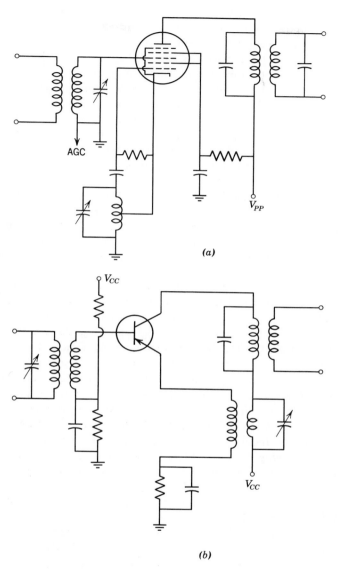

(a)

(b)

FIGURE 20.18
Converter circuits which do not use a separate oscillator.

types of special converter tubes in use, and recommended circuit diagrams are available in tube manuals and other literature. Various types of oscillator circuits, including the crystal-controlled types, may be used in conjunction with a wide variety of detector, or modulator, circuits. Thus, converter circuits may be devised to meet almost any requirement.

Both transistors and vacuum tubes generate more noise at very low frequencies and very high frequencies than they do at moderate radio frequencies. Therefore, the noise in the output of a converter may be greater than the noise in the output of a conventional amplifier because the converter includes a modulator. Hence, the low-frequency noise voltages will modulate the carrier and will cause noise side frequencies which are accepted by the output filter.

The RF amplifier shown in Fig. 20.16 may improve the signal-to-noise ratio of the superheterodyne receiver because the RF amplifier is not a good modulator. Thus, the comparatively noise-free amplification ahead of the frequency converter can materially improve the signal-to-noise ratio in the receiver. Also the RF amplifier improves the selectivity and the sensitivity of the receiver.

Diodes are sometimes used as the nonlinear device in a converter because of their low noise characteristics or because the incoming signal frequency is so high that ordinary amplifiers have practically no gain. The circuit diagram of an ordinary diode converter is given in Fig. 20.19a. This type of converter may be used in the GHz frequency range if the lumped elements are replaced by distributed elements such as wave guides and cavity resonators. Also, a tunnel diode may be used in a converter circuit as shown in Fig. 20.19b. When used in this application, the tunnel diode is biased near the zero conductance point or slightly into the positive conductance region as shown in Fig. 20.20 by point B. If the diode is biased at point A so the quiescent diode conductance is zero, the local oscillator causes the diode conductance to swing into the positive conductance region during the negative half cycle of oscillator voltage and into the negative conductance region during the positive half cycle. It can be shown that the excursion into the negative conductance region may result in conversion gain. In fact, tunnel diode converters have been constructed which provide conversion gain of more than 20 dB and a noise figure (decrease of signal-to-noise ratio) of less than 3 dB when used to convert 220 MHz to 60 MHz.[3]

The stability of the tunnel-diode converter is improved when the bias point is changed from point A (Fig. 20.20) to point B. On the other hand, the circuits may provide self-oscillation as the bias point is moved into the

---

[3] K. K. N. Chang, G. H. Heilmeier, and H. J. Prager, "Low-Noise Tunnel-Diode Down Converter Having Conversion Gain," *Proceedings of IRE*, Vol. 48, May 1960, p. 854.

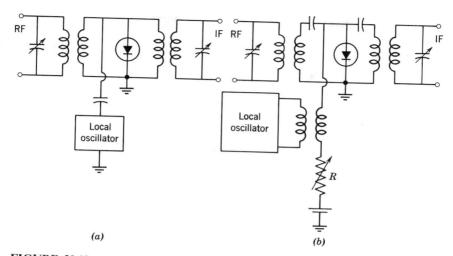

FIGURE 20.19
(a) A diode converter, (b) a tunnel diode converter.

negative conductance region to a point such as C. The separate oscillator may then be eliminated. However, since the diode does not provide isolation between the input, output, and oscillator circuits, it may be difficult to independently control the frequency of oscillation.

PROBLEM 20.11   A frequency converter is to be used to translate 225 MHz to 60 MHz. What local oscillator frequencies could be used and what would be the frequency components in the output for each local oscillator frequency?

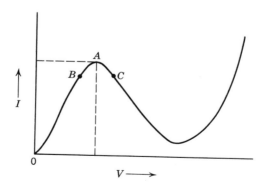

FIGURE 20.20
Bias points for a tunnel diode.

PROBLEM 20.12 A frequency converter incorporates an amplifier that has $g_m = 2000$ μmhos at the quiescent operating point. The local oscillator signal causes g_m to vary from 1000 to 3000 μmhos. Assuming that g_m is proportional to the oscillator voltage, determine the conversion transconductance of the amplifier.

PROBLEM 20.13 Draw the circuit diagram for a complete super-heterodyne broadcast band receiver. Use transistors including FETs and MOSFETs. Component values need not be determined, but specify the frequency (or frequency range) and the bandwidth of the tuned circuits.

PROBLEM 20.14 Draw the circuit diagram for a complete super-heterodyne broadcast band receiver. Use integrated circuits where possible and use as few components as possible. Specify the size of any discrete components used.

21

Frequency Modulation

Amplitude modulation was discussed in Chapter 20. Another common method of modulating a carrier signal is known as *frequency modulation*. In this type of modulation, the signal frequency is changed in accordance with the amplitude of the modulation as expressed by the following equation

$$v = A \cos[\omega_c t + M_f F_m(t)] \tag{21.1}$$

where $F_m(t)$ is the modulating signal expressed as a function of time and ω_c is the radian frequency of the carrier. The modulation index M_f relates the amplitude of the modulating signal to the variation of the carrier frequency which it produces.

Since the modulating signal can be resolved into frequency components by the Fourier series technique, it will be convenient to assume, as was done in Chapter 20, that the modulating signal is sinusoidal. Then

$$F_m(t) = B \sin \omega_m t \tag{21.2}$$

When this modulating signal is used, Eq. 21.1 becomes

$$v = A \cos(\omega_c t + M_f \sin \omega_m t) \tag{21.3}$$

The modulation index M_f absorbs the amplitude factor B of the modulating signal as will be shown below. Actually, Eq. 21.3 indicates that the term $M_f \sin \omega_m t$ must have the dimensions of an angle θ, not a frequency. Since $\omega = d\theta/dt$, the frequency associated with this angle may be obtained by differentiating the term $M_f \sin \omega_m t$ with respect to time.

$$\frac{d(M_f \sin \omega_m t)}{dt} = M_f \omega_m \cos \omega_m t \tag{21.4}$$

Then Eq. 21.3 can be written as

$$v = A \cos(\omega_c + M_f \omega_m \cos \omega_m t)t \tag{21.5}$$

It may be seen from Eq. 21.5 that the maximum frequency deviation is

$$(\Delta\omega)_{\text{max}} = M_f\,\omega_m \tag{21.6}$$

$$M_f = \frac{(\Delta\omega)_{\text{max}}}{\omega_m} = \frac{(\Delta f)_{\text{max}}}{f_m} \tag{21.7}$$

The modulation index M_f is known as the *deviation ratio*, since it is the ratio of the maximum deviation of the signal frequency to the modulating frequency.

21.1   MODULATING CIRCUITS

The characteristics of frequency modulation will be better understood after a few typical modulation circuits have been considered. Probably the simplest way to obtain frequency modulation is to connect a capacitor microphone across the tuned circuit of an *LC* oscillator as shown in Fig. 21.1. The capacitor microphone has a thin metal diaphragm stretched

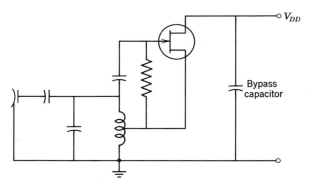

FIGURE 21.1
A simple frequency-modulation circuit using a capacitor microphone and a Hartley oscillator.

in front of a fixed metal plate. The acoustic pressure in the air causes vibration of the diaphragm and consequently variation of the capacitance between the diaphragm and the plate. This varying capacitance varies the oscillator frequency and thus causes frequency modulation. The capacitor microphone method is not very practical because the oscillator frequency would also depend on the length of the microphone cable, stray capacitance, and so on.

A varactor diode may be used to provide frequency modulation as shown in Fig. 21.2. The modulating signal varies the junction capacitance of the

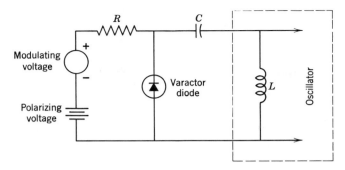

FIGURE 21.2
A frequency modulator that uses the variable capacitance of a junction diode.

varactor diode which in turn varies the frequency of the oscillator. The capacitor C should be small compared with the quiescent capacitance of the diode so the RF voltage across the diode will be small compared with the polarizing voltage. Then the diode capacitance will not vary appreciably over the RF cycle. Also the reactance of C is high compared with R at the highest modulating frequency so the modulating signals are not shunted to ground through the tuned circuit of the oscillator. In addition, the resistance of R should be large enough to prevent the modulating circuit from excessively loading the oscillator-tuned circuit. Both the polarizing and modulating voltages may be conveniently obtained from the collector circuit of a transistor (or FET) modulator.

The linearity of the varactor-diode modulator will now be investigated. Ideally, the oscillator frequency should be a linear function of the modulating voltage as shown in Fig. 21.3. However, as discussed in Chapter 3, the capacitance C_d of an abrupt-junction diode varies inversely as the square root of the barrier potential, as given below

$$C_d = K(v_B)^{-1/2} \tag{21.8}$$

where the barrier potential v_B is essentially equal to the sum of the polarizing voltage and the modulating voltage. Also, the frequency of oscillation is approximately

$$\omega = \frac{1}{[L(C_o + C_d)]^{1/2}} \tag{21.9}$$

Substituting Eq. 21.8 into Eq. 21.9, we have

$$\omega = [L(C_o + Kv_B^{-1/2})]^{-1/2} \tag{21.10}$$

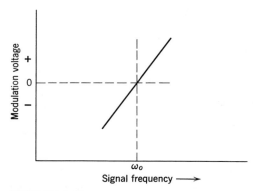

FIGURE 21.3
The desired linear relationship between the modulating voltage and the signal frequency.

It can be seen from Eq. 21.10 that the oscillator frequency ω is not a linear function of the diode voltage v_B. The nonlinear distortion for a given magnitude of v_B may be determined if the right side of Eq. 21.10 is expanded into a Taylor series. Then the techniques of Chapter 11 can be used to determine the harmonic distortion for any given excursion of the modulating voltage. The mathematics is simplified considerably if the diode provides essentially all of the tuning capacitance, or, in other words, C_o may be neglected in comparison with the capacitance of the diode. Then Eq. 21.10 reduces to

$$\omega = \frac{v_B^{1/4}}{(LK)^{1/2}} \tag{21.11}$$

using the Taylor series expansion about the polarizing voltage,

$$\frac{v_B^{1/4}}{(LK)^{1/2}} = \frac{V_0^{1/4}}{(LK)^{1/2}} + \frac{1}{4(LKV_0^{3/2})^{1/2}} (v_B - V_0)$$

$$- \frac{3}{16(LKV_0^{7/2})^{1/2}} (v_B - V_0)^2 + \cdots \tag{21.12}$$

where V_0 is the polarizing voltage applied to the diode.

When the modulator is adjusted so that the second-order term is small in comparison with the linear term, the third- and higher-order terms may be neglected.

In case the modulating signal is a sinusoid,

$$v_B - V_0 = \Delta V = V_m \sin \omega t \tag{21.13}$$

$$(v_B - V_O)^2 = V_m^2 \sin^2 \omega t = \frac{V_m^2}{2}(1 - \cos 2\omega t) \qquad (21.14)$$

Substituting Eqs. 21.14, 21.13, and 21.12 into Eq. 21.11, we obtain an expression for the signal frequency when the modulating voltage is sinusoidal. Neglecting the third- and higher-order terms, we have

$$\omega = \frac{V_O^{1/4}}{(LK)^{1/2}} - \frac{3V_m^2}{32(LKV_O^{7/2})^{1/2}} + \frac{V_m \sin \omega t}{4(LKV_O^{3/2})^{1/2}} + \frac{3V_m^2 \cos 2\omega t}{32(LKV_O^{7/2})^{1/2}} \qquad (21.15)$$

The percentage of the magnitude of the second harmonic to the magnitude of the fundamental is

$$\text{Percent 2nd harmonic} = \frac{3V_m^2/32(LKV_O^{7/2})^{1/2}}{V_m/4(LKV_O^{3/2})^{1/2}} \times 100 = \frac{3V_m}{8V_O} \times 100 \qquad (21.16)$$

Thus it may be observed from Eq. 21.16 that the second harmonic distortion may be kept within any desired limit by the proper choices of bias voltage V_O and peak modulating voltage V_m.

PROBLEM 21.1 An abrupt junction diode is used to frequency modulate an oscillator. The junction capacitance is essentially the total tuning capacitance of the oscillator circuit. When 15 V bias is applied to the diode. the oscillator frequency is 5 MHz. Determine the modulation index (deviation ratio) and the percentage of second harmonic distortion when the modulating voltage is 2 sin 6280t. Assume the third- and higher-order terms to be negligible.

Answer: $M_f = 178$, % sec $= 5$.

The situation may arise where the modulation index is not sufficiently high when the modulating voltage is at the maximum permissible level. Then the modulation index may be increased by multiplying the oscillator frequency by the use of a nonlinear amplifier, as discussed in Chapter 18.

PROBLEM 21.2 The frequency-modulated oscillator signal of Problem 21.1 is fed into a tripler. What is the carrier frequency and modulation index in the output of the tripler?

Answer: $f_c = 15$ MHz, $M_f = 534$.

21.2 A MILLER-CAPACITANCE MODULATOR

In case the carrier frequency becomes undesirably high because of the multiplication process, the carrier may be reduced to a lower frequency by the use of a heterodyne frequency converter discussed in Chapter 20. This frequency conversion will not affect the modulation index.

The *Miller capacitance* of a transistor or FET can be used to frequency modulate an oscillator, as shown in Fig. 21.4. The oscillator is basically a Colpitts circuit with the input capacitance of the FET in series with C_1 acting as the capacitance between the emitter and the collector. But the input capacitance C_i of the FET is, due to the Miller effect,

$$C_i = C_{gs} + C_{gd}(1 + g_m R_L) \tag{21.17}$$

In addition, the transconductance of the FET is a function of the gate-source voltage v_{GS} of the FET as discussed in Chapter 6.

$$g_m = g_{mo}\left(1 - \frac{v_{GS}}{V_P}\right) \tag{21.18}$$

Substituting this expression for g_m into Eq. 21.17, C_i becomes

$$C_i = C_{gs} + C_{gd} + g_{mo} R_L C_{gd}\left(1 - \frac{v_{GS}}{V_P}\right) \tag{21.19}$$

Thus C_i can be resolved into a constant capacitance plus a capacitance that varies linearly with the voltage v_{GS} and the voltage v_{GS} is approximately

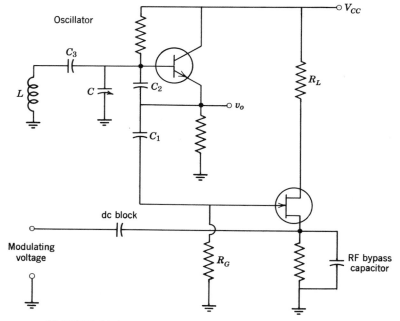

FIGURE 21.4
A frequency modulator that uses the Miller capacitance of an FET.

equal to a bias voltage plus the modulating voltage, providing C_1 is large in comparison with C_{gs}, as it should be. If C_1 is also large in comparison with C_i, the voltage gain of the transistor is approximately $X_{ci}/X_{C2} = C_2/C_i$, so C_2 may be large in comparison with C_i. Thus the oscillator frequency is primarily determined by C_i plus the adjustable capacitor C. If we use Eq. 21.19, this total tuning capacitance can then be written as

$$C_t \simeq C + C_{gs} + C_{gd} + g_{mo} R_L C_{gd} - \frac{g_{mo} R_L C_{gd} v_{GS}}{V_P} \qquad (21.20)$$

This capacitance C_t can be written as a fixed part C_o plus a component that varies with the modulating voltage v_{gs}.

$$C_t = C_o + B v_{gs} \qquad (21.21)$$

where

$$C_o = C + C_{gs} + C_{gd} + g_{mo} R_L C_{gd}\left(1 - \frac{V_{GS}}{V_P}\right) \qquad (21.22)$$

$$B = -\frac{g_{mo} R_L C_{gd}}{V_P} \qquad (21.23)$$

and

$$v_{gs} = v_{GS} - V_{GS}$$

Thus, the radian frequency of oscillation can be written

$$\omega = \frac{1}{[L(C_o + B v_{gs})]^{1/2}} \qquad (21.24)$$

The amount of harmonic distortion for a given set of conditions can be found by expanding Eq. 21.24 into a McClaurens' series.

$$\omega = [L(C_o + B v_{gs})]^{-1/2} = A_o + A_1 v_{gs} + A_2 v_{gs}^2 + \cdots \qquad (21.25)$$

The quiescent term A_o is found by letting $v_{gs} = 0$. The coefficient A_1 is found by differentiating both sides of Eq. 21.25 and then letting v_{gs} go to zero. This manipulation yields

$$A_1 = -\frac{B}{2(LC_o^3)^{1/2}} \qquad (21.26)$$

Similarly, the coefficient A_2 is found by taking the second derivative of both sides of Eq. 21.24 and then letting v_{gs} go to zero.

$$A_2 = \frac{3B^2}{8(LC_o^5)^{1/2}} \qquad (21.27)$$

Thus Eq. 21.24 can be written

$$\omega = \frac{1}{(LC_o)^{1/2}} - \frac{B}{2(LC_o{}^3)^{1/2}} v_{gs} + \frac{3B^2}{8(LC_o{}^5)^{1/2}} v_{gs}{}^2 + \cdots \quad (21.28)$$

The second harmonic distortion is found by letting $v_{gs} = V_m \cos \omega_m t$ and $v_{gs}{}^2 = (V_m{}^2/2)(1 + \cos 2\omega_m t)$. Then, dividing the magnitude of the second harmonic by the magnitude of the fundamental and multiplying by 100 yields the percent second harmonic.

$$\% \text{ 2nd har} = \frac{3B^2 V_m{}^2/16(LC_o{}^5)^{1/2}}{BV_m/2(LC_o{}^3)^{1/2}} \times 100 \quad (21.29)$$

Substituting the expression of Eq. 21.23 for B and simplifying,

$$\% \text{ 2nd har} = \frac{3g_{mo} R_L C_{gd} V_m}{8C_o V_P} \quad (21.30)$$

PROBLEM 21.3   The circuit of Fig. 21.4 is used to modulate a 10 MHz oscillator. The FET modulator has $g_{mo} = 3$ mmho, $V_P = 2.5$ V magnitude, $C_{gs} = 2$ pF, and $C_{gd} = 1.5$ pF. If $C_o = 10$ pF and $V_m = 1.0$ V (max):

    a.   Determine the value of R_L that will give 5 percent maximum second harmonic distortion.

    b.   Determine the maximum deviation ratio M_f at $f = 1$ kHz with $V_m = 1.0$ V.

    c.   Determine suitable values for the components of Fig. 21.4 not given, assuming $h_{FE} = 100$ for the bipolar transistor.

21.3   SIDEBANDS OF THE FREQUENCY MODULATED WAVE

    When the FM system was first conceived, it was hoped that the bandwidth requirements of this system might be less than that of the AM system discussed in Chapter 20. To investigate the bandwidth requirement, the equation of the frequency-modulated wave is rewritten below.

$$v = A \cos(\omega_c t + M_f \sin \omega_m t) \quad (21.3)$$

The trigonometric identity for the sum of two angles may be used to give

$$v = A[\cos \omega_c t \cos(M_f \sin \omega_m t) - \sin \omega_c t \sin(M_f \sin \omega_m t)] \quad (21.31)$$

But

$$\cos(M_f \sin \omega_m t) = J_0(M_f) + 2J_2(M_f) \cos 2\omega_m t + 2J_4(M_f)\cos 4\omega_m t + \cdots \quad (21.32)$$

and

$$\sin(M_f \sin \omega_m t) = 2J_1(M_f)\sin \omega_m t + 2J_3(M_f)\sin 3\omega_m t + \cdots \quad (21.33)$$

where $J_n(M_f)$ is the Bessel function of the first kind having order n and argument M_f. These Bessel functions have been tabulated; that is, they may be obtained from a mathematics table in the same manner as a trigonometric function. Substituting these Bessel function identities into Eq. 21.31, we have

$$v = A\{\cos \omega_c t[J_0(M_f) + 2J_2(M_f)\cos 2\omega_m t$$
$$+ 2J_4(M_f)\cos 4\omega_m t + \cdots] - \sin \omega_c t[2J_1(M_f)\sin \omega_m t$$
$$+ 2J_3(M_f)\sin 3\omega_m t + \cdots]\} \qquad (21.34)$$

Rearranging Eq. 21.34, we see that

$$v = A\{J_0(M_f)\cos \omega_c t - J_1(M_f)(2 \sin \omega_m t \sin \omega_c t)$$
$$+ J_2(M_f)(2 \cos 2\omega_m t) - J_3(M_f)(2 \sin 3\omega_m t \sin \omega_c t)$$
$$+ J_4(M_f)(2 \cos 4\omega_m t \cos \omega_c t)\cdots\} \qquad (21.35)$$

If we use the trigonometric identities, Eq. 21.35 becomes

$$v = A\{J_0(M_f)\cos \omega_c t$$
$$+ J_1(M_f)[\cos(\omega_c + \omega_m)t - \cos(\omega_c - \omega_m)t]$$
$$+ J_2(M_f)[\cos(\omega_c + 2\omega_m)t + \cos(\omega_c - 2\omega_m)t]$$
$$+ J_3(M_f)[\cos(\omega_c + 3\omega_m)t - \cos(\omega_c - 3\omega_m)t]$$
$$+ J_4(M_f)[\cos(\omega_c + 4\omega_m)t + \cos(\omega_c - 4\omega_m)t]$$
$$+ \cdots\cdots\cdots\cdots\cdots\} \qquad (21.36)$$

Observe from Eq. 21.36 that frequency modulation produces a series of sideband frequency pairs. The number of significant pairs depends on the modulation index M_f, as illustrated by Fig. 21.5. Only the first four Bessel functions are shown for the sake of clarity. It may be seen from Fig. 21.5 and Eq. 21.36 that when $M_f = 1$, two pair of significant sidebands appear in addition to the carrier. Also it may be seen that when $M_f = 2$, there are three pair of significant sidebands. If the higher order Bessel functions had been included in Fig. 21.5, it would be evident that the number of pairs of significant sidebands is one greater than the modulation index when integer values of modulation index are considered. This relationship, which may also be observed from the table of Bessel functions given in Appendix III, is expressed by the following empirical formula.

$$P_s = M_f + 1 \qquad (21.37)$$

where P_s is the number of pairs of significant sidebands.

Unfortunately, the bandwidth requirement for the amplification or transmission of a frequency-modulated wave is much greater than that of an amplitude-modulated wave unless the modulation index M_f is approximately one-half or less. In this case, there is only one pair of significant sidebands, as may be seen in Fig. 21.5. At this point in the discussion there is no reason to believe that a modulation index of one-half is not entirely adequate. This problem will be pursued in a following section.

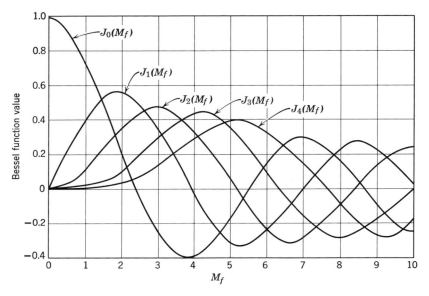

FIGURE 21.5
Value of the Bessel functions as a function of the modulation index M_f.

The frequency spectra for a frequency-modulated wave are shown in Fig. 21.6 as a function of both modulation index and modulating frequency. Observe that the bandwidth requirement is essentially proportional to the modulation index, and hence modulating voltage, as previously noted, but is almost independent of the modulating frequency. This behavior is in contrast to the amplitude-modulated case where the bandwidth requirement is proportional to the modulating frequency but independent of the modulating voltage.

PROBLEM 21.4 Verify the number and relative magnitude of the frequency components shown in Fig. 21.6a by the use of a table of Bessel functions.

PROBLEM 21.5 A 400 Hz modulating frequency has sufficient amplitude to provide a modulation index $M_f = 2$. What bandwidth would be required in order to pass the frequency spectrum of this frequency-modulated wave? What would be the effect on the required bandwidth if the modulating

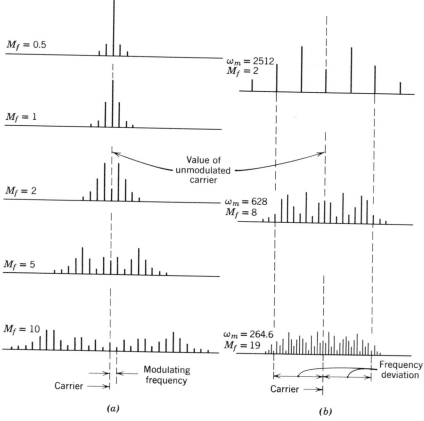

FIGURE 21.6
The frequency spectrum of a frequency-modulated wave: (a) as a function of the modulating voltage; and (b) as a function of the modulating frequency.

frequency were decreased to 100 Hz and the modulation amplitude remained unchanged? What would be the modulation index? Compare your answers with Fig. 21.6b.

21.4   INTERFERENCE TO FM TRANSMISSION

One of the knottiest problems in radio communication is the interference which undesired signals offer to the desired signal in the radio receiver. These interfering signals may be electromagnetic waves which have frequency components within the pass band of the receiver. They may be coherent signals from radio transmitters or noise signals from electric arcs such as lightning, automotive ignition systems, and other types of arc-generating machinery. The amplifiers in the receiver itself also generate interfering noise. Whatever the type of interference, the interfering signal adds vectorially to the desired signal as shown in Fig. 21.7. In this figure the desired carrier voltage v_c is used as the reference. The carrier is assumed to be unmodulated except for the interfering signal v_n which is considered to be noise. The sum of the carrier and noise voltages is v_t.

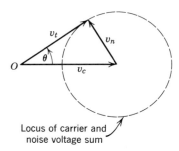

Locus of carrier and noise voltage sum

FIGURE 21.7
The phasor addition of a desired carrier voltage and an interfering noise in a radio receiver.

Figure 21.7 shows that the resultant voltage v_t varies both in phase and amplitude, compared with the carrier voltage, as the interfering signal varies in either phase or amplitude with respect to the desired carrier. In this figure, the amplitude of the interfering signal is assumed to be constant, but its relative phase varies with respect to the desired carrier. The locus of all possible values of the resultant voltage v_t forms the dashed circle shown. Observe that the amplitude modulation which results from the noise approaches 100 percent as the noise voltage magnitude approaches the desired carrier voltage magnitude. On the other hand, the maximum phase deviation θ that can result from the noise voltage will not exceed more than one radian unless the noise voltage becomes essentially equal to or exceeds the desired carrier voltage. As previously noted, the maximum phase deviation is actually equal to the modulation index of a frequency-modulated wave. Therefore, the interfering signal can cause frequency modulation, but the modulation index M_f will not exceed unity unless the interfering voltage becomes essentially equal to or greater than the desired voltage. Consequently, the following conclusions can be drawn.

    1. In a communications system which uses amplitude modulation, the modulation caused by interference will approach the modulation produced

by the desired intelligence as the magnitude of the interference approaches the magnitude of the desired carrier.

2. In a communication system which uses frequency modulation, the degree of modulation that may be produced by an interfering signal will be small in comparison to the modulation produced by the desired intelligence, providing three conditions are met. These three conditions are:

a. The modulation index produced by the desired intelligence must be large in comparison with unity.

b. The amplitude of the desired carrier must be larger than the amplitude of the interfering signal.

c. The radio receiver must be insensitive to amplitude variations of the resultant signal.

Thus, it may be seen that the frequency-modulation index must be at least 5 or more in order to provide good interference rejection. Consequently, the hope that the frequency spectrum of a frequency-modulated wave would be smaller than that of an amplitude modulated wave has completely dimmed. As previously noted, their frequency spectra are comparable providing the frequency modulation index is small (about 0.5). But then, the interference rejection capabilities of the two systems are also comparable. In fact, the AM system would be slightly superior if a high percentage of modulation could be maintained.

The interference-rejecting characteristics of the FM system may be illustrated in the following example. Assume that two communities, which are separated by only a few miles, each have FM broadcasting stations. Also assume that the two stations are operating at the same carrier frequency and their radiated powers are equal. A motorist is listening to the local FM station as he departs from one community on his way to the other. The reception is essentially free of interference until the auto is almost equidistant from the two communities. Then for a very brief time the two programs are heard with essentially equal loudness. After this brief interval, the FM station at the destination will be heard with essentially no interference from the other station. Of course, it has been assumed that both stations maintain a modulation index which is large in comparison with unity and the auto radio is insensitive to amplitude variations.

This interference-rejecting property of the FM system is a great advantage in a communication system. The chief disadvantage of the FM system is the large frequency spectrum required by a large modulation index. The resulting large bandwidth requirement of the tuned amplifiers in the transmitter and receiver is most easily attained if the carrier frequency is quite high. For example, the frequency spectrum from 88 to 108 MHz has been allotted for commercial FM broadcasting by the Federal Communications Commission. Each station is permitted to use a 150 kHz channel.

PROBLEM 21.6   An FM broadcasting station modulates with audio frequencies up to 12 kHz and maintains a modulation index $M_f = 5$ at this frequency. What is the required bandwidth of a tuned circuit if none of the significant sidebands are to be excluded?

Answer: $B = 144$ kHz.

PROBLEM 21.7   If the standard AM broadcast band (550 to 1600 kHz) were allocated to FM stations having the standards specified in Problem 21.6, how many channels could be accommodated in this band? What would be the required Q (approximately) of a single-tuned circuit which would accommodate the frequency spectrum if the carrier frequency were 1 MHz?

21.5   FM DEMODULATORS

The AM demodulator, or detector, will demodulate an FM wave providing that the detector is tuned so that the carrier frequency is on the edge of the pass band instead of in the center of the pass band. The process by which demodulation is accomplished is illustrated in Fig. 21.8. The frequency variations of the FM signal are converted into amplitude variations by the detuned circuit. The AM detector then recovers the waveform of the amplitude variations. This waveform is the same as the

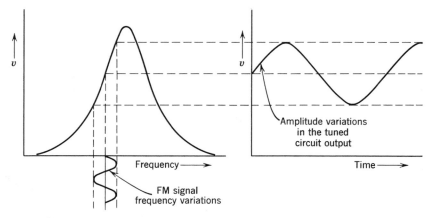

FIGURE 21.8
A detuned AM demodulator used as an FM demodulator.

waveform of the modulating voltage, providing that there is no distortion in the system. However, there will be some distortion in this demodulator because the sides of the response curve of the tuned circuit are not straight. The distortion will be small if the frequency deviation is so small that only a very small portion of the response curve is used. This system is sometimes used to receive *narrow band* FM with an AM receiver. Both the detector and the RF amplifiers of the receiver are sensitive to amplitude variations, so the conditions required for good interference rejection are not met.

An improved type of demodulator known as a Foster-Seely discriminator is shown in Fig. 21.9. In this circuit the primary voltage across the coil

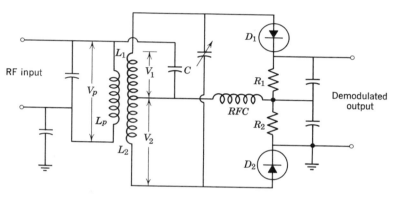

FIGURE 21.9
A Foster-Seely discriminator circuit.

L_1 is applied through the blocking capacitor C to the center tap of the inductively-coupled secondary. The relative phase of the secondary current, and hence secondary voltage, changes rapidly with the input frequency because the secondary is tuned to the carrier or center frequency of the input signal. A phasor diagram of the primary voltage V_p, the primary current I_p, and the voltage V_s induced into 21.10a. These phase relationships are easily ships $V_p = j\omega L_p I_p$ and $V_s = j\omega M I_p$. The sec the induced secondary voltage V_s when the

As the input frequency deviates above or the secondary current correspondingly lags This current phase shift may be seen w circuit of Fig. 21.10b. Since the voltages acros always lag behind the secondary current by

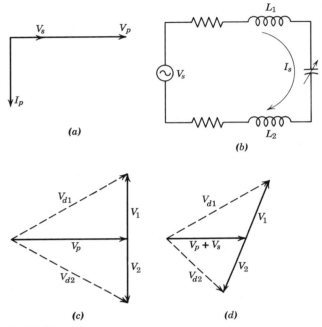

FIGURE 21.10
(a) **Phasor relationships of primary voltage, primary current, and induced voltage in the secondary of an inductively-coupled circuit,** (b) **equivalent circuit of the tuned secondary,** (c) **phasor addition of the primary voltage and the voltages across the secondary coils at the resonant frequency,** (d) **phasor addition of the primary voltage and the voltages across the secondary coils when the input frequency is above the resonant frequency of the tuned circuit.**

across each of these secondary coils is approximately 90° out of phase with the primary voltage when the input frequency is the same as the resonant frequency of the secondary circuit. This situation is illustrated in Fig. 21.10c. The phasor sum of the primary voltage V_p and the voltage V_1 across coil L_1 is applied to the upper diode and its load resistor. This voltage is V_{d1} in Fig. 21.10c. Similarly, the voltage V_{d2} is applied to the lower diode and its load resistor. The output voltage is the algebraic sum of the diode load voltages, but since the current flows in opposite ctions through these diode loads, the output voltage is actually the nce between these two load voltages. Thus, if the load resistances are output voltage is zero when the secondary circuit is tuned to the ncy.

When the input frequency is raised above the resonant or center frequency, the secondary current and hence the voltage across the secondary coils lag behind their resonant position and the phasor diagram of Fig. 21.10d results. The voltage V_{d1} applied to the upper diode and its load then becomes larger than the voltage V_{d2} applied to the lower diode and its load. The difference between these two voltages produces an output voltage of positive polarity. If the input frequency is decreased from the resonant value, the secondary becomes capacitive and the secondary current shifts phase in a leading direction. Therefore, an output voltage having negative polarity is produced. The output voltage is essentially proportional to the frequency deviation as long as the frequency remains in the flat portion of the frequency response curve of the coupled circuit. As the frequency excursions approach the edge of the pass band, the output wave becomes flattened because of the reduced amplitude of the output voltage. Thus frequency discrimination in the coupling circuit produces waveform distortion in the output voltage.

The linearity and tuning characteristics of the Foster-Seely discriminator are shown in Fig. 21.11. The circuit should be tuned to the carrier frequency ω_o. The edges of the pass band are shown at ω_H and ω_L. As illustrated, the output voltage is essentially a linear function of the input frequency providing that the total frequency deviation does not exceed about 70 percent of the pass band. This degree of linearity is attained only when

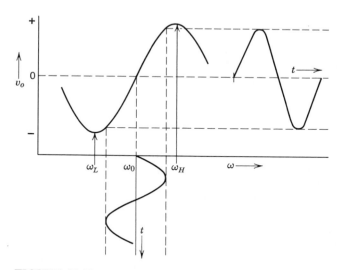

FIGURE 21.11
The demodulation characteristics of the Foster-Seely discriminator.

the discriminator circuit is double tuned as shown and when the coefficient of coupling is adjusted so that maximum flatness is obtained in the response curve.

The diode load resistance should be chosen so that the desired circuit Q is attained in the secondary circuit. This problem was discussed in Chapter 20 in conjunction with the AM diode detector. The capacitors in parallel with the diode load resistors bypass the RF and cause the load voltages to follow the peaks of the RF voltages applied to the diodes, as discussed in the case of the AM diode detector.

Although the Foster-Seely discriminator may provide good linearity, it is sensitive to amplitude variations, and therefore a *limiter* must be included in the RF amplifier if the interference-rejecting capabilities of FM are to be realized. A limiter is an amplifier which provides a constant amplitude output even though the input amplitude varies. Any amplifier will limit if the input signal is sufficiently large. A limiter, then, has small biases so that it will limit when the values of input signal are comparitively small. For example, the collector supply voltage is reduced and the base or grid bias is reduced in comparison with the conventional amplifier.

21.6 THE RATIO DETECTOR

Another type of FM demodulator which operates on the same basic principle as the Foster-Seely discriminator but which is insensitive to amplitude modulation is shown in Fig. 21.12. This demodulator is known as a *ratio detector*. The coupling circuit and the addition of the primary voltage to the center tap of the secondary coil may be identical to the corresponding portion of the Foster-Seely discriminator. However, in the ratio detector, the two diodes are connected so that their load voltages are additive rather than subtractive. In addition, a large capacitance (perhaps 20 μF to 100 μF) is connected across the series combination of the load resistors. Consequently, the total load voltage cannot follow short-term amplitude variations of the input signal and the circuit is not sensitive to amplitude variations of the input signal. Hence, the limiter previously discussed is *not* required with the ratio detector[1]. However, each individual diode load voltage must be a function of the input frequency as it is for the Foster-Seely discriminator. Thus, the ratio of the two load voltages changes with frequency even though their sum is forced to remain essentially constant, hence the name *ratio detector*. The output voltage is obtained across one of the diode load resistors and consequently the output voltage is essentially a linear function

---

[1] Limiters are often used in conjunction with a ratio detector to improve the AGC action as well as to reduce the amplitude variations to essentially zero.

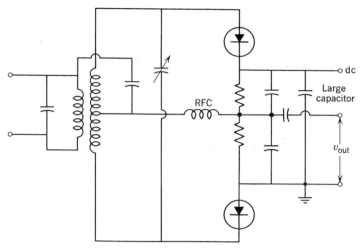

FIGURE 21.12
A ratio detector.

of the input frequency, providing that the requirements placed on the Foster-Seely discriminator are met. In contrast to the Foster-Seely circuit, the ratio detector has a dc component in the output even when the carrier is unmodulated and the circuit is properly tuned. This dc voltage may be used for automatic gain control which may be desirable when limiters are not used.

21.7   QUADRATURE DETECTOR

The FM signal can be demodulated by a simple circuit consisting primarily of a tuned circuit and a transistor, shown in Fig. 21.13. The FM signal is applied directly to the base of the transistor and the signal is also applied to the emitter through an inductively-coupled tuned circuit. The voltage across the secondary is nearly 90° out of phase with the primary when the circuit $L_2 C_2$ is tuned to the carrier frequency and the carrier is unmodulated, providing L_2 is tightly coupled to L_1 so the primary is essentially resistive. The circuit is known as a *quadrature detector* because of this 90° phase shift. Current flows through the transistor only during the times when the emitter is negative with respect to the base, which is about one-half the time when the phase shift is 90°. When the modulation increases the input frequency and the tuned circuit appears inductive to the induced voltage, the current in the tuned circuit lags the induced voltage and the phase difference between the emitter and base voltages decreases, so the transistor conducts for a shorter time during each cycle and the average collector current decreases. Conversely, when the frequency deviation is below the resonant frequency, the phase difference increases and the average collector current increases. Thus, the circuit is actually a phase detector. The capacitor C_1 in the collector circuit bypasses the RF but not the AF, so the voltage across the collector load resistor is proportional to the average collector current which is proportional to the frequency deviation.

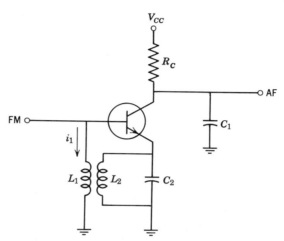

FIGURE 21.13
A single-transistor quadrature detector.

The single-transistor circuit of Fig. 21.13 is not ideal because the circuit works best when it is driven hard so the transistor behaves as a switch. However, the excess stored charge in the base that results from driving the transistor into saturation causes distortion in the output because of the recovery time.

An improved version of the quadrature detector is shown in Fig. 21.14a. The emitter-coupled amplifier is better because it cannot be driven appreciably into saturation because of the large emitter circuit resistance. Therefore it has nearly zero recovery time. Also, there is much less loading on the tuned circuit in Fig. 21.14a if the resistor R_2 is large.

The FM signal is coupled through a very small capacitor C_1 to the parallel tuned circuit. The reactance of C_1 must be large compared with the parallel resistance of the tuned circuit so the current through C_1 leads the input voltage by nearly 90° at all times. Then the voltage across the tuned circuit leads the input voltage by nearly 90° at resonance. As the input

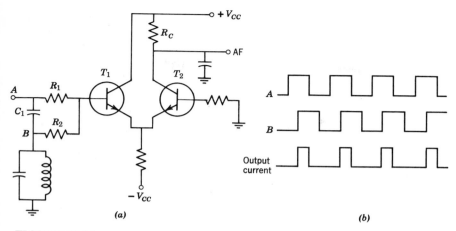

(a)

(b)

FIGURE 21.14

(*a*) **An emitter-coupled quadrature detector,** (*b*) **waveforms illustrating the detection process.**

frequency deviates above resonance, the capacitive reactance of the tuned circuit causes the voltage across it to lag the current through C_1 and the phase difference of the input and tuned circuit voltages becomes less than 90°. Conversely, when the frequency deviates below resonance, the relative phase exceeds 90°. The input and tuned circuit voltages are summed at the base of transistor T_1 and the resistors R_1 and R_2 are chosen to equalize the effect of these voltages at the summing point. The resistance of both R_1 and R_2 must be large in comparison with the input resistance of T_1, while it is conducting, in order to provide proper summing and minimize interaction between the two signals. The waveforms of Fig. 21.14*b* illustrate the detection process. Although the input signal and the voltage across the tuned circuit are essentially sinusoidal, they are usually large enough to over-drive the amplifier and cause it to operate as a switch. Therefore, square waves are used to illustrate the effect of the phase difference between the input voltages on the average value of the output current pulses. Ideally, the width of the output pulses, and hence the average output current is proportional to the relative phase shift. This condition is essentially met for the collector current of T_2 when the bias and signal levels are arranged so that transistor T_1 is driven to saturation during the time when both input signals are positive, but is otherwise cut off.

The phase shift caused by the tuned circuit is proportional to the frequency deviation only when the deviation is small compared with the bandwidth of the tuned circuit, as illustrated in Fig. 21.15. The relationship

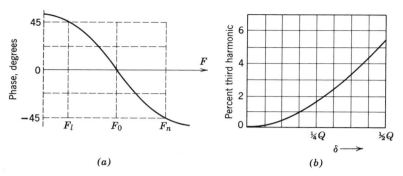

(a)   (b)

FIGURE 21.15
(a) **Phase shift of the voltage across a tuned circuit, with reference to the applied current, as a function of frequency,** (b) **percent third harmonic content of the detector as a function of** Q.

between the phase shift and the frequency deviation may be obtained from the expression of the impedance of a parallel tuned circuit found in Eq. 9.4.

$$Z_p = \frac{R}{1 + jQ[(\omega/\omega_o) - (\omega_o/\omega)]} \tag{21.38}$$

The relative phase θ of the voltage across the tuned circuit is the same as the angle of Z_p, providing the source current is controlled by the coupling capacitor as previously stated $(X_{c1} \geq 10R)$. Then

$$\tan \theta = Q\left(\frac{\omega}{\omega_o} - \frac{\omega_o}{\omega}\right) \tag{21.39}$$

Equation 21.39 can be simplified if we define

$$\delta \equiv \frac{\omega - \omega_o}{\omega_o} = \frac{\omega}{\omega_o} - 1 \tag{21.40}$$

where δ is the fractional deviation, and then write Eq. 21.39 in terms of δ as

$$\tan \theta = 2Q\delta \tag{21.41}$$

Observe from Eq. 21.41 that $\tan \theta$ is proportional to the frequency deviation. Thus the relative phase θ is proportional to the deviation only when $\tan \theta \simeq \theta$. The harmonic distortion (primarily third) is sketched as a function of δ in Fig. 21.15b. Note from Eq. 21.41 that $\theta = 45°$ when $\delta = 1/2Q$, so this value of frequency deviation extends to the edge of the pass band of the tuned circuit. Observe that the harmonic distortion will be less than 1 percent if the deviation does not exceed about 38 percent of the half bandwidth or 19 percent of the full bandwidth. Thus the tuned circuit should be designed for a bandwidth about 5.3 times as wide as the maximum deviation if the harmonic distortion is to not exceed 1 percent.

The quadrature detector is popular in integrated circuits because all the components can be integrated except the LC tuned circuit. Since transistors are easier to integrate than large value resistors, the IC version of the quadrature detector usually uses two transistors in a differential summing circuit as shown in Fig. 21.16. The resistors R_1 and R_2 may then have lower

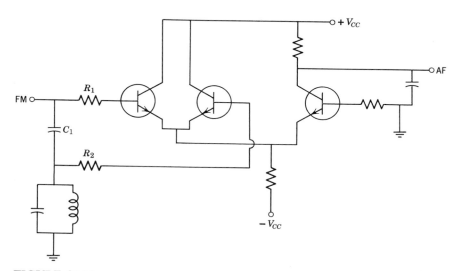

FIGURE 21.16
A typical IC version of the quadrature detector.

values than their counterparts in Fig. 21.14 for the same degree of isolation between the incoming and quadrature signals. Also, the gain of the circuit is increased somewhat because of the smaller resistance values.

PROBLEM 21.8   A standard broadcast FM has an intermediate frequency of 10.7 MHz and a maximum frequency deviation of 75 kHz. Determine suitable values for the tuned circuit, the coupling capacitor C_1, and the isolation resistors R_1 and R_2 of Fig. 21.14 if the h_{ie} of the summing amplifier is 1 kΩ. Assume $Q_o = 100$.

Several other types of FM demodulators are discussed in the literature.[2] In addition, a counter type of FM detector is given in Chapter 22. The examples included here are merely illustrative, not exhaustive. The demodulators which have been included are very commonly used, however.

Figure 21.17 is the block diagram of a typical FM receiver which incorporates a Foster-Seely discriminator.

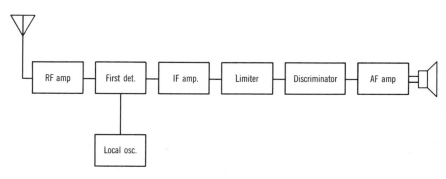

FIGURE 21.17
The block diagram of a typical FM receiver.

21.8   PHASE MODULATION

The primary problem in the generation of FM waves is the stability of the carrier frequency. In the AM system, the carrier is generated by a crystal-controlled oscillator which can easily meet the rigorous frequency tolerance requirements of the Federal Communications Commission. In contrast, the reactance-modulated oscillator cannot be crystal-controlled because the oscillator would not deviate appreciably from the resonant frequency of the crystal, as discussed in Chapter 19. However, the standard LC oscillator does not usually meet the frequency stability standards imposed by the Federal Communications Commission. This frequency stability predicament can be resolved by a technique which is known as phase modulation.

The similarity between phase modulation and frequency modulation can be seen from Eq. 21.3, repeated below for convenience.

$$v = A \cos(\omega_c t + M_f \sin \omega_m t) \qquad (21.3)$$

---

[2] For several different *IC* type FM demodulators, see *Application Considerations for Linear Integrated Circuits*, edited by Jerry Eimbinder, John Wiley and Sons, New York, 1970, pp. 97–117.

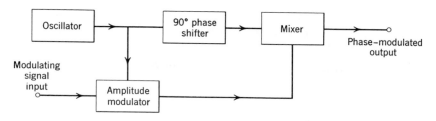

FIGURE 21.18
An elementary phase-modulating system.

As previously mentioned, the term $(M_f \sin \omega_m t)$ has the dimensions of an angle, not a frequency. Therefore, we should suspect that a modulation process could be devised wherein the phase (or relative phase angle) of the wave could be changed rather than the frequency *per se*. One of the simplest systems which may be used to accomplish phase modulation is illustrated by the block diagram of Fig. 21.18. The modulation is accomplished by adding an amplitude-modulated wave to the carrier which has experienced approximately 90° phase shift. The effect of this addition is shown by Fig. 21.19a. This figure shows that the relative phase angle θ varies as the amplitude of the modulated carrier varies. In Fig. 21.19b, the amplitude-modulated wave has been resolved into the carrier and sideband components. Only two sidebands have been included, assuming sinusoidal modulation. Since the phase-shifted carrier is used as a reference, the carrier component of the modulated portion remains stationary in the phasor diagram while the sidebands rotate at the relative angular velocity ω_m. The upper sidebands rotate in a counterclockwise direction and the lower sidebands rotate in a clockwise direction. Thus the sideband components alternately add to and subtract from their carrier component as shown, whereas the relative phase varies about the average value, θ_o, which is produced by the carrier component of the modulated wave. It may be observed from Fig. 21.19b that the increase of θ during modulation peaks

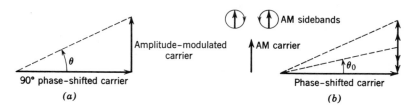

FIGURE 21.19
The relative phase shift which is accomplished by the system of Fig. 21.18.

is not as great as the decrease of θ during modulation troughs, if we assume the amplitude modulation to be distortionless. The distortion of the phase modulation θ results because $\tan \theta$ is a nonlinear function of θ.

The distortion of the phase modulation can be reduced for a given variation of θ if the carrier of the modulated wave is removed. In this case θ_o in Fig. 21.19b will be reduced to zero. The balanced modulator which was discussed in Chapter 20 can be used to suppress the carrier of the modulated wave. The block diagram of a system which incorporates a balanced modulator is shown in Fig. 21.20. In this system the output of the balanced modulator was shown in Chapter 20 to be (Eq. 20.29):

$$v_0 = \frac{BM_a}{2} \left[\cos(\omega_c + \omega_m)t + \cos(\omega_c - \omega_m)t \right] \tag{21.42}$$

where M_a is the amplitude modulation index and $BM_a/2$ is the peak amplitude of each sideband, assuming the modulating voltage to be sinusoidal of natural frequency ω_m. The phasor addition of the 90° phase-shifted carrier and the output of the balanced modulator is shown in Fig. 21.21. The 90° phase-shifted carrier is again used as the reference in this figure. Let the peak amplitude of this carrier be A. As the sideband phasors rotate, the phase modulation angle θ oscillates about the reference carrier vector at the modulation frequency ω_m. Since each sideband has a maximum amplitude $B/2$, they add together twice each modulation cycle to produce a maximum contribution B. The maximum excursion of the phase angle is then

$$\theta_{max} = \tan^{-1} \frac{BM_a}{A} \tag{21.43}$$

where A is the maximum amplitude of the phase-shifted carrier, as previously stated. But, to obtain distortionless phase modulation, the relative phase angle θ must be proportional to the sideband amplitude $B/2$. The only way this linear relationship can be obtained is by limiting θ_{max} to such

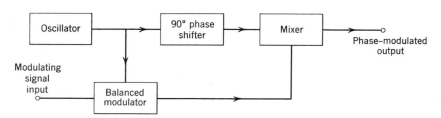

FIGURE 21.20
The block diagram of a phase-modulating system which utilizes a balanced modulator.

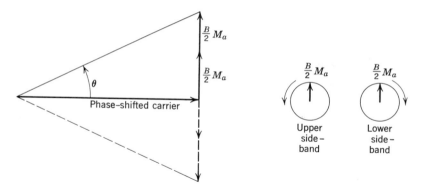

FIGURE 21.21
The relative phase shift which is accomplished by the balanced modulator system of Fig. 21.26.

small angles that the value of the angle is essentially equal to its tangent. With this restriction,

$$\theta_{\max} \simeq \frac{BM_a}{A} \tag{21.44}$$

The limit which is set on $\theta_{\max}$ will, of course, depend on the tolerable distortion. The limit $\theta_{\max} = 0.5$ rad may be appropriate in some applications. Then, assuming the upper limit of the amplitude modulation index M_a to be unity, we could obtain the value of $\theta_{\max} = 0.5$ rad by making $A = 2B$. Since A is the peak amplitude of the phase shifted carrier and $B/2$ is the peak amplitude of each sideband, it is evident that the phase shifter must also include amplification. In fact, an amplifier which has a high output impedance (such as a common-base transistor or pentode tube) and a load which is essentially a pure reactance could be used as a phase shifter.

PROBLEM 21.9 A balanced modulator is used in a phase-modulation system as discussed previously. Assuming the distortion to be primarily third harmonic, determine the distortion of the phase modulator when the modulator index is 0.5.
Answer: Dist. = 7.3%.

When the modulating signal is a single frequency (sinusoidal), as previously assumed, it is evident from Fig. 21.21 that the relative phase angle θ varies sinusoidally at the modulating frequency. Then the phase-modulated signal voltage can be written

$$v = V_m \cos(\omega_c t + \theta_m \sin \omega_m t) \tag{21.45}$$

Using Eq. 21.44, we have

$$v = V_m \cos\left(\omega_c t + \frac{B}{A} M_a \sin \omega_m t\right) \qquad (21.46)$$

Now, if a phase-modulation index $M_p = (B/A)M_a$ is defined,

$$v = V_m \cos(\omega_c t + M_p \sin \omega_m t) \qquad (21.47)$$

Since neither A, B, or M_a are functions of the modulating frequency, it is evident that M_p is not a function of the modulating frequency. On the other hand, the frequency modulation index M_f is inversely proportional to frequency, as shown in Eq. 21.7, if we assume the modulating amplitude to be constant. This difference in the character of the modulation indices is the only difference between phase modulation and frequency modulation. Viewing this difference another way, we see that the frequency deviation of the frequency-modulated wave is proportional to the modulation amplitude but independent of the modulating frequency, whereas the frequency deviation of the phase-modulated wave is proportional to the frequency of the modulating voltage as well as to its amplitude.

PROBLEM 21.10   Prove that the frequency deviation of the phase-modulated wave is proportional to both the frequency and the amplitude of the modulating voltage.

It is clear that the phase-modulating technique can be used to produce frequency-modulated waves, providing that the amplitude of the modulating voltage is inversely proportional to the modulating frequency. This inverse relationship can be obtained by including, in the modulator, an amplifier which has gain inversely proportional to the frequency.

PROBLEM 21.11   Draw the circuit diagram for an amplifier which has gain inversely proportional to frequency.

One major problem arises when frequency modulation is produced by the phase-modulating technique. As previously shown, the maximum modulation index may be of the order of 0.5. This will then be the acceptable index at the *lowest* modulating frequency. If the modulation is audio-frequency program material as in a commercial broadcast station, this lowest modulating frequency could be 30 Hz. Then, since the modulation index is inversely proportional to frequency, the modulation index at 15 kHz would be $0.5 \times 30/15,000 = 0.001$. As previously mentioned, this modulation index could be increased by frequency multiplication, but to bring this modulation index up to 5, which is considered by the FCC to be an acceptable minimum, the carrier frequency must be multiplied by 5000. The complexity of a circuit which would multiply the frequency

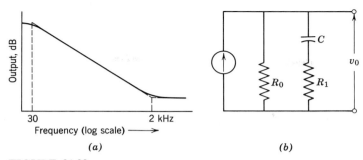

FIGURE 21.22
The predistortion and compensation characteristics of a phase-modulated
system: (*a*) predistortion characteristic, (*b*) a predistorting circuit.

by a factor of 5000 would be quite forbidding. Consequently, a compromise
is made in phase-modulated transmitters, wherein the modulation index
is required to follow the inverse relationship with frequency only up to
about 2000 Hz. Above this frequency, the modulation index is permitted
to remain independent of frequency. In this case, a compensating circuit
must be included in the output of the demodulator in the receiver. Otherwise, the frequency response characteristic would be proportional to the
frequency for frequencies above 2000 Hz.

The frequency-response characteristic of the modulator for a phase-
modulator transmitter is shown in Fig. 21.22*a*. The circuit which produces
this characteristic is frequently known as a *predistorter*. The simple *RC*
circuit shown in Fig. 21.22*b* could be used as a predistorter. In this circuit
$1/\omega C = R_0$ at 30 Hz and $1/\omega C = R_1$ at 2 kHz. Also the receiver-
compensating characteristic is shown in Fig. 21.23*a* and a typical compensating circuit is given in Fig. 21.23*b*.

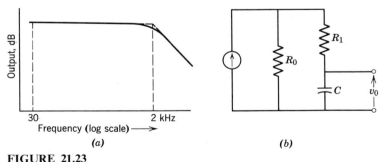

FIGURE 21.23
The predistortion and compensation characteristics of a phase-modulated
system: (*a*) receiver compensation characteristic, (*b*) compensating circuit.

PROBLEM 21.12   An FM transmitter is to have an output carrier of 50 MHz and a modulation index $M_F = 5$ above 2 kHz. The oscillator is phase modulated with a maximum phase deviation of 0.4 rad at 40 Hz. What will be the oscillator frequency if frequency multiplication is used to increase M_F?

Answer: $f_o = 80$ kHz.

Frequency-controlling crystals are not generally available for frequencies below 100 kHz. Therefore, it may be necessary to utilize frequency conversion to provide an adequate modulation index at frequencies above 2000 Hz. The block diagram of a typical FM transmitter which employs the phase-modulating technique is shown in Fig. 21.24. This system is known as the *Armstrong system* of frequency modulation because it was invented by Major E. F. Armstrong. In this diagram, a limiter follows the phase modulator because there is a small amount of amplitude modulation in the output of the phase modulator.

PROBLEM 21.13   An FM transmitter using the Armstrong system has a modulation index = 5 from 2000 to 12,000 Hz. The initial modulation index is 0.40 rad at 40 Hz. The output carrier frequency is 50 MHz. The signal-generating oscillator is crystal-controlled at 200 kHz. Assuming the block diagram of Fig. 21.24 to be used and, further, assuming the

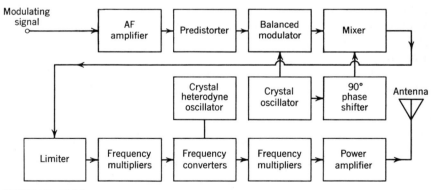

FIGURE 21.24
The block diagram of an FM transmitter which uses the Armstrong system.

frequency multiplication to be equal in the two multipliers, determine both the carrier frequency and the maximum frequency deviation following each of the blocks in the transmitter diagram. Determine the frequency of the crystal oscillator used in the frequency converter.

A vacuum tube device known as a *phasitron* has been developed for the purpose of producing phase modulation.[3] This device requires a three-phase voltage from the crystal-controlled oscillator. Modulation is accomplished by a magnetic field. The phasitron may produce relatively large phase shifts with a given degree of distortion compared with the methods previously discussed. Therefore, considerably less frequency multiplication is needed when the phasitron is used.

21.9   AUTOMATIC FREQUENCY CONTROL (AFC)

One problem that arises in a radio receiver is the detuning which results from changing circuit parameters. For example, the tuning capacitance of a tuned circuit almost always includes the inherent capacitance of the amplifying device. But the transistor junction capacitance is a function of the junction voltage, which in turn may be a function of the temperature. Also the tube interelectrode capacitances are a function of temperature. In addition, the effective capacitance between input and output circuits is a function of the amplifier gain, as previously discussed. Many other causes of detuning exist but will not be mentioned for the sake of brevity. The detuning problem in a superheterodyne-type receiver is accentuated because of the frequency drift in the local oscillator. Of course, the local oscillator can be crystal-controlled, but the receiver cannot then be continuously tunable.

One system that may be conveniently used to maintain a desirable degree of frequency stability in a superheterodyne receiver is known as *automatic frequency control* (AFC). In this system a Foster-Seely-type discriminator is used to sense a deviation from the desired intermediate frequency.[4] The rapid variations are filtered from the output voltage of the discriminator and the remaining average voltage is applied to a reactance-type frequency modulator. This reactance modulator changes the frequency of the local oscillator in a direction that will tend to reduce the average discriminator output voltage toward zero. In other words, the intermediate frequency will be changed toward the resonant frequency of the discriminator circuit. Thus, if the discriminator is tuned to the same frequency as the IF amplifiers, the AFC circuit will continually adjust the local oscillator frequency in a manner that will tend to maintain an intermediate frequency which will be very nearly the resonant frequency of the IF amplifiers. The preciseness of the frequency

---

[3] For a description of the "phasitron" see Samuel Seely, *Electron Tube Circuits*, McGraw-Hill, New York, 1950.

[4] The ratio detector circuit may be modified to produce a suitable AFC control voltage. This modification is shown in Fig. 21.25 and is discussed in connection with that figure.

control depends on the bandwidth of the discriminator and the sensitivity of the reactance modulator. Also, a dc amplifier can be used to amplify the discriminator output and thus increase the preciseness of control.

Either AM or FM receivers may employ AFC. In the FM receiver, the discriminator that recovers the modulation may also provide the control voltage. It may be observed that AFC may compensate for inaccuracy of tuning in a radio receiver. From the standpoint of adjusting the tuning of the receiver, it produces the illusion of greatly increased bandwidth. Therefore, the manual tuning of the receiver is facilitated and may be more precise if the AFC circuit is switched off during the tuning process. Some types of tuning indicators may eliminate the need for the AFC switch. The circuit diagram of an FM receiver which incorporates AFC is shown in Fig. 21.25. However, the AFC system in this receiver is different from the system described previously. In the circuit of Fig. 21.25, the center point of the ratio-detector load is grounded. Therefore, the center tap of the secondary of the ratio-detector transformer is at dc ground potential only when the circuit is in resonance. Thus, the average or dc potential at this center tap is not zero when the incoming signal is not at the resonant frequency of the

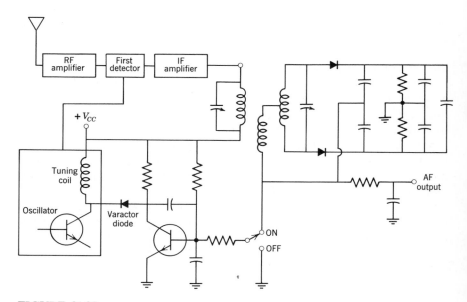

FIGURE 21.25
A partial block diagram of an FM receiver showing the details of a ratio detector and AFC system. Assume that the receiver is a standard broadcast receiver with 10.7 MHz intermediate frequency.

ratio detector. Therefore, this potential at the center tap is filtered and applied, through a dc amplifier, to a varicap diode which in turn varies the local oscillator frequency. The amplifier provides the polarizing potential for the varicap diode. There are many other uses for the AFC principle other than the solution of the frequency instability problem of a superheterodyne receiver.

PROBLEM 21.14   Select a transistor type and determine suitable component values for the amplifier in the AFC system of Fig. 21.25.

21.10   FM MULTIPLEX

The popularity of stereo music has stimulated the development of a multiplex system for standard FM broadcast transmitters and receivers. One requirement imposed by the Federal Communications Commission is that the stereo-multiplex broadcast be compatible with monaural broadcasts so the listeners with monaural receivers can receive the multiplexed signals without loss of quality compared with monaural signals. One *compatible* multiplexing system that can be used in the transmitter is shown in Fig. 21.26. The left channel L and right channel R signals are added to produce the signal needed for monaural receivers. The right channel signal is inverted and then added to the left channel signal to produce

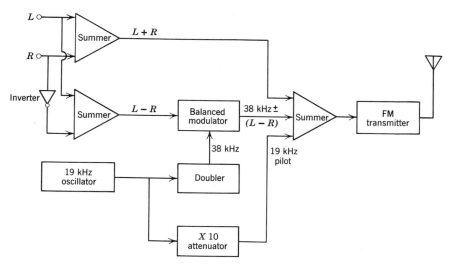

FIGURE 21.26
The block diagram of a compatible stereo-multiplex system for an FM transmitter.

an $L - R$ signal. This $L - R$ signal is used to amplitude-modulate a 38 kHz carrier. The balanced modulator suppresses the carrier and thus provides only the two sidebands of the $L - R$ signal in the output. The 38 kHz carrier is obtained by *doubling* the frequency of a 19 kHz oscillator. The 19 kHz is also attenuated and added to the $L + R$ signal and the $L - R$ sidebands to produce the *composite* signal that frequency modulates the conventional FM transmitter. The 19 kHz *pilot* signal is reduced to about 10 percent of the amplitude of the stereo signal components in the composite signal. The low-amplitude 19 kHz pilot is transmitted instead of the 38 kHz because 40 kHz is a popular bias and erase-oscillator frequency in entertainment-type tape recorders. The 38 kHz carrier would beat with these oscillator signals and audible difference frequencies would appear on the tape recordings due to intermodulation distortion.

The demodulation system in the stereo FM receiver could be the inverse of the multiplex system shown in Fig. 21.26, as shown in Fig. 21.27. However, a simpler approach to the demodulation, or decoding, is possible. It can be shown mathematically and visually verified that the 38 kHz reconstituted carrier can be used as a switch to separate the left channel and right channel information from the composite signal, as illustrated in Fig. 21.28. If the 38 kHz is of adequate amplitude and proper phase, the output of the summer appears as a 38 kHz carrier with the positive

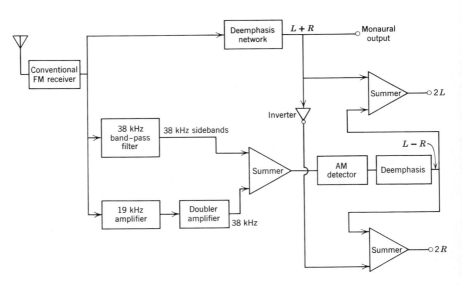

FIGURE 21.27
A stero-multiplex demodulator.

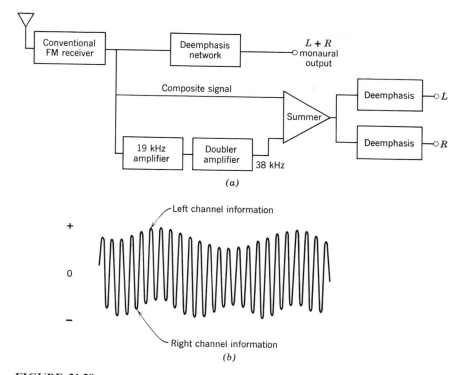

FIGURE 21.28
(a) **The block diagram of a simplified stereo decoding system,** (b) **signal waveform at the output of the summer.**

half cycles amplitude-modulated by L channel and the negative half cycles modulated by R channel. Thus, two diode detectors of opposite polarity will recover both L and R channel signals directly, as shown, without the need of additional matrixing. The 38 kHz could be a square wave as well as a sine wave, and the detector could be a synchronous detector, as discussed later, instead of the diodes.

Integrated-circuit FM stereo decoders have been developed by the semiconductor manufacturers. A circuit produced by Motorola Company will be used as an example of the special features that may be incorporated into a decoder circuit. A simplified circuit diagram of the Motorola decoder is given in Fig. 21.29. The composite signal is coupled through capacitor C_1 to the base of an emitter follower Q_4 that provides a high input impedance (about 20 kΩ) for the decoder. The composite signal at the emitter of Q_4 is passed through two resistors to another emitter follower

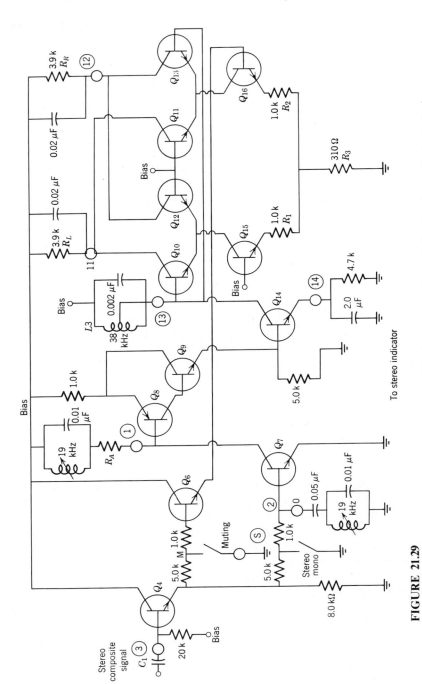

FIGURE 21.29

The simplified circuit diagram of the MC 1304 decoder. (Courtesy of Motorola Semiconductor Products, Inc.)

Q_6 and then to the synchronous detector which will be discussed later. The *muting* switch between the two resistors automatically shorts out the composite signal whenever the signal level drops below the minimum value required for satisfactory signal-to-noise ratio. Thus the receiver is quiet instead of noisy as the tuning is changed from one station to another.

The composite signal at the emitter of Q_4 is also applied through a 19 kHz band-pass filter to the base of a 19 kHz tuned amplifier Q_7. The *stereo-mono* switch between the two resistors in the 19 kHz filter provides monaural output $(L + R)$ from both decoder output terminals when the 19 kHz pilot signal is shorted to ground. This switch may be manually operated but may also be automatically operated whenever the 19 kHz pilot signal amplitude decreases below a predetermined value required for satisfactory stereo operation. A larger signal level is required for the receiver to operate with satisfactory signal-to-noise ratio in the stereo mode than in the monaural mode because the modulation index, or deviation ratio, of a 38 kHz sideband component is much lower than that produced by an AF component below 15 kHz. Thus, a weak signal that provides marginal or unsatisfactory stereo reception may be adequate for good or satisfactory monaural reception.

The 19 kHz pilot signal in the output of Q_7 is amplified by an *npn-pnp* direct-coupled amplifier (Q_8 and Q_9) and applied to the base of the doubler and 38 kHz amplifier Q_{14}. The high input impedance of Q_8 avoids loading the 19 kHz tuned circuit in the collector of Q_7.

The frequency doubler Q_{14} has small forward bias and a large input signal, thus its second harmonic distortion is high and a good 38 kHz signal is developed across the 38 kHz tuned collector circuit. The tuning coil is tapped so that only about 10 percent of the total turns are included in the collector circuit so the load impedance will be low enough (about 2 kΩ) to prevent collector saturation which would destroy the symmetry of the 38 kHz signal and thus seriously decrease the channel separation. The average collector current of Q_{14} is a strong function of the base drive; therefore, the voltage developed across its emitter resistance can be used to control a stereo indicating light. Also, this voltage can be used to operate the muting switch if the user wants his receiver to pick up *only* stereo broadcasts of adequate strength to provide good stereo reception.

The 38 kHz signal from Q_{14} and the composite signal are applied to the synchronous detector, as shown in Fig. 21.30. The composite signal is applied as a single-ended input to the lower differential pairs Q_{15} and Q_{16} and the 38 kHz signal is applied as a switching signal to the two upper differential pairs. These transistors have cross-coupled collectors which act as a matrix to separate the stereo channels whenever the 38 kHz signal is

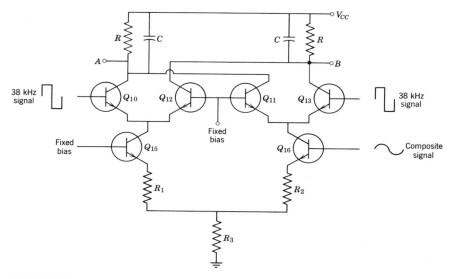

FIGURE 21.30
The synchronous detector.

present with adequate amplitude and proper phase. Observe that both Q_{10} and Q_{13} are switched on but Q_{11} and Q_{12} are switched off while the 38 kHz signal is positive. During this time the output from Q_{15} is fed to the A terminal and the output of Q_{16} is fed to the B terminal. However, during the negative half-cycle of the 38 kHz signal, the output of Q_{15} is switched to the B terminal and the output of Q_{16} is switched to the A terminal because Q_{10} and Q_{13} are cut off. The capacitors C in parallel with the load resistors R serve as deemphasis networks as well as filters to remove the high-frequency components above the audio range.

The composite signal at the collector of Q_{15} is 180° out of phase with the signal out of Q_{16} and is also reduced in amplitude because of the voltage divider action of R_1, R_2, and R_3. Let us analyze the detector circuit mathematically in order to determine the required relationship between these resistors. Assuming that the signals on Q_{16} and Q_{13} are in phase, the voltage at terminal B, neglecting the filter, is

$$v_B = [(L + R) + (L - R)\cos \omega_c t]\left(1 + \frac{4}{\pi}\cos \omega_c t\right)$$

$$- k[(L + R) + (L - R)\cos \omega_c t]\left(1 - \frac{4}{\pi}\cos \omega_c t\right) \quad (21.48)$$

where the bracketed term is the composite signal, the terms in parentheses are normalized bias plus fundamental components of the square-wave switching signals on Q_{13} and Q_{12}, respectively, and k is the ratio of the signal magnitudes on the collectors of Q_{15} and Q_{16}. The higher frequency components of the switching signal do not contribute to the audio-frequency output and are therefore omitted. Expanding Eq. 21.48 and saving only the audio-frequency components,

$$V_B = L + R + \frac{2}{\pi}(L - R) - k(L + R) + \frac{2k}{\pi}(L - R) \qquad (21.49)$$

Collecting the L (left channel) and R (right channel) terms,

$$V_B = L\left(1 + \frac{2}{\pi} - k + \frac{2k}{\pi}\right) + R\left(1 - \frac{2}{\pi} - k - \frac{2k}{\pi}\right) \qquad (21.50)$$

For perfect separation, the right channel signal should not appear. This will be accomplished if

$$1 - \frac{2}{\pi} - k - \frac{2k}{\pi} = 0 \qquad (21.51)$$

or

$$k = \frac{1 - (2/\pi)}{1 + (2/\pi)} = \frac{\pi - 2}{\pi + 2} = 0.221$$

When this value of k is used in Eq. 21.50,

$$V_B \simeq 1.56 \, L \qquad (21.52)$$

Similarly, the right channel R is received at terminal A. Motorola uses $R_1 = R_2 = 1 \text{ k}\Omega$ and $R_3 = 310 \, \Omega$ to achieve the proper value of k in their MC 1304 decoder. However, R_3 is an external resistor in the MC 1305 decoder and can be used as a channel separation control. Also, the 19 kHz and the 38 kHz filters must be properly tuned to obtain the correct 38 kHz signal phase for good channel separation.

The IC stereo decoder includes the switching circuits for the muting switch, the mono-stereo switch, and the stereo indicating light. The coils and capacitors are not included in the IC, but a complete kit of parts including a printed circuit board is available for a nominal cost. A complete circuit diagram and a discussion of the IC and auxiliary circuits is provided in an application note AN-432A published by the manufacturer.

PROBLEM 21.5 Show that the right channel only appears at terminal A, Fig. 21.30, if $k = 0.221$.

PROBLEM 21.16   Determine the optimum value of k if the 38 kHz signal applied to the synchronous detector is a sinusoid with maximum amplitude equal to the fixed bias on the switching transistors.

PROBLEM 21.17   Show that the monaural signal $L + R$ is available at both output terminals when the 38 kHz signal is missing.

PROBLEM 21.18   Draw the circuit diagram of an FM transmitter that uses a Miller capacitance modulator and provides low power output for a cordless microphone application.

PROBLEM 21.19   Draw the circuit diagram of an FM receiver of your choice that incorporates AFC. ICs are recommended where practical.

22

Switching and Pulse Circuits

The preceding chapters of this book have dealt primarily with linear circuit applications. Of course, diode circuits, class B, and class C amplifiers have been an exception.

In this chapter we shall consider applications in which transistors or tubes behave as switches. Thus, a transistor or tube that is driven into its cutoff region passes essentially no current and appears as an OPEN switch. In contrast, when a tube or transistor is driven into its saturation region, the voltage drop across this tube or transistor becomes very small and the device appears as a CLOSED switch. There is a very large variety of pulse and switching circuits,[1] so the following configurations are merely representive. However, these examples do illustrate the proper procedure for analysis and design of typical switching circuits.

22.1 TRANSISTOR OR TUBE OPERATING REGIONS

The bipolar transistor ($p\,n\,p$ or $n\,p\,n$ type transistors) has two junctions which permit four different modes of operation.

1. The *normal region* of operation occurs with the emitter-base junction forward biased and the base-collector junction reverse biased. This is the *linear region* we have studied extensively in previous chapters. The voltage from base to emitter is typically 0.6 to 0.7 V for a silicon

---

[1] For a more comprehensive selection of pulse and switching circuits, see either *Pulse, Digital, and Switching Waveforms*, by J. Millman and H. Taub, McGraw-Hill Book Company, New York, 1965, or *Wave Generation and Shaping*, Second Edition, by L. Strauss, McGraw-Hill Book Company, New York, 1970.

transistor and 0.2 to 0.3 V for a germanium transistor. The input impedance is usually about equal to h_{ie} but may differ widely from this value depending on the particular transistor configuration used. The output current is approximately a linear function of the input current. Finally, the output impedance is usually about equal to $1/h_{oe}$ but again may differ quite widely from this value depending on the exact circuit configuration used.

2.   We have also encountered the *cutoff region* in class B and class C amplifiers. In this operating region, both junctions are reverse biased. Then, the only currents flowing in the circuit are the saturation currents through the reverse-biased diodes. Thus, the output impedance and the input impedance are both very high. For example, the collector output impedance is approximately equal to r_c which we recall is equal to $h_{fe}(1/h_{oe})$. Since the emitter junction has a much higher cross-sectional area than the collector junction, the base input impedance is usually much higher than the collector output impedance when both junctions are reverse biased. For most pulse applications, sufficient accuracy is achieved by considering both base-emitter and base-collector junctions as open circuits.

3.   The third operating region is known as the *saturation region*. We have mentioned this region in the preceding chapters, but it has not been discussed fully. Thus, let us consider a circuit connected as shown in Fig. 22.1a. If the load resistor R_C is 250 Ω and V_{CC} is 25 V, the load line would be as shown in Fig. 22.1b. Now, as we begin to increase i_B above zero, i_C increases and the voltage v_{CE} decreases and the voltage v_{BE} increases. If we continue to increase v_{BE}, eventually v_{BE} will be greater than v_{CE} and *the collector junction will also become forward biased.* This condition is known as *saturation* and occurs when both junctions are forward biased. Of course, the magnitude of collector current (and consequently the magnitude of base current) which produces saturation will change if the value of R_C is changed in Fig. 22.1. However, once saturation is reached, an increase of base current produces practically no additional collector current. The device has actually saturated.

If the collector characteristics are amplified sufficiently by expanding the v_{CE} axis, each of the constant base curves follows a separate path back to the origin as shown in Fig. 22.2. However, when plotted to the scale used in Fig. 22.1b, the difference between these individual curves is so small it cannot be observed and all of the curves tend to merge into a single line back to the origin.

Notice the effect saturation has on the output impedance of the transistor. The output impedance of the transistor is equal to $\Delta v_{CE}/\Delta i_C$. As soon as saturation is reached, the characteristic falls steeply to the

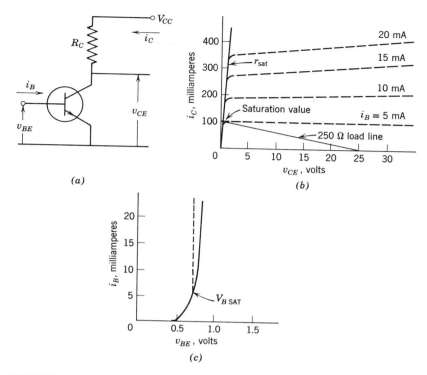

FIGURE 22.1

Effects of saturation in a transistor: (*a*) **circuit diagram;** (*b*) **collector character-istics;** (*c*) **base characteristics.**

origin. As a result, the output impedance of the saturated transistor (which is usually called r_{sat}) is very low, being that of a forward-biased junction. Many manufacturers list the value of r_{sat} in their data sheets. However, if this value is not given, the value of r_{sat} can be quickly calculated from the collector characteristic curves.

As the reader has probably surmised, the input impedance is also modified when a transistor is driven into saturation. The input characteristic for the unsaturated transistor in Fig. 22.1*b* is given in Fig. 22.1*c* as the solid line. However, as soon as the base current reaches about 6 mA (Fig. 22.1*b*), the transistor reaches saturation. Now, the base is feeding into *two* forward-biased junctions so the input impedance decreases markedly as shown by the dotted line in Fig. 22.1*c*. The slope of this new input curve (the dotted line in Fig. 22.1*c*) is roughly equal to r_{sat}. The total transistor input circuit can be approximately

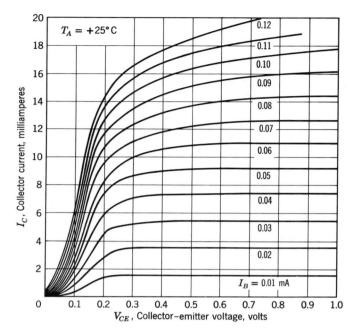

FIGURE 22.2
Collector characteristics with an expanded V_{CE} axis for a typical transistor.

represented by a battery equal to the voltage when saturation occurs ($V_{B\ SAT}$ in Fig. 22.1c) in series with a resistance r_{sat}. For many approximations, the transistor output impedance is considered to be essentially zero and the input impedance is also assumed to be zero. However, the battery $V_{B\ SAT}$ may need to be considered in some of these approximations.

4.   The fourth operating region of a transistor occurs when the base-emitter junction is reverse biased and the base-collector junction is forward biased. Transistors operating in this manner are said to be working in the *inverted region*. Actually, we are now using the emitter as the collector and the collector is now functioning as the emitter. Since the emitter junction usually has a much smaller cross-sectional area, the emitter cannot dissipate as much heat as the collector so the power dissipation must be derated when operating in the inverted mode. In addition, the reverse β_R (β in the inverted mode) is much less than the forward β_F (β in the normal mode). Similarly h_{ie} and h_{oe} are usually reduced significantly, and h_{re} is increased when operating in the inverted mode.

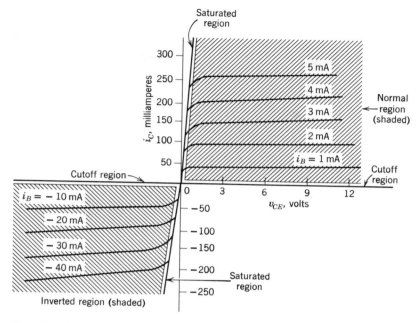

FIGURE 22.3
The four operating modes of an *n-p-n* type transistor.

The four operating regions of an *n p n* type transistor are shown graphically in Fig. 22.3. From the characteristic curves of FETs, MOSFETs, and vacuum tubes, we note that each of these devices can also be operated in the normal region, the cutoff region, or the saturation region. (Triode tubes must have a positive control grid potential to drive them into saturation.) In addition, FETs and MOSFETs can be operated in the inverted region by interchanging potentials on the drain and source. However, the substrate of the MOSFET *must* remain reverse biased with respect to the drain and source. In contrast, vacuum tubes cannot be operated in the inverted region because thermal electrons are not emitted from the cold plate.

To summarize this section, equivalent circuits can be used to represent transistors in the four operating regions. A bipolar transistor might be represented as shown in Fig. 22.4 for its four operating regions. The diagram for the normal region is different from the usual *h*-parameter configuration, because in switching applications we are usually interested in dc signal values. Consequently, the equivalent circuits for a forward-biased

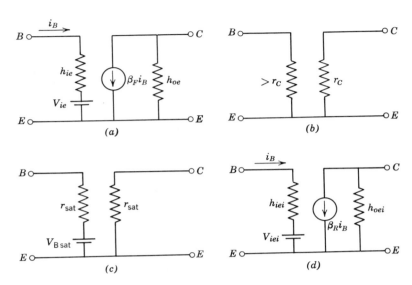

FIGURE 22.4
Equivalent circuits for bipolar transistors: (a) normal region; (b) cutoff region;
(c) saturation region; (d) inverted region.

junction, as developed in Chapter 3, is used to represent the base-emitter
junction.

PROBLEM 22.1 Determine as many of the equivalent circuit parameters
(Fig. 22.4) as possible for the transistor with characteristics given in
Fig. 22.1.

Answer: $\beta_F = 20$, $h_{ie} \simeq 50\ \Omega$, $h_{oe} \simeq 10^{-4}$ mhos, $r_{sat} = 5\ \Omega$, $v_{ie} \simeq 0.5$ V, $r_c = 200$ kΩ, $V_{BSAT} = 0.7$ (values depend on q point used).

PROBLEM 22.2 Determine as many of the equivalent circuit parameters
(Fig. 22.4) as possible for the transistor with characteristics given in Fig. 22.3.

22.2 TYPICAL TRANSISTOR SWITCHES

To illustrate the application of a transistor as a switch, let us consider
an example.

Example 22.1 A transistor is connected as shown in Fig 22.5. The
characteristic curves of the transistor are given in Fig. 22.6. If the input
signal is a sinusoid as shown, sketch the output voltage as a function of
time, assuming I_{CO} is zero. The diode D_1 is placed in the circuit to protect
the base from high reverse bias (100 V in this example).

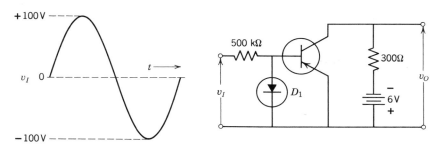

FIGURE 22.5
A transistor clipping circuit.

As a first step, the input current i_B for the transistor must be determined. Since the input resistance to the transistor is much less than the 500,000 Ω series resistor, the base current with v_I negative is

$$i_B \simeq \frac{v_I}{500,000} \tag{22.1}$$

When v_I is positive, the diode D_1 passes current. Again most of v_I is dropped across the 500 kΩ resistor with a drop of 0.6 to 0.7 V across the silicon diode. Thus, the transistor is in the cutoff mode during the positive half of the input cycle, and the base current i_B has the form shown in Fig. 22.7. With the waveform of i_B established, we are ready to determine the waveform of i_C and v_O. The load line is drawn on the characteristics for the 300 Ω load. Notice that the transistor is in the linear region from

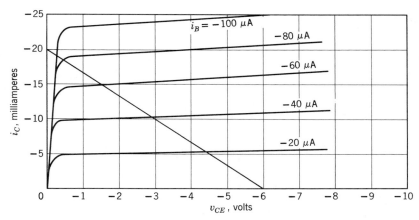

FIGURE 22.6
Characteristic curves for the transistor of Fig. 22.5.

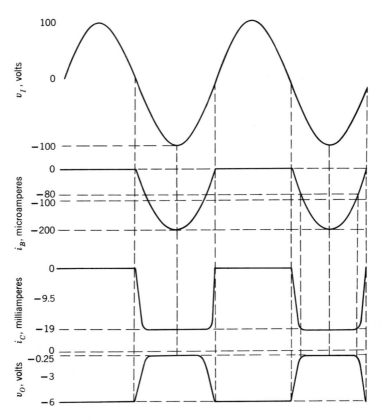

FIGURE 22.7
Waveforms for the circuit of Fig. 22.5.

$i_B = 0$ until $i_B \simeq \mu A$. Thus, i_C is a linear function of i_B over this current range. A transitional region exists between $i_B = 75$ μA to $i_B = 85$ μA. Then, the transistor enters the saturation region, and i_C does not increase above 19 mA. Thus, the waveform of i_C and v_O are as shown in Fig. 22.7. Notice that we have produced essentially a square-wave output signal from a sinusoidal input signal. This circuit is known as a *clipper circuit*.

PROBLEM 22.3   The input voltage v_I in Fig. 22.5 is changed to that shown in Fig. 22.8. Sketch the waveforms of i_B, i_C and v_O for this input waveform. List all pertinent voltage and current magnitudes.

The transistor in Example 22.1 operated in three different regions. Let us now consider an application which involves the fourth operating region. If a circuit is connected as shown in Fig. 22.9, the transistor is acting

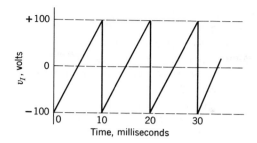

FIGURE 22.8
The form of v_I for Problem 22.3.

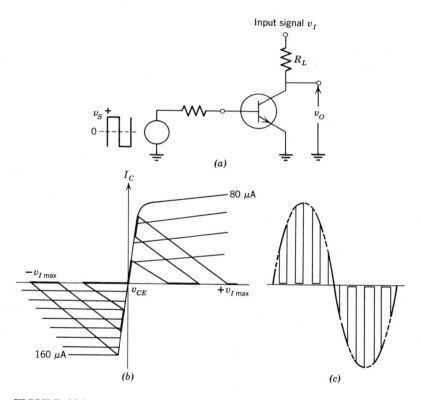

FIGURE 22.9
(a) **A simple transistor-modulating circuit,** (b) **the modulator characteristics,**
(c) **the output voltage.**

as a modulator. In this circuit, one input signal (the sinusoidal signal) is applied to the collector circuit and the second input signal (a square wave) is applied to the base circuit. When the sinusoid on the collector is positive, the transistor operates in the normal mode. However, when the sinusoid on the collector is negative, the transistor operates in the reverse mode. As noted, β_F is much larger than β_R. Consequently, one would suspect that a very nonsymmetrical output signal would result. However, if the square-wave signal is large enough to drive the transistor well into saturation in both the inverted and the normal modes, the output signal is essentially symmetrical. Notice that positive base potential is forward bias for either forward or reverse operation. Also note that the modulator circuit is in the emitter-follower configuration when v_I is negative and the collector is acting as the emitter. Therefore, the base switching voltage v_S must be larger than the peak input (collector) voltage v_I, and v_S must be an ac signal to insure driving the transistor into both saturation and cutoff each cycle. Generally, uniformly doped base transistors have higher β_R than graded-base transistors. Also, graded-base transistors have low V_{BE} breakdown ratings. Therefore, transistors with a uniformly doped base are most satisfactory for symmetric switching.

Another common application of the transistor switch is the chopper, which converts a slowly varying dc signal to an ac signal of fixed frequency, but with amplitude variations that match the amplitude variations of the original dc signal. This process is basically modulation and could be accomplished by the modulator circuit of Fig. 22.9. The input voltage v_I would then be the slowly varying dc voltage and the modulator would operate only in the forward, or normal, mode. However, a chopper is normally used to convert a *very small* dc signal to ac, so this signal can be amplified by an RC-coupled amplifier, thereby eliminating the problem of thermally generated current (I_{CO}) becoming hopelessly mixed with the signal currents. At these very small signal levels, a problem known as voltage (or current) offset causes an annoying error in the output of the simple bipolar transistor modulator. This offset voltage is illustrated in the greatly expanded collector characteristics given in Fig. 22.10a. Because of the lack of symmetry of the transistor, the collector characteristics do not converge at the point $i_C = 0$, $v_{CE} = 0$, but converge at a point $i_C = i_{Ci}$ and $v_{CE} = v_{Ci}$ instead. Only the $I_B = 0$ curve passes through the origin. Thus, the voltage v_{Ci} adds to the forward output voltage of the modulator and subtracts from the reverse output voltage. The current i_{Ci} is of the order of I_{CO} and, therefore, may be very small in silicon transistors. Thus v_{Ci} may be of the order of microvolts, which is insignificant if the signal voltage v_I is of the order of volts, but can cause a temperature-sensitive error that may be intolerable when the signal voltages are in the millivolt or microvolt

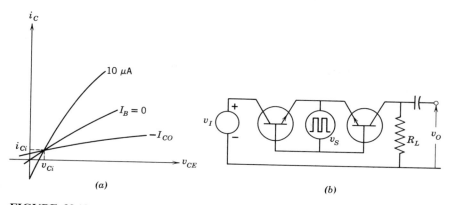

(a)   (b)

FIGURE 22.10
A symmetrical chopper (b) used to cancel the transistor offset voltage v_{Ci} shown in (a).

range. This offset voltage error may be cancelled by the symmetric chopper circuit of Fig. 22.10b. Of course, the transistors must be perfectly matched and at the same temperature for complete cancellation.

A single FET can serve as a symmetrical chopper or modulator because FETs may generally have their source and drain terminals interchanged without altering their characteristics. A simple FET circuit that can be used as a switch, modulator, or chopper is given in Fig. 22.11. An example will illustrate the basic principles and design of this circuit.

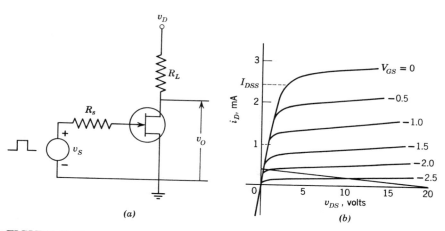

(a)   (b)

FIGURE 22.11
A switching circuit employing a FET: (a) circuit diagram; (b) drain characteristics.

Example 22.2   We assume that the drain supply voltage v_D is an ac signal voltage of varying amplitude, and the switching voltage v_S is a higher frequency square wave, so the circuit is a modulator. If the maximum drain current i_D is small compared with I_{DSS}, the voltage v_{DS} across the FET will be small compared with the pinch off voltage V_P when the FET is switched on. In Fig. 22.11 the peak v_D is assumed to be 20 V and $R_L = 50$ kΩ, so $i_{D\,max} = 0.4$ mA, which is about $I_{DSS}/7$. Smaller values of $i_{D\,max}$ may be better, but they are more difficult to illustrate. The FET is solidly switched on when $v_S = 0$, so the resistance R_s should be large enough to limit the gate current to a small value whenever v_{GS} becomes positive. In the normal (or forward) mode of operation, the maximum negative value of v_{GS} required is V_P to switch the transistor off. However, in the reverse mode when the drain and source are interchanged, the source (normally the drain) becomes negative and the base potential must become negative with respect to ground by the amount $(V_P + v_{D\,max})$ in order to switch the transistor off. In this example $v_{S\,max}$ negative must be more negative than $[-20 + (-3.0)] = -23$ V with respect to the chosen ground terminal in order to switch the FET off. Let us assume that the switching voltage v_S is symmetrical about zero and switches the transistor on with $+23$ V, with respect to ground. This would place a maximum 23 V $-(-20)$ V $= 43$ V between the gate and drain (source, while inverted) if it were not for the limiting resistance R_s. However, R_s will limit the forward-bias v_{GS} or v_{GD} to approximately $+0.5$ V, so 42.5 V must be dropped across R_s. Thus, $R_s \geq 42.5/i_{G\,max}$, and if i_G is limited to 1 mA, $R_s \geq 42.5$ kΩ.

PROBLEM 22.4   If the FET circuit of Fig. 22.11 is used as a chopper with $v_{DS\,max} = 5$ V and the gate chopping signal is symmetrical with respect to ground, what minimum values should v_S and R_s have if $I_{G\,max} = 0.5$ mA?

Answer: $v_{S\,max} > 3.0$, $R_s > 5$ kΩ.

PROBLEM 22.5   If the FET circuit of Fig. 22.11 is used as a modulator with the maximum signal applied to the drain $= \pm 5.0$ V and the gate chopping signal, usually called the carrier, is symmetrical with respect to ground, what minimum values should v_S and R_s have if $i_{G\,max} = 0.5$ mA?

Answer: $v_S > 8$ V, $R_s > 25$ kΩ.

22.3   SWITCHING TIMES

Despite the remarkable progress in achieving short switching times with transistor switches, there is always a need for higher speed switches. High efficiency as well as high speed may be achieved when a transistor switches its state from cutoff to saturation very quickly. Since power dissipation is the product of current and voltage, this dissipation is very low during cutoff because of the very low current and is also very low during saturation, providing the saturation voltage is adequately low. Thus, the dissipation is high only during the switching time. Therefore, the average dissipation may be almost proportional to the switching time, or, more accurately, to the ratio of the switching time to the switching period, which is the average time for one complete cycle of operation of the switch.

The 10 percent to 90 percent rise and fall times for an amplifier output response when a square wave is applied to the input was shown in Chapter 7 to be $2.2/\omega_h$, where ω_h is the upper cutoff frequency or bandwidth of an amplifier. A technique for either predicting the bandwidth of a given amplifier or designing an amplifier for a specified bandwidth, using a hybrid-π circuit, was also given in Chapter 7. Thus, the prediction or control of rise and fall times, or switching times, may appear to be within our present capability without further discussion. However, the techniques of Chapter 7 were developed for *linear* amplifiers and hold approximately for a switching circuit that is driven only to the edges of the cutoff and saturation regions and therefore *not* into these regions. The main purpose of this section is to show that switching times can be reduced considerably below the values predicted for a linear amplifier if the input or switching signal is capable of driving the transistor well into the saturation and cutoff regions.

We shall first investigate the effect of driving the transistor well into saturation. The circuit diagram and the collector characteristics are given in Fig. 22.12 (*a* and *b*) for convenience. The stored charge in the transistor base and the relative circuit response times are illustrated in Fig. 22.12 (*c* and *d*). We first assume that the square-wave input v_S is just sufficient in magnitude to drive the transistor to the edge of saturation, namely $i_B = 90$ μA. Then the normal charge is stored in the base region as the collector current i_C rises exponentially to almost $V_{CC}/R_L = 10$ mA, assuming the saturation voltage is negligible. The approximately exponential rise of i_C has the time constant $\tau = 1/\omega_2$, as was discussed in Chapter 7, and the rise time t_{r1} is approximately $2.2\tau = 2.2/\omega_h$, as expected. When the base driving voltage v_S returns suddenly to zero, the base stored charge decreases as the collector current decreases exponentially (approximately) toward zero with a fall time equal to the rise time.

Now let us assume that voltage v_S is increased so that the base current is doubled (180 μA) while the transistor is switched on. Then the base stored charge rises toward point 2 (Fig. 22.12c) at the emitter junction and the collector current rises exponentially toward $h_{FE} i_B = 20$ mA with time constant τ. However, the transistor saturates at $i_C \simeq 10$ mA, which occurs at about 0.7τ, so the rise time is about 0.6τ instead of 2.2τ. Thus, the rise time is reduced by a factor of about 3.6 as a result of the excess drive. The

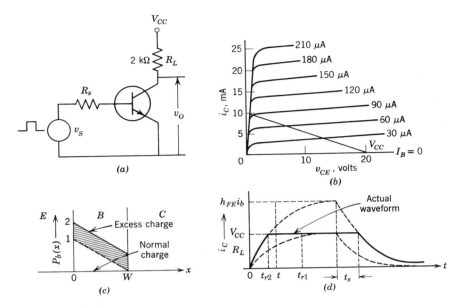

FIGURE 22.12
Illustration of the effect of base drive on switching times: (a) circuit diagram; (b) collector characteristics; (c) stored base charge; (d) response times.

ratio of actual base current to $V_{CC}/h_{FE} R_L$, which is the base current required to barely saturate the transistor, is usually called the *overdrive factor*. A plot of rise time as a function of overdrive factor is given in Fig. 22.13. Note that a small overdrive factor can cause a significant reduction of rise time.

Example 22.3 Let us assume that a transistor with $h_{FE} = 100$ is to be used as a switch with $R_L = 1$ kΩ, $V_{CC} = 20$ V and $R_s = 2$ kΩ. We wish to determine the rise time if the overdrive factor is (a) 0, (b) 1, (c) 2. First we must determine the upper cutoff frequency ω_h for the circuit as a linear amplifier, using the techniques discussed in Chapter 7. This requires additional data such as f_τ and C_{ob}. The q point chosen for this calculation is not critical. A point near the center of the load line is appropriate. We assume that ω_h has been calculated and is 10^6 rad/sec. (Of course, ω_h could be obtained from a frequency response curve.) Then, with a square-wave input of $i_B = 20$ V/1 kΩ(100) = 200 μA, $t_r = 2.2/10^6 = 2.2$ μs. If the base current drive is increased to 400 μA (overdrive factor = 1), $t_r = 0.6 \times 10^{-6} = 0.6$ μs, as shown by either the preceding work or Fig. 22.13. When the base drive is increased to 600 μA, the overdrive factor is 2 and the rise time $t_r = 0.4 \times 10^{-6}$ or 0.4 μs.

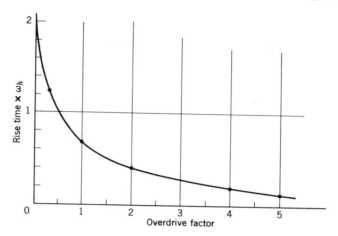

FIGURE 22.13
A plot of rise time as a function of overdrive factor.

The base-current overdrive, which improves the rise time, causes a problem called *storage time* in the transistor. Figure 22.12c showed that excess charge is stored in the base region when the transistor is driven into saturation. When both junctions are forward biased, the collector current is clamped at approximately V_{CC}/R_L, and the charge in the base region must increase until the increased recombinations in the base region use up the excess base current. At the end of the input pulse the collector current cannot decrease appreciably until the *excess* stored charge is removed from the base region because the collector junction is forward biased as long as excess minority carriers are stored next to the collector junction in the base. This phenomenon was discussed in connection with diode diffusion capacitance in Chapter 3. The time required to dissipate this excess stored charge is known as *storage time* (t_s).

If the base current is just reduced to zero at the end of the input pulse, the base stored charge decreases only because of charge recombination in the base (Figs. 22.12d and 22.14). However, if reverse base current flows because of either a low-base circuit resistance or the application of a reverse bias voltage at the end of the input pulse, the stored charge is removed from the base region more quickly. First, the excess stored charge is removed, shortening the storage time, and then the normal base charge is removed, shortening the fall time. Figure 22.14 illustrates the magnitude of improvement that can be expected for a given value of reverse base current $(-i_B)$. In this figure $-i_B$ was assumed to be about equal to the forward bias

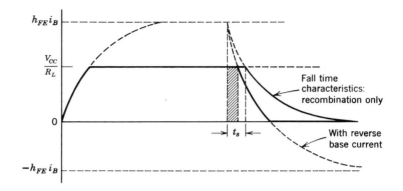

FIGURE 22.14
Illustration of the effect of reverse base drive on storage time and fall time.

just required for saturation $(V_{CC}/R_L h_{FE})$. Therefore, at the end of the forward-bias input signal, the negative base current $-i_B$ tends to cause the collector current to reverse, so the collector current decreases exponentially toward $-h_{FE} i_B$ rather than zero. Of course, the collector current does not actually reverse direction because the stored charge in the base becomes depleted at the same time the collector current reduces to about zero. In addition, the negative base current must also stop because the excess charge is the source for this negative base current. Thus, this reverse drive has the same effect on fall time as overdrive has on rise time. The reverse drive factor is the ratio of $-i_B$ to $V_{CC}/R_L h_{FE}$. Thus, Fig. 22.13 can be used for fall time as a function of reverse-drive factor as well as rise time as a function of overdrive factor.

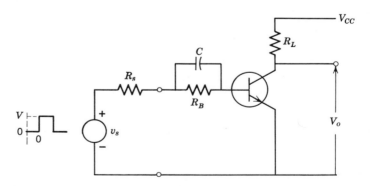

FIGURE 22.15
A typical driving circuit for a transistor switch.

We can see from the foregoing discussion that the ideal base drive for a transistor switch is a source that provides significant overdrive to turn the switch ON quickly, then reduce the drive so that the *excess* stored charge in the base region is nearly zero when the switch is to be turned OFF. This action provides very small storage time. The ideal base drive should also provide reverse drive while turning off the switch to provide fast fall time. This type of drive can be provided easily when the driving-source voltage is large in comparison with the Δv_{BE} required to drive the switch, which is usually the case. A resistor R_B is then placed in series with the driving source to limit the excess drive (Fig. 22.15), and a capacitor C is placed in parallel with R_B to provide overdrive at turn-on and reverse drive at turn-off. The charging current of capacitor C provides the extra drive at turn-on and the discharging current provides the reverse drive at turn-off. Of course, the source resistance, R_s, should be small as it limits the size of this capacitor current.

The circuit designer must determine suitable values for R_B and C (Fig. 22.15); therefore, the equivalent circuits of Fig. 22.16 are given to provide further insight into the circuit design. A hybrid-π circuit is first

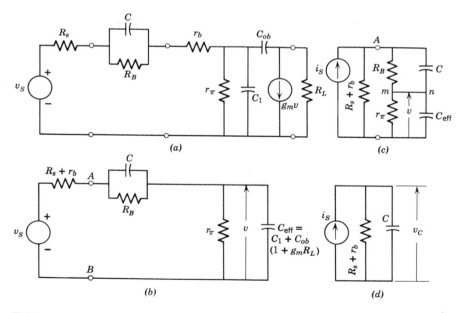

FIGURE 22.16
Equivalent circuits representing the switching circuit of Fig. 22.14: (*a*) equivalent circuit using hybrid-π; (*b*) simplified input circuit; (*c*) current-source version of (*b*); (*d*) approximation of (*c*) when $R_B C = r C_{eff}$ and $(R_B + r_\pi) \gg (R_s + R_b)$.

substituted for the transistor in Fig. 22.16a. Then r_b is lumped with R_s, since the same current must flow through both resistors, and the hybrid-π circuit is replaced by its equivalent input circuit, as was developed in Chapter 7, in Fig. 22.16b. Next, the voltage driving source to the left of points A and B is replaced by its equivalent current cource in Fig. 22.16c. We can further simplify this circuit only if we place some special constraints, or conditions, on it. Let us make the time constant $R_B C$ equal to the time constant $r_\pi C_{eff}$. Then the circuit to the right of points A and B is a balanced bridge as shown below.

$$R_B C = r_\pi C_{eff} \qquad (22.2)$$

Multiplying both sides of Eq. 22.2 by ω,

$$R_B \omega C = r_\pi \omega C_{eff} \qquad (22.3)$$

Rearranging Eq. 22.3,

$$\frac{R_B}{r_\pi} = \frac{\omega C_{eff}}{\omega C} = \frac{X_C}{X_{C\,eff}} \qquad (22.4)$$

This (Eq. 22.4) is the required relationship for a balanced bridge. Then if $R_B C = r_\pi C_{eff}$ so that the bridge is balanced, no current flows through the conductor from point m to point n, and it can be removed without altering the circuit performance. With this conductor removed, the resistive branch $(R_B + r_\pi)$ is directly in parallel with the modified source resistance $(R_s + r_b)$, and these resistors can be lumped together into a single equivalent resistance. Also, the capacitors C and C_{eff} are directly in series and can be lumped into an equivalent capacitance equal to $[CC_{eff}/(C + C_{eff})]$. This simplified equivalent circuit is given in Fig. 22.16d. However, a further simplification is indicated that may be made if our initial assumption is valid, namely that the driving-source voltage is large in comparison with Δv_{BE}. Then R_B is large compared to the other resistance in the circuit, so $(R_B + r_\pi) \gg (R_s + r_b)$ and the smaller resistance branch will primarily determine the resistance of the parallel combination. Also, if R_B is predominantly large, the capacitor C in parallel with it will be much smaller than C_{eff} and the value of the series combination is approximately equal to C, the small capacitor.

The equivalent circuit of Fig. 22.16d illustrates the advantage gained from using the comparatively large driving voltage v_S with its accompanying large value of R_B. The rise and fall times of the circuit are determined primarily by the time constant of the circuit, which is approximately

$$\tau = (R_s + r_b)C \qquad (22.5)$$

and

$$t_r = t_f = 2.2(R_s + r_b)C \qquad (22.6)$$

But since R_s is normally small in comparison with R_B, the time constant τ (Eq. 22.5) is small in comparison with $R_B C = r_\pi C_{eff}$ and the rise and fall times of the voltage v_C are small in comparison with either $2.2\,R_B C$ or $2.2\,r_\pi C_{eff}$. Since the bridge circuit (Fig. 22.16d) is balanced, the voltage v_C has precisely the same waveform and rise time as the voltage v (Fig. 22.16c) that controls the output current $g_m v$ and the output voltage $g_m v R_L$. This technique for improving rise time is precisely the same as that used in an oscilloscope probe, which converts the high capacitance of an oscilloscope input and the associated shielded input cable into a much smaller capacitance at the expense of a reduction in gain (usually a factor of 10).

The preceding technique was based on the assumption that the switching circuit can be represented by a linear equivalent circuit, which seems inappropriate and highly inaccurate. However, one requirement is that the transistor *not* be driven appreciably into saturation to avoid storage time, and if this requirement is met, the equivalent circuit is usable as a rough approximation. In fact, the time constant $r_\pi C_{eff}$ is relatively constant over a wide range of collector currents even though both r_π and C_{eff} are strongly dependent on collector current. Therefore, the equivalent circuit yields final results that are much more accurate than might be expected. The base circuit resistance R_B is chosen so that the steady-state base current during the input pulse just barely drives the transistor into saturation. Then

$$R_s + R_B = \frac{v_S - v_{BE}}{V_{CC}/h_{FE}R_L} = \frac{(v_S - v_{BE})h_{FE}R_L}{V_{CC}} \tag{22.7}$$

and

$$R_B = \frac{(v_S - v_{BE})h_{FE}R_L}{V_{CC}} - R_s \tag{22.8}$$

Equations 22.8 and 22.2 can be used to design the driving circuit and Eq. 22.6 can then be used to determine the rise and fall times. An example will illustrate the design procedure.

Example 22.4 A silicon transistor with $h_{FE} = 100$, $r_b = 100\ \Omega$, $C_{ob} = 2$ pF, and $f_\tau = 300$ MHz at $I_C = 10$ mA and $V_{CE} = 5$ V is to be used as a transistor switch with $R_L = 1$ kΩ and $V_{CC} = 20$ V. The driving source has $R_s = 2$ kΩ and provides a 5.0 V rectangular pulse. Then, the value of R_B that will just drive the transistor to saturation is $[(5.0 - 0.5)/0.2\ \text{mA}] - 2\ \text{k}\Omega = 20\ \text{k}\Omega$, where V_{BE} is assumed to be 0.5 V and $V_{CC}/R_L h_{FE} = 0.2$ mA is the base current required for saturation. At $I_C = 10$ mA, which is an average value, $r_\pi = 100/0.4 = 250\ \Omega$ and $C_1 + C_{ob} = g_m/\omega_T = 212$ pF so

$C_1 = 210$ pF (Fig. 22.16a) and $C_{\text{eff}} = 210 + 2(1 + 400) = 1012$ pF. Then, by using Eq. 22.2, $C = 250\ \Omega \times 1012$ pF$/20$ k$\Omega = 12.6$ pF. A 13 pF capacitor would probably be used in parallel with the 20 kΩ resistor R_B. This completes the design. The rise and fall times for the switch can now be calculated with the aid of Eq. 22.6, which gives $t_r = t_f = 2.2 \times 2.1 \times 13 \times 10^{-9} = 6.0 \times 10^{-8}$s or 60 ns.

We shall now check the validity of the approximations used in Eq. 22.5. The actual shunt resistance was 2.1 kΩ in parallel with 20.25 kΩ, or about 1.9 kΩ, and the actual capacitance is 13 pF in series with 1012 pF, which is very nearly 13 pF. Therefore, the rise time calculated above is about 10 percent high, which is within usually acceptable limits. You may observe that the addition of capacitor C improved the rise time approximately by the ratio $R_B/(R_s \| R_B) = (R_s + R_B)/R_s = 22$ k$/2$ k $= 11$.

PROBLEM 22.6   A circuit is connected as shown in Fig. 22.15. $V_{CC} = 12$ V and $R_L = 2$ kΩ. The signal v_S is a pulse with an amplitude of 12 V. If R_s is 1 kΩ and the transistor is silicon, determine the proper values of R_B and C. Assume $h_{FE} = 50$, $r_b = 50\ \Omega$, $C_{ob} = 5$ pF, and $f_\tau = 100$ MHz. What is the rise time and fall time for the output voltage pulse?

22.4   BISTABLE MULTIVIBRATORS

A set of devices known as multivibrators are very useful as pulse-generating, storing, and counting circuits. They are discussed here as examples of switching-type circuits. Multivibrators are basically two-stage amplifiers with positive feedback from the output of the second amplifier to the input of the first.

A *bistable* multivibrator is shown in Fig. 22.17. We first assume that transistor T_1 is cut off or nonconducting so that the base current of transistor T_2 is $(V_{CC} - V_{BE})/(R_{L1} + R_{B2})$. If R_{B2} is small enough, the base current is large enough to saturate T_2. Then the saturation voltage of T_2 is smaller than the turn-on voltage of T_1; thus, essentially no base current flows through R_{B1} and transistor T_1 is cut off, as we assumed. The multivibrator remains in this stable state until a switching pulse is applied some place in the circuit to turn transistor T_1 on. Then T_1 will saturate and its saturation voltage will be insufficient to provide forward bias to T_2, so T_2 will be turned off. The multivibrator will then remain in this second stable state until another switching pulse is applied, which will start T_2 conducting again and T_1 will cut off. Thus the multivibrator has two stable states and is therefore *bistable*. It is also known as a *flip-flop* because a trigger pulse will make it flip, or flop, from one state to the other. The trigger pulse need only start conduction in the OFF transistor, and the

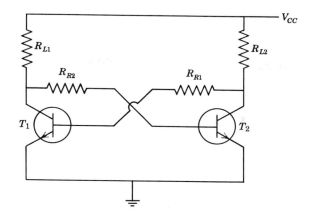

FIGURE 22.17
A basic multivibrator circuit.

circuit regeneration or feedback will carry through with the change of state because, as seen in Fig. 22.17, if T_1 begins to conduct, its collector voltage decreases and T_2 is pulled out of saturation. The collector voltage of T_2 then rises and forward bias is applied through R_{B1} to reinforce the initial trigger. This effect is cumulative, and the change of state is completed. Note that the change of state can be initiated with a very small pulse if T_2 is just barely saturated, but if T_2 is heavily saturated the trigger pulse must be large enough and remain long enough to permit T_1 to pull T_2 out of saturation. In other words, the trigger pulse must be longer than the *storage* time.

Bistable multivibrators are often used in counting circuits, since two trigger pulses are required to cycle a flip-flop through its two stable states. A typical counting circuit is shown in Fig. 22.18. This circuit includes a triggering system consisting of C_1 and the two diodes. Also, compensating capacitors C have been added in parallel with the base circuit resistors to permit high-speed switching. These circuit features are described in the following paragraphs.

The diodes D_1 and D_2 are known as *steering diodes* because they alternately apply the input trigger to the collectors of T_1 and T_2. We shall assume that T_1 is not conducting, so T_2 is in saturation. Then diode D_1 is slightly reverse biased because of the base current flowing through R_{L1}, but D_2 is heavily reverse biased because of the large voltage drop across R_{L2}. Therefore, when a negative trigger pulse is applied to the junction of the diodes, D_1 is forward biased and the trigger pulse is applied to the collector of T_1. This negative voltage tends to turn transistor T_2 off and as soon as T_2 pulls

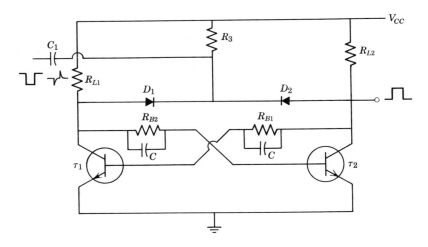

FIGURE 22.18
A binary counting circuit.

out of saturation its collector voltage rises and forward biases T_1, thus completing the switching action. The next time a negative trigger pulse arrives, diode D_2 has the small reverse bias and D_1 is heavily reverse biased, so the pulse is applied through D_2 to the collector of T_2 and the base of T_1; therefore, the multivibrator changes back to its initial state. The diode currents tend to charge the capacitor C_1 to the peak value of the input pulse; therefore, resistor R_3 is required to discharge C_1 and maintain the potential of the cathodes of D_1 and D_2 at approximately V_{CC} at the beginning of each trigger pulse. The resistor R_3 is often replaced by a diode with its cathode connected to V_{CC}. The potential of the junction of the three diodes is then clamped at approximately V_{CC}. If C_1 is too large, the pulse will remain on the bases of the two transistors until all junction charges have been removed. Then, both transistors will be cut off. As the pulse decreases, the stronger transistor will go to the ON state and the other transistor will be OFF. Under these conditions, the stronger transistor will always return to the ON condition. Thus, the value of C_1 must be carefully chosen so the trigger pulse is long enough to produce switching in the OFF transistor but the trigger pulse must not be too long or no switching will occur.

Binary counters are often connected in cascade to produce higher count-down ratios. One flip-flop produces a countdown of two (i.e., two input pulses produce one output pulse). Two cascaded flip-flops produce a count-down of four since four input pulses are required to produce one output pulse, and so on. Now, as mentioned, the capacitor C_1 (Fig. 22.18) should be small enough to produce a narrow trigger pulse compared to the duration of the rectangular input pulse. This short time-constant circuit, known as a differentiator, then produces a sharp trigger pulse at both the leading and trailing edge of the input pulse (Fig. 22.19). The positive pulses at the trailing

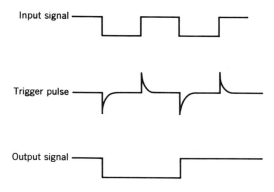

FIGURE 22.19
Input, trigger, and output voltages of a bistable multivibrator.

edges of the input signal do not trigger the circuit because they increase the reverse bias of both steering diodes and are clipped (or eliminated) by the clamping diode, if one is used. When a clamping diode is used, the time constant of the coupling circuit is

$$\tau_c = (R_s + R_L)C_1 \tag{22.9}$$

where R_s is the output resistance of the driving source and R_L is the approximate resistance to the trigger pulse on the right-hand side of the coupling capacitor C_1, assuming $R_B \gg R_L$. If R_3 (Fig. 22.18) is used instead of a clamping diode, the parallel combination of R_3 and R_L must be used in place of R_L.

The speedup capacitors C are the compensating capacitors discussed in the preceding section and can be calculated by the technique described there. If some doubt exists concerning their optimum value, it is preferable to err by making them too large instead of too small.

The resistors R_{B1} and R_{B2} must be small enough to insure saturation, otherwise there will be no stable states, and the circuit will oscillate. When a transistor is in saturation, the base and collector potentials are nearly the same. Therefore, if T_2 is assumed to be saturated (Fig. 22.18) $I_{B2}(R_L + R_{B2}) = I_{C2}R_{L2}$ and

$$R_{L1} + R_{B2} = \frac{I_{C2}}{I_{B2}} R_{L2} \tag{22.10}$$

Transistor T_2 will be in saturation if $I_{C2}/I_{B2} \leq h_{FE}$ so that

$$R_{L1} + R_{B2} \leq h_{FE} R_{L2} \tag{22.11}$$

or

$$R_{B2} \leq h_{FE} R_{L2} - R_{L1} \tag{22.12}$$

Normally, $R_{L1} = R_{L2}$ and $R_{B1} = R_{B2}$, so the subscripts can be dropped. Then the required value of R_B for saturation is

$$R_B \leq R_L(h_{FE} - 1) \tag{22.13}$$

The value of h_{FE} used in Eq. 22.13 should be the minimum expected value, from a given batch, at the minimum expected temperature.

The value of R_L is chosen in accordance with the required switching speed. As was previously discussed in this chapter and in Chapter 7, high speed, which requires short rise, fall, and storage times, is possible only with high-frequency amplifiers that require high f_τ and small R_L as well as proper size speedup, or compensating capacitors. Thus, values of R_L barely large enough to limit the transistor dissipation to safe values may be used in very high-speed circuits, while values of about 10 kΩ may be used in circuits not requiring high speed. The larger R_L is, the lower the power dissipation for a given V_{CC}, of course.

Very often the triggering circuits are noisy. In this case, reverse bias is applied to the OFF transistor of the bistable multivibrator as shown in Fig. 22.20, to prevent triggering on noise pulses. If the circuit is symmetrical, as shown, the current through R_E is essentially constant, being the emitter current of the ON transistor; the voltage drop $(-I_{E\,\text{sat}} R_E)$ across R_E reverse biases the OFF transistor. Then, the trigger pulse must be large enough to change the collector voltage of the saturated transistor by an amount greater than this reverse bias before the circuit will begin to change states. The resistors R_1 are usually large enough so that the current through them is about the same magnitude as the base current.

Example 22.5 We shall now design a moderately high-speed binary counting circuit that uses a silicon transistor with $f_\tau = 300$ MHz, $C_{ob} = 2$ pF, $h_{FE\,\text{min}} = 80$, $R_L = 1$ kΩ, and $V_{CC} = 20$ V. Let us choose the trigger desensitizing voltage $I_E R_E = 2$ V. Then, since the transistor is in saturation, $R_E/R_L = 2$ V/18 V and $R_E = 110\,\Omega$. The saturation value of $I_C = 18$ V/1 k$\Omega = 18$ mA, so the minimum saturation value of $I_B = 18$ mA/80 = 225 μA. If this value of current is allowed to flow through R_1, $R_1 = 2.5$ V/225

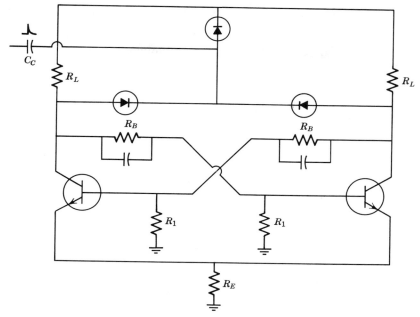

FIGURE 22.20
A bistable multivibrator with trigger desensitization.

$\mu A = 11$ kΩ. A standard value of 10 kΩ would be adequate. Since 450 μA must flow through R_B, the maximum value of $R_B = 18$ V/450 μA − 1 kΩ = 39 kΩ. Since $R_1 \gg h_{ie}$, the speedup capacitors will be calculated neglecting R_1. Since the transistor load resistance and V_{CC} are the same in this example as in Example 22.4, we shall use the time constant $\tau = r_\pi C_{eff} = 250 \times 1012$ pF determined in Example 22.4 and $C = 250 \times 1250$ pF/39 kΩ = 8 pF. A value between 8 and 20 pF would be suitable. Let us assume that the driving source has an internal resistance $R_s = 5$ kΩ and provides a 10 μs pulse duration; we want to provide a trigger pulse of approximately 1 μs at 37 percent of its peak amplitude. Then we need $(5\ k\Omega + 1\ k\Omega)C_C = 10^{-6}$s and $C_C = 167$ pF. The steering and clamping diodes need to be moderately fast diodes to accomplish the tasks well. This completes the design.

PROBLEM 22.7 Design a medium-speed binary counter that uses silicon transistors with $h_{FE\,min} = 50$, $V_{CC} = 20$ V, $R_L = 10$ kΩ, $f_t = 100$ MHz, $C_{ob} = 5$ pF, and reverse bias = 2 V. The duration of the input pulse is about 100 μs.
Answer: $R_E = 560$, $R_1 = 68$ kΩ, $R_B = 250$ kΩ, $C = 5\text{-}10$ pF.

22.5 MONOSTABLE AND ASTABLE MULTIVIBRATORS

The *monostable* multivibrator has only *one* stable state, as the name implies. It differs from the bistable multivibrator in that one of the dc coupling circuits is replaced by an ac coupling circuit as shown in Fig. 22.21. In its stable mode of operation, transistor T_1 is normally cut off and T_2 is normally in saturation. Therefore, resistor R_{B2} must be small enough to maintain T_2 in saturation or $R_{B2} \leq h_{FE}R_{L2}$. Then, the base of T_1 is at the same potential as the saturated collector of T_2 and T_1 remains cut off.

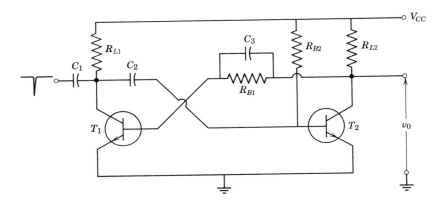

FIGURE 22.21
A monostable multivibrator circuit.

Switching action is initiated by applying a negative trigger pulse through C_1. This trigger pulse passes through C_2 to the base of T_2. The negative pulse pulls T_2 out of the saturation and both the collector of T_2 and the base of T_1 become more positive. The increase of potential on the base of T_1 permits T_1 to begin conduction. Consequently, the potential on the collector of T_1 decreases. This decrease of potential is coupled through C_2 to the base of T_2 and the conduction through T_2 is further decreased. Obviously, the action is accumulative and only terminates when T_1 is saturated and T_2 is cut off. Now, even though the collector side of C_2 is maintained at a very low potential (T_1 is saturated), C_2 is charged by current through R_{B2} until the base of T_2 becomes forward biased and T_2 begins to conduct. Then a switching action occurs and T_2 switches to its ON condition and T_1 becomes cut off. The circuit then remains in this stable condition until the next trigger pulse is applied.

To clarify these concepts and develop some useful design relationships, let us consider an example.

Example 22.6   Two silicon transistors are connected as shown in Fig. 22.21. The transistor parameters are: $h_{FE} = 50$, $v_{CE} = 0.3$ V when saturated, and $v_{BE} = 0.7$ V when saturated. In addition, $V_{CC} = 20$ V, $R_{L1} = R_{L2} = 1$ kΩ and $C_2 = 0.1$ μF. Let us analyze the behavior of the circuit if a trigger pulse is applied at time $t = 0$.

The proper value of R_{B2} is $R_{B2} \leq h_{FE} R_{L2} = 50 \times 1$ kΩ. We will use a 47 kΩ value. In addition, $R_{B1} + R_{L2}$ should also be equal to 50 kΩ or less. Again, let $R_{B1} = 47$ kΩ.

Prior to $t = 0$, transistor T_2 is saturated so $v_{B2} = 0.7$ V and $v_{C2} = 0.3$ V. In addition, T_1 is cut off with $v_{B1} = v_{C2} = 0.3$ V and $v_{C2} = 20$ V. These values are plotted in Fig. 22.22. At time $t = 0$, the negative trigger pulse is applied and T_1 becomes saturated and T_2 is cut off. Before $t = 0$, the

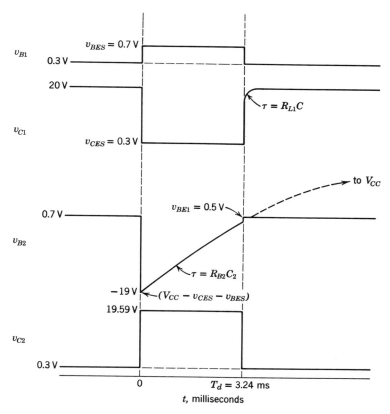

FIGURE 22.22
Voltage waveforms for the circuit of Fig. 22.21.

capacitor C_2 is charged with $v_{C1} - v_{B2} = 20\ \text{V} - 0.7\ \text{V} = 19.3\ \text{V}$ across C_2. The switching occurs so rapidly that no appreciable change of capacitor charge occurs. Consequently, immediately after $t = 0$, the circuit containing C_2 appears as shown in Fig. 22.23a. The collector circuit of the transistor T_1 is replaced by r_{sat} and the reverse-biased base circuit of transistor T_2 is replaced by an open circuit. When the circuit to the left of point x-x in Fig. 22.23a is replaced by its Thevenin's equivalent circuit, the configuration in Fig. 22.23b results. Since r_{sat} is so much smaller than R_{L1},

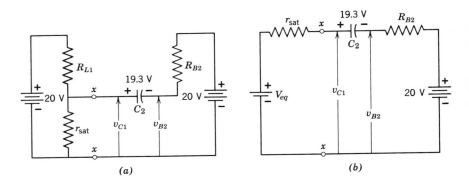

FIGURE 22.23
Equivalent circuits for Fig. 22.20 after $t = 0$: (a) actual equivalent circuit; (b) simplified equivalent circuit.

the parallel combination of R_{L1} and r_{sat} is essentially equal to r_{sat}. The voltage $V_{eq} \simeq 0.3$ as specified for v_{CE} for the saturated transistor. From Fig. 22.23b we note that $v_{C1} \simeq 0.3\ \text{V}$ after $t = 0$. Then, $v_{B2} = +0.3 - 19.3 = -19$ V immediately after $t = 0$. Notice that v_{B2} will charge toward 20 V with a time constant essentially equal to $R_{B2}C_2$, since $r_{sat} \ll R_{B2}$. The actual equation for v_{B2} is

$$v_{B2} = V_{CC} - (2V_{CC} - v_{CES} - v_{BES})e^{-t/R_{B2}C_2} \qquad (22.14)$$

where v_{CES} and v_{BES} are the saturated values of v_{CE} and v_{BE}. Equation 22.14 is only valid until v_{B2} reaches the value where T_2 begins to conduct (let us call this value v_{BEO}) and the output pulse terminates. If the output pulse duration is given by T_d, Eq. 22.14 can be written as

$$v_{BEO} = V_{CC} - (2V_{CC} - v_{CES} - v_{BES})e^{-T_d/R_{B2}C_2} \qquad (22.15)$$

Equation 22.15 may be simplified if $v_{CES} \ll V_{CC}$ and $v_{BES} \ll V_{CC}$. Then, Eq. 22.15 becomes

$$v_{BEO} \simeq V_{CC} - 2V_{CC}\, e^{-T_d/R_{B2}C_2} \tag{22.16}$$

or

$$\frac{V_{CC} - v_{BEO}}{2V_{CC}} \simeq e^{-T_d/R_{B2}C_2} \tag{22.17}$$

If we invert both sides of Eq. 22.17 and take the natural logarithm of this relationship, we have

$$\ln\left(\frac{2V_{CC}}{V_{CC} - v_{BEO}}\right) \simeq \frac{T_d}{R_{B2}\,C_2} \tag{22.18}$$

or

$$T_d \simeq R_{B2}\,C_2 \ln\left(\frac{2V_{CC}}{V_{CC} - v_{BEO}}\right) \tag{22.19}$$

The value of v_{BEO} is about 0.5 V for a silicon transistor and 0.2 V or so for a germanium transistor.

If v_{BEO} is negligible compared to V_{CC}, Eq. 22.19 can also be written as

$$T_d \simeq R_{B2}\,C_2 \ln 2 \tag{22.20}$$

or

$$T_d \simeq 0.69 R_{B2}\,C_2 \tag{22.21}$$

In this example, the value of T_d is 3.24 milliseconds if $v_{BEO} = 0.5$ V and we use the exact relationships given by Eq. 22.15. If we use the approximate relationship given by Eq. 22.21, the value of T_d is also equal to 3.24 milliseconds. (The error of neglecting v_{CES} and v_{BES} was cancelled by neglecting v_{BEO}.)

The sketch of v_{B2} as determined by Eq. 22.14 is given in Fig. 22.22. Since T_1 is saturated during the pulse duration, v_{C1} remains essentially at 0.3 V. Also v_{B1} would remain at 0.7 V. Transistor T_2 is cut off for the pulse duration because of the reverse bias of v_{B2}. However, i_{B1} must flow through R_{L2} so v_{C2} has a value of

$$v_{C2} = V_{CC} - \frac{(V_{CC} - v_{BES})R_{L2}}{R_{L2} + R_{B1}} \tag{22.22}$$

during the pulse duration. In this example $v_{C2} = (20 - 0.41)$ V or 19.59 V. All of these waveforms are plotted in Fig. 22.22.

At time $T_d\,(t = 3.24$ milliseconds) transistor T_2 turns ON and T_1 resumes its OFF state. Unfortunately, the capacitor C_2 is essentially uncharged ($v_{C1} = 0.3$ V and $v_{B2} = v_{BEO} = 0.5$ V) so a transient is present while C_2 recharges.

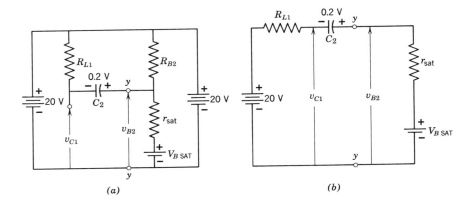

FIGURE 22.24
The equivalent circuit for the charging cycle of C_2: (a) **actual equivalent circuit;** (b) **simplified equivalent circuit.**

The appropriate equivalent circuit containing C_2 is given in Fig. 22.24a. In this case, the collector of T_1 is represented as an open circuit since T_1 is cut off. Also, the base of the saturated transistor T_2 is represented by r_{sat} and $v_{B\,SAT}$ as suggested in Fig. 22.4. (Note that $V_{B\,SAT} = v_{BES} = 0.7$ V in this example.) The initial 0.2 V charge across C_2 is also included. The circuit in Fig. 22.24a can be simplified by replacing the circuit to the right of terminals y-y by its Thevenin's equivalent circuit as shown in Fig. 22.24b. The Thevenin's equivalent resistance is the parallel combination of R_{B2} and r_{sat} which is essentially equal to r_{sat}. In addition, since $r_{sat} \ll R_{B2}$, the Thevenin's equivalent voltage source is essentially equal to $V_{B\,SAT}$.

Since r_{sat} is very much less than R_{B2}, the value of v_{B2} remains essentially at V_{BSAT} or $v_{BES} = 0.7$ V. Then, v_{C1} has an initial value of $0.7 - 0.2 = 0.5$ V and charges toward V_{CC} (20 V) with a time constant of $R_{L1}C_2$ (since $r_{sat} \ll R_{L1}$). Since $R_{L1}C_2 = 1$ k$\Omega \times 10^{-7}$ F $= 0.1$ msec, the waveform of v_{C1} would be as shown in Fig. 22.22. If r_{sat} is not negligibly small, v_{B2} will have a small positive spike at time T_d.

As noted in Example 22.6, the output voltage (v_{C2} in Fig. 22.22) of the monostable multivibrator is a rectangular pulse, often called a *gate* signal, the duration of which can be controlled by varying either R_{B2} or C_2. Often a potentiometer is provided as part of R_{B2} (as shown in Fig. 22.25) to provide adjustment of T_d. Of course, large changes in T_d are sometimes made by switching C_2 values.

Graded-base transistors usually cannot withstand the large reverse bias (Fig. 22.22) that occurs when transistor T_2 is switched off. Therefore, a diode

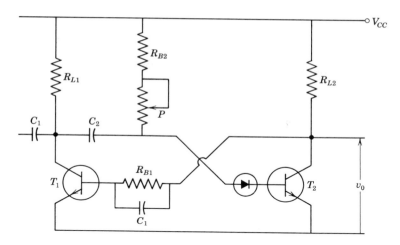

FIGURE 22.25
A monostable multivibrator with adjustable t_d and diode protection for T_2.

should be placed in series with the base of T_2, as shown in Fig. 22.25, whenever the maximum V_{BE} rating is less than V_{CC}. This diode protects the transistor from possible damage and insures the proper value of pulse duration or width.

A negative going pulse, or gate signal, can be obtained from the collector of T_1, but the fall time of this pulse is degraded as shown in Fig. 22.22 as a result of the charging current of C_2. A sharp negative gate can be achieved by inverting the signal from the collector of T_2 or by using p-n-p transistors in the monostable multivibrator.

An *astable* or *free-running* multivibrator can be constructed by capacitively, or ac, coupling both transistors, as shown in Fig. 22.26. Then there is no permanently stable state, and each transistor alternately saturates and cuts off. The *off* time for each transistor can be determined from the relationship $T = 0.7R_B C$, approximately, as was discussed for the monostable. The circuit is symmetrical if $R_{B1} = R_{B2}$ and $C_1 = C_2$, and so forth, but large assymmetry is sometimes desirable and attainable. The base circuit resistors must be chosen from the relationship $R_B \leq h_{FE} R_L$ for proper operation. The load resistors and transistor types are chosen in accordance with the frequency requirements as previously discussed. The rise times are determined by the coupling capacitor charging currents flowing through the load resistors, as discussed in connection with the monostable multivibrator. This charging current can be diverted from the load resistors by the addition of a resistance and a diode in each collector circuit, as shown in Fig. 22.27. When the transistor T_1 is driven into saturation, the

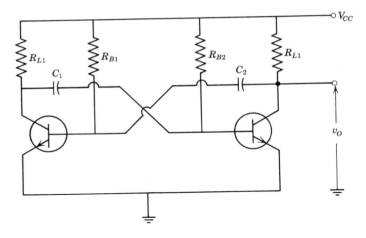

FIGURE 22.26
An astable or free-running multivibrator.

collector voltage drops quickly and the diode D_1 is forward biased and therefore has little influence on the circuit operation. However, when transistor T_1 is cut off, its collector rises very rapidly to V_{CC}, reverse-biasing diode D_1, and capacitor C_1 charges through R_1. Diode D_2 and R_2 provide the same fast rise time at the collector of T_2, as shown in the sketch of the output voltage. The resistors R_1 and R_2 must be small

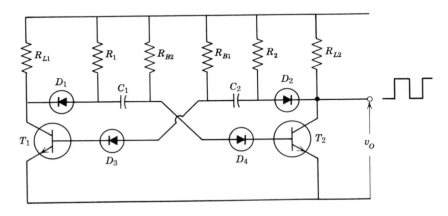

FIGURE 22.27
An astable multivibrator with diodes for transistor protection and diode-resistor combinations to permit fast rise and fall times.

enough to permit C_1 and C_2 to charge almost completely between the times of switching. An example will illustrate the design of an astable multivibrator.

Example 22.7   Assume that we need a pulse generator that will produce 100 μs pulses at the rate of 1000 pulses/sec. An astable multivibrator is a suitable generator. High-speed switching is not required so we will use $V_{CC} = 20$ V, $R_{L1} = 4$ kΩ, and $R_{L2} = 10$ kΩ. Since R_1 is in parallel with R_{L1} when T_1 is conducting and R_2 is in parallel with R_{L2} when T_2 is conducting, we tentatively assume that $R_1 = R_{L1}$ and $R_2 = R_{L2}$ for the purpose of determining the values of R_{B1} and R_{B2}. If $h_{FE\,min} = 80$, and by using $R_B \leq h_{FE}R_L$, $R_{B1} \leq 80 \times 2$ k$\Omega = 160$ kΩ and $R_{B2} = 80 \times 5$ k$\Omega \leq 400$ kΩ. Then, by using Eq. 22.19, $C_1 = 9 \times 10^{-4}/(0.7 \times 4 \times 10^5) = 3.2 \times 10^{-9}$ F and $C_2 = 10^{-4}/(0.7 \times 1.6 \times 10^5) = 880$ pF. The time constant $R_1 C_1 = 4 \times 10^3 \times 3.2 \times 10^{-9} = 12.8$ μs, which is about one-eighth of the 100 μs during which T_1 conducts. Therefore C_1 has adequate time to charge through R_1 during the 100 μs. Also $R_2 C_2 = 10^4 \times 880$ pF $= 8.8$ μs, which is very small in comparison with the 900 μs during which T_2 conducts. Thus R_2 could be much larger than 10 kΩ and R_1 could be larger than 4 kΩ.

Field-effect transistors are especially useful in monostable or astable multivibrator applications that require very long time constants, as may be seen in Fig. 22.28. Since R_{G1} and R_{G2} may be very large (many megohms), the time constants can be very large compared with those of a bipolar transistor. Since the FET switches ON when the reverse bias is reduced to V_P, the off-period of each transistor can be calculated approximately from the following relationship (compare Eq. 22.19).

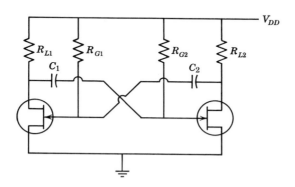

FIGURE 22.28
An astable multivibrator using FETs.

$$T_d = R_G C \ln\left(2 - \frac{V_P}{V_{DD}}\right) \qquad (22.23)$$

The base-circuit or gate-circuit resistors in an astable multivibrator are usually made adjustable by the inclusion of a potentiometer, so that the time constants can be controlled.

Multivibrators lend themselves to a wide range of uses. One interesting application is in a counter circuit as shown in Fig. 22.29. Notice that this is a conventional monostable configuration with T_1 added as a trigger input circuit. Whenever the base of T_1 becomes positive enough to permit con-

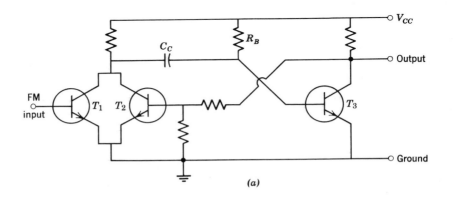

(a)

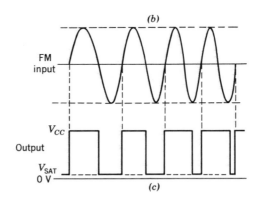

FIGURE 22.29
A counter type of FM detector: (*a*) **basic circuit;** (*b*) **exaggerated FM signal;** (*c*) **output signal.**

duction, the collectors of T_1 and T_2 begin to drop and the switching action is initiated. Thus each input cycle produces a constant pulse at the output as shown in Fig. 22.29c. Consequently the *average* value of the output signal is proportional to the frequency of the input signal. (A pulse-forming circuit may have to precede the input to T_1 so the circuit is only triggered once for each cycle.)

If the input signal in Fig. 22.29 is a pulse from an auto (or boat) ignition system, a milliameter can be used as part of the load on T_3 and the meter can be calibrated in RPM and used as a tachometer. In contrast, if the input signal to T_1 is an FM signal, as indicated in Fig. 22.29, the circuit can be used as an FM detector. Then the average value of the output signal is proportional to the frequency of the input signal. If a low-pass filter is connected to the output terminal, the signal out of the filter will be the required audio signal.

The value of resistance R_B is chosen so T_3 will be driven into saturation when T_1 and T_2 are both not conducting. Then C_C is chosen so the duration of the output pulse plus the circuit recovery time is just equal to the period of the highest frequency to be detected. An example will be used to further illustrate the performance of this circuit.

Example 22.8   An FM detector circuit is connected as shown in Fig. 22.29. The recovery time of this circuit is 5 percent of the period for a total cycle. (The total cycle is equal to the output pulse plus the recovery time of the circuit.) The value of V_{CC} is 20 V and the saturated collector voltage is assumed to be essentially zero. Let us determine the expected output signal if the maximum signal frequency is 10 MHz.

First, we note that at 10 MHz, the period of one cycle is $1/10^7 = 0.1$ μs. The pulse duration for the monostable multivibrator is (0.1×0.95) μs $= 0.095$ μs. Then the output signal will be a pulse of 20 V amplitude for 0.095 μs and 0 V for the remainder of the cycle, or 0.005 μs. The average value of this waveform is 20 V $\times 0.95 = 19$ V. As frequency decreases to zero, the average value of the output voltage also decreases to zero as shown in Fig. 22.30. Therefore, this detector will have a sensitivity of 19 V/10 MHz or 1.9 mV/kHz. Now, if the actual input signal has a frequency deviation of 40 kHz, the output signal from the detector and low-pass filter will have a peak-to-peak amplitude of 76 m V, or 0.076 V.

From the foregoing example, we note that we can define the duty cycle of the multivibrator, D_{cycle}, as

$$D_{\text{cycle}} = \frac{\text{pulse duration}}{\text{pulse duration} + \text{recovery time}} \quad (22.24)$$

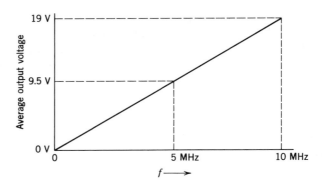

FIGURE 22.30
A plot of average output voltage versus frequency for Example 22.8.

Then, the sensitivity of the detector can be written as

$$S = \frac{D_{\text{cycle}} \, V_{CC}}{f_{\text{max}}} \qquad (22.25)$$

where f_{max} is equal to $1/(\text{pulse duration} + \text{recovery time})$ for the multivibrator. If the saturation voltage V_{SAT} of the transistors is not negligible compared to V_{CC}, the sensitivity of the detector is

$$S = \frac{D_{\text{cycle}}(V_{CC} - V_{\text{SAT}})}{f_{\text{max}}} \qquad (22.26)$$

From Eq. 22.26 we note that the sensitivity can be increased by using higher supply voltages or lower values of f_{max}. (The duty cycle is fixed for a given IC.) Since the supply voltage is limited for given transistors, the usual approach for good sensitivity is to reduce the frequency (and consequently f_{max}) of the signal which is applied to the detector. Thus, low IF frequencies are desirable for counter-type FM detectors.

PROBLEM 22.8   Determine the pulse duration and the sensitivity of a counter-type FM detector which uses an IF of 455 kHz with $f_{\text{max}} = 475$ kHz. Assume $V_{CC} = 20$ V and $D_{\text{cycle}} = 0.95$.

PROBLEM 22.9   A given silicon transistor that has $h_{FE\,\text{min}} = 50$, $f_\tau = 100$ MHz, and $C_{ob} = 5$ pF is used in a monostable multivibrator circuit with $V_{CC} = 20$ and $R_{L1} = R_{L2} = 10$ kΩ. Determine the other circuit components that will provide a 50 μs gating pulse. Draw a circuit diagram.
Answer: $R_B \leq 500$ kΩ, $C_2 = 140$ pF for $R_B = 500$ kΩ, $C_1 = 6$ pF. Diode required.

PROBLEM 22.10   Redesign the multivibrator of Example 22.7 using $R_2 = 1$ MΩ and $R_1 = 10$ kΩ to obtain the same output characteristics.
Answer: $R_{B1} = 220$ kΩ, $R_{B2} = 720$ kΩ, $C_1 = 1800$ pF, $C_2 = 610$ pF.

PROBLEM 22.11   Design a symmetrical astable multivibrator, using FET transistors with $V_{DD} = 25$ V, $V_P = -2$ V,   and $g_{mo} = 2000$ μmho at $I_{DSS} = 2$ mA, that will produce a square-wave output signal with a period of 10 s.

22.6   THE UNIJUNCTION TRANSISTOR

The unijunction transistor is a very useful device for timing or clock pulse generation. The physical construction is essentially as shown in Fig. 22.31. A bar of lightly doped n-type semiconductor material (usually

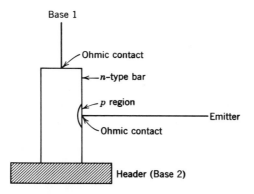

FIGURE 22.31
The physical arrangement of a unijunction transistor.

silicon) has electrical contacts or terminals connected to each end. These terminals are known as the base 1 (B_1) and base 2 (B_2) terminals. Approximately halfway along the n-type bar a region of p-type material is formed, thus producing a single junction. The terminal that is connected to this p region is known as the emitter (E) terminal.

The symbol for a unijunction transistor as well as a typical biasing arrangement is shown in Fig. 22.32a. The electrical behavior of the circuit can be visualized with the help of the equivalent circuit shown in Fig. 22.32b. The silicon bar acts as a resistor, which is tapped where the junction is formed. The resistance from the junction to the B_2 terminal will be called R_{b2} and the resistance from the junction to the B_1

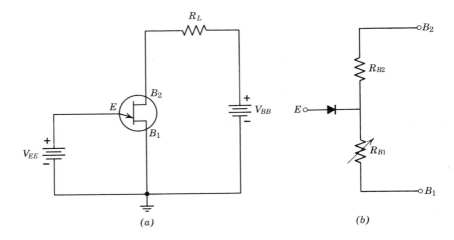

FIGURE 22.32
(a) The symbol and biasing arrangement of a unijunction transistor, and (b) an equivalent circuit for the unijunction transistor.

terminal will be called R_{b1}. The junction acts as a conventional p-n diode. Now, if B_1 is connected to ground and B_2 is connected to a positive potential V_{BB}, a current will flow through R_{b2} and R_{b1}. If the emitter terminal is open, a voltage V_{EO} will appear between the E and B_1 terminals. Because of the voltage divider action of R_{b2} and R_{b1}, this voltage V_{EO} will be a fraction η of the voltage on base 2, with respect to base 1.

$$V_{EO} = \eta V_{b2} \qquad (22.27)$$

If the emitter is biased at a potential less than V_{EO}, the p-n junction is reverse biased and only the diode saturation current flows in the emitter circuit. However, if the voltage of the emitter is increased above V_{EO}, the junction becomes forward biased. Under these conditions, holes are injected from the p material into the n bar. These holes are repelled by the positive base-two end of the bar and are attracted toward the base-one end of the bar. These additional p carriers in the emitter-to-base-one region results in a decrease of resistance for resistor R_{b1} (Fig. 22.32). The decrease of resistance R_{b1} results in a lower emitter voltage. Thus, a negative resistance effect is produced since the voltage decreases as the current increases. As more p carriers are injected, a condition of saturation will eventually be reached, as shown in Fig. 22.33, and in the current-voltage characteristics in Fig. 22.34. The similarity of these characteristics to those of the tunnel diode is at once evident. The unijunction transistor,

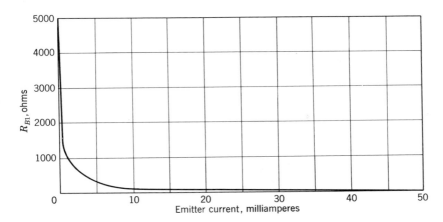

FIGURE 22.33
A plot of R_{b1} versus emitter current for a 2N492.

however, has three terminals. Therefore, to completely describe the characteristics, a family of curves (as given in Fig. 22.35) is required.

Unijunction transistors are used extensively in oscillator-, pulse-, and voltage-sensing circuits. To help illustrate some typical applications, let us consider two examples. The first of these is a circuit known as a relaxation oscillator.

Example 22.9   A unijunction transistor is connected as shown in Fig. 22.36. Determine the voltage waveforms on the emitter and on base 2 of the

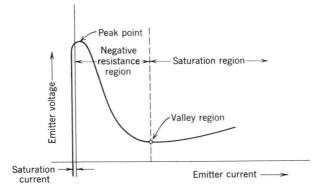

FIGURE 22.34
The emitter current-voltage characteristics of a unijunction transistor.

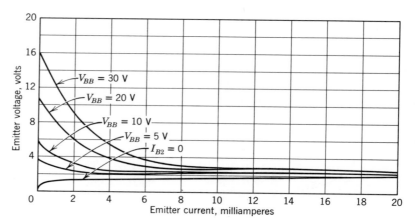

FIGURE 22.35
Static characteristic curves of a unijunction transistor.

unijunction transistor. The transistor is a 2N492 (characteristics given in Fig. 22.33) with $R_{b1} + R_{b2} = R_{bb} = 7.5$ kΩ and $\eta \simeq 0.67$.

At time $t = 0$, the switch S_1 is closed. The capacitor is initially uncharged, so the voltage V_{EO} is 0 at time $t = 0$. Current can flow through R_2 and the unijunction transistor base to ground. This current will have a magnitude of

$$I_B = \frac{V_{BB}}{R_2 + R_{bb}} \tag{22.28}$$

or $12/(2.5$ k$\Omega + 7.5$ k$\Omega) = 1.2$ mA.

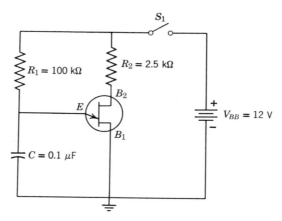

FIGURE 22.36
A relaxation oscillator.

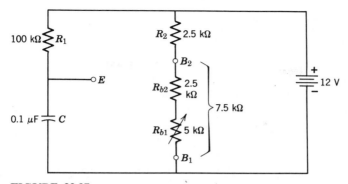

FIGURE 22.37
The equivalent circuit for Fig. 22.35 when the emitter junction is reverse biased.

The voltage on B_2 is $V_{BB} - I_B R_2 = 12 - (1.2 \times 10^{-3})(2.5 \times 10^3) = 9$ V. From Eq. 22.27, $V_{EO} = 0.67 \times 9 = 6$ V. Since the emitter voltage is initially 0, the emitter junction is reverse biased and appears as an open circuit. The circuit appears as shown in Fig. 22.37. The capacitor will charge from 0 V toward $+12$ V with a time constant $\tau = R_1 C = 10^5 \times 10^{-7} = 10^{-2}$ s. Thus, the equation for the voltage on the emitter (while the emitter junction is reverse biased) is

$$v_E = 12(1 - e^{-t/0.01}) \tag{22.29}$$

A sketch of v_E versus time is given in Fig. 22.38.

The voltage v_E follows Eq. 22.29 until v_E is approximately 6 V. The time required for v_E to reach 6 V can be found from Eq. 22.29. A

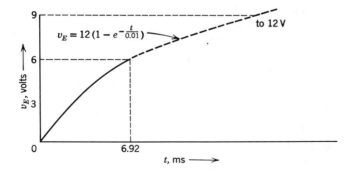

FIGURE 22.38
A plot of v_E for the circuit in Fig. 22.36.

solution of this equation with $v_E = 6$ V yields $t = 6.92$ ms. When v_E exceeds 6 V, the emitter junction becomes forward biased and p carriers are injected into the base region. As was noted previously, these carriers reduce the resistance of R_{b1}. As the resistance of R_{b1} decreases, the voltage at the junction of R_{b1} and R_{b2} decreases. This action causes the emitter junction to become more heavily forward biased and a greater number of p carriers are injected into the base region, further reducing the resistance R_{b1}. In fact, the action is cumulative and R_{b1} rapidly approaches its minimum value of approximately 50 Ω as shown by Fig. 22.33. The forward-biased junction acts almost as a short circuit, so the equivalent circuit for Fig. 22.36 (an instant after $t = 6.92$ ms) becomes that shown in Fig. 22.39a. This circuit can be further simplified (as shown in Fig. 22.39b)

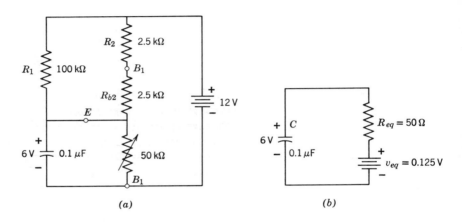

(a)   (b)

FIGURE 22.39
The equivalent circuit for Fig. 22.36 an instant after v_E is equal to 6 V: (a) equivalent circuit; and (b) simplified circuit.

by replacing all the circuitry that connects to the capacitor by a Thevenin's equivalent circuit. The equivalent resistance R_{eq} for this Thevenin's circuit is $(R_2 + R_{b2})$ in parallel with both R_1 and R_{b1}. Thus $R_{eq} = 5$ k$\Omega \parallel 100$ k$\Omega \parallel 50$ $\Omega \simeq 50$ Ω. The Thevenin's equivalent voltage is $v_{eq} = (12 \times 50)/(R_p + 50)$ where R_p is the parallel combination of $(R_2 + R_{b2})$ and R_1. Thus, $R_p = 4.76$ kΩ and $v_{eq} = (12 \times 50)/(4.76$ k$\Omega + 50) \simeq 0.125$ V.

From Fig. 22.39b, the time constant of this circuit is $R_{eq}C = 50 \times 10^{-7}$ or 5 μs. Thus, the voltage across C (which is also v_E) will decrease from 6 V toward 0.125 V with a time constant of 5 μs. In four or five time constants (20 to 25 μs) the capacitor will be essentially discharged. As the capacitor current decreases, the p carriers in the base 1 region decrease, and R_1 begins to increase. An increase of R_1 allows less current to flow, so the action again is cumulative and the emitter junction quickly becomes reverse biased. The circuit is now in essentially the same condition as at time $t = 0$, so the cycle repeats. Consequently, the emitter voltage v_E will have the form shown in Fig. 22.40.

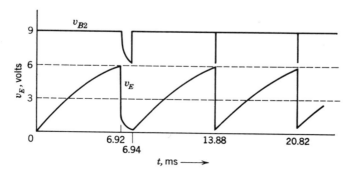

FIGURE 22.40
A plot of the voltage waveforms on the emitter and base 2 of the transistor in Fig. 22.35. The period from $t = 6.92$ ms to $t = 6.94$ ms is expanded to permit showing the detail of the waveform.

Since the emitter junction is essentially a short circuit while C is discharging, the lower end of R_{b2} will have essentially the same potential as v_E during this period. Thus, the lower end of R_{b2} will shift from 6 V to about 0.2 or 0.3 V with a time constant of 5 μs. As a result, the potential at the terminal B_2 will drop from 9 V (the value while the emitter junction is reverse biased) to a voltage $v_{B2\,min}$ while the capacitor is discharging. The minimum value for v_{B2} is

$$v_{B2\,min} = v_{E\,min} + (V_{BB} - v_{E\,min})R_{b2}/(R_{b2} + R_2) \qquad (22.30)$$

where $v_{E\,min}$ is the minimum voltage on the emitter before the junction becomes reverse biased again. If we assume $v_{E\,min} = 0.4$ V, $v_{B2\,min} = 0.4 + (12 - 0.4)2.5/(2.5\text{ k}\Omega + 2.5\text{ k}\Omega) = 0.4 + 5.8 = 6.2$ V. The waveform for v_{B2} is also shown in Fig. 22.40. Actually, since the p-n junction has a finite width, the value of R_{b2} also decreases during the firing period (the time when R_{b1} decreases in magnitude). Consequently, the value of v_{B2} decreases below that given by Eq. 22.30. If one measures the actual value of $v_{B2\,min}$, Eq. 22.30 can be used to calculate the value of R_{b2} while the transistor is firing.

From this example, we note that the unijunction can be used to produce a periodic trigger pulse (v_{B2}) when used as a relaxation oscillator.

There are two limitations to this type of circuit that should be understood. First, if R_1 (Fig. 22.36) is too large, the current through R_1 may be no larger than the saturation current flowing through the reverse-biased emitter junction. (The leakage current through the capacitor, especially if an electrolytic capacitor is used, will add to the saturation current of the diode.) Then, the charge on the capacitor remains constant and the circuit will not operate. In the other extreme, if R_1 becomes too small, the current flowing through R_1 and the emitter junction after the circuit has "fired" may be enough to keep R_1 in its saturated (or low impedance) condition. For example, if an emitter current of 10 mA is maintained in the 2N492 (Fig. 22.33), the resistance of R_{b1} will remain at about 100 Ω. Then, the emitter of the unijunction will not be turned off and the relaxation oscillator will *not* oscillate because the emitter current will remain above the valley region shown in Fig. 22.34.

PROBLEM 22.12   Repeat Example 22.9 if C is changed to 0.01 μF.

Answer: The period for one cycle = 694 μs.

PROBLEM 22.13   Modify the circuit elements in Fig. 22.36 so that the circuit can produce 1000 pulses/s.

The narrow pulses generated by the unijunction can be used to trigger other circuits such as digital ICs (to be considered in Chapter 23) or SCRs. To illustrate this capability, let us consider a last example.

Example 22.10   A phase-controlled dc power supply is connected as shown in Fig. 22.41. Analyze the performance of this circuit and determine suitable values for circuit elements to provide proper operation.

The two diodes D_1 and D_2 act as a full-wave rectifier circuit. Consequently, the voltage v_A at point A has the waveform shown in Fig. 22.42. The diode D_3 is a 12 V reference diode that acts with R_1 to limit the voltage at point B to 12 V. If D_3 is a 1 W diode, the current through D_3 should be limited to $I = W/V = 1/12 \simeq 83$ mA. In order to limit the current through D_3 to 83 mA, $R_1 = (165 - 12)$ V/0.083 A = 1850 Ω. We shall use a value of 2200 Ω for R_1.

The unijunction circuit operates in essentially the same fashion as the circuit in Example 22.9. However, we do wish to obtain a positive pulse to trigger the SCRs in this circuit. Therefore, a resistor R_5 is included in the base 1 circuit of the unijunction. To determine the value of R_5, the characteristics of the SCR must be considered. A 2N1597 is an industrial-type SCR with a peak-reverse voltage rating of 200 V and a forward current capacity of 1.6 A rms. We shall use this type of SCR in our circuit. The specifications for the 2N1597 indicate that the most sensitive gate will

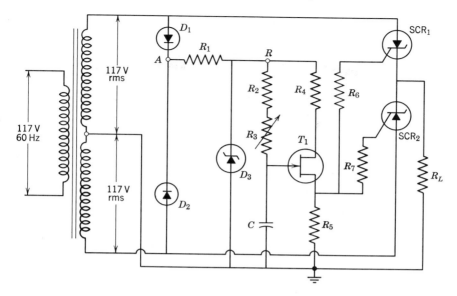

FIGURE 22.41
An adjustable dc power supply.

not fire unless at least 0.2 V is applied. In addition, the most insensitive gate will fire if 3 V is applied to the gate. Let us use the 2N492 unijunction transistor we used in Example 22.9. In that example, a 6 V pulse was generated across R_{b1} from a 12 V supply. The value of R_{b2} was about 50 Ω while the pulse was present. Therefore, if we make R_5 equal to 50 Ω, the voltage pulse across R_5 will have a value of 3 V, which is sufficient to fire the most insensitive SCR.

To insure at least 3 V across R_5, we reduce R_4 to 500 Ω. Then, the voltage from B_2 to ground will be $v_{RB2} = 12(R_{bb} + R_5)/(R_{bb} + R_4 + R_5) = 12(7550)/8050 = 11.25$ V. Since $R_5 \ll R_{bb}$, the value of $V_{EO} \simeq 0.67 \times 11.25 \simeq 7.5$ V and the voltage across $R_5 = 7.5/2 = 3.75$ V. The time required for the voltage across C to charge to 7.5 can be found from the equation

$$v_E = 12(1 - e^{-t/RC}) \tag{22.31}$$

where RC is $(R_2 + R_3)C$. When a value of 7.5 V is substituted into Eq. 22.31 for v_E, we have

$$7.5 = 12(1 - e^{-t/RC}) \tag{22.32}$$

This equation can be solved for t to yield

$$t = 0.98RC \tag{22.33}$$

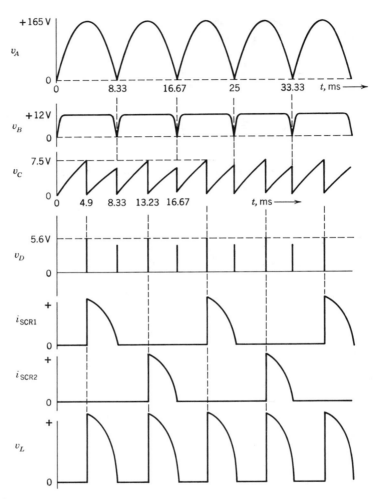

FIGURE 22.42
The voltage and current waveforms of the circuit in Fig. 22.41.

We would like the unijunction firing time to be adjustable form 0 to 8.33 ms (the period for a half cycle) from the beginning of each half cycle. However, to protect the emitter circuit of the unijunction and to prevent the saturation of its R_{b1}, we limit the current through R_2 and R_3 to 5 mA. Then when R_3 is set to zero, the value of R_2 will be $R_2 = 12 \text{ V}/5 \times 10^{-3} \text{ A} = 2.4 \text{ k}\Omega$. When R_3 is all in the circuit, the firing time should be 8.33 ms. Then, from Eq. 22.33, $(R_2 + R_3)C = (8.33 \times 10^{-3})/0.98 = 8.5 \times 10^{-3}$ s. We shall limit R_3 to a value of about 100 kΩ so the saturation current through the reverse-biased unijunction is small compared to the minimum current through R_3. Therefore, let R_3 be a 100 kΩ potentiometer. Then, C must have a value of about 0.085 μF. We shall use a value of 0.1 μF for C. The minimum firing time is $0.98 \times 2.4 \times 10^{-4} \simeq 0.24$ ms.

If the resistor R_3 is adjusted so that $(R_2 + R_3)$ is 25 kΩ, the value of t from Eq. 22.33 will be $t = 0.98 \times 2.5 \times 10^4 \times 2 \times 10^{-7}$ s $= 4.9$ ms. The voltage across the capacitor, v_c, will have the form shown in Fig. 22.42. At $t = 4.9$ ms, the unijunction will fire and the capacitor will discharge. The capacitor begins to charge again, but before the voltage rises to 7.5 V, the voltage v_B drops to zero. Since base 2 is connected to v_B, the voltage on base 2 (v_{B2}) also decreases. As v_{B2} decreases, v_{EO} decreases. When v_{EO} is reduced to the same potential as v_c, the unijunction fires and discharges the capacitor. Thus, the capacitor always begins a charging cycle at the beginning of each half cycle of the input power. Of course, each discharge produces a positive pulse across R_5. This voltage waveform is shown in Fig. 22.42 as voltage v_D. This positive pulse is applied to the gates of the SCRs. When the positive gate potential is applied to a forward-biased SCR, the SCR will fire and conduct current. The positive gate potential will have no effect on the reverse-biased SCR. Thus, one SCR will conduct for part of one half cycle and the other SCR will conduct for part of the next half cycle. The action is much the same as a full-wave rectifier except that the SCRs only conduct for part of a half cycle. The current through the two SCRs will have the form shown in Fig. 22.42.

The resistors R_6 and R_7 are used to limit the current flow into the gates. The most insensitive SCR (of this type) will fire if a gate voltage of 3 V and a gate current of 10 mA is present. The actual voltage pulse v_D has a value of 3.75 V. Hence we can drop 0.75 V (with 10 mA of current) across R_6 or R_7 and still get proper operation. Thus, $R_6 = R_7 = 0.75/0.01 = 75\ \Omega$. A value of 68 Ω for these two resistors would be sufficient.

The voltage across the load will be equal to R_L times the current through R_L. This current will be the sum of the two currents through the two SCRs. Therefore, the output voltage v_L will have the form shown in Fig. 22.42. The average output current and voltage will vary as the SCR conduction time varies. Since R_3 controls the firing time of the unijunction, it also controls the average output voltage.

If the value of R_3 is reduced to a low enough value, two or more positive pulses may be produced by the unijunction transistor within one half cycle of the applied ac voltage. In this case, the first pulse fires the SCR and any following pulses will have no additional effect. The SCR is turned off only when the anode voltage drops to zero. (As was mentioned previously, a positive pulse on the gate of a reverse-biased SCR will have no effect on this SCR.)

The pulse at the beginning of each half cycle may be large enough to trigger the SCRs. However, since the anode potentials are zero at this time, these pulses do not fire the SCRs. From this description, we see the conduction period of each SCR can be controlled from 0 ($R_3 \simeq 100$ kΩ) to a maximum value ($R_3 = 0$). The maximum dc voltage across the load is essentially equal to 0.636 times the peak value of one-half the secondary voltage (from the center tap to either end).

PROBLEM 22.14 Determine the value of each circuit element in Fig. 22.41 if the secondary winding has 220 V rms on each side of center tap. The 2N1599 has a peak reverse voltage rating of 400 V. Otherwise, its characteristics are the same as those of the 2N1597.

23

Digital Integrated Circuits

Most of the modern computers are of the digital type. With digital technology well developed, much of the modern instrumentation is also digital. In fact, digital instrumentation can be interfaced with digital computers to produce automatic data reduction systems or automatic control systems. In some instances, entire processing plants, such as oil refineries, are controlled by digital computers.

As the use of digital circuits increased, the variety of digital integrated circuits increased and the cost decreased. At the present time, a wide variety of digital *IC*s are available, and the cost is very low.

23.1 THE BINARY SYSTEM

In Chapter 22 we noted that low power is dissipated in the transistor when it is cut off (OFF) or when it is saturated (ON). As a result, we find that practically all modern computer and digital circuits are designed so the transistor is either ON or OFF. Consequently, essentially all digital computer calculations are done in the *binary* (or two-number) system. Actually, the basic manipulation of numbers can be accomplished in any base. The only reason our present base of 10 system has developed is because man had 10 fingers (or toes, if you count that way). A comparison of the first twenty numbers as written in base 10 and binary is given in Table 23.1. With a little practice, it becomes quite easy to solve mathematical problems in the binary system. To illustrate the four basic mathematical steps, consider Example 23.1.

TABLE 23.1

A Comparison of Binary and Base 10 Numbers.

Base 10	Binary (Base 2)
0	0
1	1
2	10
3	11
4	100
5	101
6	110
7	111
8	1000
9	1001
10	1010
11	1011
12	1100
13	1101
14	1110
15	1111
16	10000
17	10001
18	10010
19	10011
20	10100

Example 23.1 Let us solve the problems (a) $9 + 3 = 12$, (b) $9 - 3 = 6$, (c) $9/3 = 3$ and (d) $9 \times 3 = 27$ in the binary system.

$$
\begin{array}{llll}
(a) & \begin{array}{r} 1001 \\ +11 \\ \hline 1100 \end{array} & \begin{array}{r} 9 \\ +3 \\ \hline 12 \end{array} &
(b) & \begin{array}{r} 1001 \\ -11 \\ \hline 110 \end{array} & \begin{array}{r} 9 \\ -3 \\ \hline 6 \end{array}
\end{array}
$$

$$
\begin{array}{llll}
(c) & \begin{array}{r} 11 \\ 11\overline{)1001} \\ 11 \\ \hline 11 \\ 11 \\ \hline 0 \end{array} & \begin{array}{r} 3 \\ 3\overline{)9} \end{array} &
(d) & \begin{array}{r} 1001 \\ 11 \\ \hline 1001 \\ 1001 \\ \hline 11011 \end{array} & \begin{array}{r} 9 \\ \times 3 \\ \hline 27 \end{array}
\end{array}
$$

The main thing to remember is that $1 + 1 = 10$, conversely that $10 - 1 = 1$. It then follows that if you borrow 1 from 10 you leave 1.

The rest of the rules are the same as in the base of 10 math: $0 + 0 = 0$, $0 + 1 = 1$, $1 - 0 = 1$, $1 - 1 = 0$, $1 \times 0 = 0$, $1 \times 1 = 1$, $1/1 = 1$ and $0/1 = 0$.

The designer of a system can define which state (ON or OFF) represents the 1 in his system. Normally, a low signal (the collector on a saturated transistor) of 0.5 to 0 V represents a 0 and a high signal (the collector on a cutoff transistor) of 1 to 5 V, depending on the value of V_{CC}, represents a 1.

PROBLEM 23.1   Extend Table 23.1 to 35. Is 27 correctly written as 11011 in binary?

PROBLEM 23.2   Solve the following problems using binary numbers. (a) $6 + 7$, (b) $9 - 6$, (c) $18/6$, (d) 6×3.

23.2   DIGITAL LOGIC

Most digital logic is performed by interconnecting a few basic circuit configurations. One basic configuration is known as an AND gate. In a three-input (input signals A, B, and C) AND gate, a 1 is obtained in the output only when a 1 is present on input A *and* on input B *and* on input C. The common symbol for an AND gate is given in Fig. 23.1a. If relays or actual switches are used as the logic elements, the circuit would appear as

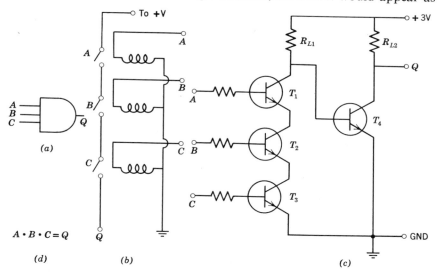

(a)

(d)

(b)

(c)

FIGURE 23.1
The AND gate: (a) symbolic representation; (b) relay configuration; (c) transistor configuration; (d) algebraic notation.

shown in Fig. 23.1b. If switches are used, the 1 position is a closed switch and the 0 position is an open switch. In the relay configuration, the 1 signal pulls the relay closed while a 0 signal lets the relay open. Notice that an output signal of 1 ($+$ V in Fig. 23.1b) only appears on the output (terminal Q) when all three input signals are 1. A transistor AND gate is shown in Fig. 23.1c. If a 1 signal (say $+3$ V) is present on all three input terminals, then T_1, T_2, and T_3 will be in saturation and T_4 will be cut off. (R_{L1} must be large enough to limit the current through T_1, T_2, and T_3 to a low value so the voltage drop across these three r_{sat} resistors is less than the turn-on voltage for the base of T_4.) Then, the output signal on terminal Q will be a 1. If one, or more, of the input signals is a 0 (essentially 0 V), that transistor will be cut off and T_4 will be in saturation. (R_{L1} should be $\leq h_{FE4} i_C$.) In this case, the output signal on terminal Q will be a 0. Algebraically, the AND function is written as $A \cdot B \cdot C = Q$, or simply as $ABC = Q$.

A second basic digital gate is known as an OR gate. The OR gate will produce a 1 on its output if one *or* more of the input signals is a 1.

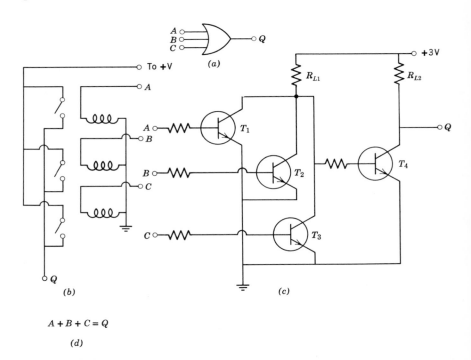

$A + B + C = Q$

(d)

FIGURE 23.2
The OR gate: (a) symbolic representation; (b) relay configuration; (c) transistor configuration; (d) algebraic notation.

The symbol for a three-input OR gate is given in Fig. 23.2a. The switch or relay configuration for an OR gate is shown in Fig. 23.2b. Again, a 1 is represented by a closed switch or relay, and 0 is represented by an open switch or relay. The transistor configuration for an OR gate is shown in Fig. 23.2c. If a 1 signal appears at one of the input terminals (A, B, or C), the corresponding transistors (T_1, T_2, or T_3, respectively) will be driven into saturation. If one or more of these transistors (T_1, T_2, or T_3) is saturated, transistor T_4 will be cut off and this signal on Q will be a 1. Only when A, B, and C are all 0 (with T_1, T_2, and T_3 cut off) will T_4 be saturated and $Q = 0$. Algebraically the OR function is written $A + B + C = Q$.

Another basic digital manipulation is produced by the NOT gate or *inverter*. The output signal from a NOT gate is always the opposite of the input signal. Thus, if the input signal A is a 1, the output signal, which is called $\bar{A}$ (read as A-not, or sometimes just A-bar), would be a 0. If A is a 0, then $\bar{A}$ is a 1. The NOT gate or inverter is represented symbolically as shown in Fig. 23.3a. The little circle at the end of the conventional amplifier

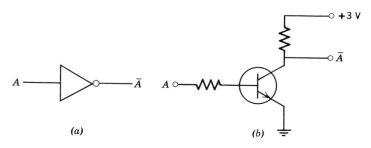

(a)   (b)

FIGURE 23.3
The NOT gate: (a) symbolic representation; (b) transistor configuration.

symbol indicates that the output signal is inverted with respect to the input signal. A simple transistor amplifier (or transistor switch since we are working in the cutoff or saturation region) can be used as an inverter as shown in Fig. 23.3b.

Two additional types of gates can be obtained by combining the NOT gate with the AND and OR gates. Thus, if the output of the AND gate is inverted, we have a NOT-AND configuration. Rather than write NOT-AND, this notation is normally reduced to NAND. Similarly, if the output of an OR gate is inverted, a NOT-OR or NOR gate is produced. The symbol for a NAND and a NOR gate is given in Fig. 23.4. Notice that these are simply AND or OR gate symbols with the little circles on the output which indicate that the output is inverted. Also note that if T_4 is omitted in Fig. 23.1c and

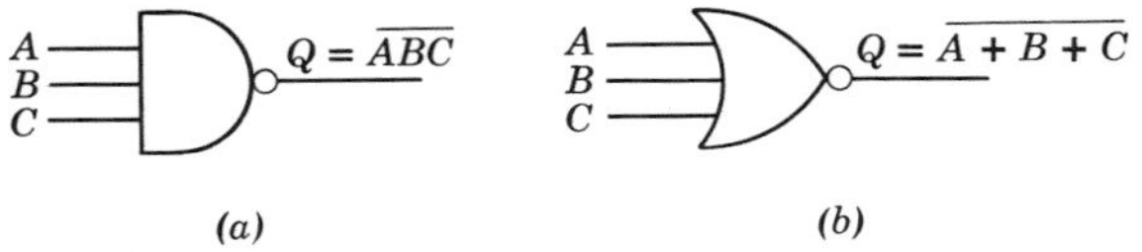

$$A - B - C \quad \boxed{\ }\!\!\!\rangle\!\!\circ \quad Q = \overline{ABC} \qquad\qquad A - B - C \quad \boxed{\ }\!\!\!\rangle\!\!\circ \quad Q = \overline{A + B + C}$$

(a)   (b)

FIGURE 23.4
The symbols for (a) a NAND gate, and (b) a NOR gate.

Fig. 23.2c and the input to T_4 is used as Q, the resulting gates will be NAND and NOR gates, respectively.

A useful method of representing the relationship between the input and output signals of a digital gate is known as the *truth table*. The truth table of a gate or digital device is simply a tabulation of all of the possible input signal combinations and the corresponding output signals. Thus, the truth table for a NOT gate would be as shown in Table 23.2. We can also

TABLE 23.2
The Truth Table for a NOT Gate.

A	$\overline{A}$
0	1
1	0

construct a combination truth table for each of the other four gates as shown in Table 23.3.

In this section, we have introduced the basic digital logic circuits. We have used three-input gates, but of course other gates could be constructed for two or more input signals. In addition, other transistor (or diode or diode-transistor) configurations could be used to produce logic gates.

TABLE 23.3
The Truth Table for AND, OR, NAND, and NOR gates.

A	B	C	AND	OR	NAND	NOR
0	0	0	0	0	1	1
0	0	1	0	1	1	0
0	1	0	0	1	1	0
0	1	1	0	1	1	0
1	0	0	0	1	1	0
1	0	1	0	1	1	0
1	1	0	0	1	1	0
1	1	1	1	1	0	0

PROBLEM 23.3   Draw the circuit diagram and write the truth table for a two-signal input NOR gate.

23.3   BOOLEAN ALGEBRA

George Boole, an English logician and mathematician, wrote a pamphlet, *Mathematical Analysis of Logic*, in 1847 and a larger work, *An Investigation of the Laws of Thought, on which are Founded the Mathematical Theories of Logic and Probabilities*, in 1854. This work of Boole forms the basis for Boolean algebra which lends itself well to the development of digital logic.

While we do not have the space or time to develop all of the rules for Boolean algebra, a few of the common relationships are given in Table 23.4. Note that many of these relationships $(A \cdot 1 = A, \; A + 0 = A,$

TABLE 23.4
Some Common Boolean Algebra Relationships.

$$A + 0 = A$$
$$A \cdot 1 = A$$
$$A + B = B + A$$
$$A \cdot B = B \cdot A$$
$$A + (B \cdot C) = (A + B) \cdot (A + C)$$
$$A \cdot (B + C) = (A \cdot B) + (A \cdot C)$$
$$A \cdot \bar{A} = 0$$
$$A + \bar{A} = 1$$
$$A + A = A$$
$$A \cdot A = A$$
$$A + 1 = 1$$
$$A \cdot 0 = 0$$
$$A + A \cdot B = A$$
$$A \cdot (A + B) = A$$
$$\bar{\bar{A}} = A$$
$$A + (\bar{A} + C) = 1$$
$$A \cdot (\bar{A}C) = 0$$
$$(A + B) + C = A + (B + C)$$
$$(AB)C = A(BC)$$
$$A + \bar{A}B = A + B$$
$$AC + \bar{A}B + BC = AC + \bar{A}B$$
$$(A + B)(\bar{A} + C) = AC + \bar{A}B$$
$$\overline{(AC + B\bar{C})} = \bar{A}C + B\bar{C}$$
$$\overline{(A + C)(B + \bar{C})} = (\bar{A} + C)(\bar{B} + \bar{C})$$
$$\left.\begin{array}{l} \overline{(A + B)} = \bar{A} \cdot \bar{B} \\ \overline{AB} = \bar{A} + \bar{B} \end{array}\right\} \text{De Morgan's theorems}$$

etc.) are valid in conventional algebra. However, some Boolean expressions $(A + 1 = 1, A \cdot A = A$, etc.) are not valid in conventional algebra. Nevertheless, if truth tables are constructed and the definitions of AND and OR functions are maintained, the validity of these expressions is evident. For example, the truth table for $A + 1$ is given in Table 23.5. Notice that since 1 is always 1, then $A + 1$ must also always be 1. All of the expressions in Table 23.4 can be verified by constructing appropriate truth tables.

TABLE 23.5
The Truth Table for
$A + 1$.

A	1	$A + 1$
0	1	1
1	1	1

A graphical representation of the Boolean expressions can be constructed as shown in Fig. 23.5. In this representation, the area where $A = 1$ is represented as the area of a circle (circle A) and the area where $B = 1$ is represented as the area of a second circle. Then, $A + B$ is the shaded

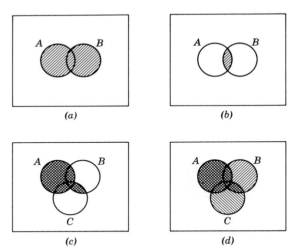

(a)   *(b)*

(c)   *(d)*

FIGURE 23.5
A graphical representation of Boolean expressions: (a) **shaded area** $= A + B$; (b) **shaded area** $= A \cdot B$; (c) **crosshatched area** $= A + (B \cdot C)$; (d) **crosshatched area** $= (A + B) \cdot (A + C)$.

area in Fig. 23.5a. If either A or B is 1, the output is also 1. However, for the expression $A \cdot B$ only the small area where both A and B are 1 (Fig. 23.5b) produces a 1 in the output. The two diagrams, Fig. 23.5c and Fig. 23.5d, show that $A + (B \cdot C) = (A + B) \cdot (A + C)$ as given in the fifth line of Table 23.4. Thus, the graphical representation can also be used to verify or prove the relationships given in Table 23.4.

It is a bit confusing to use the symbols $\cdot$ and $+$ to represent the AND and OR operations. To determine if these symbols are misleading, let us construct a truth table for A times B (or $A \times B$) where A and B are binary numbers. The truth table is given in Table 23.6 as well as the outcome for

TABLE 23.6
The Truth Table for $A \times B$ and $A \cdot B$.

A	B	$A \times B$	$A \cdot B$
0	0	0	0
0	1	0	0
1	0	0	0
1	1	1	1

$A \cdot B$. Notice that $A \cdot B$ is *identical to* the $A \times B$ operation. Therefore, if two binary numbers are to be multiplied, the binary signals representing these numbers can be applied to an AND gate. The output signal from this gate would represent the product of the two input signals. In fact, an n-input AND gate can be used to multiply n digital signals.

Now let us develop the truth table for A plus B where A and B are digital signals. This truth table is shown in Table 23.7. The column S represents the sum of A and B and the column C_o represents the carry term to be added to the next column. Notice that S is not given by any of the binary functions we have considered to this point. Thus, $A + B$ *in Boolean*

TABLE 23.7
The Truth Table for A plus B.

A	B	S	C_o
0	0	0	0
0	1	1	0
1	0	1	0
1	1	0	1

algebra is not the same as A plus B in common algebra. However, from the truth table we note that the output should be a 1 when A is 0 *and B* is 1 *or* when A is 1 *and B* is 0. This relationship can be written in Boolean form as

$$\bar{A}B + A\bar{B} = S \tag{23.1}$$

A logic configuration which accomplishes this digital manipulation is shown in Fig. 23.6. A careful study of Fig. 23.6 will confirm that a 1 appears on the

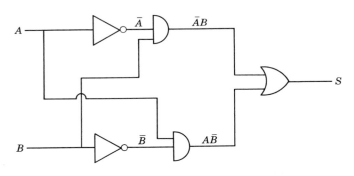

FIGURE 23.6
A configuration to solve Eq. 23.1.

S output only if A or B are 1 but not if both A and B are 1. This configuration (Fig. 23.6) occurs so often that it is given a special name, the *exclusive OR* circuit. Interestingly, the carry term C_o is obtained from an AND gate with A and B as the input signal. The configuration derived from Table 23.7 is known as a *half-adder* and is only valid for the right column of digits in a binary adder.

An example will be used to further acquaint the reader with binary logic.

Example 23.2   A *full-adder* will not only add two binary digits A and B but will also add the carry term from the adjoining column of binary digits. Let us design a full-adder circuit.

First, let us write the appropriate truth table for our adder. If C_i is the carry term to be added to A and B, and if C_o is the carry term derived from our addition, the truth table will be as shown in Table 23.8. Notice that the sum (the S column in Table 23.8) is 1 if C_i and A are both 0 *and B* is 1. If C_i and A are 0, then $\bar{C}_i$ and $\bar{A}$ will be 1. Thus we have S is one if $\bar{C}_i$ and $\bar{A}$ and B are one. In Boolean notation $S = \bar{C}_i \cdot \bar{A} \cdot B$. Actually there are four combinations of A, B and C that will cause S to be 1. The truth table (Table 23.8) states that S is one if $(\bar{C}_i \cdot \bar{A} \cdot B)$ *or* $(C_i$ and B are both 0 *and A* is 1) *or* (A and B are both 0 *and* C_i is 1) *or* (C_i and A and B are all 1).

TABLE 23.8
The Truth Table for a Full-Adder.

C_i	A	B	S	C_o
0	0	0	0	0
0	0	1	1	0
0	1	0	1	0
0	1	1	0	1
1	0	0	1	0
1	0	1	0	1
1	1	0	0	1
1	1	1	1	1

Again, this relationship can be written in Boolean form as

$$\overline{C}_i \overline{A} B + \overline{C}_i A \overline{B} + C_i \overline{A}\overline{B} + C_i AB = S \tag{23.2}$$

The logic configuration which performs this algebra is shown in Fig. 23.7. The output signal S will be 1 only if the conditions given by Eq. 23.2 are met. For any other input conditions, the output signal S will be a 0.

From Table 23.8 the output carry term C_o is 1 if the following conditions are satisfied.

$$\overline{C}_i AB + C_i \overline{A} B + C_i A \overline{B} + C_i AB = C_o \tag{23.3}$$

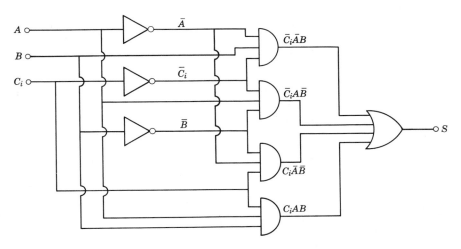

FIGURE 23.7
A logic configuration for Eq. 23.2.

Again, this logic operation (Eq. 23.3) can be performed by using three inverters, four three-input AND gates, and one four-input OR gate. However, from the Boolean algebra table (Table 23.4), we note that

$$\overline{C}_i AB + C_i AB = (\overline{C}_i + C_i)AB = AB \tag{23.4}$$

Therefore, Eq. 23.3 can be written as

$$C_i \overline{A}B + C_i A\overline{B} + AB = C_o \tag{23.5}$$

Then, the logic configuration for C_o can be drawn as shown in Fig. 23.8. Of course, the $\overline{A}$ and $\overline{B}$ signals could be taken from the inverters in Fig. 23.7. The full-adder would consist of the circuit shown in Fig. 23.7 and the circuit

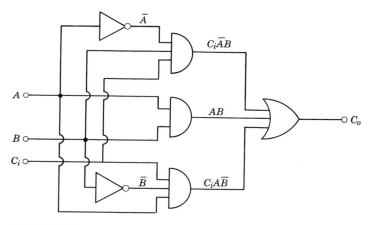

FIGURE 23.8
A logic configuration for Eq. 23.5.

shown in Fig. 23.8 with A, B, and C_i connected in parallel. The S term would be the sum of the given column of numbers and the C_o term could become the C_i term for the adjacent column of numbers.

In Example 23.2, Boolean algebra was used to simplify the circuit configuration which produced the C_o term. Often significant simplifications can be realized by the proper use of Boolean algebra. In addition, Boolean algebra can be used to modify the form of the equations and consequently the form of logic circuits used. For example, it is very difficult to extend the logic gate shown in Fig. 23.1c to many more than three input signals. Actually, it is somewhat difficult to bias three transistors so that the sum of three saturated collectors produces less voltage than the turn-on base voltage for T_4. Consequently, the configuration shown in Fig. 23.2c is preferred over that used in Fig. 23.1c. In fact, the inverting transistor T_4 in Fig. 23.2c is normally omitted to produce a NOR configuration. By using DeMorgan's theorems, configurations using AND and NAND gates can be converted to one using OR and NOR gates. Similarly, the reverse transformation can also be made by using DeMorgan's theorems. To illustrate this type of conversion, an example will be used.

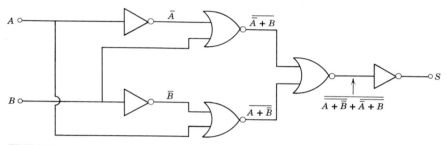

FIGURE 23.9
A circuit using NOR gates and inverters which solves Eq. 23.8.

Example 23.3   The series of digital integrated circuits we wish to use produce only NOR gates and inverters. Let us design a half-adder circuit which uses these ICs.

The equation for a half-adder was given in Eq. 23.1 and is repeated here for convenience.

$$\bar{A}B + A\bar{B} = S \tag{23.1}$$

The DeMorgan's theorem we wish to use is given in Table 23.4 as $\overline{AB} = \bar{A} + \bar{B}$. This relationship can also be written as $AB = \overline{\bar{A} + \bar{B}}$. Then, the term $\bar{A}B$ in Eq. 23.1 can be written as

$$\bar{A}B = \overline{A + \bar{B}} \tag{23.6}$$

Notice that $\bar{\bar{A}} = A$ in this equality (as noted in Table 23.4). Similarly, $A\bar{B}$ in Eq. 23.1 can be written as

$$A\bar{B} = \overline{\bar{A} + B} \tag{23.7}$$

With these two conversions (Eq. 23.6 and Eq. 23.7), Eq. 23.1 can be written as

$$\overline{A + \bar{B}} + \overline{\bar{A} + B} = S \tag{23.8}$$

The logic configuration which solves this equation is shown in Fig. 23.9. The equation for C_o in the half-adder is

$$AB = C_o \tag{23.9}$$

This equality can be written as

$$AB = \overline{\overline{A} + \overline{B}} = C_o \qquad (23.10)$$

The circuit which solves this equation is given in Fig. 23.10. The $\overline{A}$ and $\overline{B}$ signals are derived from the inverters in Fig. 23.9.

FIGURE 23.10
A circuit which solves Eq. 23.10.

By using the other DeMorgan's theorem, $\overline{A + B} = \overline{A} \cdot \overline{B}$, equations can be arranged so that they can be solved by using only AND (or NAND) gates and inverters. In this case, it may be easier to arrange this theorem as

$$A + B = \overline{\overline{A} \cdot \overline{B}} \qquad (23.11)$$

PROBLEM 23.4 Develop the equations and draw the circuit configuration for a half-adder which uses only NAND gates and inverters.

Answer: $\overline{\overline{AB} \cdot \overline{AB}} = S$ and $AB = C_i$.

PROBLEM 23.5 Develop the equations and draw the circuit configuration for a full-adder which uses only NOR gates and inverters.

PROBLEM 23.6 We wish to add two binary numbers A and B. Then we wish to multiply this sum by a binary number C. Draw the truth table and develop the equations for this logic function. Draw the circuit configuration which solves this function and uses only NOR gates and inverters.

Answer: $C\overline{A}B + CA\overline{B} = S$ and $CAB = C_o$.

23.4 DIGITAL INTEGRATED CIRCUITS

The semiconductor manufacturers produce a wide variety of digital integrated circuits. These integrated circuits are divided into several general types of circuit configurations. For example, there is a series of logic circuits which use diode-transistor circuits known as DTL (diode-transistor logic). Another series which is very popular uses transistor-transistor logic (TTL). For very fast switching circuits a series of emitter-coupled circuits known as ECL (emitter-coupled logic) have been developed. Each of these series has certain advantages and certain disadvantages which suit them to particular applications. However, if one is adept at working with one series, it is quite easy to convert to another series with a little study of the manufacturer's data sheets and literature.

In light of the foregoing statement, we will study the resistor-transistor series (RTL) in this chapter since these plastic encapsulated units are inexpensive. In addition, the RTL circuits have configurations which are very similar to the discrete circuits given in Chapter 22. Thus, the Motorola[1] MC802 has the form shown in Fig. 23.11. Notice that this circuit is very similar to the bistable multivibrator shown in Fig. 22.17. The main difference is the parallel triggering transistors T_1 and T_4 that have been added in Fig. 23.11. To understand the behavior of this circuit, let us consider an example.

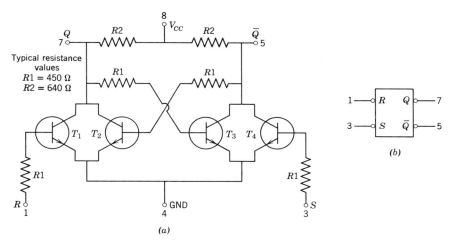

FIGURE 23.11
The RS flip-flop of the MC802: (*a*) **schematic circuit;** (*b*) **symbolic block diagram.**

Example 23.4   Assume that terminal 8 is at $+3$ V with reference to the ground terminal 4. Also assume that T_2 is in saturation and T_3 is cutoff. Then, terminal 7 will be at the collector saturation potential (0.2 V or less). The addition of a positive signal at terminal 1 will then have no effect on the circuit since T_2 is already drawing essentially all the current from V_{CC} that R_2 will allow. Of course, a zero or negative signal at terminal 1 merely drives T_1 into cutoff and in effect removes it from the circuit. If the signal on terminal 3 is zero or negative, then both T_3 and T_4 will be cutoff and the only

---

[1] We will use Motorola numbers in this section, but essentially identical units are made by other manufacturers. The Motorola RTL plastic encapsulated units are listed as the MC700P or MC800P series. Units of the MC800P series will operate through the temperature range of 0°C to 75°C. The MC700P units will operate satisfactorily for temperatures of 15° to 55°C.

current flowing from V_{CC} through R_2 toward terminal 5 will be the base current for transistor T_2. Then, the voltage on terminal 5 is

$$v_5 = V_{CC} - \frac{(V_{CC} - v_{BE2})R_2}{R_1 + R_2} \tag{23.12}$$

If v_{BE2} is assumed to be about 0.6 V and R_1 and R_2 have the nominal values given in Fig. 23.11, the value of v_5 is $3 - (3 - 0.6)640/(450 + 640) \simeq 1.6$ V.

The circuit will remain in this condition until a positive signal is applied to terminal 3. Then transistor T_4 begins to draw current. This additional current through R_2 will reduce the bias on T_2 and initiate the switching action between T_2 and T_3. The manufacturer indicates that an input voltage of 0.85 V (or more) on terminal 3 will cause the circuit to switch.

After switching, terminal 3 will have a potential of about 0.2 V and terminal 7 will have a potential of about 1.6 V. The circuit will now remain in this condition until a positive pulse (0.85 V or more) is applied to terminal 1. Thus, a positive pulse on terminal 3(S) *sets* the circuit and a positive pulse on terminal 1(R) clears or *resets* the circuit. Hence it is named the *RS flip-flop*. Observe that a positive trigger pulse applied simultaneously to inputs 1 and 3, by connecting them together, will make a binary counter out of this flip-flop. Thus, the triggering transistors T_1 and T_4 perform the same function as the steering diodes of Fig. 22.18.

Each circuit in the RTL series is constructed so that it can be directly interconnected to another circuit in the same series. Thus, the output signal from the *RS* flip-flop (Fig. 23.11) is either 0.2 V or less, or 1.5 V. If this output signal is connected to the input of another MC802 (or any circuit in the MC800 series), the 0.2 V signal will not trigger the second circuit (we have a 0) but the 1.5 V signal will (we have a 1).

The simplest circuit in the RTL series is an inverter. The inverter is simply a one-stage amplifier configuration. Usually several inverters are included in the same package. In fact, since each inverter is so simple, the number of leads to the package usually determine the number of inverters which can be included. Thus, a 14-lead flat pack can include six inverters as shown in Fig. 23.12. Due to the basic nature of the inverter configuration, the inverter can be used as building blocks to construct other configurations. For example, if terminal 1 in Fig. 23.12 is connected to terminal 13 and terminal 2 is connected to terminal 14, we have the bistable multivibrator of Fig. 22.17.

PROBLEM 23.7   Show how you could connect the inverters in Fig. 23.12 and a few external circuit elements to construct a monostable multivibrator or an astable multivibrator.

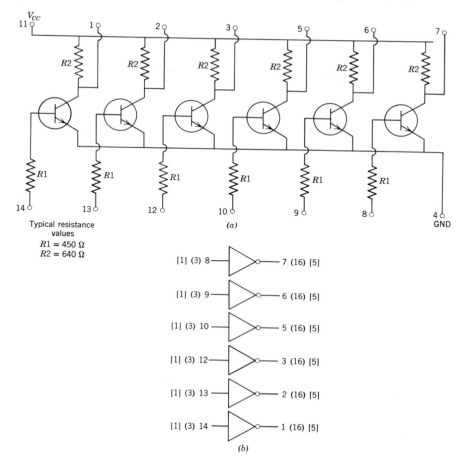

FIGURE 23.12
The MC789P/89P/889P hex inverter circuit: (*a*) schematic circuit; (*b*) symbolic block diagram.

The basic RTL logic gate is a NOR gate as shown in Fig. 23.13. Notice that this is the basic configuration shown in Fig. 23.2c with the inverter T_4 omitted. Of course, two-input gates (four in one package) or three-input gates are also constructed. Notice that by connecting terminals 10 and 12 (Fig. 23.13) together, an eight-input NOR gate is formed. (An inverter is a one-input NOR gate!)

PROBLEM 23.8 Show how inverter circuits could be connected to the four-input NOR gates of Fig. 23.13 to produce two four-input AND gates.

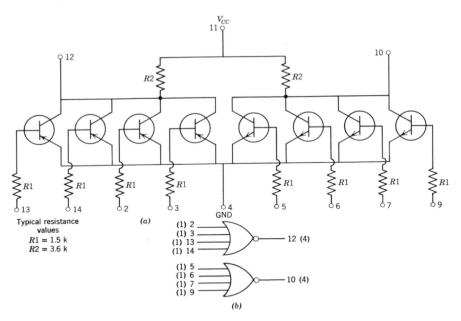

FIGURE 23.13
The MC719P/MC819P dual four-input gates: (*a*) schematic circuit; (*b*) symbolic block diagram.

PROBLEM 23.9   Show how two two-input NOR gates can be connected to produce an *RS* flip-flop as shown in Fig. 23.11.

The exact voltage on the output terminal of an *IC* (terminal 12 in Fig. 23.13 for example) is determined by the load connected to this terminal. If too many circuits are connected to a given output terminal, the voltage in the 1 state may not be high enough to trigger these circuits. To insure proper operation, the manufacturer provides data on the maximum number of *IC*s that can be connected to the output terminal and still obtain reliable operation. The information is given in terms of a *loading factor* or *fan out*. The loading factor indicates the number of "standard" input circuits that can be connected to a given output terminal. In the symbolic block diagrams, the number in parentheses near the output terminal is the loading factor for that output. In some units, the loading factors of the MC700 series and MC800 series are different. In these units, the loading factor of the MC700 series is given in parentheses, and the loading factor of the MC800 circuit is given in brackets. The actual input impedance of some circuits may be low enough to cause as much loading as two or more of the "standard" units.

Thus, a loading factor (using the same notation as on the output circuits) is also given for the input circuits. For example, the MC789P (Fig. 23.12) has an input conductance equal to three "standard" circuits.

The *buffer* is designed to have a low output impedance so that it can drive a greater number of load circuits than the basic *IC*. The circuit of two buffers is given in Fig. 23.14. Notice that these buffers also invert the signal.

Most simple multivibrators can be readily constructed by cross-coupling inverter circuits or gate circuits. However, a more complex multivibrator configuration, known as the *JK* flip-flop is especially designed for clocked operation. (The *clock* is a unit that produces a series of evenly spaced pulses

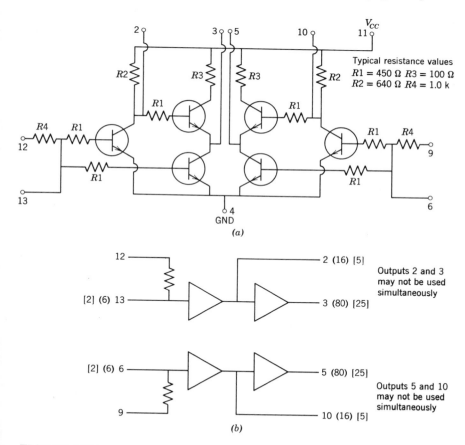

FIGURE 23.14
The MC799P/MC899P dual buffer: (*a*) schematic circuit; (*b*) symbolic block diagram.

to maintain synchronization in the system.) In the JK flip-flop, signals (1's or 0's) are applied to the input terminals but the flip-flop in the IC is not activated until the clock pulse is applied.

The diagram of a typical JK flip-flop is given in Fig. 23.15. The basic flip-flop in this circuit is formed by transistors T_1 and T_2. The input C_D (terminal 9) is a *direct clear* input. Whenever a 1 is applied to this C_D

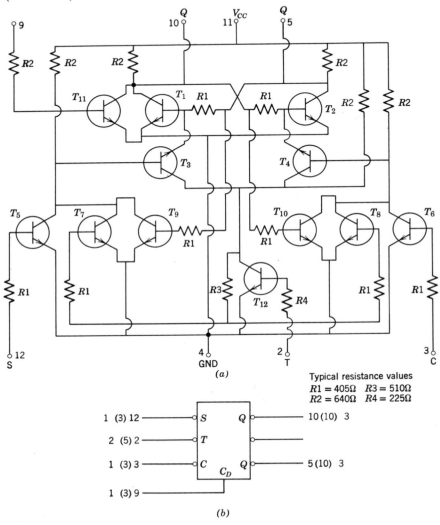

Typical resistance values
$R1 = 405\Omega$ $R3 = 510\Omega$
$R2 = 640\Omega$ $R4 = 225\Omega$

(a)

(b)

FIGURE 23.15
A JK **flip-flop (MC723P/MC816P):** (a) **schematic circuit;** (b) **symbolic block diagram.**

terminal, the circuit will be reset so that output Q (terminal 10) has a 0 for an output. The input signals are applied to terminals 12 and 3. These inputs are known as the *set* (S) and *clear* (C) inputs. The clock or *trigger* (T) pulse is applied to terminal 2.

A careful study of the circuit in Fig. 23.15 reveals that this complex circuit can be represented as an interconnection of several of the circuits already examined. For example, transistor T_{12} can be viewed as an inverter for the clock pulse. Also, notice that T_5, T_7, and T_9 are connected in parallel to form a three-input gate. An identical configuration is formed by T_6, T_8, and T_{10}. Transistors T_3 and T_4 couple the output of these gates into the multivibrator base circuit. Consequently, the JK flip-flop can be represented by the block diagram given in Fig. 23.16.

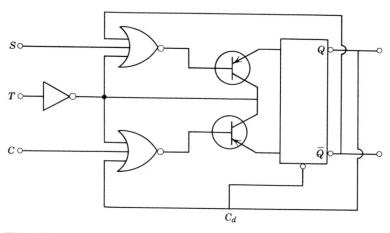

FIGURE 23.16
A block diagram of the JK flip-flop in Fig. 23.15.

Analysis of Fig. 23.15 reveals that no signal can be passed through T_3 and T_4 into transistors T_1 and T_2 when the clock pulse is 0 or 1. However, if transistors T_3 and T_4 are constructed with a longer storage time than the other transistors, an interval exists, when the clock pulse is switching from a 1 to a 0 state, when transistors T_3 and T_4 pass signals to T_1 and T_2. Thus, switching occurs in this circuit on the negative-going portion of the clock pulse. For proper action, the clock pulse fall time must be less than 100 ns. If the S input is a 1 and the C input is a 0, the signal on the Q terminal (after switching) will be a 1. The signal on the $\bar{Q}$ terminal will always be opposite to the signal on the Q terminal. If the S input is a 0 and the C input is a 1, the signal on the Q terminal (after the clock pulse) will be a 0.

If both S and C are 1s the clock pulse will not produce any change in the Q signal. In contrast, if both S and C are 0s, the clock pulse will shift the output to its opposite state (a 0 to a 1 or a 1 to a 0) whenever the clock pulse appears. The truth table for an RTL type[2] JK flip-flop would be as shown in Table 23.9. The term Q_{n+1} indicates that this is the state of output Q after the clock pulse. Q_n and $\bar{Q}_n$ indicate that these are the states of Q and $\bar{Q}$, respectively, during the clock pulse.

Two JK flip-flops may be packed as a single unit. For example, the MC790P has the block diagram shown in Fig. 23.17. The circuit of each flip-flop is similar to that shown in Fig. 23.15.

To illustrate the design techniques that can be used with these IC units, let us consider an example.

TABLE 23.9
The Truth Table for an
RTL Type JK Flip-
Flop.

C	S	Q_{n+1}
0	0	$\bar{Q}_n$
0	1	1
1	0	0
1	1	Q_n

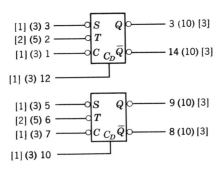

[1] (3) 3 — S Q — 3 (10) [3]
[2] (5) 2 — T
[1] (3) 1 — C C_D $\bar{Q}$ — 14 (10) [3]

[1] (3) 12 —

[1] (3) 5 — S Q — 9 (10) [3]
[2] (5) 6 — T
[1] (3) 7 — C C_D $\bar{Q}$ — 8 (10) [3]

[1] (3) 10 —

FIGURE 23.17
Block diagram of the MC790P/MC890P
dual JK flip-flop.

[2] The reader should be sure to study the truth table for the type of logic he is using. For example, the truth table for a TTL type JK flip-flop indicates that the output changes when both C and S are 1s and there is no change when both signals are 0s.

Example 23.5 Construct a circuit that will count to 10. When the count reaches 10, the circuit should reset itself and produce one output pulse that could be applied to a following identical unit.

Examination of Table 23.1 indicates that four different binary numbers must be used to represent the number 10. Consequently, we shall need four different flip-flops. We can use two MC790P units with two flip-flops in each unit. The counter could be constructed as shown in Fig. 23.18. When the reset button is pushed, a 1 signal is applied to each of the C_D inputs so that each of the flip-flops are set so that Q is 0.

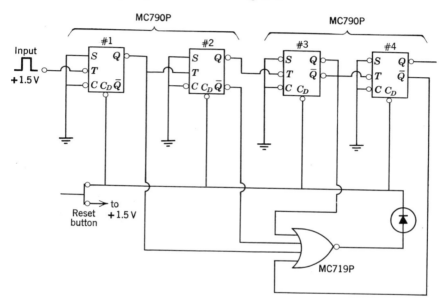

FIGURE 23.18
A simple decade counter.

Since all S and C terminals are connected to ground (0 condition), each circuit will switch whenever its clock pulse shifts from a 1 to a 0 state. For example, number 2 flip-flop switches whenever Q_1 switches from a 1 to a 0 state. Similar switching occurs in the other flip-flops. Hence, the action of the circuit can be tabulated as shown in Table 23.10.

We wish the circuit to reset itself when the tenth pulse is applied. The tenth pulse produces a 1 on Q_4, so $\bar{Q}_4$ will have a 0 for its output. Similarly, $\bar{Q}_2$ will be a 0. In addition Q_3 and Q_1 are both 0. Therefore, if we use $\bar{Q}_4$, Q_3, $\bar{Q}_2$, and Q_1 to activate one of the MC719P gates (Fig. 23.13), the output of this gate will be 1 when (and only when) the count is 10.

TABLE 23.10
Truth Table for the Circuit in Fig. 23.18.

Condition	Q_4	Q_3	Q_2	Q_1
	The Signal on the Q Terminals			
Reset	0	0	0	0
After pulse 1	0	0	0	1
After pulse 2	0	0	1	0
After pulse 3	0	0	1	1
After pulse 4	0	1	0	0
After pulse 5	0	1	0	1
After pulse 6	0	1	1	0
After pulse 7	0	1	1	1
After pulse 8	1	0	0	0
After pulse 9	1	0	0	1
After pulse 10	1	0	1	0

The 1 on the output of this gate can be applied to the C_D terminals to reset all of the Q outputs back to 0. The configuration that accomplishes this switching is shown in Fig. 23.18. The diode is included in this circuit to protect the gate when the reset button is pressed.

Note that Q_4 will switch from 1 to 0 as flip-flop 4 is reset. Therefore, the signal from Q_4 can be coupled directly to the input of the next decade counter, which can be identical to that just considered.

The decade counter in the preceding example is only one of many configurations which can be used and is known as a BCD (binary conversion to decade) counter. Actually, many different types of counters have been developed since almost each digital instrument incorporates one or more counting circuits. For example, a digital timer can be constructed which uses a clock to generate a 1 MHz pulse train. Then, a counter is used to count the number of pulses (in effect the number of μs) between two triggering pulses.

The versatility and usefulness of the digital IC units can be visualized from the foregoing material. Of course, we have just been able to introduce the basic concepts of digital ICs in this section. Nevertheless, you should be able to design some digital circuits with this background. Of course, further reading in this interesting area is strongly recommended.

PROBLEM 23.10 We wish to activate an indicator circuit from the counter in Fig. 23.18. Use buffers (MC799P) if required and NOR gates to design a circuit that will supply a 1 signal at terminal A if and only if the count is 0, a 1 signal at terminal B if and only if the count is 1, a 1 signal at terminal C if the count is 2, etc., for the 10 positions in the counter.

PROBLEM 23.11   A popular type of decade readout is the seven-segment readout shown in Fig. 23.19. By lighting the proper combination of segments, any number from 0 to 9 can be obtained. Determine the circuit which must be used with the BCD counter of Fig. 23.18 to property activate the D segment. *Hint:* It may be easier to write the equation for the conditions when segment D is *not* activated.

Answer: $F = \overline{A}\overline{B}C\overline{D} + A\overline{B}C\overline{D} + AB C\overline{D}$ or

$$F = \overline{B + C + D} + \overline{\overline{A} + \overline{B} + \overline{C} + D}$$

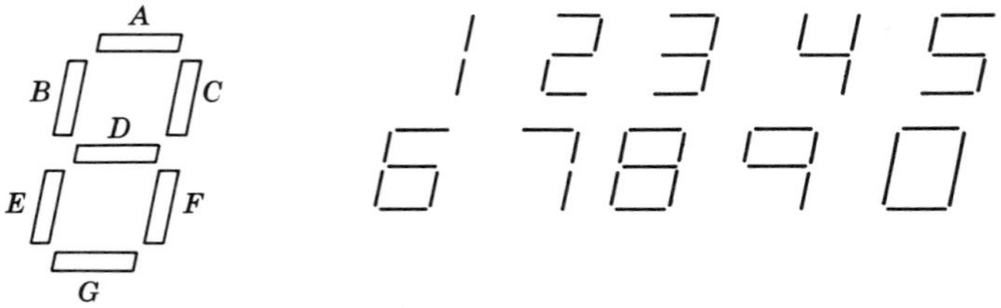

(a)   (b)

FIGURE 23.19
A seven-segment display unit: (a) the display; (b) the ten numbers used.

23.5   FET AND MOSFET DIGITAL CIRCUITS

Pulse or digital circuits can also be constructed from FET or MOSFET units. These devices have the advantages of high input impedance, lower power consumption, circuit simplicity, and extremely small sizes. Unfortunately, as we go to higher impedance circuits, the shunt capacitances in the circuits (as was already noted) have a greater effect on the output waveforms and switching speeds. Consequently, the FET and MOSFET circuits are slower speed circuits than their bipolar counterparts.

A typical MOSFET three-input NOR gate is shown in Fig. 23.20. The transistor T_4 is used as a load resistor and is biased to a conducting state when the output terminal is more positive than the V_{DD} supply. A negative pulse on any one (or more) of the input terminals will cause the associated transistor to conduct, and the output voltage will approach zero. Only if all the input terminals have a near zero signal will the output voltage approach $-V_{DD}$. In this case, the near zero signal is a 0. This arrangement

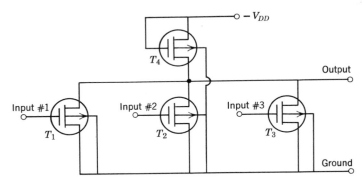

FIGURE 23.20
A typical MOSFET NOR gate.

is known as a *negative logic* system since the more negative signal is the 1 state. The circuit in Fig. 23.20 is, therefore, a NOR configuration for a negative logic system.

By using a MOSFET as a load resistance, two desirable effects are achieved. First, a MOSFET uses less area on the silicon chip than a conventional diffused resistor. Second, the effective resistance of the MOSFET is much higher than the resistance of a typical diffused resistor, as can be readily observed from Fig. 23.21. The curve for operation with $V_{GS} = V_{DS}$ (the condition of T_4 in Fig. 23.20) is shown as a solid line in Fig. 23.21.

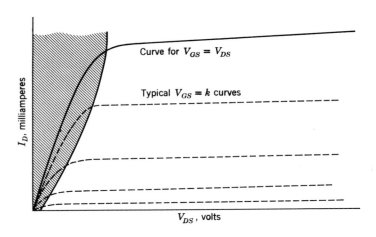

FIGURE 23.21
Operation of a MOSFET as a resistor.

The dynamic drain resistance r_d, as discussed in Chapter 6, is defined as

$$r_d = \frac{\Delta v_{DS}}{\Delta i_S} \tag{23.13}$$

where Δv_{DS} represents a change of drain-source voltage and Δi_D represents a change of drain current. Note in Fig. 23.21 that if operation in the shaded area is not permitted, a large change of voltage (along the $V_{GS} = V_{DS}$ line) produces only a small change of current. In fact, typical resistances with values of about 100 kΩ to 200 kΩ can be realized from a MOSFET. Compare these values with the typical resistors in the integrated bipolar transistor circuits ($R \simeq 1000\ \Omega$ or less).

The integrated MOSFET circuit can be very compact. For example, Robert Crawford[3] indicates the entire circuit of Fig. 23.20 can be constructed on an area 5.9 mills by 2.6 mills (one mill $= 10^{-3}$ in). Consequently, very complex digital circuits can be fabricated in a relatively small area. Thus, MOSFET devices are important contenders in the large-scale integrated (LSI) circuit area. In these LSI circuits, the digital signal processing is performed by very small MOSFETs. Larger units are used to obtain higher power levels for the output MOSFETs. A typical MOS type of LSI is given in Fig. 23.22. Over four thousand MOSFETs can be formed on a chip 1/8 of an inch by 1/8 of an inch!

Of course, MOSFET's can be used to construct multivibrators. For example, the configuration for an RS type flip-flop is given in Fig. 23.23. The similarity to the flip-flop in Fig. 23.11 is obvious. The transistors T_1 and T_2 form the load resistances for the NOR gates, which are composed of T_3 and T_4, or T_5 and T_6. By adding additional transistors, the RS flip-flop configuration can be converted to a JK type flip-flop. One LSI configuration involves a large array of MOS flip-flops arranged to serve as a section of memory for a computer. At the present time, an electronic calculator is on the market and the entire electronic circuitry (except for keyboard and readout) is contained on *three* LSI chips.

PROBLEM 23.12 Is it possible to construct a MOSFET NAND gate as an *IC*? If so, draw the proper configuration.

PROBLEM 23.13 Draw the circuit which uses MOSFET devices and solves the following equation.

$$AB + \overline{A}C = S$$

where A, B, and C are binary numbers. Use as few MOSFETs as possible.

---

[3] Robert H. Crawford, *MOSFET in Circuit Design*, McGraw-Hill Book Company, New York, 1967, p. 104.

FIGURE 23.22
A moderate size MOS-type integrated circuit. (Courtesy of General Instruments, Salt Lake City, Utah.)

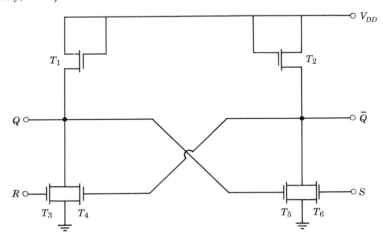

FIGURE 23.23
An *RS* flip-flop using MOSFETs.

23.6 LINEAR OP AMPS IN PULSE CIRCUITS

There are many waveshaping or pulse circuits which use linear operational amplifiers as the amplifying unit. In the interest of space, we will only present two of these applications. However, most manufacturers provide *Application Notes* which list many other configurations.

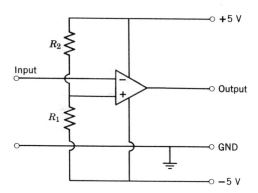

FIGURE 23.24
A voltage comparator circuit.

If an op amp is connected as shown in Fig. 23.24, we have a *voltage comparator*. The voltage level on the noninverting input is controlled by adjusting the resistance values of R_1 and R_2. Thus, if R_1 is 6 kΩ and R_2 is 4 kΩ, the voltage on the noninverting input will be +1 V (if the loading of the *IC* is negligible). If the input signal is also +1 V, the output signal will be 0 V. Now, if the gain of the op amp is 10^4, the output will be driven to saturation ($v_o = 4.7$ V or so) whenever the input signal is 0.9995 V or less. Similarly, if the input signal is 1.0005 V or more, the output will be saturated at -4.7 V or so. Thus, the output signal switches between +4.7 V or -4.7 V as the input signal switches below or above 1 V. The input switching range (1.0005 − 0.9995 = 0.001 V) decreases as the gain (open-loop) of the op amp increases. Typical input and output signals from a voltage comparator would appear as shown in Fig. 23.25.

Often, a circuit which produces a ramp signal is desirable. In Chapter 16 we found that if a circuit is connected as shown in Fig. 23.26, the output signal is the integral of the input signal. Thus, if the input signal is a square wave, the output signal will be a triangular wave as shown in Fig. 23.26. By modifying the waveform of the input signal as shown in Fig. 23.27, the output signal will be changed to a sawtooth waveform as shown. Of course, there are other ways to initiate the ramp and also to reset the circuit

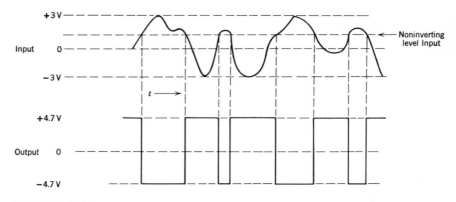

FIGURE 23.25
The typical input and output signals from a voltage comparator.

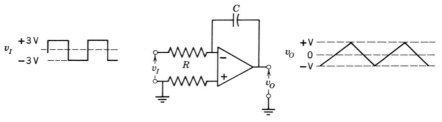

FIGURE 23.26
The integrator as a triangular wave generator.

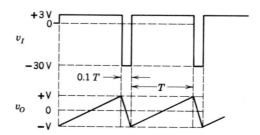

FIGURE 23.27
An input signal for an integrator which produces
a sawtooth output waveform.

at the end of the ramp, but an integrator configuration is usually incorporated in the various sawtooth signal generators.

PROBLEM 23.14 Determine the value of R and C in Fig. 23.26 which will make $+V$ of the output signal equal to 6 V and $-V$ equal to -6 V. The input signal is 1000 cycles per second.

23.7   DIGITAL INSTRUMENTATION

In developing digital instrumentation, originality and innovation are very important. However, as mentioned earlier, most digital instruments contain digital counters of some sort. Thus, a digital frequency meter can be constructed by simply converting the input signal to a square wave of one pulse per cycle (perhaps by using a signal level detector) output. Then simply count the number of pulses in one second. Of course, the number in the counter must be used to activate some type of digital display unit.

Often, a signal can be modified to be presented as a unit of time. Then by counting the number of pulses from an accurate oscillator, the unit of time can be converted to a digital signal. For example, a digital dc voltmeter could be constructed from the block diagram shown in Fig. 23.28. In this

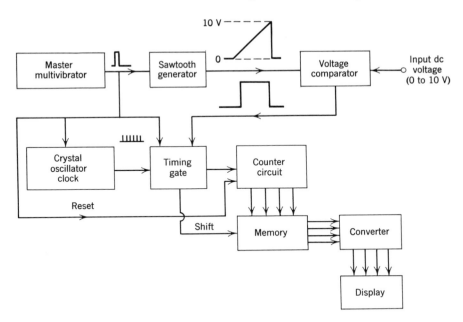

FIGURE 23.28
A block diagram for a digital voltmeter.

voltmeter, a master multivibrator (astable) is used to synchronize the meter operation. This multivibrator starts the crystal oscillator clock (an optional function), starts the sawtooth generator, and opens the timing gate so the oscillator pulses are admitted to the counter. When the output of the sawtooth generator is equal to the dc input voltage, the output of the voltage comparator changes polarity. This change of polarity is applied to the timing gate and "closes" the timing gate so no additional clock pulses are permitted through to the counter. When the timing gate closes, a signal is applied to the memory circuit and the digital information in the counter circuit is shifted into the corresponding *JK* multivibrators in the memory unit. The memory holds this information and displays it through the display units while the entire foregoing procedures are again repeated. Thus, the information in the digital readout is displayed while the counter is counting for the next value of input voltage. The information in the readout is updated once for each cycle of the master multivibrator but remains in display until newer information is available.

If the voltage signal from the sawtooth generator is linear, the pulse width from the voltage comparator will be proportional to the amplitude of the input dc voltage. Thus, the input voltage amplitude is converted to a pulse width. When the sawtooth generator output reaches 10 V, the master multivibrator generates another pulse and the procedure is repeated.

The digital unit shown in Fig. 23.28 is not very complicated when constructed with *IC* units. However, even this relatively simple circuit would be very complicated and expensive if constructed from discrete components. Therefore, we find that *IC*s permit us to construct circuits which are much more complicated than those we would build from discrete components!

PROBLEM 23.15   Draw a circuit for the timing gate in Fig. 23.28. Notice that the signal from the master multivibrator is not necessary for the proper operation of this gate.

PROBLEM 23.16   Draw the memory circuit for one decade of the memory unit in Fig. 23.28. The information from the counter (Fig. 23.18) should be transferred into the *JK* flip-flops of this memory only when a "shift" pulse is applied.

PROBLEM 23.17   The master multivibrator of Fig. 23.28 produces 100 pulses/sec, and the input voltage varies from 0 to 10 V. If we wish to read this voltage to 0.01 V accuracy, what is the required frequency of the clock? How many decades must be present in the counter, memory, and the display unit?

PROBLEM 23.18   Design a digital clock which will use the 60 Hz ac line voltage as an input and will read hours, minutes, and seconds in a digital output display.

Appendix I

Transistor and Tube Characteristics

MAXIMUM RATINGS

Collector-to-base voltage	-75 max	V
Collector-to-emitter voltage	-50 max	V
Emitter-to-base voltage	-1.5 max	V
Collector current	-5 max	A
Base current	-1 max	A
Emitter current	5 max	A

Transistor dissipation:

At mounting-flange temperatures up to 81°C	12.5 max	W
At mounting-flange temperatures above 81°C	Derate 0.66 W/°C	

Temperature range:

Operating (junction) and storage	-65 to 100	°C
Lead temperature (for 10 seconds maximum)	255 max	°C

CHARACTERISTICS

Collector-to-base breakdown voltage (with collector mA = -10 and emitter current = 0)	-75 min	V
Collector-to-emitter breakdown voltage (with collector mA = -100 and base current = 0)	-50 min	V
Base-to-emitter voltage (with collector-to-emitter V = -10 and collector mA = -50)	-0.24	V
Collector-cutoff current (with collector-to-base V = -40 and emitter current = 0)	-1 max	mA
Collector-cutoff saturation current (with collector-to-base V = -0.5 and emitter current = 0)	-70 max	μA
Emitter-cutoff current (with emitter-to-base V = -1.5 and collector current = 0)	-2.5 max	mA

Thermal resistance:

Junction-to-case	1.5 max °C/W	

In Common-Emitter Circuit

dc forward current-transfer ratio (with collector-to-emitter V = -1 and collector mA = -1000)	150	
Gain-bandwidth product (with collector-to-emitter V = -5 and collector mA = -500)	4	MHz

(Continued)

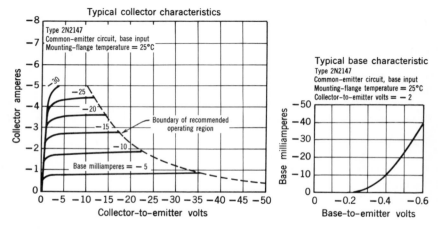

FIGURE I.1
Characteristics of the 2N2147 transistor. (Courtesy of Radio Corporation of America.)

Parameter	Test Conditions	2N3704 Min	2N3704 Max	2N3705 Min	2N3705 Max	2N3706 Min	2N3706 Max	Unit
$V_{(BR)CBO}$ collector-base breakdown voltage	$I_C = 100\ \mu A$, $I_E = 0$	50		50		40		V
$V_{(BR)CEO}$ collector-emitter breakdown voltage	$I_C = 10$ mA, $I_B = 0$ (see note 1)	30		30		20		V
$V_{(BR)EBO}$ emitter-base breakdown voltage	$I_E = 100\ \mu A$, $I_C = 0$	5		5		5		V
I_{CBO} collector cutoff current	$V_{CB} = 20$ V, $I_E = 0$		100		100		100	nA
I_{EBO} emitter cutoff current	$V_{EB} = 3$ V, $I_C = 0$		100		100		100	nA
h_{FE} static forward current transfer ratio	$V_{CE} = 2$ V, $I_C = 50$ mA (see note 1)	100	300	50	150	30	600	
V_{BE} base-emitter voltage	$V_{CE} = 2$ V, $I_C = 100$ mA (see note 1)	0.5	1.0	0.5	1.0	0.5	1.0	V
$V_{CE\,sat}$ collector-emitter saturation voltage	$I_B = 5$ mA, $I_C = 100$ mA (see note 1)		0.6		0.8		1.0	V
f_T transition frequency	$V_{CE} = 2$ V, $I_C = 50$ mA (see note 2)	100		100		100		MHz
C_{obo} common-base open-circuit output capacitance	$V_{CB} = 10$ V, $I_E = 0$, $f = 1$ MHz		12		12		12	pF

FIGURE 1.2A

Electrical characteristics of the 2N3704-2N3706 n-p-n transistors at 25°C free-air temperature. (Notes : (1) These parameters must be measured using pulse techniques; $PW = 300\ \mu sec$, duty cycle $\leq 2\%$. (2) To obtain f_T, the $|h_{fe}|$ response with frequency is extrapolated at the rate of -6 dB per octave from $f = 20$ Mc to the frequency at which $|h_{fe}| = 1$.)

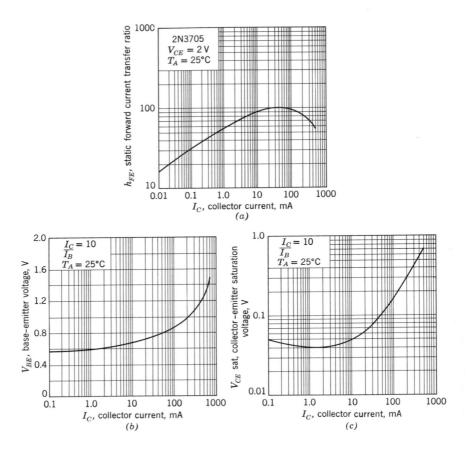

FIGURE I.2B

Characteristics of the 2N3704, 2N3705, and 2N3706 transistors. The 2N3703 is similar to the 2N3705 and the 2N3702 is similar to the 2N3706, except they are p-n-p: (a) static forward current transfer ratio versus collector current; (b) base-emitter voltage versus collector current; (c) collector-emitter saturation voltage versus collector current. (Courtesy of Texas Instruments Inc., Dallas, Texas.)

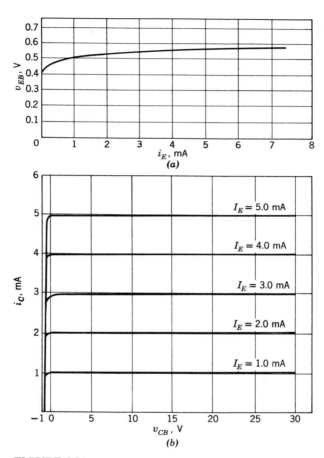

FIGURE I.3A
Common-base characteristics of the 2N3903 transistor: (a)
input characteristics; (b) collector characteristics.

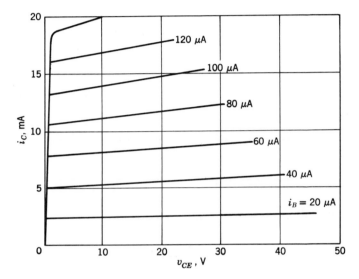

FIGURE I.3B
Collector characteristics of the 2N3903 transistor. Maximum ratings:
$T = 25°C$, $V_{CEO} = 40$ V, $I_{CO} = 0.05$ μA, $I_C = 100$ mA,
$P_{\text{diss}} = 310$ mW, **derate** 2.81 mW/°C. **Small signal characteristics:**
$I_C = 5$ mA, $v_{CE} = 10$ V, $f = 1$ kHz, $f_t = 250$ MHz, $C_{ob} = 2$ pF,
$h_{fe} = 150$, $h_{ie} = 850$ Ω, $h_{oe} = 30$ μmhos, $h_{re} = 2 \times 10^{-4}$.

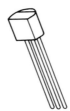

N-P-N **silicon annular transistor (MPS 6541) designed for VHF oscillator applications in television and FM receivers**

Maximum Ratings

Rating		Symbol	Value	Unit
Collector-emitter voltage		V_{CEO}	20	Vdc
Collector-emitter voltage		V_{CES}	30	Vdc
Collector-base voltage		V_{CB}	30	Vdc
Emitter-base voltage		V_{EB}	4	Vdc
Collector current		I_C	100	mAdc
Total device dissipation	$T_A = 60°C$	P_D	210	mW
	$T_A = 25°C$		310	
Junction temperature		T_J	135	°C

Thermal Characteristics

Characteristic	Symbol	Max	Unit
Thermal resistance, junction to ambient	θ_{JA}	0.357	°C/mW

(*Continued*)

Characteristic	Test Conditions	Symbol	Min	Typ	Max	Unit
Collector-emitter breakdown voltage	$I_C = 0.5$ mAde, $I_B = 0$	BV_{CEO}	20	—	—	Vdc
Collector-emitter breakdown voltage	$I_C = 10$ μAdc, $V_{BE} = 0$	BV_{CES}	30	—	—	Vdc
Collector-base breakdown voltage	$I_C = 10$ μAdc, $I_E = 0$	BV_{CBO}	30	—	—	Vdc
Collector cutoff current	$V_{CB} = 15$ Vdc, $I_E = 0$ $V_{CB} = 15$ Vdc, $I_E = 0, T_A = 60°C$	I_{CBO}	— —	— —	0.05 1.0	μAdc
Emitter cutoff current	$V_{EB} = 4$ Vdc, $I_C = 0$	I_{EBO}	—	—	1.0	μAdc
dc current gain	$V_{CE} = 10$ Vdc, $I_C = 4$ mAdc	h_{FE}	25	—	—	—
Small Signal current gain	$V_{CE} = 10$ Vdc, $I_C = 4$ mAdc, $f = 100$ MHz	h_{fe}	6	9	15	—
Output capacitance	$V_{CB} = 15$ Vdc, $I_E = 0, f = 100$ kHz	C_{ob}	—	1.3	1.7	pF
Oscillator power output	$V_{CC} = 12$ Vdc, $I_C = 4$ mAdc, $f = 257$ MHz Test circuit Fig. I.4C(a)	P_{out}	10	13	—	mW

FIGURE I.4A
Characteristics of the MPS 6541 n-p-n transistor. ($T_A = 25°C$ unless otherwise noted.) (Courtesy of Motorola Inc.)

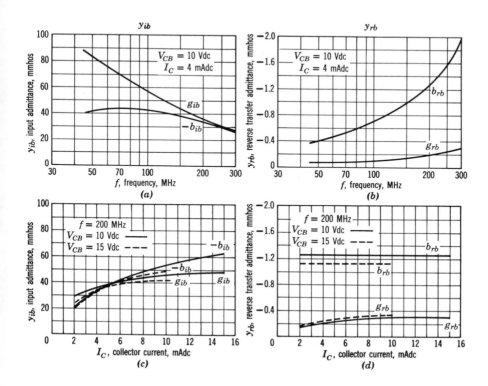

FIGURE I.4B

Characteristics of the MPS 6541 *n-p-n* transistor : (*a*) input admittance versus frequency; (*b*) reverse transfer admittance versus frequency; (*c*) input admittance versus collector current; (*d*) reverse transfer admittance versus collector current. ($T_A = 25°C$ unless otherwise noted.)

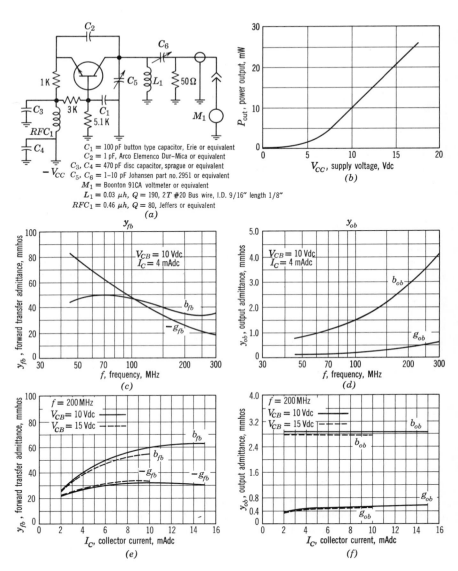

FIGURE I.4C

Characteristics of the MPS 6541 n-p-n transistor: (a) 257 MHz oscillator; (b) typical output power versus supply voltage; (c) forward transfer admittance versus frequency; (d) output admittance versus frequency; (e) forward transfer admittance versus collector current; (f) output admittance versus collector current. ($T_A = 25°C$ unless otherwise noted.)

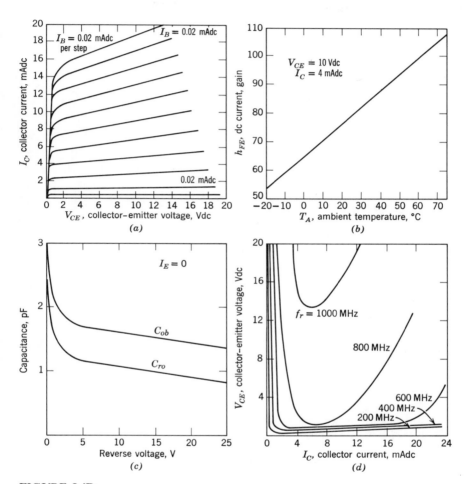

FIGURE I.4D
Characteristics of the MPS 6541 *n-p-n* transistor: (*a*) collector output characteristics; (*b*) dc current gain versus temperature; (*c*) capacitances; (*d*) contours of constant gain-bandwidth product.

For Power-Amplifier and High-Speed-Switching Applications Designed for complementary use with TIP36, TIP36A, TIP36B, TIP36C

- 125 W at 25°C case temperature
- 25 A rated collector current
- Min f_T of 3 MHz at 10 V, 1 A

mechanical data

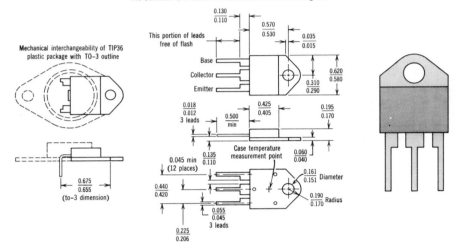

The Collector is in Electrical Contact with the Mounting Tab

Mechanical interchangeability of TIP36 plastic package with TO-3 outline

All dimensions are in inches

absolute maximum ratings

	TIP35	TIP35A	TIP35B	TIP35C
Collector-Base voltage	40 V	60 V	80 V	100 V
Collector-emitter voltage (see note 1) . .	40 V	60 V	80 V	100 V
Emitter-base voltage	←	5 V		→
Continuous collector current	←	25 A		→
Peak collector current (see note 2) . . .	←	40 A		→
Continuous base current	←	5 A		→
Continuous device dissipation at (or below) 25°C case temperature (see note 3)	←	125 W		→
Continuous device dissipation at (or below) 25°C free-air temperature (see note 4)	←	3.5 W		→
Unclamped inductive load energy (see note 5)	←	90 mJ		→
Operating collector junction temperature range	←	−65°C to 150°C		→
Storage temperature range	←	−65°C to 150°C		→
Lead temperature 1/8 inch from case for 10 seconds	←	260°C		→

(*Continued*)

796

Electrical Characteristics

Parameter	Test conditions	TIP35		TIP35A		TIP35B		TIP35C		Unit
		Min	Max	Min	Max	Min	Max	Min	Max	
$V_{(BR)CEO}$ Collector-emitter breakdown voltage	$I_C = 30$ mA, $I_B = 0$, (see note 6)	40		60		80		100		V
I_{CEO} Collector cutoff current	$V_{CE} = 30$ V, $I_B = 0$		1		1					mA
	$V_{CE} = 60$ V, $I_B = 0$						1		1	
I_{CES} Collector cutoff current	$V_{CE} = 40$ V, $V_{BE} = 0$		0.7							mA
	$V_{CE} = 60$ V, $V_{BE} = 0$				0.7					
	$V_{CE} = 80$ V, $V_{BE} = 0$						0.7			
	$V_{CE} = 100$ V, $V_{BE} = 0$								0.7	
I_{EBO} Emitter cutoff current	$V_{EB} = 5$ V, $I_C = 0$		1		1		1		1	mA
h_{FE} Static forward current transfer ratio	$V_{CE} = 4$ V, $I_C = 1.5$ A, (see notes 6 and 7)	25		25		25		25		
	$V_{CE} = 4$ V, $I_C = 15$ A, (see notes 6 and 7)	10	50	10	50	10	50	10	50	
V_{BE} Base-emitter voltage	$V_{CE} = 4$ V, $I_C = 15$ A, (see notes 6 and 7)		2		2		2		2	V
	$V_{CE} = 4$ V, $I_C = 25$ A, (see notes 6 and 7)		4		4		4		4	
V_{CEsat} Collector-emitter saturation voltage	$I_B = 1.5$ A, $I_C = 15$ A, (see notes 6 and 7)		1.8		1.8		1.8		1.8	V
	$I_B = 5$ A, $I_C = 25$ A, (see notes 6 and 7)		4		4		4		4	

(Continued)

Electrical Characteristics

Parameter		Test conditions	TIP35		TIP35A		TIP35B		TIP35C		Unit
			Min	Max	Min	Max	Min	Max	Min	Max	
h_{fe}	Small-signal common-emitter forward current transfer ratio	$V_{CE} = 10$ V, $\quad I_C = 1$ A, $f = 1$ kHz	25		25		25		25		
$|h_{fe}|$	Small-signal common-emitter forward current transfer ratio	$V_{CE} = 10$ V, $\quad I_C = 1$ A, $f = 1$ MHz	3		3		3		3		

Thermal Characteristics

	Parameter	Max	Unit
$R_{\theta JC}$	Junction-to-case thermal resistance	1	°C/W
$R_{\theta JA}$	Junction-to-free-air thermal resistance	35.7	

Switching Characteristics at 25°C case temperature

	Parameter	Test condition (see note 8)	Typ	Unit
t_{on}	Turn-on time	$I_C = 15$ A, $\quad I_{B(1)} = 1.5$ A, $\quad I_{B(2)} = -1.5$ A, $V_{BE(off)} = -4.15$ V, $\quad R_L = 2\,\Omega$,	1.2	μsec
t_{off}	Turn-off time		0.9	

FIGURE 15A

Characteristics (at 25°C case temperature) of the TIP35–TIP35C n-p-n power transistor. (Notes: (1) this value applies when the base-emitter diode is open-circuited; (2) this value applies for $t_w \leq 0.3$ msec, duty cycle $\leq 10\%$; (3) derate linearly to 150°C case temperature at the rate of 1 W/°C; (4) derate linearly to 150°C free-air temperature at the rate of 28 mW/°C; (5) this rating is based on the capability of the transistor to operate safely with inductive load energy $\approx I_A{}^2 L/2$ where $L = 20$ mH; (6) these parameters must be measured using pulse techniques—$t_w = 300$ μsec, duty cycle $\leq 2\%$; (7) these parameters are measured with voltage-sensing contacts separate from the current-carrying contacts; (8) voltage and current values shown are nominal—exact values vary slightly

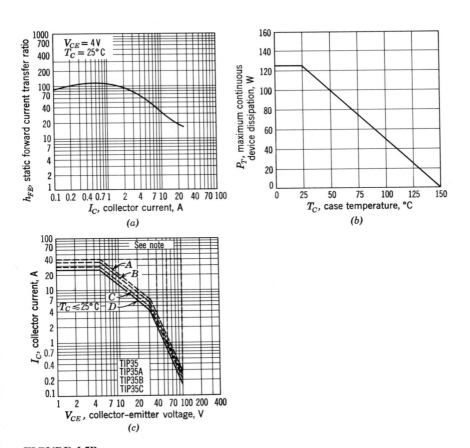

FIGURE I.5B
Characteristics of the TIP35–TIP35C *n-p-n* power transistor. The TIP36–TIP36C has essentially the same characteristics, but is a *p-n-p* type: (*a*) static forward current transfer ratio versus collector current, V_{CE} and T_C must be measured using pulse techniques ($t_w = 300$ μsec, duty cycle $\leq 2\%$) and with voltage-sensing contacts separate from the current-carrying contacts; (*b*) thermal information—dissipation derating curve; (*c*) maximum safe operating region (note: this combination of maximum voltage and current may be achieved only when switching from saturation to cutoff with a clamped inductive load). (Courtesy of Texas Instruments Inc., Dallas, Texas.)

Absolute Maximum Ratings (25°C)

Gate-drain voltage and gate-source voltage . . . . 30 V
Gate current (forward biased) . . . . 50 mA
Total device dissipation at (or below) 25°C free-air temperature (see note) . . . . 300 mW
Storage temperature range . . . . −65 to +200°C

Characteristics	2N2606			2N2607			2N2608			2N2609			Unit
	Min	Typ	Max	Min	Typ	Max	Min	Typ	Max	Min	Typ	Max	
I_{GSS} Gate-source cutoff current at: $V_{GS} = 30$ V, $V_{DS} = 0$			1			3			10			30	nA
I_{GSS} Gate-source cutoff current at: $V_{GS} = 5$ V, $V_{DS} = 0$, $T_A = 150°C$			1			3			10			30	μA
BV_{GDS} Gate-drain breakdown voltage at: $I_G = 1\ \mu$A, $V_{DS} = 0$	30			30			30			30			V
I_{DSS} Drain current at zero gate voltage at: $V_{DS} = -5$ V, $V_{GS} = 0$	−0.10	−0.17	−0.50	−0.30	−0.52	−1.50	−0.90	−1.60	−4.50	−2.00	−3.60	−10.0	mA

Symbol	Characteristic	Conditions												Units
V_p	Gate-source pinch-off voltage	at: $V_{DS} = -5$ V, $I_D = -1$ μA	1	2	4	1	2	4	1	2	4	1	2	4	V
g_{fs}	Small-signal common-source forward transconductance	at: $V_{DS} = -5$ V, $V_{GS} = 0$, $f = 1$ kHz	110	175		330	525		1000	1600		2500	3600		μmho
C_{gss}	Gate-source capacitance	at: $V_{DS} = -5$ V, $V_{GS} = 1$ V, $f = 140$ kHz	3.7	6		7	10		12	17		25	30		pF
NF	Noise figure	at: $V_{DS} = -5$ V, $V_{GS} = 0$, $f_0 = 1$ kHz, $BW = 16\%$, $R_{gen} = 10$ MΩ	0.5	3		0.5	3								dB
		$R_{gen} = 1$ MΩ	0.5	3		0.5	3								dB

FIGURE 16A

Electrical characteristics of the 2N2606–2N2609 p-channel FETs (25°C unless otherwise noted) (note: derate linearly to +175°C free-air temperature at the rate of mW/°C).

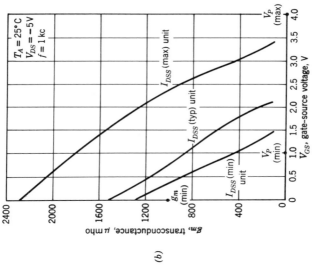

(b)

(a)

For 2N2606, 7, 9 characteristics, multiply I_D and g_{fs} scales and R_S by values shown in Table I.1.

TABLE I.1

Type	Multiply I_D and g_{fs} scales by:	Multiply R_S by:
2N2606	0.11	9.99
2N2607	0.33	3.08
2N2609	2.25	0.44

FIGURE I.6B

Typical characteristics of the 2N2608 p-channel FET: (a) drain characteristics; (b) transconductance versus gate-source voltage. (Courtesy of Siliconix Inc.)

Characteristic		Symbol	Min	Typ	Max	Unit
Off Characteristics						
Gate-source cutoff voltage		$V_{GS(off)}$				Vdc
($V_{DS} = 15$ Vdc, $I_D = 10$ nAdc)	2N5457		−0.5	—	−6.0	
	2N5458		−1.0	—	−7.0	
	2N5459		−2.0	—	−8.0	
Gate-source voltage		V_{GS}				Vdc
($V_{DS} = 15$ Vdc, $I_D = 100$ μAdc)	2N5457		—	−2.5	—	
($V_{DS} = 15$ Vdc, $I_D = 200$ μAdc)	2N5458		—	−3.5	—	
($V_{DS} = 15$ Vdc, $I_D = 400$ μAdc)	2N5459		—	−4.5	—	
On Characteristics						
Zero-gate-voltage drain current		I_{DSS}				mAdc
($V_{DS} = 15$ Vdc, $V_{GS} = 0$)	2N5457		1.0	3.0	5.0	
	2N5458		2.0	6.0	9.0	
	2N5459		4.0	9.0	16	
Dynamic Characteristics						
Forward transfer admittance		$|y_{fs}|$				μmhos
($V_{DS} = 15$ Vdc, $V_{GS} = 0, f = 1$ kHz)	2N5457		1000	3000	5000	
	2N5458		1500	4000	5500	
	2N5459		2000	4500	6000	

FIGURE I.7A

Electrical characteristics of the 2N5457, 2N5458, and 2N5459 n-channel FETs. ($T_A = 25°C$ unless otherwise noted.)

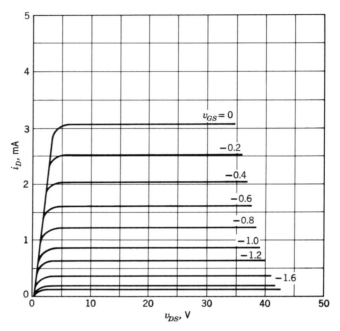

FIGURE I.7B
Drain characteristics of the 2N5457 FET.

Typical Electrical Characteristics ($T_A = 25°C$ unless otherwise noted)

Gate-source breakdown voltage	BV_{GSS}	-25	V
Gate reverse current (max)	I_{GSS}	-1	nA
($T_A = 100°C$)	I_{GSS}	-200	nA
Output conductance ($v_{GS} = 0$, $v_{DS} = 15$ V)	G_{os}	10	μmhos
Input capacitance	C_{iss}	4.5	pF
Reverse transfer capacitance	C_{rss}	1.5	pF
Maximum power dissipation	P_{diss}	310	mW
Derate 2.82 mW/°C above 25°C			

Characteristics	Symbols	Test Conditions	Limits Configuration Dual-gate			Single-gate			Units
			Min	Typ	Max	Min	Typ	Max	
Gate-to-source cutoff voltage:									
Dual-gate no. 1	$V_{G1S(off)}$	$V_{DS} = +15\ V, I_D = 50\ \mu A, V_{G2S} = +4\ V$	–	–2	–	–	–	–	V
Dual-gate no. 2	$V_{G2S(off)}$	$V_{DS} = +15\ V, I_D = 50\ \mu A, V_{G1S} = 0$	–	–2	–	–	–	–	V
Single-gate	$V_{GS(off)}$	$V_{DS} = +15\ V, I_D = 50\ \mu A$	–	–	–	–	–1.6	–	V
Gate-to-source forward breakdown voltage:									
Dual-gate no. 1	$V_{(BR)G1SSF}$	$I_{G1SSF} = I_{G2SSF} = 100\ \mu A$ $V_{G2S} = V_{DS} = 0$	–	9	–	–	–	–	V
Dual-gate no. 2	$V_{(BR)G2SSF}$	$V_{G1S} = V_{DS} = 0$	–	9	–	–	–	–	V
Single-gate	$V_{(BR)GSSF}$	$I_{GSSF} = 100\ \mu A, V_{DS} = 0$	–	–	–	–	9	–	V
Gate-to-source reverse breakdown voltage:									
Dual-gate no. 1	$V_{(BR)G1SSR}$	$I_{G1SSR} = I_{G2SSR} = 100\ \mu A$ $V_{G2S} = V_{DS} = 0$	–	9	–	–	–	–	V
Dual-gate no. 2	$V_{BR)G2SSR}$	$V_{G1S} = V_{DS} = 0$	–	9	–	–	–	–	V
Single-gate	$V_{(BR)GSSR}$	$I_{GSSR} = 100\ \mu A, V_{DS} = 0$	–	–	–	–	9	–	V
Gate terminal forward current:									
Dual-gate no. 1	I_{G1SSF}	$V_{DS} = V_{G2S} = 0, V_{G1S} = 6\ V$	–	–	60	–	–	–	nA
Dual gate no. 2	I_{G2SSF}	$V_{DS} = V_{G1S} = 0, V_{G2S} = 6\ V$	–	–	60	–	–	–	nA
Single-gate	I_{GSSF}	$V_{DS} = 0, V_{GS} = 6\ V$	–	–	–	–	–	120	nA
Gate terminal reverse current:									
Dual-gate no.1	I_{G1SSR}	$V_{DS} = V_{G2S} = 0, V_{G1S} = -6\ V$	–	–	60	–	–	–	nA
Dual-gate no. 2	I_{G2SSR}	$V_{DS} = V_{1S} = 0, V_{G2S} = -6\ V$	–	–	60	–	–	–	nA
Single-gate	I_{GSSR}	$V_{DS} = 0, V_{GS} = -6\ V$	–	–	–	–	–	120	nA
Zero bias drain current:									
Dual-gate	I_{DS}	$V_{DS} = +15\ V, V_{G1S} = 0, V_{G2S} = +4\ V$	–	10	–	–	–	–	mA
Single-gate	I_{DSS}	$V_{DS} = +15\ V, V_{GS} = 0$	–	–	–	–	3.7	–	mA

(*Continued*)

Characteristics	Symbols	Test Conditions	Limits — Configuration — Dual-gate			Limits — Configuration — Single-gate			Units
			Min	Typ	Max	Min	Typ	Max	
Forward transconductance (gate-to-drain)									
Dual-gate	g_{fs}	1 kHz	–	12000	–	–	–	–	μmho
Single-gate	g_{fs}		–	–	–	–	7000	–	μmho
Small-signal, short-circuit input capacitance (see note 1)	C_{iss}		–	6.5	–	–	11	–	pF
Small-signal, short-circuit, reverse transfer Capacitance (drain-to-gate-no. 1) (see note 2)	C_{rss}	$V_{DS} = +15$ V, $I_D = 10$ mA	–	0.20	–	–	0.54	–	pF
Small-signal, short-circuit output capacitance	C_{oss}	[Dual-gate only $V_{G2S} = +4$ V]	–	2	–	–	–	2	pF
Audio spot noise figure (see note 3)									
Dual-gate	NF	$f = 1$ kHz	–	0.46	–	–	–	–	dB
Single-gate	NF	$f = 1$ MHz	–	–	–	–	0.29	–	dB
Power gain	G_{ps}	44 MHz	–	32	–	–	–	–	dB
Conversion gain	$G_{ps(C)}$		–	24	–	–	–	–	dB

FIGURE 1.8A

Electrical characteristics of the dual-gate MOSFET-type 40841 at $T = 25°C$. (Notes: (1) capacitance between gate no. 1 and all other terminals (dual-gate); gate and all other terminals (single-gate); (2) three-terminal measurement with gate no. 2 and source returned to guard terminal (dual-gate); (3) noise figure $= 10 \log_{10}[1 + (v_n^2/4kTBWR_g)]$ where $k = 1.38 \times 10^{-23}$, $T =$ temperature in °K, $BW =$ bandwidth in Hz, $R_g =$ generator resistance.) (Courtesy of Radio Corporation of America.)

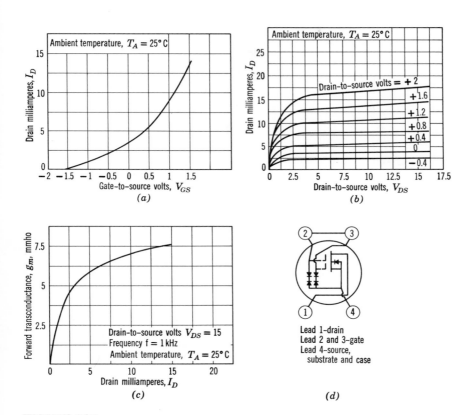

FIGURE I.8B
Characteristics of the *n*-channel MOSFET-type 40841 in the single-gate configuration:
(a) I_D versus v_{GS}; (b) I_D versus v_{DS}; (c) g_m versus I_D; (d) single-gate configuration.

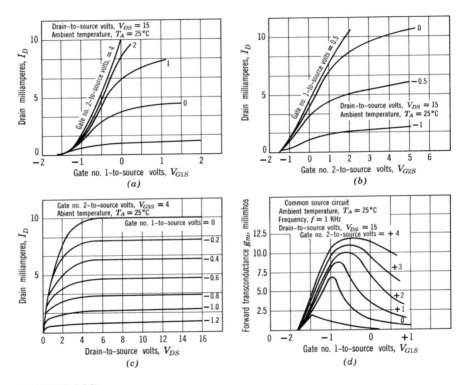

FIGURE I.8C
Characteristics of the *n*-channel dual-gate MOSFET-type 40841: (*a*) I_D versus v_{G1S};
(*b*) I_D versus v_{G2S}; (*c*) I_D versus v_{DS}; (*d*) g_m versus v_{G1S}. (**Courtesy of Radio Corporation of America.**)

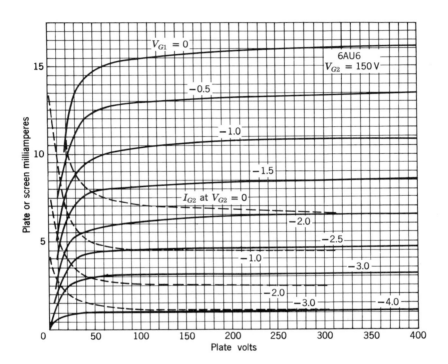

FIGURE I.9
Characteristics of the 6AU6 tube. (Courtesy of Radio Corporation of America.)

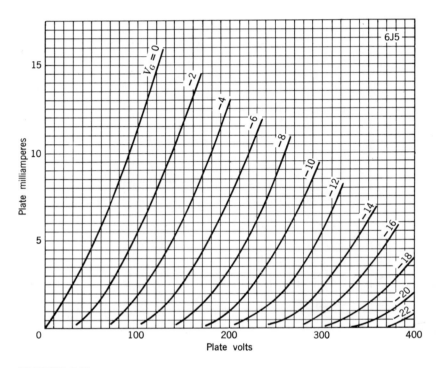

FIGURE I.10
Characteristics of the 6J5 tube. (Courtesy of Radio Corporation of America.)

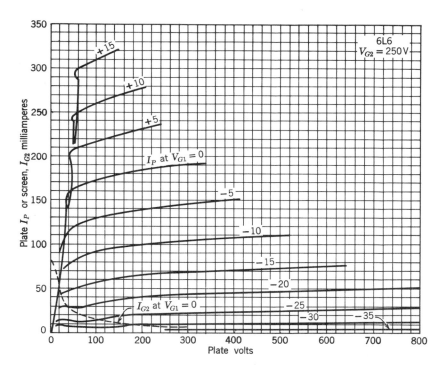

FIGURE I.11A
Characteristics of the 6L6 beam power tube. (Courtesy of Radio Corporation of America.)

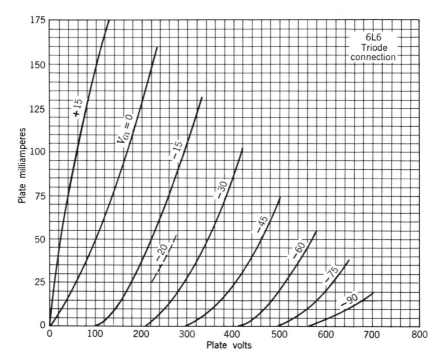

FIGURE I.11B
Characteristics of the 6L6 triode connected. (Courtesy of Radio Corporation of America.)

MONOLITHIC OPERATIONAL AMPLIFIER

This was designed for use as a summing amplifier, integrator, or amplifier with operating characteristics as a function of the external feedback components.

- Low-input offset voltage—3.0 mV max
- Low-input offset current—60 nA max
- Large power-bandwidth—20 Vp-p output swing at 20 kHz min
- Output short-circuit protection
- Input over-voltage protection
- Class AB output for excellent linearity
- Slew rate—34 V/μsec typ

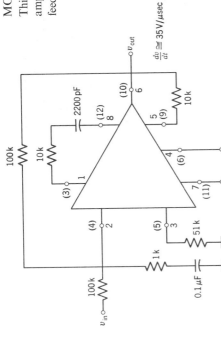

(a) High slew rate inverter-pin numbers adjacent to terminals apply to 8-pin package, numbers in parentheses apply to 14-pin packages

$$\frac{dv}{dt} \cong 35 \text{V}/\mu\text{sec}$$

(b) Output-nulling circuit

$10 \text{ k} \leqslant R_5 \leqslant 100$
$(R_3) (V^+)$

(c) Output-limiting circuit

FIGURE I.12A
Characteristics of the MC1539–MC1439 operational amplifier. (Courtesy of Motorola Corp.)

Characteristic	Symbol	MC 1539			MC 1439			Unit
		Min	Typ	Max	Min	Typ	Max	
Input bias current	I_b							μA
($T_A = +25°C$)		—	0.20	0.50	—	0.20	1.0	
($T_A = T_{low}$, see note 1)		—	0.23	0.70	—	0.23	1.5	
Input offset current	$|I_{io}|$							nA
($T_A = T_{low}$)		—	—	75	—	—	150	
($T_A = +25°C$)		—	20	60	—	20	100	
($T_A = T_{high}$, see note 1)		—	—	75	—	—	150	
Input offset voltage	$|V_{io}|$							mV
($T_A = +25°C$)		—	1.0	3.0	—	2.0	7.5	
($T_A = T_{low}, T_{high}$)		—	—	4.0	—	—	—	
Average temperature coefficient of input offset voltage ($T_A = T_{low}$ to T_{high})	$|TC_{Vio}|$							μV/°C
($R_S = 50\ \Omega$)		—	3.0	—	—	3.0	—	
($R_S \leq 10\ k\Omega$)		—	5.0	—	—	5.0	—	
Input impedance ($f = 20$ Hz)	Z_{in}	150	300	—	100	300	—	kΩ
Input common-mode voltage swing	CMV_{in}	±11	±12	—	±11	±12	—	V_{pk}
Equivalent input noise voltage ($R_S = 10$ kΩ, noise bandwidth = 1.0 Hz, $f = 1.0$ kHz)	v_n	—	30	—	—	30	—	nV(Hz)$^{1/2}$

Parameter (Test Conditions)	Symbol							Units
Common-mode rejection ratio ($f = 1.0$ kHz)	CM_{rej}	80	110	—	80	110	—	dB
Open-loop voltage gain ($V_o = \pm 10$ V, $R_L = 10$ kΩ, $R_5 = \infty$) ($T_A = +25°C$ to T_{high})	A_{VOL}	50,000	120,000	—	15,000	100,000	—	—
($T_A = T_{low}$)		25,000	100,000	—	15,000	100,000	—	—
Power bandwidth ($A_v = 1$, $THD \leq 5\%$, $V_o = 20$ Vp-p) ($R_L = 2.0$ kΩ)	p_{BW}	20	50	—	10	50	—	kHz
($R_L = 1.0$ kΩ)		—	50	—	—	—	—	kHz
Step response								
Gain = 1000, no overshoot, {$R_1 = 1.0$ kΩ, $R_2 = 1.0$ MΩ, $R_3 = 1.0$ kΩ, $R_4 = 30$ kΩ, $R_5 = 10$ kΩ, $C_1 = 1000$ pF}	t_f	—	130	—	—	130	—	nsec
	t_{pd}	—	190	—	—	190	—	nsec
	dV_{out}/dt	—	6.0	—	—	6.0	—	V/μsec (see note 2)
Gain = 1000, 15% overshoot, {$R_1 = 1.0$ kΩ, $R_2 = 1.0$ MΩ, $R_3 = 1.0$ kΩ, $R_4 = 0$, $R_5 = 10$ kΩ, $C_1 = 10$ pF}	t_f	—	80	—	—	80	—	nsec
	t_{pd}	—	100	—	—	100	—	nsec
	dV_{out}/dt	—	14	—	—	14	—	V/μsec
Gain = 100, no overshoot, {$R_1 = 1.0$ kΩ, $R_2 = 100$ kΩ, $R_3 = 1.0$ kΩ, $R_4 = 10$ kΩ, $R_5 = 10$ kΩ, $C_1 = 2200$ pF}	t_f	—	60	—	—	60	—	nsec
	t_{pd}	—	100	—	—	100	—	nsec
	dV_{out}/dt	—	34	—	—	34	—	V/μsec
Gain = 10, 15% overshoot, {$R_1 = 1.0$ kΩ, $R_2 = 10$ kΩ, $R_3 = 1.0$ kΩ, $R_4 = 1.0$ kΩ, $R_5 = 10$ kΩ, $C_1 = 2200$ pF}	t_f	—	120	—	—	120	—	nsec
	t_{pd}	—	80	—	—	80	—	nsec
	dV_{out}/dt	—	6.25	—	—	6.25	—	V/μsec
Gain = 1, 15% overshoot, {$R_1 = 10$ kΩ, $R_2 = 10$ kΩ, $R_3 = 5.0$ kΩ, $R_4 = 390$ Ω, $R_5 = 10$ kΩ, $C_1 = 2200$ pF}	t_f	—	160	—	—	160	—	nsec
	t_{pd}	—	80	—	—	80	—	nsec
	dV_{out}/dt	—	4.2	—	—	4.2	—	V/μsec

(Continued)

Characteristic	Symbol	MC 1539			MC 1439			Unit
		Min	Typ	Max	Min	Typ	Max	
Output impedance (f = 20 Hz)	Z_{out}	—	4.0	—	—	4.0	—	kΩ
Output voltage swing	V_{out}							V_{pk}
(R_L = 2.0 kΩ, f = 1.0 kHz)		±10	—	—	±10	±13	—	
(R_L = 1.0 kΩ, f = 1.0 kHz)		±13	±13	—	—	—	—	
Positive supply sensitivity (V^- constant)	S^+	—	50	150	—	50	200	μV/V
Negative supply sensitivity (V^+ constant)	S^-	—	50	150	—	50	200	μV/V
Power supply current (V_o = 0)	I_{D+}	—	3.0	5.0	—	3.0	6.7	mAdc
	I_{D-}	—	3.0	5.0	—	3.0	6.7	

FIGURE 1.12B
Electrical characteristics of the MC1539–MC1439 operational amplifier. (V^+ = +15 Vdc, V^- = −15 Vdc, T_A = +25°C unless otherwise noted.) (Notes: (1) T_{low} = 0°C for MC1439, −55°C for MC1539, T_{high} = +75°C for MC1539, +125°C for MC1439, +25°C for MC1539; (2) dV_{out}/dt = slew rate.) (Courtesy of Motorola Corp.)

Maximum Ratings

Rating	Symbol	Value	Unit
Power supply voltage	V^+	$+18$	Vdc
	V^-	-18	Vdc
Differential input signal	V_{in}	$\pm[V^+ + |V^-|]$	Vdc
Common-mode input swing	CMV_{in}	$+V^+, -|V^-|$	Vdc
Load current	I_L	15	mA
Output short-circuit duration	t_S	Continuous	
Power dissipation (package limitation)	P_D		
Metal can		680	mW
Derate above $T_A = +25°C$		4.6	mW/°C
Ceramic dual in-line package		750	mW
Derate above $T_A = +25°C$		6.0	mW/°C
Plastic dual in-line package		625	mW
Derate above $T_A = +25°C$		5.0	mW/°C
Operating temperature range			°C
MC1539	T_A	-55 to $+125$	
MC1439		0 to $+75$	
Storage temperature range	T_{stg}		°C
Metal and ceramic packages		-65 to $+150$	
Plastic package		-55 to $+125$	

(*Continued*)

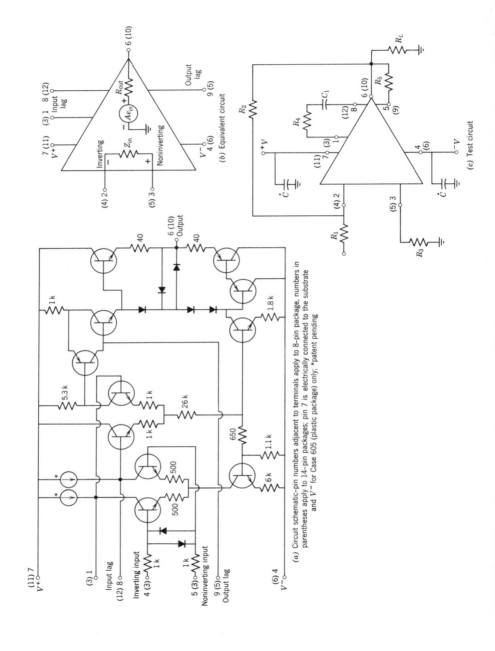

(a) Circuit schematic–pin numbers adjacent to terminals apply to 8-pin package, numbers in parentheses apply to 14-pin packages; pin 7 is electrically connected to the substrate and V^- for Case 605 (plastic package) only; *patent pending

(b) Equivalent circuit

(c) Test circuit

| Figure | Curve | Voltage | Test Conditions (Fig. 6) | | | | | |
No.	No.	Gain	$R_1 (\Omega)$	$R_2 (\Omega)$	$R_3 (\Omega)$	$R_4 (\Omega)$	$R_5 (\Omega)$	$C_1 (pF)$
	1	A_{vol}	0	∞	0	∞	∞	0
	2	1	10 k	10 k	5.0 k	390	10 k	2200
7, 8, 10, 12	3	10	1.0 k	10 k	1.0 k	1.0 k	10 k	2200
	4	100	1.0 k	100 k	1.0 k	10 k	10 k	2200
	5	1000	1.0 k	1.0 M	1.0 k	30 k	10 k	1000
	6	1000	1.0 k	1.0 M	1.0 k	0	10 k	10
13	All	1	10 k	10 k	5.0 k	390	10 k	2200
14	All	10	1.0 k	10 k	1.0 k	1.0 k	10 k	2200
15	All	100	1.0 k	100 k	1.0 k	10 k	10 k	2200
16	All	1000	1.0 k	1.0 M	1.0 k	30 k	10 k	2200

FIGURE I.12C
Characteristics of the MC1539–MC1439 operational amplifier. ($V^+ = +15$ Vdc, $V^- = -15$ Vdc, $T_A = +25°C$ unless otherwise noted.) (Courtesy of Motorola Corp.)

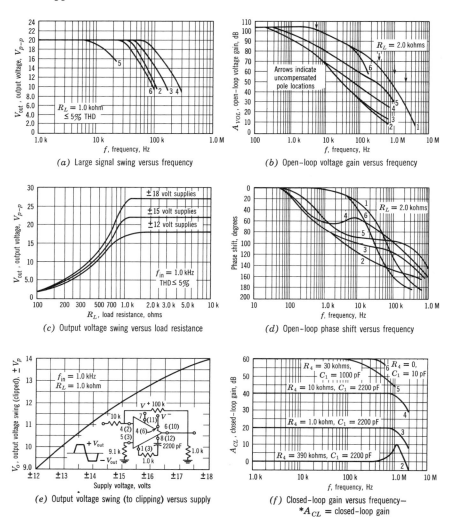

(a) Large signal swing versus frequency

(b) Open-loop voltage gain versus frequency

(c) Output voltage swing versus load resistance

(d) Open-loop phase shift versus frequency

(e) Output voltage swing (to clipping) versus supply

(f) Closed-loop gain versus frequency—
*A_{CL} = closed-loop gain

FIGURE I.12D

Characteristics of the MC1539-MC1439 operational amplifier—pin numbers adjacent to terminals apply to 8-pin package, numbers in parentheses apply to 14-pin packages. ($V^+ = +15$ Vdc, $V^- = -15$ Vdc, $T_A = 25°C$ unless otherwise noted.) (Courtesy of Motorola Corp.)

Appendix II

Feedback

Feedback circuits have the general configuration shown in Fig. II.1. As indicated in Section 19.2, the transfer function G_{13} for this type of circuit can be written as

$$G_{13} = \frac{G_{23}}{1 + G_{23}G_{34}} \qquad \text{(II.1a)}$$

The feedback is known as negative when the denominator of the total transfer function G_{13} contains a plus sign as in the case of Eq. II.1a. On

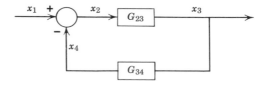

FIGURE II.1
The block diagram of a feedback system.

the other hand, when the feedback is positive, the polarities of the signals x_1 and x_4 of Fig. II.1 are alike and the total transfer function becomes

$$G_{13} = \frac{G_{23}}{1 - G_{23}G_{34}} \qquad \text{(II.1b)}$$

In the general case, the transfer functions G_{23} and G_{34} are complex functions of s. Accordingly, the transfer functions G_{23} and G_{34} can be written as

$$G_{23} = H_1 \frac{P_1(s)}{Q_1(s)} \qquad \text{(II.2)}$$

and

$$G_{34} = H_2 \frac{P_2(s)}{Q_2(s)} \qquad \text{(II.3)}$$

821

where H_1 and H_2 are constants and $P_1(s)$, $P_2(s)$, $Q_1(s)$, and $Q_2(s)$ are polynomials of s. Therefore, Eqs. II.1a and II.1b become

$$G_{13} = \frac{H_1[P_1(s)/Q_1(s)]}{1 \pm H_1 H_2[P_1(s)P_2(s)/Q_1(s)Q_2(s)]} \tag{II.4}$$

Usually the product $H_1 H_2$ is written as K. Simplifying, Eq. II.4 becomes

$$G_{13} = H_1 \frac{P_1(s)/Q_1(s)}{[Q_1(s)Q_2(s) \pm KP_1(s)P_2(s)]/Q_1(s)Q_2(s)} \tag{II.5}$$

or

$$G_{13} = H_1 \frac{P_1(s)Q_2(s)}{Q_1(s)Q_2(s) \pm KP_1(s)P_2(s)} \tag{II.6}$$

To determine the time response of this transfer function to a specific excitation, the denominator of Eq. II.6 must be factored. In other words, the roots of the denominator or poles of the transfer function must be found. The zeros of the complete transfer function (Eq. II.6) are known because the numerator is normally in factored form.[1]

A different set of poles of the complete transfer function exists for each different value of K. For example, inspection of Eq. II.6 shows that if K approaches zero, the roots or poles of G_{13} approach the roots of $Q_1(s)Q_2(s)$ which are the poles of the open-loop transfer function $G_{23}G_{34}$. In contrast, as K becomes very large the roots or poles of G_{13} approach the roots of $P_1(s)P_2(s)$ which are the zeros of $G_{23}G_{24}$. The root-locus plot is, as the name implies, a plot of the loci of all the possible poles of G_{13} as K is varied from 0 to ∞.

To obtain the root-locus plot, Eq. II.4 is rewritten as

$$G_{13} = H \frac{P_1(s)/Q_1(s)}{1 \pm KF(s)} \tag{II.7}$$

where

$$F(s) = \frac{P_1(s)P_2(s)}{Q_1(s)Q(s)} \tag{II.8}$$

---

[1] The reader should be careful to preserve the known zeros by writing the polynomials as

$$P = (s + a)(s_1 + a_2) \cdots$$

and not lose the identity of these known zeros by writing the polynomial as

$$P = s^n + b_1 s^{n-1} + b_2 s^{n-2} + \cdots + b$$

The root-locus is a plot of

$$KF(s) = \mp 1 \tag{II.9a}$$

Therefore,

$$F(s) = \mp 1/K \tag{II.9b}$$

A set of rules has been developed to simplify the plotting of Eq. II.9. Since the proof of these rules is given in the literature,[2] the rules (with no proof) will be given in this appendix.

    1.   A pole-zero plot is made of the function $F(s)$ of Eq. II.8.

    2.   Each root-locus branch departs from a pole of $F(s)$ as K increases from zero and terminates at a zero of $F(s)$ as K approaches infinity.

    3.   The number of individual paths or branches is equal to the number of poles of $F(s)$. (In all actual circuits, the number of finite poles is equal to or greater than the number of finite zeros.)

    4.   Since poles only occur on the real axis or as conjugate pairs, the root-locus plot will be symmetrical with respect to the real axis.

    5.   If the number of finite poles of $F(s)$ is n and the number of finite zeros of $F(s)$ is m, $n - m$ branches will extend to ∞ as K approaches ∞. (In addition, it is possible for a branch to extend through ∞ in going from a finite pole to a finite zero.)

    6.   (a) For a positive feedback, the portions of the real axis to the right of all finite poles and zeros and to the left of an even number of poles and zeros are branches of the root-locus. (b) For negative feedback, the portions of the real axis to the left of an odd number of finite poles and zeros are branches of the root-locus.

The foregoing six rules (and other rules to follow) are illustrated by Fig. II.2.

    7.   As the infinite-seeking branches become far removed from the finite poles and zeros of the function $F(s)$, these infinite-seeking branches approach asymptotes that intersect on the real axis. To facilitate the root-locus plots, these asymptotes must be found; $F(s)$ is of the form

$$F(s) = \frac{(s + z_1)(s + z_2)(s + z_3) \cdots (s + z_m)}{(s + p_1)(s + p_2)(s + p_3) \cdots (s + p_n)} \tag{II.10}$$

The asymptotes will intersect on the real axis at a value A, where

$$A = \frac{\sum\limits_{m=1}^{m=m} z_m - \sum\limits_{n=1}^{n=n} p_n}{n - m} \tag{II.11}$$

[2] See L. Dale Harris, *Introduction to Feedback Systems*, John Wiley & Sons, New York, 1961.

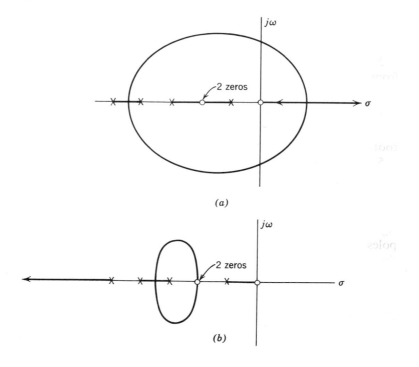

FIGURE II.2
A root-locus plot for $F(s) = s(s + 2)^2/(s + 1)(s + 3)(s + 4)(s + 5)$: (a)
positive feedback; (b) negative feedback.

8. As noted in rule 5, the number of infinite-seeking branches is $n - m$.
The angles of the asymptotes for these branches are found in the following
manner: (a) For positive feedback, the angles of the asymptotes are:

$$\text{1st angle} = 0° \tag{II.12}$$

$$\text{2nd angle} = \frac{360°}{n - m}$$

$$\text{3rd angle} = 2\frac{360°}{n - m}$$

$$\text{kth angle} = (k - 1)\frac{360°}{n - m} \tag{II.13}$$

(*b*) For negative feedback, the angles of the asymptotes are:

$$\text{1st angle} = \frac{360°}{2(n-m)} \tag{II.14}$$

$$\text{2nd angle} = \text{1st angle} + \frac{360°}{n-m}$$

$$\text{3rd angle} = \text{2nd angle} + \frac{360°}{n-m}$$

$$k\text{th angle} = (k-1)\text{th angle} + \frac{360°}{n-m} \tag{II.15}$$

The rules 7 and 8 are illustrated in Fig. II.3, where $n - m = 3$, and in Fig. II.4, where $n - m = 4$.

9. When there are at least *two* more poles than zeros in the function $F(s)$, the sum of the roots (poles of G_{13}) for a given value of K are equal to a constant which is independent of K. This constant is equal to the sum of the poles of $F(s)$. Hence, as one locus moves to the right with increasing K, one or more other branches must move to the left, as shown in Fig. II.5. Also note that as one locus increases in the $+j\omega$ direction, another locus increases in the $-j\omega$ direction (rule 4).

The angle at which a branch approaches or leaves a pole or zero can also be found. These angles are known as departure angles and will be considered next.

10. The departure angle from *single* poles or zeros on the real axis will always be along the real axis. Rule 6 determines the direction of departure.

11. The departure angle from *double* poles or zeros on the real axis will either lie along the real axis or be at right angles to the real axis. Again, rule 6 will determine the direction.

12. When a third-order or triple pole (or zero) lies on the real axis, one root-locus branch will always depart from this triple pole along the real axis. The other two branches will depart at 120° angles from this real axis branch. Rule 6 determines the direction of departure along the real axis. In general, there will be as many departing branches as there are poles or zeros at the given location. In addition, if there are k poles or zeros, the angle between adjacent directions of departure with be $360°/k$. Rule 6 will determine the directions of departure along the real axis. In any case, the directions of departures will be symmetrical with respect to the real axis (rule 4). Rules 10, 11, and 12 are illustrated in Fig. II.6.

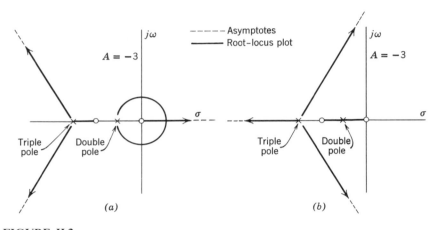

FIGURE II.3
The asymptotes for the infinite-seeking branches of $F(s) + s(s + 2)/(s + 1)^2(s + 3)^3$:
(*a*) positive feedback; (*b*) negative feedback.

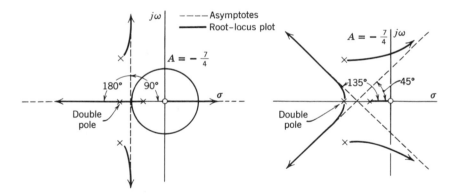

FIGURE II.4
The asymptotes for the infinite seeking branches of $F(s) = s/(s + 1)(s + 2)^2(s + 2 + j2)$
$(s + 2 - j2)$.

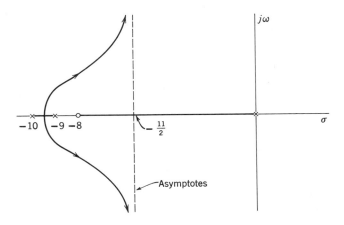

FIGURE II.5
A root-locus plot for negative feedback where $F(s) = (s + 8)/$
$s(s + 9)(s + 10)$.

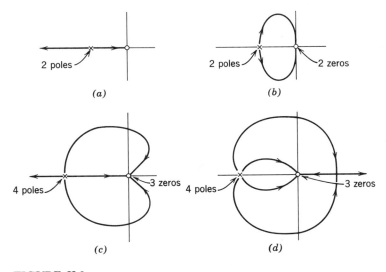

FIGURE II.6
An illustration of departure angles from the real axis: $(a$–$c)$ **negative**
feedback; $(b$–$d)$ **positive feedback.**

13. The departure angle from poles not on the real axis can be found by using Eq. II.10.

$$F(s) = \frac{(s + z_1)(s + z_2)(s + z_3) \cdots (s + z_m)}{(s + p_1)(s + p_2)(s + p_3) \cdots (s + p_n)} \tag{II.10}$$

Now, assume the required departure is from pole p_2. Let s approach $-p_2$, and rewrite Eq. II.10 as

$$F(-p_2) = \frac{(-p_2 + z_1)(-p_2 + z_2)(-p_2 + z_3) \cdots (-p_2 + z_m)}{(-p_2 + p_1)\delta/\psi(-p_2 + p_3) \cdots (-p_2 + p_n)} \tag{II.16}$$

where δ/ψ is the distance and direction from a point on the locus to p_2. Convert each of the terms $(-p_2 + z_m)$ and $(-p_2 + p_n)$, etc., to the polar form. Then Eq. II.16 becomes

$$F(-p_2) = \frac{Z_1{}^{/\phi_1} Z_2{}^{/\phi_2} Z_3{}^{/\phi_3} \cdots Z_m{}^{/\phi_m}}{P_1{}^{/\theta_1} \delta/{}^{\psi} P_3{}^{/\theta_3} \cdots P_n{}^{/\theta_n}} \tag{II.17}$$

For positive feedback,

$$\phi_1 + \phi_2 + \phi_3 + \cdots + \phi_m - \theta_1 - \psi - \theta_3 - \cdots - \theta_n = 0° \tag{II.18}$$

For negative feedback,

$$\phi_1 + \phi_2 + \phi_3 + \cdots + \phi_m - \theta_m - \theta_1 - \psi - \theta_3 - \cdots - \theta_n = 180° \tag{II.19}$$

All of the angles in Eq. II.18 and Eq. II.19 are known except ψ. The angle ψ is the departure angle from p_2. The departure angle from a typical pole is shown in Fig. II.7. The same approach is used to determine the

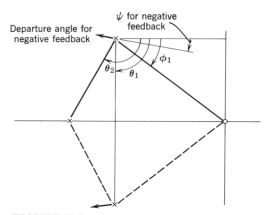

FIGURE II.7
An illustration of the departure angle from a pole not on the real axis.

departure angles from zeros not on the real axis. However, in this case, Eq. II.18 becomes

$$\phi_1 + \psi + \phi_3 + \cdots + \phi_m - \theta_1 - \theta_2 - \theta_3 - \cdots - \theta_n = 0° \quad \text{(II.20)}$$

and Eq. II.19 becomes

$$\phi_1 + \psi + \phi_3 + \cdots + \phi_m - \theta_1 - \theta_2 - \theta_3 - \cdots - \theta_n = 180° \quad \text{(II.21)}$$

14. When two branches cross the real axis or depart from the real axis, the departure angle is always 90°. The point of departure from the real axis can be found by trial and error from the following procedure: (a) Assume a point of departure from the real axis. (b) Transpose any singularities not on the real axis to an equivalent location on the real axis by the following procedure: (1) Draw a straight line (L_1) from the singularity to be transposed to the assumed departure point. (2) Draw a straight line (L_2) through the singularity and normal to the line L_1. (3) The intersection of line L_2 and the real axis is the required equivalent location for the singularity. (c) After transposing all of the singularities to their equivalent location on the real axis, apply the rule

$$\sum \frac{1}{P_{li}} + \sum \frac{1}{Z_{rj}} = \sum \frac{1}{P_{ri}} + \sum \frac{1}{Z_{lj}} \quad \text{(II.22)}$$

where P_{li} is the distance from the ith pole (left of the departure point) to the assumed departure point, P_{ri} is the distance from the ith pole (right of the departure point) to the departure point, Z_{rj} is the distance from the jth zero (right of the departure point) to the assumed departure point, Z_{lj} is the distance from the jth zero (left of the departure point) to the assumed departure point. (d) If the assumed point of departure does not satisfy

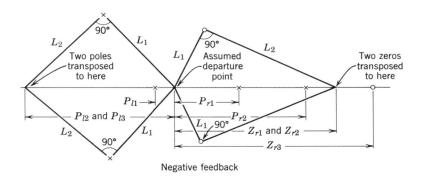

Negative feedback

FIGURE II.8
A method of determining the point of departure from the real axis.

Eq. II.22, a second trial point must be selected and the foregoing steps repeated. Fortunately, good initial trial points can be made by an experienced root-locus plotter.

Fig. II.8 will help clarify the foregoing procedure.

The above list of rules will help the uninitiated reader in simple root-locus plots. A device known as a *spirule* can be used to speed up the plotting process when accurate plots are required. However, in most situations a rough sketch is sufficient to obtain a fairly good idea of how the given feedback circuit will perform. In these cases, the foregoing rules can quickly be applied to obtain the rough sketch.

Bessel Functions

p	$J_p(1)$	$J_p(2)$	$J_p(3)$	$J_p(4)$	$J_p(5)$	$J_p(6)$	$J_p(7)$	$J_p(8)$	$J_p(9)$	$J_p(10)$	$J_p(11)$	$J_p(12)$	$J_p(13)$	$J_p(14)$
0	+.7652	+.2239	−.2601	−.3971	−.1776	+.1506	+.3001	+.1717	−.09033	−.2459	−.1712	+.04769	+.2069	+.1711
0.5	+.6714	+.5130	+.06501	−.3019	−.3422	−.09102	+.1981	+.2791	+.1096	−.1373	−.2406	−.1236	+.09298	+.2112
1.0	+.4401	+.5767	+.3391	−.06604	−.3276	−.2767	$-.0^24683$	+.2346	+.2453	+.04347	−.1768	−.2234	−.07032	+.1334
1.5	+.2403	+.4913	+.4777	+.1853	−.1697	−.3279	−.1991	+.07593	+.2545	+.1980	−.02293	−.2047	−.1937	−.01407
2.0	+.1149	+.3528	+.4861	+.3641	+.04657	−.2429	−.3014	−.1130	+.1448	+.2546	+.1390	−.08493	−.2177	−.1520
2.5	+.04950	+.2239	+.4127	+.4409	+.2404	−.07295	−.2834	−.2506	−.02477	+.1967	+.2343	+.07242	−.1377	−.2143
3.0	+.01956	+.1289	+.3091	+.4302	+.3648	+.1148	−.1676	−.2911	−.1809	+.05838	+.2273	+.1951	$+.0^23320$	−.1768
3.5	$+.0^27186$	+.06852	+.2101	+.3658	+.4100	+.2671	$-.0^23403$	−.2326	−.2683	−.09965	+.1294	+.2348	+.1407	−.06245
4.0	$+.0^22477$	+.03400	+.1320	+.2811	+.3912	+.3576	+.1578	−.1054	−.2655	−.2196	−.01504	+.1825	+.2193	+.07624
4.5	$+.0^3807$	+.01589	+.07760	+.1993	+.3337	+.3846	+.2800	+.04712	−.1839	−.2664	−.1519	+.06457	+.2134	+.1830
5.0	$+.0^32498$	$+.0^27040$	+.04303	+.1321	+.2611	+.3621	+.3479	+.1858	−.05504	−.2341	−.2383	−.07347	+.1316	+.2204
5.5	$+.0^474$	$+.0^22973$	+.02266	+.08261	+.1906	+.3098	+.3634	+.2856	+.08439	−.1401	−.2538	−.1864	$+.0^27055$	+.1801
6.0	$+.0^42094$	$+.0^31202$	+.01139	+.04909	+.1310	+.2458	+.3392	+.3376	+.2043	−.01446	−.2016	−.2437	−.1180	+.08117
6.5	$+.0^56$	$+.0^3467$	$+.0^25493$	+.02787	+.08558	+.1833	+.2911	+.3456	+.2870	+.1123	−.1018	−.2354	−.2075	−.04151
7.0	$+.0^51502$	$+.0^31749$	$+.0^32547$	+.01518	+.05338	+.1296	+.2336	+.3206	+.3275	+.2167	+.01838	−.1703	−.2406	−.1508
7.5	—	—	—	—	—	+.08741	+.1772	+.2759	+.3302	+.2861	+.1334	−.06865	−.2145	−.2187
8.0	$+.0^79422$	$+.0^42218$	$+.0^34934$	$+.0^24029$	+.01841	+.05653	+.1280	+.2235	+.3051	+.3179	+.2250	+.04510	−.1410	−.2320
8.5	—	—	—	—	—	+.03520	+.08854	+.1718	+.2633	+.3169	+.2838	+.1496	−.04006	−.1928
9.0	$+.0^85249$	$+.0^52492$	$+.0^48440$	$+.0^39386$	$+.0^25520$	+.02117	+.05892	+.1263	+.2149	+.2919	+.3089	+.2304	−.06698	−.1143
9.5	—	—	—	—	—	+.01232	+.03785	+.08921	+.1672	+.2526	+.3051	+.2806	+.1621	−.01541
10.0	$+.0^92631$	$+.0^62515$	$+.0^41293$	$+.0^31950$	$+.0^31468$	$+.0^26964$	+.02354	+.06077	+.1247	+.2075	+.2804	+.3005	+.2338	+.08501

Note: $.0^27186 = .007186$ and $.0^3807 = .000807$. Reproduced by permission from "Reference Data for Radio Engineers," copyright 1956 by International Telephone and Telegraph Corporation.

Index